Mechatronics

Ryszard Jabłoński and Tomaš Březina (Eds.)

Mechatronics

Recent Technological and Scientific Advances

Springer

Editors

Prof. Ryszard Jabłoński
Warsaw University of Technology
Faculty of Mechatronics
Institute of Metrology and
Biomedical Engineering
Sw. A. Boboli Street 8
02-525 Warsaw
Poland
E-mail: yabu@mchtr.pw.edu.pl

Prof. Tomaš Březina
Brno University of Technology
Faculty of Mechanical Engineering
Institute of Automation and
Computer Science
Technická 2896/2
616 69 Brno
Czech Republic
E-mail: brezina@fme.vutbr.cz

ISBN 978-3-642-23243-5 e-ISBN 978-3-642-23244-2

DOI 10.1007/978-3-642-23244-2

Library of Congress Control Number: 2011935381

Typeset & Cover Design: Scientific Publishing Services Pvt. Ltd., Chennai, India.

Printed on acid-free paper

9 8 7 6 5 4 3 2 1

springer.com

Preface

Mechatronics is in fact the combination of enabling technologies brought together to reduce complexity through the adaptation of interdisciplinary techniques in production.

Focusing on the most rapidly changing areas of mechatronics, this book discusses signals and system control, mechatronic products, metrology and nanometrology, automatic control & robotics, biomedical engineering, photonics, design manufacturing and testing of MEMS.

This fact is reflected in the list of contributors, including an international group of 240 academicians and engineers representing 6 countries.

The selection of papers for inclusion in this book was based on the recommendations from the preliminary review of abstracts and from the final review of full lengths papers, with both reviews concentrating on originality and quality. Finally 94 papers are included in this book.

It is hoped that the volume can serve the world community as the definitive reference source in mechatronics and more generally it is intended for use in research and development departments in academic, government and industry.

This book comprises carefully selected contributions presented at the 9th International Conference Mechatronics 2011, organized by Faculty of Mechatronics, Warsaw University of Technology, on September 21–24, in Warsaw, Poland. It is the third volume in series Recent Advances in Mechatronics, following the editions in 2007 and 2009.

We would like to thank all authors for their contribution to this book.

Ryszard Jabłoński
Conference Chairman
Warsaw University of Technology

Contents

Part I: Mechatronic Products

Part II: Metrology and Nanometrology

Part III: Automatic Control

Part VII: MEMS and Nanotechnology

Part VIII: Photonics

Part I

Mechatronic Products

Virtual Prototyping as Tool for Powertrain Development

L. Drápal*, P. Novotný, V. Píštěk, and R. Ambróz

Brno University of Technology, Technická 2896/2,
Brno, 616 69, Czech Republic

Abstract. The paper presents a use of virtual prototyping in designing, different variants comparing, and selecting of the most suitable design of crankshafts for two-stroke compressed-ignition engine with contra-running pistons and through-flow scavenging. Applying CAE methods enable a solution of different problems like torsional vibrations or loads of main bearings.

1 Introduction

The engine is primarily intended for aircraft industry, and this lays excessive demands on efficiency parameters, weight, compactness, fuel consumption and reliability.

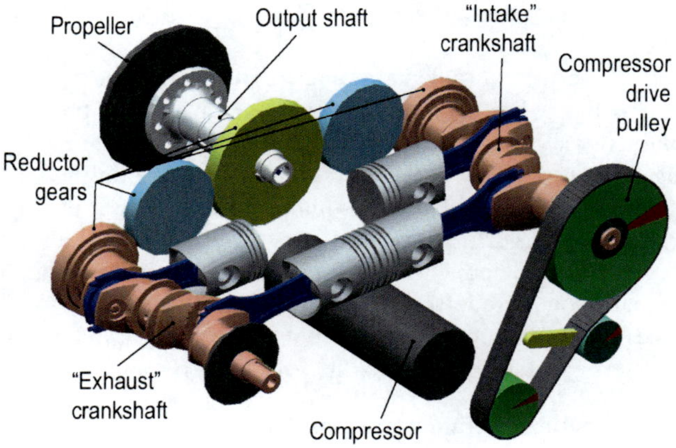

Fig. 1. 3D dynamic model of target engine cranktrain

---

* Corresponding author.

A crankshaft is crucial component of combustion engines. The crankshaft under development can be successful only if a specific limit value of the output weight is not exceeded. That is why the preferences for newly designed crankshafts are arranged in the following:

- high strength;
- low weight;
- balance of moments of the inertial forces in rotating mass.

For a small-lot production, four variants of steel and completely machined crankshafts were suggested:

- "Light unbalanced" hereinafter LU (7.521 kg);
- "Light balanced" hereinafter LB (8.687 kg);
- "Heavy unbalanced" hereinafter HU (8.277 kg);
- "Heavy balanced" hereinafter HB (9.442 kg).

Each variant has its specific weight, stiffness and level of balance. The designs include even lubricating channels for pressure engine oil supply to the connecting rod pins, whose openings are located as close to the neutral pin deflection axis as possible.

2 Powertrain Dynamic Solution

All following analyses are based on virtual engine as a dynamic computational model created in Multi-Body Dynamics (MBS), where crankshafts and output shafts are used as linearly elastic bodies. These are modally reduced Finite Element (FE) models. Other engine parts are modeled as solid bodies, except for reducer gear wheels.

The interaction between crankshafts and output shaft and the engine block is ensured via a non-linear hydrodynamic bearing model, where pre-calculated force databases obtained when solving separate hydrodynamic problem are used.

Virtual engine is excited by means of cylinder pressure defined by calculation, and by means of inertial forces from moving parts. Simulated start-up from crankshaft speed 1650 min^{-1} is carried out till the overspeed mode under 4125 min^{-1}. Such simulation is completed with all crankshaft variants and the consequential comparison is based on a predefined output obtained from MBS.

In order to evaluate and compare different crankshaft variants, suitable criteria have to be selected. For these purposes, the following ones have been used:

- level of torsional vibration in crankshafts;
- fatigue life;
- load of main bearings.

2.1 Cranktrain Torsional Vibrations

This type of vibrations in fully embedded shafts, which is caused by shaft flexibility, is the most dangerous one. Its forced form is caused also by time variability of the torsional moment. Such torsional vibrations are superimposed on the shaft oscillation that appears due to the engine run irregularity and the static torsion from tangential forces.

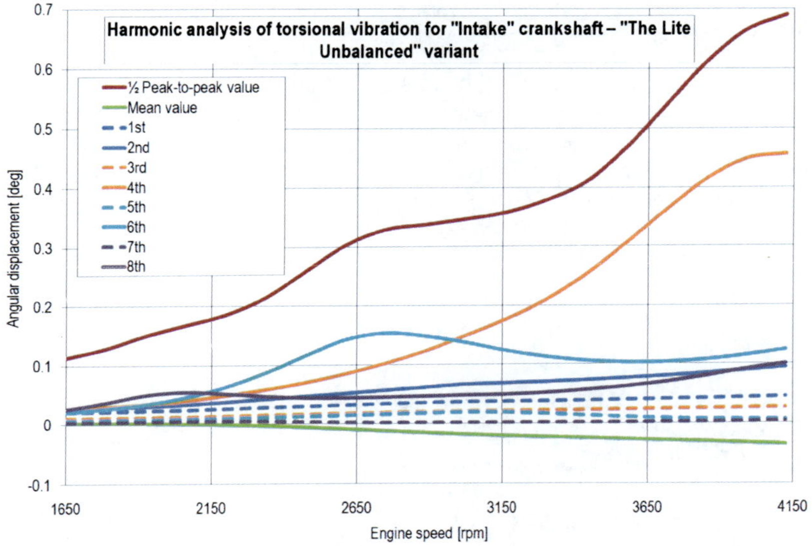

Fig. 2. Harmonic analysis of torsional vibrations of the "inlet" crankshaft of LU variant

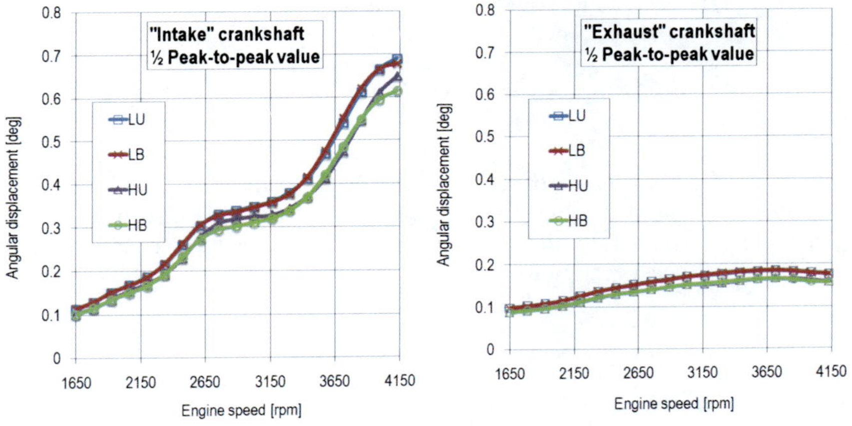

Fig. 3. Harmonic analysis of torsional vibrations of all crankshaft variants

The results obtained from harmonic analysis of torsional vibrations of the "inlet" crankshaft of LU variant are shown in Figure 2. It is clear that there is a strong impact of the fourth harmonic component of the torsional vibrations that is reaching resonance under the highest speed. Under 2700 min^{-1} the resonance from the sixth harmonic component is evident. The eighth harmonic component shows resonance too but its impact on the total shaft oscillation is considerably smaller. These analyses are carried out for all crankshaft variants and their results are shown in Figure 3.

2.2 Main Bearing Loads

Main bearing load is determined by bonding reactions to the load forces while for identical excitation by forces from gas pressure, it differs according to the stiffness and level of crankshaft balancing. It has to be considered which parameter could be suitable for comparison of individual variants. Reaction forces and relative eccentricity can be used as parameters.

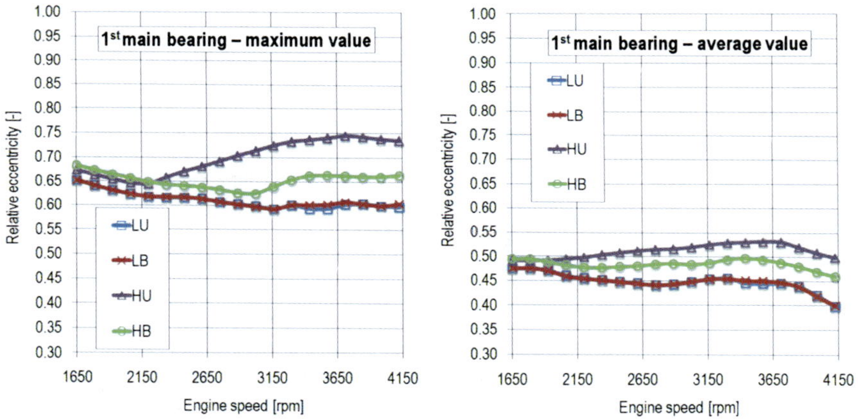

Fig. 4. Maximum and average relative eccentricity on the engine speed for all crankshaft variants

The above mentioned term of reaction force includes the size of fictitious single force which is caused by integration of function of pressure distribution in the hydrodynamic layer of lubrication. The second parameter is bearing relative eccentricity. Relative eccentricity is defined as

$$\varepsilon = \frac{2e}{D-d} \quad , \tag{1}$$

where e is an eccentricity of the pin centre against the centre of the bearing shell, D is the bearing shell diameter and d stands for the pin diameter. All is based on the assumption of an absolutely stiff pin and shell.

Newly designed variants, i.e. “inlet“ and “exhaust“ crankshafts, are then compared. Figure 4 summarizes the results obtained in analyses of maximum and average relative eccentricity on the engine speed for all crankshaft variants.

3 Conclusions

Evaluation of dynamic simulations proved that especially “inlet“ crankshaft is loaded by substantial torsional vibrations because there is a robust compressor drive connected to its front part. These vibrations are lower in “heavy” variants which show higher torsional stiffness because there are no discharge holes in the crankpins. As far as bearing load is concerned, paying attention to the impact of balancing of moments of the inertial forces in rotating mass on the bonding reactions in main bearings, all variants are convenient.

In conclusion, for the designed rated power and selection criteria, the “heavy unbalanced” variant seems to be the most suitable one.

Acknowledgements

Published results were acquired with the help of the project Powertrain Dynamic Solution considering Elastohydrodynamic Effects, No. FSI-J-11-17 granted by specific university research of Brno University of Technology.

References

[1] Novotný, P.: Virtual Engine – A Tool for Powertrain Development, Brno: Brno University of Technology, Inaugural Dissertation (2009)

[2] Novotný, P., Píštěk, V., Drápal, L., Ambróz, R.: Cranktrain Development of Aircraft Diesel Engine. In: XLI. International Scientific Conference of Czech and Slovak University Departments and Institutions Dealing with The Research of Combustion Engines (2010) ISBN 978-80-7372-632-4. 2010

Design of Fuzzy Logic Controller for DC Motor

O. Andrs*, T. Brezina, and J. Kovar

Brno University of Technology, Faculty of Mechanical Engineering,
Institute of Automation and Computer Science,
Technicka 2896/2, Brno, Czech Republic,
andrs@fme.vutbr.cz

Abstract. This contribution presents a simple and efficient approach to the fuzzy logic controller design and simulation. The proposed controller uses a Sugeno type fuzzy inference system (FIS) which is derived from discrete position state-space controller with an input integrator. The controller design method is based on anfis (adaptive neuro-fuzzy inference system) training routine. It utilizes a combination of the least-squares method and the backpropagation gradient descent method for training FIS membership function parameters to emulate a given training data set. The proposed fuzzy logic controller is used for the position control of a linear actuator which is a part of a Stewart platform.

1 Introduction

Parallel manipulators such as a Stewart platform [1] represent a completely parallel kinematic mechanism that has major differences from typical serial link robots. During the recent years, a large variety of parallel mechanisms were introduced by research institutes and by industries [2], [3], [4]. However, they have some drawbacks of relatively small workspace and difficult forward kinematic problems. Generally, forward kinematic of a parallel manipulators is very complicated and difficult to solve. The commonly used methods are usually based on a numerical iterative principle [4], [6], use of genetic algorithm [7] or for example using of extra sensors [8]. On the other hand, dynamics of a parallel manipulator, which is very important to develop a controller, tends to be very complicated. Most studies on this domain are usually based on the Lagrangian formulation [9], [10], [11] the Newton–Euler formulation [12], [13] and the principle of virtual work [14], [15]. Description of a parallel manipulator full dynamics via above-mentioned methods is usually quite complicated and numerical solution of the obtained model is too much time consuming. Even with a high-speed hardware, therefore, real-time calculation was not easy to achieve in the control system. Therefore simplifying

* Corresponding author.

suggestions shortening the computational time are often made. Thus, instead of considering the whole system, some suggest that each actuator may be controlled independently with a control law more robust than a simple PID [4].

This paper presents a parallel mechanism which is designed for testing of the backbone segments and hip joints [5], Fig. 1.

Fig. 1. The Stewart platform prototype model

The proposed mechanism makes possible to simulate the physiological movements of the human body and observe degradation processes of the cord implants. The demand for fastening the spinal segments determines the size of the mechanism. Single layer of the device consists of couple of toroidal bodies (base and platform) linked to each other. Both plates are coupled with six linear actuators which generate movement of the platform. The linear actuator consists of a ball screw with a nut, a spur gearing, DC motor Maxon RE 35 with a planetary gearbox and IRC sensor. Thus, the main aim of this paper is to describe design of the fuzzy logic controller for the linear actuator.

2 Design Methodology

Conventional proportional/integral/derivative (PID) controllers are usually used in industry due to their easy interpretation and implementation.

Meanwhile, fuzzy logic control, an intelligent control method inspired by the logical thinking of expert, can exceed some drawback of the traditional control. Conventional fuzzy PID control which combines the traditional PID control and the fuzzy control algorithm is described in [16], [17]. However, the fuzzy controllers mentioned in these studies are experimentally designed based on working conditions of the control systems and their dynamic responses. Consequently, the design of fuzzy rules depends largely on the experience of experts. There is no systematic methodology to design and examine the number of rules, scaling of input space and membership functions.

Neural network represents another powerful tool for learning static and nonlinear dynamic systems. Learning algorithms are capable of approximating nonlinear functions with desired accuracy via the back-propagation algorithm.

Advantages of both methods are used in adaptive neuro-fuzzy inference system. The neuro-adaptive learning method works similarly to that of neural networks. A network-type structure, which maps inputs through input membership functions and associated parameters, and then through output membership functions and associated parameters to outputs, can be used to interpret the input/output map. Using a given training data set, the function *anfis* (from MATLAB software) constructs a Sugeno type fuzzy inference system whose membership function parameters are tuned (adjusted) using either a backpropagation algorithm alone or in combination with a least squares type of method [18], [19]. This adjustment allows our fuzzy inference system to learn from the data generated by discrete position state-space controller.

The intended design solution is divided into three main design stages, modeling of DC motor, design of the discrete position state-space controller and design of the fuzzy logic controller. In this manner the paper extends the result of [20], [21].

2.1 Modeling of DC Motor

A simplified linear state-space model of DC motor with additional disturbance input was used (1) [21].

$$\begin{aligned} \mathbf{x}' &= \mathbf{A}\mathbf{x} + \mathbf{B}\mathbf{u} + \mathbf{E}\mathbf{z} \ , \\ \mathbf{y} &= \mathbf{C}\mathbf{x} + \mathbf{D}\mathbf{u} \ . \end{aligned} \tag{1}$$

A, **B**, **C**, **D**, **E** are in sequence the state matrix, the input matrix, the output matrix, the feedforward matrix and the disturbance matrix. The state-space model of DC motor Maxon RE 35 (1) can be written as:

$$\begin{bmatrix} \varphi' \\ \omega' \\ i' \end{bmatrix} = \begin{bmatrix} 0 & 1 & 0 \\ 0 & -k_f/J & k_m/J \\ 0 & -k_b/L & -R/L \end{bmatrix} \begin{bmatrix} \varphi \\ \omega \\ i \end{bmatrix} + \begin{bmatrix} 0 \\ 0 \\ 1/L \end{bmatrix} [u_M] + \begin{bmatrix} 0 \\ 1/J \\ 0 \end{bmatrix} [M_Z] ,$$
$$\begin{bmatrix} \varphi_M \\ \omega_M \\ i_M \end{bmatrix} = \begin{bmatrix} 1 & 0 & 0 \\ 0 & 1 & 0 \\ 0 & 0 & 1 \end{bmatrix} \begin{bmatrix} x_1 \\ x_2 \\ x_3 \end{bmatrix} + \begin{bmatrix} 0 \\ 0 \\ 0 \end{bmatrix} [u_M] , \tag{2}$$

where k_f, k_m, k_b, R, L and J are the motor parameters. The state vector $\mathbf{x}$ represents angular displacement of the rotor φ, angular velocity ω and electrical current in the rotor i, the vector of inputs $\mathbf{u}$ presents motor driving voltage u_M. The vector of model outputs $\mathbf{y}$ contains angular displacement of the rotor φ_M, angular velocity ω_M and electrical current in the rotor i_M. The disturbance vector $\mathbf{z}$ presents the DC motor loading torque M_Z.

2.2 The Discrete Position State-Space Controller Design

Implementation of the control structure with the integrator on the input without loading disturbance leads to the following control law [21]

$$\mathbf{u} = -\mathbf{R}\mathbf{x} + \mathbf{r}_i \mathbf{v} ,$$
$$\mathbf{v}' = \mathbf{w} - \mathbf{y} , \tag{3}$$

where $\mathbf{R} = [r_\varphi \; r_\omega \; r_i]$ is the vector of gains of the system's state vector, $\mathbf{r_i}$ presents gain of the integrator and $\mathbf{w}$ is reference value. Presented control law modifies the state-space model of the system to the state-space equation:

$$\begin{bmatrix} \mathbf{x}' \\ v' \end{bmatrix} = \left(\begin{bmatrix} \mathbf{A} & \mathbf{0} \\ -\mathbf{C} & 0 \end{bmatrix} - \begin{bmatrix} \mathbf{B} \\ -\mathbf{D} \end{bmatrix} \begin{bmatrix} \mathbf{R} & -\mathbf{r}_i \end{bmatrix} \right) \begin{bmatrix} \mathbf{x} \\ \mathbf{v} \end{bmatrix} + \begin{bmatrix} \mathbf{0} \\ 1 \end{bmatrix} [\mathbf{w}] . \tag{4}$$

The controller gains $\mathbf{R}$ and the integrator gains $\mathbf{r_i}$ are adjusted in such a way that state matrix of the system with the integrator

$$\begin{bmatrix} \mathbf{A} & \mathbf{0} \\ -\mathbf{C} & 0 \end{bmatrix} - \begin{bmatrix} \mathbf{B} \\ -\mathbf{D} \end{bmatrix} \begin{bmatrix} \mathbf{R} & -\mathbf{r}_i \end{bmatrix} \tag{5}$$

is solved for specified eigenvalues, providing stability of the system for the sufficient position, in allowed control values range and with given power limitations. Note, that there is very efficient to use MATLAB's function *place* and *c2d* (discrete version) to obtain real controller gains [21]. There is also possibility to extension with state space observer [22].

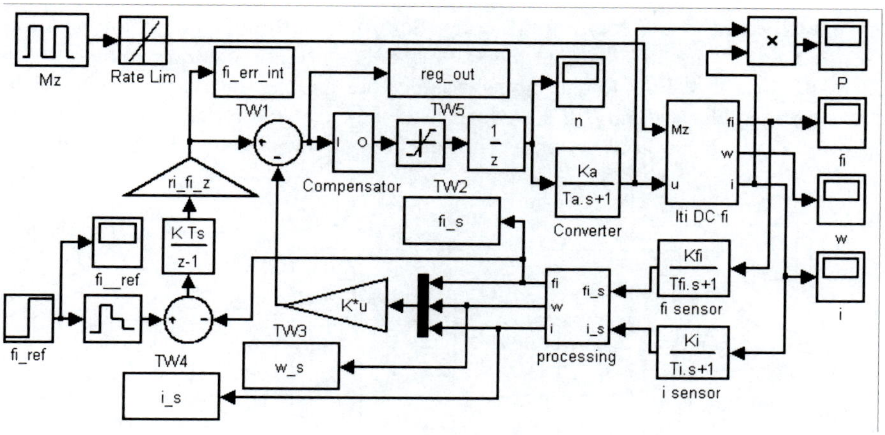

Fig. 2. The discrete state-space controller simulation model

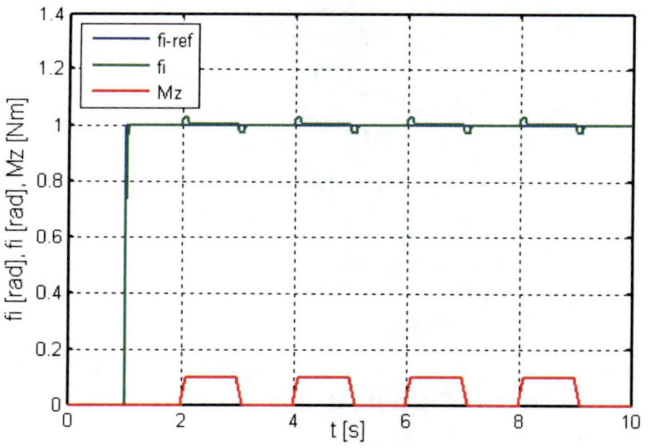

Fig. 3. The impact of the step change in M_Z on the position response φ_M

The simulation model of the position controller with DC motor model was developed in software environment MATLAB/Simulink, Fig. 2. This model is used to observe the impact of the step change in M_z on the position response (Fig. 3.) and to generate training data set for *anfis* routine. The power converter, position and the electrical current sensor were modeled as first order systems [21].

2.3 The Fuzzy Logic Position Controller Design

Consider the DC motor model and the discrete state-space controller, shown in Fig. 2. The control algorithm is formed by an inner feedback loop of state-space controller and an outer feedback loop of position error integral. Our goal is to design a Sugeno type fuzzy logic controller for controlling the motor position. Consequently, we can substitute the state-space controller for fuzzy logic controller with all inputs, Fig. 4. The input variables are the angular displacement of the rotor, the angular velocity, the electrical current in the rotor and the integral of position error. The output variable represents duty cycle of the driving voltage.

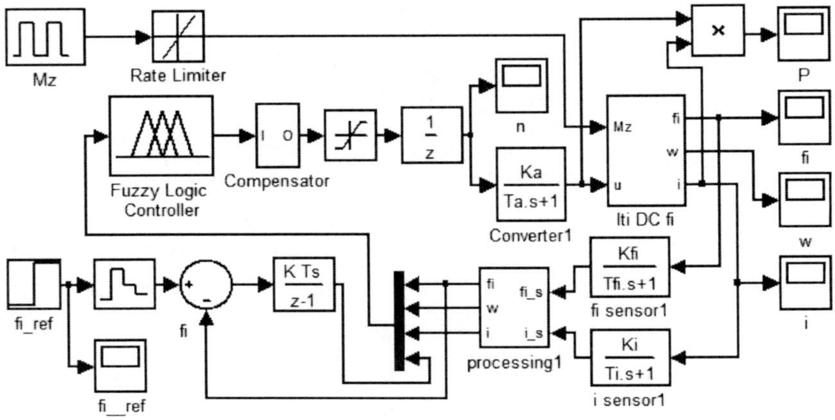

Fig. 4. The fuzzy logic controller simulation model

To determine the fuzzy relation, the input universe of each signal was partitioned into four input variables. The Π-shaped fuzzy sets were chosen with sufficient results of approximation. This partitioning led to inference system with 256 rules, Tab. 1. The fuzzy set of the output variable is inferred by weighted average of all rule outputs. In order to simplify this inference system, the number of membership functions was modified, Tab. 1. It has been verified that the performance of the optimized inference system is sufficient even when using two membership functions, Fig. 6.

The parameters associated with the membership functions and inference system changes through the learning process according data from state-space controller. The computation of these parameters is facilitated by a gradient vector. This gradient vector provides a measure of how well the fuzzy inference system is modeled. When the gradient vector is obtained, any of several optimization routines can be applied in order to adjust the parameters to reduce some error measure. The a*nfis* routine uses either the back propagation or a combination of a least squares estimation and the backpropagation for the membership function parameter estimation [19].

Table 1. The inference system parameters

Parameter	4x Π-shaped	3x Π-shaped	2x Π-shaped
InMFTypes	pimf	pimf	pimf
Number of nodes:	551	193	55
Number of linear parameters:	1280	405	80
Number of nonlinear parameters:	64	48	32
Number of training data pairs:	61425	61425	61425
Number of fuzzy rules:	256	81	16
Inputs/Outputs	[4 1]	[4 1]	[4 1]
NumInputMFs	[4 4 4 4]	[3 3 3 3]	[2 2 2 2]
InRange	[-2.4 2.4]	[-2.4 2.4]	[-2.4 2.4]
	[-30 30]	[-30 30]	[-30 30]
	[-3 3]	[-3 3]	[-3 3]
	[-8 8]	[-8 8]	[-8 8]
Error	5.7143e-006	1.5942e-006	1.7406e-007

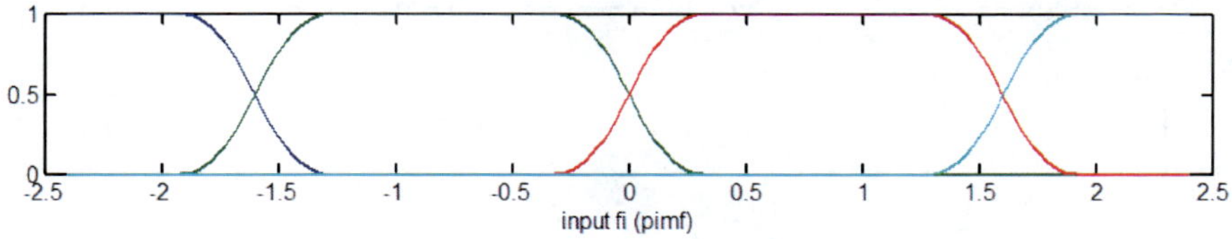

Fig. 5. An example of membership functions for input variable

The proposed method utilizes the Sugeno type fuzzy logic controller to compensate the impact of the step change in M_Z with noise on the position response, see simulation results in Fig. 6., The Sugeno type fuzzy logic controller works similar to discrete state-space controller, Fig. 7.

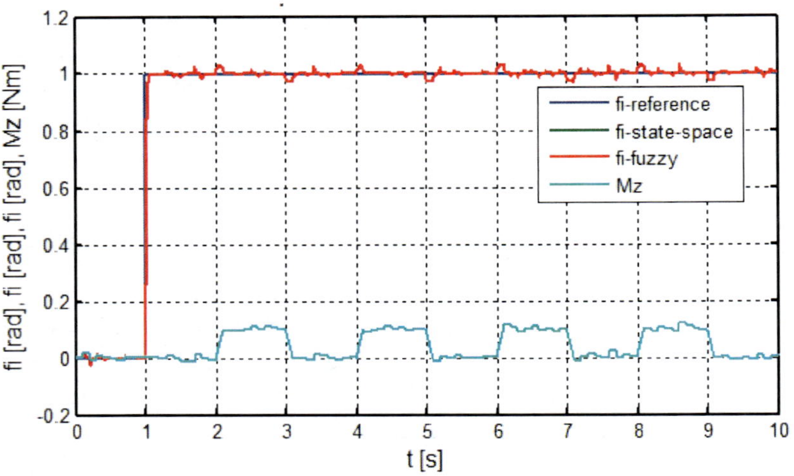

Fig. 6. The impact of the step change in M_Z with noise on the position response for the discrete state-space controller and Sugeno type fuzzy logic controller

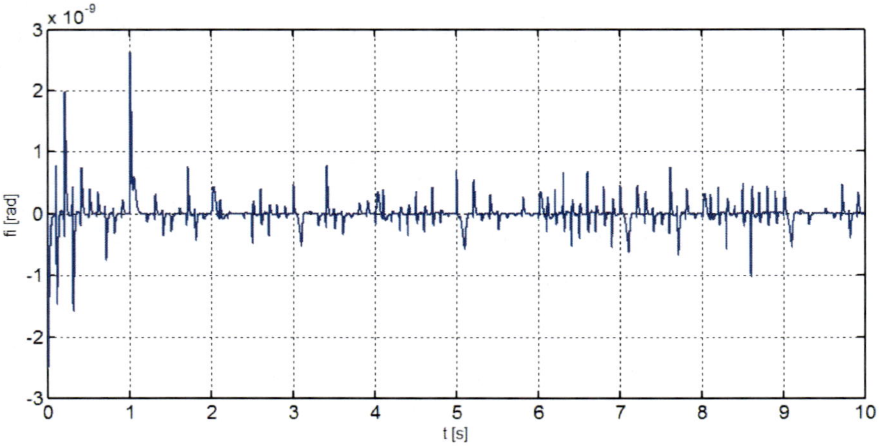

Fig. 7. The difference between the discrete state-space controller and the Sugeno type fuzzy logic controller on position responses

3 Results

In this contribution, the Sugeno type fuzzy logic controller was proposed for the DC motor position control and the effectiveness of the proposed control system is shown in simulation under the load variations. From the simulation, it can be concluded that the fuzzy controller produces no steady state error and acceptable overshoot under load disturbances. In order to show the advantage of the suggest control method, the performance comparison of this controller with the state-space controller has also been investigated. It has not been found that the performance of the test controller is more robust as compared to the state-space controller because of using a linear DC motor model. Thus, next approach will lead to changing and improving a DC motor model in order to obtain better working performance of fuzzy logic in non-linear systems.

Acknowledgements

This work is supported from research plan MSM 0021630518 Simulation modeling of mechatronic systems, FSI-J-11-39 and FSI-S-11-23.

References

[1] Stewart, D.: A platform with six degrees of freedom. In: Proceedings of the Institution of Mechanical Engineers, pp. 371–381 (1965)
[2] Gough, V.E., Whitehall, S.G.: Universal tire test machine. In: Proceedings 9th Int.Technical Congress F.I.S.I.T.A., London (1962)
[3] Gopalakrishnan, V., Fedewa, D., Mehrabi, M., Kota, S., Orlandea, N.: Parallel structures and their applications in reconfigurable machining systems. In: Proceeding of Year 2000 PKM International Conference, Michigan, Michigan, USA, pp. 87–97 (2000)
[4] Merlet, J.P.: Parallel robots, 2nd edn. Kluwer Academic Publishers, Dordrecht (2005) ISBN 1402041322
[5] Brezina, T., Andrs, O., Houska, P., Brezina, L.: Some Notes to the Design and Implementation of the Device for Cord Implants Tuning. In: Recent Advances in Mechatronics, pp. 395–400. Springer, Heidelberg (2009) ISBN 978-642-05021-3
[6] Liu, K., Fitzgerald, J.M., Lewis, F.L.: Kinematic analysis of a Stewart platform manipulator. IEEE Transactions on Industrial Electronics 40(2), 282–293 (1993)
[7] Boudreau, R., Turkkan, N.: Solving the forward kinematics of parallel manipulators with a genetic algorithm. Journal of Robotic Systems 13(2), 111–125 (1996)
[8] Inoue, H., Tsusaka, Y., Fukuizumi, T.: Parallel manipulator. In: Proceedings of ISRR, Gouvieux (1985)
[9] Wang, S., Hikita, H., Kubo, H., Zhao, Y., Huang, Z., Ifukube, T.: Kinematics and dynamics of a 6 degree-of-freedom fully parallel manipulator with elastic joints. Mechanism and Machine Theory 38(5) (2003)
[10] Cheng, H., Yiu, Y., Li, Z.: Dynamics and Control of Redundantly Actuated Parallel Manipulators. IEEE/ASME Transactions on Mechatronics 8(4), 483–491 (2003)
[11] Lebret, G., Liu, K., Lewis, F.L.: Dynamic analysis and control of a Stewart platform manipulator. J. Robotic Syst (1993)
[12] Carvalho, J.C.M., Ceccarelli, M.: A Closed-Form Formulation for the Inverse Dynamics of a Cassino Parallel Manipulator. Multibody System Dynamics 5(2), 185–210 (2001)
[13] Dasgupta, B., Choudhury, P.: A general strategy based on the Newton-Euler approach for the dynamic formulation of parallel manipulators. Mechanism and Machine Theory 34(6), 801–824 (1999)

[14] Zhang, C., Song, S.: An efficient method for inverse dynamics of manipulators based on the virtual work principle. J. Robotic Syst. 10(5) (1993)
[15] Gallardo, J., Rico, J.M., Frisoli, A., Checcacci, D., Bergamasco, M.: Dynamics of parallel manipulators by means of screw theory. Mechanism and Machine Theory 38(11), 1113–1131 (2003)
[16] Kha, N.B., Ahn, K.K.: Position control of shape memory alloy actuators by using self tuning fuzzy PID controller. In: Proceedings of IEEE International Conference on Industrial Electronics and Applications, Singapore (2006)
[17] Ahmed, R., Abdul, O., Marcel, C.: dSPACE DSP-based rapid prototyping of fuzzy PID controls for high performance brushless servo drives. In: Proceedings of IEEE International Conference, USA, pp. 1360–1364 (2006)
[18] Sugeno, M.: Industrial applications of fuzzy control. Elsevier Science Pub. Co., Amsterdam (1985)
[19] MATLAB, Fuzzy Logic Toolbox™ User's Guide (2011)
[20] Andrs, O., Brezina, T.: Control simulation of a linear actuator (In Czech). In: Automatizacia a riadenie v teorii a praxi 2010, Technicka univerzita v Kosiciach, Stara Lesna, pp.13-1–13-7 (2010) ISBN 978-80-553-0146-4
[21] Andrs, O., Brezina, T.: Controller Implementation of the Stewart Platform Linear Actuator, pp. 1357–1359 (2010); Katalinic, B.: Annals of DAAAM for 2010 & Proceedings of the 21st International DAAAM Symposium, p. 0679. DAAAM International, Vienna (2010) ISBN 978-3-901509-73-5, ISSN 1726-9679
[22] Brezina, T., Brezina, L.: Controller design of the Stewart platform linear actuator. In: Recent Advances in Mechatronics, pp. 341–346. Springer, Heidelberg (2009) ISBN 978-3-642-05021-3

Integration of Risk Management into the Process of PLC-Software Development in Machine Tools

P. Blecha[1,*] and L.W. Novotný[2]

[1] Brno University of Technology, Faculty of Mechanical Engineering, Technická 2896/2, Brno, 616 69, Czech Republic
[2] TOSHULIN, a.s, Wolkerova 845, 768 24 Hulin, Czech Republic

Abstract. Risk management is currently a key component of the design process. The designers are required to identify all hazards within the whole life cycle of a machine and, if necessary, to take appropriate measures in order to reduce the risk of these hazards to an acceptable level. To achieve overall safety of a machine, it is also very important to carry out risk management during the development of software to ensure correct function of the designed mechatronic system. The presented paper describes realization of risk management during development of PLC-software for the system of automatic tool change in a machine tool.

1 Introduction

The minimal requirements on machinery safety are given in the machinery directive 2006/42/EC [7] and in the harmonized standards issued to this directive. In the area of functional safety, the standard IEC 61508-3 [8] defines the requirements on software of the systems related to safety. However, from the aspect of machine tools development it is desirable to apply risk management not only with regard to potential damage to health or environment, but also with regard to property damage that can be caused by software malfunction (e.g. collision of fixed and moving parts of the machine).

The above-mentioned standard recommends various activities that should be realized within the software development; yet, it does not give instructions on how to realize these development activities. Many authors have dealt with the question of risk management during the development phase of a new product design. However, integration of safety into the design process usually either concentrates at a general procedure of the design proposal realization without more detailed focus on implementation of software into the developed product and verification/validation of its correct functioning in all reasonably foreseeable circumstances [1, 2, 3, 4], or the

* Corresponding author.

presented method is too scientific and therefore not suitable for application in industrial practice [5, 6]. The process of software development for a newly-designed machine is described in general in the well-known mechatronic V-model (Fig. 1), which is in accordance with the ISO/IEC 12207 [9].

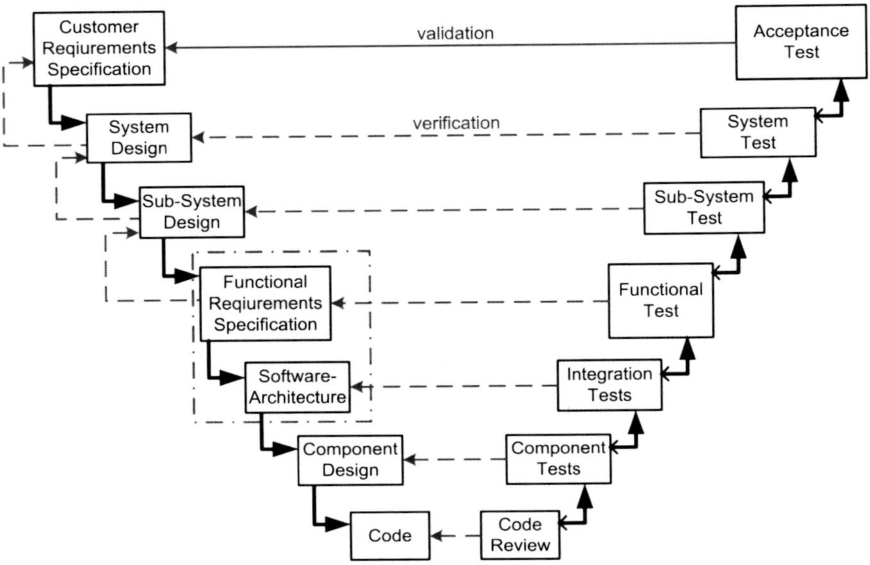

Fig. 1. V-Model according to ISO/IEC 12207

Two critical points of the software development process are "Functional Requirements Specification" and "Software Architecture". In these phases it is essential to pay attention to identification of all significant hazards and to adaptation of the software architecture to preventive measures for risk minimization in significant risks. Here, risk management blends with safety as well as with reliability of the developed system. The following chapters focus on description of the developed automatic tool change system and the process of risk management realization in the development of software for this mechatronic system of a machine tool by the TOSHULIN company.

2 Description of the Automatic Tool Change System

Automatic tool exchange system comprises several basic components (Fig. 2.) including a cartridge tool magazine (a), tool manipulator (b) and a tool loading position (e). The tool manipulator includes a tool holder (adapter) discharge block (c), able to hold six tool holders (adapters). The cleaning station (d) serves for cleaning of the tool clamping shanks and is mounted on the front pillar of the cartridge magazine (a). The tool magazine space is fenced with a wire netting protective cover to ensure safety during operation. The magazine can be reached through a swing door on the right side of the wire netting protective cover of the magazine, in its rear part. The picture shows the magazine only.

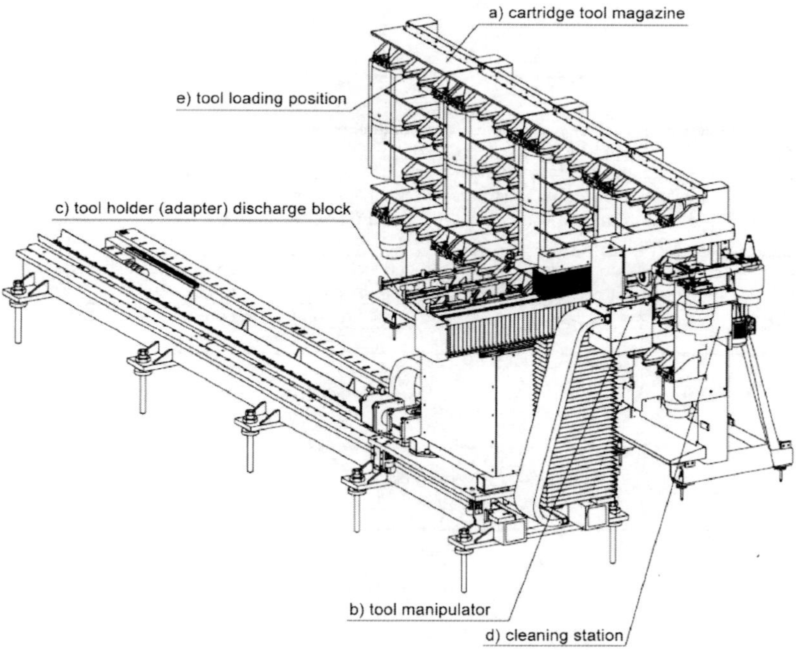

Fig. 2. Automatic tool change system

The tool manipulator including the cartridge tool magazine is separated from the machine working space by a cover with hydraulically-operated door for travel of the crossbar support into the space of the automatic tool change. The tool manipulator uses six moving axes – three linear axes (translational) controlled by digital servo-drives in position coupling and three hydraulically-controlled rotary axes. The linear axes are oriented in the directions of the axes X, Y and Z.

Operation of a mechatronic device such as the presented tool magazine requires, besides the mechanical part, i.e. the support part and the drive part, also a number of mechanical, optical and other switches and control elements to control of the end positions of the manipulator (b), identification of tool clamping, identification of the tool type and other. These sensors are necessary for correct operation of the automatic tool change in direct connection with the PLC program, as the signals from sensors activate the next lines of the program code.

3 Risk Management by Software Development

Identification of hazards is the most important part of risk management. To identify the significant hazards with success, it is necessary to divide the process of automatic tool change into individual sub-processes and to analyze them separately.

Process of automatic tool change	
Sub-process	Description
1st step	Instruction from the CNC system to change the tool (ram is empty).
2nd step	Travel of the manipulator support into the defined position in X axis, gradual rotation of the rotary axes a, b and c into the position for gripping of the holder with a tool from the magazine (rotation in axes a, b and c are written in machinery constants).
3rd step	Manipulator is gradually positioned in axes Y, Z for gripping of the selected tool (coordinates Z, Z and X are written in machinery constants).
4th step	Release of the pneumatic fixture in the arm on the side prepared for holder clamping.
etc.	etc.

Fig. 3. A list of sub-processes within the tool change process

Sub-process	Regime of machine	Potential failure	Consequence of failure	Cause of failure	Elements of risk S	A	E	W	RISK	Preventive measures for risk reduction	Elements of risk S	A	E	W	reduced RISK
1. step – Instruction from the CNC system to change the tool (ram is empty).	AUTOMATIC	CNC system does not send instruction for the tool change.	Tool change is not carried out.	Incorrect program	2	2	2	1	9	Primary checkout of the software. Protection against unauthorized changes.	1	2	2	1	3
		CNC system sends incorrect instruction for the tool change.	Tool change is carried out incorrectly. Possibility of collision of the tool with the workpiece.	Incorrect program	3	2	2	3	17	Primary checkout of the software. Protection against unauthorized changes. Check of the gripped tool.	1	2	2	1	3
	MANUAL /ADJUSTMENT	Staff can not perform tool change.	Tool change is not carried out. The machine is incapable of machining.	Insufficient knowledge of the staff	1	2	2	2	4	Description of the procedure in the instruction manual for the machine.	1	1	1	1	0

Fig. 4. Documentation of risk management during software development

In the machine in question, we have identified 66 sub-processes within the tool exchange, 26 sub-processes within the tool loading into the magazine and 23 sub-processes within the tool pick-up from the magazine. The table in Fig. 3 shows an example of documentation of the identified sub-processes.

For the identified sub-processes it is then necessary to analyze the risks in every operational mode of the machine. For this stage of risk management we have modified the FMECA form (Fig. 4); the analysis of criticality has been adapted to the requirements of the machinery directive and the harmonized standards.

Fig. 5 shows the graph of risk used for determination of the level of risk in individual hazards identified within the individual sub-processes. Categories of the individual risk components are defined: S – seriousness of possible health damage (S0 - irrelevant injury/negligible financial loss; S1 - slight injury/minor financial loss; S2 - serious injury/greater financial loss; S3 – death/considerable financial loss), A - frequency and length of threat (A1 - seldom to occasional; A2 - frequent to continuous); E - possibility of avoiding the hazard (E1 - possible; E2 - possible at certain conditions; E3 - hardly possible), W - probability of hazardous event occurrence (W1 - low; W2 - medium; W3 - high). Also, the level of risk acceptability is marked out in this risk graph.

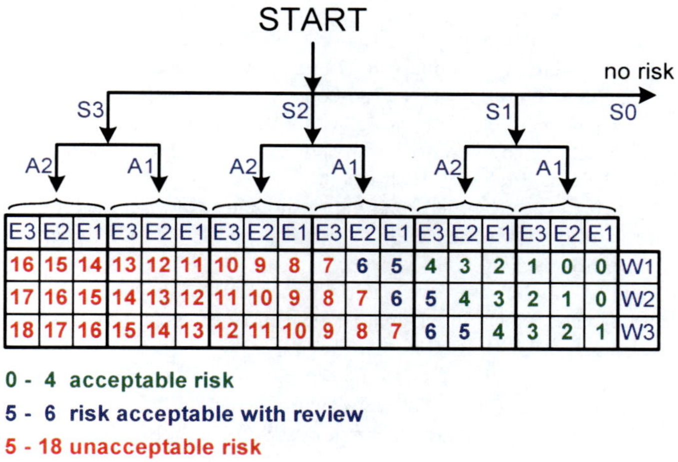

Fig. 5. Risk estimation graph

4 Conclusion

At the present, the requirements on safety of machinery are continuously increasing. The presented method of risk management integration into the process of PLC software development for the system of tool exchange is generally applicable in all mechatronic systems. If these mechatronic systems belong to machinery, the above-described procedure also enables easier fulfilment of the requirements on development of safety components of control systems.

Risk analysis applied during the design of automatic tool change not only in relation to protection and safety of operating staff, but also to its inner structure, i.e. up to the level of PLC software, has helped to detect and find weak points in inspectional activities, leading, for example, to changes in the types, number and positions of the switches and sensors.

The presented approach to risks management has proved as very useful during development of PLC software in Machine Tools and now it belongs to the company's standard within the designing process.

Acknowledgments. Research work for this contribution was financed by TOSHULIN, a.s., world leading producer of vertical lathes, and by the Ministry of Education, Youth and Sports of the Czech Republic (project 1M0507 "Research of production techniques and technologies").

References

[1] Bernard, A., Hasan, R.: Working situation model for safety integration during design phase. Annals of the CIRP 51(1), 119–122 (2002) ISSN 1726-0604
[2] Eder, W.E., Hosnedl, S.: Design Engineering - A Manual for Enhanced Creativity. CRC Press - Taylor & Francis Group (2007) ISBN 1420047655
[3] Ghemraoui, R., Mathieu, L., Tricot, N.: Design method for systematic safety integration. Annals of the CIRP 58(1), 161–164 (2009) ISSN 1726-0604
[4] Krause, F.-L., Kimura, F., Kjellberg, T., Lu, S.C.-Y.: Product Modelling. Annals of the CIRP 42(2), 695–706 (1993) ISSN 1726-0604
[5] Marek, J.: Management of risk at design of machining centres. In: Housa, J. (ed.) Proceedings of Machine Tools, Automation and Robotics in Mechanical Engineering, pp. 91–98. CVUT Praha, Praha (2004) ISBN 80-903421-2-4
[6] Marek, J.: How to go further in designing methodology of machine tools? In: Hosnedl, S. (ed.) Proceedings of AEDS 2006 Workshop, pp. 81–88. W. Bohemia University, Plzen (2006) ISBN 80-7043-490-2
[7] Directive 2006/42/EC of the European parliament and of the Council of 17 May 2006 on machinery, and amending Directive 95/16/EC (recast); Official Journal of the European Union L 157/24
[8] IEC 61508-3:2010 Functional safety of electrical/electronic/programmable electronic safety related systems – Part 3: Software requirements
[9] ISO/IEC 12207:1997 Information technology – Software life cycle processes

Experimental and Simulation Tests of Dynamic Properties of Stepping Motors

M. Bodnicki[1,*], A. Pochanke[2], K. Szykiedans[1], and W. Czerwiec[1]

[1] Institute of Micromechanics and Photonics, Faculty of Mechatronics, Warsaw University of Technology, 8 Sw. Andrzeja Boboli Str., Warsaw, 02-525, Poland
[2] Institute of Electrical Machines, Faculty of Electrical Engineering, Warsaw University of Technology, Pl. Politechniki 1, Warsaw, 00-661, Poland

Abstract. There are two kinds of frequency characteristics of stepping motors: pull-in and pull-out, both realized according to two different methods. In presented test stand torque determination of characteristics torque metering can be taken directly from motor shaft with use of an rotary torque meters or it can be processed also as a stator reaction force measurement made with use of dynamometric torque meter equipped with piezoelectric force transducer. Experiments results, despite its utilitarian use, have taken advantage in comparing of different methods of characteristics determining. Computer simulation methods give approximate numerical solution of the set of differential equations that are the mathematical model of the drive with stepping motor or model of the measuring station. Problems of obtaining the high dynamic properties of stepper motors in extremely conditions are complicated because of a character of operation of these micro-machines that introduces pulsating input functions into an flexible electromechanical structure.

1 Introduction

Designers of drives with stepping motors are interested in their work in directly nearness to characteristics illustrating dependency of the torque developed at the motor shaft on its average rotational speed or the frequency of the controlling pulses (being proportional to the speed) [1,5]. This means a maximal utilisation of the motor properties. Mentioned characteristics should be obtained for a given controlling method and a given kind of load. There are non-linear dependencies of mentioned characteristics on instantaneous values of currents of windings and design features both of the motor and mechanical devices.

* Corresponding author.

Problems of obtaining the dynamic properties of stepper motors in extremely conditions are complicated because of a character of operation of these micro-machines that introduces pulsating input functions into an flexible electromechanical structure. Interactions within the structure disturb operation of the motor by generating resonance phenomena. This problem is strongly imported during fast commutation in extreme condition of the load [7].

Information about static parameters of stepping motor is important e.g. for designer of positioning system. Measuring stations are designed for measurement of following static properties:

- step error (accuracy of positioning),
- static torque-angle characteristics.

There are following required tests during experimental analysis of dynamic properties:

- determination of dynamic characteristics - pull-in torque and pull-out torque,
- determination of transient characteristic - single-step response.

These characteristics should be determined both during operation in full step mode, as well as half and micro steps. The question is put about accuracy of process of determination of dynamic characteristics.

The appropriations of designed stand are comparative tests of measuring process realised:

- with use of different types of transducers,
- with different methods of torque measurement,
- with different procedures of determination of points of characteristics.

There is of course also utility profit from design of this kind of test stand – possibility of making investigation of various motors (e.g. with new control or supply units).

2 Frequency Characteristics of Stepping Motors

General methods and dedicated algorithms are presented in literature [4]. Synthetic graph is presented on the Fig.1. While determining pull-out characteristics the special supplementary procedures of acceleration are indispensable.

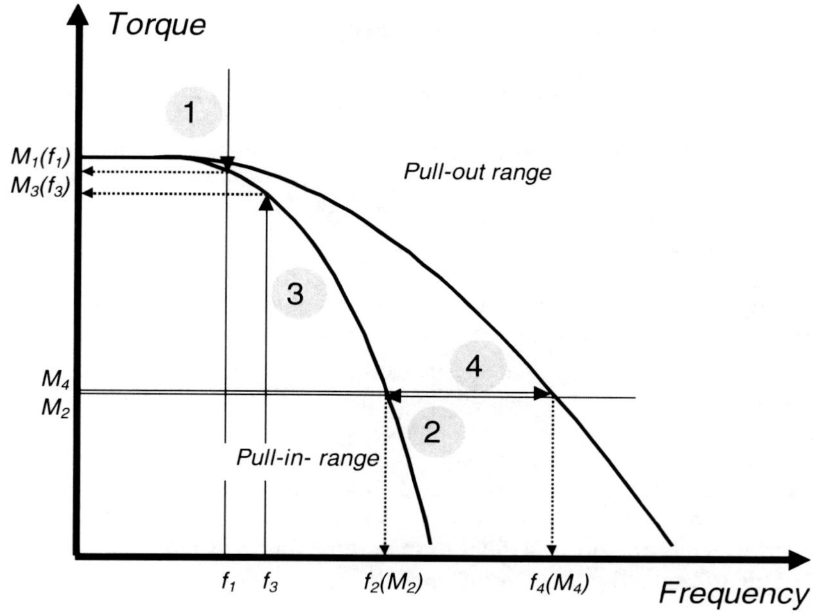

Fig. 1. Dynamic characteristics M = M(f) of stepping motor – idea of determination of following pair of values (M_i, f_i) during determination of *pull-in* (1, 2) and *pull-out* (3,4) characteristics; description of the methods in text

Pull-in characteristic

- Fixing of f_1 frequency and reducing by stages of breaking torque values to a start of rotors rotation – consequence value of torque is $M_1 = M_1(f_1)$ - *„No 1 method"*
- Fixing of breaking torque value M_2 i reducing by stages of frequency of steps to a start of rotors rotation – consequence value of torque is $f_2 = f_2(M_2)$ - *„No 2 method"*

Pull-out characteristic:

- Fixing of f_3 frequency and increasing (according to special procedure) of breaking torque values to a end of rotors rotation – consequence value of torque is $M_3 = M_3(f_3)$ - *„No 3 method"*
- Fixing of breaking torque value M_4 and increasing (according to special procedure) of control frequency values to a end of rotors rotation – consequence value of torque is $f_4 = f_4(M_4)$ - *„No 4 method"*.

3 Hardware of the Measuring Stand

There are necessary following subsystems for realization of presented procedures:

- braking unit which makes possibility of controlling of value of braking torque,
- drive circuit of testing motor;
- rotors movement sensor.

Realization of the above-mentioned procedures required application of torque applying devices and supply units for motor under test, which unable continuous changing of load torque as well as control frequency. Additionally a loading device is expected to have constant and possibly small moment of inertia of rotating parts [2].

In powder brakes the braking torque is produced by the friction between molecules of a ferromagnetic powder, placed in a space separating the rotor and the stator electromagnet. Friction couple is dependent on magnetisation rate of the powder, which is proportional to the current supplying the stator winding.

The significant advantages of such brakes are the following:

- durability (negligibly small wear of the stator and rotor surfaces),
- possibility of matching such operating conditions, when there is practically no dependence between the braking torque and the rotational speed of the rotor,
- simplicity of achieving automated control because of the existing dependence of the developed torque on the excitation current.

There are used for detection of movement of the rotor – depending on configuration of load system and torque measuring device – small tachometric generators, internal encoders of the *OPM* type rotary torque meters [3] or additional encoders (e.g. integrated with motor).

Control of the stand as well as measurement data-acquisition is realized via standard PC compatible DAC. It is equipped with analogue as well as digital outputs and inputs – what was essential in realization of measuring procedures – making possible generating signals of assumed frequencies'. The test stand is presented on fig. 2 and 3.

While testing *PM* type stepper motor we noticed danger of causing uncontrolled oscillations or even resonance of all parts used in testing station. Bellow coupling used in test equipment brings about this effect. Due to specially selected pair of bellow couplings in the discussed case parasitically oscillations were removed up to the value of *2000 Hz* and strongly reduced above this value.

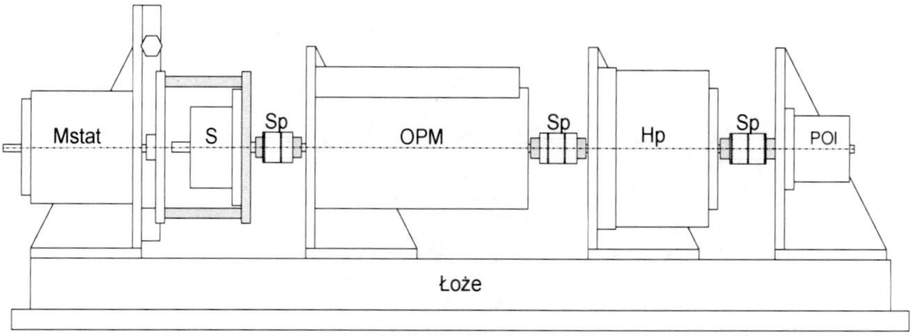

Fig. 2. Test stand in version „measurement of torque on shaft and measurement of stators reaction" with additional angular encoder for detection of shafts movement S – stepping motor under test, Hp – electromagnetic powder brake, OPM – rotary torque meter (measurement of „on shaft" torque), Mstat – stationary piezoelectric torque meter (measurement of the stators reaction torque), Sp – couplings

Fig. 3. Photo of test stand in version „measurement of torque on shaft and measurement of stators reaction" (see Fig. 2)

An exemplary course of the frequency characteristic of the PM type (permanent magnet in rotor) stepping motor registered by described system is presented on Fig. 4. Both pull-in and pull out are shown - obtained with use of piezoelectric torque meter according methods no 1 and 3 from Fig, 1. Resonance areas are similar to presented by manufacture of the motor.

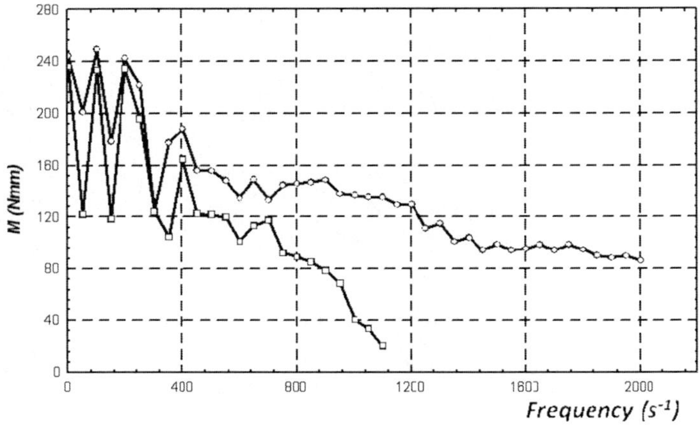

Fig. 4. Presentation of example pull-in and pull-out characteristics

4 Modeling of the Test Stands

The three mass model is proposed for test stand with two kinds of torque meters – see fig. 4. The maximum three-mass system with lumped parameters with constant values not only exactly represent the measuring system with stationary torque meter (measurement of the reaction of the motors stator) and rotary torque meter (with twisted shaft) but also gives the possibility of modelling the procedures of mechanical characteristics determination.

Mechanical system has specified dynamic properties, which according to the circumstances can significantly influence parameters of the exciting torque. In the case of the real stepper motors, the exciting torque T_e introduces a nonlinearity into the equation of motion since even while the most far-reaching simplifying assumptions accepted - what means e.g. taking into account fundamental harmonic only - it is a sinusoidal function of the angular position [1,8]:

- for a permasyn motor with p pairs of poles:

$$T_e = -T_{max} sin(p\delta) \tag{1}$$

- for a reluctance motor with Z_r number of rotor teeth:

$$T_e = -T_{max} sin(Z_r \delta), \tag{2}$$

where: δ denotes a discrepancy angle of position of the rotor axis in relation to the central electric axis of the stator (i.e. in relation to the point of stable equilibrium for unloaded motor).

Model of the test rig, described by (3-5) could be used for modelling and simulation analysis of the electromagnetic torque determination and analysis of the characteristics determination process [6].

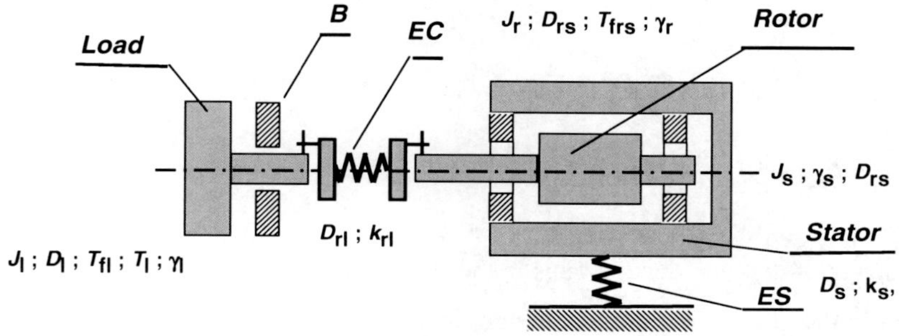

Fig. 3. Model of the mechanical unit containing a motor and two flexible connections – presented as a three-mass system *B* - outer bearing of the rotating parts, *EC* - mechanical transducer (elastic element) of the stationary torque meter, *ES* - elastic element of the stator holder (elastic element of the rotary torque meter)

Mathematical model of the three-mass system consists of differential equations:

- of the rotor:

$$J_r \frac{d^2\gamma_r}{dt^2} + D_r \frac{d\gamma_r}{dt} + T_{fr}\, sgn\left(\frac{d\gamma_r}{dt}\right) + k_{rl}\left(\gamma_r - \gamma_l\right) + D_{rl}\left(\frac{d\gamma_r}{dt} - \frac{d\gamma_l}{dt}\right) +$$
$$+ D_{rs}\left(\frac{d\gamma_r}{dt} - \frac{d\gamma_s}{dt}\right) + T_{frs}\, sgn\left(\frac{d\gamma_r}{dt} - \frac{d\gamma_s}{dt}\right) = T_e \tag{3}$$

- of the stator:

$$J_s \frac{d^2\gamma_s}{dt^2} + D_s \frac{d\gamma_s}{dt} + k_s\gamma_s + T_{fs}\, sgn\left(\frac{d\gamma_s}{dt}\right) + D_{rs}\left(\frac{d\gamma_s}{dt} - \frac{d\gamma_r}{dt}\right) +$$
$$+ T_{frs}\, sgn\left(\frac{d\gamma_s}{dt} - \frac{d\gamma_r}{dt}\right) = -T_e \tag{4}$$

- of the load:

$$J_l \frac{d^2\gamma_l}{dt^2} + D_l \frac{d\gamma_l}{dt} + T_{fl}\, sgn\left(\frac{d\gamma_l}{dt}\right) + k_{rl}\left(\gamma_l - \gamma_r\right) + D_{rl}\left(\frac{d\gamma_l}{dt} - \frac{d\gamma_r}{dt}\right) +$$
$$+ T_f\, sgn\left(\frac{d\gamma_l}{dt}\right) + T_L = 0 \tag{5}$$

under initial conditions of angular positions and velocities specified as follows:

$$\gamma_r(0)=\gamma_{r0};\quad \gamma_s(0)=\gamma_{s0};\quad \gamma_l(0)=\gamma_{l0};\quad \left.\frac{d\gamma_r(t)}{dt}\right|_{t=0}=\omega_{r0};$$

$$\left.\frac{d\gamma_s(t)}{dt}\right|_{t=0}=\omega_{s0};\quad \left.\frac{d\gamma_l(t)}{dt}\right|_{t=0}=\omega_{l0}$$

One used the following designations in the system of equations:

γ - rotation angle of the rotating element in relation to a permanent axis; J - moment of inertia of rotating masses, D - coefficients of viscous friction, k - coefficients of elasticity of the flexible coupling and holder, T_f - torque of dry friction i.e. having the sense contrary in relation to the sense of angular velocity of the load, T_l - torque of an active-type of load; i.e. having the sense independent on the angular velocity, T_e - exciting torque in case of electrical motor - electromagnetic torque. For the presented model of the stand the individual subscripts mean suitable: (s) - stator, (r) - rotor, (l) - load.

The presented system of equations is non-linear. This means that there is no general, analytic solution of such system of equations. Nevertheless, it is possible to obtain an approximate solution while applying numerical methods, provided there are known values of the parameters, initial conditions, and the form of the input function.

It is possible to derive from presented set of equations simpler forms of mathematical description, also having the practical applications. The case of ideal stiff fixing of the stator gives the two-mass model (characterised by two degree of freedom, because of $\gamma_s = 0$). However for ideal stiff both fixing of the stator and connection between rotor and load one received one-mass system with one degree of freedom ($\gamma_s = 0$ and $\gamma_r = \gamma_l$).

At least the two-mass model is enough for analysis of the stepping motor, while discrete changes of the excitation, especially with use of the elastic couplings.

Identification of presented model was presented in authors papers, e.g. [9].

As an example of the simulation tests the pull-in characteristics determination process is presented. The procedure of the simulation determination of the pull-in characteristic one works out on the basis of its definition. Co-ordinates of the point of the characteristics are defined as maximal frequency of the controlling pulses that could be apply in jump from zero to request value with the particular friction load and when the motor starts without losing the synchronism. The method of the load of the motor (kind of external torque) is substantial during determination of this characteristic in simulation experiments. Only the active load torque (T_L in eq. (5)) gives the possibility of the univocal definition of the initial conditions of the angular position of the rotor before start of simulation. The experiments shown that simulation results strongly depend on these conditions.

The comparison between characteristics obtained with change of models parameters is shown on fig. 5.

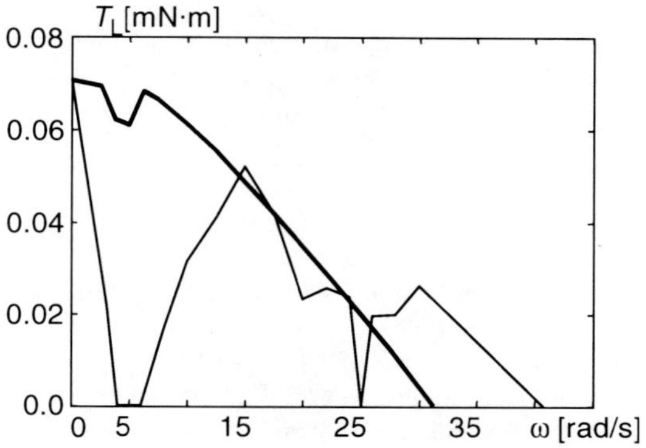

Fig. 5. Simulation of pull-in characteristics - description in text.

Thin line indicated investigations without viscous damping and thick line experiments with viscous damping in the model from eq. 3-5. The obtained characteristics confirm known perceptions about their shapes especially of occurrence of the resonance phenomena.

Computer simulation methods give approximate numerical solution of the set of differential equations that are the mathematical model of the drive with stepping motor or model of the measuring station.

5 Summary and Conclusion

The results of the tests show that the parameters of the mechanical structure of the test stand have the determined influence on the measurement of the motor torque. Model of the electromechanical structure is a basis for a complex modeling of the test station - where the following are also taken into consideration: dynamics of the system for applying load (an electromagnetic brake with a power supply unit) and processing of the signals in torque meters. This complex model makes it possible to optimize the measuring process during determination the frequency characteristics of the stepping motors.

The presented model makes it possible to foresee resonance phenomena in the test station and to limit them, e.g. by selecting appropriate design features of the torque meters. It allows us also to compare the electromagnetic torque of the motor with torque signals in the mechanical transducers of the torque meters. While determining the electromagnetic torque of the motor with application of the method based on measuring the stator reactions (when the force transducer used in a piezoelectric stationary dynamometric torque meter is characterized by linearity and high rigidity, and diameter of the stator is small) the torque of reaction will be proportional to the product of coefficient of elasticity of the sensor (being equal to the coefficient of elasticity of the stator fixing) and angular movement of the stator.

Designing and building a test rig, which makes it possible to carry out comparative studies of various algorithms of determining characteristics for various methods of measuring the torque, is considered by the authors an original contribution to the development of research techniques.

References

[1] Accarnley, P.P.: Stepping Motors: a Guide to Modern Theory and Practice. IEE and Peter Peregrinus Ltd., London (1982)
[2] Bodnicki, M.: Engineering Mechanics 8(6), 373–386 (2001)
[3] Bodnicki, M.: Journal of Theoretical and Applied Mechanics 38(3) (2000)
[4] Jaszczuk, W., et al.: Mikrosilniki elektryczne. Wyznaczanie właściwości statycznych i dynamicznych (English title: Electrical Micromotors. Determination of Static and Dynamic Properties). PWN, War-szawa, PWN (1991)
[5] Kenyo, T., Sugawara, A.: Stepping Motors and their Microprocessor Control. Clarendon Press, Oxford (1994)
[6] Pochanke, A., Bodnicki, M.: Electromotion 9(4), 210–216 (2002)
[7] Pochanke, A.: Indirectly coupled circuit-field models of brushless motors with the conditioning of feeding. In: OWPW, Warszawa (1999) (in Polish)
[8] Pochanke, A.: Przegląd Elektrotechniczny (9), 42–45 (2005)
[9] Bodnicki, M., Pochanke, A.: Proceedings of the 4th Polish-German Mechatronic Workshop 2003, Suhl (Germany), pp. 123–128 (2003)

Development of Hybrid Power Unit for a Small UGV's (Lessons Learned)

Z. Bureš, J. Mazal, and T. Túro

University of Defence, Kounicova 65,
Brno, 66210, Czech Republic

Abstract. The article is focused on problematic of hybrid power unit development and its system integration as a main power source for a small UGV (able to propel vehicle approximately up to 400kg). Based on previous experiences with combustion power sources for robotic and remotely controlled platforms were identified fundamental factors of propeller requirements, which converge to an electrical motor application. Advantage of a control ability, effectiveness, reliability and durability is hardly reachable by other type of aggregates. Hybrid system what was developed for this purpose consisted of water cooled 2-stroke combustion engine, electric generator, battery field, DC/DC converter, starting and controlling electronics. Depend on the application it gives better result than original propeller and now it is in stadium of functional demonstrator, where some areas are open for following optimization.

1 Introduction

Today's maximum attention is paid to effective propeller system which comply with future emission limits (amount of CO_2 produced per kilometer driven) and weight reduction in development of a new military technology (including UAV and UGV). The application of hybrid engine to UGVs and other systems has another positive effect, more important than CO_2 emission. The main reason is usage of electrical propeller which could be much more smoothly controlled and enables quiet mode for special (reconnaissance) application. Because of limited amount of energy stored in onboard batteries, and for real experiments, there must be applied combustion generator to extend the original (battery mode) period of operation. Using that combination, it is possible to exploit all benefits of electrical servo propeller with sufficient time period of operation also with less consumption comparing to full combustion models.

2 Effective Operation and Lower Emission

Emission reduction can be achieved by reduction of the vehicle weight by design or modification of the combustion engine itself, and electricity consumption reduction at times when the engine runs outside its optimum revolution range. Contrary to this trend, however, there are the ever-growing requirements of the on-board electricity network. Especially usage of servo engines significantly increases the electricity consumption. Modern UGVs are equipped with high-efficiency alternators and could use two electrical networks with different voltages (12V / 24V / 48V). The transition to a higher voltage brings a number of benefits arising from reduced weights of power looms and components (starter, alternator, pumps). Unfortunately, these modifications do not eliminate the high electricity drain from the alternator during moving off after the engine starts and during vehicle acceleration or during steady driving. Electricity is currently produced only by transforming the fuel's chemical energy into mechanical force by means of combusting in the engine, and this is further transformed into electrical energy by means of an alternator. The power taken from the engine by the alternator manifests itself in the engine's increased nominal consumption and in the amount of CO_2 produced, which is directly related to it. This can be partly solved by combining a combustion engine and an electric motor, i.e. by "hybrid propulsion". Even vehicles with hybrid propulsion, struggle with two limiting factors:

- the battery – a structural element with a slow charging cycle, great weight and great volume and
- the weight of the whole system (up to a 200 kg).

3 Propulsion System Used in Designed UGV

It was necessary to solve energy shortfall during the development. Original combustion engine doesn't provide required operational modes, so it was decided to modify total power propulsion. Main combustion engine was replaced by BLDC (Brushless Direct Current Motor) electro motor with two-level gearbox extended by additional auxiliary gearbox. The Handle-bar was replaced by servo motor with embedded planetary gearbox with external auxiliary gearbox. These electrical appliances require huge energy storages with high current ability. There are two methods how to solve this problem. Firstly it is possible to use big and heavy traction battery or, on the other hand, it is beneficial to use smaller battery together with generator connected with two-stroke combustion engine. We decided to use hybrid propulsion after weight analysis like is displayed on Fig. 1.

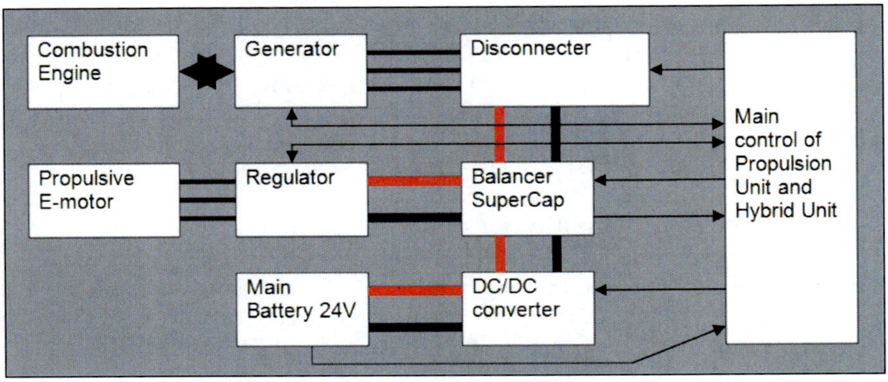

Fig. 1. Total scheme of propulsion system

Next, the two-stroke engine is directly connected with generator through the belt gearbox. Three phase generator based on BLDC motor is connected with the the disconnector through super-capacitor balancer and DC/DC converter which supplies the main battery. Propulsion is made by second three phase BLCD electro motor controlled by regulator.

The main control unit is divided into two separate parts:

- Hybrid unit
- Propulsion unit.

4 Hybrid Unit

The main part of hybrid unit is compact two stroke water cooled 86 ccm engine with embedded DC starter and generator (Fig. 4.). Combustion engine is controlled by separate unit HCU1 (Hybrid Control Unit 1). It provides high voltage for spark plug, high current source for starter and low power switch. HCU1 is controlled by superior controlling unit HCU2 (Hybrid Control Unit 2) and is connected with main evaluation microprocessor through RS232 and via signaling wires with HCU1. It enables a start or stop combustion engine, it sets its rpm by throttle valve position and finally it makes all necessary diagnostic. Basic diagnostic provides all necessary information about hybrid unit as itself (battery voltage, engine rpm, cooling system, engine operation hour). General concept of the hybrid unit is demonstrated on the next Fig. 2.

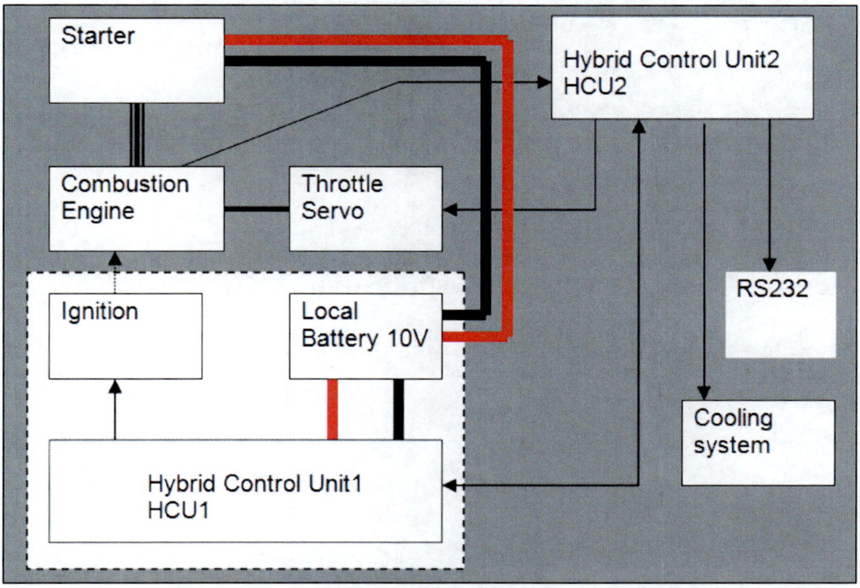

Fig. 2. Hybrid system

5 Propulsion Unit

As mentioned before, it was necessary to use an electric instead of combustion motor. It required very low speed, minimal thermal sign, low noise (or silent) and finally there were needed a lot of power for all electrical appliances. High torque BLDC electro engine was used to achieve these requirements. Three phase electro engine is driven through special bidirectional regulator. It is connected with ECU1 (Engine Control Unit). ECU1 receives command from superior controlling system, it controls regulator and it evaluates engine rpm and rear wheel rpm. Main system can evaluate slip of tires and type of surface. General scheme of the propulsion unit is illustrated on the Fig. 3.

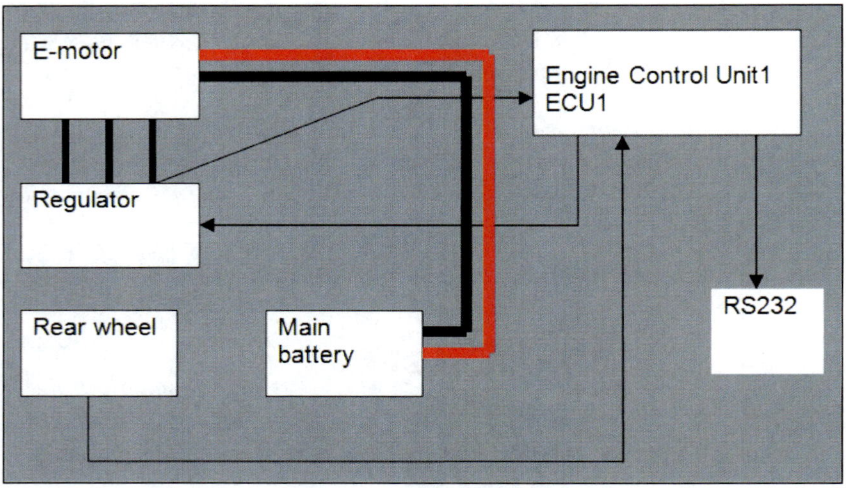

Fig. 3. Basic chart of propulsion unit

6 Power Characteristics and Application

Following Fig. 4., demonstrates the overall design of propeller, generator and starter.

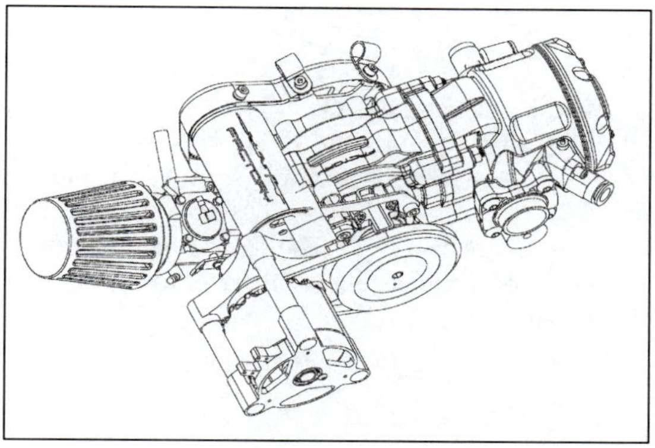

Fig. 4. Combustion engine with generator and starter

Relating to the real tests, the next Fig. 5., describes the Voltage–power characteristics which were achieved during the experiments.

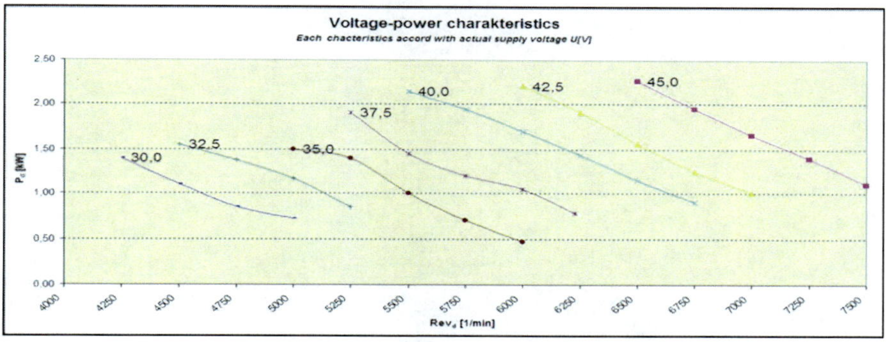

Fig. 5. Basic power characteristics of the propulsion unit

Installation of hybrid engine took less space than previous propeller and enabled application of other devices and sensors. The installation is displayed of following Fig. 6.

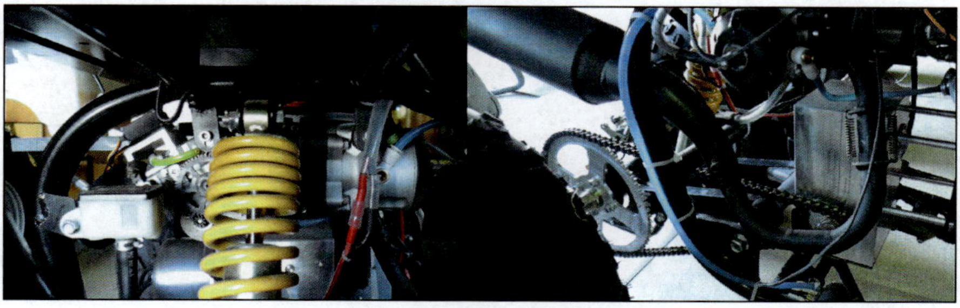

Fig. 6. Basic chart of propulsion unit

7 Conclusion

Concerning the latest information and experience, the hybrid engine is probably one of the best solutions for propelling the UGVs in various types of conditions by that time. Technical approach and optimized design was a relatively complex task with lot of harnesses but the final hybrid unit is able to meet industry (reliability and effectiveness) requirements [1],[2]. The future extended experiments in real conditions show if this approach succeeds, but, according to other advanced solutions it seems to be a right step.

References

[1] James, L., John, L.: Electric Vehicle Technology Explained. John Wiley & Sons, Chichester (2003) ISBN: 0470851635
[2] Hughes, A.: Electric Motors and Drives: Fundamentals, Types and Applications, Newnes (2005) ISBN: 0750647183

Design of an Autonomous Mobile Wheel Chair for Disabled Using Electrooculogram (EOG) Signals

Mohammad Rokonuzzaman[2,*], S.M. Ferdous[1], Rashedul Amin Tuhin[1], Sabbir Ibn Arman[1], Tasnim Manzar[1], and Md. Nayeemul Hasan[1]

[1] Islamic University of Technology,
Board Bazar, Gazipur - 1704, Bangladesh
[2] The University of Asia Pacific
Dhanmondi, Dhaka- 1209, Bangladesh

Abstract. This paper discusses the implementation of a simple, effective and low cost design of a microcontroller based wheelchair using the EOG signal collected from muscles those are responsible for the movement of the human eye. This was an exploratory research that was carried out to allow a disabled person to control the wheelchair by using only the movement of his eyes. Electrooculography (EOG) is a technique for sensing and recording the activation signal of the muscles and was used to collect and evaluate the myoelectric signal generated by the eye muscles during different movements. The main purpose of the work is to design a cost-effective, easily affordable and accessible wheel chair for the disabled general masses where advanced attachments like on board computer, digital cameras, sophisticated sensors etc. are not being used, rather concentration has been paid on designing a more simple, practical but effective system using an electronically controlled differential drive structure with only two wheels. A low cost microcontroller (ATMEGA 32) serves as the brain of the system for all types of control purposes.

1 Introduction

Over the years, the neurophysiology and biomechanics of muscle systems have been investigated quite extensively based on the research of surface EMG signal. As the measurement of given muscle activity, surface Electromyography (EMG) signals represent the electrical activity of a muscle during contraction [1]. More elaborate research had been attempted by the researchers around the world to

[*] Corresponding author.

abstract the features of EMG in motion and to classify the movement patterns. [2] [3] [4]

The surface EMG signals are complex, non stationary time sequence that can be considered as direct reflection of the muscle activity [5]. In this work EMG signals collected from the muscles responsible for the movement of human eyes are used as the controlling signal for the wheelchair movement.

In the past decade, a number of simple yet effective hands-free human machine interfaces (HMI) are brought into applications such as sip & puff switch, head infrared pointer, head touch switch buttons and these interfaces are widely used for their simplicity to control and easy implementation. At the same time, there have been some studies on developing HMIs based on human physiological signals such as electromyography (EMG), electrooculography (EOG) and electroencephalography (EEG). As can be seen in literature [1-5], HMIs developed from these signals are used for hands-free control of electrically powered Wheelchairs.

With the recent advancement of machine and computing ability, many HMIs have been developed from computer vision information [5], speech information and multi-modality phenomenon. Ju and Shin et al. introduce a vision-based intelligent wheelchair system developed for people with quadriplegic [6]. Li and Tan [5] propose a bimodal wheelchair control approach by integrating vision and speech controls. Matsumoto and Ino et al. [6] apply the recognition of head motion and eye gaze onto a locomotive wheelchair system. Ferreira and Silva et al. proposed an HMI structure to control a robotic wheelchair by scalp EMG and EEG signals. Both eye blinking movements and eye close movements are used as command movement to control a mobile wheelchair through an onboard PDA.

This paper presents a solution for those kinds of disabled people who are unable to spend a lot of money to buy a fancy or sophisticated wheelchair which does not need any on board computer or any other expensive instruments. A simple structure and user friendly control system is used to control the wheelchair navigation using only the movement of eyes. Generally this type of design would suit most to the people who are totally disabled, that means completely unable to move their hand, leg or head.

First part of the paper shows the extraction of EOG signal by elaborately analyzing the anatomy of the eye muscles and processing of those signals to make it compatible to use in conjunction with a microcontroller. Second part shows the structure and control mechanism of the wheelchair.

2 Anatomy behind the Movement of Human Eye

Movement of human eye is governed by the extra ocular muscles. A unique and exclusive combination of muscle(s) is responsible for a particular movement of the eye ball. For this reason EOG signals have been chosen as the control parameter of the system. Figure.1 shows the six different extra ocular muscles of the human eye.

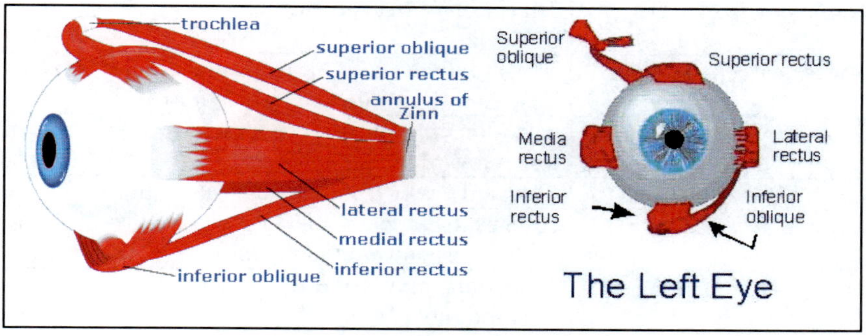

Fig. 1. Six extra ocular muscle responsible for the movement of eye

As seen from Figure.1, there are three pairs of muscles which control the movement of human eye. From Figure.2, it can be found that which muscle(s) are responsible for a particular direction of gaze.

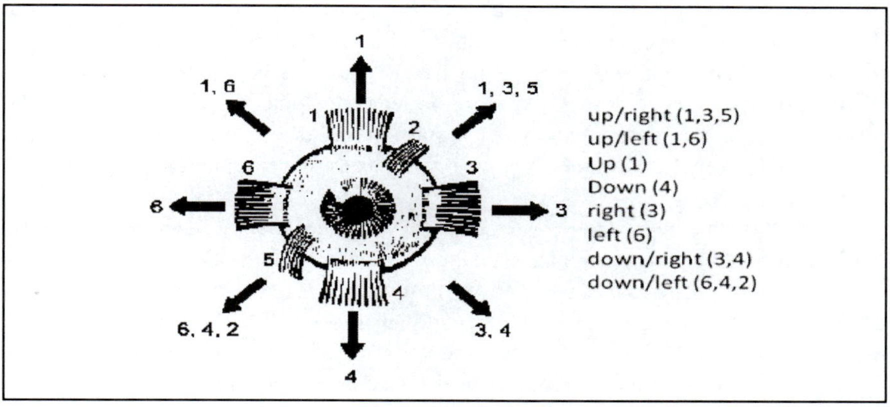

Fig. 2. Pictorial representation of the muscle combination for a particular direction of gaze

Referring to Figure.2, the muscle movements are enlisted in Table 1

Table 1. List of extra ocular muscles with associated eye movements

Name of the Muscle	**Direction of Movement**
Superior Rectus (1)	Upward (elevation), rotates the eye towards the nose (intorsion), inward (adduction)
Superior Oblique (2)	Intorsion, moves the eye downward (depression) and outward (abduction)
Lateral Rectus (3)	Outward, away from the nose (abduction)
Inferior Rectus (4)	Rotates the top of the eye away from the nose (extorsion), inward (adduction)
Inferior Oblique (5)	Extorsion, moves the eye upward (elevation) and outward (abduction)
Medial Rectus (6)	Inward, Toward the nose (adduction)

3 Sensing and Acquisition of EOG Signals

EOG is a method for sensing eye movement and is based on recording the standing corneal-retinal potential arising from hyperpolarizations and depolarizations existing between the cornea and the retina; which is commonly known as Electrooculogram[7]. This potential can be considered as a steady electrical dipole with a negative pole at the fundus and a positive pole at the cornea [refer to Fig.3 (a)]. The standing potential in the eye can thus be estimated by measuring the voltage induced across a system of electrodes placed around the eyes. The placement of electrodes are shown in figure.3 (b). As the eye moves to a direction an EOG signal is obtained which is a measurement of the electric signal of the ocular dipole.

The EOG values vary from 50 to 3500μV with a frequency range of about DC-100Hz. Its behaviour is particularly linear for gaze angle of ±30° [7].

For acquisition of EOG signal from the eye to navigate the wheel chair, four electrodes A, B, C and D are placed as shown in Fig. 3(b). For the navigation of the wheel chair according to the developed control algorithm, a total three types of eye movement is required which are – right gaze, left gaze and blinking of the eye. To clearly distinguish among the EOG signals both horizontal and vertical movement of the eyes should be detected. According to the developed algorithm the detected signals can be divided into two parts-

1. Eye movement detection (horizontal movement of eye)
2. Eye Blinking Detection (vertical and inward movement of eye)

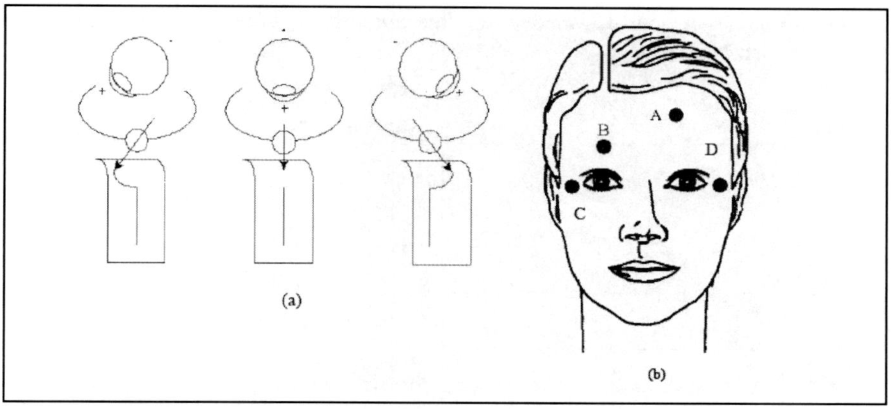

Fig. 3. (a) Ocular dipole due to movement of eye. (b) Placement of Electrodes for EOG signal acquisition [7]

A. Eye Movement Detection

To detect the horizontal movement (right/left gaze) two electrodes (C-D) are placed to the right and left of the outer canthi. A reference electrode (A) is placed on the forehead. When the subject looks at right, the voltage V_{CA} will be high. Similarly for gazing at left the voltage V_{DA} will go high. Depending on the direction of gaze i.e. right or left, the wheel chair will start to rotate (clockwise or anti-clockwise) on that particular direction. According to the direction of gaze, the wheel chair will navigate to move the person in the direction he wishes to go.

B. Eye Blinking Detection

Blinking of the eye is accomplished by the vertical and inward movement of the eyeball. For the purpose of driving the wheel chair, detection of the vertical movement is sufficient. To detect the blinking of eye (i.e. vertical movement) electrode (B) is placed on top of the right eye. When the subject blink his eyes the voltage, V_{BA} will go high. This signal will be used to turn on/off the power supply to the driving motors i.e. this signal will act as a master on/off control of the system. To activate the wheel chair the subject has to close his eyes for a certain period determined by the control algorithm. For deactivation, the same process has to be followed again.

The wheel chair will move normally move forward in straight direction with desired speed for normal gazing of the subject. Prior to that, he must close his eyes for a certain period to activate the system.

The main source of noise in EOG signal is a signal of electrical activity of face's muscles. A movement of a head or speaking can also disturb EOG signal. The rapid change of amplitude of EOG signal (shown in the Fig. 2) is caused by the saccadic movement of eye's ball. Three spikes which appeared in a vertical plane are eye's blinks. An example of single blink is presented in the Fig. 4. Blinks of eye are very well seen in a signal which is recorded in a vertical plane. A high change in the potential distribution around the eye is a result of movements of muscles of upper and lower eyelid. This potential is greater in a vertical plane than in a horizontal plane.

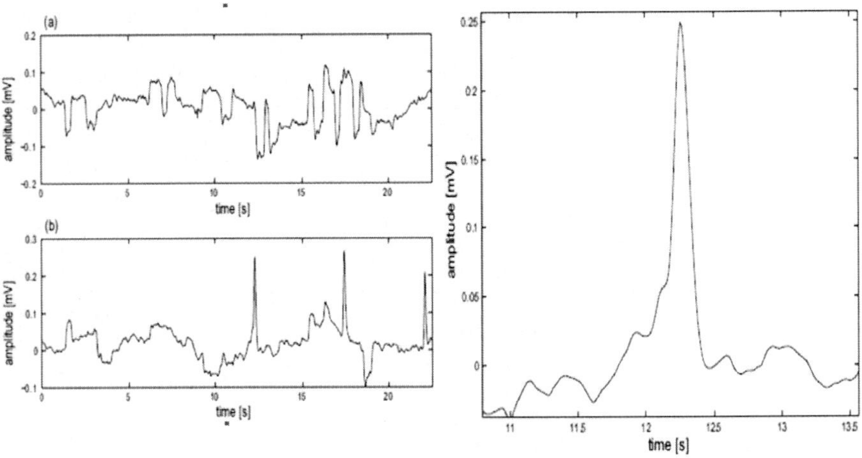

(a) (b)

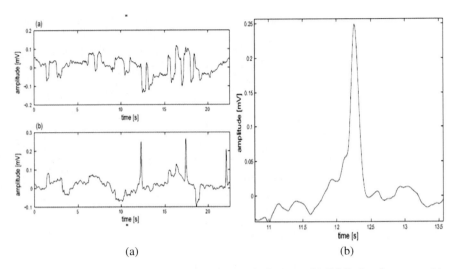

Fig. 4. (a) The EOG signal in horizontal and vertical plane. (b) EOG signal generated by blinking of eye

C. Signal Acquisition

A signal acquisition model is developed to acquire the EOG signal due to the movement of eye. Acquired signals are processed, filtered and amplified to feed to a microcontroller which in turn will produce the necessary driving pulses of the motors to navigate the wheel chair. The functional blocks of the acquisition module and steps are shown in Figure.5. With exactly same process three identical acquisition modules are used to collect signals from the eye muscles. The EOG signal changes approximately 20μV for each degree of eye movement. The EOG signal is the result of a number of factors such as- eyeball rotation and movement, eyelid movement, head movement, influence of illumination etc.

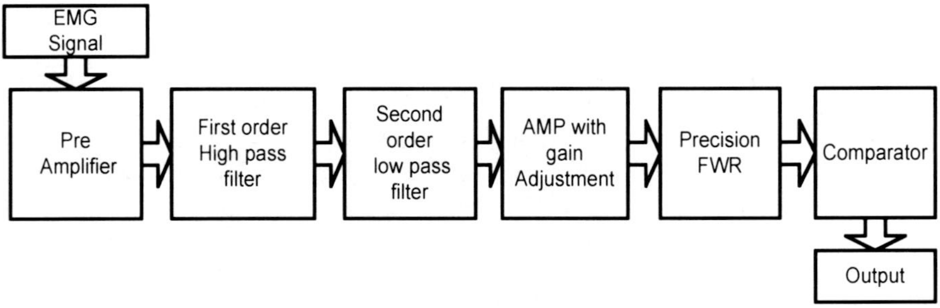

Fig. 5. Steps and components of EOG signal acquisition module

It is necessary to eliminate the shifting resting potential (mean value). To avoid this problem a high gain ac differential amplifier (1000-5000) is used along with a high pass filter with cut-off frequency of 0.05 Hz and a low pass filter with cut-off frequency at 35Hz.

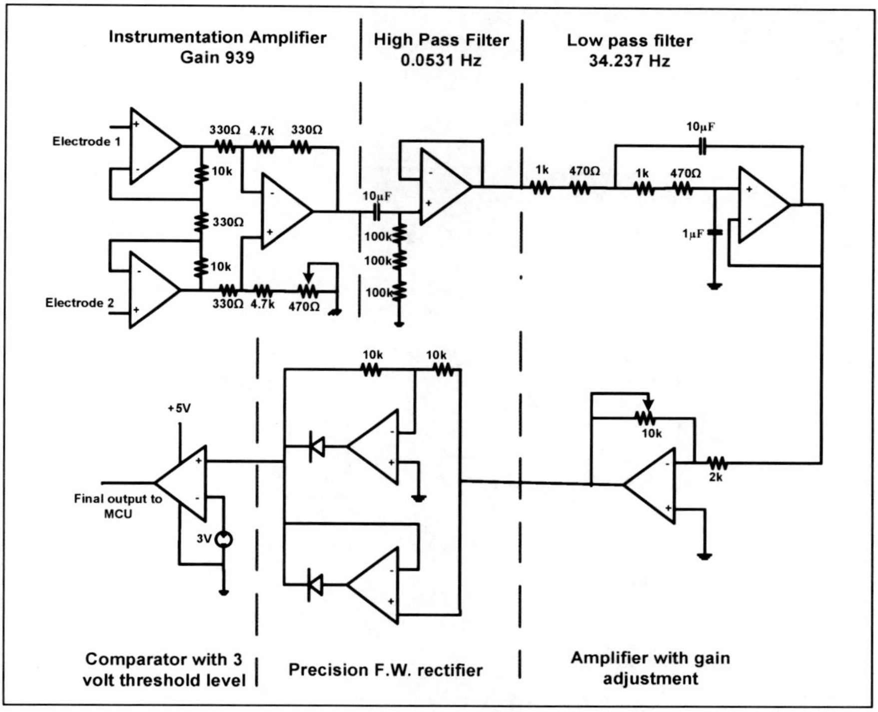

Fig. 6. Complete circuit diagram of EOG signal acquisition model.

An adjustable gain amplifier with a Full wave precision rectifier is used to produce a unidirectional signal which is then compared with a threshold voltage of 3Volts. The input port of the microcontroller will assume LH when the EOG signal crosses this threshold value. The complete circuit diagram is shown in Figure.6.

4 System Design and Development

The main part of the design is to control the motion of the wheelchair. In our system we only considered only three movements- forward, left and right movement. If any user wants to go the opposite direction he is facing than he would have to turn 180 degree by using left or right movement then go forward. The wheelchair employs a differential drive system where the motion controller transmits objective linear and angular velocities, which are in turn mapped to wheel velocities. The differential drives used have two independent drive wheels on the left and right sides, enabling the chair to move back and force with or without rotation and to turn in place. Casters on the front or back or both ends keep the chair level. The layout of the chair showing all the components are shown in Fig.7. Two 12V, 5A (60W) PMDC motor is used to drive the wheelchair where the motors are connected with a Bevel gear arrangement which are finally connected with the axis of the wheel.

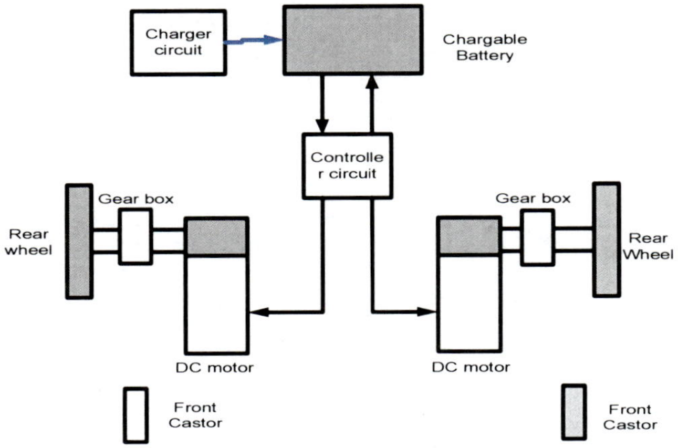

Fig. 7. Layout design of the wheelchair.

5 Wheelchair Control and Navigation

Control for the two motors in the system is carried out by using an H- bridge motor controller. The driving signals are generated by the microcontroller which produces appropriate PWM signals for appropriate movement of the chair. Depending on the direction of the gaze the associated electrode output will go high. The three different scenarios are shown below-

V_{RE} = High when looking at right

V_{LE} = High when looking at left

V_{UE} =High when blinking is detected

When V_U high for 3 seconds the system will be activated and motor will start rotating and as a result the wheelchair will start to move forward. Again if a blinking is detected for 3 seconds then the system will stop. Depending on the value of V_{UE} an output port of the microcontroller will switch its output state from LH to LL or vice versa. A transistor is connected in series with the H-bridge circuit which is being driven by this port. Now if V_R is high then the wheelchair will turn right and if V_L is high than system will turn left.

A. Motor Control Algorithm

The complete movement of the motor depending on the movement of the eye is summarized in Table.2. A simple but effective algorithm is developed and implemented using the microcontroller. Depending on the control algorithm the duty cycle of a particular motor is varied to obtain the desired response from the system.

Table 2. Control Algorithm for the movement of the motor

Right electrode output V_{RE}	Left electrode output V_{LE}	**Direction of gaze**	**Wheel chair response**	Assigned error value e_1	Duty cycle (Right motor) DC_R	Duty cycle (left motor) DC_L
0	0	Straight	Forward	0	100%	100%
0	1	Left	Rotate counter clockwise	-2	60%	0%
1	0	right	Rotate clockwise	+2	0%	60%
1	1	Invalid	No response	0	0%	0%

Based on the algorithm, next the controller is developed using the microcontroller. The control structure of the wheel chair is modeled in Figure.8. The control signals will be continuously compared with a reference value which in turn will determine the appropriate action to be taken. Based on these values set point will be fixed to drive the motor at a predetermined speed. The accuracy of the system is ensured by employing feedback signals. To obtain a smooth and fast response a PID controller is applied in the system. The error signal, e_l is generated depending on the direction of gaze and is assigned with a value enlisted in Table.2. Next the value is compared with a pre-assigned reference value which in turn will generate the necessary duty-cycle for the PWM signals. Duty-cycles are generated by following the equations (1a) to (3b). The PWM signal will operate a motor controller IC which will furnish sufficient amount of current from the power supply to run the motor.

$$V_{ref} = 5 \qquad (1a)$$

$$V_R = V_{ref} + e_l \qquad (1b)$$

$$V_L = V_{ref} - e_l \qquad (1c)$$

For Right Motor (to rotate CW/ right)

$$\text{Duty Cycle, } DC_R = 20V_R \;; \quad \text{for } 0 \le V_R < 5 \tag{2a}$$

$$\text{Duty Cycle, } DC_R = 0 \;; \quad \text{for } V_R \ge 5 \tag{2b}$$

For left Motor (to rotate CCW/ left)

$$\text{Duty Cycle, } DC_L = 20V_L \;; \quad \text{for } 0 \le V_L < 5 \tag{3a}$$

$$\text{Duty Cycle, } DC_L = 0 \;; \quad \text{for } V_L \ge 5 \tag{3b}$$

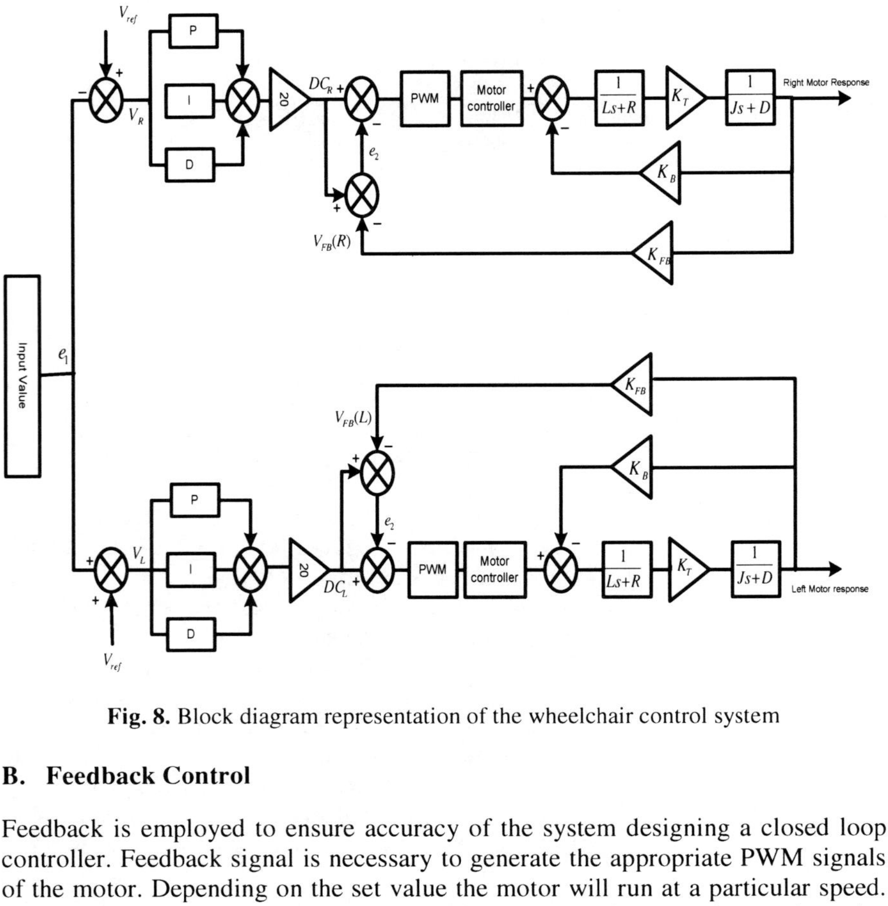

Fig. 8. Block diagram representation of the wheelchair control system

B. Feedback Control

Feedback is employed to ensure accuracy of the system designing a closed loop controller. Feedback signal is necessary to generate the appropriate PWM signals of the motor. Depending on the set value the motor will run at a particular speed. The speed of the motor is sensed by an Infra red transmitter- receiver and a pulse generating circuit. These pulses are sent to a frequency to voltage converter which is calibrated to produce 5 volts when the input signal has a frequency of 25 Hz (i.e. 25 pulses per second). The output voltage of an F/V converter is fed to micro-controller where it is sampled and stored using Analog to Digital converter. The

Feedback signal, V_{FB} is then generated using the following equation implemented by the MCU.

$$V_{FB} = (25/128) \times (\text{ADC value}) \quad (4)$$

The total process of feedback signal acquisition is enlisted in Table.3.

Table 3. Feedback signal acquisition and generation of V_{FB}.

PWM signal (reference of set value)	RPM of the motor	Voltage generated by the F/V converter.	ADC value (generated by the MCU)	Respective value assigned according to ADC	Feedback voltage V_{FB}
0%	0	0	0	0	0
20%	300	1	102	1	20
40%	600	2	205	2	40
60%	900	3	308	3	60
80%	1200	4	410	4	80
100%	1500	5	512	5	100

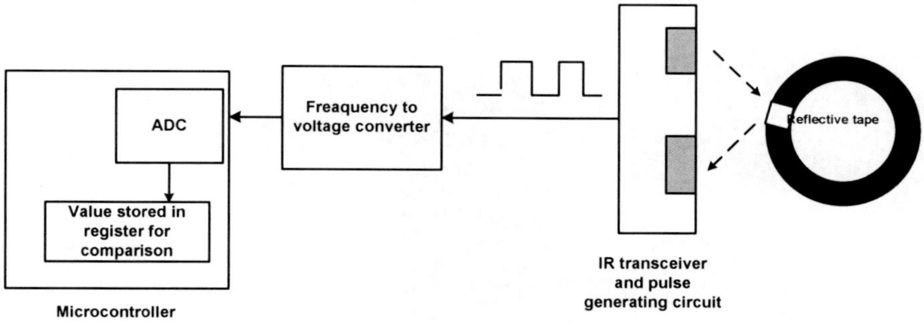

Fig. 9. Block diagram representation of feedback signal acquisition

6 Simulation and Result

Before implementation, the whole project is simulated on computer to check the functionality of the design. The software used for simulation is Proteus 7.4 sp3. The Signal acquisition part is simulated on ISIS.

The demo EMG signal we have used in our simulation was a wave audio file. It was a collection of EMG data collected from a human face to analyze facial expression.

The specification of the demo signal is as follows:

Filename: left.emg.wav and right.emg.wav
Channel: Dual channel Stereo
Sampling rate: 2000 Hz
Duration: 10 minutes

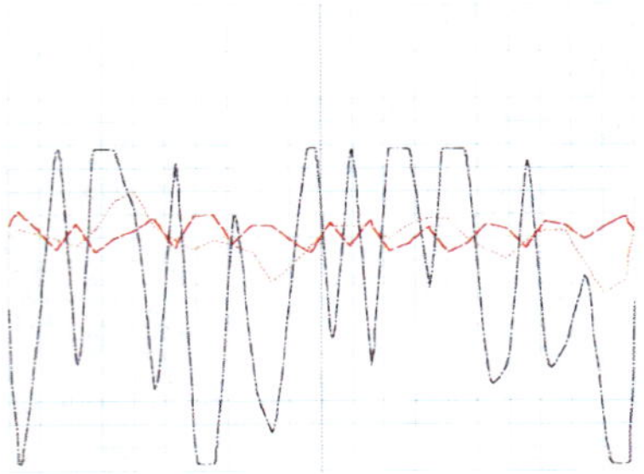

Fig. 10. Amplified EOG signal before filtering

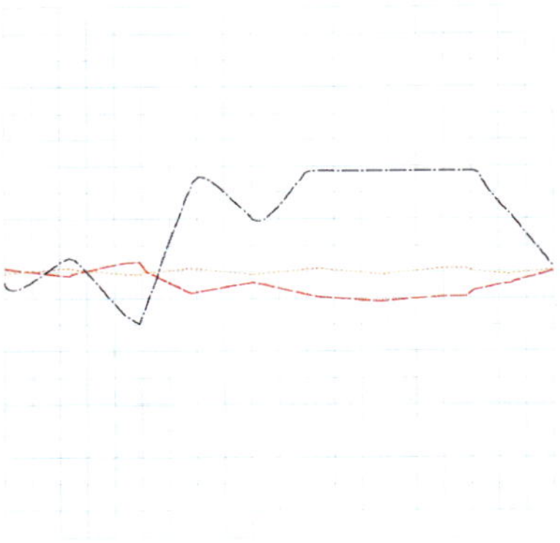

Fig. 11. EOG signal after filtering

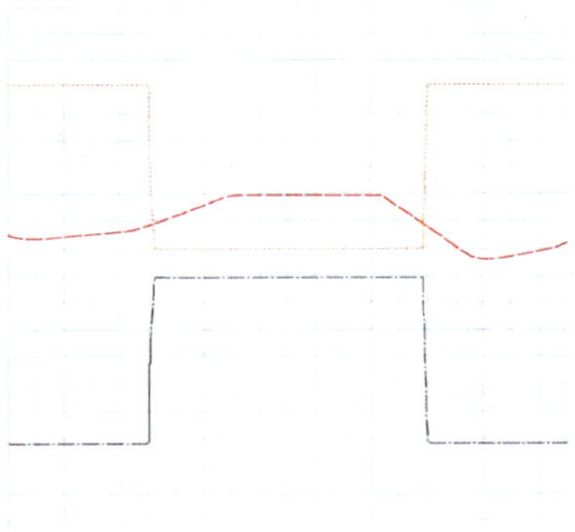

Fig. 12. EOG signal after amplification and rectification

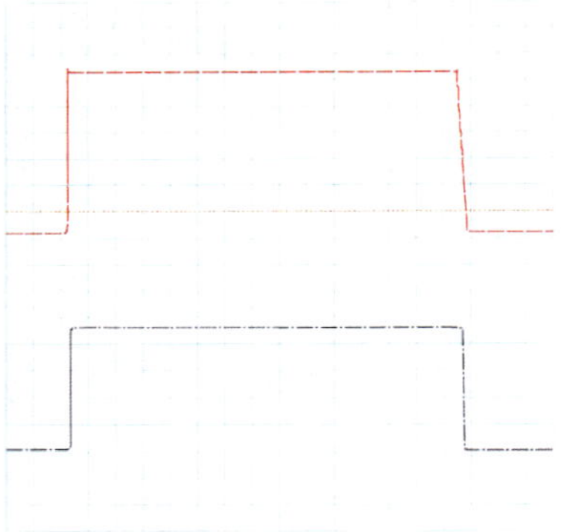

Fig. 13. Final output to be fed to the micro-controller obtained at the Output of the comparator

7 Conclusion and Future Work

The design of the wheel chair along with its simple but effective algorithm suggests that, it can be a very good method to obtain system assisted mobility for the disabled. The accuracy and performance of the system depends greatly on the signal acquisition. So, the signal acquisition and processing module should be designed with great care. The simplicity of the system makes it a perfect candidate for practical implementation. In spite of its effectiveness still there is a lot of room for improvements. Intelligent control system like neuro-fuzzy controller, adaptive control may be introduced in the system to obtain a better performance from the system and make the system more versatile, dynamic and fast.

References

[1] Lowery, M.M., Stoykov, N.S., Taflove, A., Kuiken, T.A.: A Multiple-Layer Finite-Element Model of the Surface EMG Signal. IEEE Trans. on Biomed. Eng. 49(5), 446–454 (2002)

[2] Alemu, M., Kumar, D.K., Bredley, A.: Time- Frequency Analysis of SEMG- with special Consideration to the Interelectrode spacing. IEEE Trans. on Neural System and Rehabilitations Eng. 11(4), 341–345 (2003)

[3] Bonato, P., Member IEEE, Boissy, P., Croce, U.D., Roy, S.H.: Changes in the EMG Signal and the Biomechanics of motion During a repetive Lifting Task. IEEE Transaction on Neural System and Rehabilitation Eng. 10(1), 38–47 (2002)

[4] Bogey, R.A., Perry, J., Gitter, A.: An EMG-to-Force Processing Approach for determining Ankle Muscle Forces During Normal Human Gait. IEEE Trans. on Neural System and Rehabilitations Eng 13(3), 302–310 (2005)

[5] Gao, Z., Lei, J., Song, Q., Yu, Y., Ge, Y.: Research on the Surface EMG Signal for Human Body Motion Recognizing Based on Arm Wrestling Robot. In: Proceedings of the 2006 IEEE International Conference on Information Acquisition, Weihai, Shandong, China, August 20-23 (2006)

[6] Moon, I., Lee, M., Chu, J., Mun, M.: Wearable EMG-based HCI for electric-powered wheelchair users with motor disabilities. In: Proceedings of IEEE International Conference on Robotics and Automation, pp. 2649–2654 (2005)

[7] Barea, R., Boquete, L., Mazo, M., Lopez, E.: System for assisted mobility using eye movements based on electrooculography. IEEE Transactions on Neural Systems and Rehabilitation Engineering 10, 209–218 (2002)

Dependence of Functional Characteristics of Miniature Two Axis Fluxgate Sensors Made in PCB Technology on Chemical Composition of Amorphous Core

Piotr Frydrych[1], Roman Szewczyk[1,2], Jacek Salach[1], and Krzysztof Trzcinka[2]

[1] Institute of Metrology and Biomedical Engineering, Warsaw University of Technology, sw. A. Boboli 8, Warsaw, 02-525, Poland
[2] Industrial Research Institute of Automation and Measurements PIAP, Al. Jerozolimskie 202, Warsaw, 02-486, Poland

Abstract. Fluxgate sensors are common used in week and low frequency magnetic field measurements. Recently rapid development of miniature fluxgate sensors made in PCB technology is observed. Until now single core layer sensors were unable to measure magnetic field for two directions. Moreover no modelling results of influence of amorphous alloy chemical composition on sensor properties were available.

This paper presents results of tests of functional characteristics of miniature single tape ply fluxgate sensors with two kinds of chemical compositions of amorphous alloys cores: $Fe_{67}Co_{18}Si_1B_{14}$ and $Co_{66}Fe_4Ni_1Si_{15}B_{14}$.

Sensors were made with seven layer printed board. Cores, which are produced using photolithography technology from single tape ply, has frame shape. Thus they are able to measure independent two perpendicular direction of magnetic field. Excitation and pick up coils are printed on board.

Experimental results shown strong dependence of sensor characteristics on chemical composition of core material. It proves that further investigation of material influence on sensor properties and optimisation of cores are strong desired.

1 Introduction

Low level magnetic fields measurements are used in geology, geodesy, medicine, nondestructive testing, and many others industry branches. For that purpose different types of magnetometers are used: SQID magnetometers, magnetoresistive sensors, or magnetoopic sensors [1]. Fluxgate sensors are also often used because of their ability to work in wide range of temperatures, low production cost and continuous process of their miniaturization. Miniature fluxgate sensors are also low energy consuming and has low weight. Usage of frame shaped single layer ribbon cores enables further miniaturization. This kind of cores are able to measure magnetic field independent in two directions. Thus only one core is needed to achieved features of two sensors [2]. In this paper comparison of functional characteristics of sensors with different kinds of core material is presented.

2 Tested Cores

Rapid Quenched amorphous alloys are produced by spreading liquid metal on rotating reel [3]. Thanks that method metal can not crystallised. Effect of faster cooling on edges and slower in the middle of ribbon is large magnetostrictive anisotropy energy [4]. For that reason it was no attempted to build two-axis frame shaped sensors.

During test it was observed, that anisotropy energy is no big enough to make two direction measurements impossible [2]. Tested cores has high magnetic permeability, what can help to increase sensitivity of sensors. Disadvantage of that material is high fragility. Thus it is necessary to use photolithography method in production process.

Ribbons were coated with photosensitive lacquer and exposure on UV light across stencil. Than ribbons were dipped in developer. Back side was coated with insulating lacquer. After that they were digested in sodium persulfate solution. In that way equilateral frame with internal dimension 6 mm and outer dimension 8 mm. Two kinds of alloys were used: $Fe_{67}Co_{18}Si_1B_{14}$ and $Co_{66}Fe_4Ni_1Si_{15}B_{14}$.

3 Production Technology of Two-Axis Magnetic Sensors

Fluxgate sensors produced in PCB technology consisted core which is placed in middle layer. Wires connected between layers surrounding core creates excitation and pick up coils. In case of one axis sensors, three layers PSBs were used [5]. In this case seven layers are needed to placed additional excitation and pick up coils for second direction. Dimensions of sensors are 40x40x1mm (fig. 1.). On first and seventh layer pick up coils for X direction were placed, on second and sixth layer for Y direction, on third and fifth excitation coils for both directions. On fourth layer core is placed.

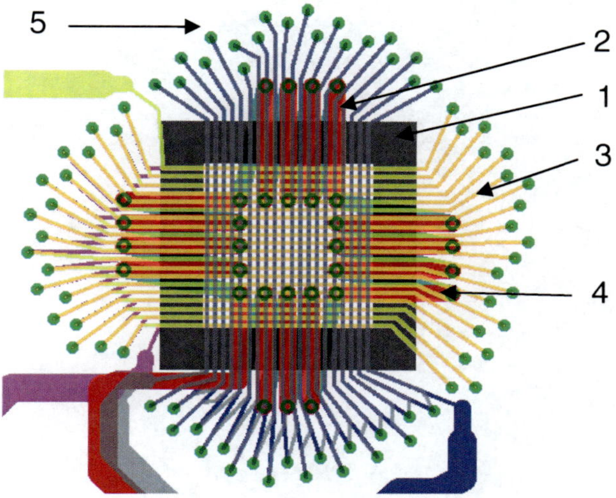

Fig. 1. Developed miniature two-axis fluxgate sensor: 1 – core, 2 –excitation coil in *X* direction, 3 – pick up coil in *Y* direction, excitation coil in Y direction, 5 – pick up coil in *X* direction

4 Experimental Setup

Excitation coils were supplied with 1kHz frequency sinusoidal current signal and up to 0,53A amplitude, which was generated by functional generator AFG3021B from Tectronics and voltage-current transducter BOP36 type from Kepco. Signal from pick up coils was amplified and measured by selective nanovoltometer set up for 2kHz, what is second harmonic frequency. Helmholz coils were used to apply constant magnetic field. For given magnetic field value voltage was measured. Test were carried out for different values of amplitude of excitation current.

5 Measurements Results

Using Helmholz coils constant magnetic field up to 62 A/m was applied. Sensor with $Fe_{67}Co_{18}Si_1B_{14}$ core was tested for excitation current amplitude measured in current efficiency value I_Z from 0.16 to 0.48A. For every amplitude value eight voltage values were recorded for increasing magnetic field value. Before changing excitation current amplitude core was demagnetised by sinusoidal 62A/m amplitude current. For measured characteristics for every excitation current amplitude value sensitivity was calculated. For that material sensitivity decreased for higher amplitude for both directions. Sensor achieved maximum sensitivity 0.47mV/A/m for X direction and 0.39mV/A/m for Y direction. In case of $Co_{66}Fe_4Ni_1Si_{15}B_{14}$ core measurements were possible only for excitation current efficiency value

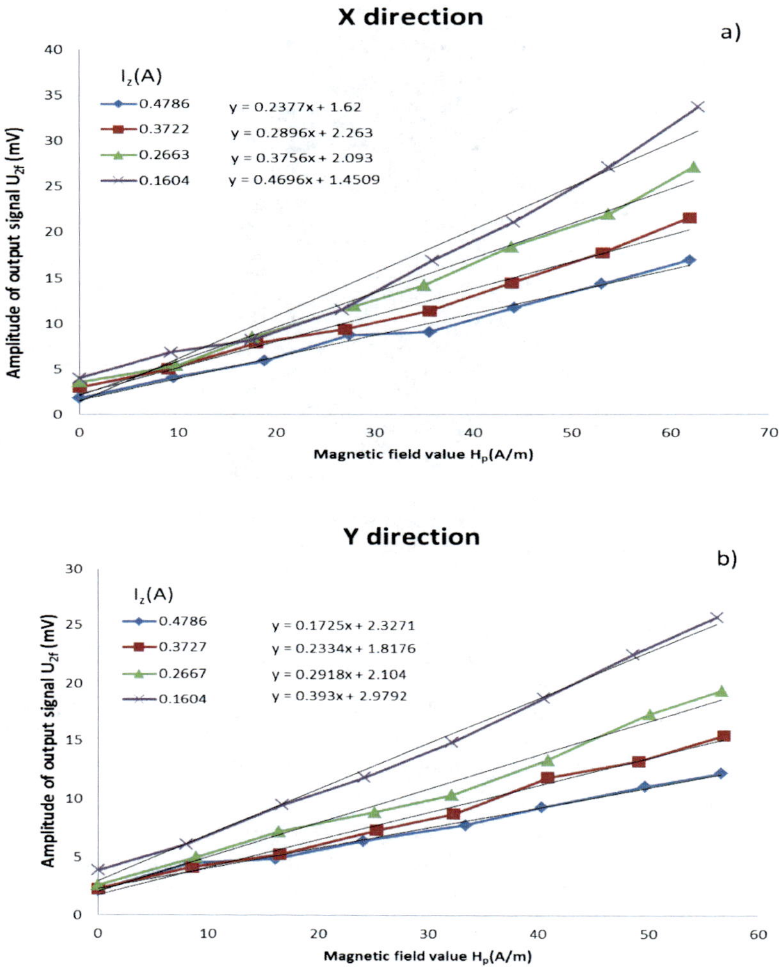

Fig. 2. Functional characteristics of developed fluxgate sensor with $Fe_{67}Co_{18}Si_1B_{14}$ core: a) *X* direction, b) *Y* direction

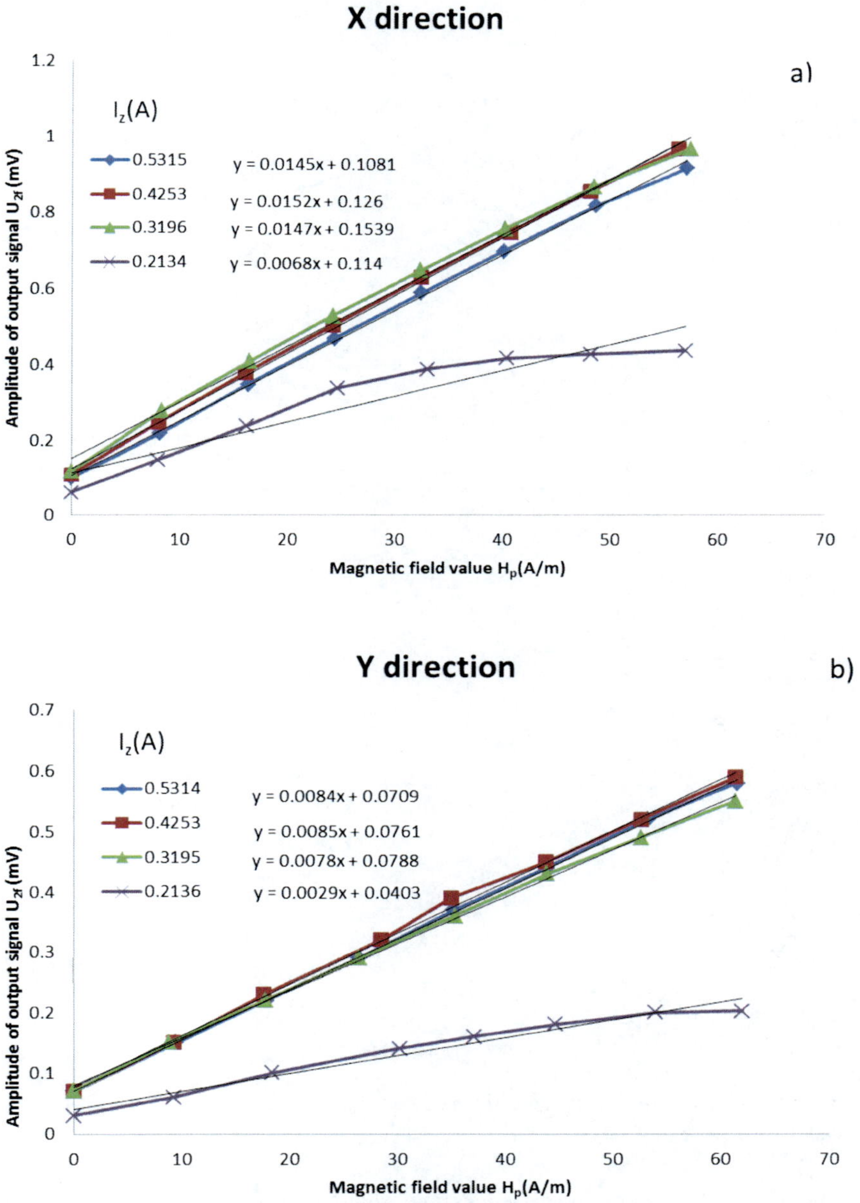

Fig. 3. Functional characteristics of developed fluxgate sensor with $Co_{66}Fe_4Ni_1Si_{15}B_{14}$ $Fe_{67}Co_{18}Si_1B_{14}$ core: a) *X* direction, b) *Y* direction

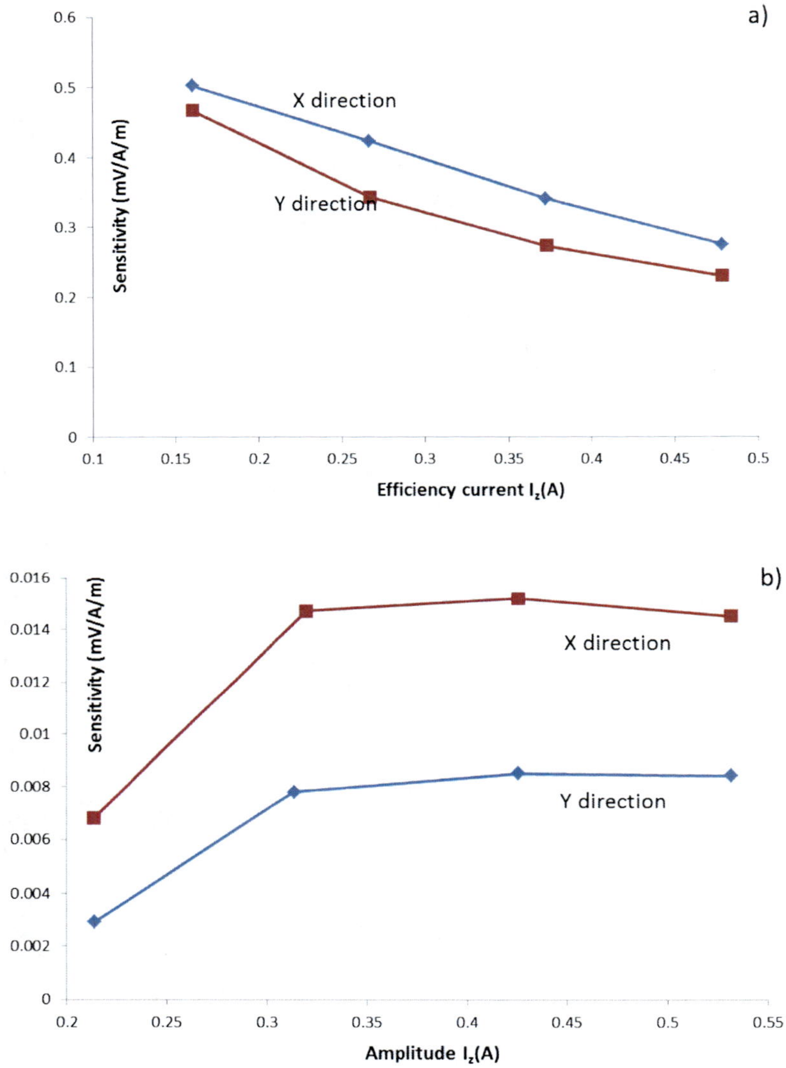

Fig. 4. Sensitivity dependence on efficiency current value I_z. a) for $Fe_{67}Co_{18}Si_1B_{14}$ b) for $Co_{66}Fe_4Ni_1Si_{15}B_{14}$

higher than 0.21A. For lower values any changes of output signal was not observed. Maximum excitation current amplitude was limited to 0.53A value like in previous case by wires resistance. For that kind of material only for 0.21A amplitude sensitivity was lower than for others amplitude values, for whom was almost constant equal to 0.0152mV/A/m for X direction and 0.0085mV/A/m for Y direction. Results show that sensor with $Fe_{67}Co_{18}Si_1B_{14}$ core has higher sensitivity, but its characteristic are less linear. Therefore its accuracy can be worst, because sensitivity coefficient will not be constant for whole range. Sensor with prevalent cobalt content has high linearity for higher excitation current amplitude values. Limit of sensor range is saturation induction causes by external magnetic field. When excitation signal causes induction amplitude is less than saturation induction of the core it can be observed remanence magnetisation in the core. This will cause errors. For that reason it is better to keep maximum values of induction not less than saturation value. 0.53A current efficiency value is enough to produce saturation in core.

6 Conclusion

Presented results show important differences in functional characteristics between fluxgate sensors with two kinds of cores. Sensor with $Fe_{67}Co_{18}Si_1B_{14}$ core has higher sensitivity but poor linearity. Output signal from sensor with $Co_{66}Fe_4Ni_1Si_{15}B_{14}$ core is very weak, but it can be amplified more than hundred times because of its good accuracy. Most prevailing feature of sensor is linearity. Thus cobalt alloy is better for fluxgate sensor, especially only second harmonic value of output signal is measured, what means that it has high resistance for noise. Presented research show, that testing and modelling of influence of chemical composition on sensor properties is very important for further development of fluxgate sensors.

References

[1] Gordon, D., Brown, R., Haben, J.: IEEE Trans. Magn. 8, 48 (1972)

[2] Frydrych, P., Szewczyk, R., Salach, J., Trzcinka, K.: PAR 2, 766–773 (2011)

[3] Kubo, Y., Kamijo, A., Yoshitake, T., Igarashi, H.: IEEE Trans. Magn. 2, 649 (1987)

[4] Tremolet, E.: Magnetostriction. CRC Press, London (1992)

[5] Vcelak, J., Petrucha, V., Kaspar, P.: Sensors. In: 5th IEEE Conference on Sensors, Daegu, p. 859 (2006)

Mechatronic Model of Anti-lock Braking System (ABS)

V. Goga[1,*], T. Jediný[1], V. Královič[1], and M. Klúčik[2]

[1] Slovak University of Technology, Faculty of Electrical Engineering and Information Technology, Institute of Power and Applied Electrical Engineering, Department of Mechanics, Ilkovičova 3, Bratislava, 812 19, Slovakia
[2] Slovak University of Technology, Faculty of Electrical Engineering and Information Technology, Institute of Control and Industrial Informatics, Department of Robotics and Artificial Intelligence, Ilkovičova 3, Bratislava, 812 19, Slovakia

Abstract. In this paper we propose a newly designed quarter-model that describes the vehicle braking process and dynamics. The model is based around a main scheme with ABS that can be fully modified for various adhesive conditions to simulate and further optimize the functionality of the ABS system.

1 Introduction

Nowadays, the ABS is a basic auto feature that it improves its overall controllability when subject to an abrupt braking. Therefore, the knowledge of its function is a necessity for students of automotive mechatro-nics. The proposed model has been designed for students to create various control schemes and investigate their models.

2 The Braking System with ABS Model

Breaking is a process of a wanted slowing down of the vehicle. Most of passenger vehicles have friction brakes whose principle is displayed in Fig. 1. The force applied via the driver's brake pedal is transmitted through the brake booster onto the main brake piston. The hydrostatical pressure in the brake fluid in the main brake piston is further distributed through a series of pipes towards a system of wheel cylinders. Above the main brake piston there is a fluid container. The braking mechanisms can be sub-divided into disc brakes and shoe brakes with two jaws.

In the two-circuit brake system design, the front wheels are inhibited independently on the rear ones while the construction and mechanism of both is identical.

* Corresponding author.

An enhanced effect can be achieved when a brake-power controller is implemented. Dual master cylinder enables independent breaking of both axles [1].

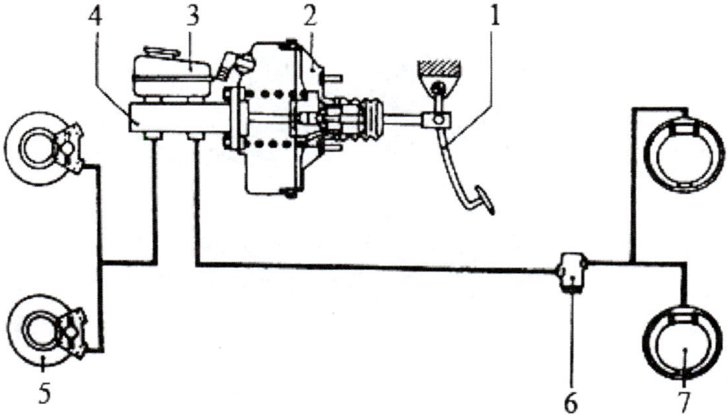

Fig. 1. A two-circuit brake system design of a passenger vehicle [2];
1 - brake pedal, 2 – brake booster, 3 – main brake piston, 4 – fluid container, 5 – disc brake, 6 – brake-power controller, 7 – shoe brakes

Load regulation is capable of a brake power control and division among axles as a result of an upright/vertical forces applied, however, it is unable to evaluate and react to different tyre surface adhesion; thus is it unable to prevent wheel locking.

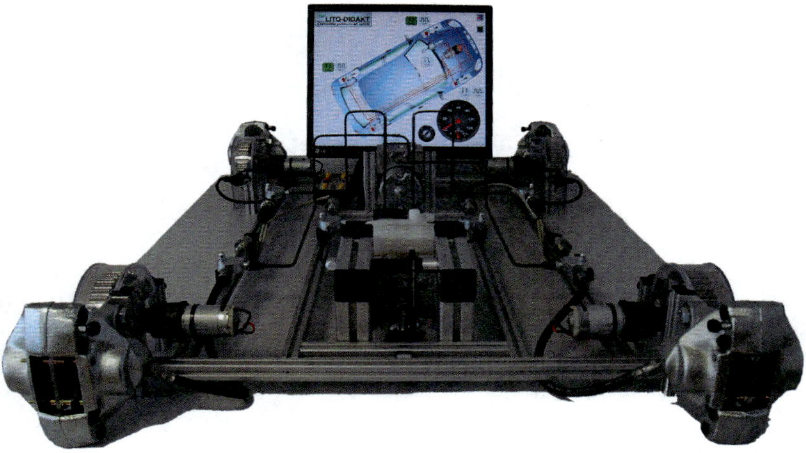

Fig. 2. The braking system with ABS model

2.1 Braking Process

The braking power is not constant. Its rundown is displayed in Fig. 3. t_r is the so-called reaction time between the driver sights the obstruction until he pushes the braking pedal. The response time is t_p. This refers to a lapse between pushing of a pedal and the braking effect taking place. The rise time of braking is t_n. In praxis, this is the time until which the braking power reaches certain levels. The time t_1 is made up of $t_p + t_n$. Time t_2 is the time until a full slowing-down effect is reached. Summing up of all time lapses, t_c is the total brake time (till stop), t_u - effective braking lapse, t_z - time for complete stopping (from realizing until a complete stop). b_p is the maximal level of a deceleration.

In Fig. 3., a brake force-time dependency is shown. In this graph, the following notation is used:

- t_{Fo} – regulatory force rise time (pressure),
- t_p – braking time, i.e. time that elapses between the beginning and at the time when releasing the brake when the braking force ceases to operate,
- F_p – maximum of braking force (pressure).

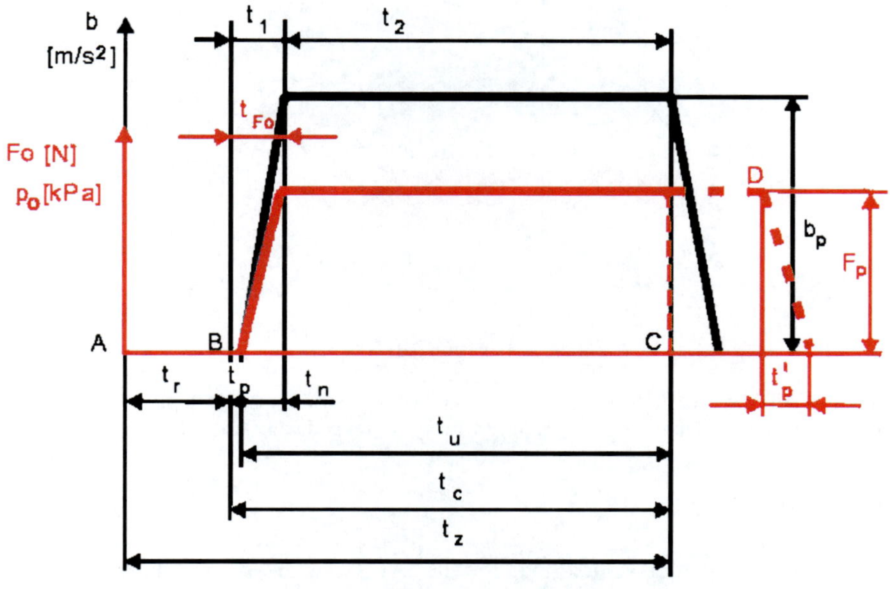

Fig. 3. A real vehicle braking lapse [3]

2.2 ABS Principle

In the process of braking, for a tyre it is necessary to not only employ the maximum brake power but also to make use of side forces to enable a directional control of the vehicle. The braking moment causes a stretch in the tyre contact and front area. As a result, the wheel travels a longer distance. This phenomenon it often referred to as tyre slip and can be described as (1) [4]:

$$\lambda = \frac{(v - \omega \cdot r)}{v}, \omega \neq 0 \tag{1}$$

where λ is the relative tyre slip, v is the car speed, ω is the tyre rotation speed and r is the tyre rolling radius.

Fig. 4. shows that the biggest longitudinal forces occur at 15–25% of the relative tyre slip. In this area, there's still a substantial amount of lateral force for a sufficient directional control. The ABS, therefore, is designed to keep the braking relative slip in this range.

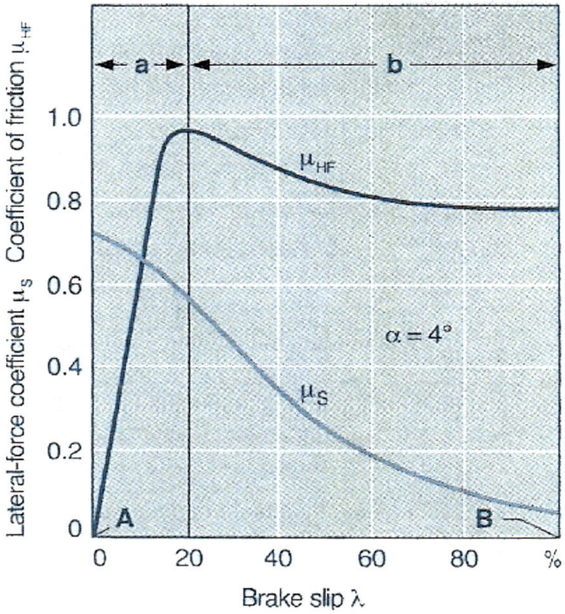

Fig. 4. Coefficient of friction and lateral-force coefficient relative to brake slip [5], a – stable zone, b – unstable zone, A – rolling wheel, B – locked wheel, α – slip angle

3 Model ABS

The proposed quarter-vehicle-model with ABS has been programmed in the MSC.Adams View. It consists of a surface element, tyre, brake, axle, speed and wheel revolution sensor and quarter vehicle mass element (Fig. 5.). The friction coefficient between the tyre and surface varies in the range of <0,1>. The control scheme has been written in MATLAB/Simulink (Fig. 6., Fig. 8.). The control input is the vehicle speed and wheel revolution. The outputs are the braking forces on discs. The maximum braking is considered. The control block lessens the brake pressure losing the wheel in order to avoid wheel locking. The friction coefficient is calculated from the result of the actual velocity and wheel revolutions. Its value is then being sent to the model.

The braking force models have been obtained from experiments (Fig. 7.) on a physical model (Fig. 2.).

The control is based on a switch-scheme; the control block lessens the brake-pipe pressure resulting in brake force reduction, until the relative slip reaches 0.25.

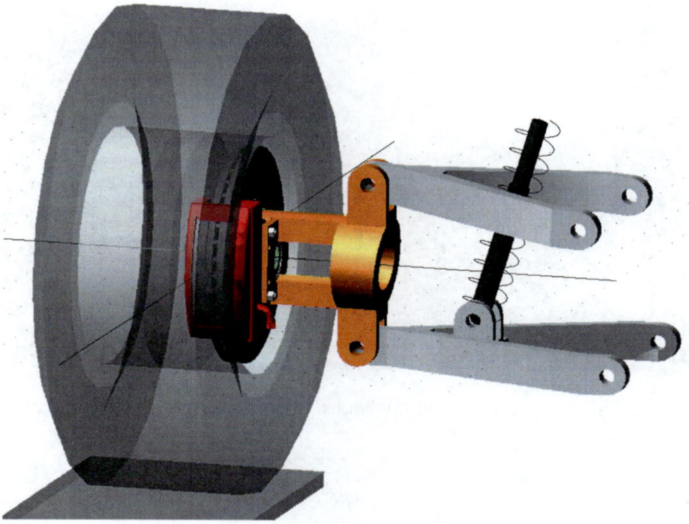

Fig. 5. The MSC.Adams brake model

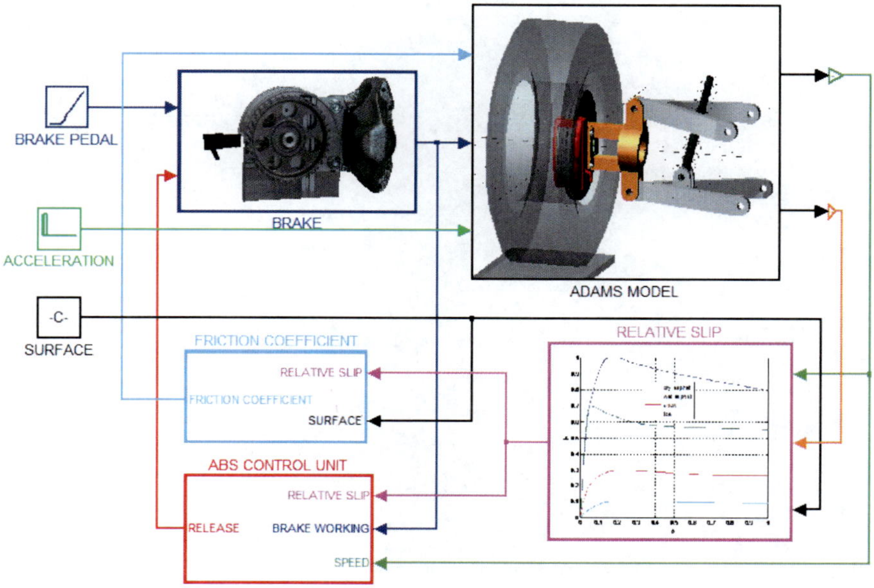

Fig. 6. ABS block scheme

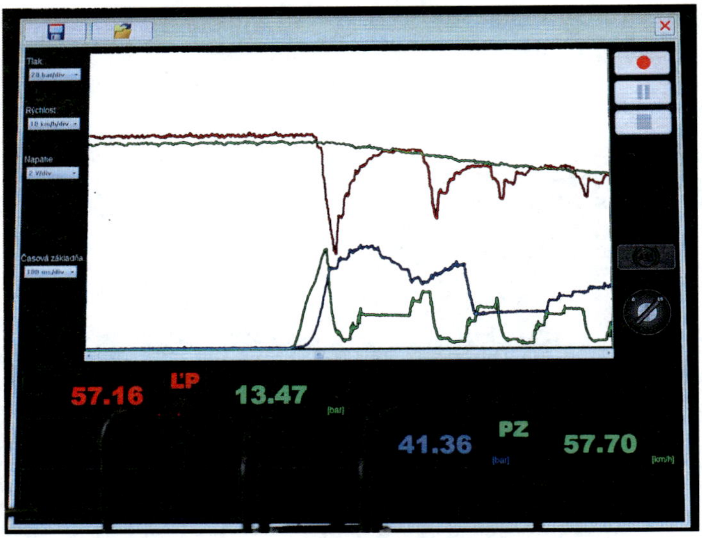

Fig. 7. Experimental results: wheel revolution / brake-pipe pressure

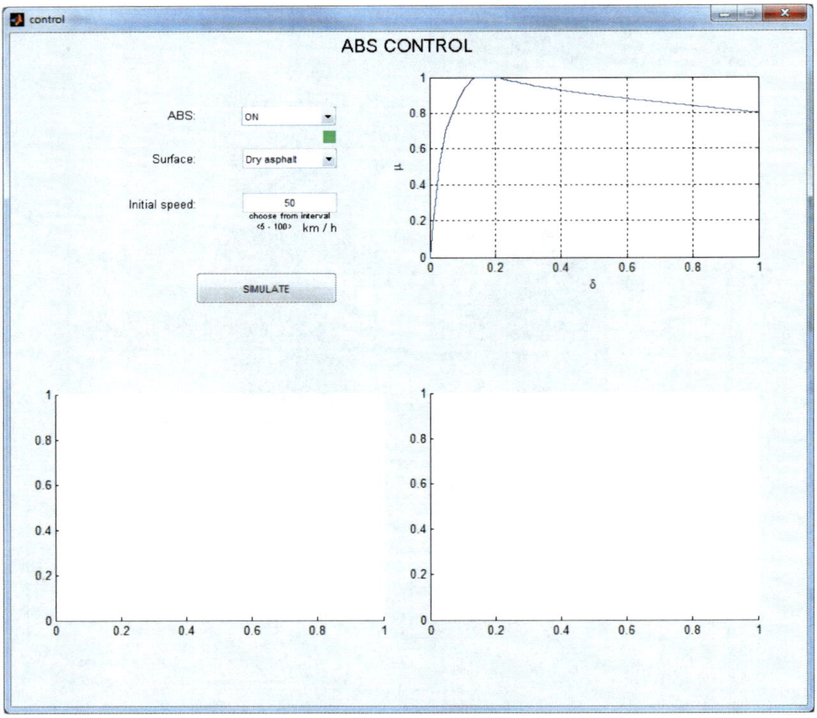

Fig. 8. ABS control panel

4 Simulations

We provided several simulations on different models of surfaces with activated and deactivated ABS. Initial speed was set to 50 kmh^{-1}. Dependence of friction coefficient based on relative slip is shown in Fig. 9. We modeled friction coefficient for these four surfaces: dry asphalt (dark blue), wet asphalt (green), packed snow (red) and ice (light blue).

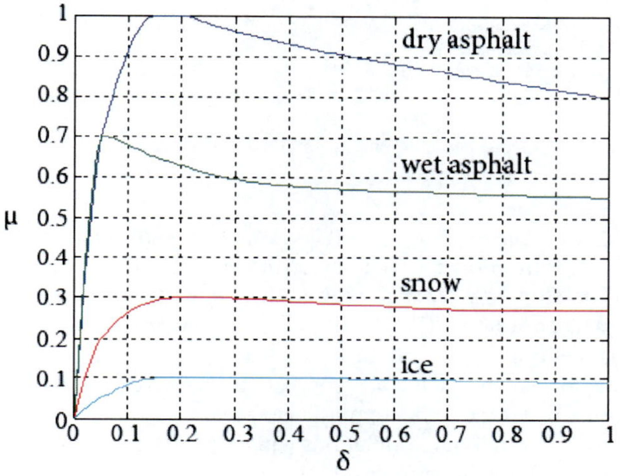

Fig. 9. Dependency of the friction coefficient on relative slip for different surfaces

There is a significant difference between braking with ABS turned ON and OFF shown in Fig. 10. and Fig. 11. With an activated ABS (Fig. 10.) control unit, the pressure in brakes decreases avoiding tyre locking, relative slip varies between 20-30% and the friction coefficient between the tyre and surface is almost on its maximum.

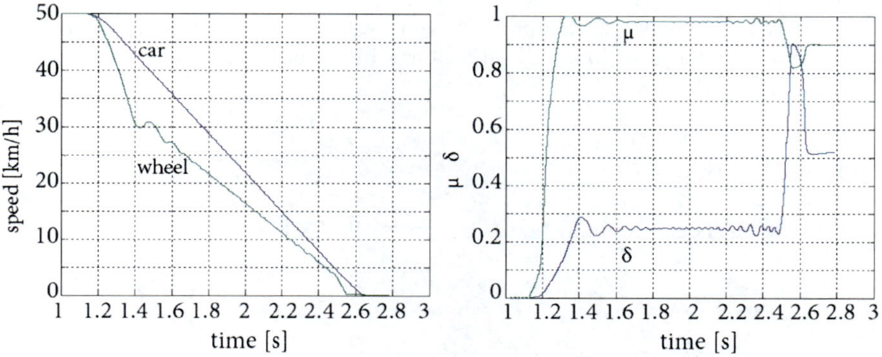

Fig. 10. Left: car and wheel speed; Right: time course of relative slip and friction coefficient; ABS – ON, surface – dry asphalt

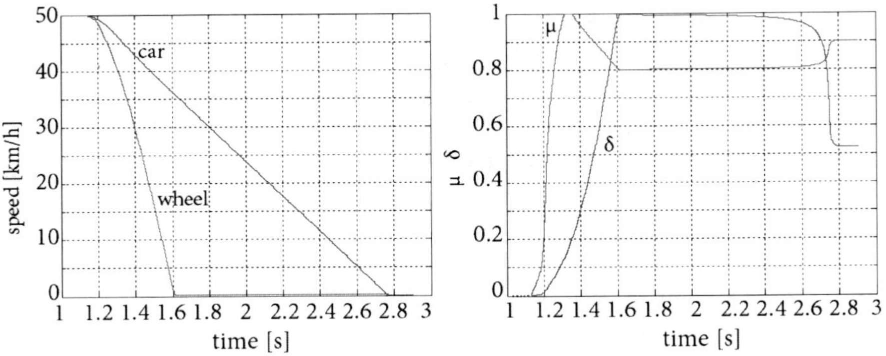

Fig. 11. Left: car and wheel speed; Right: time course of relative slip and friction coefficient; ABS – OFF, surface – dry asphalt

When the ABS is deactivated, the tyre becomes locked, relative slip increases rapidly to 100%, the friction coefficient is smaller and the effect of braking far from optimal. The stopping distance with activated ABS is shorter by about one meter. However, according to Fig. 4. the most important contribution of ABS is that the tyre is directionally controllable.

This phenomenon is similar for all types of surfaces. According to all simulations there are only differences in stopping distances.

Results obtained through our simulations can help optimizing some of the braking control system features in various vehicles.

5 Conclusion

The process of vehicle braking is an important aspect in car design. In this paper we discussed and showed some selected results of simulations carried out using our quarter-vehicle-model. Results showed that a very simple control unit can also improve the characteristics of braking.

Further analyses and model optimization is desired to enhance some of the features of this model to include new algorithms in the braking system for a variety of boundary conditions.

Acknowledgement

This article has been accomplished under grant VEGA 1/0093/10.

References

[1] Vlk, F.: Motor vehicles chassis, Brno (2006)
[2] Vlk, F.: Motor vehicles 1, Brno (2002)

[3] Information portal about the braking and braking systems of present cars (2006), http://www.brzdnevls.wz.cz/index.html
[4] Žilák, P., Smeja, M., Belavý, C.: Analysis and synthesis of anti-lock braking system of a vehicle in Matlab-Simulink. In: Conference Technical Computing, Prague (2006)
[5] Bosch, R.: GmbH, Safety, Comfort and Convenience Systems, Postfach (2006)

Composite Controller for Electronic Automotive Throttle with Self-tuning Friction Compensator

R. Grepl

Mechatronics Laboratory, ISMMB, FME, Brno University of Technology,
grepl@fme.vutbr.cz
www.mechlab.cz

Abstract. The automotive throttle is a highly nonlinear servo system with significant parameter changes caused by temperature, wear and lubrication during the vehicle operation. This paper describes the controller structure with basic heuristically tuned (fixed) PID controller coupled with tuned model-based feedforward spring and feedback error-based friction compensator. The extended Kalman filter with augmented state vector is used for parameter and state estimation based on nonlinear model and noisy position measurement. Simulation results of controller behavior and friction estimation are presented.

1 Introduction

An electromechanical position servo (Fig. 1) often called "electronic throttle" is nowadays a part of every produced passenger car. However, it is still an attractive research area due to strong nonlinearity, strict performance requirements and high parameters uncertainty. Active research is focused on modeling [3] as well as control. The heuristic-based adaptive controller is presented in [1] and [7]. This paper deals with the simulation study of feedback and feedforward compensator designed according to the estimated nonlinear model of the plant. Basic estimation is performed offline (e.g. using characteristics shown in Fig. 2). During the operation (online), slow parameter variations caused by temperature, vibration, wear, lubrication and other disturbances are estimated.

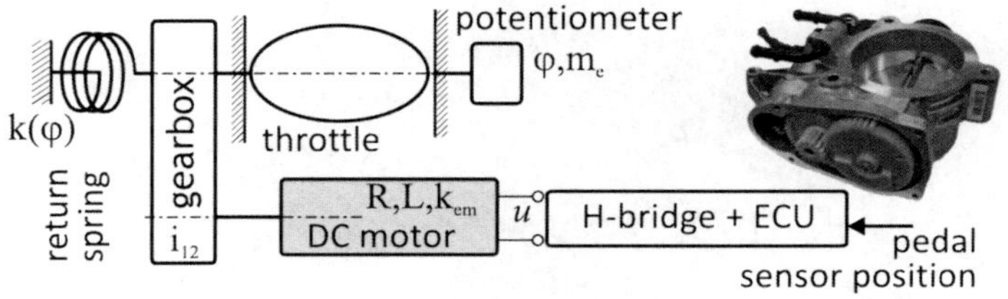

Fig. 1. Schema of electromechanical throttle

2 Modeling of Electromechanical Throttle

Modeling is based on practical experiments and described in details in [5,6,9]. The dynamics of electrical part (rotor winding inductance) can be neglected as well as clearance and stiffness in gears and then the system can be considered as 1dof mechanical model:

$$I\ddot{\varphi} = u - b\dot{\varphi} - u_{\mathrm{S}}(\varphi) - u_{\mathrm{F}}(\dot{\varphi}), \tag{1}$$

where I is system inertia, b is reduced viscous damping (includes also motor EMF), u_{F} is dry friction (eq. 3) and u_{S} piecewise linear spring model:

$$u_S = \begin{cases} k_{\mathrm{LH}}(\varphi_{\mathrm{LHO}} - \varphi_{\mathrm{LH}}) + k_{\mathrm{OP}}(\varphi - \varphi_{\mathrm{LHO}}), & \varphi > \varphi_{\mathrm{LHO}} \\ k_{\mathrm{LH}}(\varphi_{\mathrm{LHC}} - \varphi_{\mathrm{LH}}) + k_{\mathrm{CL}}(\varphi - \varphi_{\mathrm{LHC}}), & \varphi < \varphi_{\mathrm{LHC}} \\ k_{\mathrm{LH}}(\varphi - \varphi_{\mathrm{LH}}), \text{ otherwise.} \end{cases} \tag{2}$$

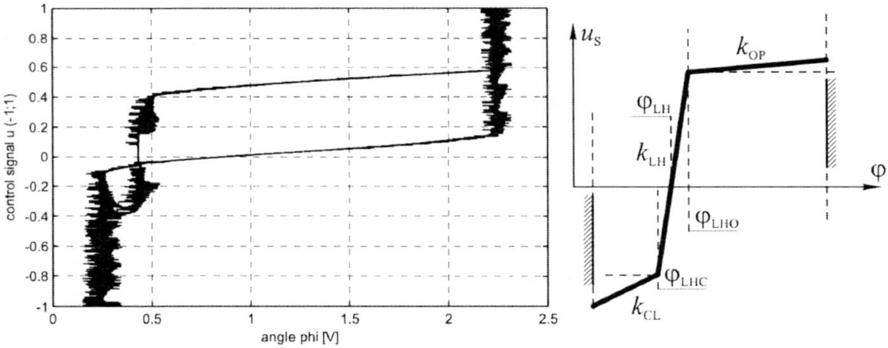

Fig. 2. Measured quasi-static characteristic (left) [6] and static model of the spring (right)

According to series of estimation experiments [6,8,9], the static friction model cannot fully represent the measured data; the dynamic model gives significantly better results. Instead of common LuGre model, the comparatively simpler Reset Integrator model [8] is used:

$$u_{\mathrm{F}} = \frac{(1 + a(p))u_{\mathrm{kin}}\, p}{p_0} + \beta \dot{p} \tag{3}$$

$$a(p) = \begin{cases} a & \text{if } |p| < p_0 \\ 0 & \text{otherwise} \end{cases} \tag{4}$$

$$\dot{p} = \begin{cases} 0 & \text{if } (\dot{\varphi} > 0 \wedge p \geq p_0) \vee (\dot{\varphi} < 0 \wedge p \leq -p_0) \\ \dot{\varphi} & \text{otherwise} \end{cases} \tag{5}$$

where p is friction state (can be understood as bending of virtual bristle), a represents the increase of friction force in low velocity and β is a damping coefficient which effects the stability of computation. Simulation results presented in this paper are obtained using the described model.

3 Controller Structure and EKF

The controller shown in Fig. 3 consists of PID, feedback friction compensator (FFBC) and feedforward spring compensator (FFC).

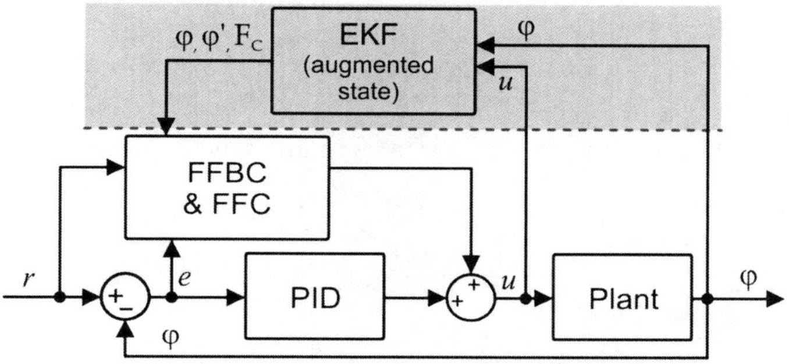

Fig. 3. Block diagram of proposed controller (PID, nonlinear feedback friction compensator, non-linear feedforward spring compensator)

The PID is tuned heuristically, e.g. using Ziegler-Nichols and its constants are invariable during the system operation. The FFC includes the nonlinear spring model (2) with φ substituted by reference r. Parameters are roughly identified offline [6]. The FFBC implements the error-based friction compensation [9]

$$u_{FFBC} = \begin{cases} 0, & |e| < \varepsilon_F \\ \operatorname{sgn}(e)F_K, & \text{otherwise} \end{cases} \tag{6}$$

The value of kinetic friction F_K is crucial for the compensator performance and is also very dependent on the throttle temperature, wear and lubrication. First, F_K is estimated offline [9]. Next, during the operation, the value is continuously estimated using EKF with augmented state vector. The computed signal u is saturated at (-1,1).

The discrete-time implementation of EKF is based on the continuous model (1) and the Euler integration formula. Although the dynamic friction model (3-5) must be used in simulation of the plant to have a good relevance to experiments, a simpler static model with stiction [8] can be successfully used in EKF:

$$u_{FE} = \begin{cases} \operatorname{sgn}(\dot{\varphi})F_K, & |\dot{\varphi}| > \varepsilon \\ F_e & |\dot{\varphi}| < \varepsilon \wedge |F_e| < F_S \\ \operatorname{sgn}(F_e)(1+a)F_K & \text{otherwise} \end{cases} \quad (7)$$

$$F_e = u - b\dot{\varphi} - u_S \quad (8)$$

Finally, the nonlinear state space model with the augmented state vector $\mathbf{x} = [\varphi, \dot{\varphi}, p, T]^T$ can be formed:

$$\dot{\mathbf{x}} = \begin{bmatrix} \dot{\varphi} \\ 1/I(F_e + u_{FE}) \\ 0 \end{bmatrix}. \quad (9)$$

The related Jacobian matrix used in EKF algorithm is of form:

$$\mathbf{J} = \begin{bmatrix} 0 & 1 & 0 \\ -J_S/I & -b/I & -J_F/I \\ 0 & 0 & 0 \end{bmatrix}, \quad (10)$$

where

$$J_F = \begin{cases} \operatorname{sgn}(\dot{\varphi}), & |\dot{\varphi}| > \varepsilon \\ 0 & |\dot{\varphi}| < \varepsilon \wedge |F_e| < (1+a)F_K \\ \operatorname{sgn}(F_e)(1+a) & \text{otherwise} \end{cases} \quad (11)$$

$$J_S = \begin{cases} k_{\mathrm{OP}}, & \varphi > \varphi_{\mathrm{LHO}} \\ k_{\mathrm{CL}}, & \varphi < \varphi_{\mathrm{LHC}} \\ k_{\mathrm{LH}}, & \text{otherwise.} \end{cases} \quad (12)$$

4 Simulation Results and Conclusion

The proposed controller and relevant behavior of EKF have been tested on the simulation model with parameters set according to a real throttle [9]. Fig. 4 and 5 show the quality of position control, velocity estimation and convergence of friction parameter F_K estimated by EKF. Also the proportion between PID and compensator actions is shown in Fig. 4. Simulation experiments prove the ability of adaptive controller scheme to compensate slow parameter changes caused by temperature, wear, and lubrication during operation.

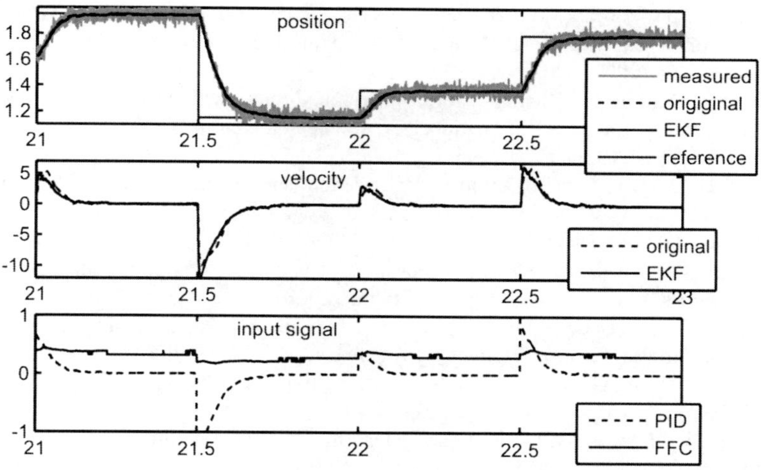

Fig. 4. Simulation results: position control of throttle, estimated velocity and input values (comparison of PID and FFBC & FFC)

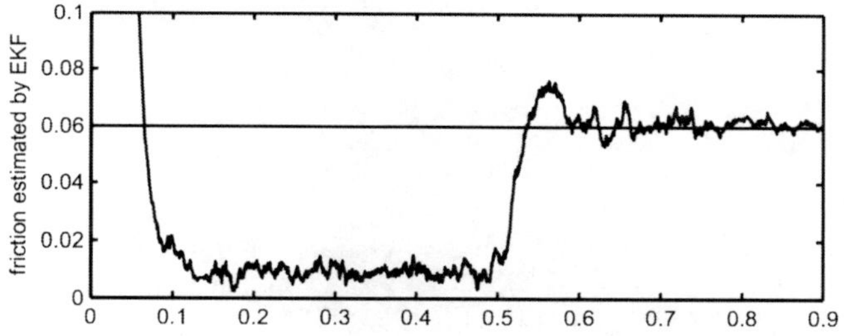

Fig. 5. Simulation results: estimation of friction F_K online using EKF

Acknowledgement

The work presented in this paper was supported by project FSI-S-11-15 "Design, testing and implementation of control algorithms with use of nonlinear models of mechatronics systems".

References

[1] Deur, J., Pavkovic, D., Jansz, M., Peric, N.: Automatic Tuning of Electronic Throttle Control Strategy. In: 11th Mediterranean Conference on Control and Automation MED 2003, Rhodes, Greece, June 18-20 (2003)

[2] Stence, R.W.: Digital By-Wire Replaces Mechanical Systems in Cars. In: Electronic Braking, Traction, and Stability Controls, pp. 29–36. Society of Automotive Engineers, Inc., USA (2006)

[3] Vasak, M., Baotic, M., Petrovic, I., Peric, N.: Hybrid Theory-Based Time-Optimal Control of an Electronic Throttle. IEEE Transactions on Industrial Electronics 54, 1483–1494 (2007)
[4] Zhang, P., Yin, C., Zhang, J.: Sliding Mode Control with Sensor Fault Tolerant for Electronic Throttle. In: International Conference on Automation Science and Engineering, Shanghai, China (2006)
[5] Grepl, R., Lee, B.: Modelling, identification and Nonlinear Control of Electronic Throttle. In: The KSAE Daegu Gyungbuk Regional Conference, pp. 45–53. Korean Society of Automotive Engineers, Daegu (2008)
[6] Grepl, R., Lee, B.: Modelling, identification and control of electronic throttle using dSpace tools. Technical Computing Prague (2008)
[7] Pavkovic, D., Deur, J., Jansz, M., Peric, N.: Adaptive Control of Automotive Electronic Throttle. Control Engineering Practice 14, 121–136 (2006)
[8] Olsson, H., et al.: Friction models and friction compensation. Eur. J. Control 4(3), 176–195 (1998)
[9] Grepl, R., Lee, B.: Modeling, parameter estimation and nonlinear control of automotive electronic throttle using a Rapid-Control Prototyping technique. Int. J. of Automotive Technology 11(4), 601–610 (2010)

Non Invasive Identification of Servo Drive Parameters

R. Neugebauer, A. Hellmich*, S. Hofmann, and H. Schlegel

Chemnitz University of Technology, Faculty of Mechanical Engineering,
Institute for Machine Tools and Production Processes,
Reichenhainer Str. 70, 09126 Chemnitz, Germany
wzm@mb.tu-chemnitz.de

Abstract. For the tuning of servo controllers as well as for monitoring functions, significant parameters of the controlled system are required. In contrast to identification methods with determined input signals, the paper focuses on the problem of identification with regular process movements (non invasive identification), leading to a lack of power density in some frequency ranges. A nonlinear Least Squares (LS) approach with single mass system and friction characteristic is investigated regarding the accomplishable accuracy and necessary constraints. The proposed method is applicable on industrial motion controllers and has been carried out with a multitude of input sequences. To verify the performance of the approach, achieved experimental results for the model parameters are exposed.

1 Introduction

With an increasing number of installed position controlled servo drives in production machines and machine tools, the importance of parameter identification grows as well. The biggest object of identification techniques is to provide significant parameters or a model transfer function of the controlled systems in the servo drive. These identification results form the basis for automated ways of controller tuning, such as tuning rules or optimization strategies [1,2]. Another aspect of parameter identification is to generate models as a basis for monitoring functions, such as wearout detection or control loop performance monitoring [3,4,5]. Especially for machines with a high number of drives, the optimal setting and adoption of controller parameters requires automatic strategies, which again need informative models. Consequently, a wide range of identification methods from control engineering was adjusted to the needs of drive control [6, 7].

This paper focuses on the field of non invasive (i.e. not interfering with the process) identification and explores the possibility of using only typical movements of the servo drives, so called "natural excitations" [8].

* Corresponding author.

Hence, a non invasive, online capable identification approach is presented after an analysis of identification techniques and model structures with regard to the application with natural excitation. Furthermore, experimental results are shown in section 4. The paper is finalized by conclusions and an outlook.

2 Non Invasive Parameter Identification in Drive Control

The choice of a suitable excitation signal is of primary importance for the identification of an accurate plant model. Only if a minimal power density is present in the whole spectrum, a decision for a convenient model structure can be made. Furthermore, only in this case all model parameters are estimated in a good quality. Excitation signals, which meet these demands, are e.g. pseudo random binary signals (PRBS) or on-off-type-signals (relay signals). Consequently, they are applied in currently in use identification algorithms of industrial drive systems [9, 10] as well as in many parameter identification approaches [2, 8, 11, 12].

In contrast to identification with determined test signals, there is a need for the identification of plant parameters during regular operation of machine tools or production machines. This is the case if changes of the plant parameters are possible and expected more often, than maintenance breaks are feasible.

Consequently, established identification methods have to be explored, whether they provide convenient results despite the insufficiency of only natural excitation. Another fact to consider is the online capability if the approaches for industrial motion controllers or numerical controllers. These problems can be narrowed down to the core issues of choosing an identification technique and using this approach with a certain model structure.

2.1 Identification Techniques

For the example of the velocity control loop of a cascaded position control (Fig. 1), [7] and [8] compare the applicability of extended Kalman Filters and Basis Function Networks. It is stated, that single mass mechanical systems can be identified with natural excitation, if certain acceleration is present.

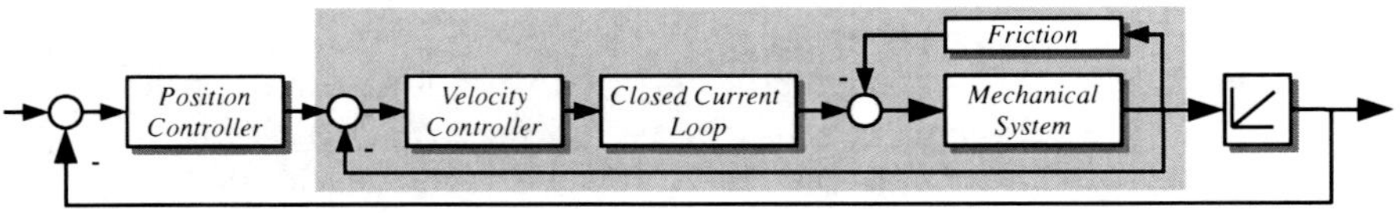

Fig. 1. Cascaded position control with velocity loop (grey)

Furthermore, [11] presents an identification approach for mechanical systems, where a Least Squares (LS) method is used. Although it is combined with a PRBS excitation in [11], it seems promising to consider this technique for non invasive identification as well. [13] compared several approaches to estimate the total moment of inertia of a position controlled servo drive based on different test signals (sinus-, ramp-, rectangular- signals) and thus provides an overview of their suitability for the us with natural excitation, too.

2.2 Model Structures

Generally it is not necessary to take high model orders into account for non invasive identification after a model structure choice and complete parameter estimation during commissioning. In truth it is worthwhile to narrow the models down to a few significant parameters. This concurs with several publications, where only a single mass system (characterized by a total moment of inertia J_{tot}) and a friction model is taken into account for the chosen example [11,14,15]. According to [8] there is limitation in the non invasive identification quality for higher order models, which can only be shifted by applying additional excitation to the process. Furthermore, filters to damp significant eigen frequencies are common in the velocity loop which entails a further deterioration of the identifiability.

Hence, the paper focuses on a single mass system with a friction characteristic, consisting of a constant friction moment (M_{RC}) and a speed depending friction coefficient (μ_R).

The resulting parametric model entails three parameters as follows:

$$\dot{n}_{act} = \frac{1}{J} \cdot (M_{act} - M_{fric}) = \frac{1}{J} \cdot (M_{act} - M_{RC} \cdot \mathrm{sign}(n_{act}) - \mu_R \cdot n_{act}) \tag{1}$$

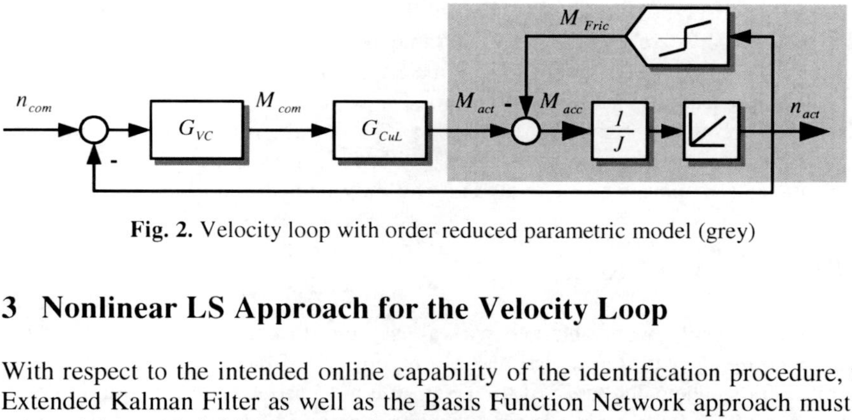

Fig. 2. Velocity loop with order reduced parametric model (grey)

3 Nonlinear LS Approach for the Velocity Loop

With respect to the intended online capability of the identification procedure, the Extended Kalman Filter as well as the Basis Function Network approach must be rejected due to their high complexity. The methods from [13] do not consider friction influences and thus lead to inaccurate results. Hence, the introduced parametric model will be combined with the Least Squares method. Due to the nonlinear friction characteristic, the LS problem becomes nonlinear as well. For the realization, equation 1 has to be discretizised with the sample time T_{sample}. The resulting difference equation (eq. 2) and is further transformed to match the nomenclature of [6] (eq. 3), where input values are denoted with u(k) and output values with y(k). Notice that the parameters a_1 and b_1 still depend on J_{tot} and μ_R directly:

$$n_{act}(k) + a_1 \cdot n_{act}(k-1) = b_1 \cdot \left[M_{act}(k-1) - M_{RC} \cdot \mathrm{sign}(n_{act}(k-1))\right]$$

$$\text{with: } a_1 = -e^{\frac{-\mu_R \cdot T_{sample} \cdot 60}{J_{tot} \cdot 2\pi}} \qquad b_1 = \frac{1}{\mu_R} \cdot \left(1 - e^{\frac{-\mu_R \cdot T_{sample} \cdot 60}{J_{tot} \cdot 2\pi}}\right) \tag{2}$$

$$y(k) = a_1 \cdot -y(k-1) + b_1 \cdot \left[u(k-1) - M_{RC} \cdot \mathrm{sign}(y(k-1))\right] \tag{3}$$

This leads to the description of the LS problem with a data vector ψ (eq. 4) and a parameter vector Θ (eq. 5):

$$\boldsymbol{\psi}^T(k) = \left[-y(k-1) \quad u(k-1) \quad \mathrm{sign}(y(k-1))\right] \tag{4}$$

$$\boldsymbol{\Theta}^T = \left[a_1 \quad b_1 \quad b_1 \cdot M_{RC}\right] \tag{5}$$

With this definition, the LS problem is carried out recursively [1]. The calculation is feasible on industrial controllers due to the relatively small order of the resulting matrices (max. 3).

Due to the orientation towards non invasive identification, sufficient excitation to the process needs to be detected. Firstly, the identification procedure should be started in a phase of acceleration or deceleration of the drive, when a certain value of torque is present. This is to prevent the LS algorithm from diverging due to a bad signal to noise ratio. Secondly, suitable results can only be expected, if a minimum power density is exceeded. Some experimental results to exemplify this fact will be presented in section 4.

Furthermore, it is stated in literature [6, 8, 11], that the filtering of the input and output signals with the same filter helps to improve the accuracy. Consequently, the identification approach was extended by a moving average filter.

4 Experimental Results

Experimental results were obtained with a test rig with known moment of inertia, which is equipped with an industrial motion controller SIMOTION with a sample time of 500 µs. The calculation of the LS problem is carried out in the motion controller as well as in Matlab whereas the concordance of both variants has been proven.

Fig. 3. Test rig with known moment of inertia

Despite the aspired non invasiveness of the presented identification approach, test signals were used to investigate and develop the method. The experimental results in this chapter are based on variety of input sequences with two frequencies in different velocity ranges and three exemplary velocity controller (VC) settings as displayed in Table 1.

Table 1. Variety of input sequences

Input sequence	Time period [ms]	Magnitude [1/ min]	Offset	Controller setting G_{VC}		
				Nr.	K_P [Nm s/rad]	T_N [ms]
Rectangle	100	50	0	1	1.309	6.85
Stairs		250	2 x Magn.	2	0.8	20
Sinus	300	500	3 x Magn.	3	0.3	50
Trapezoid						

On the one hand, this is done to evaluate the performance of the identification technique and to locate minimum requirements, detached from a restricting case of application. On the other hand, it allows a comparison of the accuracy of the approach with other invasive identification methods.

For all experiments the LS calculation is started in a phase of acceleration or deceleration, when a minimum value of torque (20% of the nominal torque) is present. Consequently some experiments do not lead to any results due to the lack of excitation. The main focus is the determination of the total moment of inertia J_{tot}, which is displayed in the following figure.

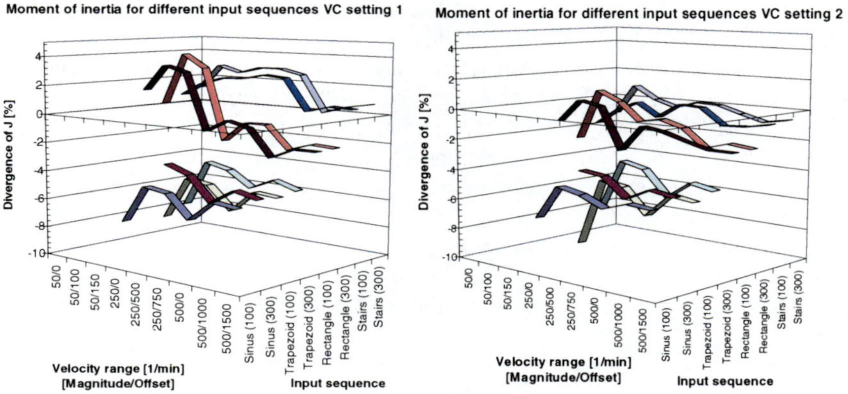

Fig. 4. Identification results for the moment of inertia J_{tot}

The figures show results for the velocity controller settings 1 and 2 (Table 1) after filtering the input and output values by using a moving average filter with the length of 5 elements. It is noticeable, that the shape of the input sequence as the biggest influence to the accuracy, followed by the present velocity controller setting. Another apparent result is that all experiments with a velocity offset unequal to zero lead to discrepancies of less than five percent. The biggest divergence is about 10% while 86% of the experiments provide results with a divergence less than ±5%, which is a typical value of discrepancies for invasive identification as well.

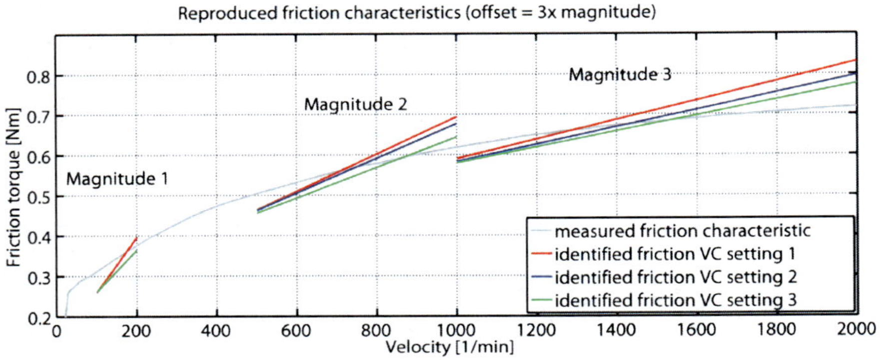

Fig. 5. Identification results for the friction parameters

Results for the identification of friction parameters are displayed in figure 5. The presented curves result from the reproduction of the friction characteristic based on the estimated parameters M_{RC} and μ_R. Notice, that the identification results for those values are only relevant for the specific velocity range of the process movement. For a better illustration three experiments with rectangular input sequences and the introduced magnitudes (1-3) were combined in figure 5. Additionally, a measured friction characteristic was included for verification. The identified characteristics match the measured curve in the linearization points, but vary towards lower or higher velocity values. The influence of different velocity controller parameters is comparatively small. A more precise image of the friction in the mechanical system could be issued by carrying out further movements for different velocity ranges.

Furthermore, the multitude of experimental results provides a basis for the detection of suitable process excitation. For this purpose, the power density spectra of all input sequences were computed (eq. 6 and 7). One example is displayed in figure 6.

$$\boldsymbol{\Phi}_{uu}(\boldsymbol{\tau}) = \left\{ \frac{1}{N} \cdot \sum_{k=1}^{N-\boldsymbol{\tau}} u(k) \cdot u(k+\boldsymbol{\tau}) \right\} \quad \text{for } 1 \leq \boldsymbol{\tau} \leq N \tag{6}$$

$$S_{uu}(j\boldsymbol{\omega}) = FFT(\boldsymbol{\Phi}_{uu}(\boldsymbol{\tau})) \tag{7}$$

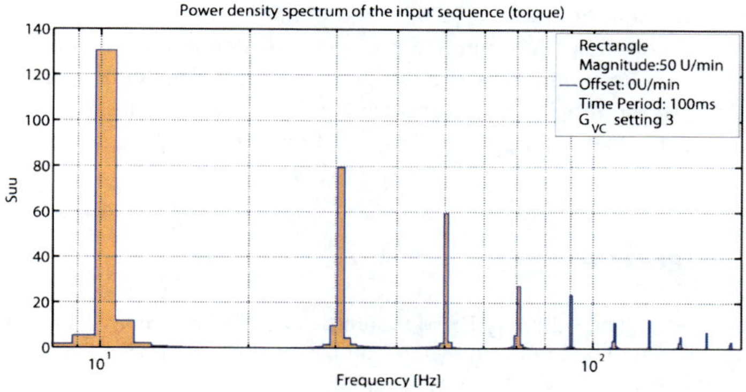

Fig. 6. Power density spectrum of a rectangular input sequence

In the next step, the area under the power density spectrum was calculated to represent the level of excitation (orange area in figure 6). The results were proportioned with the identified moment of inertia of equation 3 (red area in figure 7). Identification results with a divergence of less then ± 10% are rated as sufficient (green area) and were achieved for high excitation. Hence, a minimum size of the excitation representative is about 200. Whether this statement can be generalized for other drives and thereby serve as excitation detection, future investigations will show.

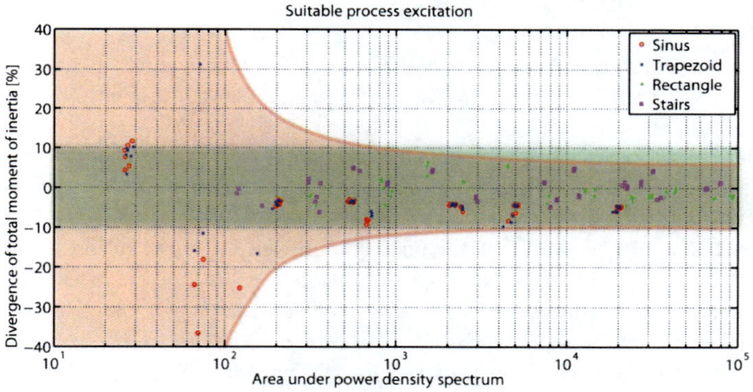

Fig. 7. Divergence of moment of inertia over power density area

5 Conclusions and Outlook

An identification method for the estimation of an order reduced parametric model was presented. The approach is online capable on industrial motion controllers and has been implemented as an automatic application. Experimental results were presented based on a multitude of different input sequences to prove the independency on a specific type of test signal and thus the suitability for non invasive identification. More investigations will be done regarding the detection of a minimum power density of the input signal to guarantee adequate identification results. Furthermore it seems promising to explore different filtering strategies for signal processing.

Acknowledgement

The Cluster of Excellence „Energy-Efficient Product and Process Innovation in Production Engineering" (eniPROD®) is funded by the European Union (European Regional Development Fund) and the Free State of Saxony.

Europa fördert Sachsen.
EFRE
Europäischer Fonds für regionale Entwicklung

References

[1] O'Dwyer, A.: Handbook of PI and PID Controller Tuning Rules. Imperial College Press (2003)

[2] Åström, K.J., Hägglund, T.: Advanced PID Control. ISA, Research Triangle Park (2006)

[3] Villwock, S.: Identifikationsmethoden für die automatisierte Inbetriebnahme und Zustandsüberwachung elektrischer Antriebe, Dissertation Thesis, University of Siegen (2007)

[4] Bebar, M.: Regelgütebewertung in kontinuierlichen verfahrenstechnischen Anlagen anhand vorliegender Messreihen, Dissertation Thesis, Ruhr-Universität Bochum (2005)

[5] Neugebauer, R., et al.: Überwachung und Bewertung von Antriebsregelungen bei Verzicht auf zusätzliche Sensorik. In: VDI-Berichte 2092, pp. 117–120 (2010)

[6] Isermann, R.: Identifikation dynamischer Systeme 1. Springer, Heidelberg (1992)

[7] Beineke, S., et al.: Comparison of parameter Identification Schemes for Self-Commissioning Drive Control of Nonlinear Two-Mass Systems. In: IEEE Industry Applications Society, Annual Meeting New Orleans, pp. 493–500 (1997)

[8] Beineke, S.: Online-Schätzung von mechanischen Parametern, Kennlinien und Zustandsgrößen geregelter elektrischer Antriebe. VDI Fortschritt-Berichte 816 (2000)

[9] Siemens, A.G.: SINAMICS S120/150 Function Manual (2008)

[10] Bernecker, Rainer: Industrie Elektronik GmbH. Autotuning ACP 10 Software (2006)

[11] Schütte, F.: Automatisierte Reglerinbetriebnahme für elektrische Antriebe mit schwingungsfähiger Mechanik, Dissertation Thesis, Shaker Verlag (2003)

[12] Doenitz, S.: Verfahren zur Minimierung des Einflusses von Störkräften in lagegeregelten Vorschubantrieben, Dissertation Thesis, Technische Universität Darmstadt (2003)
[13] Burkhard, H.: Applikation von Algorithmen zur Bestimmung von Massenträgheitsmomenten im Umfeld von Bewegungssteuerungen, Diploma Thesis, Chemnitz University of Technology (2010)
[14] Wertz, H., Beineke, S., Fröhleke, N.: Computer aided commissioning of speed and position control for electrical drives with identification of mechanical load. In: IEEE Industry Applications Conference (1999)
[15] Mink, F., Bähr, A., Beineke, S.: Self-commissioning feedforward control for industrial servo drive. Elektromotion (2009)

Optimal Design of a Magneto-Rheological Clutch

P. Horváth and D. Törőcsik

Széchenyi István University, Egyetem tér 1,
Győr, 9026, Hungary

Abstract. Appliances using magneto-rheological fluid (MRF) can find application in shock absorbers, brakes and clutches. Until now the spread of MRF clutches was limited by the moderate torque they can transmit. This paper summarizes the factors influencing the torque transmitted. During the design process of a prototype MRF clutch its most important geometric data have been determined by means of simulation of the magnetic circuit.

1 Introduction

Beneficial features of MRF clutches can be summarized as follows: short response time (range of milliseconds), simple mechanical construction without moving parts, little wearout, easy controllability, moderate current and voltage needs, insensitivity to soiling, operation features' independency of temperature. The only disadvantage of application of MRF is its not secured resistance of ageing. The applied MR fluids are suspensions of ferromagnetic particles in a carrier fluid that can drastically change their apparent viscosity from liquid to semi solid described by yield stress in some milliseconds under a magnetic field. In literature there are some clutches available. Bansbach [1] suggested a double-plate and a multi-plate MRF clutch to be placed between the car engine and the differential. Lampe [2] studied the transitional state of a brake between liquid and solid states. Kavlicoglu [3] at University of Nevada built a prototype double-plate MRF clutch of 7,9 Nm torque. Later their group increased the torque up to 240 Nm applying 43 plates. Gratzer et al. at MAGNA [4] prepared a multi-cylindrical MRF clutch of 700 Nm torque with a magnetization unit rotating together the shaft. Its current was supplied with slip rings.

2 Designing a Prototype MRF Clutch

Our goal was to design a prototype limited slip MRF clutch with the following requirements: disc type, easy to manufacture construction, a plate of 45mm outer radius (torque about 10 Nm) and non-rotating magnetization coil to avoid slip rings. The clutch developed consisted of three well- distinguishable units. The driving shaft (1), the two ferromagnetic flux guide rings (2) attached to each other by a copper tube (3) and the right hand side cap (4) formed the first unit. The driven shaft (6), as the second unit was supported inside the first unit by sealed deep-groove ball bearings (7). The ferromagnetic plate (5) was attached to the driven shaft by screws. The third stationary assembly involving the electromagnetic coil (8) was wound around a core. The coil took place inside a ferromagnetic casing (9). Driving assembly is supported in this stationary casing by deep groove ball bearings, so no slip rings needed to activate the coil.

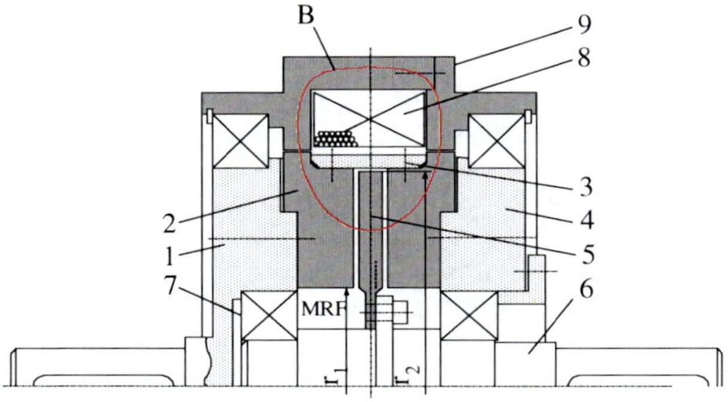

Fig. 1. Schematic drawing of the MRF clutch modeled

3 Optimization of Flux Guide Rings

The most important factor of the magnetic circuit is the inner radius r_1 of flux guide rings, because it determines the effective area of the driven plate. Anticipatory simulations showed that B is almost constant over the gap. Assuming $J=4\cdot10^6$ A/m^2 constant magnetization current density and $h=1$ mm gap thickness, the magnetic flux density was calculated over the MRF gap at inner radii of 16, 20, 24, 28 and 32 mm, respectively. For flux guide parts the outstanding ferromagnetic material P900 by Böhler AG. was used. MRF manufacturers supply the yield stress (τ_y) vs. magnetic field (H) and magnetic flux density (B) vs. magnetic field characteristics. Since in calculations the yield stress vs. magnetic flux density curve can be better applied, we constructed it in Fig.2 for a commercially available MR 132 LD fluid of the leading manufacturer LORD Corp. Prior to saturation the magnetic flux density can be approximated by the

$$\tau_y = 54.2 \cdot 10^3 \cdot B \quad \text{(Pa), if } B \le 0{,}8 \text{ (T)} \tag{1}$$

linear expression. In designing MR clutches the most important issue is the appropriate intensity and distribution of the magnetic field over the fluid gap.

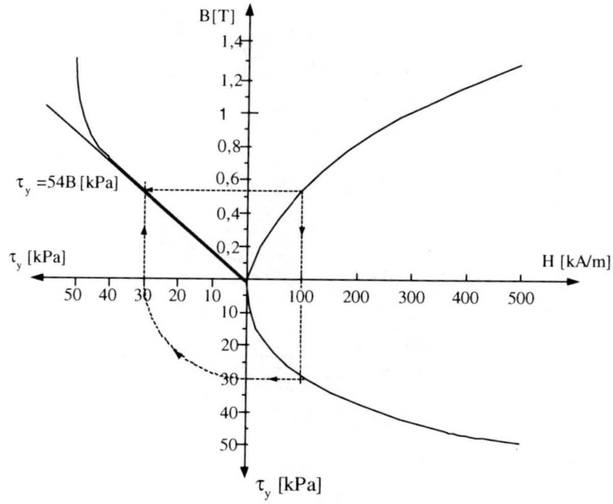

Fig. 2. Yield stress vs. magnetic flux density curve

In Fig.3a one can see the relevant parts of the magnetic circuit where the darkness of picture is proportional to the intensity of magnetic flux density. COMSOL Multiphysics software has the opportunity to visualize the flux density distribution in the centerline of the gap (Fig. 3b). Simulations showed that B is almost constant over the radius and its mean value depends only on the inner radius by the

$$B_{\text{mean}}(r_1) = 4.5 \cdot 10^{-4} r_1^2 - 7.8 \cdot 10^{-3} r_1 + 0.5873 \tag{2}$$

second order regression function. As the magnetic flux density did not exceed saturation limit at any point, the (1) relationship can be applied for torque calculation in the

$$\begin{aligned} T(r_1) &= \int_{r_1}^{r_2} \tau_y(r_1) 2\pi r^2 dr = \\ &= 113.45 \cdot (4.5 \cdot 10^{-4} r_1^2 - 7.8 \cdot 10^{-3} r_1 + 0.5873)(r_2^3 - r_1^3) \end{aligned} \tag{3}$$

expression. The relationship between the inner radius of the plate and the torque transmitted by one gap can be seen in Fig. 4. From constructional point of view the shaded area can not be taken into consideration because this space is reserved for the shaft. A maximum value of torque transmitted can be found at inner radius of about 22 mm. As both sides of plate are active, the transmissible torque of this construction is about 11.5 Nm. Power consumption of the clutch is 25.2W. The torque of the presented clutch is generally not enough for industrial applications, especially in automobile transmission systems. However the torque transmitted can be multiplied by increasing the number of plates but that construction needs of course higher magnetization power.

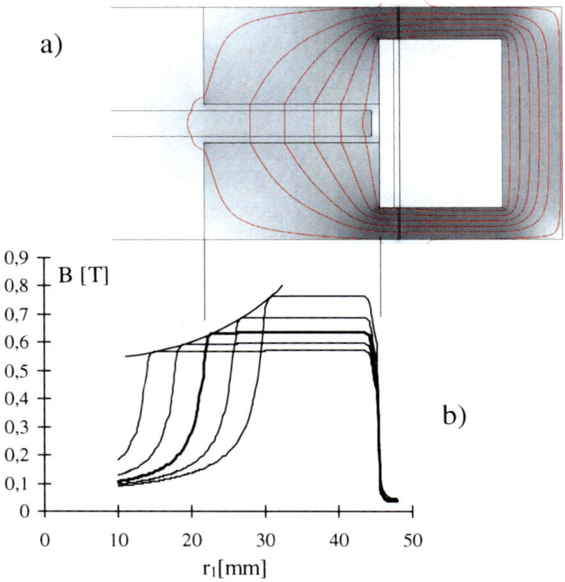

Fig. 3. Magnetic flux density distribution over the gap

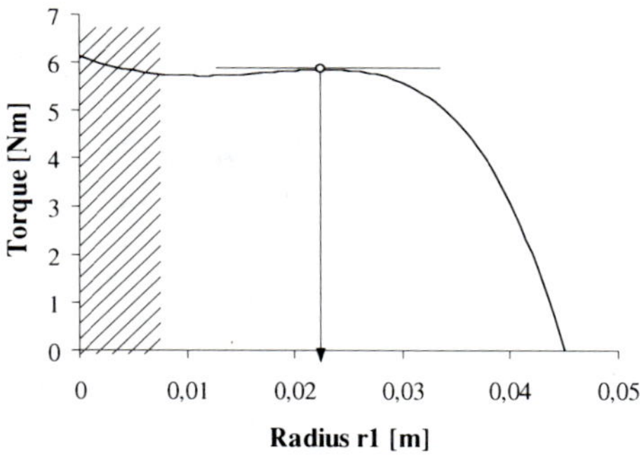

Fig. 4. Torque vs. inner radius relationship

4 Conclusions

This paper has dealt with some construction issues of a limited slip magneto-rheological clutch. The investigated clutch transmits 11.5 Nm torque at 25.2 W magnetization power. Simulation results of the magnetic circuit proved the existence of an optimal inner radius of the rotating plate. Further works will include the increase of torque by optimizing the number of plates.

Acknowledgement

Our research was supported by the project „TAMOP-4.2.1/B-09/1/KONV-2010-0003: Mobility and Environment: Research in the fields of motor vehicle industry, energetics and environment in the Central- and Western-Transdanubian Regions of Hungary". The Project is supported by the European Union and co-financed by the European Social Fund.

References

[1] Bansbach, E.E.: Torque Transfer Apparatus Using Magnetorheological Fluids. US Patent, No: 5,779,013 (1998)
[2] Lampe, D.: Anwendung von Magnetorheologischen Fluiden in Kupplungen, `http://www.donnerflug.de/publikationen/antrieb.pdf`
[3] Kavlicogu, B.M., Gordaninejad, F., Evrensel, C.A., Cobanoglu, N., Liu, Y., Fuchs, A.: A high-torque magneto-rheological fluid clutch. In: Proceedings of SPIE Conference on Smart Materials and Structures, San Diego (2002)
[4] Gratzer, F, Steinwender, H, Kusej, A: Magnetorheologische Allradkupplungen. ATZ, pp. 902–909 (Jahrgang 10, 2008)

Contribution to Environmental Operation of Industrial Robots

Z. Kolíbal* and A. Smetanová

Faculty of Mechanical Engineering, Institute of Production Machines, Systems and Robotics, Technická 2896/2, Brno, 61600, Czech Republic

Abstract. At the Institute of Production Machines, Systems and Robotics at Brno University of Technology there was mathematical model performed and after that the possibility of economical run-up and run-out of kinetic functions of industrial robots was tested. The results of experiments demonstrated validity of mathematical model and so they have shown one of possibilities of environmental operation by industrial robots. In case of multiple practical applications significant energetic savings can be found.

1 Introduction

Saving of energy is one of important conditions for preservation of permanently sustainable development. That is why ecological design started to be developed as new scientific discipline. You can meet ecological design in machinery especially in the field of new automatic CNC machine tools design [5; p.368-376]. In particular case of unique non-conventional milling centre design TAJMAC-ZPS Zlín following is mentioned in final report of project IMPULS: "Improvement of energetic requirements for movement in axes by identical dynamic parameters as well as improved machine accuracy is expected from non-conventional design". Comparison of energetic requirements of original and new machine shows positively the advantages of new solution [9; p, 18-19].

The engineering and non-engineering applications of industrial robots are extremely sophisticated for their dynamic properties by complicated run-up and run-out movements. These dynamic relations and local extremes related to consumption of required energy were tested at the Institute of Production Machines, Systems and Robotics FSI Brno for development of ecological design.

2 Local Extremes of Energy Consumption

2.1 Steady Linear Movement

The total energy is sum of energy needed for moving the object, energy for supply of driving unit and energy for the friction force of ambience.

* Corresponding author.

It is necessary to assign some simplifications. Object is moved along given linear path with steady linear or steadily accelerated / decelerated movement. Object is carried over the base through the air and friction forces are not evaluated. We evaluate the friction force of ambience. The control unit input is constant. The other influences are disregarded including energy consumed for robot components movement – there is a large amount of robot configurations and measurement can be arranged so that we are able to evaluate exactly the energy consumed for these conditions given.

$$dW = Fvdt + UIdt + F_0 vdt \tag{1}$$

The friction force of ambience is deduced from the Bernoulli equation (2) where for given object shape and size and for given ambience the constant K can be simplification.

$$F_0 = \tfrac{1}{2} CS\rho v^2 = Kv^2 \tag{2}$$

For steady linear movement acceleration is zero, velocity is constant and the movement starts with zero velocity and zero initial consumed energy.

$$W = \int_{t_0}^{t} Fvdt + \int_{t_0}^{t} UIdt + \int_{t_0}^{t} Kv^3 dt \tag{3}$$

$$W = Fvt + Pt + K\frac{s^3}{t^3}t \tag{4}$$

After substitutions and modifications we can get from the equation (3) the equation (4).

For this type of movement there are only present the driving system supply and the force necessary for overcoming of ambience friction (5):

$$W = Pt + K\frac{s^3}{t^2} \tag{5}$$

$$\frac{\partial W}{\partial t} = \frac{\partial}{\partial t}(Pt + Ks^3 t^{-2}) = 0$$

The local extreme of this function can be found:

$$t = s\sqrt[3]{\frac{2K}{P}}$$

By calculating the second derivation according time and making substitution of variable t we can find out that there is local minimum:

$$\frac{\partial^2 W}{\partial t^2} = \frac{\partial^2}{\partial t^2}(Pt + Ks^3 t^{-2}) = \frac{\partial}{\partial t}(P - Ks^3 t^{-3})$$

$$\text{and} \quad t = s\sqrt[3]{\frac{2K}{P}} > 0$$

for the equation of energy consumption in the form

$$dW = Fvdt + UIdt + F_0 vdt$$

which was converted to

$$W = Pt + K\frac{s^3}{t^2}. \tag{6}$$

We can say that the local minimum of energy consumed for object movement in dependence on the time of movement can be found. That also means that there exists minimum velocity that causes minimum energy used by robot movement.

2.2 Steadily Accelerated Movement

Similar calculation can be performed for steadily accelerated movement (7). Simplifications are the same as for steady linear movement.

$$dW = Fvdt + UIdt + F_0 vdt \tag{7}$$

$$F_0 = \tfrac{1}{2} CS\rho v = Kv^2$$

$$W = \int_{t_0}^{t} Fvdt + \int_{t_0}^{t} UIdt + \int_{t_0}^{t} Kv^3 dt$$

$$v = f(t), v = at, a = konst$$

For steadily accelerated movement acceleration is constant and the same is valid for the force F (8).

$$W = \int_{t_0}^{t} Fatdt + \int_{t_0}^{t} UIdt + \int_{t_0}^{t} Ka^3 t^3 dt \tag{8}$$

$$W = ma^2 \frac{t^2}{2} + Pt + Ka^3 \frac{t^4}{4} \tag{9}$$

The equation (9) was calculated for energy consumption by manipulation with an object for steadily accelerated movement.

The positive or negative value of the result shows to local minimum or maximum similar as it was shown above.

$$\frac{\partial W}{\partial t}=\frac{\partial}{\partial t}(ma^2\frac{t^2}{2}+Pt+Ka^3\frac{t^4}{4})=0 \tag{10}$$

$$Ka^3t^3+ma^2t+P=0 \tag{11}$$

This function (11) has two imaginary roots that are not reflected and one real root – solved with help of program Matematica5.

$$t \to -\frac{\left(\frac{2}{3}\right)^{1/3} \mathrm{ma}^2}{\left(-9\,\mathrm{Ka}^6\,\mathrm{P}+\sqrt{3}\,\sqrt{4\,\mathrm{Ka}^9\,\mathrm{ma}^6+27\,\mathrm{Ka}^{12}\,\mathrm{P}^2}\right)^{1/3}}+\frac{\left(-9\,\mathrm{Ka}^6\,\mathrm{P}+\sqrt{3}\,\sqrt{4\,\mathrm{Ka}^9\,\mathrm{ma}^6+27\,\mathrm{Ka}^{12}\,\mathrm{P}^2}\right)^{1/3}}{2^{1/3}\,3^{2/3}\,\mathrm{Ka}^3}\}$$

The equation of energy consumption by robot movement and manipulation with an object for steadily accelerated movement in the form of:

$$W=ma^2\frac{t^2}{2}+Pt+Ka^3\frac{t^4}{4} \tag{12}$$

This equation (12) has the local extreme (13) from which we can find out if it is local minimum or local maximum which help of substitution of its value into the second derivation of the function:

$$\frac{\partial^2 W}{\partial t^2}=\frac{\partial^2}{\partial t^2}(ma^2\frac{t^2}{2}+Pt+Ka^3\frac{t^4}{4})$$

$$\frac{\partial^2 W}{\partial t^2}=\frac{\partial}{\partial t}(ma^2t+P+Ka^3t^3)$$

$$\frac{\partial^2 W}{\partial t^2}=ma^2+Ka^3t^3>0 \tag{13}$$

Because both parts of the inequality are bigger then zero:

$$ma^2>0 \qquad Ka^3t^2>0$$

$$K-\frac{1}{2}CS\rho>0$$

$a^3>0$ for accelerated movement $t^2>0$.

It is not necessary to substitute the value counted with help of the program Matematica5 into the equation (13) because for all P, K, and M the root of this equation takes positive value.

There exists optimal time of movement for given path by steady linear and also steadily accelerated movement where minimum energy is needed for operation.

Therefore there also exists optimal speed respective optimal acceleration for the same types of movement from the point of view of energy consumption.

3 Experiment Implementation

The mathematical model mentioned above was verified with help of the robot type IRB 4400/60 which is placed in the laboratory of the Institute of Production Machines, Systems and Robotics at Brno University of Technology.

Fig. 1. Robot IRB 4400/2

Electric energy consumption was measured with digital wattmeter Yokogawa WT 1600 S. It is a four-canal wattmeter that makes possible measuring of voltage, current, output and electrical energy consumption on base of instant values integration measured with high frequency. The measure results were gained with the period of 200 ms.

4 Measure Results

4.1 Speed Influence

The path selected for this measure was 500 mm in y axis in the framework of the base. The following five graphs show the results of measuring in axis y with the load of 0 to 20 kg. The points are chained with help of regression curve in form of polygon with the fourth degree. This degree was the lowest one which could be used. The equations of these polygons were computed by program Excel.

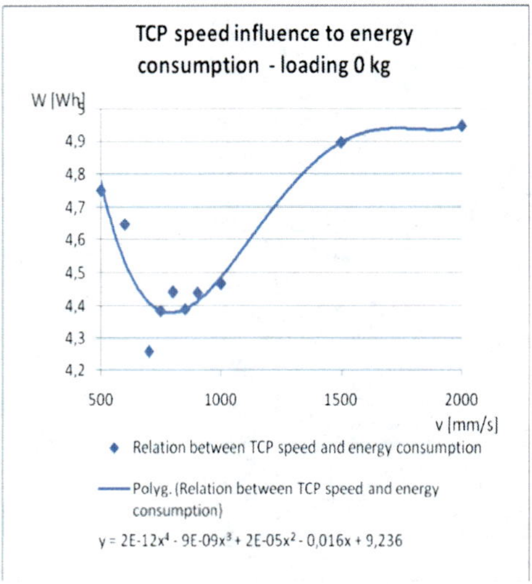

v [mm/s]	W [Wh]	σ [Wh]
500	4,571	0,0084
600	4,648	0,0426
700	4,259	0,0193
750	4,384	0,0235
800	4,442	0,0223
850	4,390	0,0219
900	4,441	0,0159
1000	4,467	0,0222
1500	4,898	0,0597
2000	4,947	0,0543

Fig. 2. Graph and table TCP speed influence to energy consumption, loading 0 kg

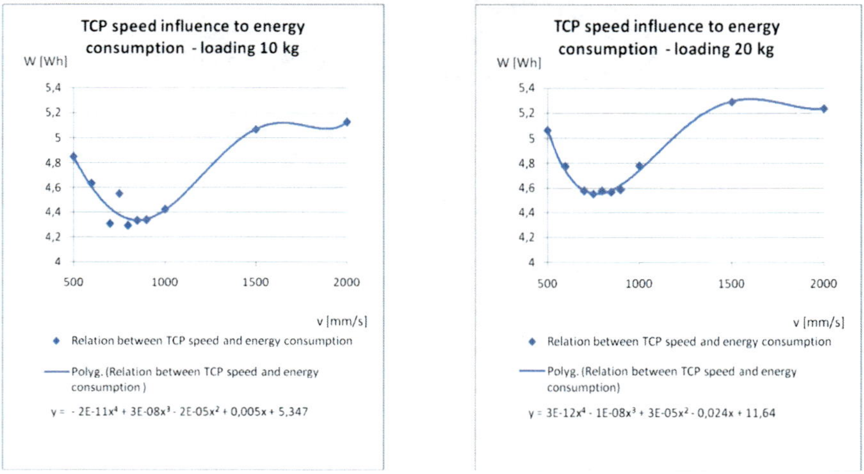

Fig. 3. TCP speed influence to energy consumption with loading 10 and 20 kg

The other graphs are included without their tables because of lack of space. In the graphs the existence of optimum speed in relation with the consumption of energy by robot operation can be seen.

4.2 Acceleration Influence

The drives run during starting with maximal acceleration that comes out from given movement, loading and maximal output of drives. The acceleration can be reduced with function AccSet which has two parameters setting up the reduction of maximal acceleration and the value jerk.

For this measure steady linear movement in axis x in the framework of the base with length 1 000 mm was chosen, TCP speed was v =2000 mm/s. With help of the function AccSet the maximum acceleration was reduced to 10-20-30-40-50-60-70-80-90% of maximal available one. The following five graphs show the results of measuring with the load of 0 to 20 kg. The points are also chained with help of regression curve in form of polygon with the degree number four which were computed by program Excel.

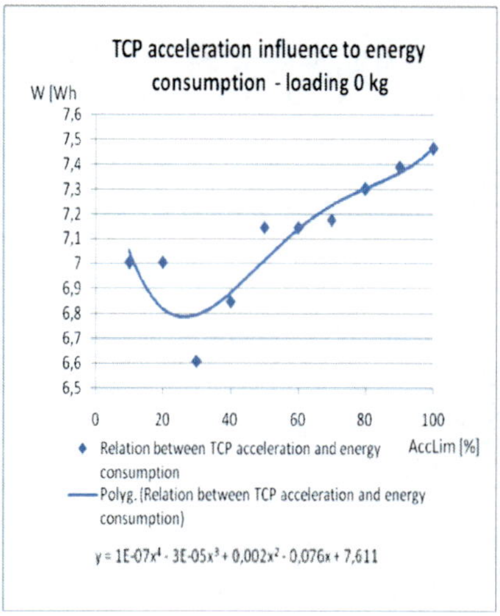

Zrychlení Reduce [%]	W [Wh]	σ [Wh]
10	7,005	0,0047
20	7,008	0,0086
30	6,610	0,0264
40	6,850	0,0124
50	7,147	0,0117
60	7,147	0,0048
70	7,178	0,0113
80	7,302	0,0085
90	7,387	0,0145
100	7,464	0,0117

Fig. 4. Graph and table acceleration reduction influence to energy consumption with loading 0 kg

The minimum energy consumption was in all cases by a = 30% of maximal available value. The other graphs are included without their tables because of lack of space. The conclusion of all these results is that there exists optimum reduction of acceleration and its use reduces the consumption of energy by robot operation.

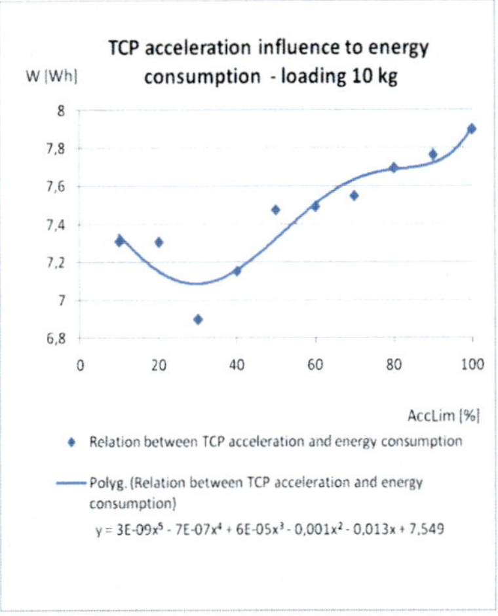

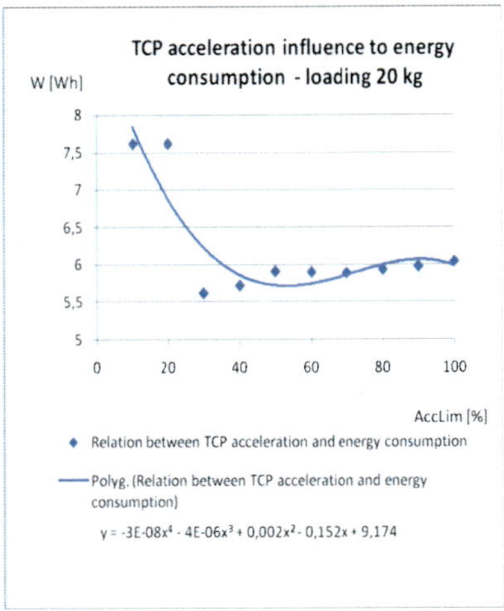

Fig. 5. Acceleration reduction influence to energy consumption, load. 10 and 20 kg

If we reduce for example acceleration of movement running for about 20 seconds from 100% to 70% , then the difference of energy consumption is (according table 2):

7,464 – 7,178 = 0,286 Wh

By continuous operation it is after 24 hours:

(0,286 /20) . 3600 . 24 = 1 235,52 Wh

In case of mass use of for example 100 robots in car production can we save during one year:

1,23552 . 100 . 365 = 45 096 kWh, that is about 45 MWh.

5 Conclusion

In case of setting larger amount of robots into production e.g. production of cars, it is possible to reach significant saving of electric energy by appropriate optimisation of their run-up and run-out movement control and so contribute to required permanently sustainable society development. This method can be also interesting by limited energy sources for example by research of distant areas or by emergency operation etc. In these cases functionality of device can be distinctly prolonged with setting of suitable parameters.

References

[1] ABB Robotics AB, Rapid Reference Manual, System Data Types and Routines, 3HAC 774-1, Vasteras, Sweden (1999)
[2] ABB Robotics AB, User's Guide, IRB 4400 Industrial Robot, 3HAC 9299-1, Vasteras, Sweden (1999)
[3] Karpíšek, Z.: Mathematics, Statistics and Probability. CERM, Brno (2002) ISBN 80-214-2055-3
[4] Kolíbal, Z., Smetanova, A.: Experimental Implementation of Energy Consumption by Robot Movement. In: RAAD 2010 (2010)
[5] Marek, J. a kol.: Konstrukce CNC obráběcích strojů. MM publishing (2010) ISBN 978-80-254-7980-3
[6] Schmidt, D.: Control and Regulation for Machinery and Mechatronics. Europa-Sobotales, Praha (2005) ISBN 80-86706-1
[7] Smetanova, A.: Optimization of Energy by Robot Motion, Doctoral Thesis, Brno University of Technology, Faculty of Mechanical Engineering, Institute of Production Machines, Systems and Robotics, Brno (2009)
[8] Smetanova, A.: Optimisatin of Energy by Robot Motion. MM Science Journal, Praha (December 2009) ISSN 1803-1269
[9] TAJMAC-ZPS Zlín, Seismicky vyvážený obráběcí stroj. Final report of project MPO FI-IM5/081 (January 22, 2011)
[10] Valný, M.: Optimization of 6 Axes Robots Control and Framework, Doctoral Thesis, Brno University of Technology, Brno (2006)

Performance of dsPIC Controller Programmed with Code Generated from Simulink

V. Lamberský and J. Vejlupek

Brno University of Technology, Faculty of Mechanical Engineering,
Technická 2896/2, Brno, 616 69, Czech Republic

Abstract. In this paper, the performance of the controller dsPIC33F CPU is tested with code generated directly from Simulink. We have created systems with different orders and multiple inputs to demonstrate controller performance with basic control algorithms. We have summarized computing times in result section for simple PID, PID controller with anti-windup protection, state space controller with constant gain and Kalman gain. Computing efficiency using different data types (doubles and integers) is evaluated. We have concluded that using fixed point arithmetic instead of doubles will significantly improve computing speed.

Keywords: dsPIC33F, Matlab Embedded Coder, Simulink, performance, controller, PID, LQ, LQG.

1 Introduction

The standard approach to develop a new controller requires a lot of time. It involves modeling, testing controller behavior, and writing a code for the device [1], [2]. Furthermore, if some changes need to be performed, this cycle is repeated.

The development cycle can be significantly reduced using higher programing languages like Simulink [3]. Using additional toolboxes [4], the code for embedded processor can be generated directly from Simulink model.

Main advantages of Embedded Coder are high speed code generation, easy implementation of changes. The code is generated directly from Simulink block scheme; hence there is no need to know C language. Replacing one block is much faster compared to rewriting the extensive block of C code.

An automatically generated code has also some disadvantages. Unlike the C language, Simulink language was not designed to straightforward translation to CPU instructions. Therefore, the generated code is not as efficient as the hand written C code [5], [1]. This problem can be minimized using special block structure and optimization functions.

When generating the code for embedded CPU, among other things, blocks for peripheral handing and CPU settings need to be added. If the CPU or some peripheral is not supported we need to create an appropriate block by ourselves. A special TLC language is used to describe how to generate the C code based on the

Simulink model. It is not necessary to do this as the dsPIC family is supported by Kerhuel Blockset [6].

There are several studies that benchmark computing efficiency of very basics [7] or special function [8] using different CPU type [9] available, but they do not provide an overall idea of computing performance for given application like a system controller.

Our aim is to use a cheap dsPIC controller instead of standard Freescale MPC, which is supported by Matlab tools and verify its performance on basics control algorithms. The CPU is programmed with the code generated directly from Simulink and the time required for executing one computing period is measured.

The controller sampling period plays one of key roles in system stability. Therefore knowledge, of computing period for given system complexity, is essential.

The CPU dsPIC33F family has the same 16 bit ALU (Arithmetic Logic Unit) with following features. It operates at a maximum clock frequency of 10 MHz, with 16-bit wide data. It is not equipped with FPU (Floating Point Unit), therefore using float data types does not lead to efficient code.

Tests using different data types and controllers demonstrate how the computing time can be reduced transforming simulation with double data types into fixed point arithmetic. Measured computing times are summarized in the result section.

2 Simulation Model

For controller design, different controlled plants have been modeled. A linear system, consisting of particles - *m* each with one degree of freedom connected with springs - *k* and dampers – *b,* whose force magnitude is computed according to (1) and (2), was used for SISO (Single Input Single Output) controller design.

$$F_k = k \cdot x \tag{1}$$

$$F_b = b \cdot \dot{x} \tag{2}$$

Varying the number of mass points will change the system order. Force acting upon the first mass point is input; position of the last mass point is the output. See Fig. 1.

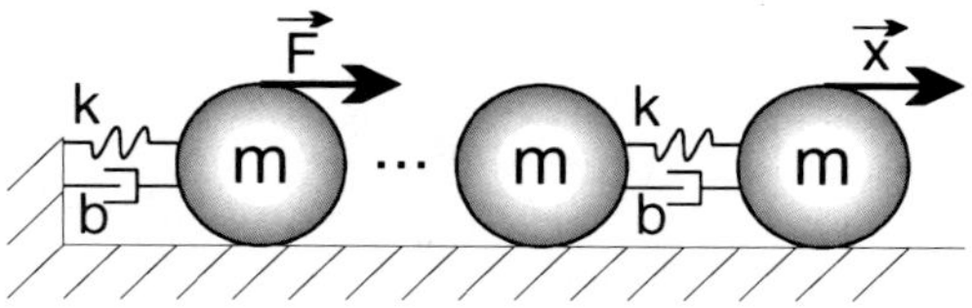

Fig. 1. Scheme of the controlled SISO plant (F – input, x – output).

For testing controller performance on complex MIMO (Multiple Input Multiple Output) systems, the previous model has been modified. There were more inputs (forces) acting upon masses and outputs (measured mass point positions) added. This system is illustrated in Fig. 2.

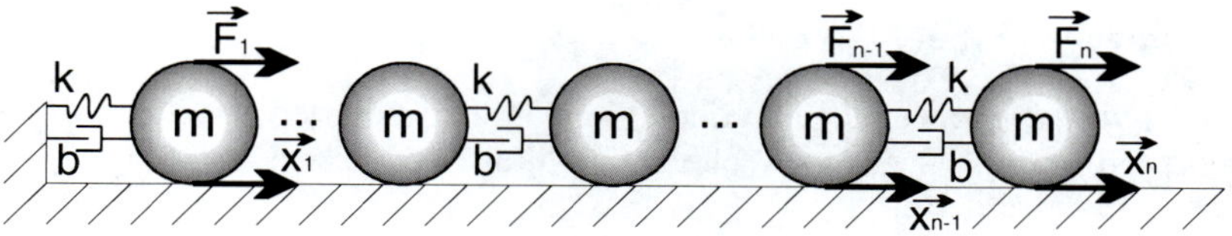

Fig. 2. Scheme of the controlled MIMO plant (F_i – input, x_i – output).

A model of this system was used in Kalman filter and the observer of state space controller.

3 Controller Design

We have tested two basics controller types. The PID type controller is not suitable for complex systems; therefore it was tested only with SISO system. The state space type controller was tested on booth system types (SISO and MIMO).

The controller was designed and tested in simulation with controlled plant and then moved to another, which was set up for code generation. This process is illustrated in Fig. 3.

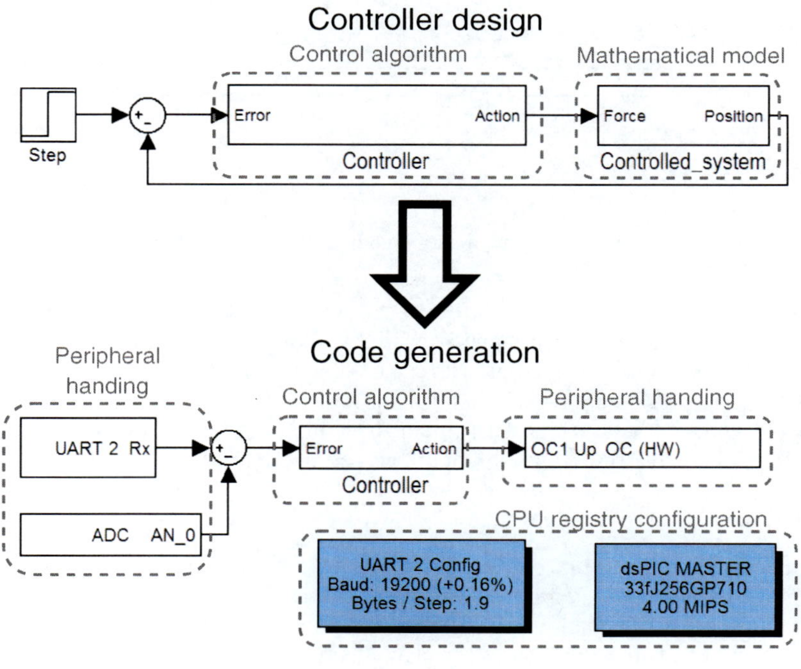

Fig. 3. Simulation scheme for designing controller and direct code generation.

PID Controller Type

The PID was implemented in a simple version and modified version with "anti-windup" protection. Their block schemes are illustrated in Fig. 4 and Fig. 5.

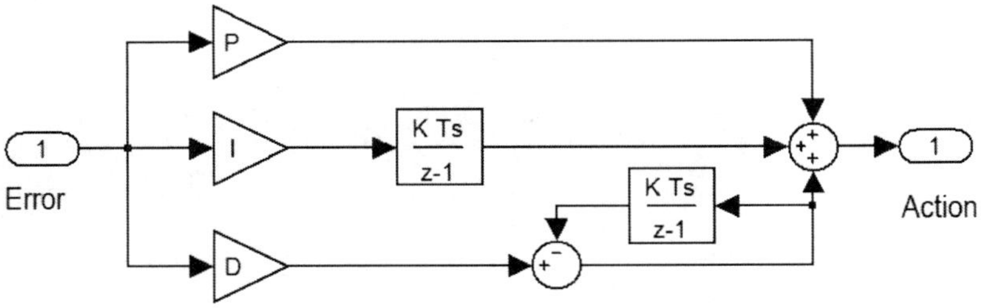

Fig. 4. Simple PID controller design

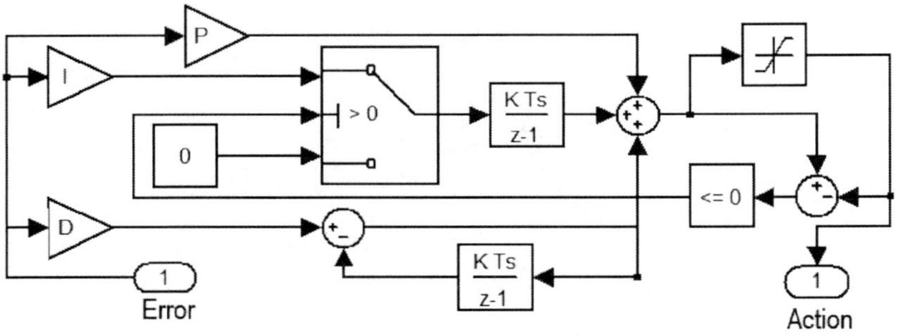

Fig. 5. PID with anti-windup protection design

By default, the floating point arithmetic, operating with double data type (real), is used. To get optimal performance on CPU without FPU we have manually converted this model to Fixed-point arithmetic. This was performed by changing the data type in Block Parameters dialog from double to fix. We have used 16 bit variable with 5 fraction bits. Performance of this controller, implemented using floating point and fixed point arithmetic, was compared.

State Space Controller Type

The second group of controllers, we have tested, is state space based controller. We have modeled LQ controller with a "Constant gain" and "Kalman gain" observer [10]. The "Kalman gain" observer is an observer which uses Kalman filter to reconstruct state vector based on a mathematical model and measured data, considering disturbances entering measurements and state variables. Booth variants of observers have an integrator placed on its input. This generally serves to compensate eventual steady state error.

The model with a "constant gain" observer is illustrated in Fig. 6. The observer matrix *K* is generated using Pole placement method [10]. The controller is implemented in both floating point and fixed point arithmetic. The fixed point arithmetic is implemented changing block signal type parameter. The same data precision as for PID controller was used.

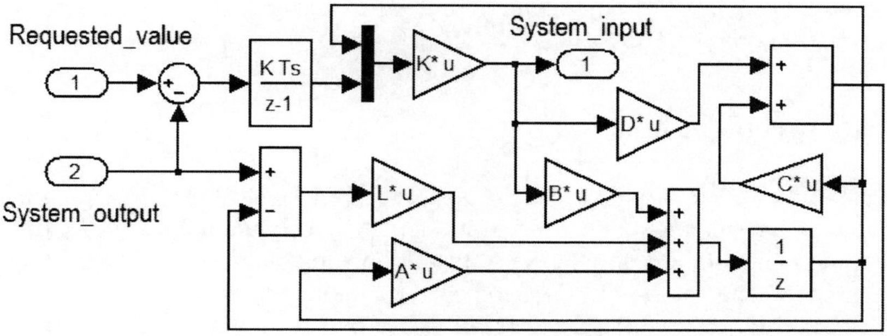

Fig. 6. State space controller with constant gain observer.

The "Kalman gain" observer was implemented using Embedded Matlab Function (Fig. 7). Simulation scheme remains similar to Fig. 6, but the state vector is computed in different way.

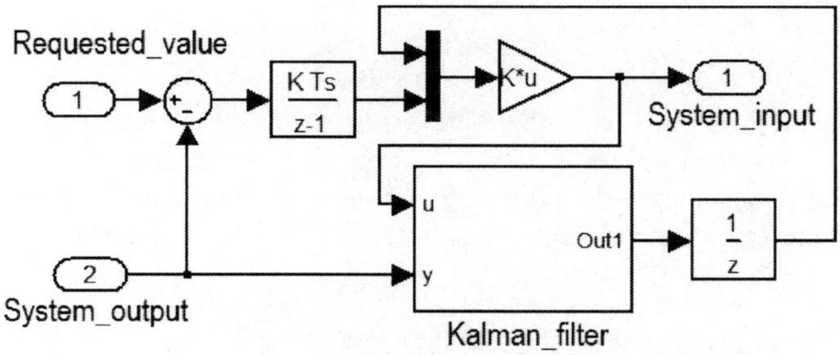

Fig. 7. State space controller with Kalman filter observer

The Kalman filtration algorithm [10] is implemented in double (3) and fixed point arithmetic (4). Matlab *fi* object does not support matrix inverse operation, therefore we have computed it using the QRD – MGS algorithm [11]. The *fimath* object - *F*, with 32 bit multiplication and accumulation word length and 5 bit fraction length, is used for arithmetic operations.

```
xk = A*xkn+B*u;     Pk = A*Pkn*A.'+Q;

xkn = xk + (Pk*C.')*((C*Pk*C.'+R)\(y-C*xk));
Pkn = Pk - (Pk*C.')*((C*Pk*C.'+R)\C*Pk);
```
(3)

```
xk = A*xkn+B*u;    Pk = A*Pkn*A.'+Q;

Kg=fi((Pk*C.')*QRD_MGS(C*Pk*C.'+R)*(C*Pk),1,16,5,F);
xkn = xk + Kg*(y-C*xk);
Pkn = Pk - Kg*C*Pk;
```
(4)

For this controller type, computation time of systems with different order and inputs was recorded.

4 Measurement Method

The experiments were done on dsPIC33FJ256GP710 CPU. The time period was computed from timer associated with one step. When simulation starts, the timer is reset, after finishing one simulation step, value from counter is read. Knowing the counter speed, the one step time period can be calculated.

The C30 Microchip compiler was set to use basics optimization level (mcpu = 33fJ256GP710 -O3 –fschedule-insns –fschedule-insns2). The inline parameters option was set for code generation from Simulink.

The simulation was performed using PIL method. That means controller inputs and outputs are connected to signal transmitted from computer, on which the simulation model was running instead of physical system. The connection between Computer and controller was realized using RS232 interface.

5 Results

In the following graph in Fig. 8, there are summarized the results from PID controller used to control a simple SISO system. For comparison, we have also designed this controller using integer data types. Since the integer is a special type of fix data type (it has zero fraction bits), computation time should be similar to fix data type.

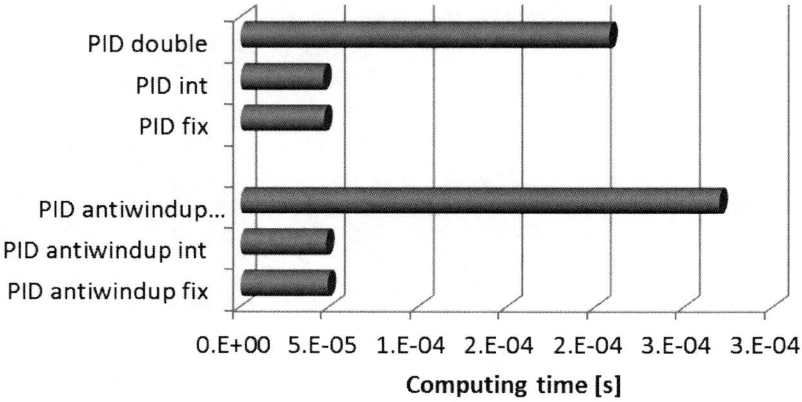

Fig. 8. PID Computation times for different data types

These computation times are summarized in table below (Table 1). Controller performance for systems with different order, one output and different data types is shown in Fig. 9.

Table 1. Measured computational times of PID type controllers

PID double	2.08E-04 s	PID antiwindup double	4.84E-04 s
PID int	4.67E-05 s	PID antiwindup int	2.70E-04 s
PID fix	4.72E-05 s	PID antiwindup fix	4.93E-04 s

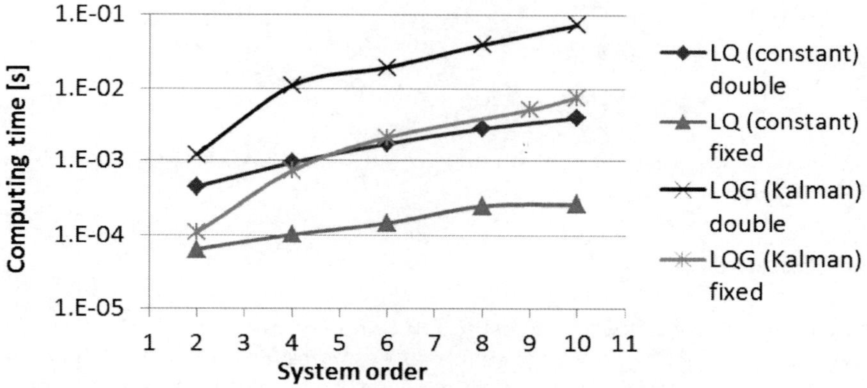

Fig. 9. Measured computational times of state space controller for system with one input using different observer gain types and different arithmetic

Computing times for system with different order and varying number of outputs were measured. The number of outputs was set to be same as the number of inputs and the system order is at least twice greater or equals the number of inputs. These results are shown in Fig. 10 and some of them are noted in Table 2.

Table 2. Measured computational times of LQG type controllers with different system order and different number of inputs

LQG double		LQG fixed	
6-th order 2 inputs	0.0261 s	6-th order 2 inputs	0.0024 s
10-th order 2 inputs	0.0894 s	10-th order 2 inputs	0.0077 s
6-th order 3 inputs	0.0336 s	6-th order 3 inputs	0.0030 s
10-th order 3 inputs	0.1070 s	10-th order 3 inputs	0.0086 s

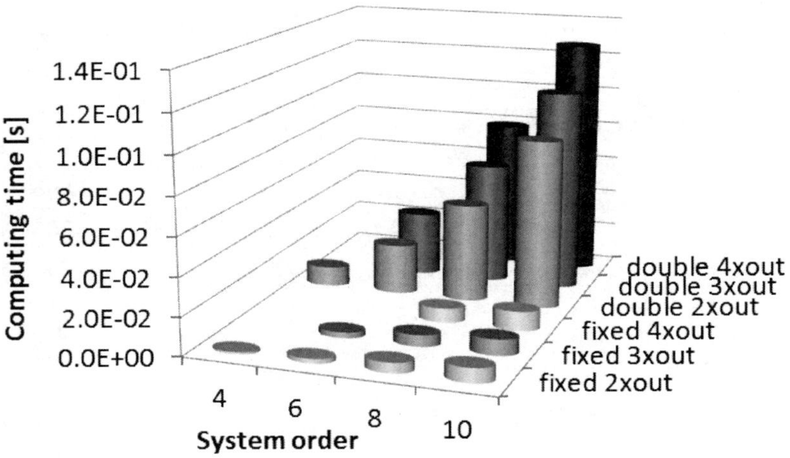

Fig. 10. Computational times of LQG controller with different system order and number of outputs

6 Conclusions

Based on the measurements, the dsPIC33F controller can be used for controlling very fast (0.1 ms time constant) simple processes using PID control algorithm. Generally, only slower systems, with time constant near one second, with higher order, and multiple inputs can be controlled using this controller.

Using fixed point arithmetic is strongly recommended as it significantly improves performance of micro controller. System converted to fixed point arithmetic with 16 bit data precision is in some cases nearly ten times faster compared to code which uses double data types, as can be seen from Fig. 9 and Fig. 10.

Acknowledgment

The work presented in this paper has been supported by research project FSI-S-11-15 *"Design, testing and implementation of control algorithms with use of nonlinear models of mechatronics systems"*.

References

[1] Vlachy, D., Zezula, P., Grepl, R.: Control unit architecture for biped robot. In: International Conference on Mechatronics 2007. Recent Advances in Mechatronics, Warsaw, pp. 6–10 (2007)

[2] Zezula, P., Vlachy, D., Grepl, R.: Simulation modeling, optimalization and stabilisation of biped robot. In: International Conference on Mechatronics 2007. Recent Advances in Mechatronics, pp. 120–125. Warsaw Univ Technol., Warsaw (2007)

[3] Grepl, R., Vejlupek, J., Lambersky, V., Jasansky, M., Vadlejch, F., Coupek, P.: Development of 4WS/4WD Experimental Vehicle: platform for research and education in mechatronics. In: IEEE International Conference on Mechatronics (ICM 2011), Istanbul, April 13-15 (2011)
[4] The MathWorks Inc. Real-Time Workshop 7 Embedded Coder Users Guide
[5] Roscoe, A.J., Blair, S.M., Burt, G.M.: Benchmarking and optimisation of Simulink code using Real-Time Workshop and Embedded Coder for inverter and microgrid control applications. In: 2009 Proceedings of the 44th International Universities Power Engineering Conference (UPEC), September 1-4, pp. 1–5 (2009)
[6] Kerhuel, L.: Simulink block set embedded target for microchip devices (2010), `http://www.kerhuel.eu/wiki/Simulink_-_Embedded_Target_for_PIC`
[7] Microchip Technology Inc. The Embedded Control Solutions Company® dsPIC33F Reference Manual
[8] Grepl, R.: Real-Time Control Prototyping in MATLAB/Simulink: review of tools for research and education in mechatronics. In: IEEE International Conference on Mechatronics (ICM 2011), Istanbul, April 13-15 (2011)
[9] Netland, Ø., Skavhaug, A.: Adaption of MathWorks Real-Time Workshop for an Unsupported Embedded Platform. In: 2010 36th EUROMICRO Conference on Software Engineering and Advanced Applications (SEAA), September 1-3, pp. 425–430 (2010)
[10] Phillips, C.L., Nagle, H.T.: Digital Control Systems Analysis and Design. Prentice-Hall, Upper Saddle River (1990)
[11] Irturk, A., Mirzaei, S., Kastner, R.: An Efficient FPGA Implementation of Scalable Matrix Inversion Core using QR Decomposition. CS2009-0938 (2009)

Stirling Engine Development Using Virtual Prototyping

V. Píštěk* and P. Novotný

Brno University of Technology, Technická 2986/2,
Brno, 616 69, Czech Republic

Abstract. A successful realization of Stirling engines is conditioned by its correct conceptual design and optimal constructional and technological mode of all parts. Initial information should provide computation of real cycles of the engines. The information form base engine dimensions like piston bore or piston stroke. After that more sophisticated methods can be used for the development. The paper presents calculation models and results of dynamics and thermodynamic cycles of the external heat supply engines. High-level FE (Finite Element), MBS (Multi Body System) or CFD (Computational Fluid Dynamics) models arising from the description of real processes which run in an external heat supply engine are used for virtual prototype of 3 kW Stirling engine with Rhombic mechanism.

1 Introduction

A prototype of γ-modification Stirling engine with an electric plant with 500W power output has been realized in previous stages. Within the frame of this pilot project 1D computational models [1], [4], and subsequently 3D computational models of virtual prototype levels [2], [3] have been developed and applied.

These computational models are based on CFD in the case of thermodynamics and on MBS in the case of structural dynamics of Stirling engine. The computational models have been verified by experiments on the Stirling engine prototype.

The application of the virtual development on Stirling engine with rhombic mechanism (β-modification) for an electric plant up to 3000 W is the aim of the next stage. Figure 1 shows parts of the virtual development of the 3000W Stirling engine. A result of the virtual development is a prototype design of the engine with specified parameters.

* Corresponding author.

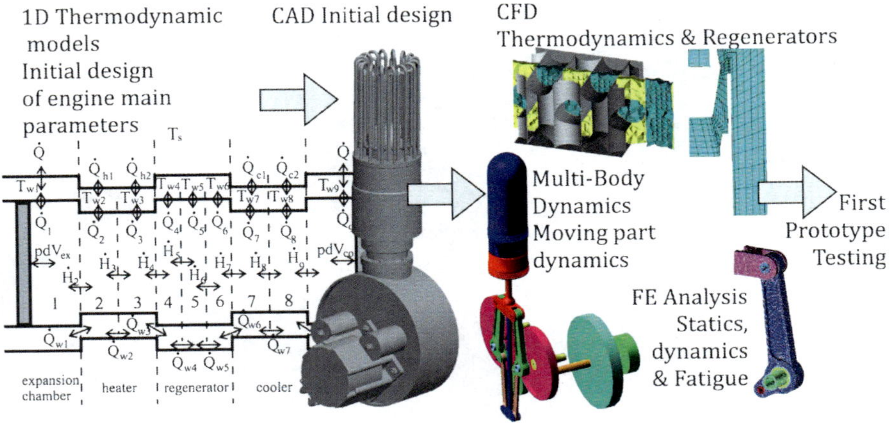

Fig. 1. Virtual development process of Stirling engine

2 Thermodynamic Cycles

In this particular case, the operating gas volume is sealed with fixed walls and two moving pistons. The engine consists of an expansion chamber, a heater, a regenerator, a cooler and a compression chamber. Individual chambers can be further split into a suitable number of zones. The volumes of the expansion and compression chambers change periodically with movements of the pistons. Temperatures, pressures and gas flow speeds in different zones can be assumed as constant for now or can be approximated between the zone limits using a linear or another suitable function [1].

In particular energy transport members include the volume work of both pistons, heat flows between the engine walls and the operating gas, enthalpy flows through the limits of the nine zones, heat flows through the walls separating different chambers and the heat flows between the engine walls and the ambient room [1].

Balance equations for the mass, energy and momentum can be designed for a generalised Stirling engine volume element. The following one-dimensional balance applies for this volume element:

Mass:

$$\frac{dm}{dt} = \dot{m}_i - \dot{m}_j \quad . \tag{1}$$

Energy:

$$\frac{dU}{dt} + \frac{1}{2}\frac{d(mw^2)}{dt} = \dot{m}_i\left(h_i + \frac{1}{2}w_i^2\right) - \dot{m}_j\left(h_j + \frac{1}{2}w_j^2\right) + \dot{Q} + P \,. \tag{2}$$

Momentum:

$$\frac{d(mw)}{dt} = \dot{m}_i w_i - \dot{m}_j w_j + \sum F \quad , \tag{3}$$

The w symbol presents velocity, t is time, $\dot{Q}$ is heat flow, P is power, U is internal energy and F means force. Forces in the equation (3) are pressure forces and friction forces, which affect the operating gas in the generalised volume.

The following applies for the power P led in/out in the equation (2):

$$P = p\dot{V} = p\frac{dV}{dt} \ , \tag{4}$$

Where V , T and r are volume, temperature and gas constant respectively. Together with the status equation for the ideal gas,

$$pV = mrT \quad , \tag{5}$$

these equations form a single-dimensional computational model of flows inside the Stirling engine.

The above set of differential equations must be supplemented with additional computational models for:

- calculation of pressure losses in the Stirling engine,
- calculation of heat transfer in the Stirling engine,
- calculation of Stirling engine operating chamber volume changes.

3 Optimizing of Driving Mechanism

A rhombic driving mechanism has been applied for the proposed Stirling engine. While the issue of the dynamic balance of the driving mechanisms has been well elaborated and is based on the fact that both the geometrical and mass parameters of individual cylinder units are identical in internal combustion engines, these prerequisites are usually not met in case of the Stirling engine.

MBS can be applied as effective tools for solving Stirling engine dynamics. Multi-body systems enable solving different dynamic issues of complex systems combining rigid and flexible bodies. In the case of Stirling engine mechanisms, they can be used to find the optimum alternative for balancing the driving mechanism [1].

In general, the driving mechanism transfers forces to attachments. These forces should be as minimal as possible. Therefore, the reaction force is selected as a target function. A balance weight static moment and an angular position of balance weight are used as parameters for optimization. Figure 2 presents results of a design process and determines global minimum of driving mechanism reaction force.

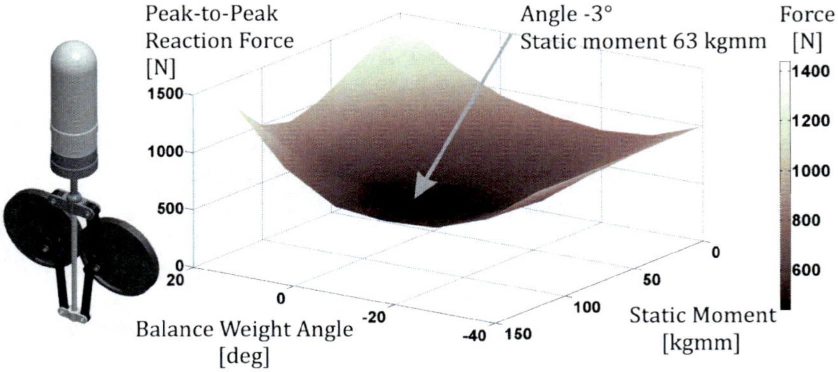

Fig. 2. Design results of driving mechanism balance weights

4 Dynamic Solution of Driving Mechanism

Driving mechanism dynamics considering loadings (e.g. combustion pressure forces) is also solved in MBS. Body deformations, velocities and accelerations and forces between different bodies are results of the MBS calculation model.

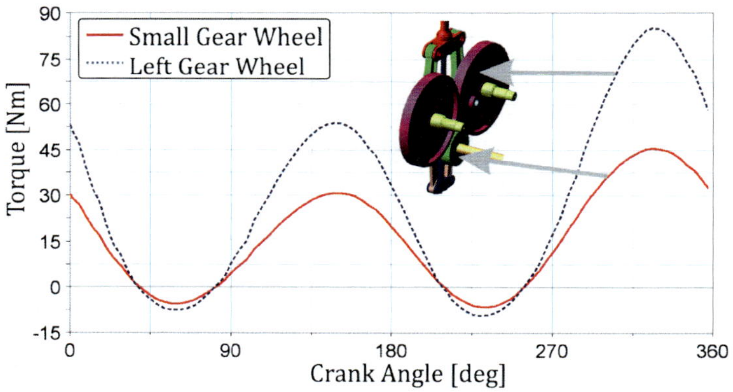

Fig. 3. Computed gear wheel torques

The engine speed of 1500 rpm is chosen for the central shaft. Practically, the values of the actual engine speed of the central shaft vary about this value in dependency on body inertia moments and torsional stiffness of some parts.

Stirling engine concept uses mainly roller bearings for interaction between the components. Slide bearing are used only for small connection rod bearings. Loadings of the bearings or gear wheels (reaction forces vs. crank angle) are determined from MBS calculation results and used for the design of the bearings. Figure 3 presents the calculated torque of a left gear wheel and a small gear wheel placed on the central shaft.

5 Conclusions

The virtual prototypes based on CFD, MBS and FEM enable to speed up the development process of new Stirling engine solutions significantly. Simultaneously they enable to decrease the costs required for the production of expensive prototypes. The documentation for production of Stirling engine components is produced on the basic of numeric simulations.

Acknowledgements

Published results were acquired using the subsidization of the Ministry of Education, Youth and Sports of the Czech Republic, research plan MSM 0021630518 "Simulation modelling of mechatronic systems" and with the help of the project FSI-S-11-8 granted by specific university research of Brno University of Technology.

References

[1] Píštěk, V., Novotný, P., Kaplan, Z.: Virtual Prototypes of External Heat Supply Engines. In: Simulation Modelling of Mechatronic Systems I (2005)
[2] Novotný, P., Píštěk, V.: CFD Approach to Stirling Engine Virtual Design Process. In: Simulation Modelling of Mechatronic Systems II (2006)
[3] Píštěk, V., Novotný, P.: Multidimensional Simulation of Stirling Engine Regenerator. In: Simulation Modelling of Mechatronic Systems III (2007)
[4] Sikora, M., Vlach, R.: The Numerical Model of α-Stirling Engine Modification. In: Simulation Modelling of Mechatronic Systems III (2007)

Complex Mechanical Design of Autonomous Robot Advee

T. Ripel and J. Hrbáček

Institute of Solid Mechanics, Mechatronics and Biomechanics,
Faculty of Mechanical Engineering, Brno University of technology,
Technicka 2896/2, Brno, 616 69, CZ

Abstract. This paper describes mechanical design of the autonomous mobile robot Advee which serves as an advertising tool. The robot provides the client with a unique kind of propagation using various fitted devices as a touch monitor or a thermal printer. Complex mechanical construction of this information portal or an electronic hostess, as it can be presented, has major effect on final characteristics of the whole robotic system. Low power consumption, stability and robust safety precautions are the most important attributes directly bound to mechanical design of the body. The outer look also plays important role, as the operation environment of the robot is indoor public sector and its purpose is to attract people's attention.

1 Introduction

Mechanical design of an autonomous mobile robot designated for commercial utilization in indoor public environment is a complex task where a robust, safe and well-developed construction stands next to attractive and quality outer design. Fusion of these two features is a base for a reliable and attractive robotic device. Although mobile robotics has taken a big step forward recently, commercial utilization of mobile robots is still rare. Necessity of long operational time, reliability, safety precautions and eye-catching outer design puts high demands on mechanical, electronic and control design of the robots and makes it a real challenge for both academic and industrial field.

This paper describes complex mechanical design of the autonomous mobile robot Advee which has been developed by Bender Robotics s.r.o. in cooperation with Faculty of Mechanical Engineering Brno University of Technology. Robot Advee is a device providing several types of advertisement in a very unique way. The robot is designed for indoor public environment. It operates as a mobile information portal which is able to orient in space, detect obstacles and avoid physical contact with persons filling the space around it.

The main goal is to design a robust and stable mobile platform integrating all the necessary equipment. The robot is intended to operate in public indoor space, which makes mechanical safety precautions essential. Also the operating performance (speed, acceleration) have to be adjusted to ensure safe performance of the robot. Attractive appearance is a must when an industrial device is intended to be successful on the market. Special attention is therefore paid to the design of the outer shell, its connection with the mechanical platform and implementation of sensory equipment.

2 Chassis Concept

The design of the chassis directly affects several features of the mobile robot. Among the most important ones belong maneuverability and operational time depending on energy efficiency.

According to the technical task defined by operating conditions and purpose of the robot, proper selection of the chassis is essential. There are several different chassis concepts usable for a mobile robot [1]. Among the most utilized concepts belong differential, omnidirectional, hybrid and Ackerman chassis types. Each one has specific features, which make it suitable for different tasks. Former experience with design of the robotic platforms (testing and racing robotic platform Bender I and II [2,3]) reduced the selection to Ackerman or hybrid chassis which are convenient for the high energy efficiency. Hybrid chassis with independent control and actuation of all drives is a complex task to both mechanical design and motion control [4]; on the contrary, Ackerman chassis requires only one propulsion drive with differential gear and one steering actuator. Even with the cost of limited maneuverability resulting in more complex motion planners due to nonholonomic constraints, the Ackerman steering still proves as the best variant.

Presentation robots require rigid construction especially in vertical direction. Steering suspension commonly used in combination with Ackerman chassis lowers the rigidity and makes the mechanical design complex. To keep the chassis stable enough and ensure continuous contact with the surface, the swinging axle is utilized. This concept allows overcoming minor obstacles while keeping the robot rigid with no special demands to the mechanical design.

Mechanical design of the platform serving as an information portal needs to take into account the ergonomics and sufficient space for all necessary equipment. The main dimensions of the platform result from an optimal position of the touch screen required to be comfortably approached. On the figure 1c the simplified 1:1 model is depicted, from which the optimal angle and position of the monitor was determined.

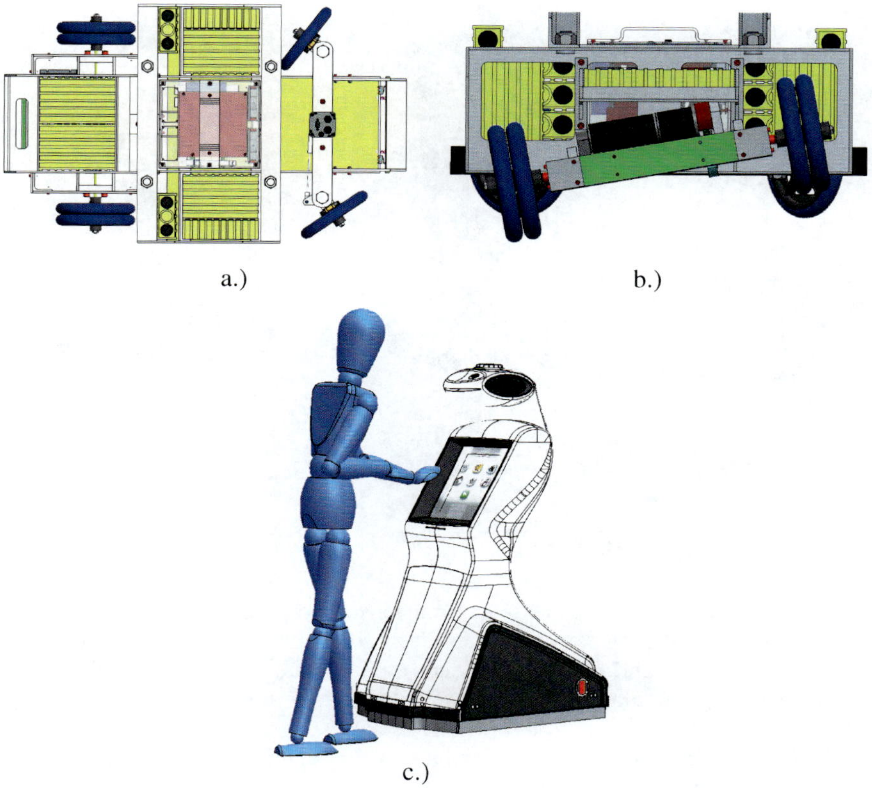

Fig. 1. a) Top view – the Ackerman platform, b) Function of the swinging axle, c) Touch screen position study

3 Frame Design

The frame of the robot is made of aluminum bars for its sufficient weight to rigidity ratio. The most important aspects defining the shape of the chassis are its purpose, proportions of all necessary equipment and the design of outer shell. The platform consists of three parts where the front part represents the main body, in the rear part the actuator with control unit is embedded and the upper part positions electronic equipment into proper locations (see Figure 1a).

Considering the complexity of the frame and weight of the all the equipment stress and strain analysis was performed using FEM. Most significant modifications were performed in the upper part, where even indiscernible displacement could cause damage of the outer shell mounted to both front and upper parts of the frame. The final modified shape of the frame is shown on Figure 2b. Maximum displacement under full static load of equipment is 0.13mm.

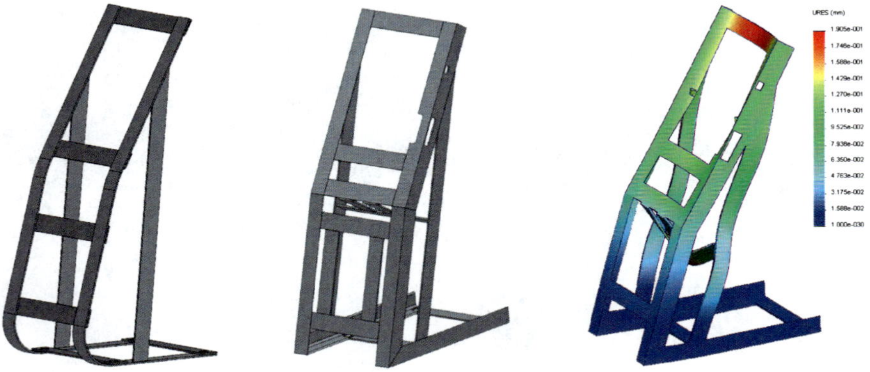

a.) b.) c.)

Fig. 2. Upper frame design a) Original b) Adjusted c) FEM study

4 Actuators Calculation

Proper selection of actuator has major impact to energy consumption and riding qualities of the platform. Power system design depends on many parameters (as for example the weight, maximum speed, type of wheels and the maximum ascent). Following calculation shows how to determine the optimal power of the actuator.

The power of the mechanical rotary movement is generally defined as a torque multiplied by an angular speed. Angular speed depends on known parameters (maximal desired velocity of the robot and wheel diameter) on the contrary to torque which is defined as an unknown action force multiplied by wheel diameter. Determination of the action force (1) is the main task while designing proper actuator.

$$F = F_v + F_a + F_s \tag{1}$$

The action force consist of three parts where F_v is a the traction force (2), F_a is the force necessary to overcome inertial factors (3) and F_s is the force comprising motion on inclined plane (4).

$$F_v = G\frac{\xi}{r} \tag{2}$$

For determination of the traction force (2) it is necessary to estimate the weight of the robot from which the gravity G can be defined. Result is multiplication of G by quotient of rolling resistance ξ and wheel diameter.

$$F_a = ma_{\max} \tag{3}$$

Desired maximal acceleration a_{max} of the robot multiplied by its weight define minimal force necessary to overcome inertial factors.

$$F_s = G \sin \alpha \tag{4}$$

The force acting against the robot movement while moving on inclined plane (loading platform) depend on the apparent gravity G and angle of ascent α.

Based on the calculation comprising all the physical influences the actuator Maxon RE 50 with 24V/200W DC motor, planetary gear 26:1 and incremental sensor has been chosen.

The determination of sufficient power of the steering mechanism is difficult task, as the calculation would depend on very precise acquaintance with mechanical model simulation and used friction model, that can vary for different types of surfaces the robot is intended to move on. Therefore the required power was determined on the base of the measurement of forces actions in the testing platform Bender II steering mechanism, adjusted to the wheel load of Advee, resulting in Berger Lahr actuator with planetary gear 40:1. The measurement scheme is shown on Figure 3. and the torque required on different surfaces is shown at Tab. 1.

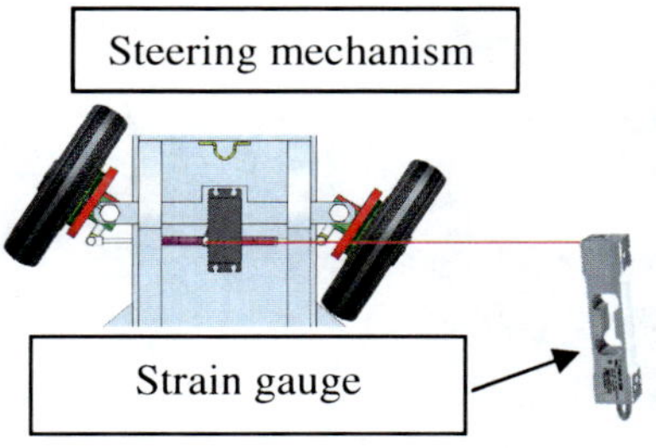

Fig. 3. Scheme of measurement

Table 1. Steering power measurement

Surface	**Force**	**Torque**
Bituminous road	35 N	2,1Nm
Sheet pavement	28 N	1,68Nm
Linoleum	26 N	1,56Nm
Carpet (long pile)	60 N	3,6Nm

5 Safety Precautions

The utilization of Advee in public spaces puts substantial demands on its safety precautions. One part of the chassis design forms a mechanical safety device directly connected to the frame. It represents the last of three levels of safety precautions, however the most important one. In case of failure of both software and electronic systems, the only way to avoid the collision with an obstacle is a mechanical bumper producing a stop signal fed to the power control.

Realization of the bumper is shown on Figure 4. The bumpers are embedded through springs to the front and rear face of the frame. While pressed (see Figure 5a) the bumper activates the circuit and the control unit immediately stops the actuator without reference to higher level control system. After release of the pressure the bumper returns to its original position.

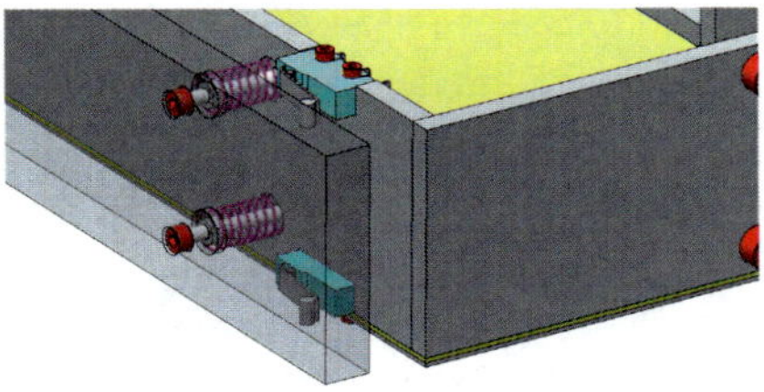

Fig. 4. The bumper model study

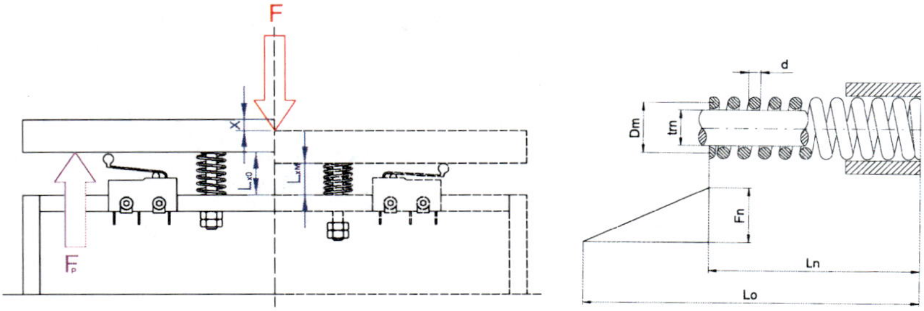

a.) b.)

Fig. 5. a) Scheme of action of force in the bumper b) The spring dimensions

$$F_p = n\left(L_0 - L_{x0}\right)c \tag{1}$$

$$F = n\left(L_0 - L_{xm}\right)c \tag{2}$$

While designing the mechanical bumper, estimation of the pressing force is crucial. Too small force would cause unexpected interruption of the robot movement, on the other side an inadequately high force could cause even injury in case of collision with a human. The sensitivity of the bumper is defined by characteristics *c* and quantity *n* of the compression springs. The total force is controllable by prestressing of the sprigs while changing its operational distance (L_0-L_{x0}). Prestress is defined in equation (1), total maximal force necessary to activate the bumper in (2). The bumpers in Advee are set to prestress of 9 N and maximal force necessary to interrupt its movement is less than 18 N. The bumper is fitted in four spots which ensure full functionality without relation to locality of bumping force action. The springs are centered by a screw connection to the chassis which allows the bumping force to act in a wide range of directions.

6 Outer Shell

To ensure attractive appearance of the robot, special attention was paid to the outer design. The outer shell is made of composite sandwich to maximize rigidity with the lowest possible weight. The body and the head form the main part of the shell with several minor pieces like bumpers or covers. Position and shape of covers is determined by service requirements. Connection of the outer shell to the mechanical platform is realized by several screw joints.

Various sensors necessary for safe autonomous motion are embedded directly to the outer shell to a specific imposition depending on its function. There are 16 infra red (IR) receivers in the head cover, 16 ultrasonic sensors in the bottom part of the body and the camera in the eye. Fusion of the data gained from sensors enables fully controlled motion of the platform with monitoring of obstacles and orientation in space [5]. Infra red receivers require to be positioned at the top of the robot to stay in visual connection with IR beacons [6] positioned high enough to avoid shielding by humans. On the contrary ultrasound sensors serve as obstacles detector and need to be placed in the lower sector of the platform. Figure 4 shows detailed shape of the outer shell with sensors position.

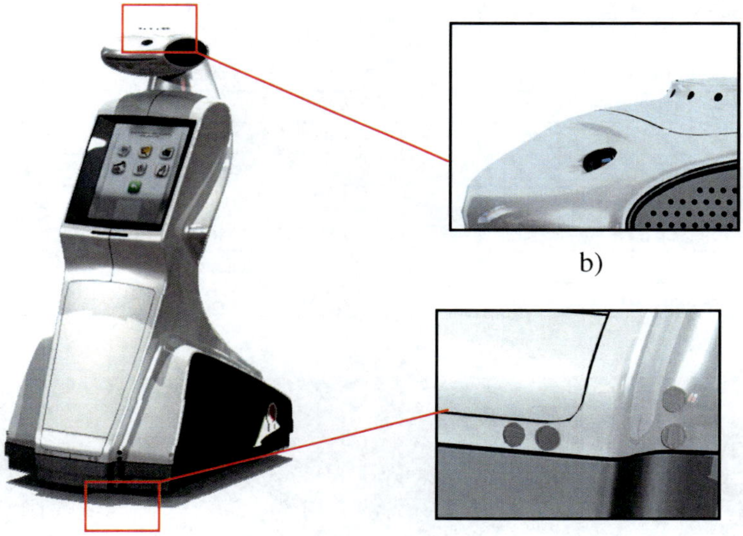

b)

a) c)

Fig. 6. a) Outer shell b) IR receivers c) Ultrasonic sensors

7 Conclusion

This paper describes complex mechanical design of the autonomous mobile robot Advee. The 1600 mm high robot proves excellent behavior while meeting requirements of an advertisement portal. Even with the weight over 80kg it can offer over 8 hours of autonomous service which shows high energy efficiency of the platform. The basic parameters of the robot are presented in Table 2.

Robot Advee is currently commercially used (Figure 7) and after 500 hours of operation no mechanical defects have appeared.

Table 2. Basic parameters of autonomous robot Advee

Basic parameters	
Main dimensions	1100x660x1600mm
Weight	80kg
Total wheel base	570mm
Wheel truck	500mm
Charging and accumulators	
Accumulator	8 cells LiFePO4
Charging	2500Wh
Continuous run time	8h
Voltage	28V
Riding qualities	
Main motor	DC motor Maxon RE 50 200W
Steering motor	DC motor Berger Lahr IclA DC024
Drive wheel diameter	900mm
Maximal speed	1m/s
Traversable sill max.	50mm
Sensors	
Incremental sensors	US Digital 2" 1024 ppr
Ultrasonic sensors	SRF08 Ultra sonic range finder
IR sensors	16 IR receivers
Touch sensors	Microswitch 250V/5A P-B172C

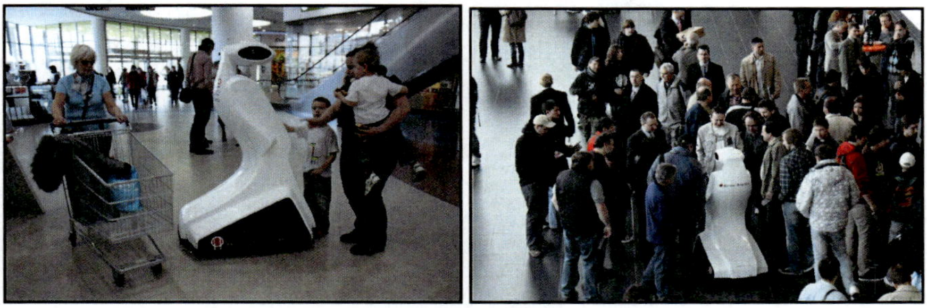

Fig. 7. Commercial utilization of Advee

Acknowledgement

Published results were acquired with the support of project FSI-S-11-15 "Design, testing and implementation of control algorithms with use of nonlinear models of mechatronics systems".

References

[1] Siegwart, R., Nourbakhsh, I.: Introduction to Autonomous Mobile Robots. MIT Press, Cambridge (2004)

[2] Věchet, S., Krejsa, J., Ondroušek, V.: The Development of Autonomous Racing Robot Bender. Engineering Mechanics 14(4), 277–287 (2007) ISSN 1802-1484

[3] Hrbáček, J., Ripel, T., Krejsa, J.: Ackermann mobile robot chassis with independent rear wheel drives. In: Proceedings of EPE- PEMC 2010, Skopje, Republic of Macedonia: 2010, s. T5-46 (T5-51 s.) (2010) ISBN: 978-1-4244-7854- 5

[4] Grepl, R., Vejlupek, J., Lambersky, V., Jasansky, M., Vadlejch, F., Coupek, P.: Development of 4WS/4WD Experimental Vehicle: platform for research and education in mechatronics. In: IEEE International Conference on Mechatronics (ICM 2011), Istanbul, April 13-15 (2011)

[5] Věchet, S., Krejsa, J., Ondroušek, V.: Sensors Data Fusion via Bayesian Filter. In: Proceedigns of EPE-PEMC 2010 Conference, Republic of Macedonia, pp. T7/29–T7/34. IEEE cat num: CFP1034A-DVD (2010) ISBN 978-1-4244-7854-5

[6] Krejsa, J., Věchet, S.: Odometry-free mobile robot localization using bearing only beacons. In: Proceedigns of EPE-PEMC 2010 Conference, Republic of Macedonia, pp. T5/40–T5/45. IEEE cat num: CFP1034A-DVD (2010) ISBN 978-1-4244-7854-5

Analysis of Mechanisms of Influence of Torque on Properties of Ring-Shaped Magnetic Cores, from the Point of View of Magnetoelastic Sensors Sensitivity

Jacek Salach, Adam Bieńkowski, and Roman Szewczyk

Institute of Metrology and Biomedical Engineering,
Warsaw University of Technology
św. A. Boboli 8, 02-525 Warsaw, Poland

Abstract. The paper presents analysis of consequences of the shearing stresses generated by torque in the ring-shaped magnetic core. In this analysis both compressive and tensile stresses generated by torque are considered, from the point of view of magnetoelastic Villari effect. As a result, changes of magnetic flux density in the core are predicted, on the basis of stress induced easy axis direction. These changes are particularly important from the point of view of development of magnetoelastic torque sensors based on ring-shaped magnetoelastic cores.

1 Introduction

The magnetoelastic effect can be observed as changes of the flux density B as well as permeability μ_a (for given magnetizing field H_m) under influence of mechanical stresses [1]. The most significant changes of magnetoelastic-based effect can be observed when participation of magnetoelastic energy in the total free energy of the material is the highest [2]. Therefore, amorphous materials without magneto-crystaline anisotropy, are most suitable for development of the magnetoelastic sensing elements, including torque sensors [3]. Moreover, due to minimization of the demagnetization energy (which reduced magnetoelastic sensitivity), cores with closed magnetic circuit should be used [4].

On the other hand, development of magnetoelastic sensors requires a mechanical analysis of stress generation in the sample, which also covers the analysis of stress distribution. For uniform distribution of stresses in the ring-shaped magnetic core three following cases may be distinguished:

- uniform distribution of compressive stresses [5],
- uniform distribution of tensile stresses [6],
- uniform distribution of shearing stresses [3] generated by the torque.

From mechanical point of view the uniform distribution of shearing stresses in the core may be considered as a coincidence of compressive and tensile stresses distribution. This assumption is a starting point for analysis of torque's influence on changes of the flux density B in the ring shaped magnetoelastic core, as presented in this paper.

2 Theoretical Analysis of the Influence of Torque on the Ring-Shaped Core

Detailed analysis of the mechanisms of the influence of axial stresses on magnetic hysteresis loop were presented before [1, 7]. The stress induced anisotropy energy E_σ, generated by stresses σ, is described by the equation (1):

$$E_\sigma = \frac{3}{2}\lambda_s \sigma \cdot \sin\phi^2 \tag{1}$$

where λ_s is the saturation magnetostriction of material and ϕ is the angle between direction of stresses σ and direction of magnetization on the sample.

As a result, the dependence connecting small changes of flux density dB, under small changes of stresses $d\sigma$, with small changes of magnetostriction $d\lambda$ under small changes of magnetizing field dH, may be determined by the equation (2) [2]:

$$\left(\frac{dB}{d\sigma}\right)_H = \left(\frac{d\lambda}{dH}\right)_\sigma \tag{2}$$

This analysis leads to generalization of the influence of the axial stresses σ (both compressive marked as negative, and tensile marked as positive) on the flux density B, achieved in the magnetoelastic core, for the given value of magnetic field H. Results of this analysis are presented graphically in figure 1.

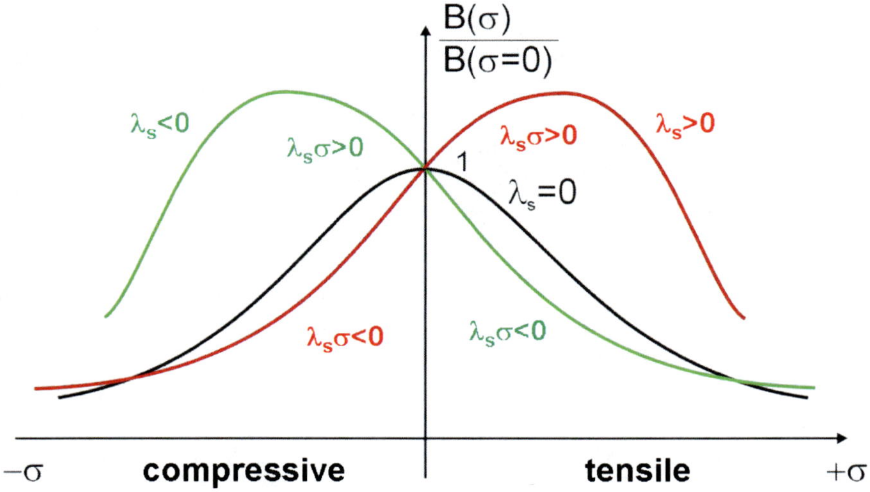

Fig. 1. Influence of both compressive and tensile stresses σ on value of flux densi-ty B (for given value of magnetizing field H_m) diversified from the point of view of saturation magnetostriction λ_s of the magnetoelastic material [5]

It should be indicated, that accordingly to equation (2), for small value of stresses σ, when value of the product of λ_s and σ is positive, increase of the flux density B may be observed, whereas for negative values of the product $\lambda_s \bullet \sigma$ decrease of the flux density B appears.

Increase of the flux density B under stresses is limited by Villari point. This point may be described by given value of stresses σ, for which the value of flux density B starts to decrease. The decrease is connected with changes sign of saturation magnetostriction λ_s under stress (as it was experimentally proven in [8]). As a result, it is in close agreement with dependence presented by the equation (2).

Figure 2a presents the method of application of the torque T_s to the ring-shaped core. Figure 2b presents the compressive $-\sigma$ and tensile $+\sigma$ stresses, generated by the shearing stresses in the elementary part of the ring-shaped core.

As it is presented in figure 3, for both directions of the torque's vector (clockwise and counter-clockwise) compressive $-\sigma$ and tensile $+\sigma$ stresses are always directed with the 45^0 angle to magnetizing field H_m and to the ribbon of the ring-shaped core.

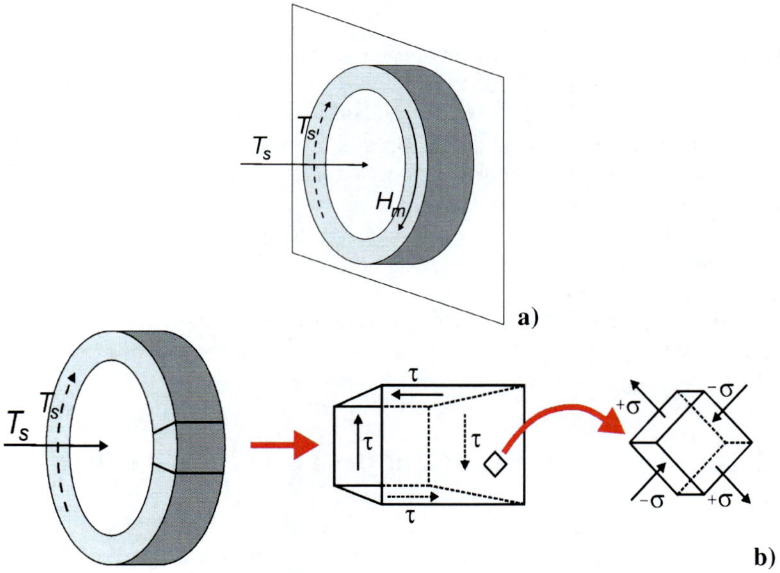

Fig. 2. Application of torque T_s to the ring-shaped sample: a) method concept, b) generation of compressive $-\sigma$ and tensile stresses $+\sigma$ due to the shearing stresses τ

A schematic diagram of the magnetic domain configuration of the elementary part of the core, subjected to torque T_s, is presented in figure 3b this elementary element is corresponding with elementary element in figure 2b.

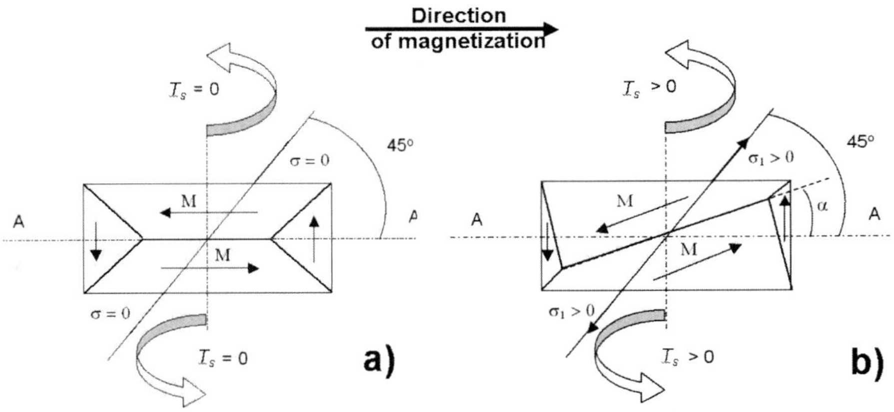

Fig. 3. Schematic diagram of the magnetic domain configuration in sample subjected to torque T_s: a) $T_s = 0$, b) $T_s > 0$.

Taking into consideration the equation (1) and mechanical analysis presented in figure 2 it should be indicated, that for both directions of torque (clockwise and counter-clockwise), as well as for both positive and negative values of the saturation magnetostriction λ_s, mechanical stress induced the easy axis of manetisation in the ring-shaped magnetic core. This axis tends to be located with the 45^0 angle to the direction of magnetization of the core (direction of the ribbon). As a result, for nearly isotropic ring-shaped core material, torque will always cause decrease of the value of flux density B, reached for given value of amplitude of magnetizing field H_m. Only for cores with pre-induced anisotropy, perpendicular to the direction of the ribbon (such as cores presented in [9]), application of torque may cause the increase of flux density B, due to reduction of the pre-induced anisotropy [10], perpendicular to the ribbon.

3 Experimental Verification of the Theoretical Analysis

To confirm the theoretical analysis, a specialized mechanical device was developed, for application of torque to the ring shaped cores, as it is presented in figure 2a. The device is presented in figure 4.

In the presented device, the ring-shaped sensing core (1) is mounted on the base planes of the special nonmagnetic backings (2), and the torque T_s is applied in the direction of axis of the ring. The nonmagnetic backings (2) have radial grooves, which enable core to be winded by magnetizing and sensing windings (3). As a result, the changes of sensing element's parameters, under the influence of the torque T_s, can be measured.

The experimental verification was performed for the ring-shaped core made of $Fe_{40}Ni_{38}Mo_4B_{18}$ amorphous alloy, annealed in 380°C for 1 hour. The ring-shaped core is 8 mm height, and it's outside diameter equals 32 mm, while inside diameter is 25 mm. The magnetic characteristics $B(H)_{Ts}$ and $B(T_s)_{Hm}$ were measured using digitally controlled hysteresisgraph HBPL-10. The torque was generated by the calibrated set of pulleys and weights. Figure 5a presents the influence of torque T_s on the shape of hysteresis $B(H)_{Ts}$ loop. Figure 5b presents the influence of torque T_s on the maximum value of flux density B_m in the core, reached for given value of amplitude o magnetizing field H_m.

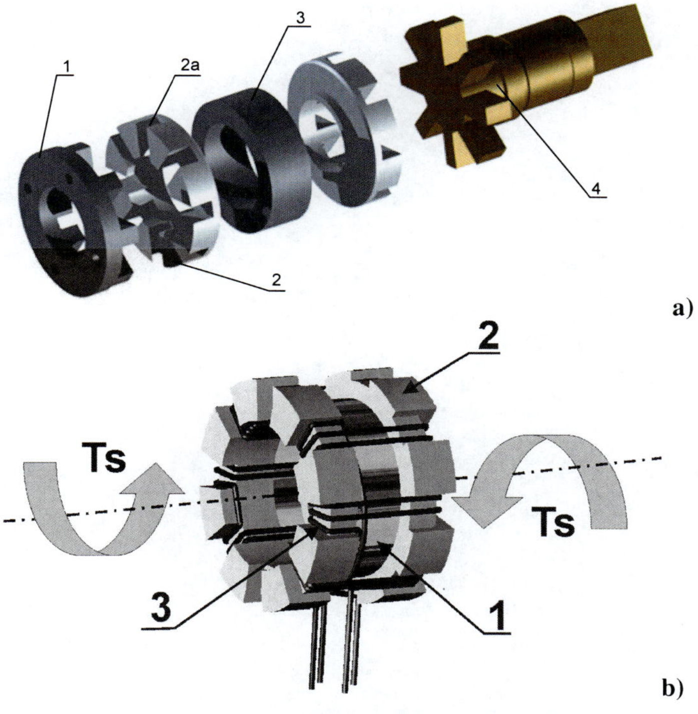

Fig. 4. Device for application of the torque T_s to magnetoelastic ring-shaped core [3]: a) elements of the device: 1 – base backing, 2 – epoxy backings, 2a – grooves for winding, 3 – sensing core, 4 – shaft, b) key element of device with winding: 1 – magnetoelastic sensing core, 2 – non-magnetic backings with grooves for winding, 3 – magnetizing and sensing windings

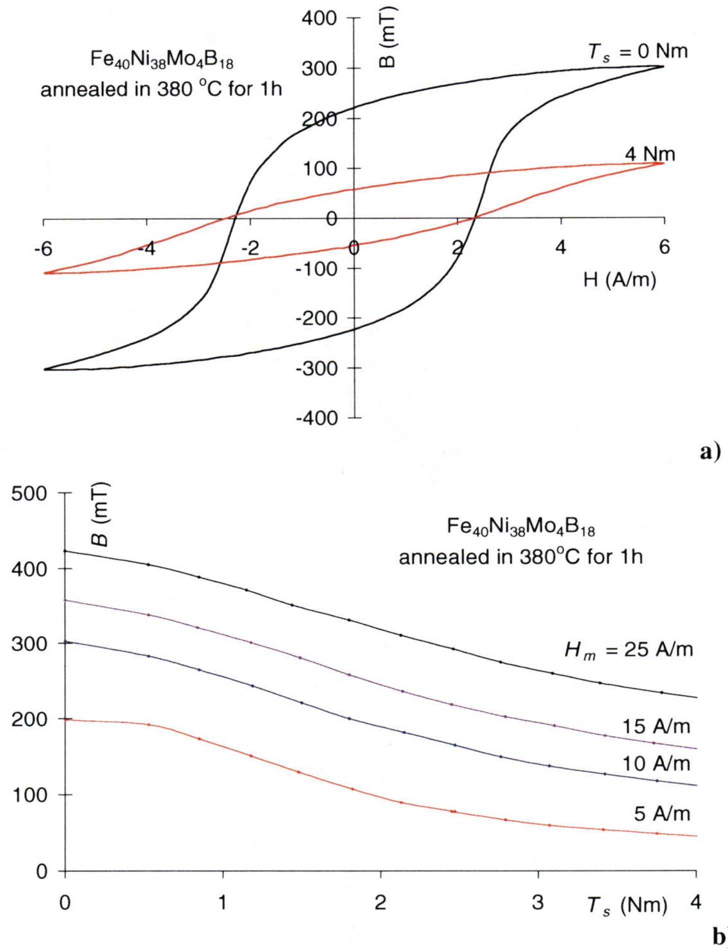

Fig. 5. The influence of torque T_s on magnetic characteristics of the core made of $Fe_{40}Ni_{38}Mo_4B_{18}$ amorphous alloy, annealed in 380°C for 1 hour:
a) influence on the shape of hysteresis loop $B(H)_{Ts}$,
b) influence on the maximum flux density B reached for given value of amplitude H_m of magnetizing field

Experimental results confirm that for amorphous magnetic cores, without anisotropy pre-induced perpendicularly to the direction of the ribbon, torque T_s causes decrease of the flux density B reached in the core. This decrease is caused by the stress-induced easy magnetisation axis, created in direction of 45^0 angle to the direction of magnetizing field.

4 Conclusion

The theoretical analysis presented in the paper indicates, that application of torque to the ring-shaped core creates the stress-induced easy magnetization axis, which is located with the 45^0 angle to the magnetizing field direction. The angle between easy axis and magnetizing field direction is equal to 45^0, for both clockwise and counter-clockwise application of torque, as well as for materials with both positive and negative saturation magnetostriction.

Such stress-induced easy magnetization axis causes a significant decrease of value of flux density B in all cores, which do not have the pre-induced perpendicular anisotropy. In the case of $Fe_{40}Ni_{38}Mo_4B_{18}$ amorphous alloy, annealed in 380°C for 1 hour, a decreasing value of the flux density B, of up to 75%, was observed for torque T_s up to 4Nm.

This work was partially supported by Polish Ministry of Science and Higher Education within research grant N505 465338.

References

[1] Bozorth, R.M.: Ferromagnetism. D. van Nostrood, New York (1956)
[2] Sablik, M.J., Kwun, H., Burkhardt, G.L., Jiles, D.C.: J. Appl. Phys. 61, 3799 (1987)
[3] Salach, J., Bieńkowski, A., Szewczyk, R.: J. Magn. Magn. Mater. 316, e607 (2007)
[4] Jiles, D.C.: Introduction to Magnetism and Magnetic Maerials. Chapman & Hall, London (1998)
[5] Bieńkowski, A., Szewczyk, R., Salach, J.: Electrical Review 1, 74 (2007)
[6] Salach, J., Szewczyk, R., Bieńkowski, A., Kolano-Burian, A., Falkowski, M.: Acta Physica Polonica A (in printing)
[7] Sablik, M., Rubin, S., Riley, L., Jiles, D., Kaminski, D., Biner, S.: J. Appl. Phys. 74, 480 (1993)
[8] Bieńkowski, A., Kulikowski, J.: J. Magn. Magn. Mater. 101, 122 (1991)
[9] Hasegawa, R.: J. Magn. Magn. Mater 100, 1 (1991)
[10] Andrei, P., Caltun, O., Stancu, A.: IEEE Trans. Magn. 34, 231 (1998)

Analyses of Micro Molding Process of the Thermoplastic Composition with Ceramic Fillers

Andrzej Skalski, Dionizy Bialo, Waldemar Wisniewski, and Lech Paszkowski

Warsaw University of Technology, Institute of Metrology and Biomedical Engineering, sw. A. Boboli 8, 02-525 Warszawa, Poland
A.Skalski@mchtr.pw.edu.pl

Abstract. The article discusses the issues related to molding of micro-elements from powder materials and it covers the first stage – injecting to the mold cavity and filling the micro-channels. The material, which constituted the composition of a special thermoplastic binder and ceramic powder, was injected. The binder consisted of paraffin, polyethylene, wax and stearic acid. The nanometer powders from Al_2O_3 ceramic material with granularity of 660 nm and 135 nm and irregularly shaped particles were used. Different loading of composition by powder were used. The analyses of impact of injection parameters such as the mold's temperature and the temperature of the material, on the quality of filling the micro-mold cavity with the material, were presented. The special molding insert with micro-channels was made to perform the analyses to that effect with width from 50 to 1000 µm. The presented results of filling the micro-channels indicate considerable influence of the mold's temperature and the cross-section of micro-channels. Slight impact of the material's temperature was observed, however this factor does not have a considerable influence on filling the channel. The obtained information was used in the experiments of injection the samples for bending tests and tensile tests, and the shapes in the form of toothed wheels. The analyses are part of the tests of injection molding of micro-elements with use of powders. The second part of this article will discuss the impact of pressure and filling of the composition with powder.

Keywords: injection molding, micro-elements, ceramic powders.

1 Introduction

In the recent decade we have observed intensive development of processes involving production of micro-products and micro-elements related to development of Micro Technologies Systems (MTS). The Micro Technologies Systems combine micro-electronics with many other micro technologies e.g. mechanical, optical, chemical technologies, etc. The objective is to fully concentrate various functions in a single miniaturized product. It should be noted that the dimensions of the produced structures or their components are expressed in micrometers. Previously used methods of production of micro-elements are limited to selected groups of materials. This pertains to the following processes: LIGA, laser processing, erosion, etching, etc [1,2,4,5]. A large problem involves adaptation of well-known technologies from the macro scale to the micro scale. It turns out that this is not always possible or does not give the desired effects because the dimensions of the elements are less than 1 mm. One of the more promising processes is the process of molding micro-elements through injecting. Such molding method allows to produce micro-elements with complex shapes, with high accuracy, in large series, and in a manner that is efficient and competitive compared to other production methods.

The micro injection molding is based on the currently used method of making products from thermoplastic polymers through injection. It should be emphasized that the first analyses and applications focused on micro-elements manufactured form such materials [3,5,6,7].

The micro-elements injected with metal and ceramic powders are designated for operation in harsher thermal and mechanical conditions than elements made from plastics. In this technology, the problem involves not only making the injection micro–mold but also selecting the parameters for the entire injection process. No tests have been developed which would allow to unequivocally determine the material's suitability for production of micro-elements by the injection method. Such test, which is known as the spiral test, exists for macro-elements.

The presented article concerns the analyses of the first stage of micro-element production from ceramic powders. The entire process of micro-element production includes preparation of the mass consisting of powder and special thermoplastic binder, injection, binder removal (debinding) and sintering [8,9,10]. The results of filling the channels and the micro-cavity of molds were presented along with the examples of the obtained shapes. In this part of the article, the impact of temperature of the injected mass and the mold's temperature was discussed.

2 Materials

The injection molding uses the masses which are the compositions consisting of the thermoplastic binder and the specified micro-powder. In the presented analyses, Al_2O_3 ceramic micro-powders with granularities of 0.66 μm and 135 nm were used.

The description of the individual powders is included in table 1 and the photographs of the selected powders are presented in figure 1.

Table 1. Powder description

Powder type	Symbol	Manufacturer	Average granularity	Particle shape
Al_2O_3	M	Martoxide	0.66 μm	Irregular
Al_2O_3	TM	Tamei Chemicals Co, Ltd	135 nm	Irregular

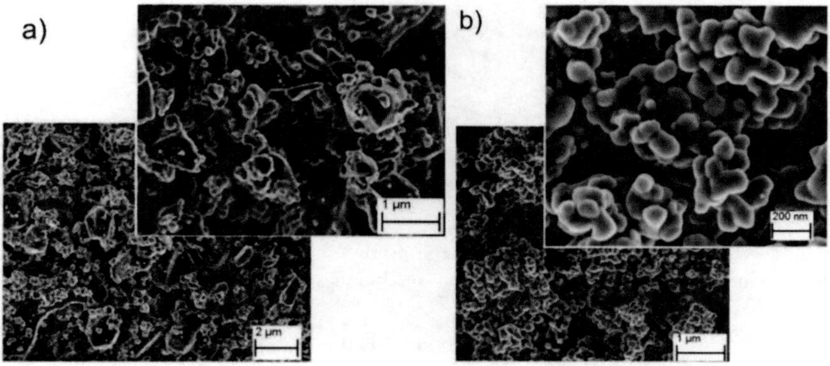

Fig. 1. Al_2O_3 ceramic powders: M, 660 nm (a) and TM, 135 nm (b)

The composition of the binder is as follows:

LD polyethylene	20%
paraffin	69%
Carnauba wax	10%
stearic acid	1%

Preparation of the injection mass consisting of powder and binder was carried out in the type 2Z mixer with the heating mantle in the temperature of 125 °C for the time of 1 hour. Such period of time was sufficient to prepare a homogeneous mass.

For tests involving injections and filling the micro-channels, the mass with the powder content of V_p = 50 and 55% in terms of volume was used. Such high V_p values are necessary to produce micro-elements with complicated shapes and very small structural details.

3 Tests

Special injection mold with micro-channels with cross-sections of 0.033; 0.077; 0.13; 0.21; 0.3; 0.54 and 0.84 mm2 was designed for the analyses. This was necessary because there are no standardized tests for analyzing the micro-molding, and the spiral test, which is commonly used in molding the macro-elements from thermoplastics, cannot be used in the case in question. The mold has a built-in heater with the temperature regulation system as well as the cooling system. The molding insert is presented in figure 2. The mold's injection nozzle directed the mass centrally to the molding insert, from where it continued to flow to micro-channels located along the radii. Dimensions of the channels are presented in table 2.

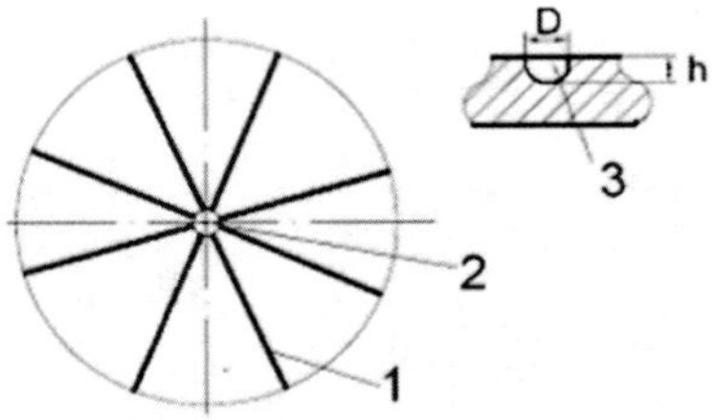

Fig. 2. The molding insert for evaluating the inflow distance of the mass in the mold. 1 – micro-channel, 2 – injection point, 3 – cross-section view of the channel.

Table 2. Micro-channel dimensions

Width	D, mm	0.2	0.3	0.4	0.5	0.6	0.8	1
Depth	h, mm	0.17	0.27	0.35	0.44	0.54	0.71	0.9
Cross-section	S, mm2	0.033	0.077	0.13	0.21	0.3	0.54	0.84

The mold, which was designed in such way, allows to simultaneously obtain a series of data from one injection cycle because it gives the information on the process of filling several micro-channels with different cross-sections at the same time.

The following technological parameters of the injection process were used:

Temperature of the mass T_w – 125, 150 and 170 °C
Temperature of the mold T_f – 25, 40, 50, 60, 70 and 80 °C
Pressure p – 60 MPa
Powder content in the mass V_p – 50 and 55% by vol.

4 Results

Figure 3 presents the chart describing the dependencies, which are typical of the discussed analyses, between the micro-channel filling process (mass inflow distance, L) and the temperature of the mold T_f. The chart pertains to the mass containing the ceramic powder with granularity of 0.135 μm with powder content in the mass of $V_p = 55\%$.

The obtained results show that as the temperature T_f increases, the flow distance of the material increases.

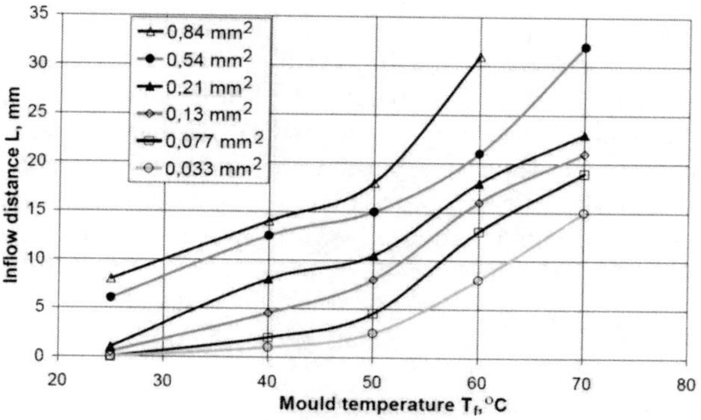

Fig. 3. Dependency among the inflow distance L, the temperature of the mold T_f and the cross-sections of micro-channels for mass with TM powder (135 nm) with V_p = 50% and T_w =125 °C.

The chart shows the temperature threshold above which it is easier for the material to fill the channels. For larger micro-channels, this temperature is approx. 45 °C, for smaller ones it is slightly higher. The cross-section of the channel is an important factor. As it increases, the inflow distance considerably increases. In such case, the stream of material flowing in the channel has relatively higher volume and therefore higher weight. Consequently, it cools off at a slower pace in contact with the mold which is colder. According to literature [6], under extreme conditions, when the length of the feeding channels is large and the micro-molds have complicated shapes, in order to facilitate the flow of mass the so-called "Variotherm" process is used during the injection, in which the temperature of the mold reaches up to 100 °C.

Figures 4 and 5 present dependencies between the inflow distance L (mm) and the temperature of the mold T_f for various cross-sections of micro-channels and both injected materials.

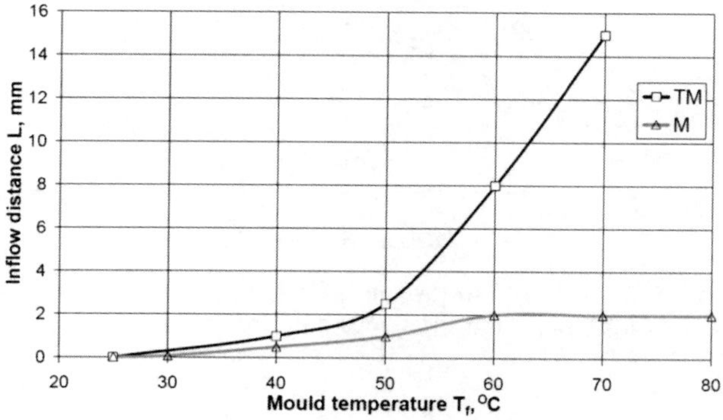

Fig. 4. Dependency between the inflow distance L of mass in the micro-channel with cross-section of S = 0.033 mm^2 and the temperature of the mold T_f with V_p = 50% and T_w =125 °C.

Figure 4 concerns the filling of the micro-channel with cross-section of 0.033 mm^2. The dependency between the inflow distance and the temperature of the mold is also visible here. If the temperature is lower than 50 °C, the inflow distance in the micro-channel is similar for the individual materials and it amounts to 1 and 2.5 mm. After this value is exceeded, the impact of the type of material is observed. The mass with finer powder fills the micro-channels more efficiently, and as the temperature increases the inflow distance rapidly increases. The mass with powder M with granularity of 0.66 μm fills the micro-channel less efficiently. As the mold temperature grows, the inflow distance increases to the specified level, after which it has approximately the same value despite increase of temperature T_f.

Figure 5 presents the dependency between the mold temperature and the inflow distance for micro-channel with cross-section of 0.21 mm^2. This dependency is the same as in figure 4. Increasing the cross-section of the micro-channel resulted in extension of the inflow distance, and the curves in figures 4 and 5 for the individual granularities show the same trend.

As T_f increases, the mass with the nanometer powder (TM) has higher inflow distance. After increasing the cross-section of the channel almost six times, the mass additionally increases the inflow distance and it begins to fill the channel at a lower mold temperature (room temperature). As the cross-section increases, the slope of the curve in relation to the mold temperature axis decreases. This is the evidence for non-proportional growth of the inflow distance in relation to increase of the channel's cross-section, as well as the smaller impact of T_f. In the case of powder M with larger granularity, the curves for various cross-sections of micro-channels display similar dependency. In other words, after exceeding certain temperature threshold (approx. 60 °C), the impact of the mold temperature is not observed.

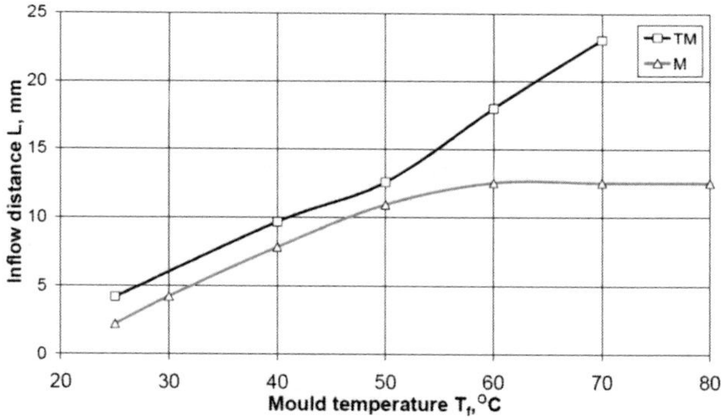

Fig. 5. Dependency between the inflow distance L of mass in the micro-channel with cross-section of S = 0.21 mm^2 and the temperature of the mold T_f with V_p = 50% and T_w =125 °C.

The next technological parameter, which may result in improvement of parameters of the produced elements, is the temperature of the injected material (T_w). Figure 6 presents the inflow distance figures as a function of micro-channels'

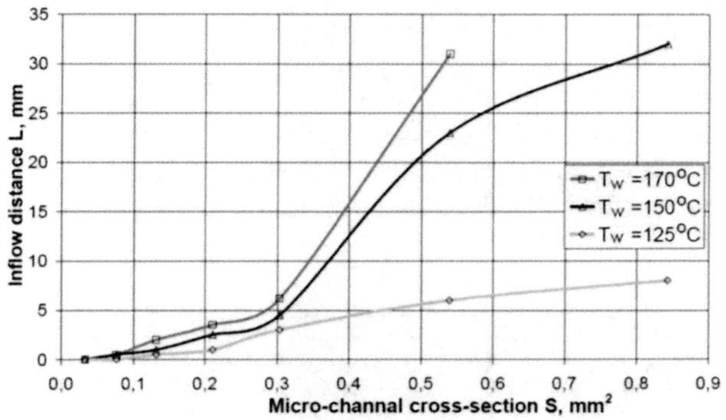

Fig. 6. Dependency of the inflow distance L of the mass on the cross-section of the micro-channel for mass with powder TM with $V_p = 50\%$ for mold temperature of T_f = 25 °C and mass temperature of T_w =125, 150 and 170 °C.

cross-sections for the temperature of the injected material of 125, 150 and 170 °C. The injected mass contained the ceramic powder with granularity of 135 nm. It may be observed that, as it was predicted, the increase of the material's temperature will improve the efficacy of filling the micro-channel and also filling the mold. In such situation, it is possible to reduce the temperature of the mold while preserving the satisfactory efficiency of filling it – figure 7.

Simultaneous impact of temperature of the mass and temperature of the mold on the inflow distance in channels with various cross-sections is presented in figure 7. The obtained curves confirm that the conclusion is right. Increase of mass temperature T_w results in increasing the inflow distance, especially for larger cross-sections of micro-channels. In addition, increasing the mold temperature facilitates the flow of mass, which is very important for small cross-sections of micro-channels when the increase of T_w itself does not have a sufficient impact on filling the micro-channels or micro-molds.

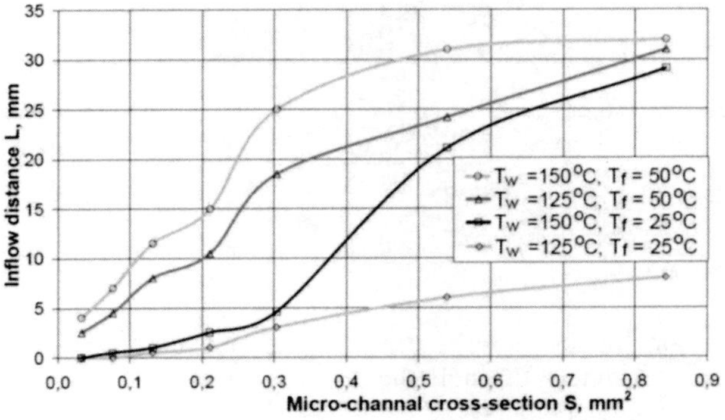

Fig. 7. Dependency of the inflow distance L of the mass on the cross-section of the micro-channel for mass with powder TM with $V_p = 50\%$ for mold temperature of T_f = 25 and 50 °C and mass temperature of T_w = 125 and 150 °C.

It should be remembered that the temperature of the material and the mold should not be excessively increased due to possibility of degradation of the mass before its injection into the mold cavity.

The presented results of the analyses were verified through production of micro-elements in the form of bars for bending tests and samples for tensile tests as well as miniature toothed wheels. The examples of such elements are presented in figures 8 and 9.

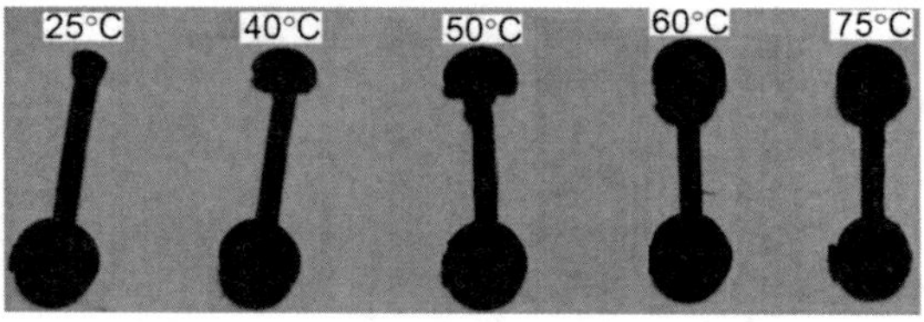

Fig. 8. Micro-samples for tensile tests molded at various temperatures of the mold and with the following injection conditions: T_w =115 °C, p = 60 MPa and V_p = 60%

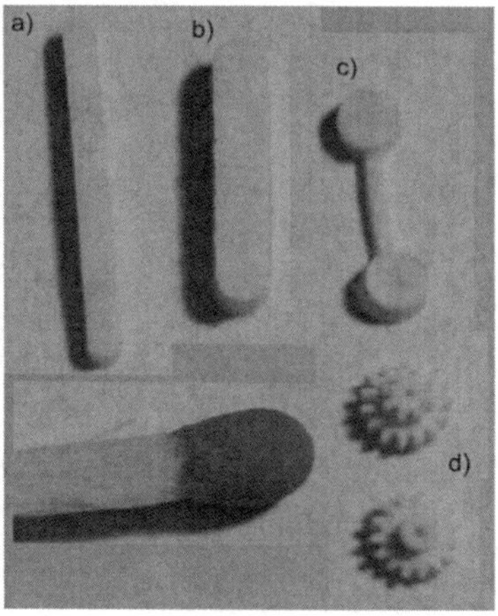

Fig. 9. Examples of molded micro-elements before sintering:a – bars for bending tests: 1x1x10, b – bar for bending tests: 2x2x12 mm, c – sample for tensile tests 0.5x0.5x5 mm, d – toothed wheel

5 Conclusions

The following conclusions may be drawn on the basis of conducted analyses of filling the micro-channels through the process of injection molding of micro-elements from ceramic powders:

- The proposed analysis method, which uses a special molding insert with many channels with different cross-sections, has turned out to be very useful in evaluating the behavior of the masses consisting of ceramic powder and binder during the process of micro-injecting.

- The greater the cross-section of the micro-channel, the longer the inflow distance of the mass with the set T_w, T_f and p.
- The most important parameter, which determines the course of filling the micro-channels, is the mold temperature. It has to be considerably higher than in the case of molding of macro-elements with ceramic macro-powders.
- The finer the powder particles, the easier it is for the mass to flow to micro-channels.

References

[1] Oczoś, K.E.: Kształtowanie mikroczęści – charakterystyka sposobów mikroobróbki i ich zastosowanie. Mechanik 5-6, s. 309–s.327 (1999)

[2] Mrugalski, Z., Rymuza, Z.: Mikromechanizmy. Pomiary, Automatyka i Kontrola 6, s.4–s.9 (1998)

[3] Gad-el-Hak, M.: The MEMS Handbook, 2nd edn. Virginia Com. University, USA (2005)

[4] Poiter, V., et al.: Micro Powder Injection Moulding. In: EURO PM 2000, Munich, Germany, October 18-20, vol. PIM, pp. 259–264 (2000)

[5] Hasselbach, J., et al.: Investigation on the International State of Art of Micro Production Technology. In: Euspen Int. Topical Conf., Aachen, Germany, May 19-20, pp. 11–18 (2003)

[6] Piotter, V., et al.: Micro Injection Molding of Components for Microsystems. In: 1st Euspen Topical Conf. on Fabrication and Metrology in Nanotechnology, Copenhagen, May 28-30, vol. 1, pp. 182–189 (2000)

[7] Biało, D., Skalski, A., Paszkowski, L.: Selected Problems of Micro Injection Moulding of Microelements. In: Pr.zbiorowa pod red J. Jabłońskiego pt. Recent Advances in Mechatronics, pp. s. 370-s.374. Springer, Heidelberg (2007)

[8] Benzler, T., et al.: Fabrication of Microstructures by MIM and CIM. In: PM World Congress PIM, Granada, Spain, vol. 3, pp. 9–14 (1998)

[9] Zauner, R., Korb, G.: Micro Powder Injection Molding for Microstructured Components. In: PM Plansee Seminar, Reute, Austia, vol. PL 5, pp. 59–68 (2005)

[10] Biało, D.: Metoda formowania wyrobów z proszków poprzez wtrysk. In: Konferencja, N.-T.: Postępy w Elektrotechnologii, Szklarska Poręba, września 14-16, r. t. 1, pp. s. 235-s. 240 (1994)

The Application of MR Dampers in the Field of Semiactive Vehicle Suspension

Z. Strecker[1] and B. Růžička[2]

[1] Institute of Machine and Industrial Design, Brno University of Technology, Technická 2896/2; 616 69, Brno; CZ
streckerz@gmail.com
[2] Institute of Machine and Industrial Design, Brno University of Technology, Technická 2896/2; 616 69, Brno; CZ
bronislav.ruzicka@trw.com

Abstract. Electrorheological (ER) and magnetorheological (MR) dampers are becoming to be used in the adaptive wheel suspension systems of serial-produced cars. If the adaptive system is fast enough (it can respond on high frequencies), it can be called semiactive. If the adaptive damper is supposed to work as semiactive, its response time should be approximately 10 times smaller than the reciprocal resonant frequency of the controlled system. From this point of view, the ER dampers are appropriate for semiactive suspension of both sprung and unsprung mass. MR dampers with standard low-voltage current drivers are too slow for semiactive suspension of unsprung mass. However the MR damper time response is mainly caused by coil induction and connected current driver. The time response of MR damper with the current driver with higher voltage should be fast enough even for semiactive suspension of the unsprung mass.

1 Introduction

The main goals of car suspension are isolation of passengers from vibrations (comfort function) and maximizing friction between tyre and road (safety function). Standard car suspension doesn't allow changing of the damping characteristics. However this is not ideal because the car parameters (especially sprung mass) alter during usage. Especially for small cars with big weight rate of full loaded and empty car it is difficult to ensure good friction while maintaining ride comfort. That is the reason why adaptive systems are starting to be used. They are able to improve both ride comfort and wheel friction. The adaptive systems based on damping characteristics change can be divided into two groups – adaptive passive and semiactive systems. Adaptive passive systems can change long-term damping characteristics while semiactive systems can respond to individual bumps.

Recently, the ER and MR dampers are becoming to be used as adaptive dampers.

The principle of the MR damper is on Fig. 1. The main difference between an ordinary and MR damper is the coil around the piston of the MR damper and the fill – a MR fluid. The MR fluid is a colloid suspension consisting of carrier fluid and ferromagnetic microparticles (2-5 μm). When electric current is applied, the magnetic field is generated in the gap through which the MR fluid flows and this field causes a concatenation of particles and an apparent increase of the fluid viscosity.

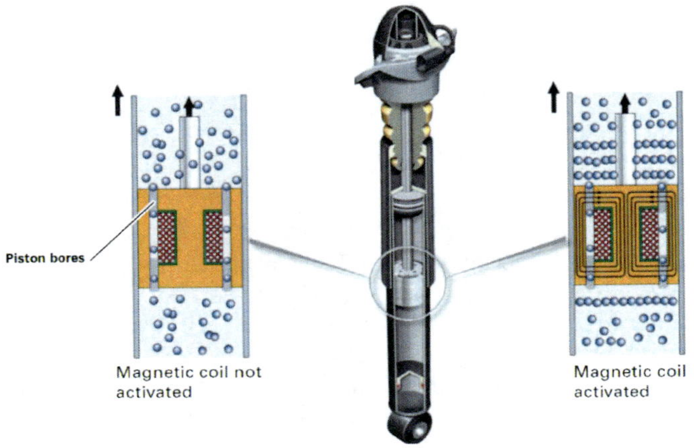

Fig. 1. MR damper [1]

The principle of the ER damper (Fig. 2.) is similar. The content of the ER damper is an ER fluid which consists of carrier fluid and electric-polarisable microparticles (in most cases PUR). The ER fluid flows through the gap between piston and cylinder where the electric field is generated (one electrode is the piston and the second electrode is the cylinder). Again, the electric field causes concatenation of particles and an apparent increase of the fluid viscosity.

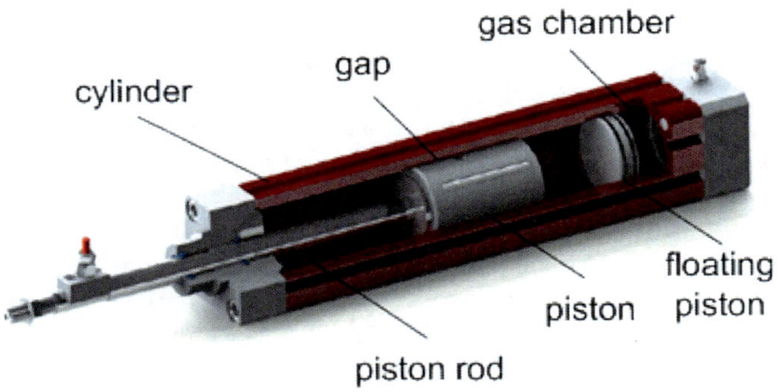

Fig. 2. ER damper [2]

If the damper is supposed to work in semiactive mode, it must be able to respond to the control signal fast enough. The next chapter examines the usage of ER and MR dampers from the time response point of view.

2 Methods and Results

As a simplification for car suspension modelling, the quarter car model is often used. This model is a 2-DOF model consisting of sprung mass M, unsprung mass m, damper with damping coefficient c_s, spring with stiffness k_s and tyre stiffness k_t. There are 2 peaks on the frequency characteristic of car wheel suspension (Fig. 3) – the sprung mass resonant frequency (the lower one) and the unsprung mass resonant frequency (the higher one). The wheel suspension should be able to suppress especially these frequencies. When the suspension is semiactive, the damper is used as a force generator. For semiactive suppression of sprung mass vibration, the ability to generate force at least on frequencies of sprung mass is necessary, for semiactive suppression of unsprung mass, the damper must generate force at least on frequency of unsprung mass.

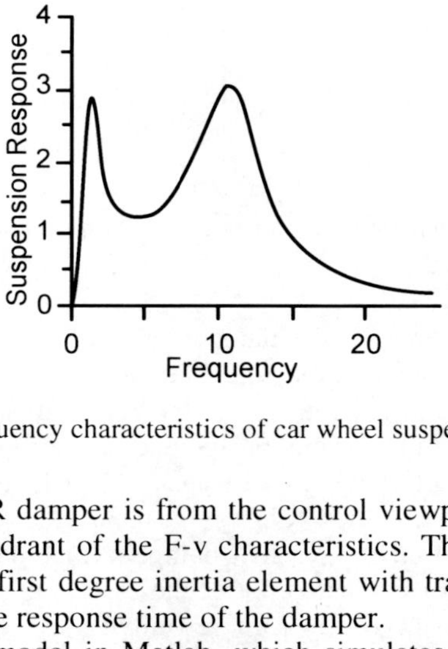

Fig. 3. Frequency characteristics of car wheel suspension [4]

Semiactive MR or ER damper is from the control viewpoint a force generator in the first and third quadrant of the F-v characteristics. The damper's behaviour can be approximated as first degree inertia element with transfer function $F_t(p) = 1/(Tp + 1)$, where T is the response time of the damper.

The authors made a model in Matlab, which simulates the behavior of force generators with different response time. In case the force generator had the response time $T = 1/f_r$ where f_r is the resonant frequency of the system, the current necessary for the required force amplitude would be twice as big in comparison with necessary current for the same force amplitude but lower frequencies, moreover the generated force would be 45° phase delayed from required waveform. For higher frequencies, the phase would grow fast (Fig. 34 dashed). The control would be very inefficient.

For efficient controlling, the force generator's response time must be at least 10 times shorter than the time constant $1/f_r$ of the controlled system. The phase delay on the f_r is only about 6° (Fig. 4 solid).

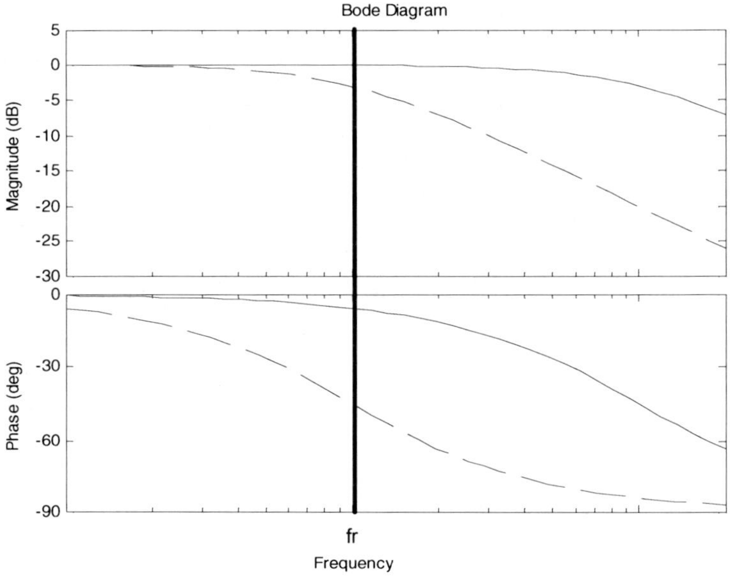

Fig. 4. Frequency characteristics of the damper with response time 1/fr (dashed) and 1/(10.fr) (solid)

The usual response times of ER dampers including driver are between 1 - 2 ms [5, 6], they should be capable of semiactive control up to 50 – 100 Hz. The maximum electric field is limited mainly by dielectric breakdown voltage. With the voltage increase, the sink current grows and the efficiency of the damper decreases. Also even the little amount of water in the ER fluid causes the increase of the sink current. In addition the ER effect weakens at low temperatures [3].

The usual response times of the MR dampers (including current driver) are between 15 - 75 ms [6]. In semiactive mode the MR dampers are capable of controlling forces up to 1 – 5 Hz. But in comparison with ER dampers, the MR dampers have bigger dynamic range (the ratio of forces with maximum current or voltage to forces without voltage or current at constant piston speed). The maximal magnetic field in the gap is limited by the number of turns, maximal permitted coil current and cross-section area of parts of a magnetic circuit. The response time of the MR fluid itself is around 1 – 3 miliseconds [7]. The authors' measurements of the current step response to the coil of commercial Delphi MR damper show that the main part of the MR damper time response is caused by the coil inductance, which determines the current flow in dependence on time and voltage according to formula (1):

$$i(t) = \frac{U_G}{R_1 + R_2}(1 - e^{-\frac{(R_1+R_2)t}{L}}) \tag{1}$$

Where U_G is the driver voltage, $R_1 + R_2$ is the coil and series resistance, L is the coil inductance and t is time.

This formula shows that it is possible to achieve a steeper current slope by increasing the supply voltage U_G during the time when the requested current is higher than the real one. In comparison with commercially used drivers, which usually use low voltage (on-board voltage – 12V), the driver with higher voltage (100-200 V) should improve the time response of the MR damper to only a couple of miliseconds.

Table 1. Suitability of adpative dampers for semiactive damping

	response time [ms]	**limiting parameter**	**appropriate for semiactive sprung mass damping**	**appropriate for semiactive unsprung mass damping**
ER damper with driver	1 - 2	fluid response time	yes	yes
MR damper with low-voltage driver	15 - 75	driver response time	yes	no
MR damper with high voltag(100V) driver	3 - 5	fluid response time	yes	yes

3 Conclusions

From the response time viewpoint, the ER dampers are capable of semiactive control of both sprung and unsprung mass. However compared to MR dampers, ER dampers have smaller dynamic range and smaller range of operating conditions.

The current MR dampers with current drivers have response time between 15 – 75 ms. Therefore they can't be used for semiactive damping of unsprung mass. But they can be used for the semiactive control of sprung mass or for adaptive passive damping systems. The time response of the MR damper is mainly caused by the current drivers with low output voltage. With an appropriate driver the MR damper time response should be reduced to a couple of miliseconds – the main limiting parameter will be the time response of MR fluid itself, which is between 1 - 3 ms. With such fast time response, the MR damper will be suitable even for semiactive control of the unsprung mass.

References

[1] Audi TT Coupé 2007 - Suspension System" Audi AG, Ingolstadt (2006)
[2] Fludicon RheDamp 16,25,32 datasheet, Fludicon (2010)
[3] Gurka, M., et al.: Journal of Physics 149, 012008 (2009)
[4] Vlk, F.: Dynamika motorových vozidel, Prof. Ing. František Vlk DrSc., nakladatelství a vydavatelství, Brno (2005)
[5] Gavin, H., Hoagg, J., Dobossy, M.: Workshop on Smart Structures for Improved Seismic Performance in Urban Regions 1, 225 (2001)
[6] Koo, J.H., Goncalves, F.D., Ahmadian, M.: Smart Materials and Structures 15, 351 (2006)
[7] Goncalves, F.D., et al.: Smart Materials & Structures 15, 75 (2006)

Modelling of a Linear Actuator Supported by 3D CAD Software

J. Wierciak*, K. Szykiedans, and A. Binder – Czajka

Warsaw University of Technology, Faculty of Mechatronics,
Institute of Mechatronics and Photonics, ul. Św. A. Boboli 8
Warszawa, 02-525, Poland

Abstract. Modelling is considered to be an inherent part of design process of mechatronic devices and systems, in particular when solving problems of dynamics and accuracy of actuators and sensors. Mathematical software packages such as *Matlab/Simulink* are commonly used for this purpose. Engineers and designers who integrate mechanical components of devices under design, usually employ special computer software known as *3D CAD* for creating three dimensional images. Typically it is used to generate technical drawings and layouts but its usability is more extensive. The paper presents ability of such software to support simulation where a linear stepping actuator is an example of a device under design.

1 Introduction

Modelling is recognized as one of characteristic features of mechatronic systems and devices designing. However the term "modelling" is a general one and it covers various techniques referring to various domains and parts of a system under design. The authors attempt to address those techniques to particular phases of design process as they were previously defined. The analysis has shown that the following phases of the design process could be identified [6]:

- developing of a system structure,
- identifying control and measuring channels,
- designing of channels,
- designing of subsystems: electromechanical, electronic and software,
- supervising of making a prototype,
- trials with modifications of a system in any layer.

* Corresponding author.

Mathematic models used in the design are commonly formulated as equalization of accumulated masses, energies or strokes, their complexity (fig.1) depends on described effects, pointed aims of tests and sometimes on available data or software.

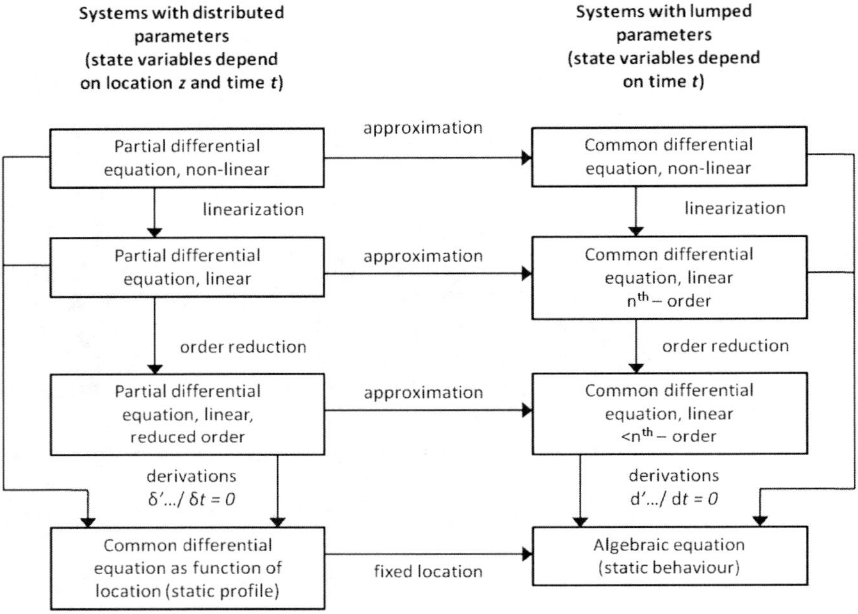

Fig. 1. Classification of models with respect to a type of state variables, by [3]

2 Determination of User's Needs and Analysis of a Main Function of the System

Designing should be started from the definition of client's requirements. It has to consists of following parts [6]:

- *system functions*, commonly presented as operating algorithms or lists of operations that device has to do with numerically defined characteristics,
- *description of a system structure* including mechanical, electric and informatics subsystems,
- *description of system environment* with physical and legal conditions, also possible presence of other systems and user

In this phase, when user's needs are reviewed, *2D* or *3D* geometrical models of the device under design can be generated in order to obtain the client's approval for either its look or layout (Fig.2). It can be extended by modelling some activities when using the device or system – sometimes technical models are used. When analysing a main function of a system, time relations between particular operations are crucial.

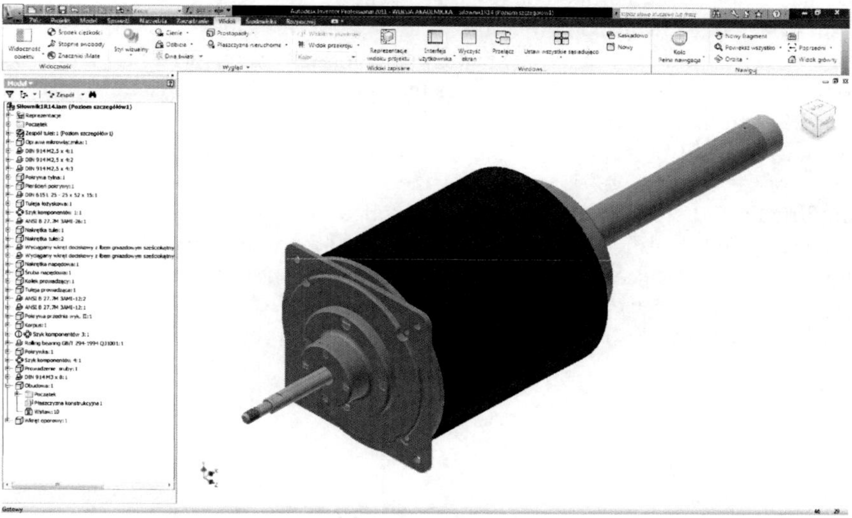

Fig. 2. Virtual three dimensional model of an actuator developed with use of *Autodesk Inventor* . Use of such a model allows presenting desired dimensions, movement ranges or displacement of mounting holes.

At the phase of main function analysis, the functional concept of a device is being developed. Upon this concept a list of necessary actuators and sensors is formulated. User's requirements are translated into technical requirements for each unit, either actuating or measuring. 3D software is not useful for this stage of project. Block diagrams are the most common form to illustrate these works.

3 Developing of Actuators and Sensors

A result of this stage of a design should be a proposal of technical solutions of particular actuators and sensors identified at previous stage. In this phase modelling is very important, it allows to determine an optimal solution with use of tests that could be applied to virtual models [6].

In this phase typical modelling and simulation of dynamical systems is performed in order to obtain time responses, which supports decisions upon technical solutions of actuators and sensors.

3.1 Modelling with CAD 3D Software

Following stage of a design can be aided with 3D CAD software helping to make a visualization or virtual models tests. A possibility to create multiple creation of virtual models of many system configurations allows to make a multiple variant test at low cost. Definition of physical parameters such as material properties, friction factors, stiffness or elastics, ensures that simulation conditions are so (?) close to reality that in most cases results are satisfactory.

Simulations of object dynamics can be proceeded on final model of an assembly or on demonstrative, draft one. In the second option parts will be made in simplified way to illustrate only their main functions, essential for results of simulation (Fig.3). Simplified models allow to proceed more tests and simulations with bigger number of possible configurations. That lead up to choosing optimal solution. After that and making a fully detailed model of chosen configuration final simulation with conditions specific description can be proceeded.

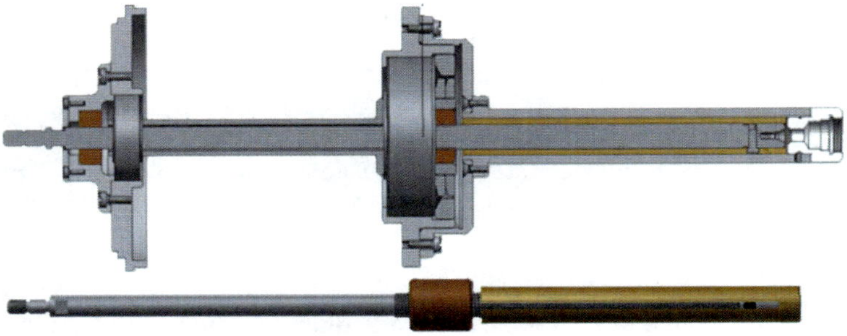

Fig. 3. Screw–nut transmission unit, virtual model capable of preliminary simulations

The following example of simulation of a helical joint in a linear actuator will present possibilities of 3D CAD software use. It will be illustrated at Autodesk Inventor base. Previously made model of a component should be connected into assemblies and prepared to be use in a dynamic simulation (Fig. 4)

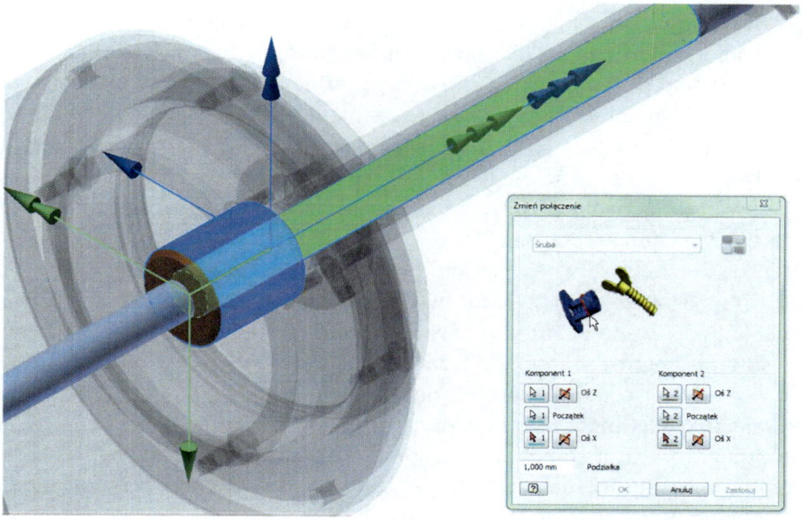

Fig. 4. Selection of components that are the parts of a helical joint

When mechanical properties of a joint are defined, actuating forces and loads can be applied. Some limitations of course occur, exemplary simulation data (Fig.5) show a simplified actuating torque and change of velocity and position.

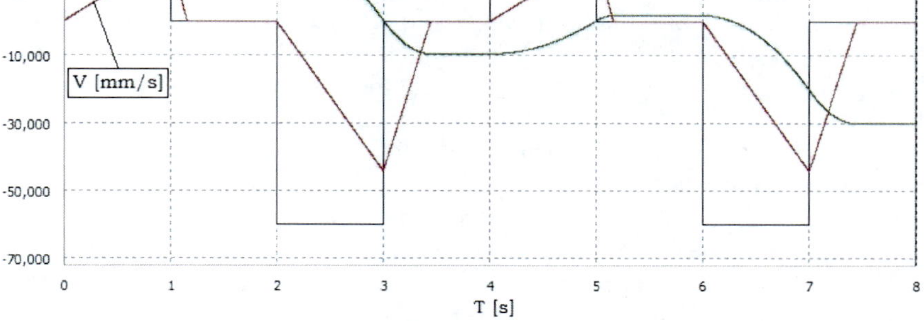

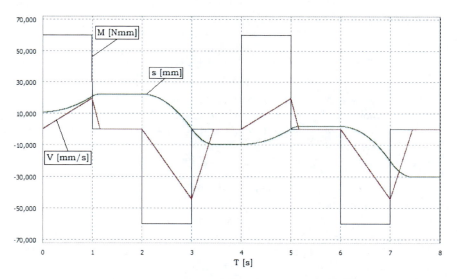

Fig. 5. Simulation of a helical joint dynamic with applied load of 30 N; actuating torque M (navy blue line), rod velocity V (red line), rod position s (green line).

The actuating torque that drives a nut has been simulated as a step function. It is very rough simplification that not describes torque routes in a real stepper motor. Velocity and position curves show large influence of load on a device way of work. While load force is opposite to rod move (0-1 s), velocity grows slowly (acceleration 19,5 mm/s^2) and after torque disappearance (1-2 s) rod-screw stops after 0,16 second. As the screw is naturally retarding, there is no movement under axially applied force. When the torque changes and the rod movement is consistent to the load force vector (2-3 s), acceleration is more than twice bigger (43 mm/s^2) so the rod travel is bigger and the rod does not take starting position after torque disappearance (3-4 s). Next, between 4th and 8th the second torque route is repeated. Velocity and position routes are similar. Finally after two sequences the rod position is approx. 40 mm different from start into the sense of load force vector. It shows significant influence of load on the mechanism.

Presented modelling and simulations results give faultless conclusions about helical joint work. But some types of load or actuating force are very difficult or impossible to apply in such a model. It leads to use different ways of modeling and simulation to receive more precise results.

3.2 Modelling with Software for Multidomain Simulation of Dynamic Systems

When modelling actuators, the procedure starts with either rotational or linear movement model, depending upon the kind of movement realised by a drive used. If rotational movement takes place, then the classical equation of torque equilibrium makes the model

$$\left(J_m + J_l\right)\frac{\mathrm{d}^2\gamma}{\mathrm{d}t^2} + K_D\frac{\mathrm{d}\gamma}{\mathrm{d}t} + M_{lf}\,\mathrm{sgn}\left\{\frac{\mathrm{d}\gamma}{\mathrm{d}t}\right\} + M_l = M_m\,, \qquad (1)$$

where: J_l – moment of inertia of rotating elements reduced to the motor shaft [kgm^2], J_m – moment of inertia of motor rotor [kgm^2], K_D. – coefficient of viscous damping [Nm/rad/s], M_m – motor torque [Nm], M_l – active load torque reduced to the motor shaft [Nm], M_{lf} – frictional load torque reduced to the motor shaft [Nm], γ – angle of rotation of rotor [rad]. It is a base for the whole model, which is created by developing models of all components included into the above equation. In Fig. 6 simulation model of the actuator developed in Matlab-Simulink software environment is presented.

The role of *CAD 3D* programs can be quite extensive at this stage for it is a source of reliable mechanical data to be included in the model, such as: masses, moments of inertia, location of gravity centres and geometrical parameters. Special modules of *3D* software are used for this purpose.

Typical software for simulation of dynamic systems produces time responses of the system output quantities as well as quantities that are not visible or not measurable in a real device, for instance discrepancy angle in a stepping motor (Fig. 7).

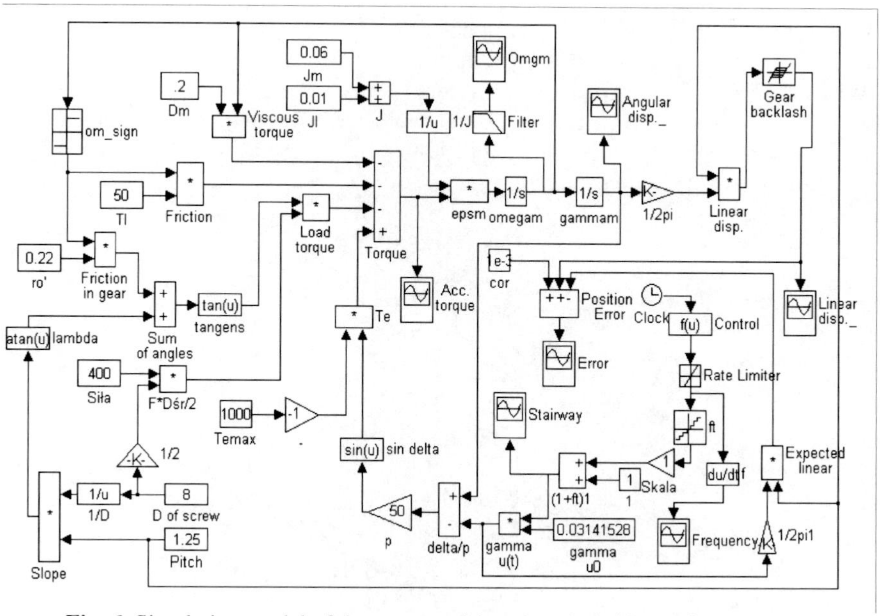

Fig. 6. Simulation model of the actuator driven by a hybrid stepping motor

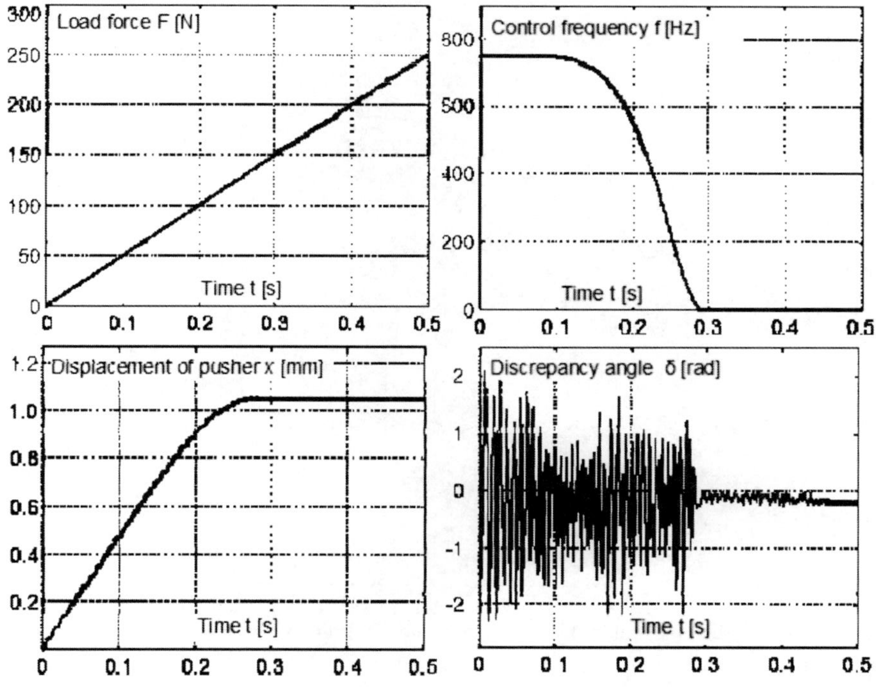

Fig. 7. Results of simulation of a linear actuator driven by stepping motor

4 Designing of a Mechanical Subsystem

When designing a mechanical subsystem, it becomes a common practice to develop its 3D model aimed at avoiding geometrical collisions between parts and used for analysing various space dependent states e.g. thermal, strength (Fig. 8), position of gravity centre etc.

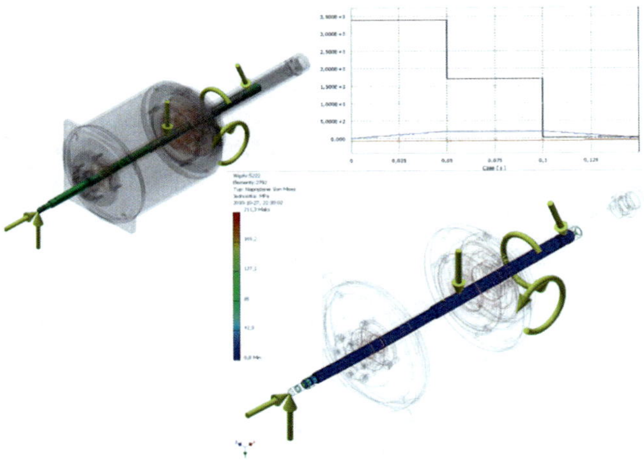

Fig. 8. Dynamic analysis results can be used to determine loads in FEM strength analysis

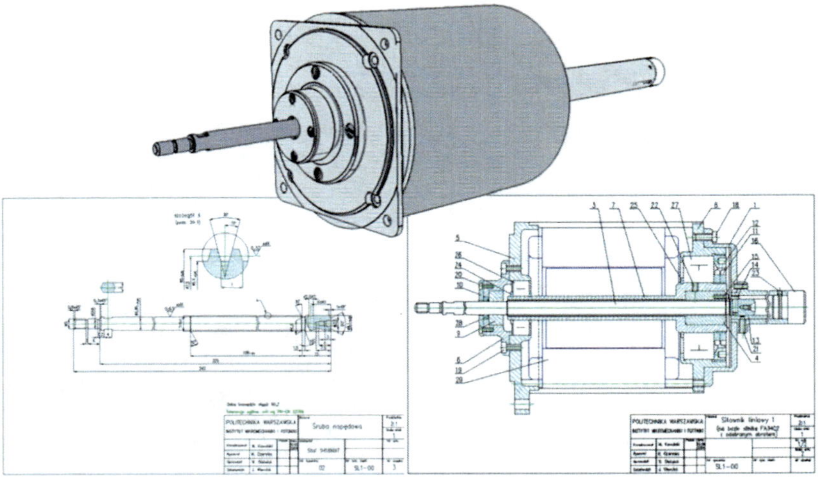

Fig. 9. Three dimensional models of parts and assemblies are used to create technical drawings

Design process, after final conception of a device/system is taken, has to be ended with technical drawings of a whole device. When 3D CAD software is used, virtual prototype of a device will be created. This prototype will represent geometry, material and physical properties that a real device will have. Besides allowing simulations and model test, it will be used to assure design correctness.

5 Summary and Conclusions

The software for various kinds of modelling becomes still more and more compatible, giving opportunities to exchange data between different packages. Analysis of conducted modeling and simulations allows authors to end with some conclusion upon reliable use of such tools in order to keep engineers conscious of actual role of modelling.

First of all, it is worthful that possibility of data exchange between engineers software constantly spreads. That makes extending model simulations easier and their results more accurate.

During designing mechanical subsystems of mechatronic devices, significant role plays three dimensional modelling bounded both with geometry of designed assemblies and with other greatness depending on placement in space.

Development of actuators and sensors requires in most cases time domain simulations. Models of lumped parameters described by ordinary differential equations are commonly used.

References

[1] Gawrysiak Mechatronika, M.: i projektowanie mechatroniczne. Politechnika Białostocka. Rozprawy Naukowe nr 44. Białystok (1997)

[2] Heimann, B., Gerth, W., Popp, K.: Mechatronika. Komponenty, metody, przykłady". Wydawnictwo Naukowe PWN. Warszawa (2001)

[3] Isermann, R.: Mechatronic Systems – Fundamentals. Springer, Heidelberg (2005)

[4] Mrozek, B., Mrozek, Z.: Matlab i Simulink. Poradnik użytkownika. Wyd. Helion, Gliwice (2004)

[5] Pelz, G.: Mechatronic systems. Modelling and simulation with HDLs. John Wiley and Sons Ltd., Chichester (2003)

[6] Wierciak, J.: An algorithm for designing mechatronic systems. In: Rohatyński, R. (ed.) Design Methods for Industrial Practice, pp. s.255–s.260. Oficyna Wydawnicza Uniwersytetu Zielonogórskiego, Zielona Góra (2008)

[7] Wierciak Modelowanie, J.: elektrycznych układów napędowych urządzeń precyzyjnych". In: XV Sympozjum Modelowanie i Symulacja Systemów Pomiarowych, Krynica, września 18-22, pp. 239–247 (2005)

Hardware-in-the-Loop Testing of Control Algorithms for Brushless DC Motor

Jiří Toman*, Zdeněk Ančík, and Vladislav Singule

BUT Faculty of Mechanical Engineering, Technická 2,
Brno, 61669, Czech Republic

Abstract. The aim of this article is to inform readers about new trends in the development of control algorithms. One of the trends currently gaining popularity is a Model Based Design (MBD) approach. MBD consists of a sequence of several steps, is the sequence being defined by the V-cycle [1] [2]. This new development method plays a role not only in science and research, but also increasingly often in the commercial development of industry applications.

The following article deals with the hardware-in-the-loop (HIL) development phase, where the control algorithms for brushless DC (BLDC) motors are tested and tuned on the dSPACE simulator.

1 Introduction

The development of the V-cycle is shown in Fig. 1. This new development life cycle enables verification, validation and testing during each phase of development. Detection and correction of errors is possible at the beginning of the development cycle. This is much cheaper and more effective than redesign of a device as usually happens during traditional development procedures during the prototyping phase. The other advantages that model based design allows are: reduction of development time and cost, easier certification, faster and easier testing, support with accompanying documents, reduction of failures, etc.

A further important aspect is higher safety during testing. It is not necessary to test the control algorithms on the real system, which is often dangerous. First, the new code is tested on the HIL simulator and then, if no hazardous state occurs, it can be applied in the target system.

* Corresponding author.

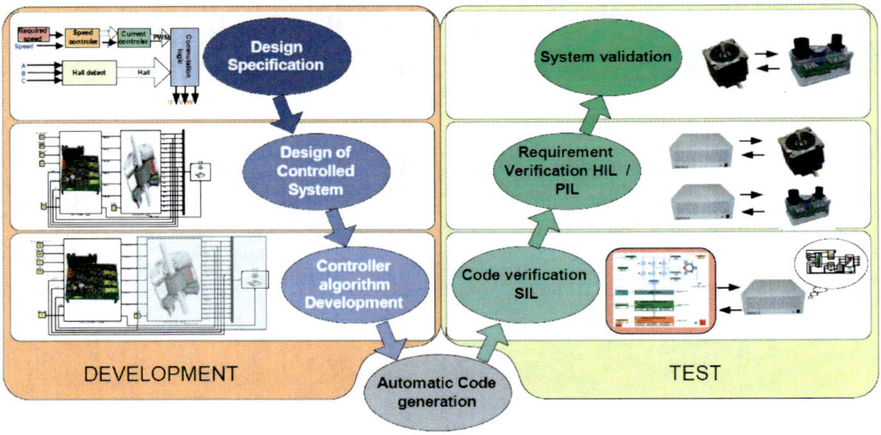

Fig. 1. Diagram of the V-cycle

2 Model Based Design – MBD

The first step of the MBD, the principles of which are shown in Fig. 2, involves the device specification and the requirements of each part. Development continues with analysis of the incoming parameters and specification of the future hardware and software. It is important to choose a suitable software environment for model based design.

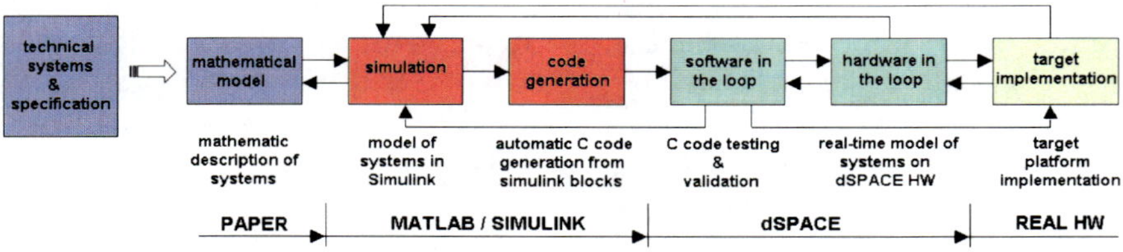

Fig. 2. Diagram of the Model Based Design principles

Complex software environment, which allows MBD technology, is i.e. MATLAB & Simulink. Simulink enables composition of the real system according to its mathematical and physical description. For such a model of the controlled system it is possible to design and realize any control algorithm and simulate it.

Simulation allows designers to correct view on the behavior and dynamics of the whole system. This phase is marked as Model-in-the-loop (MIL) in the MBD.

The next phase in the cycle is Software-in-the-loop (SIL). This part consists of automatic source code generation from the Simulink model. Real-Time-Workshop, an integral part of the Simulink libraries, is used for this purpose. Testing of the automatically generated code is provided at the end of this phase. Results from the source code testing are compared with the model's behaviour. If the results correspond to those of the mathematical model, the next phase continues.

Hardware-in-the-loop is the next part of the design. The main aim is again to test the generated source code, but now on the real simulator. The code is uploaded to the appropriate HIL hardware simulator with various set of peripherals. The dSPACE system, which is compatible with MATLAB/Simulink, can be used.

The last step of the MBD is the code generation and linking to the target platform (i.e. several embedded system). The Real-Time-Workshop-Embedded-Coder is used for this step in the Simulink environment. This software package generates source code, which is optimized for the selected target platform (i.e. Freescale, Microchip microcontrollers).

3 Conception and Requirements for the HIL Testing

In this case, the controlled system is the BLDC motor produced by Anaheim Automation (BLY342S-24V-3000). The BLDC motor is connected to the power electronics, which are controlled by the dSPACE system. The control algorithm from the Simulink environment is uploaded to this system and executed. The required peripherals for the dSPACE system are 5 analog inputs for voltage and current sensing from the motor windings, 5 digital inputs for Hall sensors and Fault state sensing, a digital output and 6 synchronous PWM channels. The connection diagram is shown in Fig. 3.

The Power Electronics consists of interface circuits and full 6 step inverter with appropriate protection components. The interface board has several power supplies and circuits for optical interconnection of the logic signals from the dSPACE and power drivers in the inverter. The inverter is a commercial evaluation kit by Microchip. It consists of power MOSFETs with appropriate integrated drivers. The maximum switching current is 30 Amps and the maximum DC bus voltage is 50 Volts. The power electronics are also equipped with an electrical brake, which is now not used. The electronics enable measurement of any analog values from the electric drive and transforms ranges for dSPACE acquisition cards.

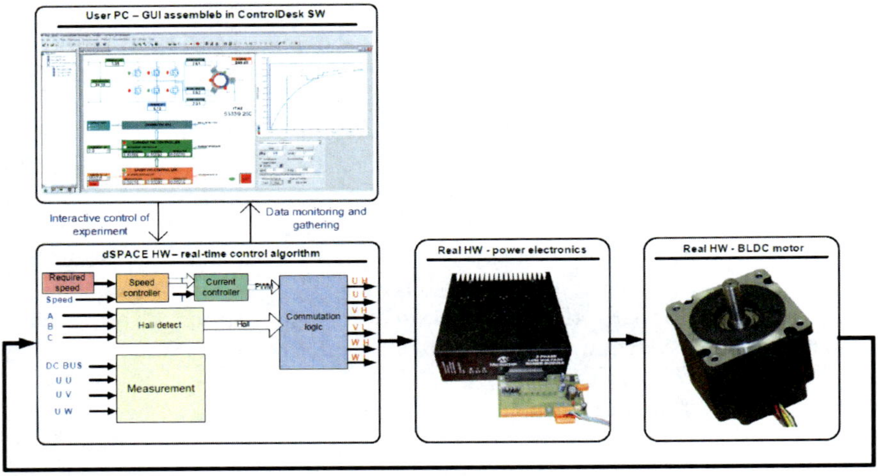

Fig. 3. Connection diagram of the whole configuration

Because the dSPACE real-time system is quite sensitive and expensive equipment, is the manufacturer recommends protection of all digital and analog inputs and, if possible, provision of high-quality optical isolation.

The last described part of the HIL system is the interface board. This part consists of three connectors used for Hall sensors from BLDC, dSPACE system and Power electronics.

The main purpose of the board is the optical isolation and protection of the dSPACE system, which could be damaged by over-voltage or over-current faults. Electrostatic protection of the dSPACE inputs is also important. The main connector on the board (dSPACE connector) consists of 6 PWM logic outputs for the Power electronics, 3 logic signals from the Hall sensors and a RESET signal for the Power electronics. Because the dSPACE system cannot directly drive the power electronics' inputs, optical isolation is also used for this purpose.

4 Simulink Model Development

Two fundamental groups of blocks are used for Simulink model development. The first group of blocks consists of standards from Simulink libraries. The second one is blocks from the Real-Time-Interface (RTI) library, which is used for dSPACE system peripherals settings. Settings and common work with the RTI blocks are similar as to the others blocks in Simulink. All blocks are interconnected by standard links on Simulink.

The Simulink model is divided into three main parts – *Measurement, Control PWM* and *Evaluation of control.* The schematic diagram is shown in Fig. 4. The Digital I/O Setup block is used for setting the digital and analog input/output ports into the dSPACE system. The BIT IN and BIT OUT blocks are for *Fault* detection and for the *Reset* function of the power electronics. For setting the current and speed controller, the ON/OFF CONTROLER block is used.

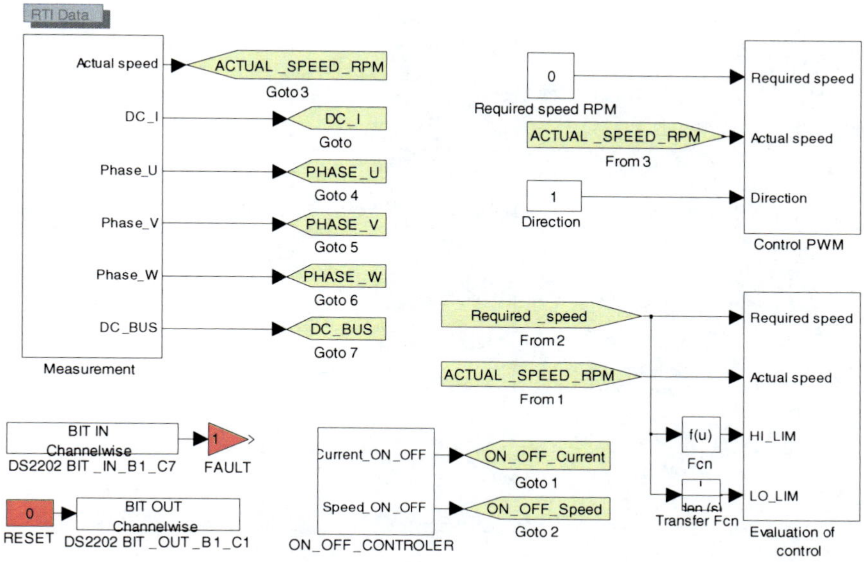

Fig. 4. Schematic diagram of the Simulink model

4.1 Measurement Part

This part is used for measurement of all analog values from the power electronics and BLDC motor. This part is consists of the 5 analog inputs and actual rotational speed. The Frequency block is connected to a Hall sensor in the BLDC motor and senses the rectangular signal frequency. This value is then recalculated to revolutions per minute (rpm) according to a simple equation. The Measurement parameters sub-block allows calibration and setting of the analog input of the dSPACE system.

4.2 Control PWM Block

This part of the model consists of a closed control loop, input and detection of the Hall sensor, commutation logic and PWM setting blocks of the dSPACE system. It is an essential part of the overall model and is used for commutation and control. The controller Switch block can be used for fast switching of various control loops according to the designer's requirements. The switching pin is ported out from the structure for easy control from the dSPACE environment. The Control PWM block is shown in Fig. 5.

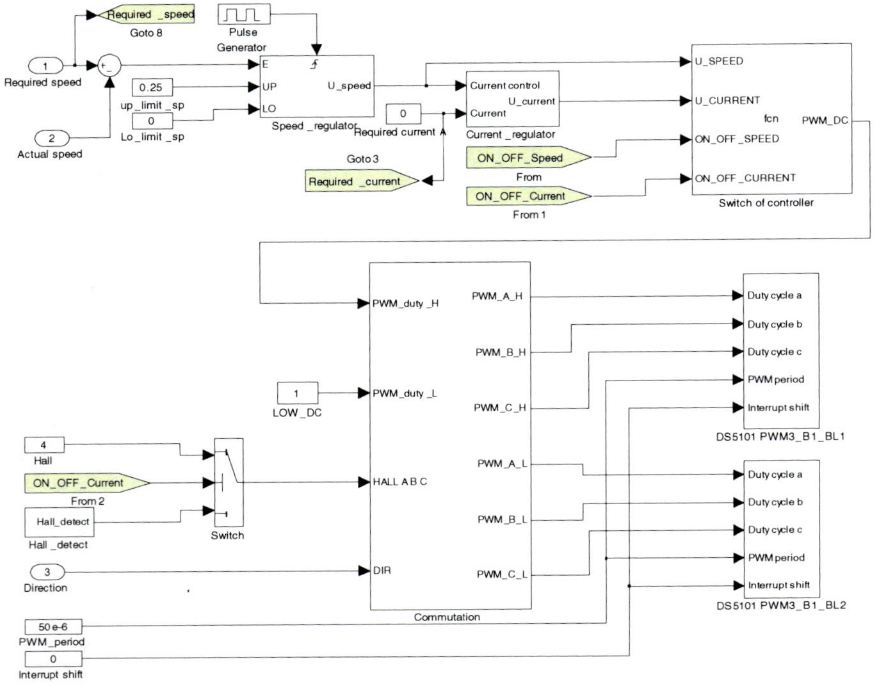

Fig. 5. Schematic diagram of the Control PWM block

The control loop consists of current and speed controllers. These are connected to the so-called cascade control loop. This is a speed control loop with a current loop which is subordinate. Inputs to the speed controller are difference between demanded and actual speed and 2 constants that constrains the maximal and minimal output value of the controller. The output value is the input for the second current controller. Firstly, it is also made a difference between input and actual value of the motor current. The control deviation leads to the current controller. The output controlling quantity is again constrained to a maximal and minimal value. This output is directly a PWM duty for PWM channels.

The controllers' conception is generally defined as a discrete PSD [6] controller. Up to three controller components may be used depending on the designer and requirements. The speed controller is sampled with the period of 10e-3 seconds and the current controller must be ten times faster (i.e. 10e-4 seconds). Structure of the PSD controller is shown in Fig. 6. The whole structure is optimized for easy implementation into a target platform.

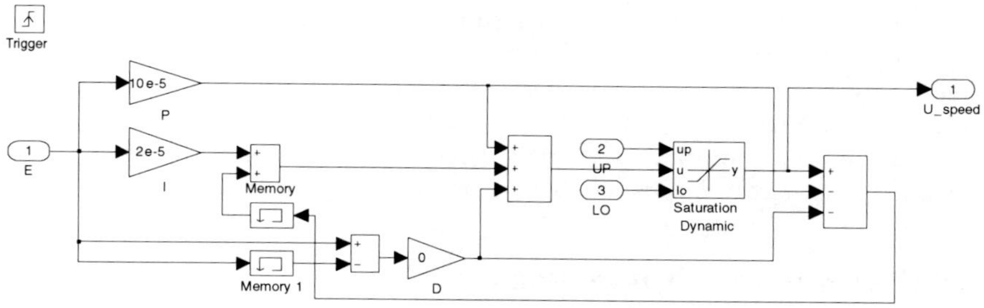

Fig. 6. The internal structure of a PSD controller

The Hall sensors are connected to three digital inputs of the dSPACE system. Combinations of all three inputs are converted by simple multiplication to the decimal system and added together. There is a unique decimal number for each commutation state, which determines with the switching combination of the appropriate power MOSFETs. This number is sent to the Commutation block, where the logic (according to a Hall state) determines which transistors are going to be on.

The Commutation block consists of the decision logic, which determines the correct combination of the output MOSFETs. It is made up of standard Simulink blocks and a lookup table. The main idea of the logic is to control two transistors at the same time in complementary mode so that in each time period a high side transistor is on and the appropriate low side has to be off. The PWM duty cycle is driven to the three high side switches directly from the current controller after multiplying with the values from the lookup table. The low side MOSFETs, or the low side PWM channels, are set with a constant PWM duty cycle. This means they are permanently on or off.

4.3 Evaluation of Control Block

The last part of the Simulink model is the block used for evaluation of the control algorithms. It determines the quality and requirements of the controllers. There are several input signals to the block – actual rotational speed, requested speed and definition of the area for the actual speed. The upper limit is set to 110 percent of the requested speed. The lower limit determines a requested dynamic of the system, so it is set as a transfer function

$$F(s) = \frac{1}{0.8s + 1} \tag{1}$$

Inside this block are the integral of time-multiplied absolute value of error (ITAE) criteria, which evaluate the quality of the control [4] [5]. ITAE criterion is defined by the equation

$$Q_{ITAE} = \int_{0}^{\infty} \left| e(t) - e(\infty) \right| \cdot t \cdot dt , \tag{2}$$

where e is control error and t is time.

This function counts until the requested speed is reached. The result is a value proportional to the difference between requested and actual speed. The lower the value is, the better and superior is the control algorithm used. The record length and calculation time of the evaluation criteria is set by the Time constant variable. The block is made up of the Enable subsystem, which allows enabling of the evaluation criteria calculation only once at any motor speed change.

5 dSPACE System Environment

The model as described above is ready for compilation. Real-Time-Workshop (RTW) builds the application, and the produced C-code (application) is loaded into real-time HW (dSPACE). So that we can control the experiment/validation, we have to prepare a "graphical user interface" in ControlDesk. This GUI provides a link between the application running in real-time on the dSPACE HW and a PC connected to this HW. Carrying out real-time experiments is simplified bythe ability to interactively control, visualize or gather almost any data from the model. Various GUIs can be made for various experiments depending on requirements. This user interface is designed and executed on a common personal computer (PC). The design of the motor control interface is shown in Fig. 7.

The control interface was designed for the convenience of the user. As the controllers are probably the most important part for testing and tuning, they are made graphically. It is possible to easily switch the appropriate controller on or off by means of common check button. For parameter tuning there are several edit fields on all blocks.

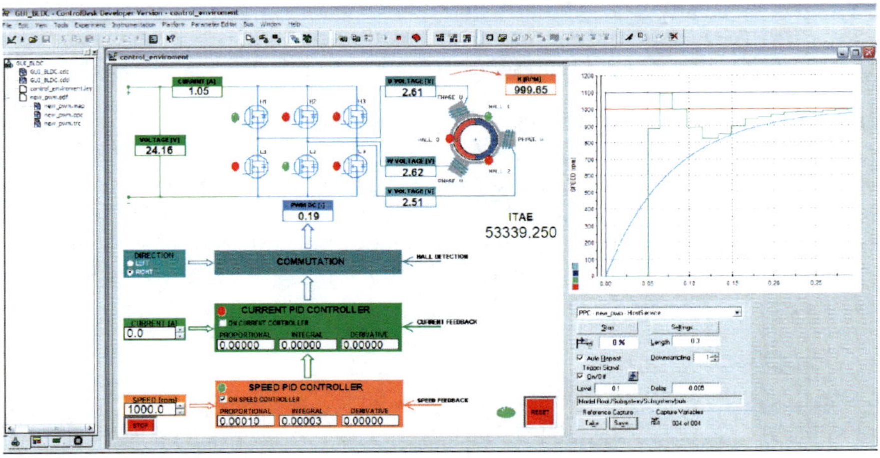

Fig. 7. The GUI for control BLDC motor in ControlDesk

The response of the system may be watched immediately in the graph window. All collected courses may be saved and redrawn in any mathematical environment. This real-time behaviour enables analysis of all electrical signals from the connected power electronic, e.g. phase voltages, currents, state of all transistors, etc. In the graphs it is possible to immediately verify whether or not the actual speed is still in specified range. The evaluation function ITAE may then be checked to determine whether or not the response of the system is acceptable.

6 Conclusion

The project involves the development of control algorithms and electronics to improve run-up performance and reliability of the BLDC motor in actuating devices for safety critical applications.

Because of these strict requirements, simulation and modeling tools are widely used to accelerate the development of control algorithms. The simulation results are tested on real hardware in the dSPACE environment. This type of development is called Model Based Design and is an integral part of every development department.

The main advantages of this mathematical modeling are its complexity, integrity and its applicability to any higher system without the need to understand the principle of the BLDC motor.

Acknowledgements

This work was supported by project No. MSM 0021630518 "Simulation modelling of mechatronics systems" carried out at the Faculty of Mechanical Engineering, Brno Technical University. Hardware and software development, production of physical modules, and project support were provided by UNIS, a.s.

References

[1] Hadas, Z., Singule, V., Věchet, S., Ondrusek, C.: Development of energy harvesting sources for remote application as mechatronic system. In: 14th International Power Electronics and Motion Control Conference, pp. 13–19 (2010) ISBN 9989-9785-1-4

[2] Březina, T., Houška, P., Singule, V., Březina, L., Hadaš, Z.: Mechatronic Education and Research Activities at Brno University of Technology. In: 11th International Workshop on Research and Education in Mechatronic, pp. 83–88 (2010)

[3] Hubík, V., Toman, J., Kerlin, T., Singule, V.: Testing of the EC motor control algorithms in the dSPACE system, pp. 20–28 (2009) ISBN 978-80-214-3974-0

[4] Awouda, A., Bin Mamat, R.: Refine PID tuning rule using ITAE criteria. In: ICCAE 2010, Singapore, pp. 171–176 (2010)

[5] Stanley Shinners, M.: Modern control system theory and design, New York (1998) ISBN 0-471-24906-8

[6] Leonhart, W.: Control of Electrical Drives, 3rd edn. Springer, Berlin (2001)

Thermal Modelling of Powerful Traction Battery Charger

R. Vlach

Brno University of Technology, Technicka 2896/2,
Brno, 616 69, Czech Republic

Abstract. In this paper, the author presents survey of the modern approaches in the thermal analysis of power electronic systems. The improvement and the new methods are applied to the real engineering problem. The thermal analysis based on lumped parameters thermal network (LPTN) and finite elements method (FEM) is considered in this paper. The both methods were compared in order to highlight the advantages and defects of each. In addition, an overview of the problems linked to the thermal parameter determination and computation is proposed and discussed.

1 Introduction

The thermal analysis of power electronic systems have receives less attention than electric design [1]-[4]. It tends to cause frequent problems which usually occur after realization and testing. At present is trend to minimize size of the engineering devices including power electronic systems. The effective dissipation of heat losses from semiconductors becomes a limiting factor of a design these devices. Therefore, it is now necessary to analyze the thermal phenomena to the same extent as the electric design. In fact, there should be an interaction between the electric and thermal designs.

A powerful traction battery charger will be constructed in TU of Brno. Useful modular conception of the power part was chosen. The modules can be serial-parallel reconfigured - for various battery voltage levels. Single-module output parameters are 160V/16kW. The charger is based on a DC/DC transistor converter using CoolMOS transistors and SiC diodes. A non-usual topology of the power circuit, especially construction of power pulse transformers and output choke coils is performed. The electric circuit of powerful battery charger is shown in Fig. 1. The scheme in Fig.1 is only a principle. In the real construction were used 3 diodes and 3 transistors parallel in each switching branch.

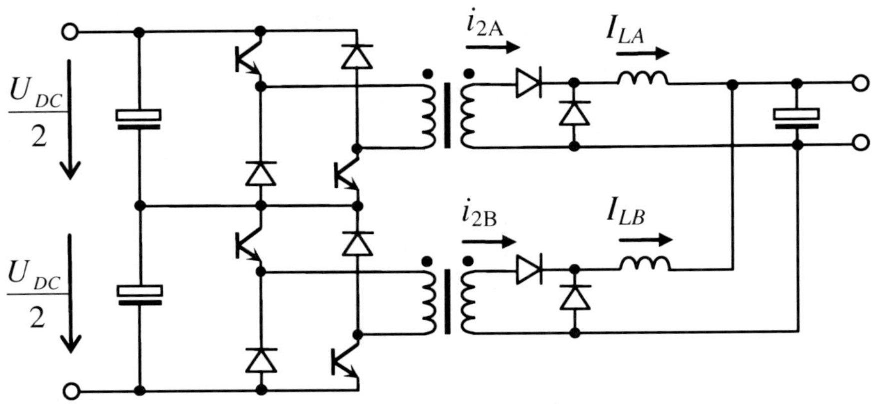

Fig. 1. Scheme of powerful battery charger electric circuit

The thermal analysis is focused on the most important parts of the power battery charger. These parts are transistors, diodes and rectifiers which are placed on common heat sink. The heat sink consists of two parts of extruded heat sink profiles. The both extruded heat sinks are connected so that the whole forms a channel for cooling air flow. The radial fan is used to create air flow through the heat sink channel with fins.

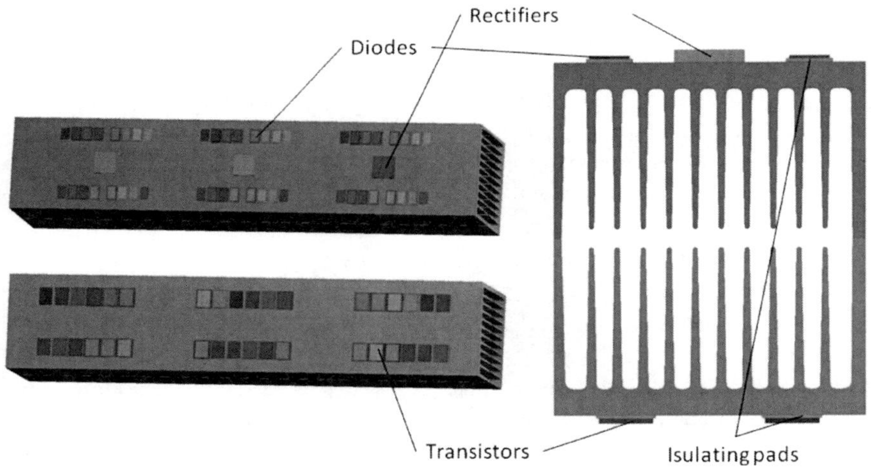

Fig. 2. Semiconductors layout on the heat sink

Semiconductors layout on the heat sink corresponds to semiconductors layout on the printed circuit boards (PCB).

2 Thermal Modelling

A thermal state of the semiconductors depends mainly on the air flow quantity passing through the heat sink channel with fins. Therefore, it is necessary to design the ventilation system for a correct cooling of semiconductors. Therefore, the ventilation system is necessary to propose for a correct cooling of semiconductors.

2.1 Ventilation System

The ventilation system has to provide the required air flow for the correct cooling. A simple estimate of the minimal air flow quantity can be calculated by the equation:

$$Q \geq \frac{P_z}{C \cdot \Delta\vartheta} = \frac{3444}{1170 \cdot 30} \cong 0{,}1 \frac{m^3}{s} \tag{1}$$

where P_z are heat losses in all semiconductors, C is thermal capacity of air and $\Delta\vartheta$ is maximal heat rise of cooling air.

The ventilation calculation is based on equivalent hydraulic circuit [5]-[7]. The total air flow is divided into three parallel streams, as shown in Fig. 3. Two streams describe air flow through choke coils which are not solved in this paper. The middle stream is air flow through the heat sink channel.

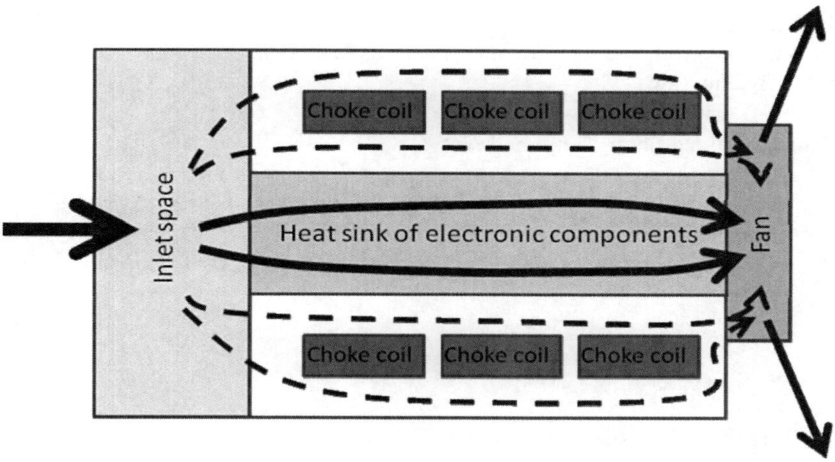

Fig. 3. Ventilation system of powerful battery charger

The radial fan was selected as pressure source which has to compensate hydraulic losses of air flow. The operating point of ventilation is shown in fig. 4.

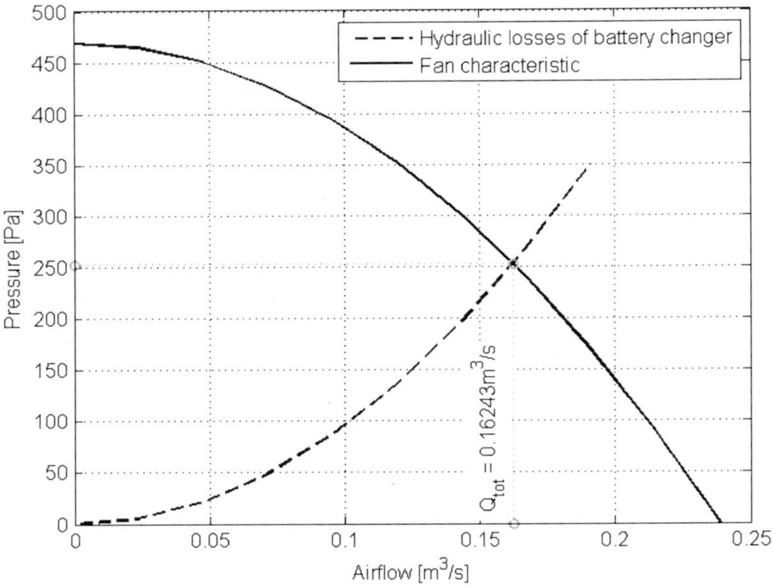

Fig. 4. Operating point of ventilation

The maim part of the total air flow goes through heat sink channel. The air flow through heat sink channel is 0.12 m^3s^{-1}. It satisfies sufficient condition which is described by equation (1).

2.2 Lumped Parameters Thermal Network

In general there are three basic heat transfer mechanisms: heat conduction, convection and radiation [2], [8]. The equation for heat transfer is analogous to the relation for electric systems well known as Ohm law. The relation between heat flux q and temperature differences $\vartheta_1-\vartheta_2$ is:

$$q = \frac{\vartheta_1 - \vartheta_2}{R_{th}} \tag{2}$$

where R_{th} is thermal resistance. Thence, the thermal resistance for conduction of the homogenous one-dimensional isolated wall of the thickness δ and area S can be expressed as:

$$R_{cond} = \frac{\delta}{k \cdot S} \tag{3}$$

where k is thermal conductivity of wall material.

Heat transfer from external surface A can be expressed by surface thermal resistance as:

$$R_{surf} = \frac{1}{h \cdot A} \tag{4}$$

where h is heat transfer coefficient which describes a common heat transfer by natural convection and radiation or only forced convection. The heat transfer coefficient in heat sink channel (Fig.2) using Colburn equation for Nusselt number [8] can be expressed as:

$$h = \frac{k}{D_h} \cdot 0.023 \cdot \mathrm{Re}^{0.8} \cdot \mathrm{Pr}^{0.3} \tag{5}$$

where Re is Reynolds number and Pr is Prandtl number. The hydraulic diameter of heat sink channels can be calculated from heat sink geometry as:

$$D_h = \frac{4 \cdot S_c}{p} \tag{6}$$

where S_c is cross-section and p is perimeter of heat sink channel.

The important thermal resistance in cooling of power electronic systems is description of contact between semiconductors and heat sinks. It can be expressed as [8]:

$$R_{cont} = \frac{1}{h_c \cdot A_{cont}} \tag{7}$$

where A_{cont} is apparent interface area and h_c is called thermal contact conductance which depends on the many factors (roughness, surface condition, contact pressure etc.).

The last thermal resistant describes heat flow in the coolant [6], [7] and can be expressed as:

$$R_Q = \frac{1}{2 \cdot C \cdot Q} \tag{8}$$

where C is thermal capacities of coolant and Q is coolant air flow quantities.

The Lumped parameter thermal network is shown in Fig. 5. The thermal network takes into account the length of heat sinks profile because influence of cooling air heating cannot be neglected as show calculation results. The heat sink of battery charger was divided to three parts. The each part includes one rectifier, sixteen diodes, twelve transistors and of course third of the heat sink. The thermal model consists from thermal network of solid parts (transistors, diodes etc.) and thermal network of coolant flow. This enables include the effect of gradual heating of cooling air which flows through heat sink channel.

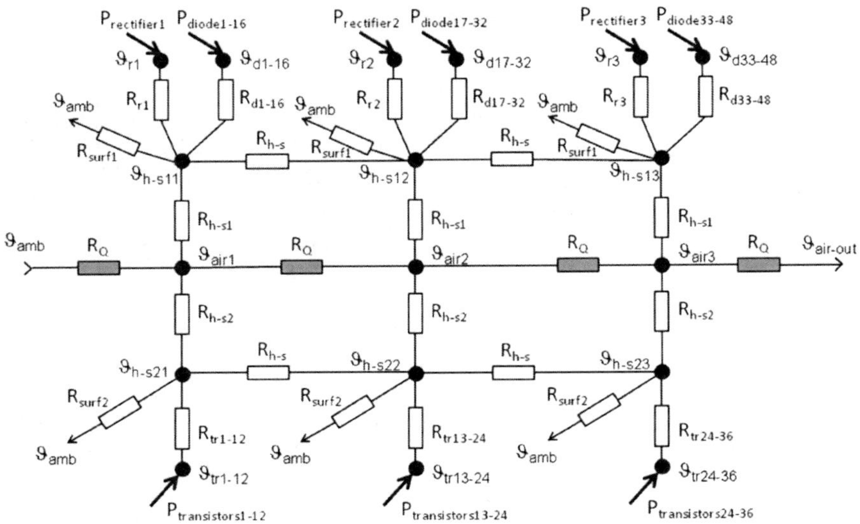

Fig. 5. Lumped parameters thermal network of power battery charger

The thermal resistance between the semiconductor junction and heat sink is composed from serial connection of several thermal resistances which describe heat flow from junction to case and contact effect. The thermal resistance of insulating pad and two contacts effect is necessary to considered in the case of total thermal resistances diodes and transistors.

The thermal model (Fig.5) is possible to be described by system of linear equation:

$$G \cdot \vartheta_i = P \tag{9}$$

where G is matrix of thermal conductivities and P is vector of heat losses in node i and heat fluxes to ambient. The thermal balance of the thermal network of coolant flow nodes is more complicated because it has to respected direction of air flow. It can be described by [6], [7]

$$\vartheta_i\left(\frac{1}{R_Q}+\sum_j \frac{1}{R_{ij}}\right)-\vartheta_{(i-1)}\left(\frac{1}{R_Q}-\sum_j \frac{1}{R_{(i-1)j}}\right)-\sum_j \vartheta_{ij}\frac{1}{R_{ij}}-\sum_j \vartheta_{(i-1)j}\frac{1}{R_{(i-1)j}}=0\cdot \tag{10}$$

The result of calculation is summarized for clarity directly in the thermal network as shown in Fig. 6. The influence of coolant air heating is evident as show increasing temperature of individual semiconductors along the heat sink.

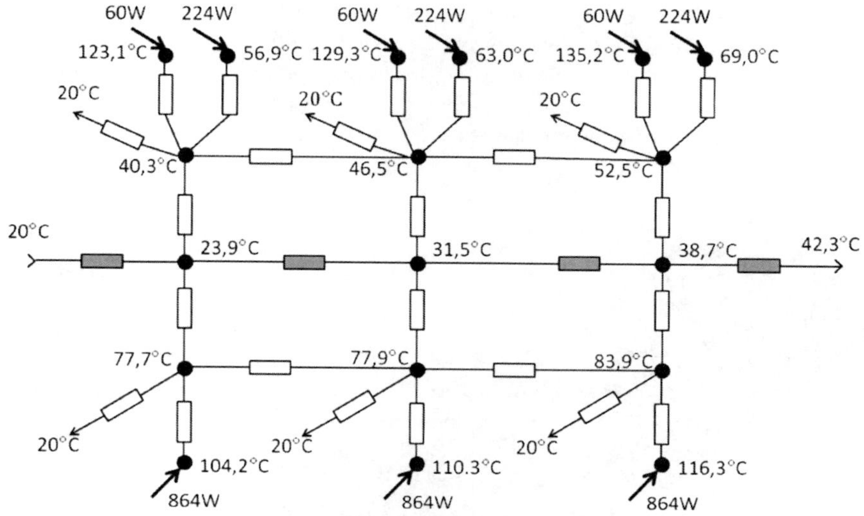

Fig. 6. Calculated temperatures in thermal network nodes of power battery charger

The heat flows thermal balance shows that the majority generated heat losses in the individual semiconductors is dissipated to air flow. Therefore, only heat losses of 131.5W is dissipated to ambient air by natural convection and radiation and the heat losses of 3312.5W is diffused to air flow from total heat losses of 3444W.

2.3 Finite Elements Method Thermal Model

The system ANSYS/Workbench 13 was used for thermal analyses of the power battery charger. This system has many modules which allow solving different physical problems [9], [10]. In this paper, the steady-state thermal module was tested for verification its applicability in thermal analyses of power electronic systems.

The three dimensional model (Fig. 7) contains semiconductors, insulating pads and heat sinks profiles was imported to Workbench from Autodesk Inventor where it was created. The Geometry model of insulating pads and heat sinks profile nearly corresponds to real geometry. The differences are in holes for screw in the insulating pads and heat sinks profiles which are not considered. In addition, the fins surface is considered as smooth. However, there are miniature triangular fins on the fins surface. It was compensated by the boundary conditions. The semiconductors were modeled as the rod corresponding to size of semiconductor case. The equivalent thermal conductivity of semiconductors was calculated using equation (3) where the thermal resistance value (junction-case) and the size parameters were used from the semiconductor datasheets.

The finite elements model and the boundary conditions which were set to model are shown in Fig.7. The mainly hex dominant and sweep method were used for mesh creation. The total number of elements is 424314. The generated heat losses in the individual semiconductors were placed on the upper surface of the volume representing them. The heat transfer coefficient of $69 Wm^{-2}K^{-1}$ which was calculated using equation (5) was set on the heat sink channel surfaces. Likewise, heat transfer coefficient of $10 Wm^{-2}K^{-1}$ representing the common heat transfer by natural convection and radiation to ambient was set on all external surfaces with the exception of heat sink channel surfaces.

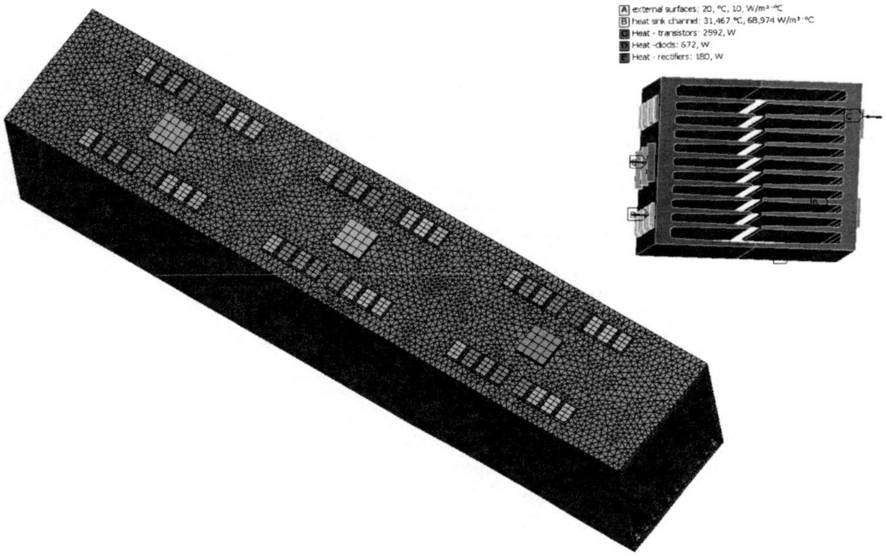

Fig. 7. Finite elements model and boundary conditions applied to the thermal model

The simulation result is distribution temperature in the individual parts of the thermal model. The main results of simulation are shown in Fig. 8. The temperatures on the upper surface of semiconductors bodies are junction temperatures. The result shows temperature distribution on the heat sink fins. The temperature profile over fins can be used for analyses of them efficiency.

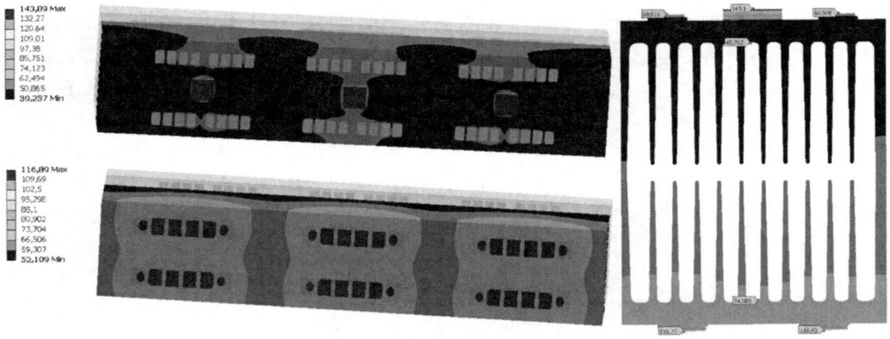

Fig. 8. Temperature distribution in semiconductors and heat sinks of power battery charger

As in the case the lumped parameters thermal network, the heat flows thermal balance was made. Thence, the heat flow to ambient air by natural convection and radiation is 202.8W, and the heat flow to air flow in the heat sink channel is 3241.2W.

3 Conclusion

The LPTN is the most used and friendly solution for a fast and low-computation time-consuming thermal analysis. The accuracy of this method is strongly dependent on the thermal parameters. As show results, a thermal network of coolant flow used in the LPNT improve the calculation credibility.

The FEM thermal model is more detailed as well as gives better idea of temperature distribution in the individual bodies. The FEM approach requires a quite long modeling and computation time with respect to the thermal network. The main disadvantage of steady-state thermal module (Workbench) is impossibility respected coolant heating. The bulk temperature depending on the length of the heat sink cannot be set to respected coolant heating.

Temperatures calculated using both methods are comparable because differences are smaller than 15%. Maximal junction temperatures of semiconductors are below safe limits of 150°C

In this paper, a summary of the current state of thermal analysis of power electronic systems was given. The most common methods used for thermal analysis were compared. Their advantages and disadvantages were discussed. This paper has given useful in information for researchers involved in thermal analysis of power electronic systems.

Acknowledgment

This research has been supported by the European Commission under the ENIAC CA-E3Car-2008-120001 E^3CAR - Nanoelectronics for an Energy Efficient Electrical Car project.

References

[1] März, M., Nance, P.: Thermal Modelling of Power-Electronic Systems, Published at `http://www.iisb.fhg.de/en/arb_geb/powersys_pub.htm`

[2] Drofenik, U., Kolar, J.W.: Teaching Thermal Design of Power Electronic Systems with Web-Based Interactive Educational Software. Interactive Power Electronics Seminar (iPES), at `http://www.ipes.ethz.ch`

[3] Boglietti, A., Cavagnino, A., Staton, D.A., Shanel, M., Mueller, M., Mejuto, C.: Evolution and Modern Approaches for Thermal Analysis of Electric Machines. IEEE Transactions (March 2009)

[4] Newell, W.E.: Transient Thermal Analysis of Solid-State Power Devices – Making a Dreaded Process Easy. IEEE Transactions on industry Applications IA-12(4), 405–420 (1976)

[5] Idelčik, I.E.: Handbook of Hydraulic Resistance, 3rd edn., New York, US (2006)
[6] Vlach, R., Grepl, R., Krejčí, P.: Control of Stator Winding Slot Cooling by Water Using Prediction of Heating. In: ICM 2007 4th IEEE international Conference on Mechatronics, Kumamoto, Japan, pp. 5–9. IEEE, Los Alamitos (2007) ISBN 1-4244-1184-X
[7] Vlach, R.: Predictor for Control of Stator Winding Water Cooling of Synchronous Machine. In: Recent Advances in Mechatronics, Mech., pp. 190–194. Springer, Berlin (2007) ISBN 978-3-540-73955-5
[8] Yunus, A., Turner, H., Cimbala, J.M.: Fundamental of Thermal-fluid Sciences, 3rd edn. McGraff Fill, Anstralia (2008)
[9] ANSYS (cit. 2011-02-11), `http://www.ansys.com`
[10] ANSYS examples (cit. 2011-01-20), `http://ansys.net/`

Part II

Metrology and Nanometrology

Measurement of the SMA Actuator Properties

M. Dovica*, T. Kelemenová, and M. Kelemen

Technical University of Košice, Faculty of Mechanical Engineering,
Letná 9, 042 00 Košice, Slovak Republic
miroslav.dovica@tuke.sk, tatiana.kelemenova@tuke.sk,
michal.kelemen@tuke.sk

Abstract. The paper deals with experimental measurement of properties of the shape memory alloy (SMA) wire actuator. A measurement stand is designed for experimental examinations of the actuator properties. Statics and dynamic characteristic is determined via experimentally way. Step response for activation and deactivation of actuator has been obtained. Experimental stand has been designed for educational purposes in mechatronic study program.

1 Introduction

Shape Memory Alloys (SMA's) are famous materials which are able to return to a predetermined shape when heated. The SMA, which is cooled below its transformation temperature, has low yield strength and it can be easy deformed into any whatever new shape. After heating above the transformation temperature, it changes crystallographic structure and it returns to the original shape. The SMA crystallographic structure changes between two phases, the low temperature (martensite) and high temperature (austenite) phases (Fig. 1). It is able to produce extremely forces from the viewpoint of volume to force ratio. SMA could be used also as force sensors and after that again as actuator. Consequently, the SMA belongs into group of smart materials. The most famous SMA material is Nitinol, which is an alloy of nickel and titanium. It has been discovered in Naval Ordance Laboratory in sixty years of twentieth century. Phenomenon of SMA is occurs in more then 20 alloy types. The SMA actuators are made in wire, spring or ribbons shape.

The shape memory effect - SME described above, whereby only the parent austenitic phase is remembered by the alloy, is known as the one-way SME. It is also possible to make an alloy remember both the parent austenite phase and the martensite shape simultaneously. This is known as the two-way SME. In this case, the alloy has two stable phases: a high temperature austenite phase, apparent on heating and a low-temperature martensite phase, apparent on cooling. Although SMA of two-way SME provides contractive and tensile forces, its tensile force is much smaller than contraction force and recoverable strain normally less than half that

* Corresponding author.

of one-way SME type. Thus SMA actuators of oneway SME are more attractive in mechatronic applications and usually preferred. The oneway SMA needs a reverse-bias force to return the wire to its original length. Bias forces can be created by many methods: gravitational pull, spring, magnetic force, opposing SMA wire [1, 2, 3].

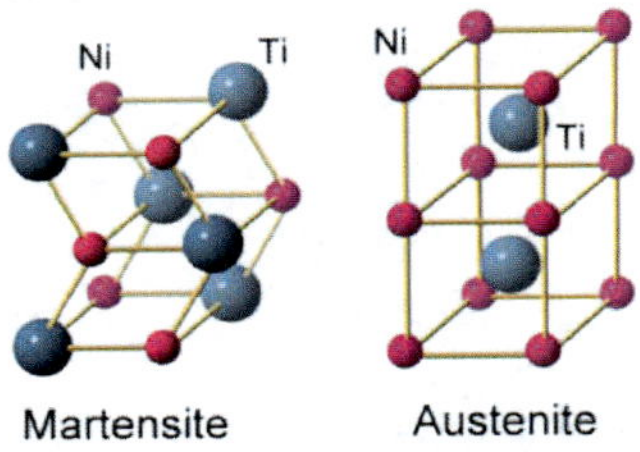

Fig. 1. The SMA crystallographic structure [4]

Thermal activation of the SMA can be easily driven by electrical current via Joule heating. Cooling of the SMA can be realized via heating radiation into surroundings at the room temperature. Other methods of improved cooling are to use: forced air, heat sinks, peltier elements, increased stress (this raises the transition temperature and effectively makes the alloy into a higher transition temperature wire), and liquid coolants. Combinations of these methods are also effective.

Hysteresis and nonlinearities cause the control troubles. Heat losses during the phase transformation phases (owing to internal friction or structural defects) cause hysteretic behaviour of SMA as shown in Fig 2.

In the [5] has been presented mathematic model of the SMA behaviours as equation:

$$\frac{dR}{R} = \pi_e \cdot d\sigma + K_\varepsilon \cdot d\varepsilon + \alpha_{RT} \cdot dT \text{ [5]} \tag{1}$$

where, R is resistivity, π_e - piezoelectric coefficient, σ - stress, K_z - coefficient of shape sensitivity, ε - strain (deformation), α_{RT} - coefficient of thernal expansion, T - temperature.

The model (1) shows the resistivity changes dependence on temperature changes. Piezoelectric coefficient and coefficient of shape sensitivity are very often unknown parameters of the SMA material. Consequently the model (1) is not so practically usable. More effective would be a model based on identification from transient response exploration, which is mentioned in fig. 11 [6].

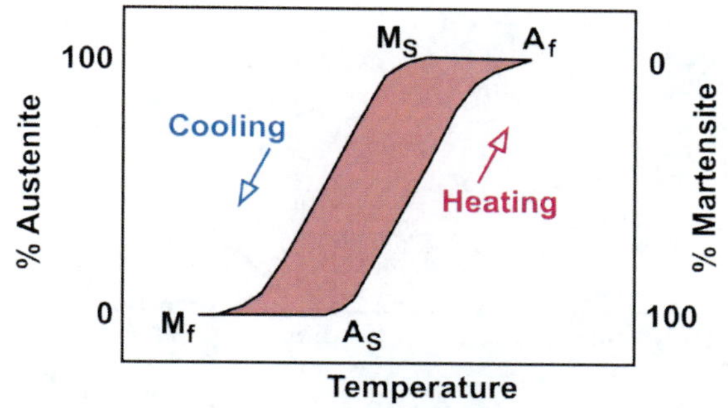

Fig. 2. The SMA hysteresis and nonlinearities [1, 2]

2 Measurement of the SMA Wire Actuator

Nitinol wire actuator named as FLEXINOL 250LT with 0.25mm in diameter has been tested in this paper. Recommended activation temperature is 70°C. Recommended recovery force is 9,3N. Activation electric current is defined under the 1000mA. Maximum stress (pull force) is 172MPa [3].

There is a standardized Test Method for Determination of Transformation Temperature of Nickel-Titanium Shape Memory Alloys by Bend and Free Recovery [7].

In this paper, the contraction and pull recovery test stand has been developed (fig. 3).

Fig. 3. Contraction and pull recovery test stand

Scheme of the measurement stand arrangement is shown on fig. 4. One end of the Flexinol wire is attached and second free end is connected via nylon wire with bias weight. Nylon wire is guided with two pulleys to the place for hook with bias weight. There is a reference point (fig. 5) from permanent magnet placed on the nylon wire. Deformation of the Flexinol wire is represented with the reference point (permanent magnet) and position of the magnet is measured via Hall position sensor. Consequently, output sensor voltage represents the deformation of the Flexinol wire. It is necessary to do calibration of the hall position sensor with length etalons (fig. 6). The calibration process associates the sensor output voltage with Flexinol wire deformation.

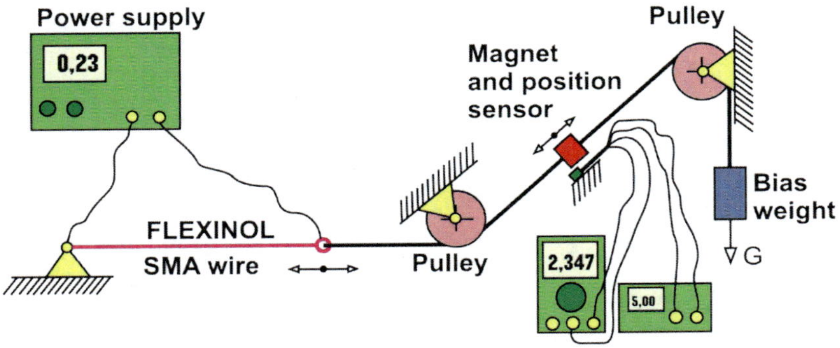

Fig. 4. Arrangement of the contraction and pull recovery test stand

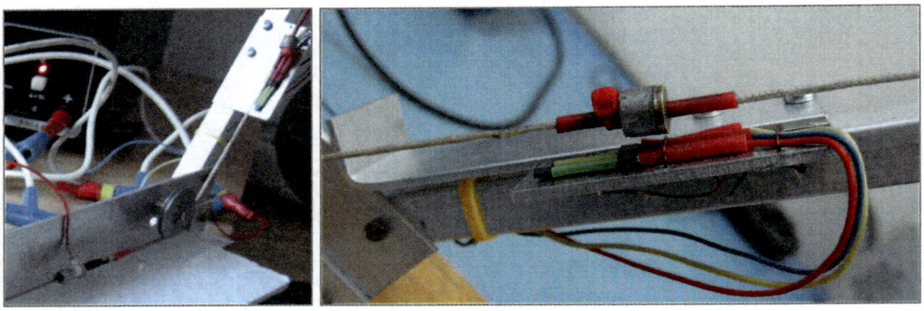

Fig. 5. Reference point for Flexinol deformation measurement

Results from calibration process are shown on figure 7. Output sensor voltage has nonlinear behavior which corresponds to Flexinol free end position. Relation between the free end position and output sensor voltage has been approximated via polynomial model mentioned on figure 7.

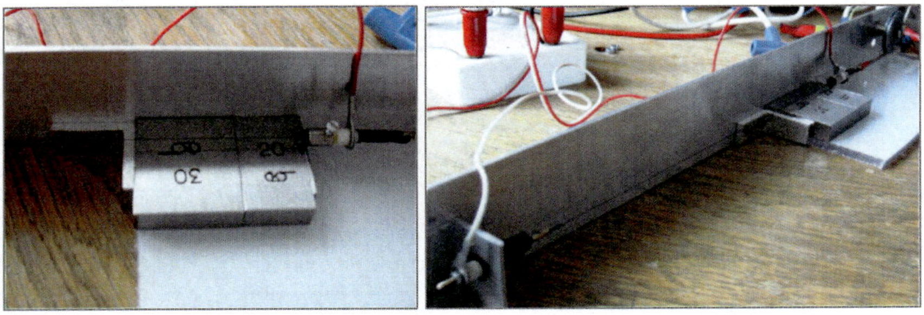

Fig. 6. Calibration process of the Flexinol deformation measurement

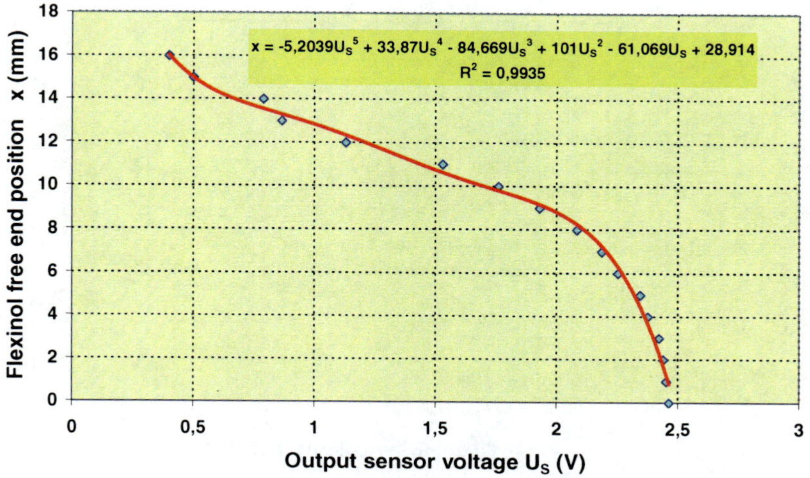

Fig. 7. Calibration process of the Flexinol deformation measurement

Heating and cooling of the Flexinol causes change of the Flexinol free end position. Fig. 8 shows this dependence of free end Flexinol position on value of electric current, which shows highly nonlinearity and hysteresis behavior. The characteristic shown on fig. 8 has been measured with bias weight 1kg, which corresponds to maximum pull stress.

The pull stress overload of the Flexinol has been also tested. Flexinol works correctly and without damaging. Results of this testing (fig. 9) shows decreasing of the relative deformation with increased value of additional bias weight. Activation value of electric current has been 1000mA for this testing. Inconsistency of data is caused with uncertainty measurements.

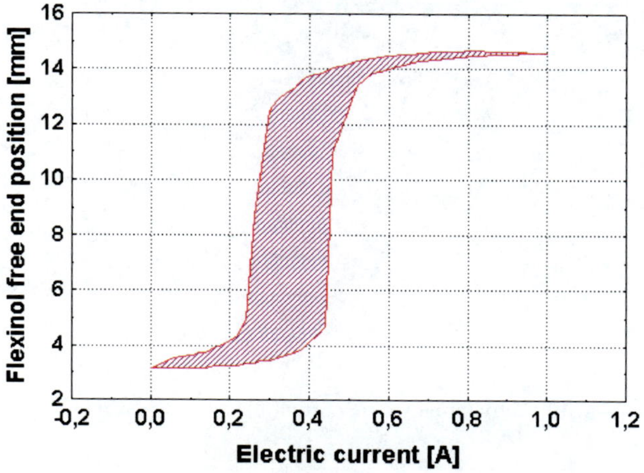

Fig. 8. Dependence of the free end Flexinol position on the value of electric current

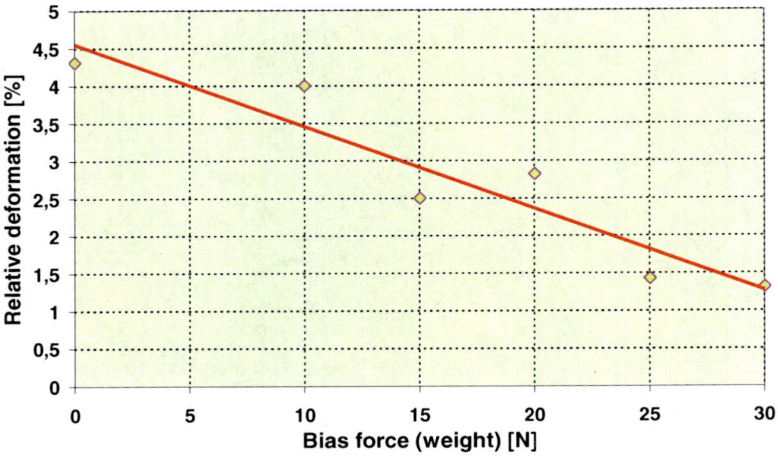

Fig. 9. Dependence of the relative deformation on additional bias weight (force)

Dynamic characteristic has been tested through the step response testing. Excitation electric current is shown on fig. 10 and it has been controlled via microcontroller. Step response has been measured via measuring adapter with personal computer. Mathematic model obtained from calibration process has been used for obtaining of the step response as time dependence of the free end Flexinol position.

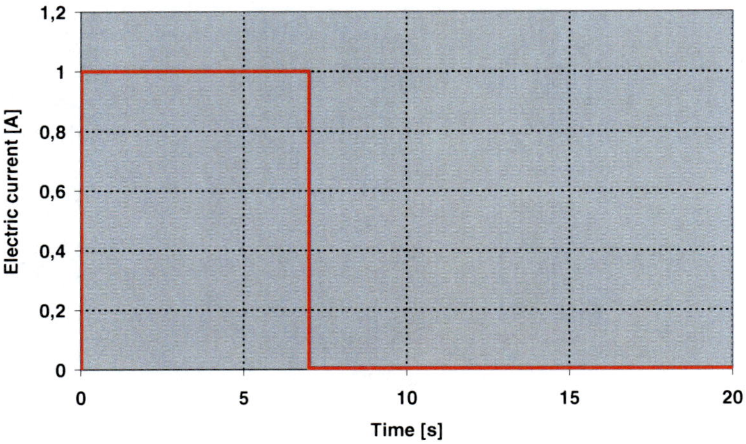

Fig. 10. Activation electric current for step response testing

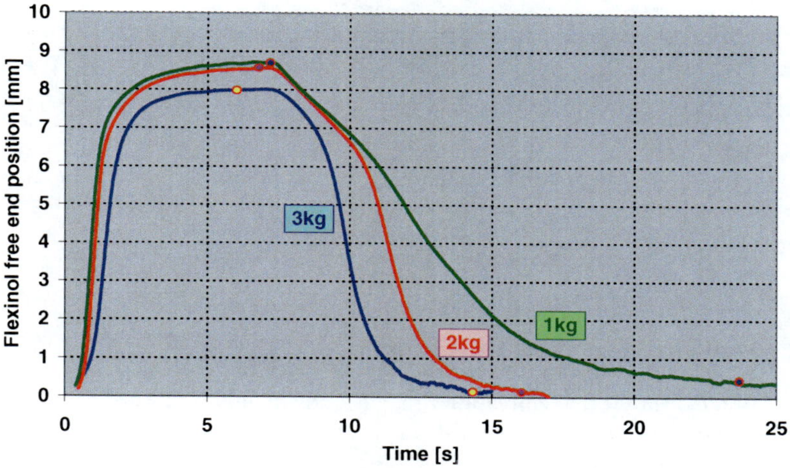

Fig. 11. Flexinol Step response for various bias weight

It is possible to determine activation and deactivation time for Flexinol which are corresponds with heating and cooling time. Figure 11 shows that cooling is slower then heating.

Very often SMA actuator is overloaded in construction. Actuator producer recommends bias weight 930g. It is important to know about overloading influence to actuator behaviors. For this reason step response is tested also for overloaded mode (for 2kg and 3kg) and it is possible to say that overloading causes a decreasing of the heating and cooling time. Overload in heating process causes decreasing of actuator deformation and that is the reason of shorter heating time. Cooling time is shorter because higher stress helps to short relaxation time.

Overloading has no degradation influence to functionality of actuator, but it is possible that overload mode may decrease lifetime. Lifetime tests will be also executed in the next future.

Activation time can be also described as ratio of the energy necessary heating and electric power consumption used for actuator activation [8].

$$t_a = \frac{E_h}{P_{el}} \tag{2}$$

Where energy necessary heating E_h and electric power consumption used for actuator activation P_{el} can be expressed in form:

$$E_h = c_v \cdot m \cdot \Delta T = c_v \cdot \rho_{Density} \cdot S_{NiTI} \cdot L \cdot \Delta T \tag{3}$$

$$P_{el} = I^2 \cdot R = \frac{I^2 \cdot \rho_{\mathrm{Re}\,sist} \cdot L}{S_{NiTi}} \tag{4}$$

Where
c_v – specific heat constant, ΔT – temperature difference of crystallization phases, m – actuator weight, $\rho_{Density}$ - density of the actuator, ρ_{Resist} - electrical resistivity of actuator, S_{NiTI} - cross section area of actuator, L – actuator length, I – heating electric current.

Activation time ta is possible to express from the equations (2, 3, 4) in form:

$$t_a = \frac{A_{Material} \cdot S_{NiTi}^2 \cdot \Delta T}{I^2} \tag{5}$$

Where
$A_{material}$ is material constant for SMA actuator.

Equation (5) shows that activation time depends mainly on its cross section and also on value of electric current used for heating. Length of actuator wire has no influence to activation time.

It is possible to obtain theoretical activation time after substitution of known actuator parameters. Theoretical activation time is 5,3s. Real time of activation (fig. 11) is varied in range from 6,1s to 7,4s (it depends on bias weight). Mentioned equation has no information about heat looses into environment. It depends on heating and cooling condition.

Activation time should be extended for this reason, which is possible to express in the form:

$$t_a = \frac{E_h + E_p}{P_{el} - Q} \tag{6}$$

where
Q is actuator heat looses into environment and
E_p is an energy needed for phase transformation.

Consequently, activation time is extended. It is very complicated to identify these additional energy parts. Experiment is the best way how to identify them.

3 Conclusion

Shape memory alloy has a lot of advantages as clean, silent and spark free operation, high biocompatibility and excellent corrosion resistance. They are also free of parts such as reduction gears and do not produce dust particles. Actuators based on this principle are very often used in robotic and mechatronic application [9, 10, 11, 12, 13].

Acknowledgment

The authors would like to thank to Slovak Grant Agency – project VEGA 1/0454/09 "Research on mechatronic systems imitating snake locomotion in confined and variable area" and VEGA 1/0022/10 „A contribution to the research of measuring strategy in coordinate measuring machines ". This contribution is also the result of the project implementation: Centre for research of control of technical, environmental and human risks for permanent development of production and products in mechanical engineering (ITMS:26220120060) supported by the Research & Development Operational Programme funded by the ERDF.

References

[1] Andrianesis, K., Koveos, Y., Nikolakopoulos, G., Tzes, A.: Experimental Study of a Shape Memory Alloy Actuation System for a Novel Prosthetic Hand, Cismasiu, C. (ed.), pp. 81-105. InTech (2010) ISBN: 978-953-307-106-0

[2] Severinghaus, E.M.: Low Current Shape Memory Alloy Devices. U.S. Patent US6969920B1, Monodotronics (November 29, 2005)

[3] Dynalloy Inc., Flexinol Technical Characteristics, [online]. Document Version 2011/01/02 T15:31:00Z 2011 (cit. 2011-01-02), `http://www.dynalloy.com`

[4] Hoi-Yin, S.: Metal has memmory??, [online]. Document Version 2011/01/02 T17:45:00Z 2011 (cit. 2011-01-02), `http://www.phy.cuhk.edu.hk/phyworld/iq/memory_alloy/memory_alloy_e.html`

[5] Drahoš, P., čIčáková, O.: Measurement of SMA Drive Characteristics. Measurement Science Review 3, Section 3, 151–154 (2003)

[6] Pirč, V., Ostertagová, E., Sedláčková, A.: Approximative solution of some differential equations. In: Buletinul Stiintific al Universitatii din Baia Mare: Dedicated to Costica Mustata on his 60 th anniversary. Fascicola Matematica Informatica, North University of Baia Mare, Rumania, pp. 73–80 (2002) ISSN 1222-1201

[7] ASTM F 2082 – 01 Standard Test Method for Determination of Transformation Temperature of Nickel-Titanium Shape Memory Alloys by Bend and Free Recovery

[8] Janke, L., Czaderski, C., Motavalli, M., Ruth, J.: Applications of shape memory alloys in civil engineering structures - Overview, limits and new ideas. Materials and Structures 38, 578–592 (2005)

[9] Vitko, A., Jurišica, L., Kľúčik, M., Duchoň, F.: Context Based Intel-ligent Behaviour of Mechatronic Systems. Acta Mechanica Slovaca. Roč. 12, č. 3-B, pp. 907-916 (2008) ISSN 1335-2393

[10] Vitko, A., Jurišica, L., Kľúčik, M., Murár, R., Duchoň, F.: Embedding Intelligence Into a Mobile Robot. AT&P Journal Plus.: Mobile robotic systems 1, 42–44 (2008) ISSN 1336-5010

[11] Olasz, A., Szabó, T.: Direct and inverse kinematical and dynamical analysis of the Fanuc LR Mate 200IC robot. In: XXV. International Scientific Conference microCAD, pp. 37–42 (March-April 2011)

[12] Lénárt, J., Jakab, E.: Machine vision used in robotics. In: OGÉT 2010, XVIII. International Conference on Mechanical Engineering, pp. 272–274 (in Hungarian)

[13] Antal, D.: Dynamical modelling of a path controlled vehicle. In: XXIV. Micro CAD, International Scientific Conference 2010, pp. 1–6 (2010)

Design of Mass Spectrometer Control in NI LabVIEW and SolidWorks

L. Ertl and P. Houška

Faculty of Mechanical Engineering, Brno University of Technology,
Technická 2896/2, 616 69, Brno, Czech Republic

Abstract. This paper deals with designing and testing new software functions for control of a mass spectrometer, which is type of TOF-MALDI and for depositing interface. The functions are tested on virtual motion analyses of 3D CAD model. Then they are implemented in final application for real system. Applications and functions are developed in graphical development system National Instruments LabVIEW and 3D models of devices are made in SolidWorks.

1 Introduction

Mass spectrometry is a method for measuring and analyzing chemical composition of samples. We measure the mass of particles depending on their time of flight. This mass spectrometer and deposition interface are built in chemical laboratory of Masaryk University in Brno [1].

Method of V-cycle [2] was used for development new software. This method was very useful because we created functions and we tested them with virtual models. This is much safer and cheaper then, developing prototype of hardware.

First we created virtual 3D model of device. This was useful for right imaging of devices' functions. We set the software requirements based on these models and started to prepare for designing of the applications. In the next step we developed code for stage control like move with stages and code for taking pictures by camera. We developed a new toolkit for dynamic characteristics, which allowed us to identify the system and to design the control.

2 Motion Analysis

Stage model was made in SolidWorks. There is possibility to use "motion analysis" for simulation of movement. Here we can set different parameters and sensors for each drive. This creates a virtual axis of movement. The rotation drives and the ratio to linear moves were set here. The sensors for limit switches were set at the ends of each axis. These stop the movement, if the stages reach the edges.

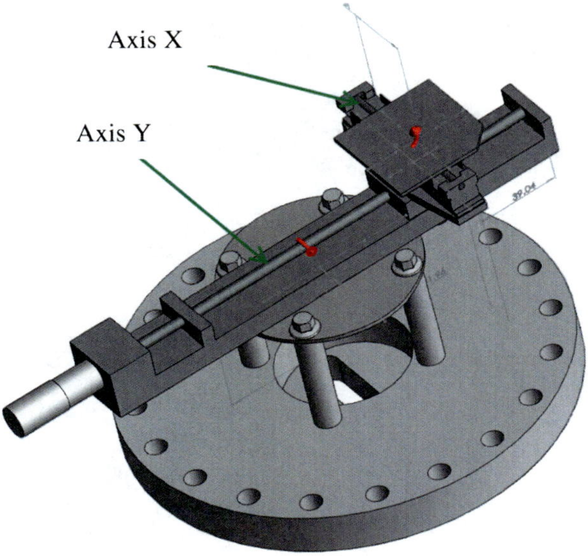

Fig. 1. 3D model of stages for mass spectrometer

The control application for simulation is created in NI LabVIEW SoftMotion. The assembly from SolidWorks is linked in to LabVIEW project [3]. Also the axis are inserted the same way and there are mapped the sensors for reading the states of them [4].

The "Scan Engine" provides connection between model and simulation. This LabVIEW tool connects itself to solver of SolidWorks. This is the way how SolidWorks and LabVIEW are synchronized and data are sent to control application.

Some functions for control mass spectrometer were developed with this simulation. For example the calculation of trajectory for movement with stages, grabbing picture by camera and setting the coordinates for moving stages from this picture. Also the function for distribution of spots on plate which is based on starting point, ending point and spacing between each other point was designed.

Described simulation does not support control using dynamic like force or torque. It is only kinematic control and it is useful for visualization and collision testing.

This approach can be used only for control verification, check the collisions, but cannot be used for driving dynamics controlled assembly. The Dynamic Toolkit was developed based on knowledge application program interface (API), this toolkit open up the dynamic studies from SolidWorks in LabVIEW.

3 Simulation with Dynamics

The minimal position step of both axes can be 4 μm and the number of samples can be 10 000 or more. From this reason we need high precision position control and maximal speed. It is necessary to know dynamic characteristics for identify the drives and design their control.

The main thought for creating this new toolkit was in creation of accessible dynamic simulations from environment used by mechanical systems developers (SolidWorks) in environment, which allows us to implement real-time control (LabVIEW).

Model of mechanical system is designed with dynamic characteristics (force, torque) as the inputs and kinematic (position, velocity, acceleration) as outputs. We are getting the dynamics from simulation of mathematical model which represents the actuator. We can simulate the real system behavior in this way. Linear forces or moments are defined in place of electro-mechanic actuators.

Model of mechanical system respects its dynamics, it is implemented like motion study in this system. There are inserted the actuators in the system and there is also specified the friction in individual mates. The electromechanical actuators are not easy to model in SolidWorks. They were replaced by their force/torque interaction and the simulation in LabVIEW contains mathematical (state-space) model of electromechanical system [5]. The inputs of mathematical model of actuator are kinematic characteristics read from SolidWorks at specific part of assembly and the outputs are the requested forces/torques, which are transferred back to SolidWorks.

Dynamic co-simulation consists of some important phases. First phase is a dynamic study initialization. It starts by opening the appropriate motion study in SolidWorks and connecting from LabVIEW to motion analysis in SolidWorks. Then we connect to the linear forces or torques defined in motion study with LabVIEW, which creates interface for co-simulation. The callbacks are registered for individual forces/torques these are generated by solver in SolidWorks. The events of solver are generated in individual calculation steps and the solver read from them actual values of forces and torques.

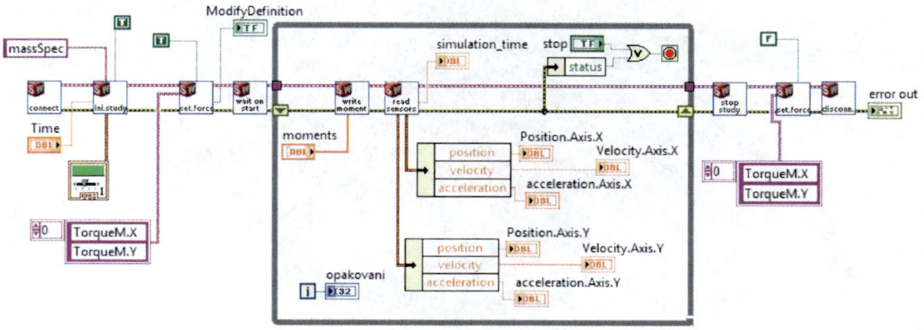

Fig. 2. Dynamic Toolkit interface for use of dynamic in simulation

Next phase is starting the simulations and their synchronization. The synchronization is driven from LabVIEW and it depends on steps of solver from SolidWorks. SolidWorks generates the events during the simulation, from these events are read the actual values of forces/torques. The actual requested kinematics of SolidWorks model is read in LabVIEW via calling the API functions. These acquired data are used as feedback for mathematical model of electro-mechanical actuator.

In next phase we do the calculations. Here are written the forces or torques and the kinematic data are read from sensors. As long as the calculation is running, we have always to check solver events. There is the problem with sample period of solver from SolidWorks. Its steps are variable and depended on complexity of the calculation in motion study. The sample period for controller is constant, so we need to resample the kinematic data read from motion study. We choose the method, when we check the length of increment, if this is smaller than sample period for mathematical model, we don't use this one. We use the first one which is in length of controller sampling.

At the end, when the calculations are finished or there is generated an event for interruption, we have to free the resources. This means stop the motion study, stop the force interaction and break the link with SolidWorks application.

This new toolkit allowed us to read dynamics, but there was a problem with solver and synchronization of calculations. We made verification for our model because the real system has some properties which are hardly to describe and we needed the most accurate model. We made identification for describing of friction and backlash in gears. We used these data for model of the system and controller. Now we can move with high accuracy and with top speed in real system.

Fig. 3. The real stages for mass spectrometer

Very important was also the optimization of system control. The stages are working in deep vacuum and the drives are specially designed for these conditions. They are very hard to maintain so we need the high reliability and repeatability.

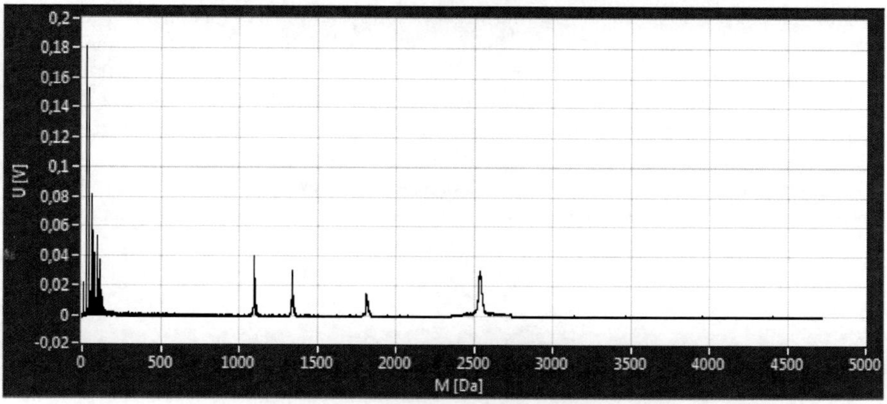

Fig. 4. Measured masses for peptides from sample

4 Conclusion

The NI LabVIEW SoftMotion for SolidWorks toolkit was useful for simulation without dynamics. It is beneficial for visualization and verification of kinematics. We could design the movement on desired trajectories with this instrument and in simulation inspect trajectory and possible collision detect.

The design of controller needs dynamic characteristics therefore we developed new Dynamic Toolkit which allows us to use dynamics and via this environment identify the system and design controllers for drives.

The designed code and controllers were used with real device for measuring of samples. We measured masses of four peptides which were in the sample as results. These facts are shown in the figure 5.

This procedure was very effective and economically beneficial because there was no need of any hardware during the tests of new functions and so it could not be damaged.

Acknowledgement

This work is supported from research plan MSM 0021630518 Simulation modeling of mechatronic systems and FSI-S-11-23.

References

[1] Moskovets, E., Preisler, J., Chen, H.S., Rejtar, T., Andreev, V.: High-Throughput Axial MALDI-TOF MS Using a 2 kHz Repetition Rate Laser. Anal. Chem. 78, 912–919 (2006)

[2] VDI 2206. Design methodology for mechatronic systems. Beuth Verlag, Berlin (2004)

[3] National Instruments. Getting Started with LabVIEW [online] (June 2010), http://www.ni.com/pdf/manuals/373427g.pdf [cit. 2011-05-09]

[4] National Instruments. Getting Started with NI SoftMotion for SolidWorks [online] (June 2010), http://www.ni.com/pdf/manuals/372876c.pdf (cit. 2011-05-09)

[5] Brezina, T., Brezina, L.: Controller Design of the Stewart Platform Linear Actuator. In: 8th International Conference on Mechatronics, Luhacovice, Czech Republic, pp. 341–346 (2009)

Shape Deviations of the Contact Areas of the Total Hip Replacement

V. Fuis[1,*], M. Koukal[2], and Z. Florian[2]

[1] Centre of Mechatronics – Institute of Thermomechanics AS CR and Faculty of Mechanical Engineering BUT, Technická 2, Brno, 616 69, Czech Republic

[2] Institute of Solid Mechanics, Mechatronics and Biomechanics, Faculty of Mechanical Engineering BUT, Technická 2, Brno, 616 69, Czech Republic

Abstract. One type of total hip replacement function loss is acetabular cup loosening from the pelvic bone. This article examines manufacture deviations as one of the possible reasons for this kind of failure. Both dimensions and geometry manufacturing deviations of ceramic head and polyethylene cup were analyzed. We find that deviations in the variables analysed here affect considered values of contact pressure and frictional moment. Furthermore, contact pressure and frictional moment are quantities affecting replacement success and durability.

1 Introduction

The paper examines a number of the biomechanical problems associated with total hip replacements (THR). The manufacture deviations [3], [12] and impacts of femoral head and acetabular cup upon the failure of function of THR. THR failure is typically accompanied with severe pain and necessitates reoperation and prolonged convalescence. THR function loss is associated with a combination of several risk factors – use of inappropriate prosthesis materials, geometry, surface finish treatment, ... A recent clinical study of 600 THR's with ceramic-ceramic contact between femoral head and acetabular cups showed there was 20% failure in period 1977 – 1989 [4]. One type of THR functionality loss is loosening of the acetabular cup from pelvic bone (Fig. 1).

* Corresponding author.

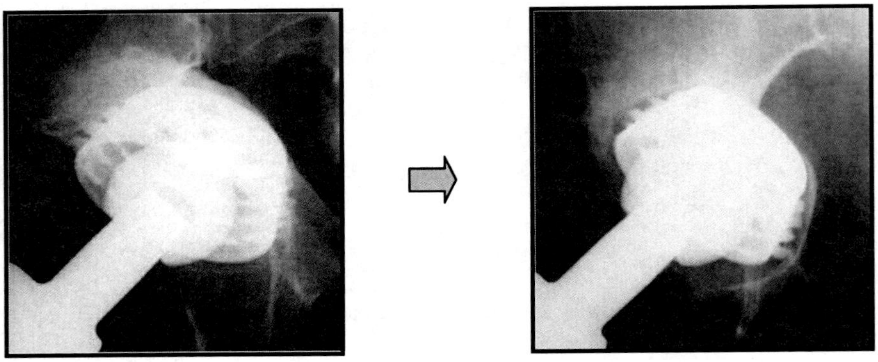

Fig. 1. Continuity of acetabular cup loosening from pelvic bone [1]

The Faculty Hospital Brno (FHB) has an ongoing project examining the clinical impact of manufacture deviations in hip prostheses [1]. During a five year period, a failure rate of X% in THR due to loosening was at the FBH (Tab. 1).

Table 1. % of failed THR [1]

Producer / Type of the cup	Failed THR due to loosening of the cup
Johnson&Johnson / Mercing	58 %
Balgrist / AlloPro	2 %
Walter-Motorlet (Fig .2)	26 %

Fig. 2. Walter-Motorlet Co. [1]

Loosening of the cups is thought to be caused by the deterioration of thepoly-ethylene, worn particles dislodged from the prosthesis then damage the tissue around the cup [2]. Amount of polyethylene wear is influenced by contact pressure and frictional moment between head and cup [8], [11]. Contact pressure and frictional moment is thought to be influenced by contact surfaces inaccuracies created during the fabrication process.

The objective of this study was therefore to computationally model the influence of fabrication tolerances on contact conditions between the acetabular cup and the femoral head in THR. The considered variables were firstly the contact pressure between acetabular cup and femoral head and secondly the frictional moment needed to be overcome to turn the femoral head in the acetabular cup.

2 Computational Modelling

A model of the hip joint (Fig. 3) was created using FEM system ANSYS in conjunction with standard Newton-Raphson's iterative scheme. The contact conditions between the femoral ball head and the acetabular cup were achieved by applying contact elements and using the Augmented Lagrange Multiplier technique.

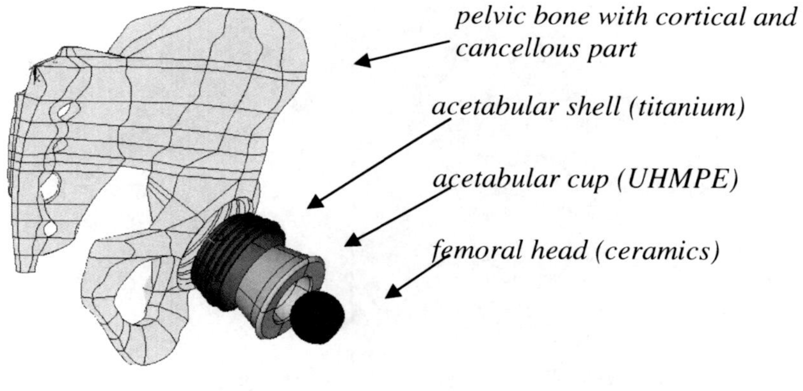

Fig. 3. Hip joint parametric model

This model was created as parametric in order to study influences of deviations of fabrication tolerances on the contact conditions. The investigated parameter is inner cup diameter.which was examined in the interval of 31.7 mm (the interference fit of 0.3 mm) to 32.4 mm (clearance fit of 0.4 mm).

Linear isotropic material models were used for all components with characteristics summarized at the Tab 2. Values of ceramic and titanium were taken from material lists, other characteristic from literature research.

Table 2. Material characteristics

Model part - material	Modulus of elasticity E [MPa]	Poisson`s ratio μ [-]
Femoral head - ceramics Al_2O_3	3.9×10^5	0.23
Acetabular cup - UHMWPE [5]	1.0×10^3	0.40
Acetabular shell - titanium	1.0×10^5	0.30
Pelvis - cancellous / cortical bone [7]	2.0×10^3 / 1.4×10^4	0.25 / 0.30

Values of modulus of elasticity of UHMWPE polyethylene varies in published literature (Tab.2). It ranges from 600 MPa to 1200 MPa. Therefore sensitivity analysis was done for this parameter with results listed in 4.1.

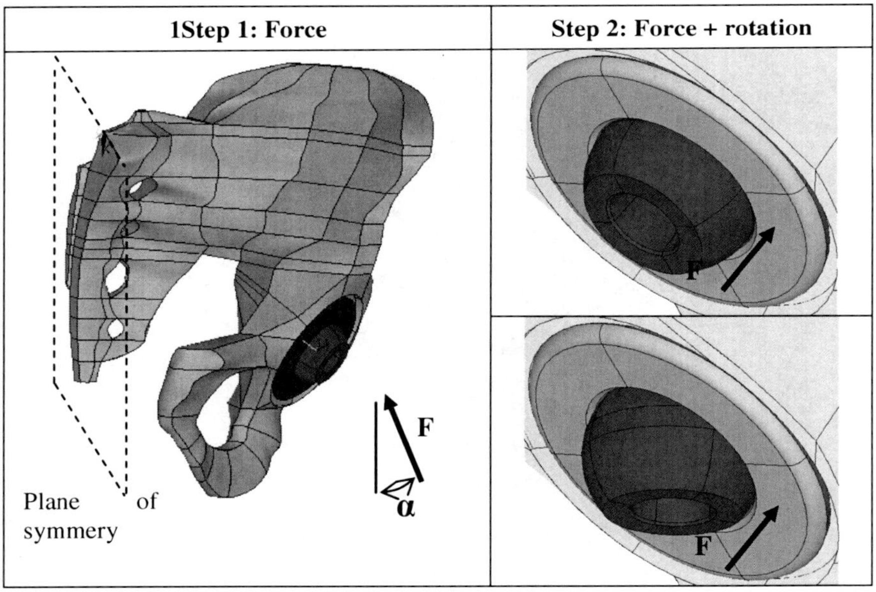

Fig. 4. Load steps of the system and boundary conditions

The first contact pair was modelled by creating contact elements between the femoral head and the acetabular cup, with with friction coefficient of 0.1 (Coulomb model used). The second contact pair was created between the acetabular cup and the shell. This pair was not further evaluated in this study. Titanium shell was bonded into the pelvic bone.

In the finite element model two boundary conditions were considered, loading applied to the centre of the femoral ball head and bonding of the model at the plane of symmetry (Fig. 4). The loading of the model was performed in two steps. In the first step a constant loading force of 2500 N was applied in a direction as shown in Fig. 4, Step 1. In the second step loading force of the same magnitude was applied while the femoral head was being rotated by 30° as shown in Fig.4, Step 2. Magnitude and direction of the loading force F were established from equilibrium of forces of a man standing on one foot. Input for the calculation was an average weight 80 kg of a human body. Resulted value of the loading force F was 2500 N and the direction of the loading force α was 13.5°.

2.1 Measuring of the Deviations

In this study both used and unused acetabular cups and femoral heads (all made by Walter-Motorlet) were studied and measured (Tab. 3). We found, that the inner spheric surface of the used cups showed non-uniform abrasion caused by wear. This was due to the non-uniform contact pressure caused by geometry and deviations. Two types of head and cup contact surface deviations were observed. First dimension type (different diameters of femoral head and acetabular cup) and second geometry type (roundness deviations of spherical surfaces). Founded dimension type of deviation evokes interference type of fit between head and cup (Fig. 5).

Table 3. Deviations measured [1]

Cup / head	**Diameter [mm]**	**Roundness deviation [mm]**	**Used / unused**
Cup 1	31.97	0.117	Unused
Cup 2	31.88	0.090	
Cup 3	31.83	0.064	
Cup 4	31.79	0.087	
Cup 5	***32.11***	***0.357***	***Used***
Head 1	32.02	0.011	Unused
Head 2	32.04	0.009	

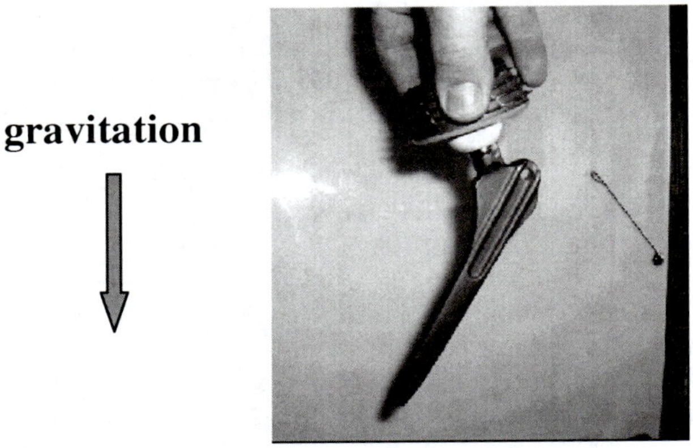

Fig. 5. Due to interference fit, the femoral stem holds in acetabular cup in spite of direction of gravitation [1]

3 Results

In order to model the dimension deviations the inner diameter of the acetabular cup was varied. This resulted in change of the fit between the femoral head and the acetabular cup (from interference fit to clearance fit). Values for the contact pressure (Fig. 6) and the frictional moment (Fig. 7) were considered.

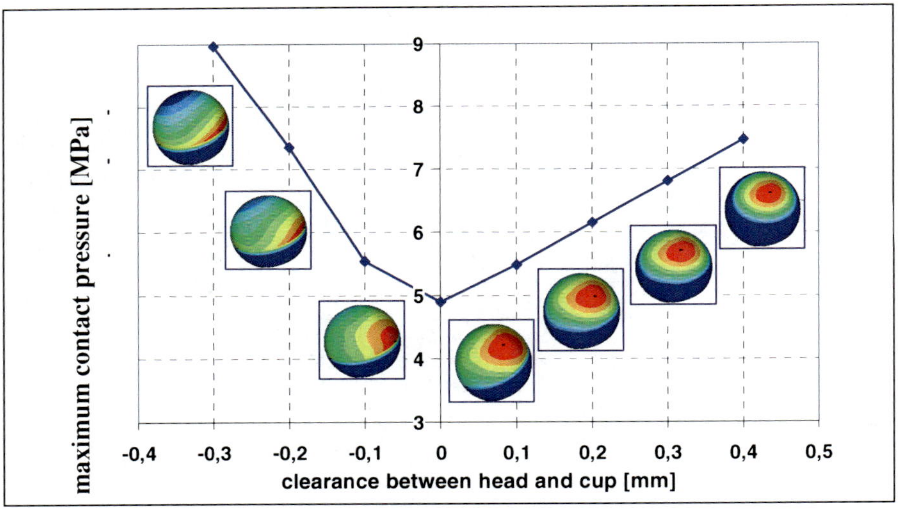

Fig. 6. Influence of the head – cup clearance on the contact pressure

We found that by increasing the clearance starting from 0.0 mm (fine fit) the value of contact pressure raised linearly within the investigated interval (Fig. 6). Increasing the clearance between the head and the cup by 0.1 mm the contact pressure raised by ~10%. On the contrary the value of the frictional moment slightly decreased (Fig. 7) when increasing the clearance. This was due to a smaller size of the contact surface.

Reducing the clearance below 0.0mm (fine fit), we found that the value of contact pressure raised with approximately double speed than for the positive clearance values (Fig. 6). However the value of the frictional moment steeply increased (circa 10%) when reducing the clearance by 0.1mm (Fig. 7). This was caused by the increase of circumferential tension due to negative clearance (interference).

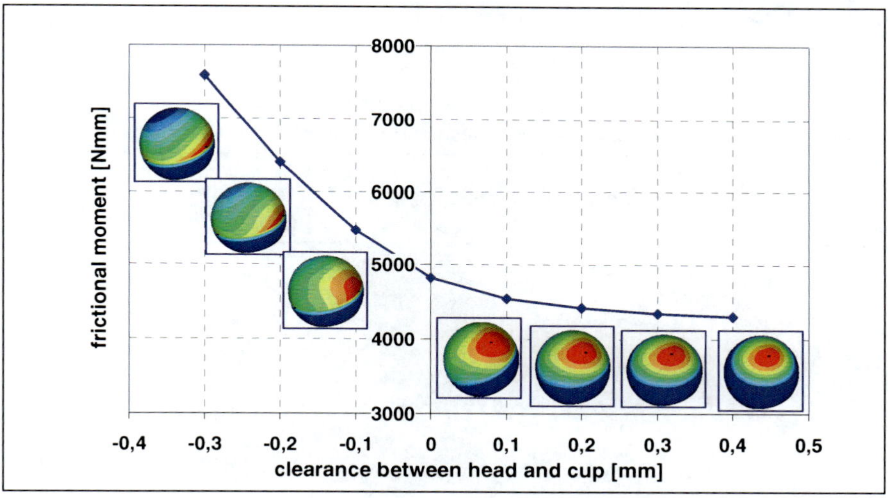

Fig. 7. Influence of the head – cup clearance on the frictional moment

3.1 Sensitivity Analysis

To study the influence of the elasticity modulus of the polyethylene cup the modulus of elasticity value varied from 600 MPa to 1200 MPa. Values of contact pressure (Fig. 8) and frictional moment (Fig. 9) were considered. Results shows that if the value of elasticity modulus of cup decrease, than the values of contact pressure decrease due to increase of contact area for all types of fit. From the same reason, the frictional moment decrease in the case of clearance fit. With decrease of modulus E value in the interference type of fit, the effect of equatorial tension decrease, and the frictional moment decrease too. In the case of clearance fit, value of frictional moment is almost the same for all values of elasticity modulus.

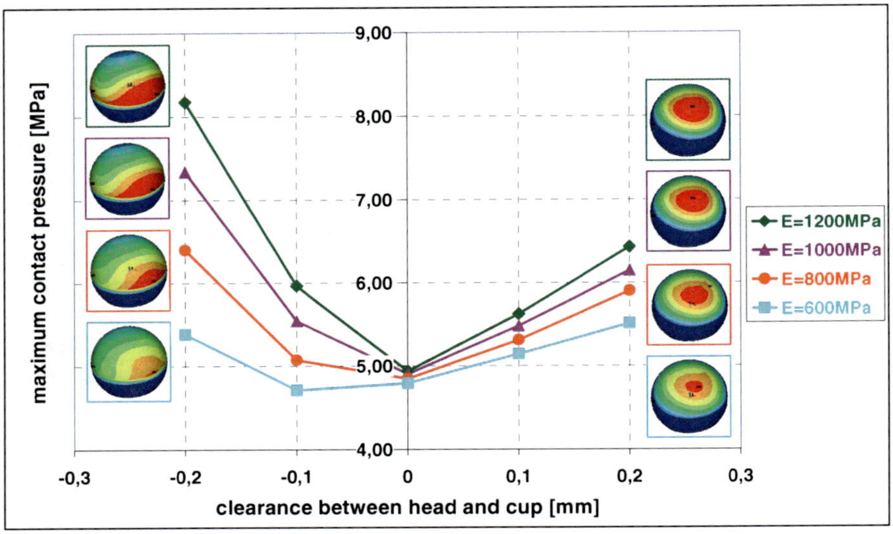

Fig. 8. Acetabular cup modulus of elasticity vs. contact pressure

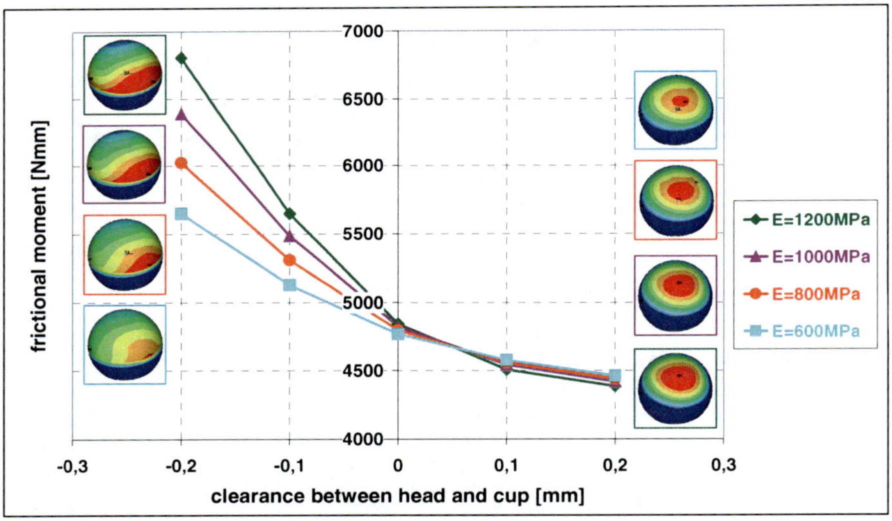

Fig. 9. Acetabular cup modulus of elasticityvs.frictional moment

4 Conclusion

In this study parametric model of the hip joint with total hip replacement was created. The influence of fabrication tolerances on contact conditions between the acetabular cup and the femoral head was studied. Values of contact pressure and frictional moment were considered.

We discovered, that if the value of deviations increase, the contact conditions deteriorate. We also found, that the best type of fit was fine fit, it means without clearence and interference. As this type of fit is hard to fabricate we recommend to fit head and cup together with clearance value ranging 0 mm to 0.05mm. In this clearance interval the values of considered quantities varied in less than ten percents. We do not recommend using interference type of fit because of strong deterioration of the contact conditions.

Both contact pressure and frictional moment values affect the success and durability of the replacement, therefore it is recommended to minimize their values. The relationship between contact pressure and wear is analyzed at [6], [8], [9] and [10].

Acknowledgment. The research has been supported by the following projects AV0Z20760514 and MSM0021630518.

References

[1] Krbec, M., Fuis, V., Florian, Z.: Loosening of the uncemented cups of total hip replacement from pelvic bone. In: V. National congress ČSOT, Prague, p. 108 (2001)

[2] Charnley, J., Kamangar, A., Longfield, M.: The Optimum Size of Prosthetic Heads in Relation to the Wear of Plastic Socket in Total Replacement of the Hip. Medical & Biological Engineering 7, 31–39 (1969)

[3] Fuis, V.: Tensile Stress Analysis of the Ceramic Head with Micro and Macro Shape Deviations of the Contact Areas. In: Recent Advances in Mechatronics: 2008-2009, pp. 425–430. Springer, Heidelberg (2009)

[4] Gualtieri, G., Calderoni, P., Ferruzzi, A., Frontali, P., Calista, F., Gualtieri, I.: Twenty Years Follow-up in Ceramic-Ceramic THR at Rizzoli Orthopaedic Institute. In: 6th International BIOLOX Symposium, pp.13–17 (2001)

[5] Návrat, T., Florian, Z.: Stress-strain analysis of surface replacement of the hip joint. In: Applied Mechanics 2005, Hrotovice, pp. 77–78 (2005)

[6] Teoh, S.H., Chan, W.H., Thampuran, R.: An Elasto-plastic Finite Element Model for Polyethylene Wear in Total Hip Arthroplasty. J. Biomech 2002 35, 323–330 (2001)

[7] Dunham, C.E., Takaki, S.E., Johnson, J.A., Dunning, C.E.: Mechanical Properties of Cancellous bone of the Distal Humerus. In: ISB XXth Congress - ASB 29th Annual Meeting, 564, Cleveland, Ohio (2005)

[8] Mak, M.M., Jin, Z.M.: Analysis of Contact Mechanics in Ceramic-on-Ceramic Hip Joint Replacement. Journal of Engineering in Medicine 216(4) (2002)

[9] Ipavec, M., Brand, R.A., Pedersen, D.R., Mavcic, B., Kralj-Iglic, V., Iglic, A.: Mathematical Modelling of Stress in the Hip During Gait. Journal of Biomechanics 32, 1229–1235 (1999)

[10] Genda, E., Iwasaki, N., Li, G., MacWilliams, B.A., Barrance, P.J., Chao, E.Y.S.: Normal Hip Joint Contact Pressure Distribution in Single-Leg Standing – Effect of Gender and Anatomic Parameters. Journal of Biomechanics 34, 895–905 (2001)
[11] Saikko, V., Calonius, O.: An Improved Method of Computing the Wear Factor for Total Hip Prostheses Involving the Variation of Relative Motion and Contact Pressure with Location on the Beraing Surface. Journal of Biomechanics 36, s.1819–s.1827 (2003)
[12] Fuis, V.: Stress and reliability analyses of ceramic femoral heads with 3DD manufacturing inaccuracies. In: 11th World Congress in Mechanism and Machine Science, Tianjin, People R. China, pp. 2197–2201 (2004)

Identification of Geometric Errors of Rotary Axes in Machine Tools

M. Holub, J. Pavlík, M. Opl, and P. Blecha

Brno University of Technology, Technická 2896/2,
Brno, 616 69, Czech Republic

Abstract. Due to the ever-increasing requirements on higher productivity of machines with simultaneous preservation, or even increase, of their working accuracy, considerable attention is paid to functional nodes of machines, such as linear and rotary axes. This paper describes a new concept of a universal measuring stand, developed to fulfil the requirements on monitoring of the headstock geometric properties in relation to the changing loading conditions. Equally important is the thermal stability, which can markedly affect the accuracy of the future workpiece. The problem of thermal stability of machine tools and their components is a continuously studied topic, especially the possibilities how to minimize these influences either by constructional changes or by a suitable compensation method.

For this reason, a methodology of measurement and evaluation of geometric errors of rotary axes has been proposed and, with regard to a long-term character of measurement of the heat transfer between the machine components, a universal measuring SW module has been developed.

1 Introduction

Working accuracy of machine tools is affected by many factors. Due to the ever-increasing requirements on machine tools, including mainly their productivity, working accuracy and price, manufacturers are forced to apply the latest technology in the development of new machine tools. Subsequently, this trend leads to implementation of compensatory methods into the structure of the machine.

The compensatory methods, having the same goal – to achieve the required level of the specific characteristics in a given tolerance zone – can apply different principles. In this case, it is previously used equipment for collection and processing of the necessary information. Application of new equipment and algorithms for processing of the measured data may lead to reduction of errors occurring in machine tools. In order to achieve such goals it is necessary to suitably describe the behaviour of the monitored unit and to make up a compensatory method on the basis of the obtained information. With appropriate application of the compensation (either of thermal influences, dynamic events or static deformations) on the machine, it is then possible to choose a more simple construction of the machine tool. In other words, to use conventional materials or more simple construction that would lower the total costs of the machine tool.

The main elements affecting the working accuracy of a machine tool are its functional parts including linear-motion and rotary axes and headstocks. Marek [1, 2] describes the impact of thermal load of individual bearings on rotary and translational motion.

The experimental equipment, described in this paper, serves precisely for monitoring of the geometric properties of headstocks in relation to the changing loading conditions.

2 Proposal of the Testing Equipment

2.1 Requirements on the Measuring Stand

A headstock with a spindle is a part of the machine tools for both rotary and non-rotary tools. The main requirements are stiffness in both radial and axial direction and low circular and face runout. As these elements form the main cutting movement, they are strongly affected by the technological process: in milling machines by the type of the tool used and by the forces generated by machining; in turning machines by the size of the workpiece and by the generated cutting forces. The solution focuses on description of the geometric properties of the machine rotary axes in relation to changing technological parameters (change of revolutions, static load). The main objective was to find out the axial and radial runout, spindle angle change or spindle shift under different loading conditions. The work on the project included a proposal of methodology and its testing on both a milling machine and a vertical lathe.

2.2 Construction Solution

Based on the requirements on the spindle running accuracy, a measuring unit has been designed, suitable for application on a machine tool during different loading cycles. In order to enable further application of these results, for example in comparison of similar machines, this test has been developed in agreement with the tests of the ISO 230 series, that is, without the action of the forces from machining.

The requirements on monitoring of the rotary axis geometric properties, including circular runout, face runout and angle change, have been incorporated in the proposal shown in Fig. 1. Due to the demandingness on the accuracy of the testing workpieces and applicability only for vertical lathes, it was not advantageous to proceed in this direction.

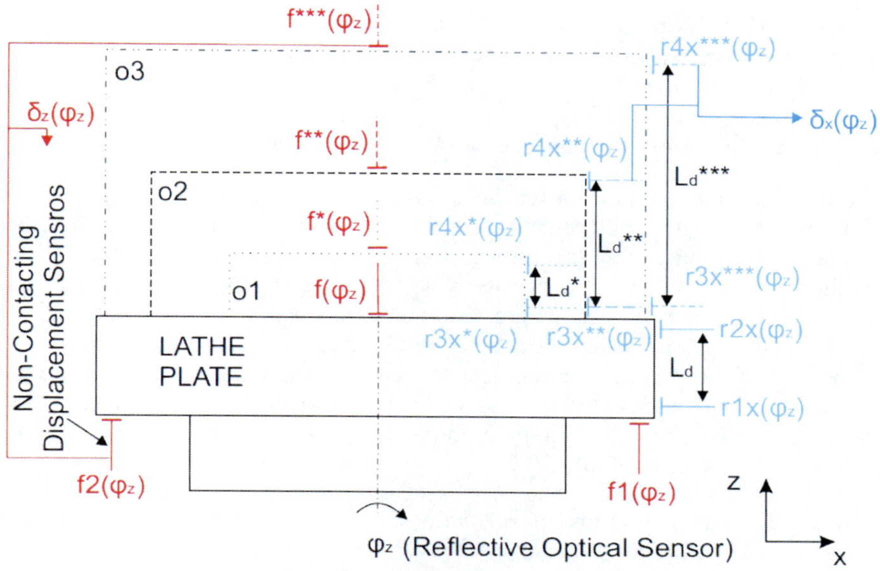

Fig. 1. Principle of rotary axes measurement

In order to achieve more universal applications, a principle of geometric error detection on a measuring pin has been proposed [3]. The minimum dimensions, material of the pin and the surface quality were chosen in accordance with the requirements on induction sensors. A 3-D model and demonstrations of a fixture measurement with a measuring pin and sensors are shown in Fig. 2.

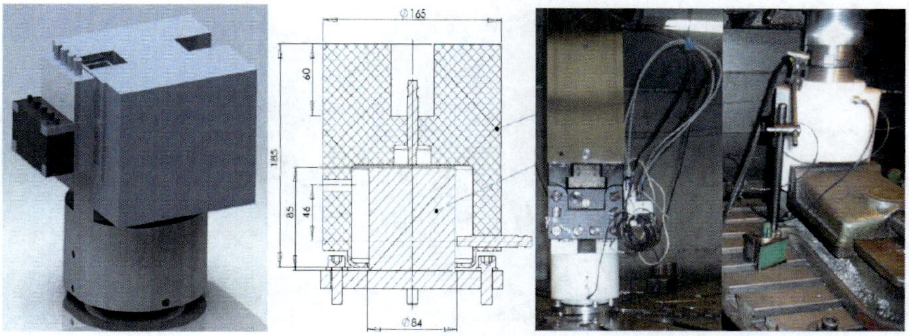

Fig. 2. Principle of rotary axes measurement

A complete identification of rotary axis geometric deviations within one measurement requires application of five sensors: two for Y-axis deviations, two for X-axis deviations and one sensor for detection of axial shift (Fig. 1). In order to capture the dependence of all shifts on the angle change of the plate φ_z, it was necessary to obtain the information on the position from the machine. As it was not possible to connect to the machine's control system, an optical sensor CNY70 was used for this purpose.

2.3 Acquisition and Processing of the Measured Data

Acquisition of the values from sensors s1 (Non-Contacting Displacement Sensor), s2 (Non-Contacting Displacement Sensor) and opt (Reflective Optical Sensor) was again performed with the software LabVIEW. The data obtained for various revolutions and loads were further processed in Matlab software. Rough data are shown in Fig. 3, however, they require further adjustment for subsequent processing. The values of strain from s1 and s2 must be further re-calculated according to calibration of the specific measuring pin material. The data from opt. serve for determination of the revolutions and subsequent re-calculation to spindle angle change φ_z. So far, the maximum revolutions tested were 200 min $^{1-}$ showing sufficient accuracy of repeatability of the measured values.

Evaluated errors, such as runout or changed position of the rotary axis, are mutually compared at different revolutions and spindle load (Fig. 4). As the setting-up of a fixture itself introduces errors of geometry into the system, reference geometry must be suitably determined. This reference geometry is defined just from the technological quantities (revolutions, load). Measurement is performed at constant revolutions and only from loading (with own weight of the workpieces and machine parts), loads from the machining are not imported here.

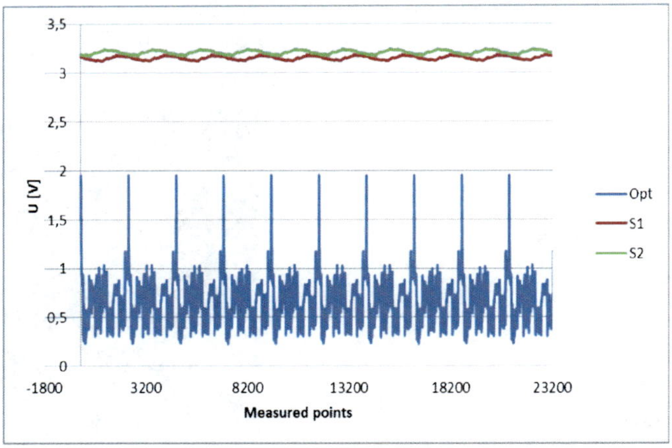

Fig. 3. Measured data (Voltage, Points) from sensors S1, S2, Opt from the vertical lathe SKIQ30, TOSHULIN

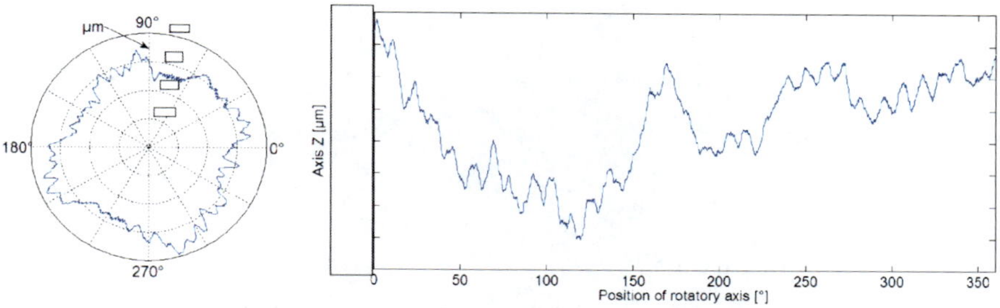

Fig. 4. Processed data errors in X, Y, Z axes from the vertical lathe SKIQ30, TOSHULIN

3 Conclusion

The measuring stand and the respective SW module were developed for the possibility of monitoring of geometric errors in rotary axes, primarily for the spindles of turning machines.

Based on the evaluation, valuable information on the behaviour of the machine are obtained, describing the behaviour of the rotary axis in relation to the specific testing conditions, that may correspond with the future revolutions during turning. In this way it is possible to evaluate the detected geometric deviations and thus to predict the effect of the machine's geometric properties on the future workpiece. Besides, these results may be further used as a source of valuable information for the feedback to construction and development of the tested machine tools.

Acknowledgements

This work is supported by Brno University of Technology, Faculty of Mechanical Engineering, Czech Republic (Grant No. FSI-J-10-39 and CZ.1.07/2.3.00/09.0162).

References

[1] Marek, J., Mareček, J.: Vliv uložení posuvového šroubu na tepelnou stabilitu. Acta Mechanica Slovaca 7(3), s.355–s.360 (2003)

[2] Marek, J., Mareček, J.: Generování tepla v hlavním ložisku svislého soustruhu. Acta Mechanica Slovaca 8(3-A), s.483–s.486 (2004)

[3] TOSHULIN, firm publication

Mechatronic Approach to Absolute Position Sensors Design

P. Houška, O. Andrš, and J. Vetiška

Faculty of Mechanical Engineering, Brno University of Technology,
Technická 2896/2, 616 69, Brno, Czech Republic
houska.p@fme.vutbr.cz

Abstract. The contribution deals with design and implementations of absolute angular position sensors for serial manipulator Mini-Swing. Manipulator design including sensors and control software was developed according to V-cycle methodology. Selected methodology brings successive improvement of design stages which leads to accomplishment of all requirements. Designed sensors are based on Hall effects.

1 Introduction

The absolute position sensors are design for manipulator Mini-swing. Mini-swing (fig. 1) represents a small planar manipulator with two controlled axis and one rotary end effector. This manipulator was designed as universal laboratory manipulator intended for automatic positioning and manipulation. Modular mechanical design of manipulator is suggested so that it is possible to change lengths of arms and thus workspace dimension can be changed. Typical applications for this manipulator are sensor manipulation and positioning [1], testing of energy harvesting devices [2] and control development.

The next reason to develop this manipulator is feasibility study of usability these kinds of manipulators for special measurement projects. Measurement of temperature field and air flow in air-conditioned areas is a first target projects. Its goal is a data collection for air-conditioning models improvement. Manipulator will be used for sensor positioning in area with dimensions about 2x1m and positioning accuracy ±1mm. Second target project is a manipulator for final inspection of gearboxes. In this case the acoustic power is measured and evaluated in defined positions. The acoustic power is measurement by microphone that is positioned by manipulator. Measured area has similar dimension with accuracy ±5m. Weight and simplicity of manipulator are keys parameters for both projects.

2 Manipulator Description

Mini-swing (fig. 1) is serial manipulator with three controlled axis. Axis A and B are axis of manipulator arms and axis C is rotational of end effector. Axis drives consist of DC motor, multi-stage planetary gearbox and incremental encoders. Encoders are fitted to motor shaft and are used as feedback for drive control. Manipulator dimension are derived from arms lengths that are 100mm.

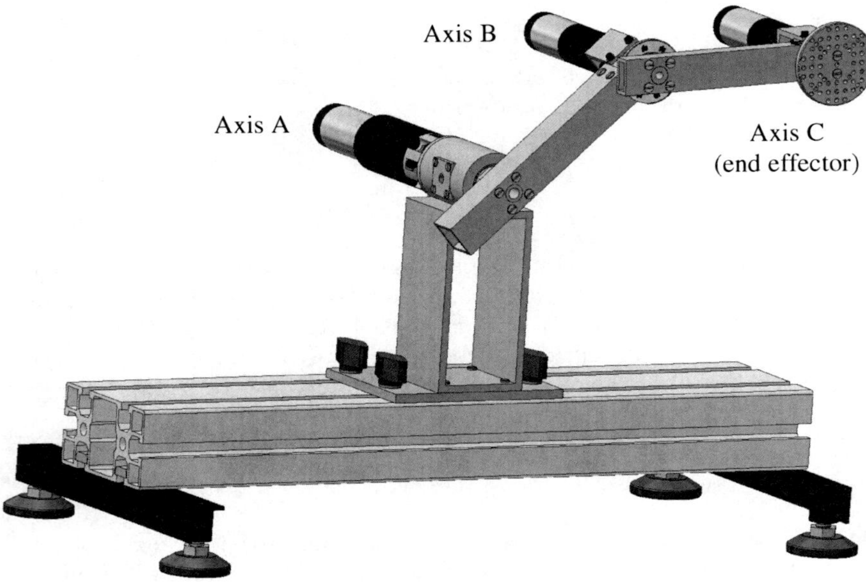

Fig. 1. 3D model of manipulator Mini-Swing

Manipulator control system is developed in LabVIEW. The control system is divided in two levels. Low level is a real-time control system of drives which collects data from sensors and checking manipulator states. Low level system is running on PAC NI CompactRIO. Interface on PC is connected with low level over Ethernet which presents second level of control system.

3 Manipulator Sensor System

Sensor system of manipulator is divided into two groups. First group consist of sensors needed for manipulator control. Second group consist of sensors attached to the end effector of manipulator. Their choice depends on solved problem including sensors connections and positioning methods.

Group of sensors for manipulator control consist of sensors for the control of individual axes of manipulator. Each axis must have sensed absolute (angular) position of the arm to its joint. Since the drives on axes are mounted with planetary gearboxes with high gear ratio (132:1 and 84:1) it is necessary to use an absolute position sensor placed between arm and axis joint. Feedback control of motors requires use of sensors mounted on motors shaft. We are using incremental encoders with resolution 1000CPR (axis A) and 512CPR (axis B and C). Using a resolver or absolute encoder on motor shaft is improper in this case.

Due to the location must be absolute position sensors incorporated into a frame of manipulator so as not significantly enlarge the size limitation of the manipulator and its workspace. Because any size increasing leads to the use of drives with higher power and generally this leads to a significant cost increasing it is necessary find a sensor that does not too affect dimension of manipulator. Because the sensor with the required parameters could not find on market we start deal with possibility of development of its own sensor.

Using of experiences with design and realization absolute position sensors for linear drive of Stewart platform [3], [4] we have design suitable sensors that meets our requirements.

4 Absolute Position Sensors Concept

Conception of sensors is based on permanent magnet position sensing by Hall IC (detector). DAQ and data processing will be running on microcontroller with A/D converter and analog inputs for detector output connecting. Measured position will be available over communication interface. The same conception was used for linear absolute position sensor on Stewart platform [3].

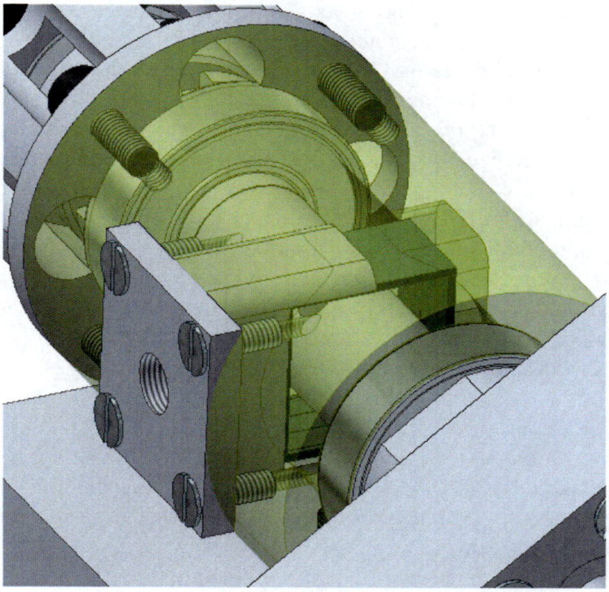

Fig. 2. 3D model of axis A with absolute sensor

Places for embedding of absolute angular position sensors were chosen on manipulator body and fundamental sensor dimensional analysis were done.

Place for magnet on axis A was chosen on axis shaft in centre between bearings. Sensor body surrounds shaft and it is embedded in bearing housing (fig 2.). Electronics with microcontroller will be clamped to bearing housing of axis A.

Axis B and C have significantly smaller dimensions than axis A. The best sensors positions were chosen on flange of bearing housing (fig. 1). Sensors magnets are set on arm/end effector and detectors are placed around annulus clamped to flange.

The next components were chosen according to experiences from solving previous projects. Hall effects detectors A1301 by company Allegro Microsystems is chosen. Sensitivity of detector is 25mV/mT, supply is 5V/10mA. Material NdFeB type N35 is chosen for permanent magnet. Magnet dimensions (Ø3x2mm) were chosen with respect to manipulator size and magnet geometrics axis is perpendicular to magnetic axis.

5 Sensor Design

Mathematics model of pair detector – magnet, is need for design of sensor geometry and simulation verification. Real characteristic of pair detector-magnet were measured by an existing linear sensor [3]. Characteristic were measured for two configurations. First one – axis of magnet is in direction with sensor movement (fig. 3 – both magnet poles are detected). It means that magnet axis is perpendicular to detector axis. Second one – axis of magnet is perpendicular to sensor movement direction (fig. 3 – detected only one pole of magnet). It means that axis of magnet and axis of detectors are parallel. Mathematical models of detector-magnet are obtained from acquired data from detectors. Model inputs are distance between detector axis and magnet and distance between detector front and magnet. Model output is normalized electric voltage proportional to magnetic induction.

Next phase of sensor design was a sensor geometry optimization and detectors placement optimization. These optimization criteria were selected: minimal dimension, minimal count of detectors (minimal power consumption) and maximal resolution. In sense of reliability increasing by our experiences extra criteria was selected – minimally three detectors must detect magnet/s in any sensor working position. Its means that minimally three detectors outputs are in range at 8% to 95% of positive output range (for absolute value) in single magnetic pole configuration and in range at 25% to 95% of positive output range (for absolute value) in two magnet poles configuration (fig. 3).

Sensor optimization passed over few steps of V-cycle by above mentioned criteria.

Detectors placement and working direction between detector front side and magnet are optimized in first step. For optimization were used mathematical models of detector-magnet pair.

In second step were made mathematical models of sensor with geometry proposed in previous step and sensitivity analysis was performed on these models. Attention is paid to output at least of three detectors is valid (in range) for every sensor working position.

Sensor dimension optimization and sensor feasibility study were done in third step.

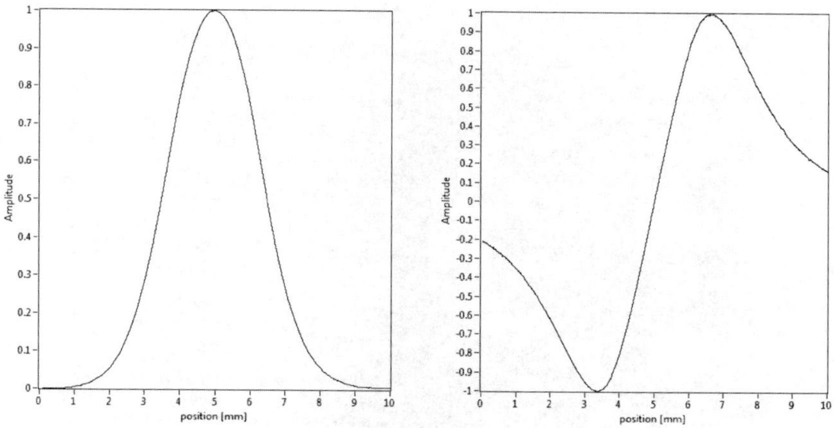

Fig. 3. Normalized typical detectors outputs (one and both magnetic poles)

Sensors variant consists of two magnets with parallel axis with detectors and opposite polarization were proved to better than second variant. Resulting sensors designs have 7 detectors for axis A (fig. 4) and 9 detectors for axis B (fig. 5) and C. Expected sensors power consumption is 80mA/5V for axis A and about 100mA/5V for axis B and C.

6 Sensing Algorithms and Calibrations

Sensor algorithm is implemented on microcontroller C8051F530A. This microcontroller has 16 general purpose pins that are configurable like digital I/O or analog input. Analog inputs are connected to the 12 bit A/D converter with programmable preamplifier. Detectors (Hall IC) outputs are connected to the analog inputs of microcontroller. For this purpose of application 11 analog inputs are used. Detector voltage output is ratiometric to supply voltage 5V and quiescent output is a half of supply voltage. Detector supply voltage is connected to next analog input of microcontroller. Communication (UART) and programming (C2) interfaces are connected to microcontroller pins too.

Developed measuring algorithm consists of these steps: preamplifier gain setup, detectors scan mode, rough guess of position and choice of active detectors, measuring of active detectors and precision position calculation.

Simply measuring algorithm is described as follows – detectors supply voltage is measured and by voltage level the preamplifier gain is trimmed in first step. Gain trim is continuing until measured value match about 95% of A/D convertor range. All detectors outputs are measured (scanned) in next step. Active detectors are selected from scanned values and rough guest of position is made. Outputs of active detectors are sampled and filtered. With help of rough position guest are identified indexes to the calibration table. From sampled outputs with help of calibration table position is evaluated. Offset addition to the position is last step of algorithm. Resulting position is accessible over communication interface.

Fig. 4. Realized absolute angular position sensor for axis A

Precision of designed and implemented sensors (fig. 4, 5) depends on precision of calibration and precision of reference sensor. In this case incremental encoders on motors shaft were used as reference sensors. The calibration process is influenced by a planetary gearbox and encoder manufacturing inaccuracies.

Fig. 5. Realized absolute angular position sensor for axis B

Over calibration process are collected data about real positions and raw outputs of active detectors. For this purpose "get raw data" command is implemented in communication interface. After that data are collected, calibrations are calculated and over communication interface are written to the microcontroller FLASH memory.

Accuracy evaluation of implemented sensors is relatively complicated due fact that the sensors are part of the manipulator structure. As a reference (calibration)

sensors for calibration and accuracy evaluation of realized absolute position sensors were used incremental encoders which are used for feedback control of drives. Due to the presence of planetary gearboxes we are obtaining sensor resolution about 0.01deg on arm. This solution is not entirely accurate but sufficient to calibrate and verify sensor. Example of measurement errors of the realized absolute position sensor is on fig. 6. This is actually a sensor calibration error including errors of incremental encoder and gearbox. Measured errors of all three realized sensors are in range of ±0.15 deg.

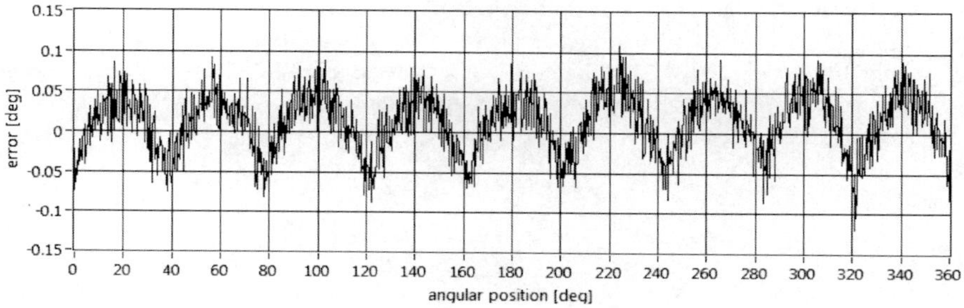

Fig. 6. Measurement errors of the realized absolute (angular) position sensor on axis B

7 Conclusion

The article describes design methodology of absolute angular position sensors for axis of robotic manipulator. Design and implementation of compact sensors system bring quite small influence on designed device dimensions and have no influence on functionality. It is the primary benefit of described approach.

Described sensors are made and fulfil desired requirements (fig. 4, 5). But sensor has two disadvantages. First disadvantage is needs for sensor calibration. Precision of calibration process has main influence on sensor accuracy. Second sensors disadvantage is higher power consumptions due high current flow over detector (Hall IC). Reduction of power consumption is possible by inactive detectors switching off, but this solution is little complicated because of we need microcontroller with higher pins count and need switch for every detectors. Next problem is start up time of detectors and its influence on detector output. Second method for reduction of power consumption is possibility of replace Hall detectors by Magneto-resistive detectors. But using of magneto resistive detectors is little complicated that using of Hall detectors and it leads to higher sensor dimensions in our case.

Acknowledgement

This work is supported from research plan MSM 0021630518 Simulation and modeling of mechatronic systems and FSI-S-11-23.

References

[1] Brezina, T., Vetiska, J., Blecha, P., Houska, P.: Design of the Controller for Elimination of Self-excited Oscillations. In: 8th International Conference on Mechatronics, Luhacovice, Czech Republic, pp. 395–400 (2000)

[2] Hadas, Z., Ondrusek, C., Singule, V.: Power sensitivity of vibration energy harvester. Microsystem Technologies 16(5), 691–702 (2010)

[3] Houska, P., Brezina, T., Brezina, L.: Design and Implementation of the Absolute Linear Position Sensor for the Stewart Platform. In: 8th International Conference on Mechatronics, Luhacovice, Czech Republic, pp. 347–352 (2009)

[4] Andrs, O., Brezina, T.: Controller Implementation of the Stewart Platform Linear Actuator 1357–1359 (2010); Katalinic, B. (ed.): Annals of DAAAM for 2010 & Proceedings of the 21st International DAAAM Symposium, p. 0679. DAAAM International, Vienna (2010)

Procedure for Calibrating Kelvin Probe Force Microscope

M. Ligowski[1,2,*], Ryszard Jabłoński[1], and M. Tabe[2]

[1] Faculty of Mechatronics, Warsaw University of Technology, Boboli 8, Warsaw, 02-525, Poland
[2] Research Institute of Electronics, Shizuoka University, Johoku 3-5-1, Hamamatsu, 432-8011, Japan

Abstract. The paper presents novel method for calibrating Kelvin Probe Force Microscope. The method is based on measuring the surface potential of a reference sample and comparing it with the results obtained by KFM. Proposed method offers a calibration possibility in the whole measuring range of the microscope both in terms of absolute values and dynamic behavior. In this novel method the surface potential of a reference sample used for calibration can be adjusted to desired value instead of being defined by physical parameters of the sample. Presented results obtained with proposed method prove its validity.

1 Introduction

Even though KFM [1] is an established technique and allows the detection of potential difference features as small as 5mV [2] which makes detection of individual charges possible [3], no standard calibration method is proposed so far. There are many possible ways to calibrate the KFM and various accuracy levels can be achieved accordingly to employed method. There is a number of papers addressing this topic[4]. The selection of a standard or standards is the most visible part of the calibration process. Ideally, the standard has less than 1/4 of the measurement uncertainty of the device being calibrated. When this goal is met, the accumulated measurement uncertainty of all of the standards involved is considered to be insignificant when the final measurement is also made with the 4:1 ratio. Maintaining a 4:1 accuracy ratio with modern equipment is difficult. The test equipment being calibrated can be just as accurate as the working standard. In papers [5,6], a KFM calibration procedure using gold-plated Ohmic contact as a reference has been introduced. The calibration procedure described in this paper has been designed to provide sufficient accuracy while being also convenient enough to allow for fast calibration. As a reference sample a gold-coated bulk silicon wafer was specially prepared.

The method is based on investigating the reference sample with KFM and applying external voltage to the sample while measuring its surface potential.

* Corresponding author.

Results obtained with KFM are then compared with the applied bias. KFM can be adjusted to assure match between detected and reference values being in the range of assumed measurement tolerance.

In presented method, a reference step of surface potential is required , to perform a calibration. By measuring this sudden change of the surface potential and comparing it with the value shown by KFM it is possible to evaluate and adjust KFM settings. Therefore to perform a calibration the reference surface potential step has to be clearly defined and its absolute value has to be known. Moreover, to assure the linear characteristic of the KFM the reference step should be measured at various surface potential levels. The time response of the KFM can be investigated by measuring the reference surface potential step and comparing obtained profile of surface potential change with reference step change (Fig. 1).

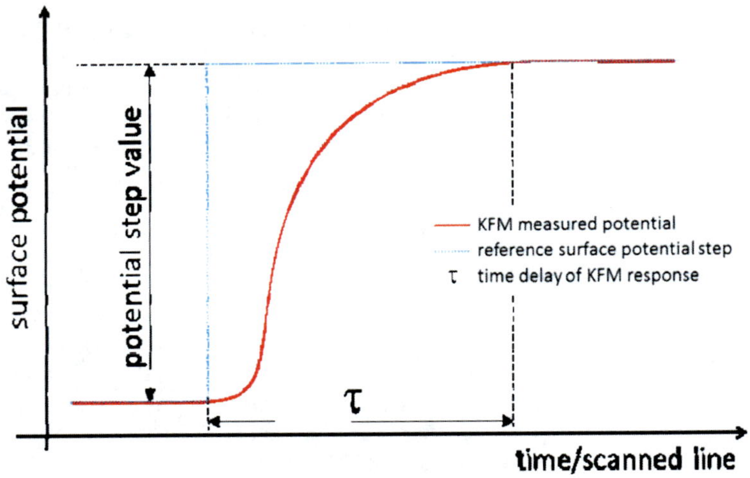

Fig. 1. The proposed method of KFM calibration including response time in reference to the step change of surface potential.

This way it is possible to estimate the time needed for KFM to reach the actual value of the surface potential. This time strongly depends on the settings of PI controller present in the KFM feedback loop. It is crucial to apply a proper feedback loop settings like P and I in PI controller or gain in Lock-In Amplifier in order obtain reliable results.

2 Reference Sample

One of the possible ways to obtain a surface potential step needed for calibration is by fabricating sample which contains two different materials. This can be for example a p-n junction or metal-semiconductor contact system. Two different conductive materials create a surface potential step due to different Fermi energy level. This method however has many drawbacks, like: difficulties in defining precisely the potential step and also the transition region can be smooth - different from step (abrupt) change which is expected due to physical configuration of atoms. Finally in such a sample obtained potential step has only one fixed value which doesn't allow for calibration in full measuring range unless a number of samples has been prepared. Thus we propose a different approach, where a surface potential of perfectly conductive sample is adjusted with external voltage source. By changing the bias suddenly during the scan the surface potential steps can be effectively achieved. The big advantage of this method is the possibility of obtaining various steps at various surface potential levels. This allows performing calibration in the whole measuring range of the KFM and adjusting the dynamical characteristic of KFM.

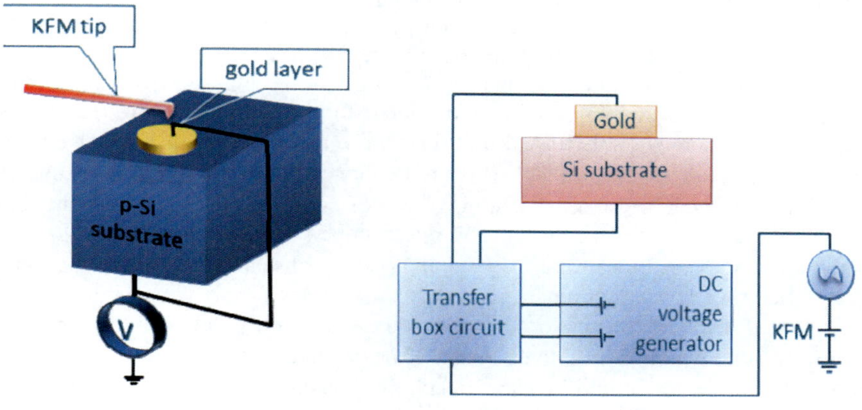

a) b)

Fig. 2. Schematic view of the a reference sample(a) and measurement circuit (b)

During calibration process gold-coated bulk silicon wafer was fabricated to serve as a reference sample. The reason for using gold is its property of non-oxidizing (self-grown oxide layer at the top of the reference sample may influence the results shown by KFM). The schematic view of the reference sample is shown in Fig. 2a. The wiring scheme, is presented in Fig. 2b.

In presented setup the resistance of the wiring between the voltage source and the sample is in the range of less than 100Ω, which compared to resistivity of bulk Si (at least thousands of MΩ) gives the negligible voltage drop at the power cables. Thus it can be assumed that the gold surface directly reflects applied voltage and the accuracy of the reference samples is defined by the accuracy of the voltage generator. For the voltage generator used, Agilent E3648A this is (podać jaka to niedokładność, na jakim poziomie)10mV. The values of reference potential were also checked with the reference voltage analyzer (Agilent Semiconductor Device Analyzer) with the accuracy and resolution orders of magnitude smaller than the reference voltage generator. Thus it can be assured that the type A uncertainty of the reference potential sample is better than 10mV.

3 KFM Measuring Mode Used for Calibration

There are few ways of measuring the surface potential using the KFM that can be used for calibration:

- live observation of surface potential detected by KFM using an external AC/DC converter;
- the observation of KFM potential within one line of the KFM image (horizontal direction potential profile obtained from KFM resultant image);
- the observation of KFM potential within few lines (vertical direction potential profile obtained from KFM resultant image);

The first choice (using a separate A/D converter) is the most straightforward approach. It provides the real time characteristics of the surface potential during the scan. KFM microscope is dividing the obtained results (forward and backward scan) in lines and save them as separate two dimensional images. However the use of internal KFM measuring circuit is expected to give more accurate results and therefore it is more suitable for calibration. The main problem of using KFM internal A/D converter and internal software is connected with the response time (τ) of KFM. If τ is bigger than the scanning time of one line, then it is more difficult to observe full time dependence. KFM microscope provides two separate surface potential images resulting from forward and backward scan. If the τ is in the range of few lines we have to compare the consecutive lines from two separate images which is a problem. Moreover there is some time delay between forward and backward scan which is hard to estimate. Thus depending on the time τ one of two different approaches should be used: checking the single line (horizontal direction potential profile) or checking the scan profile along vertical direction.

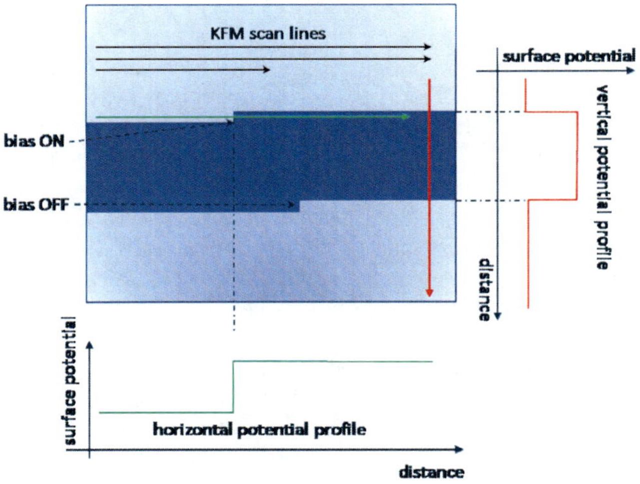

Fig. 3. Various approaches to check KFM response time depending on its the actual value.

In Fig. 3 a graphic representation of the reference image obtained during the calibration can be observed. The top of the KFM image marked with dark gray color represents surface potential of the sample in low bias condition (for example 0V). In the middle of the image, the brighter region represents biased condition (for example -1V). Two approaches to check the surface potential step obtained by KFM depending on the response time τ are represented by green and red arrows.

KFM image is obtained by scanning the surface in a raster scan. Consecutive scan lines are indicated by the black arrows at the top of the image. To check the shape of the potential step in case of small response time τ it is preferable to investigate a horizontal profile of surface potential. This profile is easily obtainable from any surface metrology. However, if the KFM response time is longer than the time which KFM takes to scan one line it is better to investigate a vertical profile (indicated by red color in Fig. 3).

Using above described method, it is possible to perform calibration of KFM in full measuring range (including linearity, uncertainty, response time and others).

4 Calibration Results

All results presented in this paper are obtained using a Low Temperature Kelvin Probe Force Microscope produced by UNISOKU Co, Ltd. At first, in order to check the accuracy of the microscope, the test revealed, that the settings of PI controller and Lock –In amplifier, which are part of feedback loop in tested KFM, have a strong impact on obtained results. When finding their appropriate settings, the balance between noise and response time has to be attained. Depending on the setup of feedback loop parameters the response time strongly varies. Wrong setting of PI controller and lock-In amplifier can cause too slow response and thus can influence the results. For higher gain of the electrostatic force the response time is shorter but at the same time the noise increases.

Further tests proved that the influence of particular tip on the results is negligible. This was proved by series of measurement with a usage of various cantilevers (chosen cantilevers had a different resonant frequency). No significant difference between the shape of the surface potential due to cantilever replacement was observed.

After the series of tests and adjustments of the lock-in amplifier and PI controller settings, the characteristic shown in Fig. 4 has been achieved.

The results show linear characteristic within -5 to 4 V range of applied bias. In this range despite of the 0.5 V shift of the measured potential respectively to applied bias, microscope shows real values of surface potential. It can be assumed, however, that since sample was fabricated by depositing thin layer of gold on the Si wafer, the 0.5 V shift is caused by the work function difference within the structure of the reference sample. Also the noise level of ~100mV is acceptable.

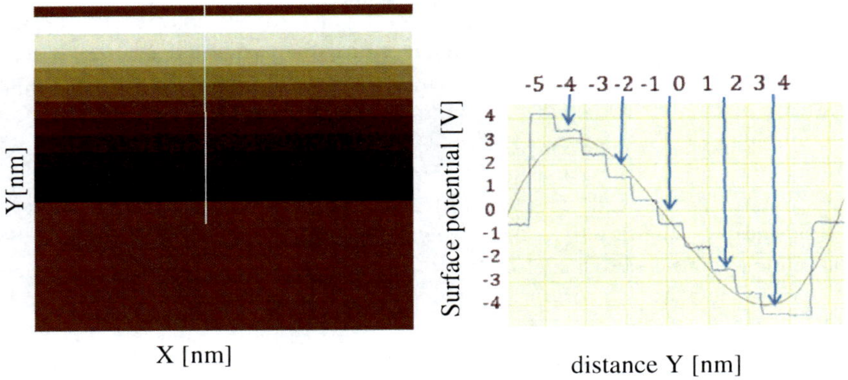

Fig. 4. The KFM characteristic after calibration.

In Fig. 5 the results of the measurements of the sample surface potential for 7 different reference voltage values (Vref from -3V to 3V) can be observed. Each line is the actual line measured at the sample surface. As the gold sample is close to perfect conductor, which do not oxidize in time, we can assume that the whole sample surface had a uniform potential distribution equal to the value setup on the DC voltage generator. It can be seen, that obtained values are close to the expected ones.

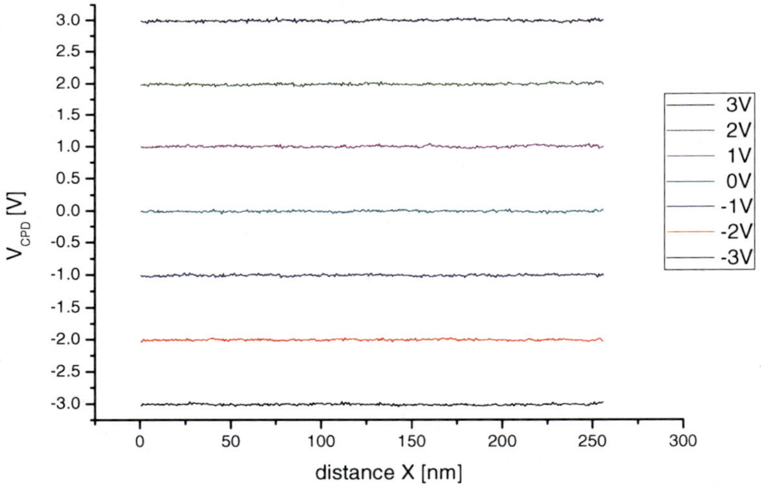

Fig. 5. KFM calibration curve

For all above measurement points the mean value and the standard deviation for each value of the V_{ref} was calculated. Table 1 presents obtained values. It can be seen that obtained values are close to the expected ones. The values of mean

potential vary from the V_{ref} between 10 and 20 mV depending on the case. In the last row the standard deviations are given .
Fig. 6 presents the calibrationcurve.

Table 1. Calibration results

position	V_{ref}=-3V	V_{ref}=-2V	V_{ref}=-1V	V_{ref}=0V	V_{ref}=1V	V_{ref}=2V	V_{ref}=3V
0.5	-3.023	-2.027	-1.008	-0.016	0.990	1.975	2.981
1.0	-3.026	-2.043	-1.022	-0.005	0.986	1.976	2.979
1.5	-2.997	-2.010	-1.025	-0.004	0.976	1.978	2.969
2.0	-3.005	-1.996	-1.019	-0.023	0.986	1.981	2.966
2.5	-3.010	-2.015	-1.018	-0.020	0.981	1.988	2.969
3.0	-3.008	-2.029	-1.019	-0.017	0.976	1.968	2.976
3.5	-3.015	-2.016	-1.035	-0.022	0.988	1.979	2.964
4.0	-3.001	-2.001	-1.032	-0.020	0.975	1.972	2.940
4.5	-2.998	-2.009	-1.017	-0.016	0.984	1.953	2.957
5.0	-3.007	-2.024	-1.022	-0.002	0.980	1.968	2.992
5.5	-3.012	-2.020	-1.036	0.005	0.986	1.994	2.987
...	...	...	...	...	...	...	...
255	-2.984	-2.030	-0.998	0.002	0.967	2.004	2.971
255.5	-2.981	-2.031	-0.992	0.002	0.988	1.996	2.978
256	-2.970	-2.037	-1.007	-0.002	0.981	1.968	2.974
mean	*-3.010*	*-2.010*	*-1.009*	*-0.016*	*0.991*	*1.979*	*2.977*
cor-rected	-2.996	-1.996	-0.995	-0.002	1.005	1.993	2.991
st. dev	*0.013*	*0.012*	*0.013*	*0.013*	*0.015*	*0.014*	*0.015*

It can be seen, that it is almost linear within the measuring range -3V to 3V, however systematic error can be found. This error can be easily corrected by adding 14mV to all results. The value of 14mV was calculated as a mean of all systematic errors obtained for each calibration point. Obtained results revealed, that after correction in the most of measuring range the deviation from reference value is smaller than 5 mV, while in whole measuring range it is smaller than 9mV. Also, it can be seen that the noise in obtained results is low. The standard deviation calculated for each of calibration points does not exceeded 15mV. The corrections in the whole measuring range do not exceed 20mV. Fig. 7 presents central part of the calibration curve (given in Fig. 6) with the 9mV correction and standard deviation δ=14mV for -1V reference value.

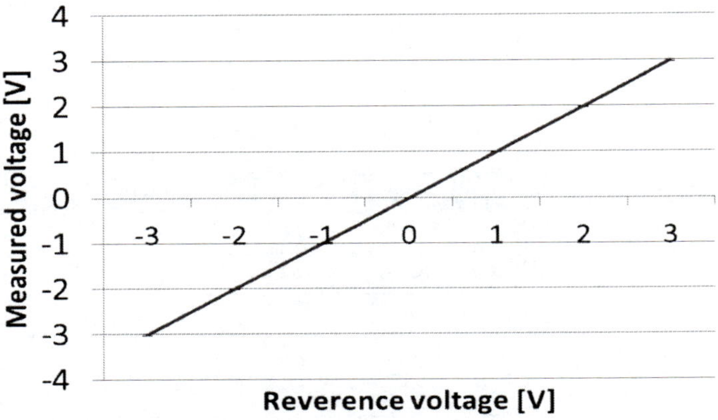

Fig. 6. Calibration curve

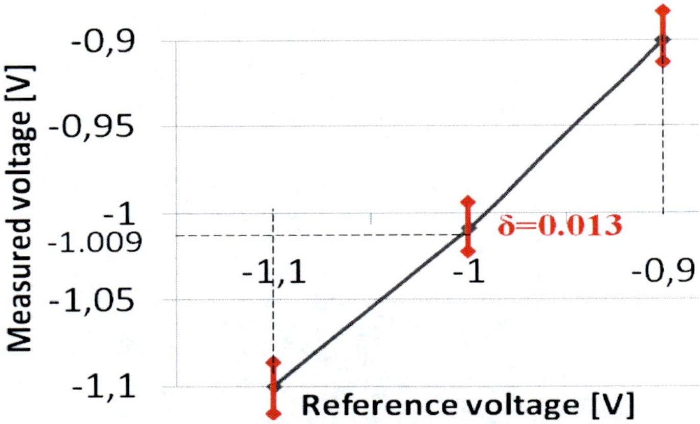

Fig. 7. Central part of calibration curve

It should be noticed that in case of KFM measurement the whole sample is artificially biased for the measurement purpose. Thus the systematic error may be easily corrected even during the measurement.

5 Summary and Conclusions

In this paper the original and novel calibration method for the KFM microscope has been proposed. As a reference sample a special gold-coated bulk silicon wafer was realized. Using this method both the time characteristics and the measurement accuracy can be adjusted by a proper setting of PI controller and Lock-In amplifier gains. With an increase of proportional gain the measuring range decreases, but obtained resolutions increases.

The paper also presents the results of the calibration of UNISOKU LT-KFM with the use of reference sample (which has been designed and fabricated). Obtained results proves that KFM can be properly calibrated using described method. The calculated standard deviation is less than 15mV. Taking into consideration also systematic error, we conclude, that microscope can measure in a whole measuring range with the uncertainty of 20mV (it corresponds to 0.3% of relative uncertainty). In reference to the present requirements, concerning KFM calibration (the standard should has less than 1/4 of the measurement uncertainty of the device being calibrated) – the proposed method is very effective.

References

[1] Nonnenmacher, M., et al.: Applied Physics Letters 58, 2921–2923 (1991)
[2] Zerweck, U., et al.: Phys. Rev. B 71, 125424 (2005)
[3] Ligowski, M., et al.: Appl. Phys. Lett. 93(14), 142101 (2008)
[4] Jacobs, H.O., Leuchtmann, P., Homan, O.J., Stemmer, A.: Journal of Applied Physics 84(3), 1168–1173 (1998)
[5] Barbet, S., et al.: Appl. Phys. Lett. 93, 212107 (2008)
[6] Freland, J.W., et al.: Appl. Phys. Lett. 93, 212501 (2008)

Multi-sensor Acoustic Tomograph for Gas Mixture Content Monitoring

W. Ostropolski, J. Wach, and A. Jedrusyna

Wroclaw University of Technology, Wybrzeze Wyspianskiego 27,
Wroclaw, 50-370, Poland

Abstract. Helium is a common coolant in the modern superconducting magnets construction. One of the most important problems related to such systems is the protection of the human personnel against the uncontrolled spilling of the helium leading to oxygen deficiency syndrome. The detection of a helium leak is possible either by slow electrochemical detectors or by modern fast acoustic detectors. An acoustic measurement uses relation between the helium content and sound velocity in gas mixture. The oxygen content can be evaluated through helium content measurement. The tomographic measurement can produce the spatial distribution of helium and temperature. In the paper a prototype of the multi-sensor acoustic tomograph was presented. The instrument uses an array of ultrasonic transducers and it is capable of reproduction of spatial distribution of helium stream.

1 Introduction

One of the most important problems encountered during the operation of large cryogenic systems is the potential hazard to the personnel related to the uncontrolled release of the liquid helium. Such a condition may lead to an oxygen deficiency syndrome [1]. The helium discharge could be detected either indirectly by electrochemical oxygen detectors or directly by helium detectors. Unfortunately, the detection of helium introduces significant technical problems. Helium content can be measured by electrothermic instruments measuring changes in thermal conductivity of the gas mixture under test and comparing it to the thermal conductivity of a reference gas housed in a sealed cell [2]. Such a detector can be used for local testing of gas properties, but it is unable to monitor helium content inside a large confined space (for instance a test chamber or control room). Both electrochemical and electrothermic measurements are also inherently slow. Therefore a need for a faster measurement method has emerged.

2 Acoustic Measurement

One of the possible solutions could be an acoustic method. In this case the helium content is derived from the changes in sound velocity. This velocity is strongly related to the helium concentration; a similar relation is also true for many other gases (see Fig.1).

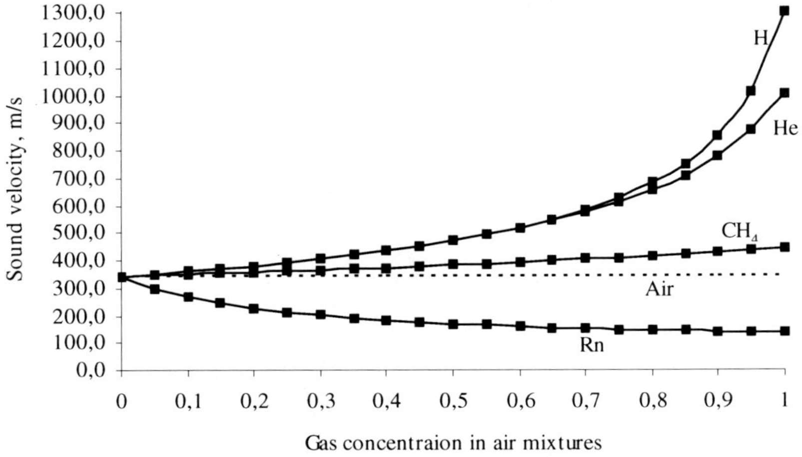

Fig. 1. Sound velocity for different gas/air mixtures [3]

Some gases like air or helium under some circumstances can be considered as perfect gases. Taking it into account and applying perfect gas law for the mixtures to the equation for sound velocity one obtains the following equation:

$$a = \sqrt{\kappa RT} = \sqrt{\left(\frac{1}{\sum \frac{z_i}{\kappa_i - 1}} + 1\right) \cdot \frac{\overline{R}T}{\sum z_i M_i}} \tag{1}$$

Where κ is specific heats ratio of the mixture, T is temperature and R is gas constant of the mixture, $\overline{R}$ is universal gas constant and z_i and M_i are respectively molar concentration and molar weight of i-th component. From (1) one can see that sound velocity in gas mixtures depends on: temperature T, molar weight M and specific heats ratio κ.

3 Experimental Setup – Acoustic Tomography

The fundamental idea of an acoustic tomography is time-of-flight measurement in a medium. During the measurement one ultrasonic transducer act as a transmitter, another as a receiver. The time-of-flight (TOF) of an ultrasonic burst between the transmitter and the receiver has to be measured. For the medium velocity v much smaller than sound velocity a, and the variation of a is much smaller than the absolute value of a the path of a sound wave can be approximated by straight line. Thus the sum of propagation times of acoustic wave from point A to B and B to A is given by equation:

$$\tau = \int_A^B \frac{2}{a} ds \tag{2}$$

A set of transceivers fixed in wall around a channel in selected plane provide in each measurement cycle a set $\boldsymbol{\tau}$ of propagation times taken between every pair of transceivers. The spatial sound velocity distribution $a(x,y)$ across the plane can be calculated by using some tomographic algorithm. Here the ART algorithm [3] has been used.

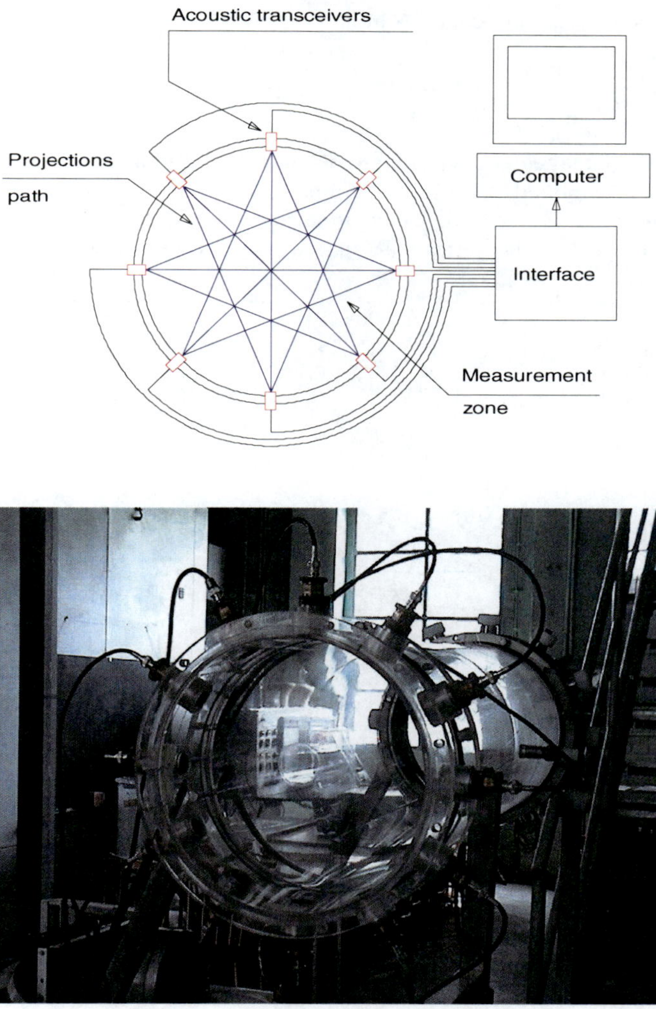

Fig. 2. Acoustic tomograph setup scheme and its general view

An acoustic tomograph system has been mounted in channel of diameter 0.3m housed inside a tube made of an acrylic glass (PMMA). Twelve piezoelectric transducers covered with layer matching wave impedance to the air have been evenly fixed around the channel. Each transducer transmits and receives acoustic wave and works at frequency of 108 kHz. Measurement unit contains 12 linear amplifiers and comparators – one for each transducer – and 5 digital counters plus an electronic switching matrix acting as a multiplexer.

The whole unit is controlled by 8-bit microcontroller that switches the transducers and transmits captured data to computer via standard serial port. The system includes also temperature sensors that can be used for corrections of the measured results of helium concentration.

4 Test Results

In order to validate theoretical calculations made for helium and depicted on Fig.1 a measurement of sound velocity in helium/air mixtures for different helium contents has been carried out using one pair of transducers. The results are given on Fig. 3. As one can see theoretical calculation are very close to experimental ones.

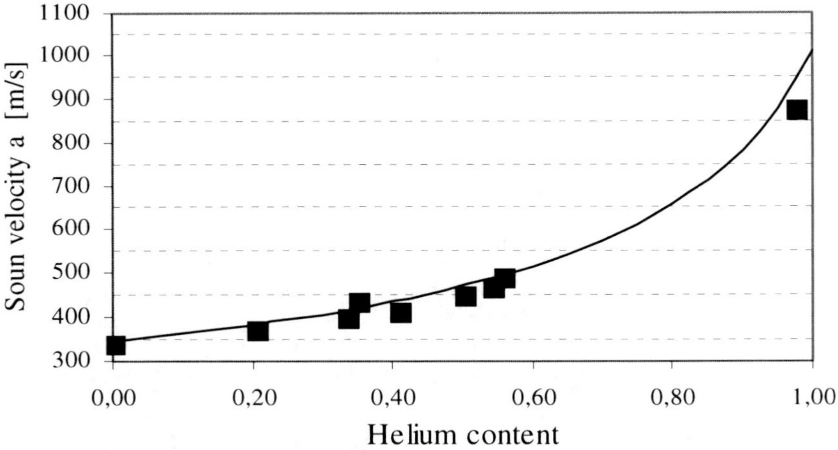

Fig. 3. Sound velocity in He/Air mixtures vs. He content for T=21°C. Line shows theoretical calculations and square marks show results of the acoustic measurements [3]

The system presented could be used for the reconstruction of spatial distribution of the helium content. In Fig.4 an image of helium jet flowing from a circular nozzle inside the test chamber have been shown.

Another application of acoustic tomography is the reconstruction of temperature spatial distribution inside a confined space. An example of such reconstruction (an effect of a candle placed in the middle of the test chamber) has been shown in Fig.5.

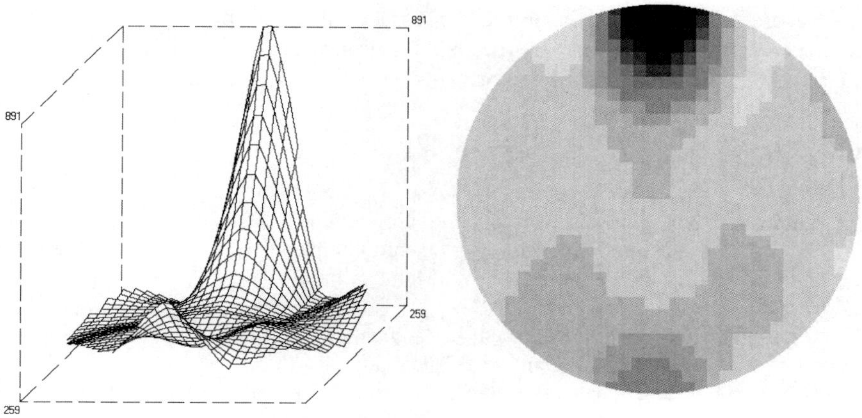

Fig. 4. Reconstruction of *a* profile for He stream flowing out from circular nozzle (Φ=0.09m) placed near to the wall of channel

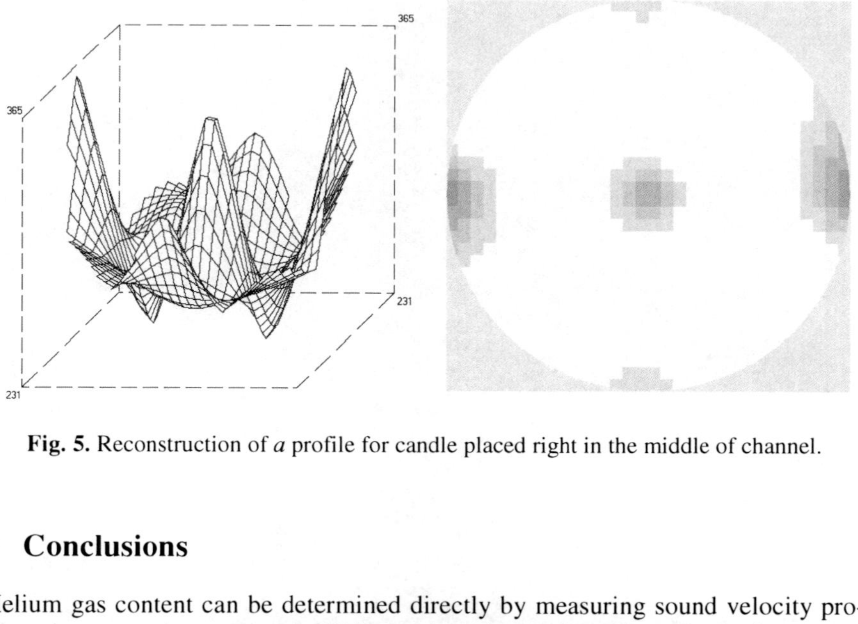

Fig. 5. Reconstruction of *a* profile for candle placed right in the middle of channel.

5 Conclusions

Helium gas content can be determined directly by measuring sound velocity profile using acoustic tomograph. This non-invasive and quick method of measurement presents an interesting alternative to electrochemical oxygen sensors, which work slowly and in presence of helium manifest some error. The results presented confirm theoretical dependency of sound velocity versus helium content in He/air mixtures. The tomographic system was able to locate precisely the region of helium release to the atmosphere; moreover, it could be also used for remote temperature measurement.

The precision of the measurement could be enhanced by application of more advanced signal processing algorithms and by improvements in hardware part of the ultrasonic generator and detector.

References

[1] Chorowski, M., Konopka, G., Riddone, G.: Helium discharge and dispersion in the LHC accelerator tunnel in case of cryogenic failure. In: Narayankhedkar (ed.) Proc. 18th Int. Cryogenic Engineering Conf., Mumbai, India (2000)

[2] http://www.atomox.com

[3] Chorowski, M., Gizicki, W., Wach, J.: Application of acoustic tomography in helium-air mixture content monitoring. In: Baguer, G.G., Seyfert, P. (eds.) Proceedings of the Nineteenth International Cryogenic Engineering Conference (ICEC 19), Grenoble, France, July 22-26 (2002)

Measurement Stand for Basic Parameters of LCD Displays

B. Kabziński[1], R. Barczyk[1], D. Jasińska-Choromańska[1], and A. Stienss[2]

[1] Warsaw University of Technology, Faculty of Mechatronics, 8 Sw. A. Boboli Street, Warsaw, 02-525, Poland
[2] Instytut Edukacji Interaktywnej estakada.pl 80/82 Grzybowska Street Warsaw, 00-844, Poland

Abstract. This article is about realization of the laboratory stand used for measuring parameters of LCD displays. The stand together with additional accessories (spectrophotometer and measuring probes) as well as software allow to measure a wide range of parameters (colour gamut, response time, contrast coefficient and its irregularity, luminance of black colour and its irregularity, luminance of white colour and its irregularity, changes of luminance depending on the viewing angle). This paper also show method and results of measurements of differences in contrast ratio and luminance uniformity between particular pieces of the same model of a display screen (research has been carried out on 5 pieces of the same model of a display). The presented results show that the measured value of the viewing angle may significantly differ from the value provided by producers of displays.

1 Introduction

This article is about realization of the laboratory stand used for measuring viewing angles of LCD displays. There is a need to build such stands in order to check whether the parameters of displays reflect the ones provided by their producers. ISO norms guidelines as regards tests of displays (ISO 9241-303:2008 „Ergonomics of human-system interaction - Part 303: Requirements for electronic visual displays" and ISO 9241-305:2008„ Ergonomics of human-system interaction - Part 305: Optical laboratory test methods for electronic visual displays") have been used to make sure the results of such research are reliable. The stand together with additional accessories (spectrophotometer and measuring probes) as well as software allow to measure a wide range of parameters (colour gamut, response time, contrast coefficient and its irregularity, luminance of black colour and its irregularity, luminance of white colour and its irregularity, changes of luminance depending on the viewing angle). In this article, the construction of a stand has been described as well as the method of measuring changes of luminance depending on the viewing angles, which is one of the most important parameters in the situation when there are several people looking at a certain screen. This paper also show method and results of measurements of differences in contrast ratio and luminance uniformity between particular pieces of the same model of a display screen (research has been carried out on 5 pieces of the same model of a display). The presented results show that the measured value of the viewing angle may significantly differ from the value provided by producers of displays.

2 Reasons for Independent Measurement Methods

Theoretically all LCD monitors seem to have parameters allowing to use them almost in any application. According to manufacturers specifications LCD monitors have wide view angles (over160^{o}), contrast ratio at least 500:1, and 16 million colour, so average user should not be able to observe any difference between view angle of TN matrix and PVA or MVA matrix. Measurements used in technical specifications are usually made by LCD matrix manufacturers, not final monitor manufacturers. During such measurement procedure LCD matrices are placed on special platforms with selected light source powered by high quality power supply. Final result is when two different monitors equipped with the same kind of LCD matrix, but different light source and electronics, have different technical parameters.

3 View Angle – Luminance Changes Depending on View Angle

According to standards maximal view angle is defined for picture where contrast ratio in central point of monitor drops to value 10:1. However picture is distorted while contrast ratio drops to value 100:1, so real loss of picture quality is visible at smaller angles. Moreover some producers accepted 5:1 as maximal contrast ratio to allow them expand theoretical view angle by approximately 20 degrees.

Measurement Method

In order to get information about changes in observed picture while moving observer around the monitor, a measurement stand was designed and build. Measurement of view angle is based on contrast ratio factor:

$$CR = \frac{L_{max}}{L_{min}} \tag{1}$$

where: L_{max} – maximal luminance display point, L_{min} – minimal luminance display point.

Compatibility of Measurement Stand with ISO Standard

To make our results comparable with values given by manufacturers, an assumptions according to ISO standards was made:

- Optimal distance of monitor observation according to ISO 9241-303:2008 [1] is between 300 and 750mm from central point of the screen.
- According to ISO 9241-305:2008 [2]:
  - Monitor which is being tested should be isolated from external light sources
  - Measurement device should be is placed perpendicular to screen axis going through center of LCD matrix, measurements are made every 10 degrees or less
  - Monitor displays 100% white plane
  - After measurements are done, luminance uniformity of measurement points is calculated using equation:

$$uniformity = \frac{L_{\min}}{L_{\max}} \cdot 100\% \tag{2}$$

Measurement Stand

Digital SLR camera is mounted on a tripod, which allow to change height of camera from lower to upper edge of monitor. Camera with tripod is able to move around the monitor on previously set radius (range ±90°from perpendicular position to matrix), motion is realized manually. Before measurements, according to ISO-9241-305:2008, few conditions should be fulfilled:

- Remove all dust from monitor matrix
- Warm up monitor – luminance is stable after approximately 20 minutes
- Preliminary parallel position of monitor and camera matrices
- Monitor calibration with spectrophotometer
- Setting monitor to native resolution (physical matrix resolution)
- External condition in closed premises, humidity 25-85%, temperature 20-50°C, atmospheric pressure 860-1060 hPa

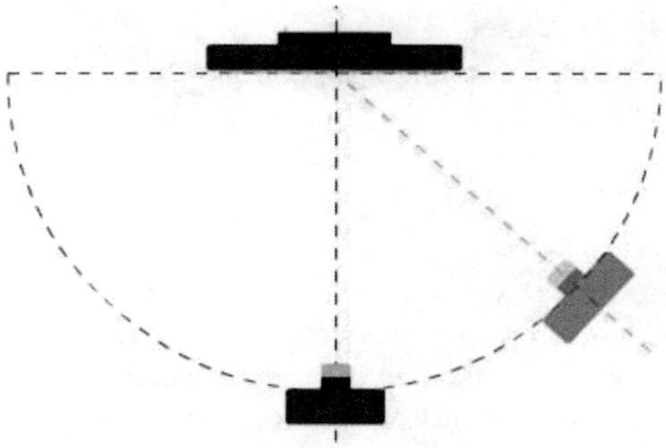

Fig. 1. Measuring stand – measuring angle's range

Measurements

Analysis of acquired data is based on finding points of minimal and maximal luminance, but most important is average luminance of whole matrix (or selected region). Measurements of horizontal and vertical angles were made for 5 monitors (same manufacturer and model), to reveal differences between theoretically identical copies of the same monitor. Charts below show average values of whole monitor matrix.

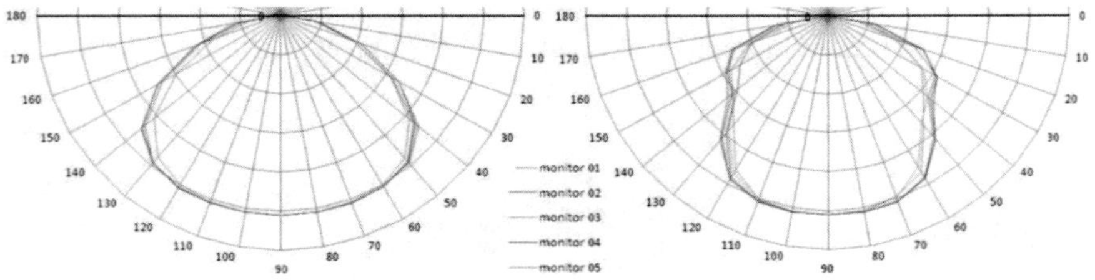

Fig. 2. Chart of the dependence of grayscale level on horizontal (left) and vertical (right) viewing angle

4 Uniformity of Contrast Ratio

In contrast ratio measurement procedure and its uniformity a set of two checkerboard. In place of white rectangles on one board, black ones are on the second.

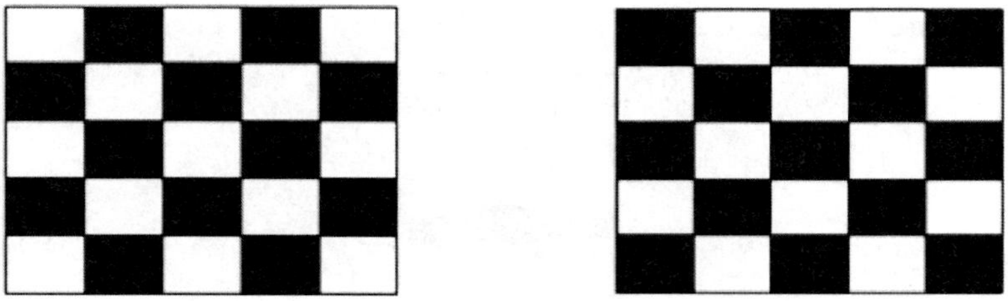

Fig. 3. Test board

Measurements were made using electronic circuit OPT101 consisting of photodiode with transimpedant amplifier in one structure. Electric current is proportional to light stream. Procedure is simple, but need accuracy.

First step is to display board on left, and measure with probe each white field (measurement on center of rectangle), next step is to display second board and repeat measurement on black fields. Result is measurements in 13 points for each board, measurement were repeated five times to make estimation more accurate and certain.

Measurements

Results in percents show distortion in uniformity of contrast ratio while setting backlight for 100% and 50% for group of 5 monitors.

Table 1. Contrast Ratio and contrast uniformity difference measurement results

	Monitor 1		Monitor 2		Monitor 3		Monitor 4		Monitor 5	
Param [%]	100	50	100	50	100	50	100	50	100	50
Min CR	68,9	57,9	67,8	69,9	63,8	64,3	63,7	63,8	58,0	50,0
Max CR	127,9	112,3	105,5	100,0	113,5	105,4	115,3	122,2	110,3	100,6
ΔCR	59,0	54,4	37,7	30,1	49,7	41,1	51,6	58,4	52,3	50,6
diff. CU	53,8	51,5	64,3	70,0	56,2	61,0	55,2	52,2	52,6	49,7

Contrast Uniformity:

$$CU = \frac{C_{\min}}{C_{\max}} \cdot 100\% \qquad (3)$$

where C_{min} i C_{max} are minimal and maximal measured value of CR.

Significant difference in contrast uniformity is observed, but average value of difference in uniformity is about 50%. It cannot be discussed without looking at results of contrast ratio for whole surface of matrix.

Table 2. Contrast Ratio values

	Monitor 1	Monitor 2	Monitor 3	Monitor 4	Monitor 5
CR	40:1	43:1	46:1	38:1	38:1

5 Conclusions

Measurement stand allow to measure LCD monitor parameters representative for image reproduction quality and comparing to values given by manufacturers. Our research show that our results of viewing angle as well as contrast ratio strongly differ from that in technical specifications published by manufacturers. Contrast uniformity and luminance uniformity are parameters that seem to be relevant in describing quality of monitor and are not included in official specifications. Significant differences in uniformity of white luminance are main source of differences in contrast ratio.

References

[1] ISO 9241-303:2008: Ergonomics of human-system interaction - Part 303: Requirements for electronic visual displays

[2] ISO 9241-305:2008: Ergonomics of human-system interaction - Part 305: Optical laboratory test methods for electronic visual displays

Flows Out of Silos – Vector Fields of Velocities

J. Malášek

Brno University of Technology, Faculty of Mechanical Engineering, Technická 2896/2, Brno, 616 69, Czech Republic

Abstract. The specific flowing areas of non-homogeneous materials can be found in flows in silos - flows out of silos or bunkers. Many possibilities of flows between mass flows and core flows contain various interesting extremes. These flowing and deformed areas of non-homogeneous materials will be identified by means of mathematical descriptions of pictures with results of vector fields of velocities.

1 Introduction – Flows Out of Silos

The important requirement for the form – design of silos against stored non-homogeneous materials (powdery materials, particular materials, bulk solids) is mass flow.

Mass flow – the entire contents of the silo is in motion - no or only minimal static areas. The surface of silo contents sinks almost evenly. Non-homogeneous material constitution of filling mass flow is very similar to constitution of extracting mass flow. The abrasion on the wall of silo increases at mass flow. The silo geometry for mass flow is typical for the steep funnel and the smaller discharge opening.

Core flow – materials flows exist only in the centre of the silo. Static areas of non-homogeneous materials exist along the walls of silos. The surface of silo contents sinks unevenly with respect to the core of flow location. Non-homogeneous material constitution of filling core flow is not similar to constitution of extracting core flow – reduction of homogeneity and other physical and mechanical characteristics. Possibilities of formation of permanently static areas of material contents entail reduction of the active silo volume. The silo geometry for core flow is typical for the flat funnel – this is a problematic advantage to the built height. Many design adjustments of silos exist to lessen tendency to core flow.

2 Vector Fields of Velocities

Flows and deformations in non-homogeneous materials are processes proceeding in time. Therefore many in time successive pictures are needed. We repeatedly need the sequences of three successive pictures. In the first picture we select a square element of the area with dimensions Δx , Δy in its first position

$(x_1;y_1)$ with a suitable limited number of pixels. In the first time-successive picture we search for the corresponding square element of the area with dimensions Δx, Δy in its second position $(x_1+dx_1;y_1+dy_1)$ by means of minimal square differences of the brightness level $I=f(x,y)$ of the corresponding pixels. In the second time-successive picture we search for the corresponding square element of the area with dimensions Δx, Δy in its third position $(x_1+dx_1+dx_2;y_1+dy_1+dy_2)$ or $(x_2+dx_2;y_2+dy_2)$ by means of minimal square differences of the brightness level $I=f(x,y)$ [1] of the corresponding pixels. For the time Δt_1, eventually the time Δt_2 we search for shifts dx_1, dy_1, eventually shifts dx_2, dy_2 of the square element of the area of non-homogeneous material by means of mathematical description of picture.

Following intervals of number of pixels are chosen for the equations (1), (2):

$\Delta x = 10$, $\Delta y = 10$, $x \in <-5;+5>$, $y \in <-5;+5>$,

$dx_1 \in <-30;+30>$, $dy_1 \in <-30;+30>$,

$dx_2 \in <-30;+30>$, $dy_2 \in <-30;+30>$.

$$\arg\min \sqrt{\sum\sum [I_1(x_1+x),(y_1+y) - I_2(x_1+dx_1+x),(y_1+dy_1+y]^2} \qquad (1)$$

$dx_1 \in <-30;+30$

$dy_1 \in <-30;+30>$

$x \in <-5;+5>$

$y \in <-5;+5$

$$\arg\min \sqrt{\sum\sum [I_2(x_2+x),(y_2+y) - I_3(x_2+dx_2+x),(y_2+dy_2+y]^2} \qquad (2)$$

$dx_2 \in <-30;+30$

$dy_2 \in <-30;+30>$

$x \in <-5;+5>$

$y \in <-5;+5$

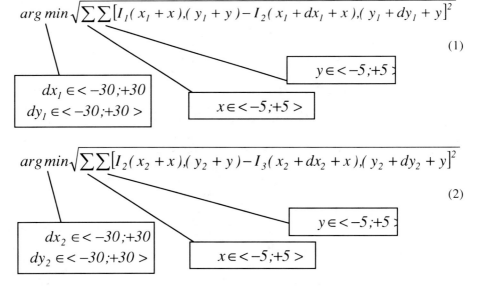

The vector of velocity relevant to the square element of the area of non-homogeneous material in location $(x_1+dx_1;y_1+dy_1) \equiv (x_2;y_2)$ [2]:

$$\overrightarrow{v_{1,3}} \equiv \left\{ \frac{dx_1 + dx_2}{\Delta t_1 + \Delta t_2}, \frac{dy_1 + dy_2}{\Delta t_1 + \Delta t_2} \right\} \tag{3}$$

This hereinbefore stated principle of calculation must be repeated many times to get results of experiments as a mathematical description.

3 Results of Experiments

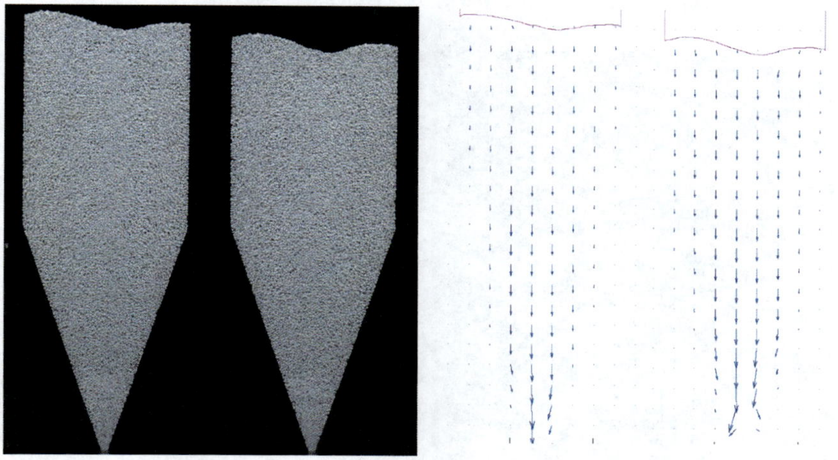

a) b)

Fig. 1. Mass flows out of silo: a) realities, b) vector fields of velocities.

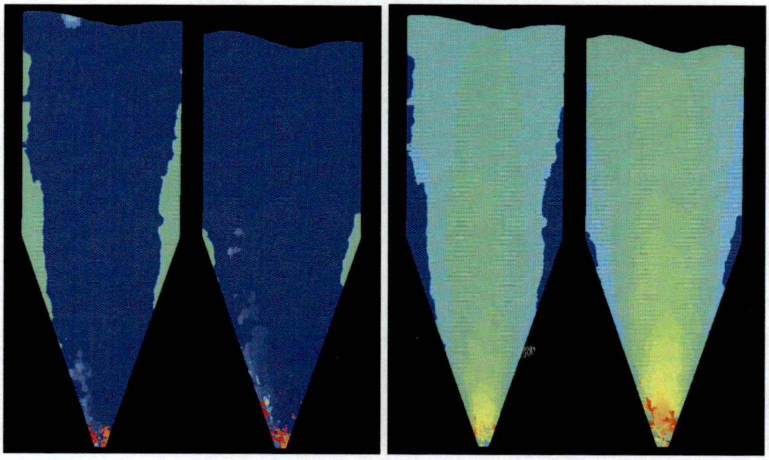

a) b)

Fig. 2. Mass flows out of silo: a) directions, b) velocities.

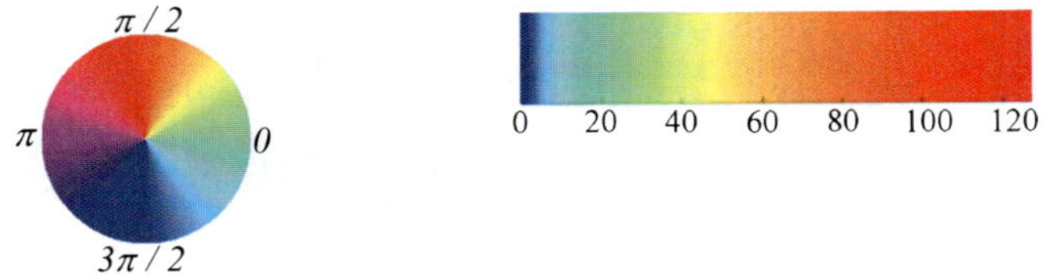

a) b)

Fig. 3. The scales: a) direction scale, b) velocity scale $[\ mm.s^{-1}\ ]$ [3].

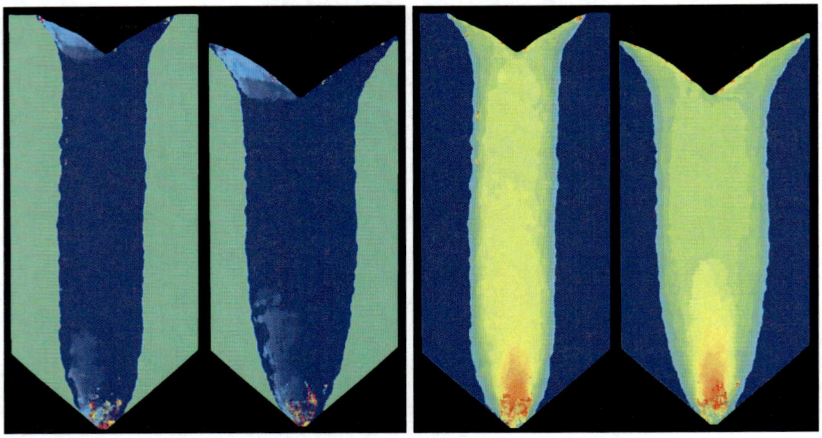

a) b)

Fig. 4. Core flows out of silo: a) directions, b) velocities.

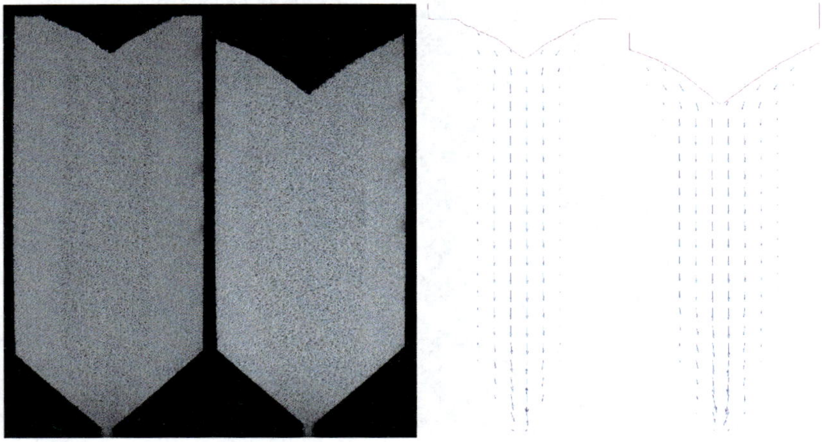

a) b)

Fig. 5. Core flows out of silo: a) realities, b) vector fields of velocities.

4 Conclusion

The flowing areas of non-homogeneous materials change its structural composition, its size and shape according to the possibilities of its physical and constitutive characteristics to the form with minimal external forces effects. These changes of sizes and shapes cause simultaneous changes of physical and constitutive characteristics in time. The pictures of transformed areas of non-homogeneous materials in a form of vector fields of velocities are chosen as the carrier of information.

Acknowledgements

The published results have been achieved with the help of the project No. FSI-S-11-8 granted by specific university research of Brno University of Technology.

References

[1] Hlaváč, V., Sedláček, M.: Zpracování signálů a obrazů. ČVUT v Praze (2000) ISBN 978-80-01-03110-0
[2] Malášek, J.: Identifikace napjatosti, deformace a proudění v nehomogenních materiálech (2008) ISBN 978-80-214-3738-8
[3] Malášek, J.: Vector fields of velocities of flows and deformations in non-continuous materials. In: CHISA 2008 (2008) ISBN 978-80-02-02050-9

Problems of the Influence of the Quality and Integrity of Input Data on the Reliability of Leak Detection Systems

Mateusz Turkowski[1] and Andrzej Bratek[2]

[1] Institute of Metrology and Biomedical Engineering,
Warsaw University of Technology św. A. Boboli 8, 02-525 Warsaw, Poland
[2] Industrial Research Institute for Automation and Measurements Al. Jerozolimskie 202, 02-486 Warsaw, Poland

Abstract. The authors describe the problems they have met during implementation of leak detection and localization systems for liquid and gas pipelines. The main problems are discontinuities of the data acquired from telemetry systems, damaged or bad calibrated measuring transmitters, drift of measuring instruments, bad balance between fluid entering and leaving the system and problems of the correct installation of temperature transmitters. The proposals how to overcome these problems have been presented.

1 Introduction

All stages of pipeline building and operating must be fulfilled the regulations and recommendations of numerous standards and regulations, whose purpose is to provide long-lasting operation of pipeline system. However, even if the pipeline has been designed and built very carefully, there is always a potential of leaks.

The leak brings always large losses of various kind: suspending the product transport, the cost of reparation of the damage and loss of transported product. In case of explosive or/and flammable or/and dangerous to environment media (e.g. petroleum and other petrochemical products), the leak causes hazard for safety of the people and the equipment as well as an environmental contamination. Events like that induce high social and financial costs, which are proportional to the intensity and duration of the leak.

If a leak happened, its effects can be minimized only by fast detection and localization of the leak point enabling quick dispatcher reaction (stopping pumping, closing the valves, organizing provisional damage repair etc.)

Pipeline leak detection systems play therefore a key role in minimization of the leaks probability and impact. The vast assortment of technologies is available today, the background information has been presented in [1].

Most of the leak detections methods are analytical, internal methods based on comparison of the pipeline system model with the real measurement data obtained from the telemetry or SCADA systems.

For the purposes of the leak detection and localization systems the following parameters can be permanently measured:

- pressure at the inlet, at the end, and at several/a dozen or so points (at the valve stations, metering and regulating stations, terminals),
- flow rate at the inlet and at the outlet of the pipeline,
- temperature usually at the same places that pressure measurement,
- sometimes density of the transported liquid at inlet of the pipeline.

Usually the procedures of leak detections are based on continuous measurements of pressure and flow-rate, measurements of other variables (temperature, density) are only auxiliary although they can influence the calculation results strongly.

Measured pressure p is compared with calculated pressure p_{cal}, in the simplest case (static model) the following formula can be used for liquids:

$$p_{cal} = p_0 - \lambda \frac{\rho w^2}{2} \frac{L}{D} \tag{1}$$

where p_0 - pressure at inlet of the pipeline, λ - friction coefficient (function of Reynolds number and pipe roughness), ρ – liquid density, w – average velocity of liquid, L – pipeline length, D – pipe internal diameter,
and for gases:

$$p_{cal} = \sqrt{p_0^2 - \frac{1.62114 q_{st}^2 \rho_{st}^2 ZRT\lambda L}{D^5}} \tag{2}$$

where q_{st} and ρ_{st} – flow rate and density at standard conditions, Z – compressibility coefficient, R – gas constant, T – absolute temperature.

In most cases, however, much more complicated, but more accurate dynamical moodels are used [2, 3].

Apart of the physical models some neural, additive, fuzzy, swarm particles models are used, sometimes supported by data mining methods [4]. They do not need precise information about the pipeline, so they are suitable in case when the leak detection system has to be installed in old pipelines.

Almost all of these methods are based on the constant comparison of the measured and calculated data. If the differences (residua) override a certain limit the alarm is generated and localization procedures are started.

The algorithms of leak detection and localization are very sensitive to the fluctuations of measuring signals, resulting from instruments noise, uncertainty [5] and systematic errors of the instruments. The quality of the measurement data is therefore of greatest importance. It is evident, that bad quality of the data will strongly influence the leak detection systems and generate false alarms which are sometimes as dangerous as the leak itself.

2 Discontinuity of the Data and Calibration Problems

The data acquired from the telemetry or SCADA system usually contains a lot of discontinuities, noises and systematic errors. The discontinuities are particularly frequent in the systems based on the GPRS connection, usually when the network is overloaded or the distance from the nearest antenna is great. Fig. 1 presents the data (pressures) from gas metering and regulating stations installed in the gas pipelines system. The number of discontinuities is important, about 30 to 35 % of all measurements. In case of data lack, error or significant noise the data from telemetry system has to be replaced by the approximated values. Approximation can be fulfilled by the linear approximation – the straight line is calculated from the previous data with the use of least squares method.

In case of long data absence the reconstruction is suspended and the given signal is no more taken into account by the leak detection system.

The serious problem is also bad calibration of the pressure transmitters. Sometimes the offset of the transmitter signal may caused that positive changes of pressure along the pipeline are observed, what is contradictory to the fundamentals of physics. In such cases the transmitter characteristics are corrected with the use of the model data.

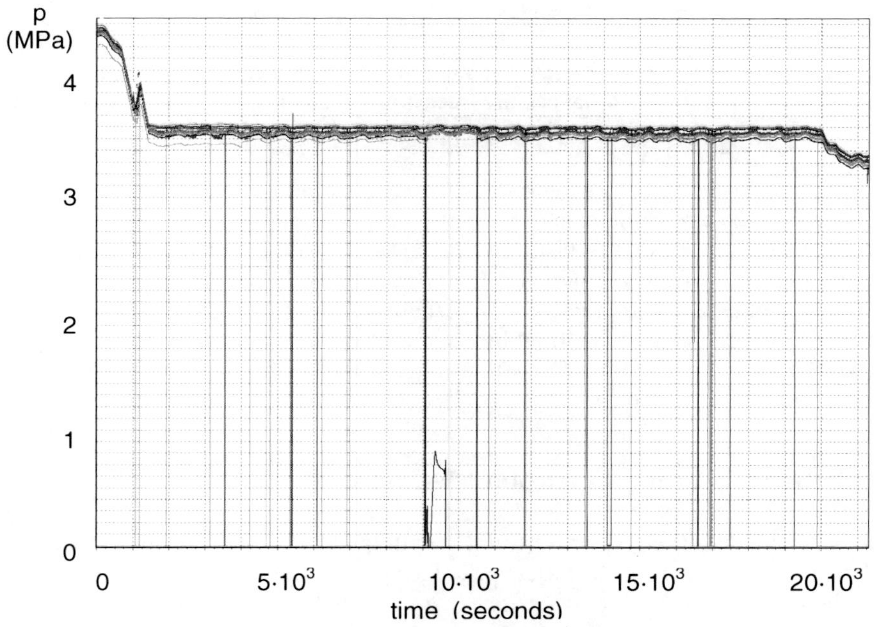

Fig. 1. Measurement signals (pressures) from the gas pipelines system

Another example, for liquid pipeline, is presented in Fig. 2.

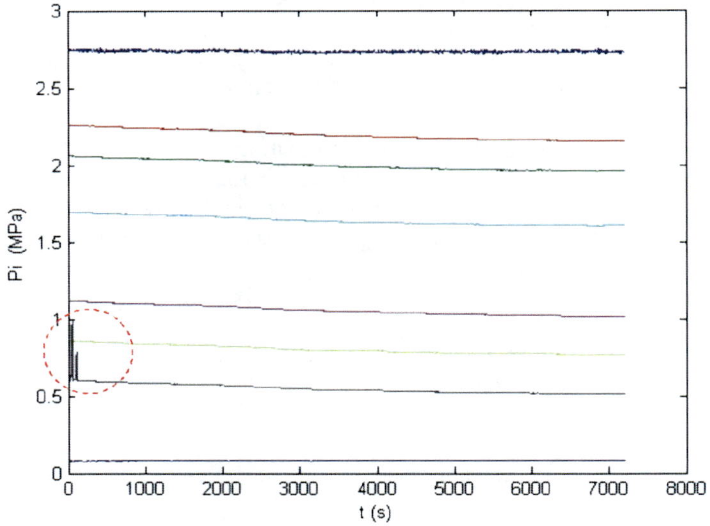

Fig. 2. Measuring signal disturbance at the pressure transmitters in liquid pipe.

This is the example of dedicated (not public, as in case of GPRS) telemetry system. The temporary disturbances are occasional, but even the short-lived disturbance of the signal used by the system can induce false alarms or leak localization faults.

The measuring signals recognized as wrong, are eliminated from the leak detection process. The identification of wrong signals is based on the observation of the momentary changes of the measured parameters ant their comparison with the values calculated from the model. The parameters with the values out of typical for the pipeline parameter ranges are rejected, as well as the values which speed of change is to high and not justified by the operational changes of pipeline parameters. The exclusion of the wrong transmitters can be fulfilled automatically or manually.

3 Efficiency of the Filtration Methods

Before the measuring signals are utilized through the system they have to be filtered to decrease the noise influence to the system efficiency.

The low pass filters were investigated and also the time averaging of the signals at various time constants. The effects of such data processing is presented in fig. 3.

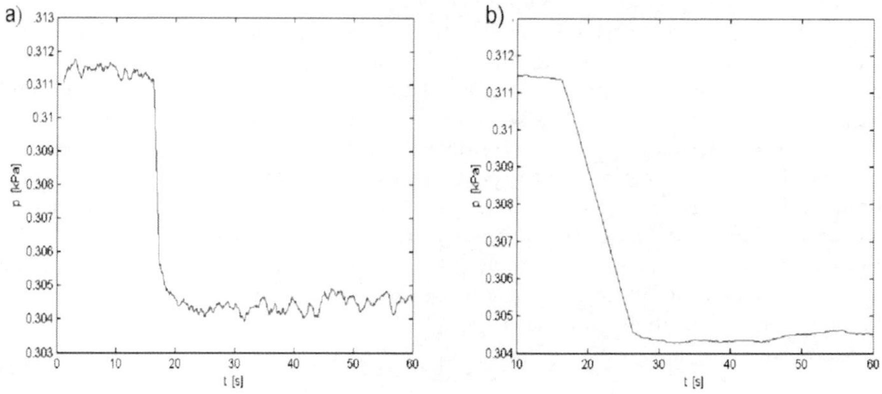

Fig. 3. Diagram of the pressure at a measurement point. Averaging with time constant 1 s (a) and 10 s (b)

The most effective method was the double filtration with time averaging with time constant 1 s and 10 s.

Lower time constant is used to detect the occurrence of the leak fast and the higher – to stabilize measurement results for the purpose of precise leak localization.

4 Tunning of the Model

The changes of the operational conditions i.e. change of the pumped medium, switching the sending/receiving tank, start or stop of the compressor in gas compressor station change the parameters of the pipeline system. It can have negative influence on the performance of the leak detection system. The system need therefore permanent tunning to assure consistency between model and the real pipeline system. The method of tunning can be described as follows:

a) In stable state of the pipeline (constant flow, no technological operations) the values of the friction coefficient λ should be permanently computed from the formula (3). They should be calculated for subsequent pipe sections between pressure measurement points.

$$\lambda_i = \frac{2D}{\rho w^2 L_i}\left[\left(p_i - p_{i+1}\right) + \rho g\left(h_i - h_{i+1}\right)\right] \tag{3}$$

where D – internal pipe diameter L_i – length of pipeline section, g – acceleration due to the gravity, ρ – density, p – pressure, w – fluid velocity, h – height, subscript i – number of the pipeline section.

b) For the calculation of the friction coefficient the mean values of the pressures and velocities for the period equal about the time of stabilization of the disturbances propagation in the pipeline should be used.
c) Periodically, during stable conditions, the consistency between model data and measured data for subsequent pressure measurement points should be checked.
d) In case of recognition in the section i the difference between calculated and measured pressure greater then the accepted threshold value dP, the friction coefficient λ in the i-th section should be corrected. The threshold value should be determined taking into account the normally existing pressure variations.
In order to do not overlook the real leak in each step only the small part of the correction should be introduced, according to the recursive formula with the coefficient ζ which can be equal i.e. 0.01

$$\lambda_i = \varsigma \lambda_{i-1} + (1-\varsigma)\lambda_i \tag{4}$$

During such procedure of tunning, despite only the friction coefficient λ is changed, all the slowly changing parameters i.e. instrumentation drift or change of pipe roughness (i.e. due to corrosion), are compensated.

5 Imbalance

Parallel to the analysis of the pressure distribution in the system the balance of the fluid flowing into and out of the system should be monitored. For this purpose some index proposed in [6] should be calculated; in case of gas according to the formula (5):

$$\tau(t) = \Delta V_{st,in}(t) - \sum_{i=1}^{n} \Delta V_{st,out}(t) - V_{st,acc}(t) \tag{5}$$

The variable $\tau(t)$ can be called corrected flow imbalance at the moment t. This is the difference between the volume of gas flowing into the pipeline system $\Delta V_{n,in}(t)$ and the volume that has flown out of the system $\sum_{i=1}^{n} \Delta V_{st,out}(t)$ (n is the number of output stations), minus the volume of the gas accumulated in the pipeline $V_{st,acc}(t)$; subscript st denotes standard conditions. The term $V_{st,acc}(t)$ depends on the gas temperature, pressure and composition and can be calculated with the use of the formula

$$V_{n,acc}(t) = V_g \rho_{st} \frac{p_{avg} T_{st}}{p_{st} T Z} \tag{6}$$

where V_g is the geometrical volume of the pipeline system, T absolute temperature. and p_{avg} , average pressure in the pipeline section can be calculated as

$$p_{avg} = \frac{2}{3}\left(p_i + \frac{p_{i+1}^2}{p_i + p_{i+1}} \right) \tag{7}$$

Calculation of the $\tau(t)$ can give important information about the leak intensity, which can be used as input parameter in some procedures of leak localization. This parameter fluctuates about some medium value m, mainly due to the instruments drift, gas meters systematic errors (changing with the flow rate) or uncontrolled temperature changes. These fluctuations can be characterized with the variance σ^2. Let us denote the momentary deviation from the mean value m, as Δm. Than, the cumulative sum $\alpha(t)$ given by the formula (8) can constitute one of the criterions of alarm generation when it excess some level [6]. This level can vary depending of the state of the pipeline, in steady state it can be rather low, and during the technological operations generating instabilities, transients etc. it can be set to the higher value.

$$\alpha(t) = \alpha(t-1) + \frac{\Delta m}{\sigma^2}\left(\tau(t) - m - \frac{\Delta m}{2}\right) \tag{8}$$

During the research on the leak detection systems the authors have encountered difficulties trying to make the use of the formulae (5) and (8). Both the differences between the incoming and outgoing flow rate and the cumulative sum exceeded significantly the expected values (fig. 4).

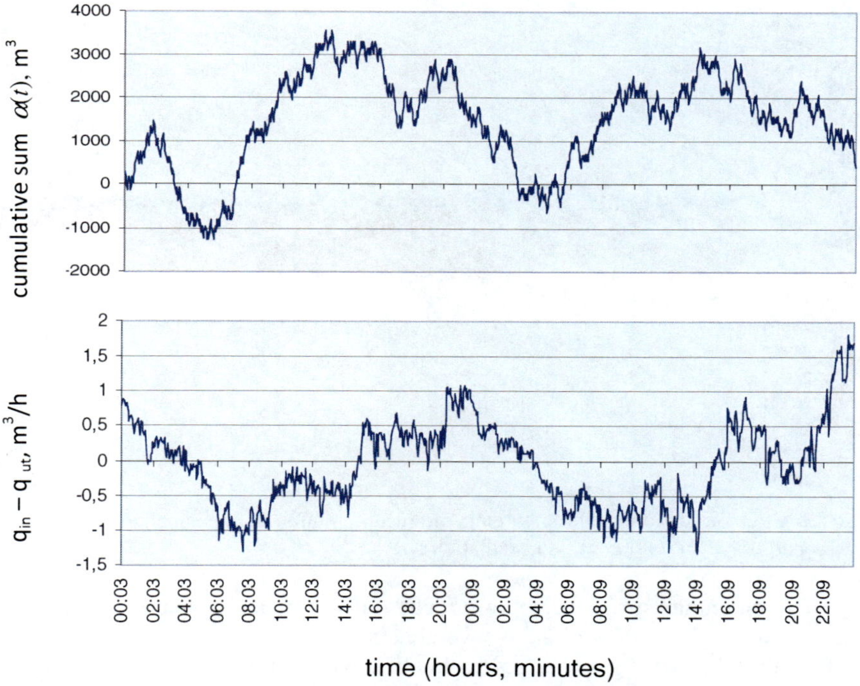

Fig. 4. Cumulative sum (upper diagram) and difference between incoming and outgoing flow rate of gas (lower diagram)

The excessive values were attributed to the improper method of temperature measurement, because they are strongly correlated with the day/night cycle.

6 Temperature Measurement

Temperature influences the viscosity, so Reynolds number and in consequence friction coefficient λ. It also influences the inventory volume of the fluid in the pipeline.

The problem is the right localization of temperature sensor. The natural place is the valve system, usually situated above the ground level. Especially for gas it can be source of significant errors. Because of low gas specific heat the measured temperature in winter is lower than the actual temperature of gas, and in summer – higher.

The simulation with the use of Computational Fluid Dynamics (CFD) is presented in Fig. 5.

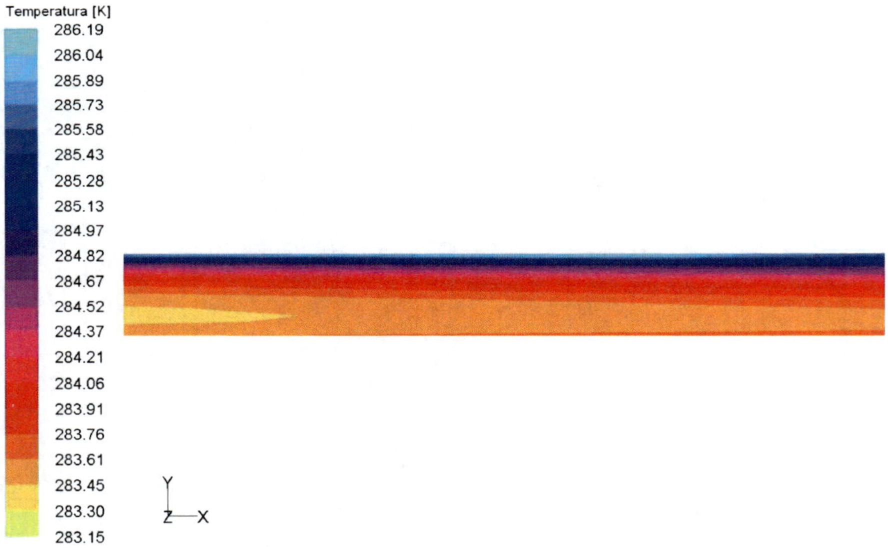

Fig. 5. CFD simulation of the gas temperature distribution in the pipeline heated in the upper part to $t = 40$ °C (i.e. by solar radiation).

The temperature difference between gas flowing under ground can differ from that measured above the soil up to several degrees.

Probably the excessive changes of the difference between incoming and outgoing gas presented in fig. 4 are due to the improper temperature sensor installation. Underground installation of the temperature sensors may be a solution, as shown in fig. 6.

Fig. 6. Installation of the temperature sensor on the gas pipeline surface

The sensors have been installed during upgrading the measurements instruments of a gas pipeline to make it possible installation of the leak detection system. Because it would be very costly to install the thermowell in the pipe under pressure (gas transport can not be interrupted), the surface temperature sensors were chosen. The sensors must be isolated, both against the moisture and thermally.

7 Conclusions

This paper can seem discouraging for the potential user or designer of the leak detection and localization systems, but the following conclusions can help to avoid failures during implementing such systems.

In most of cases the existing, dedicated for the routine maintenance, measurement and telemetry system have to be retrofitted to comply with the needs of the system.

The short discontinuities of the signal can be eliminated by extrapolation of the previous data, the longer discontinuities, however, demand the exclusion of the measured parameter from the system.

The bad calibration of the measuring transmitters can be corrected in some extent, with the use of model data. The operator of the Leak Detection System must however close cooperate with pipeline operator, it concerns mainly the procedures of calibration of field transmitters.

The slowly changing parameters, as instrumentation drift, changes of pipe roughness etc. can be compensated by the tunning of the model.

The manner of the installation of temperature sensors can have important influence on the system performance. It is preferable to install the temperature sensors underground, because the temperature measured above ground may be not representative, it concerns mainly gas pipelines.

Acknowledgements

This work was partially supported by Polish Ministry of Science and Higher Education within research project nr O R00 0013 06.

References

[1] Turkowski, M., Bratek, A., Słowikowski, M.: Methods and Systems for leakage detection in long range pipelines. Journal of Automation. Mobile Robotics and Intelligent Systems 3(07), 39–46 (2007)

[2] Turkowski, M., Bratek, A., Słowikowski, M.: The improvement of pipeline mathematical model for the purposes of leak detection. Recent Advances in Mechatronics. In: 7th International Conference Mechatronics, Warszawa, pp. 573–577 (2007)

[3] Sobczak, R., Turkowski, M., Bratek, A., Słowikowski, M.: Mathematical modeling of liquid flow dynamics in long range transfer pipelines. Part 1. Problemy eksploatacji, ITE, Radom (2007)

[4] Kościelny, J.M., Syfert, M., Tabor, Ł.: Sequential Residual Design Method for Linear Systems. In: Conference on Control and Fault-Tolerant Systems SysTol 2010, Nice (2010)

[5] Bogucki, A., Turkowski, M.: The influence of the measuring instruments accuracy on the reliability of the leak detection and localization systems. In: Academic Network Event at EGATEC 2011, The European Gas Research Group, Copenhagen, Denmark, May 12-13 (2011)

[6] Beuhausen, R., et al.: Transient leak detection in crude oil pipelines. In: Proceedings of International Pipeline Conference, October 4-8, Calgary, Alberta, Canada (2004)

Part III

Automatic Control

Simulation and Modelling of Control System for Turbo-Prop Constant Speed Propeller

Z. Ancik* and J. Toman

Brno University of Technology, Technicka 2896/2,
Brno, 616 69, Czech Republic

Abstract. The aim of this paper is to show the simulation and modelling of the behaviour of a constant speed propeller control system in Matlab/Simulink software. The constant speed propeller is a subsystem of the turbo-prop engine which is a highly complex nonlinear mechatronics system. The model is created on the basis of a mathematical description provided by the propeller's producer. The model is designed with a view to its testing in real-time application on dSPACE hardware. The aim of the model is to simulate the behaviour of this system for limit operational conditions to avoid destruction of a real device under test.

1 Introduction

The simulation and modelling of turbo-prop propeller control systems [1] is necessary for their optimal design and development. The aim of the model is to identify and predict behaviour of a real system in operational and limit conditions. Equipment essential to safe aircraft operation must be subjected to very strict and expensive tests in real operating conditions. Simulation and modelling can lead to savings in money and time during the development and certification process.

The model is based on mathematical description of the real system characterized by a system of algebraic and differential equations, transfer functions and state space [2]. These characteristics are determined on the basis of mathematical and physical phenomena and experimental measurement. The constant speed propeller is a subsystem of the turbo-prop engine which is a highly complex nonlinear [3] mechatronics system. Knowledge of independent technical branches such as thermodynamics, aerodynamics, hydrodynamics and electronics are involved.

2 Description of Control System

The constant speed propeller is a type of aircraft propeller which maintains constant engine speed by adjusting the propeller blade angle to vary the load on the engine in response to changing flight conditions [4].

---

* Corresponding author.

Engine power is changed according to flight conditions such as altitude, temperature and flight speed [7]. These variations cause changes in engine speed and need to be compensated for in order to maintain efficiency and aircraft steerability [5]. The constant speed propeller ensures a constant speed over a wide range of operational conditions. The control of pitch angle is performed by a hydro-mechanical subsystem shown in Fig. 1.

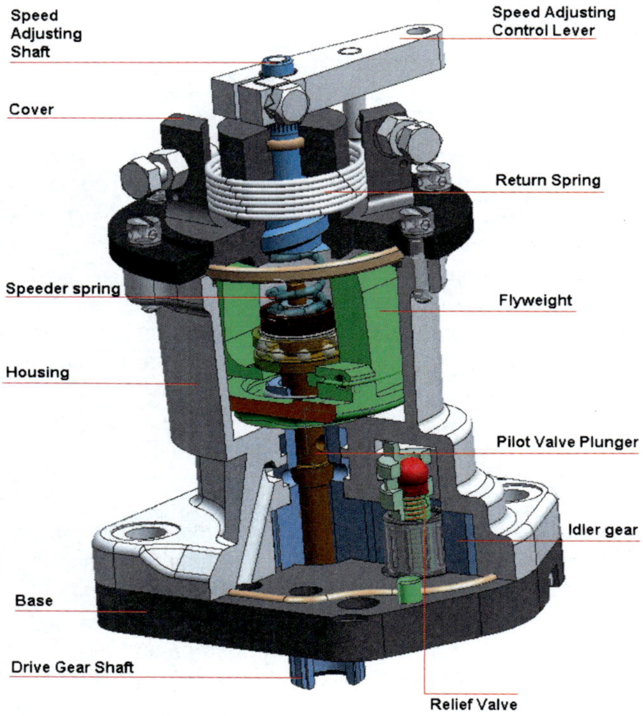

Fig. 1. Hydro-mechanical subsystem

In the hydro-mechanical subsystem a piston and cylinder are linked to the propeller blades so that when oil under pressure is pumped to this cylinder, the piston and the blades are forced to move. Movement of the propeller piston rotates the propeller blades of an uncounterweighted propeller in the increased pitch direction and the blades of a counterweighted propeller in the decreased pitch direction. The single acting propeller thus uses oil pressure to change pitch in one direction and the centrifugal force of propeller counterweights, air charge or the natural twisting movement of the blades to change the pitch in the other direction. The oil flow into the piston is controlled by a proportional electromagnet, so that movement of a slider is proportional to control current - *I20*. The block diagram of a propeller control system is shown in the following Fig. 2.

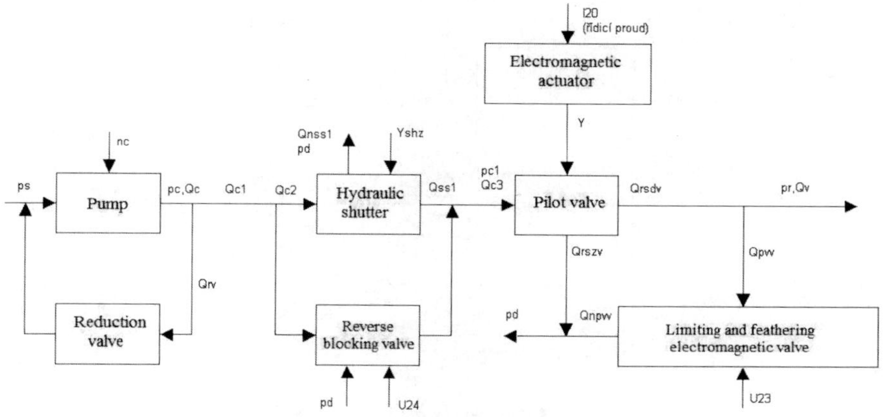

Fig. 2. Block diagram of propeller control system.

3 Modelling of Control System

The mathematical description of individual subsystems is performed by the propeller manufacturer. The subsystems are characterized by a system of algebraic equations including individual constants and characteristics. These constants were computed, measured and verified by testing. The model is created in Matlab/Simulink software.

The model is designed with a view to its testing in a real-time application on dSPACE hardware. There are a few limitations related to automatic generation of source code by Real Time Workshop toolbox (RTW) [6] for the target dSPACE platform. The "algebraic constraint" block is not supported in RTW and is therefore written in C code and implemented to Simulink model by the "S-function" block. The algebraic loop is solved by numeric Newton's iterative method. A proper choice of iteration function and initial approximation are essential for solution convergency.

The model is described in two parts. The first part is the algebraic loop subsystem intended for determination of oil flow in a propeller head - *Qv [l/h]* and pressure behind the pump – p_r *[MPa]*. Input variables to this model are slide-valve motion - *Y [mm]*, oil pressure in propeller head - p_r *[MPa]*, propeller speed - n_p *[r.p.m.]* and piston motion - *u [mm]*. The inputs to numeric solutions are calculation accuracy - *delta*, maximal number of iteration - *step* and *interval*. A model of the algebraic loop is shown in Fig. 3.

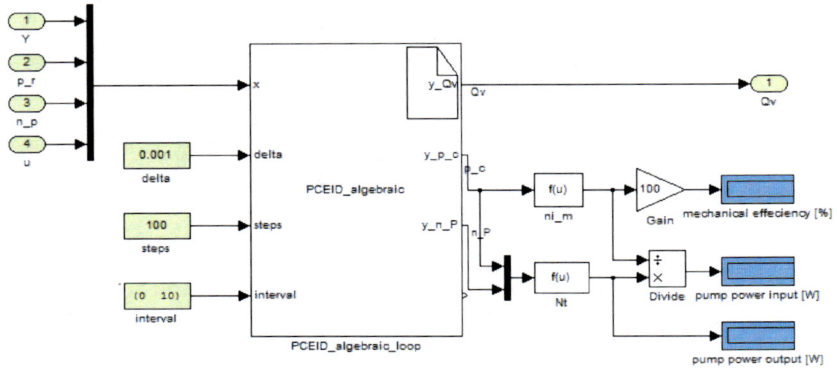

Fig. 3. Model of algebraic loop

The second part is the overall system of propeller control. Input values are propeller speed - n_p *[r.p.m.]* and control current - *I20* which is converted into slide-valve motion - *Y [mm]*. The output is blade angle φ *[°]*.

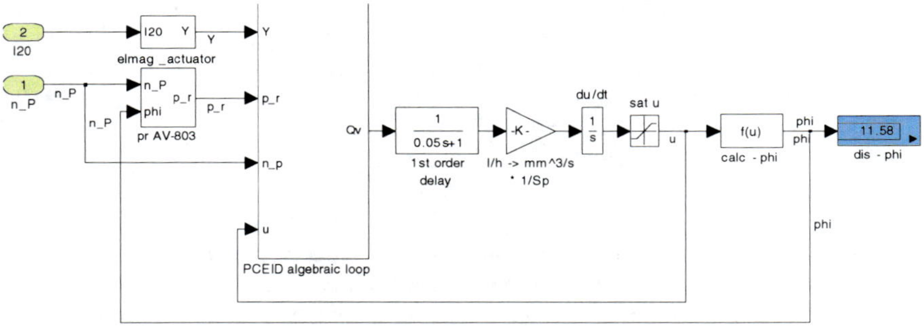

Fig. 4. Model of propeller control system

The algebraic loop gives only static characteristics. Dynamic effects of inertial forces are represented by the "First order transfer function" block where the time constant is determined by the engine producer. Dynamic effects of the piston are included in the equations below. The oil flow in a propeller head - *Qv [l/h]* is recount into the blade angle φ *[°]*.

$$\frac{du}{dt} = \frac{Q_v}{S_p} \tag{1}$$

where: piston surface - S_p *[mm^2]*, piston position – *u [mm]*. Effects of propeller friction and inertia are not included because of much higher hydraulic forces and very slow speed of motion. The blade angle φ *[°]* is determined by the kinematic equation below.

$$\varphi = 35.84 - \left(arcsin\left(\frac{17.434 - u}{40} \right) \right) \cdot \frac{180}{\pi} \tag{2}$$

This equation is specific for each type of propeller. These characteristics are given by engine producer for propeller type AV803.

4 Results of Simulation

The calculation of algebraic loop realized by "S-function" and "algebraic constraint" blocks are compared in following figure. The "S-function" approach reduces computational demands to 60% and gives better results for initial values (see Fig. 5).

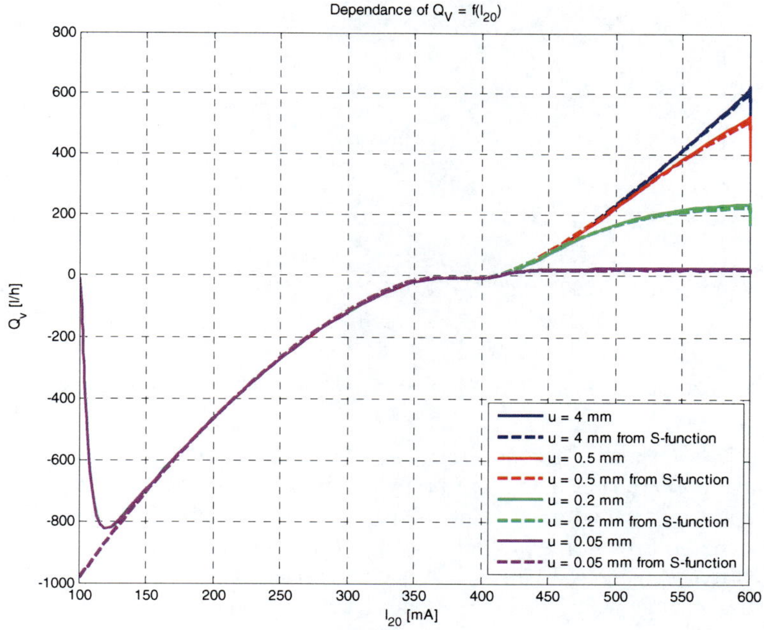

Fig. 5. The comparison of algebraic loop computed by "S-function" and "algebraic constraint" blocks

Dependence of the oil flow in a propeller head - Q_v on the control current - *I20* is shown in Fig. 6. The simulation is run for different degrees of piston motion – *u*. The beta valve which prevents crossing the minimum pitch value depends on the piston motion – *u*. The value of the control current - *I20* affects the direction of oil flow. If the value of control current is lower than 400mA, the oil flows out of the propeller head. If higher, the oil is pumped into the propeller head.

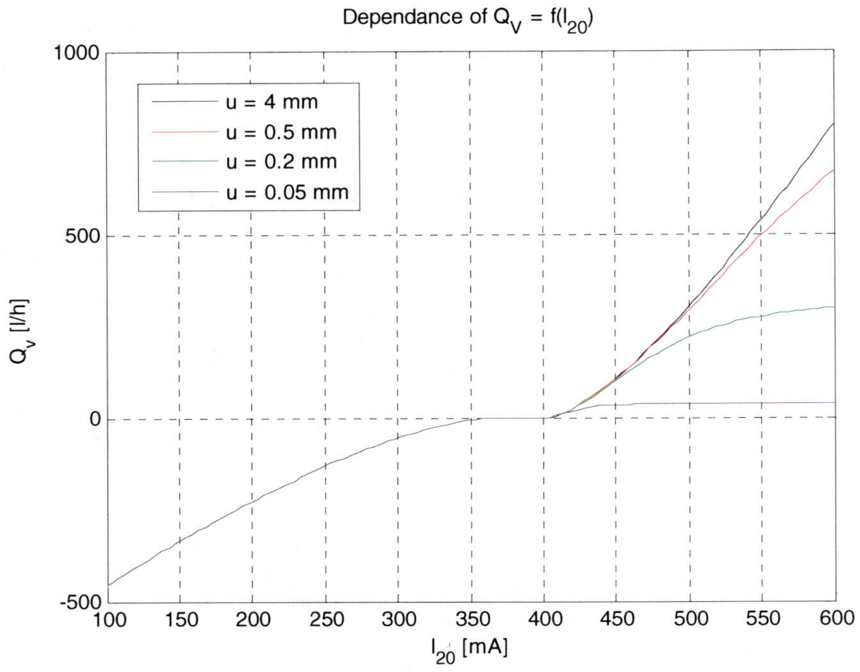

Fig. 6. Dependence of the oil flow in a propeller head - Q_v on control current - *I20*

The system response of control current - *I20* step change is shown in Fig. 7. The simulation is run for a constant propeller speed - $n_p=2150rpm$. At the beginning of simulation the nominal value of control current is $I20=400mA$ which does not affect the oil flow in the propeller head - Q_v or the blade angle φ. After 1s the control current is increased which leads to increased oil flow and decreased blade angle. The absolute value of blade angle depends on the initial value of piston motion – *u* which is set in the particular integrator. The piston motion is $u=5mm$ for this simulation.

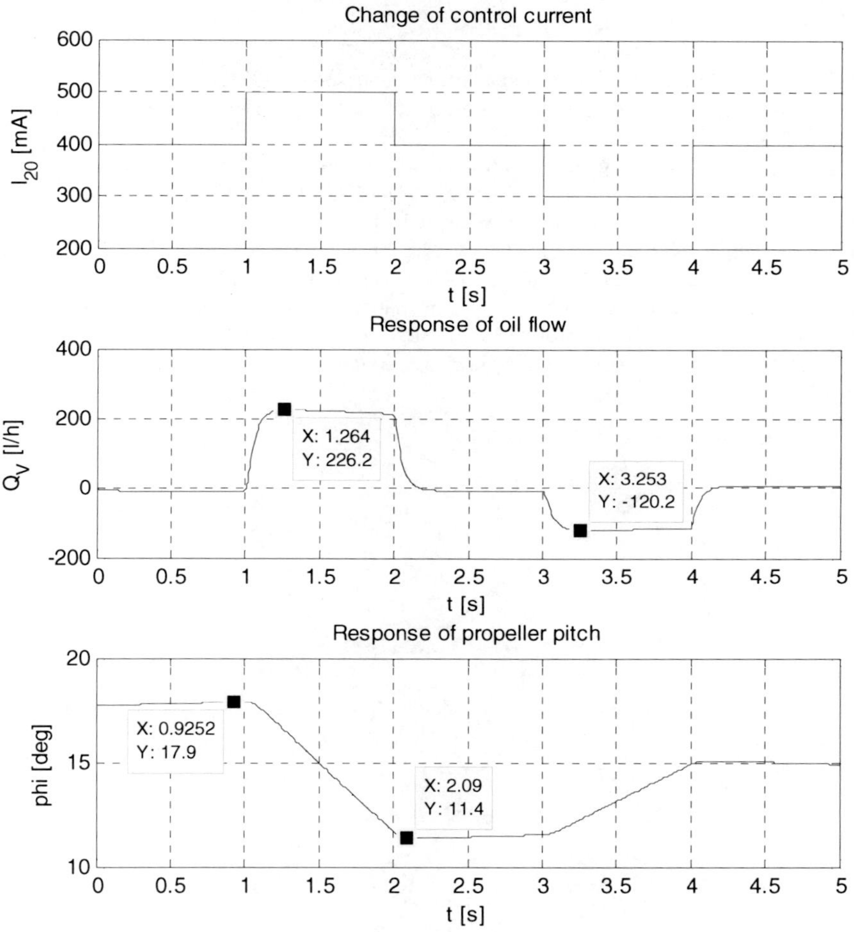

Fig. 7. Control system response of control current - *I20* step change

5 Conclusion

The model of the propeller control system created in Matlab-Simulink gives an accurate result of system behaviour. The characteristics shown in Fig. 6 were compared to measurement data from a real unit with maximal error ±10%. The model enables real-time testing on the dSPACE platform. The two approaches of the algebraic loop calculation are used and compared. The implementation of C code by the "S-function" block reduces computational demands to 60% and gives better results for initial values.

References

[1] Delp, F.: Aircraft Propellers and Controls, Jeppeson (1979) ISBN 0-89100-097-6
[2] Torenbeek, E.: Synthesis of Subsonic Airplane Design, Delft University Press (1976) ISBN 90-298-2505-7
[3] Grepl, R.: Modelling and Control of Electromechanical Servo System with High Nonlinearity, pp. s.34–s.48. In-Tech, Croatia (2009) ISBN: 978-953-7619-46- 6.
[4] Farokhi, S.: Aircraft Propulsion. Wiley, Chichester (2008) ISBN 978-0-470-03906-9
[5] D'Epagnier, K.P.: AUV Propellers: Optimal Design and Improving Existing Propellers for Greater Efficiency. In: OCEANS 2006 (2006) ISBN: 1-4244-0114-3
[6] Netland, Ø., Skavhaug, A.: Adaption of MathWorks Real-Time Workshop for an Unsupported Embedded Platform (2010) ISBN: 978-1-4244-7901-6
[7] http://www.propellergovernor.com/how-it-works/

Control of Oscillations and Instabilities in a Single Block Rocking on a Cylinder

Ardeshir Guran

Institute of Structronics, 275 Slater Street, Ottawa, K1P-5H9 Canada
Ardeshir.guran@mail.mcgill.ca

Abstract. In this paper, Nonlinear Controllers have been designed to control undesired oscillations and instabilities in a single rigid block rocking on a cylinder. Fractional controller and active controller are applied to control this system. The simulations are implemented by using SIMULINK based on frequency domain approximation. The results show that fractional controller is beneficial in certain range of bifurcation parameter while Active controller works for entire range.

Keywords: Active Control, Fractional Controller, Nonlinear Dynamics, Rigid rocking block, Unstable symmetric bifurcation.

1 Introduction

Stability of rigid blocks has been studied extensively by many authors including Psycharis [1], Guran and Rimrott [2-5], Hogan [6], plaut [7], Augusti and sinopoli [8], spanos [9], Yim et. al. [10], Housener [11], Feilder et. al. [12], virgin et. al. [13], and Kovaleva [14]. It has long history, primary associated with earthquake engineering in consideration of seismic vulnerability of structures. A similar model can be used in study of motion of structures under wind or wave loading. It also serves as a paradigm to study the rich nonlinear behavior of mechanical systems. Although the nonlinear behavior of rocking blocks is well studied, only in recent years few authors dealt with control of rigid blocks, e.g. Menini and Tornambe [15], Kovaleva [16] and Lenci and Rega [17]. In [17] a method for control of nonlinear dynamics by Lenci and Rega roughly it consist first in detecting a homo/hetroclinic bifurcation responsible for nonlinear dynamical response of the investigated system, if any, and then shifting it in parameter space.

In the recent years, emergence of effective methods in differentiation and integration of non-integer order equations makes fractional-order systems more and more attractive for the systems control community. The fractional PID controller [18], the TID controller [19], the CRONE controllers [20–22] and the fractional lead-lag compensator [23, 24] are some of the well-known fractional-order controllers. In [25], the authors proposed a simple fractional-order controller to control unstable systems. This controller can change the chaotic motion to the regular

oscillations or even eliminate them. In this paper, we proposed this Controller to control unwanted oscillations and instabilities of rigid block system.
The present paper is organized as follows:

In the next section, we explain the Physical model and derive the bifurcation diagram of the system. In section 3, a fractional control method is presented to control undesired motion of the rigid block. In section 4, an Active Controller is applied on this system. Finally, the paper will be closed by some concluding remarks in section 5.

2 Physical Model of the System

Figure 1, shows the configuration of a block rolling on the cylindrical surface. In general, the motion of a rigid body in a plane has three degrees of freedom. Considering the fact that there is no sliding this imposes two non-holonomic constraints so the whole dynamics of the system can be expressed only by one generalized coordinate, namely θ.

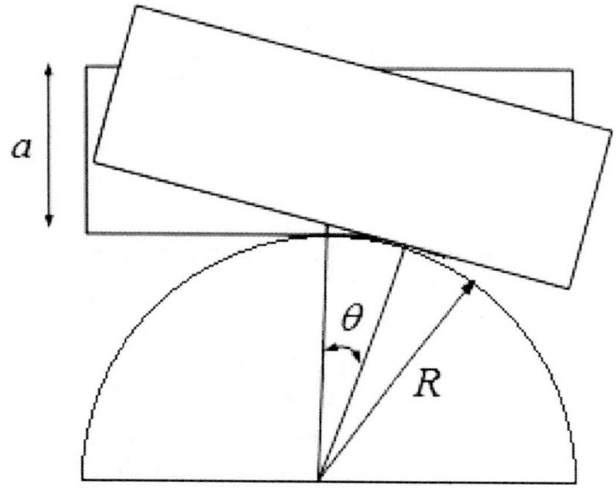

Fig. 1. configuration of the open system

The potential energy of the block is:

$$V = mg[(\frac{a}{2}+R)\cos\theta+R\theta\sin\theta] \tag{1}$$

Using Lagrange equation, this leads to [2]:

$$\ddot{\theta}-\lambda\sin\theta+\theta\cos\theta=0 \tag{2}$$

where, $\lambda=\dfrac{a}{2R}$ is the bifurcation parameter of the system.

Note that the full nonlinear Dynamics of the rocking block system including the effects of imperfections is treated in ref. [3].

From equation 2, the equilibrium paths of the block are given by:

$$\lambda = \frac{\theta}{\tan\theta} \tag{3}$$

Figure 2 shows the bifurcation diagram of the system, the stability or instability of equilibrium paths are determined by examining the second variation of the potential energy of the block. The bifurcation diagram is unstable symmetric. For values of $\lambda < 1$ there is one center and two saddle points for $\lambda \geq 1$ there is only one saddle point.

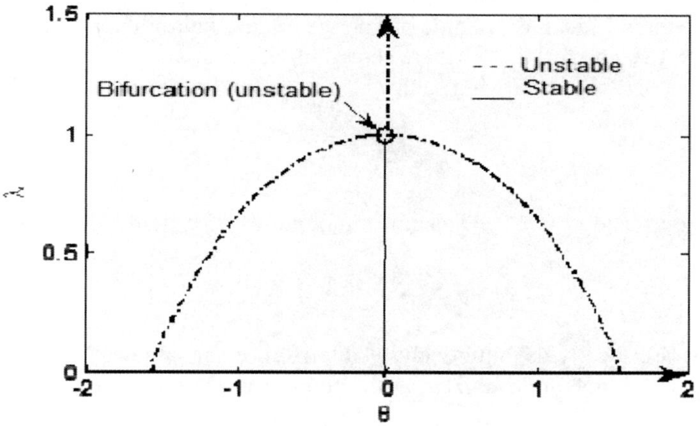

Fig. 2. Unstable symmetric bifurcation diagram of the open system

3 Fractional Controller

3.1 Stability Analysis of Fractional Order Systems

A fractional order linear time invariant (LTIFO) system can be represented by the following state-space format:

$$\begin{cases} D^{\alpha}x = Ax + Bu \\ y = Cx \end{cases} \tag{4}$$

where, $x \in R^{n}$, $u \in R^{r}$ and $y \in R^{p}$ denote states, input and output vectors of the system. The appropriate coefficients will be shown by $A \in R^{n \times n}$, $B \in R^{n \times r}$ and $C \in R^{p \times n}$ respectively whilst α is the fractional commensurate order. Fractional order differential equations are at least as stable as their integer orders counterparts. This is because; systems with memory are typically more stable than their memory-less alternatives [26]. It has been shown that an autonomous dynamic $D^{\alpha}x = Ax$, $x(0) = x_0$ is asymptotically stable if the following condition is met [27]:

$$|\arg(eig(A))| > \alpha\pi / 2 \tag{5}$$

where, $0 < \alpha < 1$ and $eig(A)$ represents the eigenvalues of matrix A. This means each component of states decays towards 0, like $t^{-\alpha}$.

In addition, this system is stable if $|\arg(eig(A))| \geq \alpha\pi / 2$. Those critical eigenvalues which satisfy $|\arg(eig(A))| = \alpha\pi / 2$ have geometric multiplicity of 1. The equality holds when the dynamic has not geometric multiplicity of 1.

The stability region for $0 < \alpha < 1$ are shown in Figure 3.

Now, consider the following autonomous commensurate fractional order system:

$$D^{\alpha}x = f(x) \tag{6}$$

where, $0 < \alpha < 1$ and $x_2 \in R^n$. The equilibrium points of system (3) are calculated when:

$$f(x) = 0 \tag{7}$$

These points are locally asymptotically stable if all eigenvalues of Jacobian matrix $J = \partial f / \partial x$, which are evaluated at the equilibrium points, satisfy the following condition [25, 26]:

$$|\arg(eig(J))| > \alpha\pi / 2 \tag{8}$$

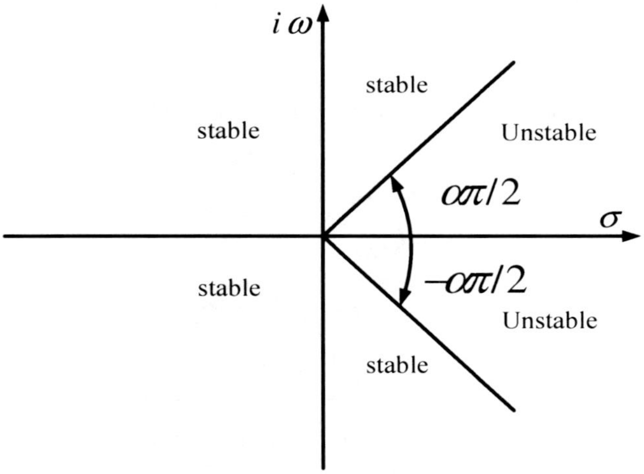

Fig. 3. Stability region of the FO-LTI system with fractional order, $0 < \alpha < 1$

3.2 Basic Idea of Fractional Controller

A controller is often used to stabilize the system or improve the performance indices. A classic stabilizing treatment may be found by a pole placement approach. This may be achieved when a state feedback controller assuming availability of the state is applied. An alternative fractional controller [24] , expands the effective range of stability according to equation (8). In this case, there is no need to locate the poles in other places explicitly. A normal integer type controller will be substituted with a fractional derivative. In Figure 4 a schematic diagram of the proposed closed loop fractional controller is shown.

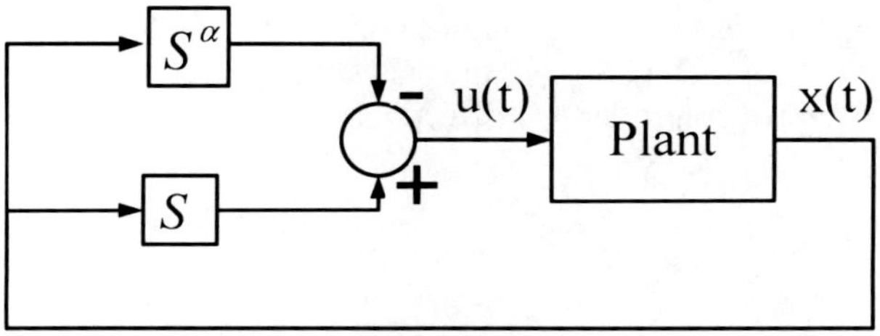

Fig. 4. schematic diagram of closed loop fractional controller

Consider a chaotic system can be represented by the following dynamic:

$$\dot{x} = f(x) + u \tag{9}$$

A fractional controller alters the dynamics in equation (9) into the following autonomous dynamics:

$$D^{\alpha} x = f(x) \tag{10}$$

In order to stabilize the overall system, a proper interval for α , should be established. If λ_i shows the ith eigenvalues of the system, an upper bound for α can be written as [25-26]:

$$\tan(\alpha\pi/2) < \left|\frac{\mathrm{Im}(\lambda_i)}{\mathrm{Re}(\lambda_i)}\right| \Rightarrow \alpha < \frac{2}{\pi}\tan^{-1}\left|\frac{\mathrm{Im}(\lambda_i)}{\mathrm{Re}(\lambda_i)}\right| \tag{11}$$

In the following, this controller implement on a rigid block system.

Let us consider rigid block system as follows:

$$\begin{cases} \dot{x}_1 = x_2 \\ \dot{x}_2 = \lambda \sin(x_1) - x_1 \cos(x_1) \end{cases} \tag{12}$$

In order to investigate the stability of the system, Jacobian matrix and corresponding Eigenvalues have to be found. The appropriate Jacobian matrix is as follows:

$$J = \begin{bmatrix} 0 & 1 \\ (\lambda - 1)\cos x_{1e} - x_{1e} \sin x_{1e} & 0 \end{bmatrix}$$

The corresponding Eigenvalues at each equilibrium points are evaluated as:

$(+ia, -ia), (+b, -b)$

where, a and b evaluate due to variation of λ.

It is seen that second set of eigenvalues show instability of the system and therefore instability of the corresponding equilibrium point i.e. the origin. Therefore, according to equation (11), $\alpha < 1$ must satisfy to maintain the stability of the system. A fractional controller for $\alpha = 0.8$ together with initial states equal to $x_1(0) = 0.05$ rad, and $x_2(0) = 0$ rad, is chosen.

This controller is applied on rigid block system. Numerical simulations have carried out using the SIMULINK based on the frequency domain approximation [26]. Corresponding results are shown in Figure 5 for $\lambda < 1$. The control action is triggered at t=2 s. As it can be seen, the states are got settled and stabilized soon after and the designed controller was effectively able to Control this system.

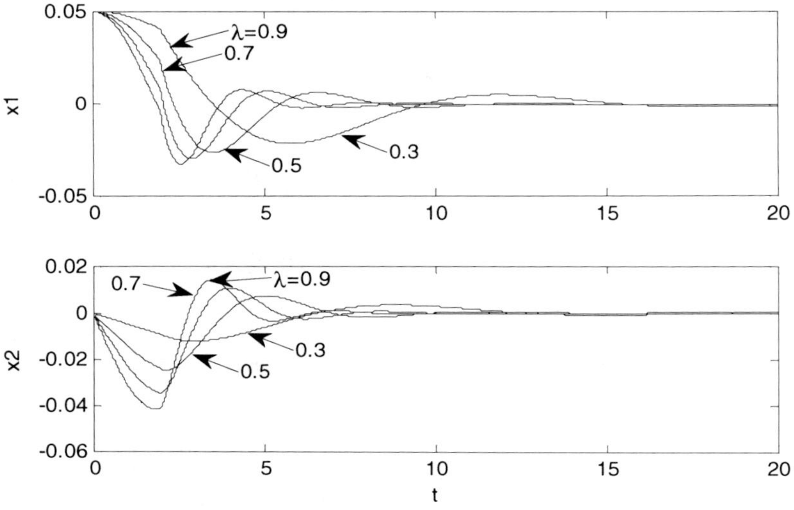

Fig. 5. simulation results for $\lambda < 1$

Now, we apply this controller for $\lambda > 1$. As shown in Figure 6, the states converge to the equilibrium points, i.e. x1e = 4.17 and x1e = 4.17 that is not feasible. To overcome this problem we apply an Active control in the next section.

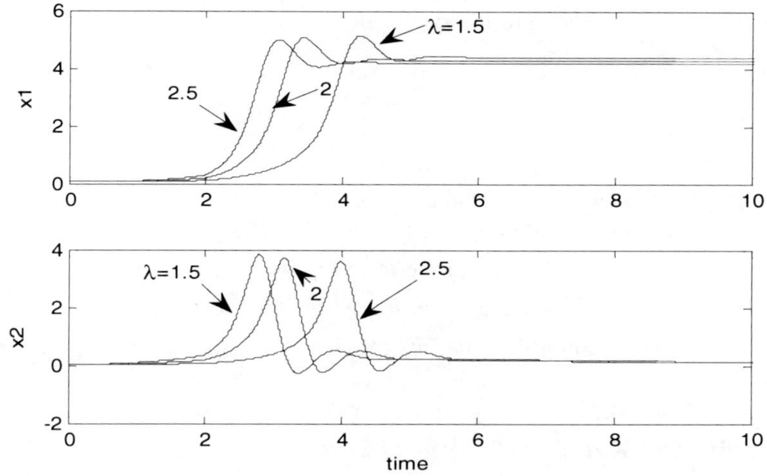

Fig. 6. simulation results for $\lambda > 1$

4 Active Control

In the active control method, the control signals are directly added to the system dynamics:

$$\dot{x} = f\ (x)\ + u\ (t). \tag{13}$$

The control signals u constructed from two parts. The first part is considered to eliminate the nonlinear part of the (13). The second part V (t) acts as external input in (13) and is design to stabilize the system

$$u(t)\ =\ -f\ (x)\ + v\ (t). \tag{14}$$

where,

$$v\ (t)\ = -Ax \tag{15}$$

Design of the controller in this method is to determine the matrix A in such a way that the dynamics in (13) become stable. In the following, relevant matrices and relations are given for Control of rigid block systems.

Let us reconsider equation 11.Now, from equation (14), the control input is determined as:

$$u(t) = \begin{bmatrix} -x_2 \\ -\lambda \sin(x_1) + x_1 \cos(x_1) \end{bmatrix} - AX \tag{16}$$

Choosing, $A = \begin{bmatrix} 1 & 0 \\ 0 & 1 \end{bmatrix}$, controlled system determines as:

$$\dot{X} = \begin{bmatrix} -1 & 0 \\ 0 & -1 \end{bmatrix} X \tag{17}$$

Simulation results in Figure (7) indicate significance of this method. Controlled states have been shown, in this figure. It should be noted that control of $u(t)$ has been activated in $t = 2s$. As we can see from Figure (7) for bifurcation parameter $\lambda < 1$ the block start to oscillate while for $\lambda \geq 1$ the block starts to topple over. At time $t = 2s$ the controller returns the block to its original opiate position $\theta = 0.0$.

5 Concluding Remarks

In summery the following results are obtained: 1-The nonlinear behavior of a rocking block rolling on the surface of cylinder is studied. It is shown that the bifurcation is of unstable symmetric type. 2-The Fractional controller works satisfactorily for $\lambda < 1$ it may leads to an equilibrium point but for $\lambda > 1$ it may lead to an equilibrium point which is not feasible.

3-For the whole range of the λ, Active Control is designed to suppress unwanted oscillations and to prevent toppling of the block.

4-Other control techniques can be applied to this fundamental problem. Rocking block system can represent a benchmark nonlinear design problem to study classical as well as modern control.

Acknowledgement

This work was completed during my stay as an Invited Professor in Universitaet der Bundeswehr Muenchen, Fakultaet fuer Luft und Raumfahrttechnik. I am indebted to Professor Joachim Gwinner (Director, Institut fuer Mathematik und Rechneranwendung) for his hospitality, stimulating suggestions, and encouragement.

References

[1] Psycharis, I.N.: Dynamic Behaviour of Rocking Two-Block Assemblies. Earthquake Engineering and Structural Dynamics 19(4), 555–575 (1990)

[2] Guran, A., Rimrott, F.P.J.: Stability of a Rocking Block on a Cylinder including the effect of Initial Imperfections of Arbitrary Magnitude. Archiv of Applied Mechanics 59(15), 390–399 (1989)

[3] Guran, A., Rimrott, F.P.J.: Dynamic Instability and Imperfection Sensitivity of a Block Rolling on a Cylinder. Archiv of Applied Mechanics 60(8), 500–506 (1990)

[4] Guran, A., Rimrott, F.P.J.: Some Remarks Concerning the Determination of Stability Boundaries for a System of Rocking Blocks. Journal of Applied Mathematics and Mechanics (Zeitschrift fur Angewandte Mathematik und Mechanik) 72(2), 141–142 (1992)
[5] Guran, A.: Static Stability of a Triple Rocking Block System. Journal of Applied Mathematics and Mechanics (Zeitschrift fur Angewandte Mathematik und Mechanik) 74(7), 284–286 (1994)
[6] Hogan, S.J.: The Many Steady State Responses of a Rigid Block under Harmonic Forcing. Proc. Roy. Soc. Lond. A. 425, 441–476 (1989)
[7] Plaut, R.H.: Rocking Instability of a Pulled Suitcase with two Wheels. Acta Mechanica 117, 165–179 (1996)
[8] Augusti, G., Sinopoli, A.: Modelling the dynamics of large block structures. Meccanica 27(3), 195–211 (1992)
[9] Spanos, P.D., Koh, A.S.: Rocking of rigid blocks due to harmonic shaking. J. Eng. Mech. ASC 110, 1627–1642 (1984)
[10] Yim, C.S., Chopra, A.K., Penzien, J.: Rocking response of rigid blocks to earthquakes. Earthqu. Eng. Struct. Dyn. 8, 565–587 (1980)
[11] Housner, G.W.: The Behaviour of Inverted Pendulum Structures During Earthquakes. Bull Seism Soc. Am. 53, 403–417 (1963)
[12] Fielder, W.T., Virgin, L.N., Plaut, R.H.: Experiments and simulation of overturning of an asymmetric rocking block on an oscillating foundation. Eur. J. Mech. A/Solids 16, 905–923 (1997)
[13] Virgin, L.N., Fielder, W.T., Plaut, R.H.: 'Transient motion and overturning of a rocking block on a seesawing foundation. Journal of Sound and Vibration 191, 177–187 (1996)
[14] Kovaleva, A.: Random rocking dynamics of a multidimensional structure. In: Ibrahim, R.A., Babitsky, V.I., Okuma, M. (eds.) Vibro-Impact Dynamics of Ocean Systems and Related Problems. LNACM, vol. 44, pp. 149–160. Springer, Heidelberg (2009)
[15] Menini, I., Tornambe, A.: Tracking of Admissible Trajectories for a Rocking Block. In: Guran, A. (ed.) Proceeding of the Second International Symposium on Impact and Friction of Solids and Structures, Montreal (1999)
[16] Kovaleva, A.: Control of Random Dynamics of a Rigid Rocking Block. International Congerss on Theoretical and Applied Mechanics (2004)
[17] Lenci, S., Rega, G.: Optimal Control and Anticontrol of the Nonlinear Dynamics of a Rigid Block. Philosophical Transaction of the Royal Society 364(1846), 2353–2381 (2006)
[18] Podlubny, I.: Fractional-Order Systems and PIλDμ -Controllers. IEEE Trans. on Autom. 44(1), 208–213 (1999)
[19] Lune, B.J.: Three-parameter tunable tilt-integral derivative (TID) controller, US Patent US5371 670 (1994)
[20] Oustaloup, A., Sabatier, J., Lanusse, P.: From fractal robustness to CRONE control. Fract. Calculus Appl. Anal. 2(1), 1–30 (1999)
[21] Oustaloup, A., Moreau, X., Nouillant, M.: The CRONE suspension. Control Eng. Practice 4(8), 1101–1108 (1996)
[22] Oustaloup, A., Mathieu, B., Lanusse, P.: The CRONE control of resonant plants: to a flexible transmission. Eur. J. Control 1(2), 113–121 (1995)
[23] Raynaud, H.F., Zergaïnoh, A.: State-space representation for fractional order controllers. Automatica 36, 1017–1021 (2000)

[24] Monje, C.A., Feliu, V.: The fractional-order lead compensator. In: IEEE International Conference on Computational Cybernetics, Vienna, Austria, August 30-September 1 (2004)

[25] Tavazoei, M.S., Haeri, M.: Chaos control via a simple fractional-order controller. Physics Letters A 372(6), 798–807 (2008)

[26] Ahmed, E., El-Sayed, A.M.A., El-Saka, H.A.A.: On the fractional-order logistic equation. J. Math. Anal. Appl. 325(1), 542 (2007)

[27] Matignon, D.: Stability result on fractional differential equations with applications to control processing. In: IMACS-SMC Proceedings, Lille, France, pp. 963–968 (1996)

Modeling and Control of Three-Phase Rectifier of Swirl Turbine

T. Brezina[1,*], T. Hejc[1], J. Kovar[1], and R. Huzlik[2]

[1] Faculty of Mechanical Engineering Brno University of Technology, Technicka 2896/2, 616 69 Brno, Czech Republic
[2] Faculty of Electrical engineering and Communication Brno University of Technology, Technicka 3058/10, 616 00 Brno, Czech Republic

Abstract. The paper deals with a three-phase rectifier controller design for monobloc configuration of a small Swirl turbine. The aim is phasing of the system to electrical power network. A cascade controller was used for the control: lower loop provides a current control and upper loop a voltage control. Symmetric optimum method and robust loop shaping control synthesis are employed for the controller design. Results are verified and compared by simulations applying original three-phase rectifier model taking into account high level of disturbances. Resulting system meets with reserve desired behavior.

1 Introduction

Swirl turbine represents a new type of a water turbine and it is determined especially for very low inclines (less than 3 meters) and high speed flows [1]. Construction of this water turbine is based on a monobloc configuration of the turbine, permanent magnet generator, active three-phase rectifier and three-phase inverter, shown in Fig.1.

The most common switching technique in switch-mode voltage source converters (VSC) is Pulse Width Modulation (PWM). The PWM control can not only manage the active power, but reactive power also, allowing this type of rectifier to correct power factor [2]. The mathematical model and control of three-phase rectifier was presented in several ways [3, 4, 5]. The main approach to design of the controller is modeling and simulation [7, 8, 9 and 10].

This paper focuses on voltage source controlled by PI controllers and alternatively robust controllers (RC) based on [6, 11] to regulate the voltage and the current of a three-phase PWM AC-DC converter. The proposed control strategy of a three-phase active rectifier is based on the load voltage and AC input current conversion into d-q frame.

* Corresponding author.

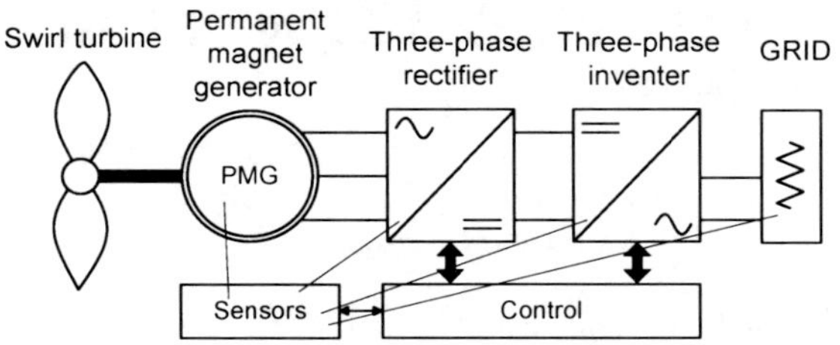

Fig. 1. Monobloc configuration of the Swirl turbine

2 The Mathematical Model

A three-phase PWM active rectifier is represented as a block diagram, shown in Fig. 2. The mathematical model for the three-phase rectifier without the neutral connection and neglecting the resistance of the power switches was derived in [3] as

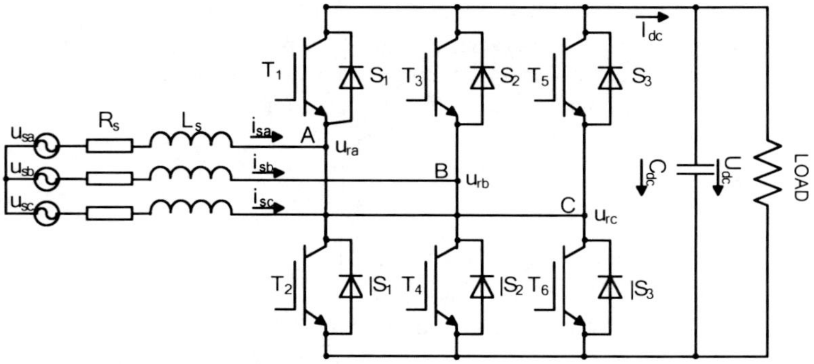

Fig. 2. Main circuit of three-phase rectifier of the Swirl turbine

$$C_{dc}\frac{dU_{dc}}{dt}+i_{dc}=\sum_{j=1}^{3}i_{sj}S_j \tag{1}$$

$$L_s\frac{di_{sj}}{dt}=u_{sj}-Ri_{sj}-u_{dc}\left(S_j-\frac{1}{3}\sum_{n=1}^{3}S_n\right) \tag{2}$$

$$\sum_{k=3}^{3}u_{sj}=\sum_{k=1}^{3}i_{sj}=0 \tag{3}$$

where is

j	index of the three-phases (a, b, c);
C_{dc}	capacitance of the DC bus;
R_s	equivalent resistance of the loop;

L_s	equivalent inductance of the loop;
U_{dc}	bus voltage;
I_{dc}	bus current;
S_j	switching functions;
i_{sj}	input AC grid current;
u_{sj}	input AC grid voltage;

It can be made a block diagram of the three-phase rectifier dynamics based on equations (1), (2) and (3), shown in Fig. 3 [4].

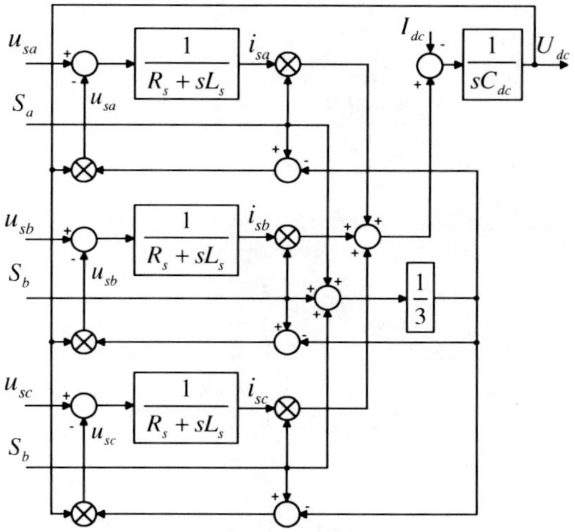

Fig. 3. Block diagram of VSC PWM rectifier in natural three-phase coordinates

3 Control of the Rectifier

The block diagram for the cascade control of the three-phase rectifier [12] is shown in Fig. 4. The lower loop is used for the current control and the upper loop for the voltage control.

AC side current is transformed to d-q frame through Clarke and Park transformation. The reference angle θ is generated by Phase Lock Loop (PLL) and synchronizes phase shift of AC-DC side. The reference voltage U_{ref} for the controller generates signal for the lower current control loop. Command $i_{sd\ ref}$ is set to zero for unity power factor. The output signals u_{sd}, u_{sq} coming from the current controller are transformed into a, b, c signal and entered to the PWM generator. Two ways of design and control of this circuit are compared in this paper.

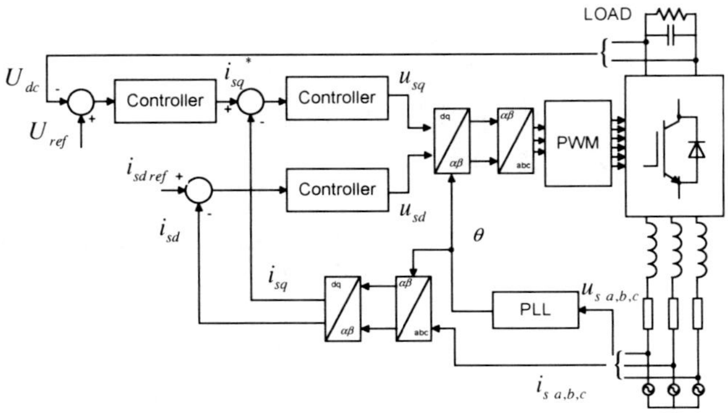

Fig. 4. Block diagram of the control of VSC PWM rectifier

3.1 PI Control of Circuit

The first way employs the lower current control loop which consists of a sampler, PI controller, PWM and coil transfers, shown in Fig. 5.

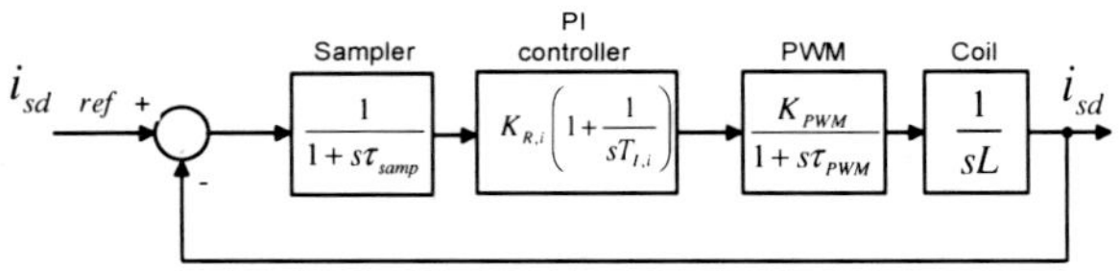

Fig. 5. Diagram of d axis current control loop

Symmetric optimum method (SO) is used for the current controller design with standard third order open loop transfer (4).

$$F_o(s) = \frac{1+4s\tau_\sigma}{8\tau_\sigma^2 s^2 (s\tau_\sigma + 1)} \tag{4}$$

The transfer of the controlled system is given by (5), where τ_{PWM}, resp. τ_{samp} are time constants of PWM, resp. sampler and K_{PWM} is a gain of PWM. Based on equations (4), (5) the transfer of controller can be then obtained from the equation (6) for $\tau_\sigma = \tau_{\sigma,i} = \tau_{TWM} + \tau_{samp}$.

$$F_{o,i}(s) = \frac{K_{PWM}}{1+s\tau_{PWM}} \cdot \frac{1}{s\tau_{samp}+1} \cdot \frac{1}{sL} \tag{5}$$

$$F_{R,i}(s) = F_o \cdot \frac{1}{F_{o,i}} \tag{6}$$

The upper control loop of the voltage is shown in Fig. 6. It contains sampler, PI controller, lower current loop and capacitor transfers.

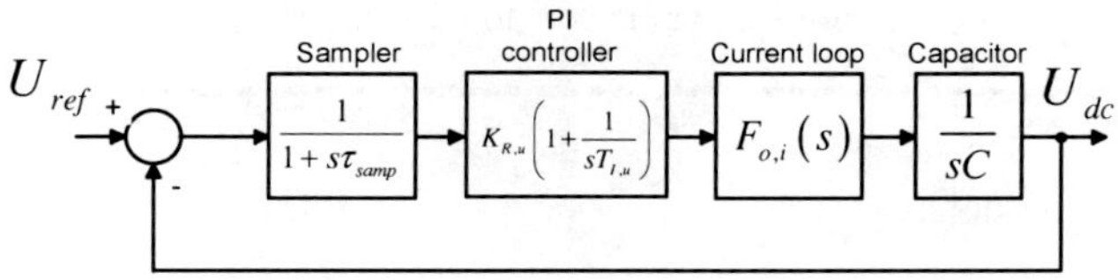

Fig. 6. Block diagram of dc bus voltage control loop

Spare transfer of the lower current loop is described as (7). Now the transfer function (8) of the system to be controlled can be derived from Fig. 6. The voltage controller is then simply given by (9) with $\tau_\sigma = 2\tau_{\sigma,i} + \tau_{samp/2} + \tau_{samp/2}$.

$$F_I(s) = \frac{1}{1+4\tau_\sigma s} \tag{7}$$

$$F_{o,u}(s) = F_I \cdot \frac{1}{s\tau_{samp}+1} \cdot \frac{1}{sC} \tag{8}$$

$$F_{R,u}(s) = F_o \cdot \frac{1}{F_{o,u}} \tag{9}$$

3.2 Robust Control of the Circuit

The cascade control concept was used again for this approach. The controllers design exploits transfer functions (5), (8) for optimal H-infinity loop-shaping controller synthesis. Design of both controllers was realized through the simplest possible desired loop shape $F_{d,i}$ = 10000/s. Similarly the voltage controller was design through the $F_{d,u}$ = 15/s. Obtained current resp. voltage controllers were of the 7th resp. 14th order. It was possible to reduce them to the 6th resp. 3rd order controllers with keeping the dynamics of the control practically without changes.

4 Results and Conclusions

The structure of the main circuit is shown in Fig. 4. The main parameters are then in Table 1. Two methods for controllers design of three-phase rectifier are compared in the paper. Simulation results of the first of them (SO method), presented by Fig. 7., show good system dynamics performance – overshoot of U_{dc} for used PI controllers is about 110V with 0.05s settling time. The controllers designed by the second mentioned method (H-infinity loop-shaping controller synthesis) lead on better behavior which is demonstrated in Fig. 8 – with overshoot of U_{dc} of about 70V with 0.02s settling time. Better is also the process of AC current side. The simulations were done using Matlab/Simulink (and employing the function loopsyn – Robust Control Toolbox for H-infinity loop-shaping controller synthesis).

The second presented solution seems to be sufficient for the stage Swirl turbine construction.

Table 1. – Parameters for simulations

Source voltage	80V/50Hz (phase-to-ground)
Switching frequency	5kHz
DC voltage	230V
Capacitance of the dc bus	1mF
Inductance of the line reactor	1.5mH

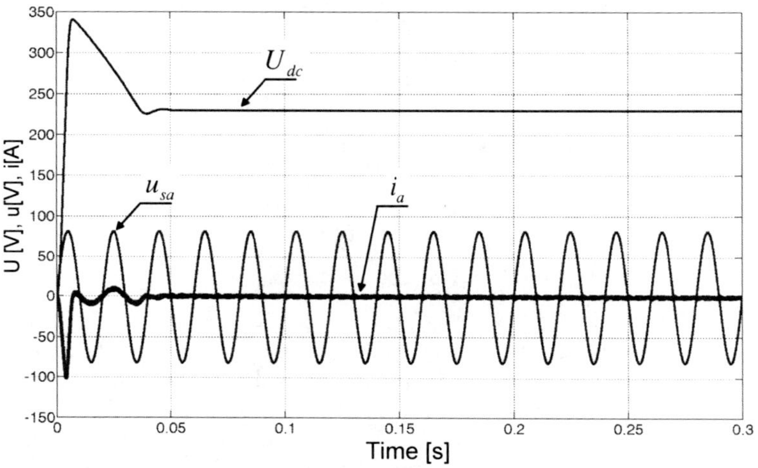

Fig. 7. Controlled by (SO) method of voltage U_{dc} to load current from AC side

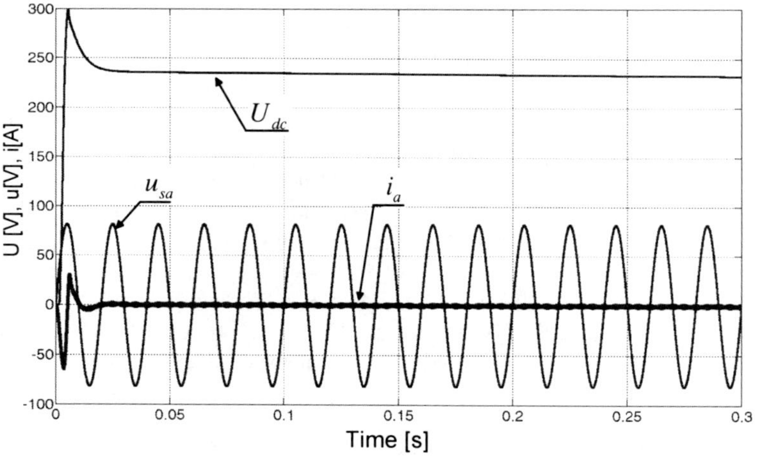

Fig. 8. Controlled by (RC) method of voltage U_{dc} to load current from AC side

Acknowledgment

This work is supported from research plan MSM 0021630518 Simulation modeling of mechatronic systems, FSI-J-11-39 and FSI-S-11-23.

References

[1] Haluza, M., Rudolf, P., Pochyly, F., Sob, F.: Design of a new Low-Head Turbine. In: Proceedings of the XXI-st IAHR Symposium Hydraulic Machinery and Systems, vol. 1, pp. 29–34 (September 2002)

[2] Ooi, B.T., Salmon, J.C., Dixon, J.W., Kulkarni, A.B.: A three-phase controlled-current PWM converter with leading power factor. IEEE Trans. Ind. Applicat., 78–84 (January 1987)

[3] Ye, Y., Kazerani, M., Quintana, V.H.: A novel modeling and control method for three-phase PWM converters. In: IEEE Power Electronics Specialist Conference, pp. 102–106 (June 2001)

[4] Blasko, V., Kaura, V.: A new mathematical model and control of a three-phase AC-DC voltage source converter. IEEE Trans. Ind. Eletron., 116–123 (January 1997)

[5] Wu, R., Shashi, B.D., Gordon, R.S.: Analysis of a PWM ac to dc voltage source converter under the predicted current control with a fixed switching frequency. IEEE Trans. Ind. Applicat., 756–764 (1991)

[6] Hejc, T., Brezina, T., Huzlik, R.: Modeling and control of three-phase inverter's RMS of swirl turbine by using system identification. In: Annals of DAAAM, Zadar, pp. 1361–1362 (2010)

[7] Hadas, Z., Singule, V., Ondrusek, C., Kluge, M.: Simulation of vibration power generator. In: Recent Advances in Mechatronics, Warsaw, pp. 350–354 (2007)

[8] Maga, D., Hartansky, R., Manas, P.: The Examples of Numerical Solutions in the Field of Military Technology. In: Advances in Military Technology, vol. 1(2), pp. 53–66 (2006) ISSN 1802-2308

[9] Fabo, P., Pavlikova, S.: Gsim - Software for Simulation in Electronics. In: Proceedings of EPE-PEMC 2010 - 14th International Power Electronics and Motion Control Conference, pp. S414–S417 (2010) ISBN 978-142447854-5

[10] Fabo, P., Siroka, A., Maga, D., Siroky, P.: Interactive Framework for Modeling and Analysis. Przeglad Elektrotechniczny 83(11), 79–80 (2097) ISSN 0033-2097

[11] Minárik, V.: Realization of three-phase active rectifier using dsp56f803. In: Proceedings of the 14th Conference Student EEICT 2008, Brno, pp. 187–191 (2008)

[12] Maga, D., Sitar, J., Bauer, P.: Automatic Control, Design and Results of Distance Power Electric Laboratories. In: Recent Advances in Mechatronics 2008-2009, pp. 281–286. Springer, Heidelberg (2009) ISBN 978-3-642-05021-3

Identification of Deterministic Chaos in Dynamical Systems

M. Houfek, P. Sveda, and C. Kratochvíl

Faculty of Mechanical Engineering Brno University of Technology,
Technická 2896/2, Brno 616 69, Czech Republic

Abstract. The existence of chaotic systems in control systems is a serious problem of engineering practice. Unfortunately, many of identification practices, permitting their identification is very difficult. The article is an example of some simple practices and their contexts, permitting the emergence of chaos to predict.

1 Introduction

In the field of mechatronic drive systems, aimed at examining the physical processes in these ongoing, is examining the current problem of non-standard operating conditions in the area of macro-and microscale. One of the main challenges is the ability to obtain predictions about future behavior and the development of dynamic behavior. Unfortunately, the systems operating in a changing environment may be located in areas of chaos, as is known, prevents reliable prediction of their behavior. We must not forget that chaos may be harmful in all circumstances, may be a symptom of a new, stronger code (steady state at a higher level).

2 Manifestations of Deterministic Chaos in His View

The chaotic state of a dynamic system can be seen on display response time mileage, when you view in the phase plane using the Poincare views on the amplitude of resonance characteristics. On this view, you can find typical features:

- Chaotic systems generate typical aperiodic oscillations, characterized by "rapid" growth in early exponential type (unstable phase). Followed by "rapid" growth arrest caused by non-linearity and boundedness energy (stabilization phase). This process is repeated intermittently.
- For chaotic systems are characterized by irregular alternation of states of instability and dissipative. These irregularities are the essence of the transition generator of chaos.
- Poincare images show a typical structured infinite value set, consisting of clusters of points and relative terms (smooth transitions)

- When you are in the phase plane to meet with the so-glorious attractors (strange attractors), representing a globally bounded but locally unstable display non-periodic oscillations
- Frequency spectrum of an isolated set of distinct frequencies, among which are "dense" areas of local noise.

3 Identification of Chaos

The analytical analysis of deterministic chaos are the main tools invariant set, the study of attractors and in particular the phenomenon non-linearity and instability of dynamical systems.

We consider a class of dynamic systems described by the equation

$$\dot{\boldsymbol{x}}_{(t)} = \boldsymbol{f}_{(x,t)}, \qquad \boldsymbol{x}_t \in X \subset \boldsymbol{R}^n \tag{1}$$

Chaotic behavior of dynamical systems is determined by the occurrence attractor. For their characterization and classification with regard to the complex topological structure, we proceed in two ways.

- We can use Ljapunov´s factors that enable "geometric" classification of attractors
- We use the term "dimension" as a static level of the attractor, chaos or rate the output process
- We analyze the phase portraits

Lyapunov´s coefficients are counted quantitative measure that characterizes the local stability properties of trajectories. Lyapunov´s coefficients of the system (Equation 1) can determine if the function f (....) differentiable. Let D is a subset of the state process x N belongs to the nth. D is called the attractor prodigious, if is bounded and all trajectories starting in D are unstable. Equation 1 (...) represents a chaotic system has at least a strange attractor.

The criteria are based on chaos approximal linearization system (1) and the use and Bolov Ljapunov´s exponents [1]. The equation representing the approximate linearization of the system 1 according to a trajectory that has the form

The criteria are based on chaos approximal linearization system (1) and the use and Boelov Ljapunov´s exponents [1]. The equation representing the approximate linearization of the system 1 according to a trajectory that has the form

$$\dot{\boldsymbol{x}}_{(t)} = \boldsymbol{A}_{(t)} \cdot \boldsymbol{x}_t, \qquad \boldsymbol{x}_t \in \boldsymbol{R}^n \tag{2}$$

Suppose that the solution of this equation is given by

$$\boldsymbol{x}_t = \Phi_{(t,t_0)} \cdot x_{t_0}, \quad \forall\ t, t_0 \in \boldsymbol{R}^n \tag{3}$$

Lyapunov´s coefficients form a sequence whose members are defined by the relation [2].

$$\lambda_i = \lim_{t\to\infty} \frac{ln\|\Phi_{(t,t_0)}\cdot x_{t_0}\|}{t-t_0} \qquad i = 1,2, \dots n \tag{4}$$

Where Φ (t, t_0) is the fundamental matrix lineralizovaného system (2).

Ticks Ljapunov´s coefficients provide information about the existence of chaos and the absolute value of an estimate of the "intensity of chaos.". Lyapunov coefficients are always real and have a global character (evaluates the behavior of the system along the entire trajectory approximal). If at least one coefficient is positive, the system is chaotic. Unfortunately, numerical calculations of their values require different approximations and averaging are often quite problematic.

Fractal dimension is the second concept, which allows us to estimate the potential for chaos. Let´s is a set of n-dimensional space, then we can define the fractal dimension D_s relationship

$$D_s = \lim_{\varepsilon\to 0} \frac{\log n(\varepsilon)}{\log(\varepsilon^{-1})} \tag{5}$$

Where n(ε) number of N-dimensional cube with side length, which is needed to cover the set S. attractor, whose dimension is not an integer is quite attractor, which is an attendant feature of chaos. There is as yet not fully provable relationship between the coefficients Ljapunov´s λ_i (4) and fractal dimension D_s (5). [2].

$$D_s = j + \frac{\sum_{i=1}^{j} \lambda_i}{-\lambda_{i+1}} \tag{6}$$

Where j is the largest integer for which

$$\lambda_1 > \lambda_2 > \cdots + \lambda_j > 0$$

When the Lyapunov´s coefficients are usually arranged, $\lambda_1 > \lambda_2 \dots \lambda_n$, for example, a system of differential equations of dimension 3 (a necessary condition for the emergence of chaos) we get $\lambda_1 > 0 > \lambda_3$ and (6) that $D = 2 - \frac{\lambda_1}{\lambda_3}$

To calculate the real dimensions of D, then we can proceed by direct calculation according to equation (5) or using equation (6). For the existence of chaos is another important condition. The system of equations (1) will be chaotic if the condition: [4].

$$\sum_{i=1}^{n} \frac{\delta f_i}{\delta x_i} < 0 \tag{7}$$

The analysis of phase portraits is another way to tell if the system is close to chaotic states. According Holmise [3] is a precondition of the existence of penetration chaos stable and unstable branches separatrix. Homoklinics in so-called values (Fig. 1).

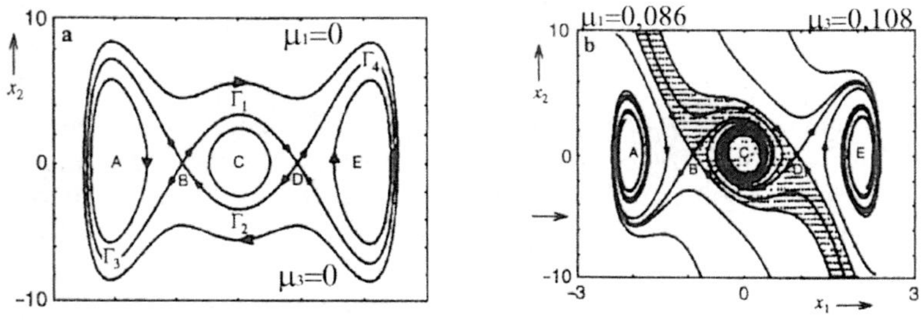

Fig. 1. samples of possible chaotic state for the parameters

Also, branches of periodic solutions immediately prior to the emergence of chaos [4].

4 Conclusion

The article gives a brief example of the criteria by which to determine whether dynamic systems are chaotic. They are aware of other approaches, such as the relationship between Feigenbaum sequence of bifurcations and intersections of the Mandelbrot set with the real axis, but they are specialized and do not have much practical significance.

Acknowledgment

Published results were acquired using the subsidisation of the Ministry of Education, Youth and Sports of the Czech Republic, research plan MSM0021630518.

References

[1] Štork, M., Kol, A.: Nelineární obvody a systémy s chaotickým chováním, slaboproudý odpor, 64 (2008)
[2] Hortel, M., Kol, A.: Analýza nelineárních časově heteronomních pohonových soustav, Praha (2010)
[3] Holmes, P.: A nonlinear oscilator with a strange attractor. Phylosofical Transactions of the Royal society, London (1997)
[4] Kratochvil, C., Heriban, P.: Dynamický systémy a chaos, Brno (2010)
[5] Nayfeh, A., Balachandran, H.: Applied nonlinear dynamics, New York (1995)

Data Fusion from Avionic Sensors Employing CANaerospace

R. Jalovecky*, P. Janu, R. Bystricky, J. Boril, P. Bojda, R. Bloudicek, M. Polasek, and J. Bajer

Dept. of Aerospace Electrical Systems, University of Defence, Kounicova 65, Brno, 662 10, Czech Republic

Abstract. The paper describes findings and results of the defence research project "Innovation of the modern aircraft sensors data processing technology and the fusion of the data in the NEC environment", whose main objective and result was an implementation of up-to-date Aircraft Electronic System (AES). The system is based on communication network CAN with CANaerospace protocol. Every module represents a compact instrument, which process a data from selected avionic sensors. Twenty individual modules were designed and implemented during the project. Ten of them fully communicate on CAN with CANaerospace in time-triggered mode.

1 Introduction

Information technology is a one of the most important factors during the establishment of Network Enabled Capability (NEC) [1] in NATO structure as well as in the professional Czech Armed Forces. The basis of the successful NEC creation and its following using are data which carry logically ordered information. In the area of aviation and avionic sensors signal processing no comprehensive requirements on data structure intended for its utilization in NEC environment have been defined, yet. However, data fusion from all aircraft on-board systems is a dominant trend in modern avionic systems.

That was the reason to begin the project, whose objective and result is manufactured and experimentally verified Aircraft Electronic System (AES), which provides a fusion of relatively large amount of information from various avionic sensors and instruments. Acquired data are continuously transmitted on the on-board communication bus CAN with CANaerospace [2] protocol, which forms a basis of the whole system.

The aim of the project was also a verification of the proposed conceptions in terms of its suitability, dimensioning, dependability, availability and usability for prospective future implementation on specific aircraft or UAS (Unmanned Aerial System).

---

* Corresponding author.

2 Modular Architecture

System architecture was conceived as a modular. The basis of the system forms CAN bus with CANaerospace protocol. All modules are connected to the CAN and continuously transmitting obtained data to it. Basic assumption is the consequent interconnection of on-board CAN to the NEC environment via RF datalink. However, this interconnection was not solved in our project.

Particular modules are designed like an intelligent sensors or on-board instruments connected to CAN. Modules of intelligent sensors are based on 8-bit microcontrollers Atmel AVR. More complex instruments, which carry out more complex signal processing, are based on FPGAs. Each module is galvanically isolated from communication bus and is designed for proper functionality under the wide range of supply voltage from 6 to 36 V DC.

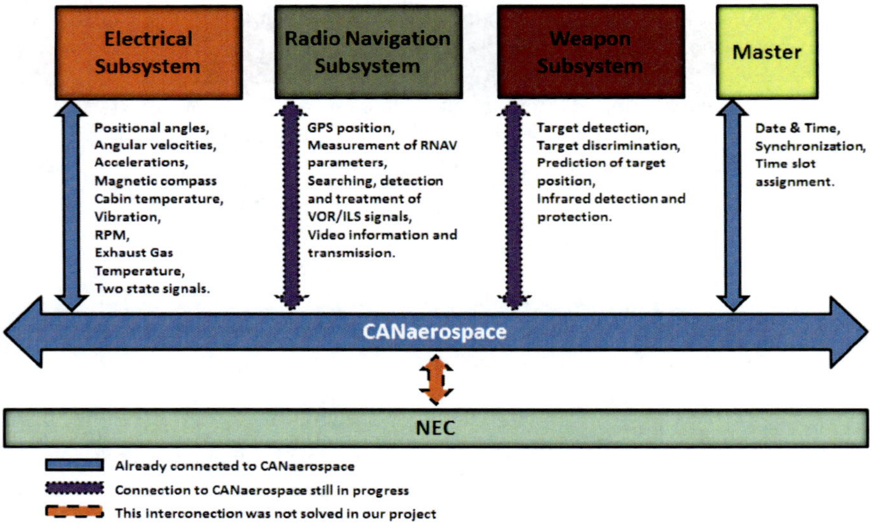

Fig. 1. AES architecture and modules implemented

According to the kind of data, which each of modules provides, are all modules divided into basic categories on electrical subsystem, weapon subsystem and radionavigation subsystem. A several representants were selected from each subsystem, which were implemented during the project and serve for experimental tests. Resulting structure of implemented system is shown in Fig. 1. Some modules have not been connected to CAN bus, yet. Their interconnection is in Fig. 1 depicted by dotted arrows.

3 Communication Timing and Control

In the avionic system it is necessary to ensure data transmissions in certain exactly defined time instants. Therefore the time-triggered method (TTCAN), which is defined by ISO standard 11898-4 [3], was chosen for communication among individual subsystems through the CAN with CANaerospace protocol [2]. TTCAN method de facto changes communication control to the TDMA (Time Division Multiple Access). While this method of the communication control is applied the transmission of individual messages is provided according to so-called Cycle matrix shown in Fig 2. The Cycle matrix contains particular messages with corresponding identifiers CAN IDs, which carry data important for the proper function of an aircraft or intended for NEC (e.g. altitude, Mach number, true airspeed, vertical airspeed, total temperature, exhaust gas temperature, oil pressure, fuel pressure, oil temperature, acceleration, angular velocity etc).

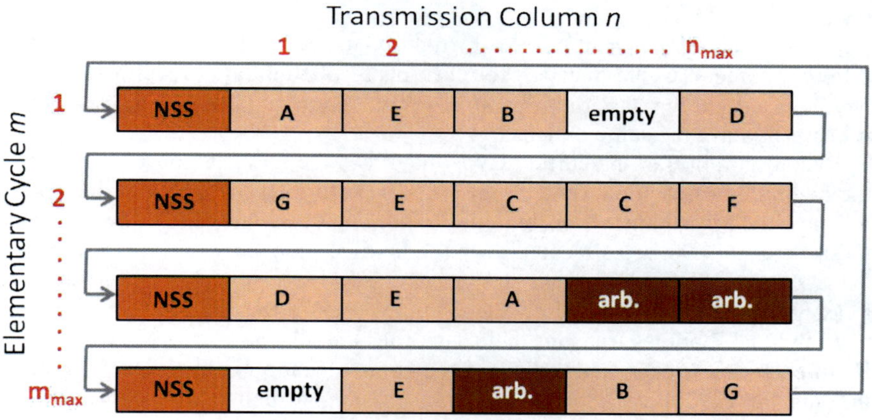

Fig. 2. Cycle matrix – transmission schedule

Each participant on the bus or better to say each unit, intelligent sensor or module has precisely defined time interval (Time-slot), in which should transmit its data. Each recipient of these data then exactly knows the moment in time when the data are expected. The network MASTER node then carries out the task of communication synchronization by transmitting of reference messages. This function is called Network Synchronization Service (NSS). In case of the MASTER block failure, synchronization messages broadcasting must be assumed by some of the other network participants, because the whole system must be equipped by such mechanism. Thus it is possible to ensure data transmission in the proper time instants by this way, but at the expense of reduced flexibility of the system.

Cycle matrix (Fig. 2) consists of elementary cycles, which always begin by the synchronization message, followed by data messages (carrying aircraft flight parameters), either cyclically repeated in every elementary cycle as e.g. the message E or individual for the entire Cycle matrix as e.g. the message G. Free time-slots may also occur here, reserved for the future modification of the Cycle matrix. There is also space allotted to so-called arbitration slot, which is used for messages of event-triggered communication control. The entire schedule of message transmissions is usually predefined and remains fixed during the full system operation.

4 Dynamic Time-Slot Assignment

Aircraft flight always consists of several phases. There are not needed or desirable to keep always the same transmission schedule during all phases of flight. Some data are specific and necessary for certain phase of flight, whereas for the other phase of flight are totally unnecessary, unreasonable or even the system is not able to provide this data at all. For this reason and also in order to increase the flexibility of the entire system a novel communication control method, so-called Dynamic time-slot assignment [4], was proposed during the solving the project. In proposed method is each phase of flight matched with the Cycle matrix of different content. Cycle matrixes for each phase of flight are predefined and may be modified during the full operation of the system. These matrixes do not necessarily have the same dimension.

Reconfiguration of the current Cycle matrix to the new one must be ensured at the beginning of each phase of flight. This is performed within the framework of the so-called assignment or reconfiguration process, in which the particular message CAN IDs, are allocated to the individual time-slots of the Cycle matrix. Time-slot Assignment Message (TAM) was newly defined message type intended to assigning the time-slots.

Individual modules transmit their data after the allocation process of the CAN ID to each time-slot is finished. Thus the communication is time-synchronized to the instant of Reference message transmission, which introduces each row (elementary cycle) of the Cycle matrix.

Command for the reconfiguration to the new needed matrix may come from the Network Enabled Capability (NEC) integrated environment, thereby ground personnel or headquarters require what data are important for them in certain moment. Cycle matrixes can be completely different for particular phases of flight, but some phases of flight can have very similar Cycle matrixes composition.

Mentioned command for Cycle matrix reconfiguration is ensured by newly defined message MMS (Matrix Matching Service). This message is asynchronous, therefore it can occur anywhere in the time communication schedule and always carries a name of the Cycle matrix, which is assigned to the flight phase, to which the reconfiguration should proceed. Current elementary cycle is completed after the reception of the MMS message and then the process of reconfiguration begins. The reconfiguration process is a very important factor and the optimal accomplishment can play significant role. Reconfiguration process optimization is described in details in [5].

Cycle matrix schedule is possible to change as needed also individually while the system is running and is not always necessary to reconfigure the whole matrix. This adjustment is performed again by TAM message.

5 Conclusion

The paper deals with results of the project aimed at the development of aircraft electronic system, which main task is an acquisition of data from various avionic sensors and their sharing through the CAN with CANaerospace protocol. The proposed system manufactured and laboratory tested during the project, for purpose of verification of the general conception. Final solution can serve as a starting point for the design and implementation of concrete solution for real aircraft or drone.

Acknowledgement

The work presented in this paper has been supported by the Ministry of Defence of the Czech Republic (Project No. OVUOFVT200802 defence research).

References

[1] NEC outline concept, Part 1 – Background and programme of work, May 2, vol. (2.0), Dstl/IMD/SOS/500/2 (2003)

[2] Stock, M.: CAN aerospace - Interface specification for airborne CAN applications, Germany, vol. 1.7, p. 58 (2006)

[3] ISO 11898-4:2004 (2004)

[4] Bajer, J., Janu, P., Jalovecky, R.: Controller Area Network based On-board Data Acquisition System on Military Aircraft. In: Concepts and Implementation for Innovative Military Communications and Information Technologies, Military University of Technology, Warsaw, Poland, pp. 589–598 (2010), ISBN 978-83-61486-70-1

[5] Janu, P., Bajer, J.: Dynamic Time-slot Assignment method applied on CAN with CAN aerospace protocol during the aircraft phase of flight transitions. In: Jalovecky, R., Stefek, A. (eds.) Proceedings of the International Conference on Military Technologies, ICMT 2011, University of Defence, Brno, pp. 619–625 (2011), ISBN 978-80-7231-787-5

Assessment of the Stub-Shaft Sleeve Bearing Chaotic Behavior in Automotive Differential

M. Klapka and I. Mazůrek

Brno University of Technology, Faculty of Mechanical Engineering, Technická 2896/2, Brno, 616 69, Czech Republic

Abstract. This article is concerned with capabilities of analysis of a chaotic behavior of the stub-shaft sleeve bearing inside the automotive differential. Analytical methods based on phase-trajectory diagrams and fractal dimensions are used. Effect of the chaotic behavior on automotive differential noise is examined as well. The purpose of the article is to give an overview of practical use of the above mentioned methods. The potential advantages, disadvantages are discussed as well as expected contributions of application of the bifurcation analysis to solving the automotive differential journal bearing noise problem.

1 Introduction

Fighting the noise emitted by a running vehicle into the environment and into the passenger area is a long-lasting problem. The development department of the Škoda Auto manufacturer has dealt with this phenomenon recently. The sleeve bearing inside automotive differential was identified as one of the noise sources. The stub shaft is inserted into the differential case through sleeve journal bearing which requires some radial clearance to be operational, see Figure 1 (b). Therefore, there is a mechanical looseness in the journal bearing which in combination with vibrations of the engine may result in an undesirable noise emission. The VW Group attempted to eliminate the problem by adjusting the radial clearance of the bearing with a tapered centring ring and an axial compression spring, see Figure 1 (a). However, new kinds of undesirable noises of the differential were observed under a different set of operating conditions compared to the previous case. The noise phenomenon related to the sleeve bearing arrangement without compression spring was labeled as "*wummern*" (perceived as drumming noise) and the noise of the bearing with the compression spring was labeled as "*schnarren*" (perceived as rough chattering noise). Our main goal was to identify the causes of the noise and to develop a method for identification of phenomena types. Therefore, an analytic method capable of determining the type of noise phenomenon was created [1]. The method processed measured values of the stub shaft center displacement relative to the center of the bearing. Although the described method proved as functional in practical applications we were looking for other ways to improve it or to perform even deeper analysis of the bearing operation.

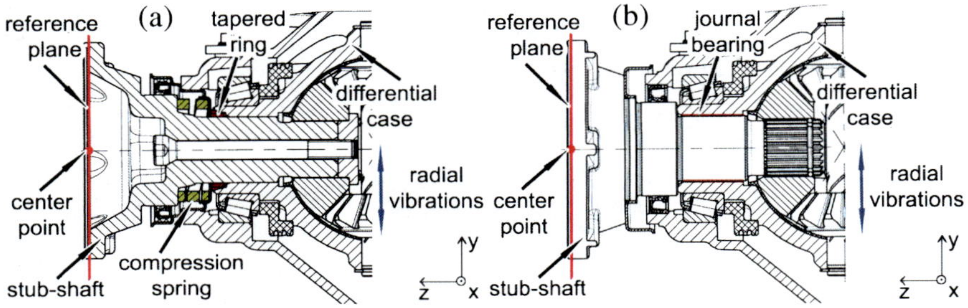

Fig. 1. Bearing arrangement with (a) and without compression spring (b) cut-away.

Achieved results proved that chaotic behavior is tightly related to overall vibroacoustic emission of examined bearing. Thus, the analysis of the bifurcation diagrams of the bearing was applied at first. However, obtained results were unsatisfactory and lack of expected information about behavior of the bearing. The analysis of the phase-trajectory diagrams and correlation fractal dimensions was suggested. Such methods were widely described in professional literature by authors like Muszynska and Goldman [2], Adiletta, Guido and Rossi [3] or Ing et al. [4]. It could be remarked that most of the articles deals with simulated models without experimental verifications on real machines. Study of the chaotic behavior of the stub-shaft sleeve bearing under real operational conditions is necessary to obtain deeper knowledge of relations between chaotic behavior and vibro-acoustic emission of the bearing. Such analysis is important for proper mechanical design of similar sleeve bearings and noise elimination and diagnostics as well.

2 Examined Bearing Arrangements, Operational Conditions and Vibro-Acoustic Behavior of the Bearing

The bearing arrangement with adjustment of the radial clearance by tapered ring and compression spring (Fig. 1 (a)) and an arrangement without the compression spring and tapered ring (Fig. 1 (b)) were experimentally examined.

The differential case was driven by controlled electromotor and was continuously slowed down from the speed of 3000 RPM to standstill. The speed of the differential case was proportional to the exciting vibrations which simulated vibrations of the vehicle engine. The exciting vibrations continuously decreased from 230 Hz to 10 Hz (corresponding to the maximum speed of 6900 RPM and the minimum speed of 300 RPM of the vehicle engine respectively, these values were also determined by technical parameters of the electrodynamic exciter). The amplitude of the exciting vibrations was 0,1 mm. The results of the experiments analyzed using the method of power summation of the stub shaft displacement are summarized in Table 1, where the greater value determines the type of noise emission.

Table 1. Parameters of the "wummern" and "schnarren" phenomena

Design of the bearing arrangement	wummern	schnarren
with compression spring, exciting amplitude 0,1 mm	0,222	0,553
without compression spring, excitation amplitude 0,1 mm	0,474	0,357

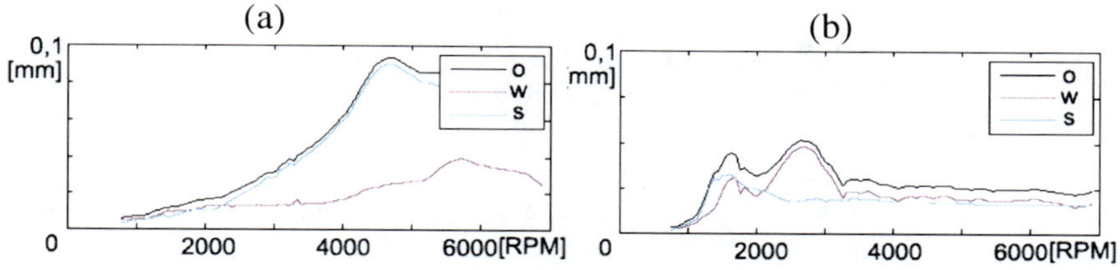

Fig. 2. Amplitude response of the stub-shaft in arrangement without compression spring (a) and with compression spring (b) proportional to engine speed.

An amplitude response of the stub-shaft oscillation is depicted in Figure 2 proportionally to speed of engine, thus frequency of exciting vibrations. The overall amplitude response describes curve „*o*". Curve "*w*" describes portion of amplitude response determining "*wummern*" vibro-acoustic emission. Curve "*s*" describes portion of amplitude response determining resonant vibro-acoustic emission called "*schnarren*". The following results were obtained from combined analysis of the values in Table 1 with curves of amplitude response in Figure 2. The sleeve bearing arrangement with the compression spring suffers from resonant vibro-acoustic emission at critical speed about 4600 RPM. On the contrary, the bearing arrangement without compression spring suffers mostly from chaotic vibro-acoustic emission "wummern" at critical speed about 2800 RPM.

3 Phase-Trajectory Diagrams and Correlation Dimensions

A phase-trajectory diagrams of the center point radial vibrations (see Fig. 1) were created for every single revolution of the differential case. It is obvious that the bearing subjected to combination of exciting vibrations and rotation of differential case is unstable in the whole range of observed operational speeds from the phase trajectories in Figures 3 and 4, irrespective of the bearing arrangement. Therefore, mostly chaotic attractors were observed. However, the bearing arrangement with compression spring tends to stable behavior for speeds above 4000 RPM and exciting vibrations amplitude of 0,1 mm (see Figure 3). Such behavior could be related to the resonance of the bearing assembly at 4600 RPM. It seems that the effect of displacement in direction of the exciting vibrations is strengthened by resonance against the effect of rotation and the system tends to be more stable. However, there are more steel-on-steel impacts which result in undesirable noise.

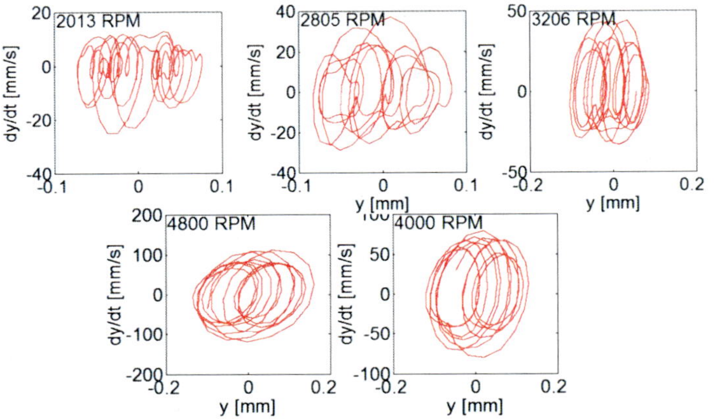

Fig. 3. Phase trajectories of the bearing arrangement with compression spring, amplitude 0,1 mm

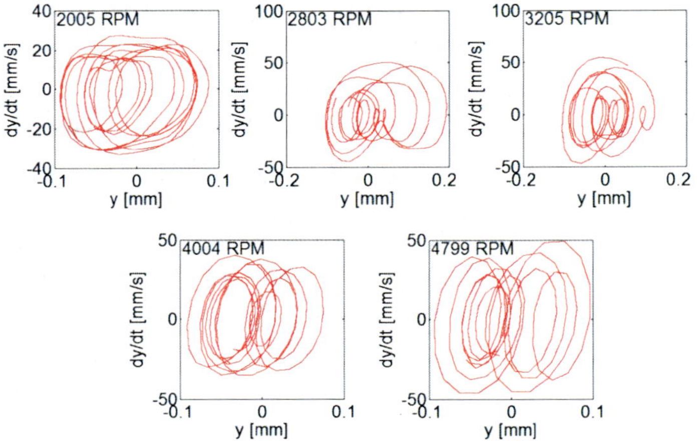

Fig. 4. Phase trajectories of the bearing arrangement without compression spring, amplitude 0,1 mm

The phase-trajectory diagrams analysis seems to be unsuitable for the objective of distinguishing between examined noises phenomena because only chaotic phase trajectories were observed. Therefore, calculation of correlation dimension values was added to analytical methodology to describe amount of chaotic behavior of the examined element. Values of the correlation dimensions D_c are calculated for each phase-trajectory. Calculated values proportional to operational speeds are depicted graphically afterwards, see Figure 5.

Curves in the diagrams in Figure 5 are capable to describe amount of chaotic behavior of the bearing where integral multiples (1,2,...) of values D_c indicates stable behavior. High values of correlation dimension at low operational speeds (under 2000 RPM) are caused by problems with regulation of the experimental rig hence they are omitted.

In case of bearing arrangement with compression spring there are some peaks between 2000 RPM and 3000 RPM which are probably caused by non-lienar characteristic of the compression spring. There is also noticeable increase of the values slightly before resonant speed at 4600 RPM and then again at 5500 RPM. Overall tendency of the bearing behavior is to be stable though. On the contrary there are significant peaks of the values of correlation dimension in case of the bearing without compression spring at 1800 RPM and particularly at critical speed 2800 RPM. Oscillations of the D_c values are more significant than in previous case as well and bearing behavior is therefore more chaotic. We could presume mostly resonant vibro-acoustic emission in the first case and chaotic vibro-acoustic emission in second examined case according obtained results.

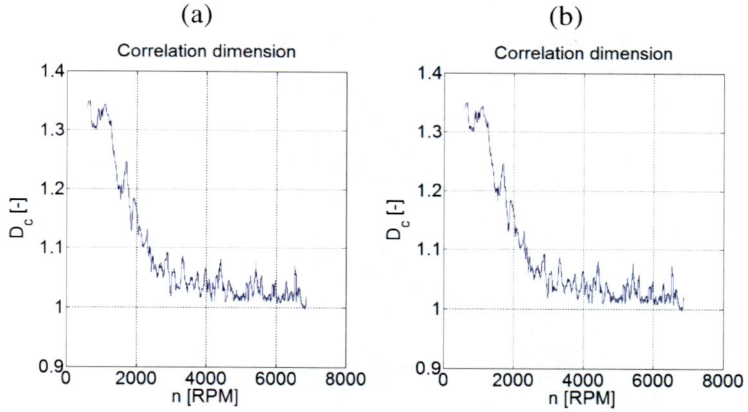

Fig. 5. Correlation dimension values proportional to operational speed for bearing arrangement with compression spring (a) and without compression spring (b), amplitude of exciting vibrations 0,1 mm

4 Conclusions

Achieved results proved relations between chaotic behavior of the machine and its overall vibro-acoustic emission. Calculation of the correlation dimension values proportionally to operational speeds appears as most suitable analytical method in particular case of the stub-shaft sleeve bearing inside automotive differential and experimental data processing. Such analysis allows to determine whether a chaotic oscillation or a resonant oscillation of the stub-shaft has significant effect to overall vibro-acoustic emission. Obtained knowledge are important to proper construction design of the bearing aimed to undesirable noise elimination. Results of analysis could be useful for verification of mathematical models or for regulation of chaotic oscillation of the sleeve bearings with mechanical looseness.

Acknowledgement

The results were acquired with support of GACR 101/10/P429.

References

[1] Klapka, M., Mazůrek, I., Roupec, J.: Mecca 46, 293 (2010)
[2] Muszynska, A., Goldman, P.: Chaos. Solit. Fract. 5, 1683 (1995)
[3] Adiletta, G., Guido, A.R., Rossi, C.: Nonlin. Dyn. 10, 251 (1996)
[4] Ing, J., et al.: Phys. D: Nonlin. Phen. 239, 312 (2010)

Using LabVIEW for Developing of Mechatronic System Control Unit

P. Krejci and M. Bradac

Institute of Solid mechanics, mechatronics and biomechanics, BUT,
Faculty of mechanical engineering, Technicka 2896/2, 61669 Brno,
Czech Republic
krejci.p@fme.vutbr.cz, y100948@stud.fme.vutbr.cz

Abstract. The paper deals with using LabVIEW graphical programming language for design and prototyping of mechatronic system controller. The paper describes possibility of using LabVIEW Real-Time and LabVIEW ARM module for control of electronic automotive throttle valve and its controller development. The advantage of this approach is using high-level graphical design tools for programming of RISC processors instead of using low level programming techniques. This allows to engineers quick transitions from hardware and software development to deployment.

1 Introduction

Design of electronic control unit (ECU) requires knowledge of several specializations especially in case of nonlinear systems. Based on system dynamical behavior the controller should be chosen. Than it is necessary to set and proof controller parameters and its behavior. Finally, the controller algorithm has to be implemented to a microcontroller in final ECU. The process of controller design can be accelerated by using of single software platform for all these steps.

The paper deals with using LabVIEW software platform for design of automotive electronic throttle valve controller which is nonlinear system because of friction and mainly because of return spring characteristic.

The National Instruments real-time computer Compact RIO with LabVIEW Real-time Module and FPGA module was used for real-system characteristic measurement, design and controller parameters optimization and its testing with real automotive throttle valve. The LabVIEW ARM embedded module for ARM microcontroller together with Luminary Micro ARM Cortex-M3 evaluation board (Fig. 1) was used for developing and testing controller on ARM microcontroller. The advantage of the evaluation board is possibility of programming ARM microcontroller directly from graphical code of LabVIEW instead of low level programming.

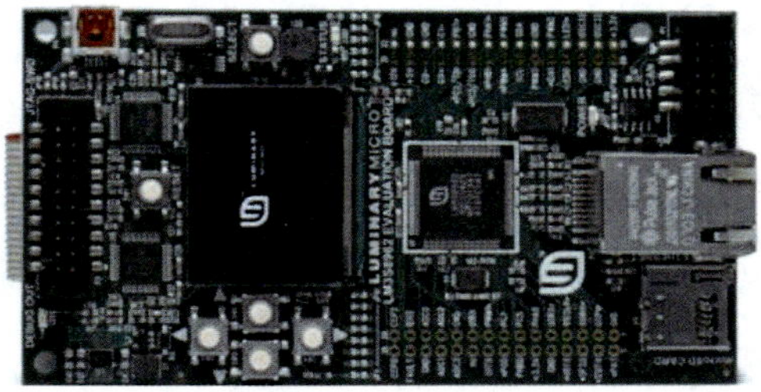

Fig. 1. Luminary Micro ARM Cortex-M3 evaluation board

Overall process of ECU development is graphically shown in Fig. 2

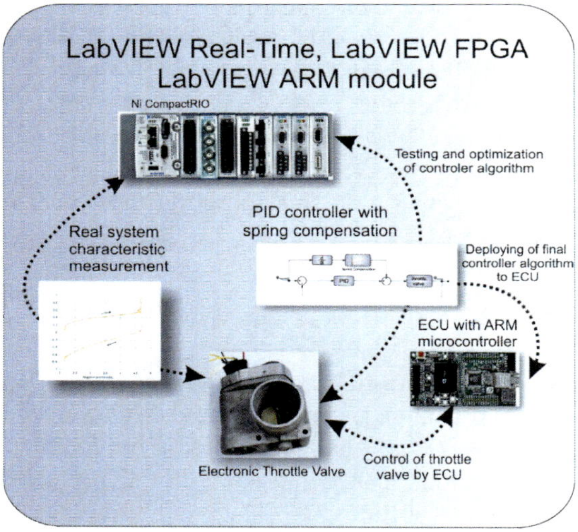

Fig. 2. Schematic of work Flow

2 Automotive Throttle Valve

The electronics throttle valve is based on drive by wire concept where car contains electronic speed pedal with position sensor instead of throttle cable. The electronic throttle valve is electromechanical system with DC machine and gearbox and is controlled by electronic signals. The throttle contains also return spring for safety reasons and actual position sensor. Due to spring an also due to friction the throttle valve is nonlinear system.

It is necessary to provide fast position control of throttle valve for proper car engine operation. Due to throttle nonlinearity, classical PID controller may fail or can be insufficient in accuracy. Therefore the PID controller with spring compensation should be used for proper operation (see [1] for more details). The nonlinear characteristic of valve is shown in Fig. 3 as relation between angular position of valve (position is shown in volts because of nature of position sensor signal) and duty cycle of DC motor PWM driver.

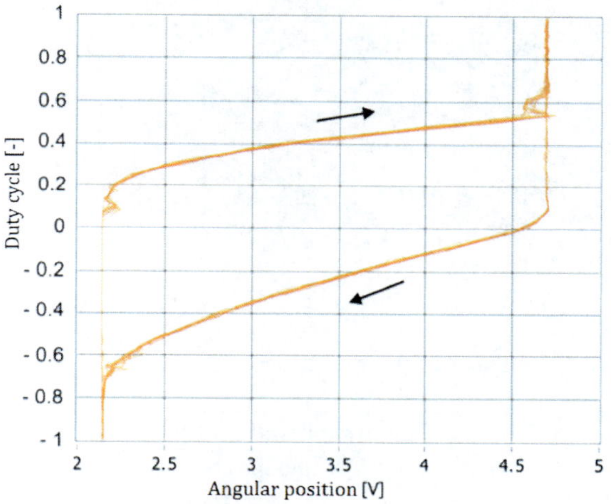

Fig. 3. Nonlinear spring characteristic of electronic throttle valve

The throttle valve used for testing is shown Fig. 4. It is Magneti Marelli C146 throttle valve and is usually used in FIAT cars. Important parameters of DC motor are shown in Table 1. For more information about automotive electronic throttle and its modeling see [1,2].

Fig. 4. Magneti Marelli C146 electronics throttle

Table 1. Throttle DC motor parameters

Parameter	Value
Supply Voltage	24V
R	3.6 Ω
L	1.2 mH

3 Programming of Controller in LabVIEW

The design of throttle valve controller with nonlinearity compensation is described in [1]. The controller consists of classical PID controller with nonlinear spring characteristic compensation (see Fig. 5). The LabVIEW graphical code is shown in Fig. 6.

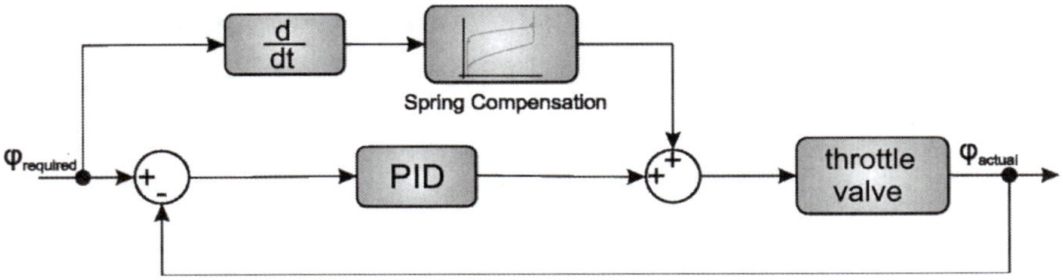

Fig. 5. PID controller of electronic throttle valve with spring nonlinearity compensation

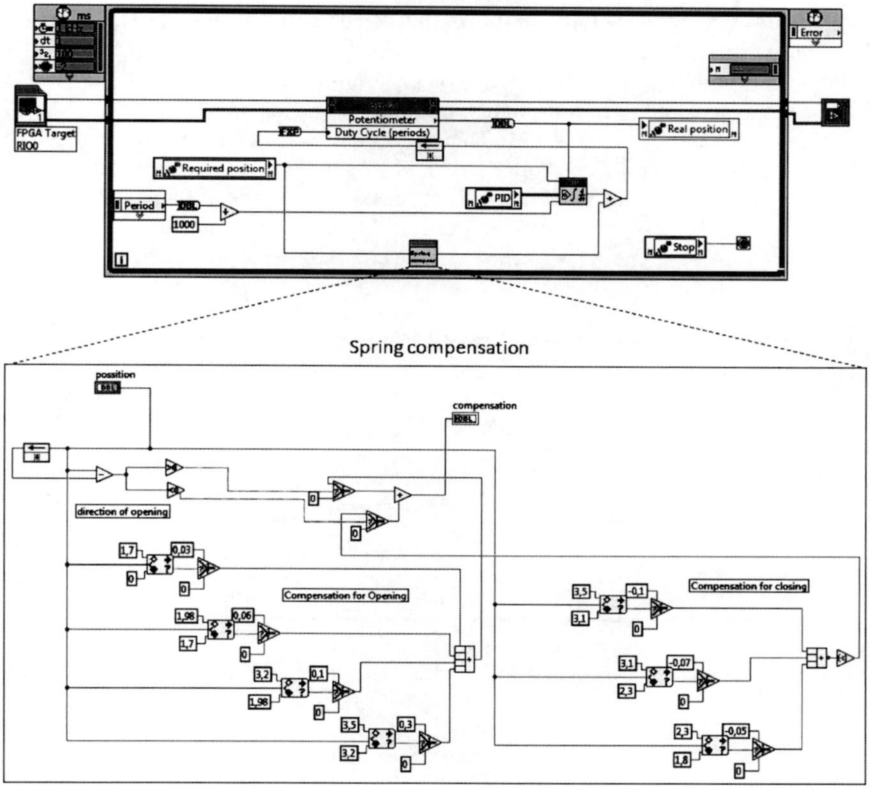

Fig. 6. LabVIEW Graphical code of PID controller with compensation of spring nonlinearity for CompactRIO Target

The Real-Time computer Compact RIO was used for development and tuning of controller because of its universality in signal manipulation and measurement and of course because of its easy debugging capabilities. The Compact RIO controller uses for input and output signals a field-programmable gate array (FPGA) which allows to program fast signal processing functions like filtering. In our case, however, FPGA was used only for transmitting signals to the control computer.

4 Deployment of Control Algorithm to ARM Controller

The final control algorithm developed on Compact RIO real-time computer was finally transferred to ARM microcontroller which will be used for final design of ECU.

The transformation of LabVIEW Graphical code from Compact RIO target to ARM target was done through LabVIEW embedded module for ARM microcontroller. The module allows easy way for programming of supported 32bit ARM microcontrollers. The access to inputs and outputs signals is for CompactRIO different from access to IO for ARM. Therefore there was necessary to modify LabVIEW Graphical code of controller to ARM target. The CompacRIO uses FPGA for I/O signals, while the ARM uses direct access to I/O pins of the controller.

Modified LabVIEW graphical code for ARM microcontroller is shown in Fig. 7.

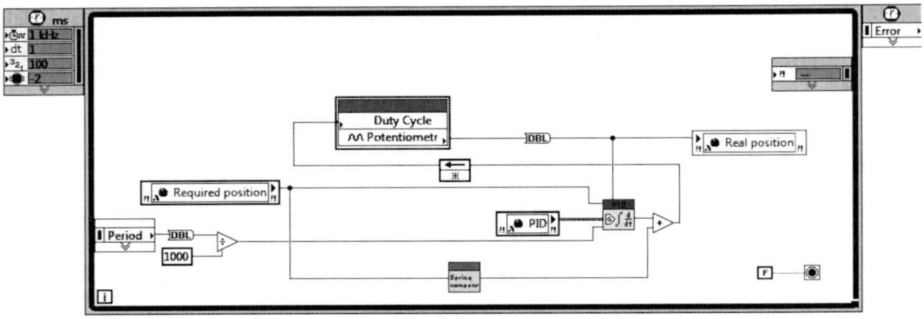

Fig. 7. LabVIEW Graphical code of PID controller with compensation of spring nonlinearity for ARM Target

The Fig. 8 shows process of throttle valve position regulation for changing of required position of throttle valve opening. The chart shows influence of spring compensation. The Fig. 9 shows Compact-RIO controller during test.

The comparison between Compact-RIO platform and ARM target in terms of control loop speed is shown in Tab. 2. The speed shown in table can be understood as duration of control loop inner code (PID, input and output, spring compensation). On the other hand the comparison can be inaccurate because of different time management of input and output on both platforms. The result also shown the possibility of increasing of control loop speed to more than 1ms which was used in our case.

Table 2. Speed of inner part of control loop

Platform	duration
Compact -RIO	26us
ARM	128us

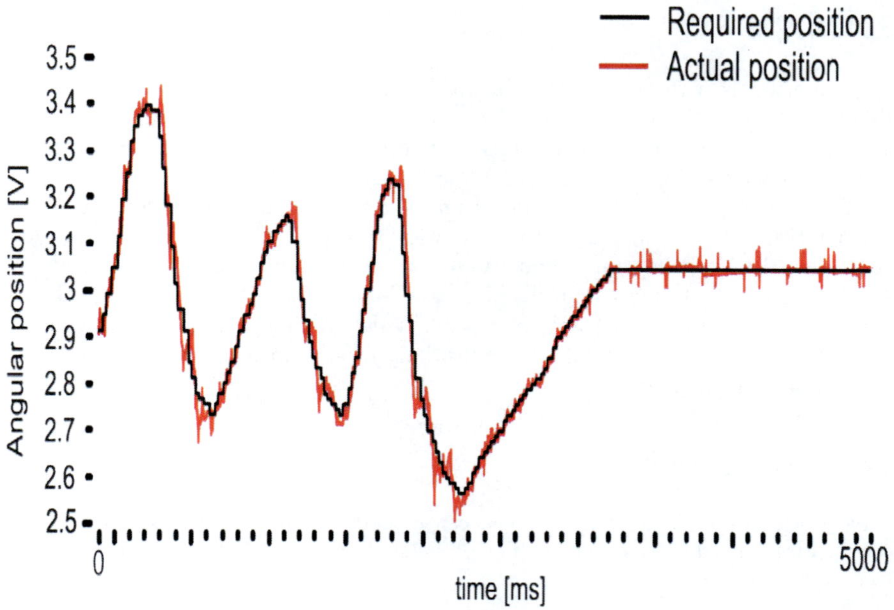

Fig. 8. Result of regulation

Fig. 9. NI CompactRIO real-time hardware with power electronics and automotive electronics throttle valve

5 Conclusion

The paper deals with possibility of using LabVIEW for design, optimization and final development of mechatronics system controller. In the paper the electronic automotive throttle valve was used as mechatronics system. Real-time computer Compact-RIO (National Instruments) was used for adjustment of throttle valve nonlinear regulator and for measuring and signal manipulation of throttle valve characteristics which was used for setting of controller parameters. The proven controller (PID with spring compensation) from Compact-RIO was than easily transferred to ARM microcontroller which will be used for final ECU. The advantage of this solution is in single platform for developing and final deploying of controller algorithm to ARM Microcontrollers, which can be program using high-level graphical design tool.

Acknowledgement

Published results and presented work has been supported by research project FSI-S-11-15 "Design, testing and implementation of control algorithms with use of nonlinear models of mechatronics systems".

References

[1] Grepl, R., Lee, B.: Model Based Controller Design for Automotive Electronic Throttle. In: Recent Advances in Mechatronics: 2008-2009, 8th International Conference on Mechatronics, Luhacovice (2009)

[2] Bradac, M.: Implementace vybraných řidicích algoritmů pro FPGA. Brno university of technology, Faculty of mechanical engineering. Diploma thesis (2011)

[3] Grepl, R.: Real-time control prototyping in matlab/simulink: review of tools for research and education in mechatronics. In: IEEE ICM 2011, Istanbul, Turkey (2011)

Simple Instability Test for Linear Continuous-Time System

J.E. Kurek

Institute of Automatic Control and Robotics
Warsaw University of Technology
ul. Boboli 8, 02-525 Warszawa, Poland

Abstract. There is presented simple test for checking if a linear system is unstable, if it has poles in the right hand half plane of complex variable *s*. Presented test can be easily used for testing system instability as well its non-minimum phase property.

1 Introduction

Stability is one of the most important problems in analysis of dynamical systems. Asymptotic stability criterion for continuous-time linear system is very simple: system is asymptotically stable if it has all poles in the left hand half plane of complex variable *s*. However, if the system is not asymptotically stable it is more difficult to find if the system is unstable or stable but not asymptotically.

There is a number of stability criterion which enables ones to test if the system is asymptotically stable, e.g. [1], [2], [3], but there is rather no tests which enable distinguish if the system is unstable or stable not asymptotically. One can use for such testing the Lyapunov stability theorem [2]. This approach is rather simply but requires some calculations. In this note we present more simple theorem for testing if the system is unstable, if it has poles in the right half plane of the *s* plane.

2 Main Result

Consider linear system described by the transfer function:

$$G(s) = \frac{L(s)}{M(s)} = \frac{b_w s^w + \ldots + b_1 s + b_0}{a_n s^n + a_{n-1} s^{n-1} + \ldots + a_1 s + a_0} \tag{1}$$

Stability of the system can be easily tested using for instance the Hurwitz criterion [1], [3], [4], or the Lyapunov criterion [2], [3], [4]. We can obtain clear answer if the system is asymptotically stable. Unfortunately, we can not recognize if the system is unstable or stable non-asymptotically. Solving this problem we propose the following theorem which gives sufficient conditions for system instability.

Theorem 1

System (1) has poles in the in the right hand plane half of complex variable s if some coefficients a_i of the transfer function denominator are positive and some are negative.

Proof

Denominator of the transfer function (1) can be presented as follows

$$M(s) = a_n s^n + a_{n-1} s^{n-1} + \ldots + a_0 = a_n (s - s_1)\ldots(s - s_n)$$

where s_i denotes its roots, i.e. poles of the system. If all coefficients a_i of the denominator are real the poles can be real or complex conjugate. For real or poles s_i, $s_j \le 0$ or complex conjugate $s_i = \bar{s}_j$ such that Re s_i = Re $s_j \le 0$ one easily finds that all coefficients of the polynomial

$$(s - s_i)(s - s_j) = s^2 - s(s_i + s_j) + s_i s_j$$

are real and positive or equal to zero. Thus, we easily find that denominator $M(s)$ of the transfer function has some poles s_k such that Re $s_k > 0$ if some its coefficients a_i are negative and some are positive.

Remarks

1. Clearly, if condition formulated in the theorem is satisfied – denominator has negative and positive coefficients it has roots with positive real part, but it can also can have roots with negative or equal to zero real part.
2. Theorem gives only sufficient conditions. Thus, system can have poles with positive real part and all denominator coefficients of its transfer function can have the same sign i.e. all are positive or all are negative. □

Proposed theorem simply checks if some roots of polynomial

$$a_n s^n + a_{n-1} s^{n-1} + \ldots + a_0 = 0, \quad a_n \ne 0 \tag{2}$$

have real part positive. It can be also used for checking roots of the transfer function (1) nominator $L(s)$, i.e. if some zeros of the system have positive real part. It enables ones simply check if the system is non-minimum phase one. This knowledge can be important if one design controller for the system.

Example

The n-th order Padé approximation of the dead time system has the following form [5]

$$e^{-T_0 s} \approx \frac{1 - a_1 T_0 s + a_2 T_0^2 s^2 \ldots + (-1)^n a_n T_0^n s^n}{1 + a_1 T_0 s + a_2 T_0^2 s^2 \ldots + a_n T_0^n s^n} = \frac{1 + b_1 T_0 s + b_2 T_0^2 s^2 \ldots + b_n T_0^n s^n}{1 + a_1 T_0 s + a_2 T_0^2 s^2 \ldots + a_n T_0^n s^n}$$

where all coefficients a_i are real and positive, $a_i > 0$.

It is well known that the approximation is asymptotically stable. However, based on the theorem one also easily finds that every Padé approximation is a non-minimum phase system, i.e. some roots of the approximation nominator have positive real part since some nominator coefficients b_i are positive and some are negative. □

3 Concluding Remarks

There is presented simply theorem for testing if the polynomial (2) has roots with positive real part. It can be easily used for checking sufficient conditions if a continuous-time linear system (1) is unstable or is a non-minimum phase one.

References

[1] Hurwitz, A.: Über die Bedingungen unter welchen eine Gleichung nur Wurzeln mit negativen reelen Theilen besitzt. Math. Ann., 273–284 (1895)

[2] Lyapunov, M.A.: Dissertation, Kharkov (1892); see also Stability of Motion, English translation of Russian edition. Academic Press, New York (1966)

[3] Takahashi, Y., Rabins, M.J., Auslander, D.M.: Control and Dynamic Systems. Addison-Wesley Publishing Company, Inc., Reading, Massachusetts (1970)

[4] Ogata, K.: State Space Analysis of Control Systems. Prentice-Hall, Englewood Cliffs (1967)

[5] Baker Jr., G.A., Graves-Morris, P.: Padé Approximants. Addison-Wesley, Reading, Massachusetts (1981)

Extended Kalman Filter and Discrete Difference Filter Comparison

M. Laurinec

Brno University of Technology, Faculty of Mechanical Engineering,
Institute of Automotive Engineering, Technická 2896/2, Brno, 616 69, Czech Republic

Abstract. In this paper we focused our attention on the mathematical background of the Extended Kalman Filter and its comparison to the Discrete Difference filter. Both of the filters are capable to estimate states of nonlinear systems but each one has its advantages and drawbacks we would like to outline. In addition to the mathematical derivation, we will show also the details of software implementation in Matlab.

1 Introduction

The main purpose of this text is to look deeply into the mathematical principles of filters used for vehicle dynamics analysis presented in [1]. Detailed algorithms together with software implementation will be presented.

First we will focus our attention on the nonlinear systems and its general description followed by basic principles of state estimation. In the next sections we will show the use of these principles in both of the filters with their algorithm development. Finally we will show a graphic comparison of both the filters used for a simple road test maneouvre.

2 Nonlinear Systems and Estimation

A usually used method for state estimation is the Kalman filter derived in 1960 by R.E. Kalman [2]. This approach is applicable in case of linear transformation. But by considering nonlinearities the classic Kalman Filter becomes unavailable.

2.1 Nonlinear System

$$\begin{aligned} \mathbf{x}_{k+1} &= \mathbf{f}(\mathbf{x}_k, \mathbf{u}_k, \mathbf{v}_k) \qquad & \mathbf{v}_k &\sim \mathrm{N}(\bar{v}_k, \mathbf{Q}_k) \\ \mathbf{y}_k &= \mathbf{g}(\mathbf{x}_k, \mathbf{w}_k) \qquad & \mathbf{w}_k &\sim \mathrm{N}(\bar{v}_k, \mathbf{R}_k) \end{aligned} \tag{1}$$

The equations (1) represent a state-space model of a dynamic system with generally nonlinear function in both state and measurement equation. The first part represents an evolution of the state $\mathbf{x}_k$ over time depending on an input $\mathbf{u}_k$ disturbed by a

Gaussian random variable $\mathbf{v}_k$ called process noise with mean $\overline{\mathbf{v}}_k$ and covariance matrix $\mathbf{Q}_k$. The second part express a relation between the state and the measurement (output) which is also disturbed with a Gaussian random variable $\mathbf{w}_k$ called measurement noise with mean $\overline{\mathbf{w}}_k$ and covariance matrix $\mathbf{R}_k$.

2.2 Estimation Principle

The main objective is to find a state estimate based on information contained in the measurements and information about uncertainty in the state-space model (1). Moreover this estimate is in every step minimizing its mean square error. As can be seen in [3] such an estimate is equal to the conditioned expectation:

$$\overline{\mathbf{x}}_k = E\{\mathbf{x}_k \mid \mathbf{Z}^{k-1}\}, \tag{2}$$

where $\mathbf{Z}^j$ is a matrix of all measurements up to time .

The measure of quality of this estimate is given by its covariance:

$$\overline{\mathbf{P}}_k = E\{(\mathbf{x}_k - \overline{\mathbf{x}}_k)(\mathbf{x}_k - \overline{\mathbf{x}}_k)^T \mid \mathbf{Z}^{k-1}\} \tag{3}$$

By the estimation procedure is the actual (a posteriori) estimate $\hat{\mathbf{x}}_k$ with its covariance $\hat{\mathbf{P}}_k$ transformed using the equations (1) which give us the prediction of the state in time $k+1$ exploiting the measurement up to time k, called as a priori estimate $\overline{\mathbf{x}}_{k+1}$ with its covariance $\overline{\mathbf{P}}_{k+1}$. Consequently the measurement in time $k+1$ is obtained and used for the a priori estimate correction in order to get the a posteriori update $\hat{\mathbf{x}}_{k+1}$ and $\hat{\mathbf{P}}_{k+1}$. For this purpose it is possible to use different relationship between the a posteriori update and measurement. But for the reason of difficulties in evaluation of conditioned expectation and covariance we usually work with an estimate which is a linear function of measurement (does not require linearity in equations (1)):

$$\hat{\mathbf{x}}_k = \overline{\mathbf{x}}_k + \mathbf{K}_k(\mathbf{y}_k - \overline{\mathbf{y}}_k) \qquad \hat{\mathbf{P}}_k = \overline{\mathbf{P}}_k - \mathbf{K}_k \mathbf{P}_{y,k} \mathbf{K}_k^T \tag{4}$$

Where $\mathbf{K}_k$ is usually called the Kalman gain:

$$\mathbf{K}_k = \mathbf{P}_{xy,k}\mathbf{P}_{y,k}^{-1} \qquad \begin{aligned} \mathbf{P}_{xy,k} &= E\{(\mathbf{x}_k - \overline{\mathbf{x}}_k)(\mathbf{y}_k - \overline{\mathbf{y}}_k)^T \mid \mathbf{Z}^{k-1}\} \\ \mathbf{P}_{y,k} &= E\{(\mathbf{y}_k - \overline{\mathbf{y}}_k)(\mathbf{y}_k - \overline{\mathbf{y}}_k)^T \mid \mathbf{Z}^{k-1}\} \end{aligned} \tag{5}$$

This method is essentially the same for all the filters introduced in next sections. There are differences in the evaluation method of the a priori estimate and appropriate covariance, which is determinant for quality of each filter.

3 Extended Kalman Filter

First and wide used extension of the Kalman filter to the nonlinear dimension is the Extended Kalman Filter (EKF), which uses the well known Taylor series [4] for approximating nonlinearities in equation (1). Just the first-order Taylor approximation is used in EKF, although other higher-order filters were derived. But because of their evaluation complexity they are rarely used. For more detailed information see [3] and [5].

3.1 Filter Algorithm

According to the estimation principle described in the section 2.2 and the approximation described above the whole algorithm of EKF can be written as follows.

A Priori Update

$$\begin{aligned}\bar{\mathbf{x}}_{k+1} &= \mathbf{f}(\hat{\mathbf{x}}_k, \mathbf{u}_k, \bar{\mathbf{v}}_k)\\ \bar{\mathbf{y}}_k &= \mathbf{g}(\hat{\mathbf{x}}_k, \bar{\mathbf{w}}_k)\\ \bar{\mathbf{P}}_k &= \mathbf{A}_k\hat{\mathbf{P}}_k\mathbf{A}_k^T + \mathbf{B}_k\mathbf{Q}_k\mathbf{B}_k^T\end{aligned} \tag{6}$$

As can be seen, the a priori update (state and output) is a result of a simple transformation of the a posteriori estimate. The appropriate covariance matrix of the a priori state estimate is utilizing the Jacobi matrices computed in the best available estimate, which in this part of algorithm is the a posteriori estimate.

A Posteriori Update

$$\begin{aligned}\mathbf{K}_k &= \bar{\mathbf{P}}_k\mathbf{C}_k^T[\mathbf{C}_k\bar{\mathbf{P}}_k\mathbf{C}_k^T + \mathbf{D}_k\mathbf{R}_k\mathbf{D}_k^T]^{-1}\\ \hat{\mathbf{x}}_k &= \bar{\mathbf{x}}_k + \mathbf{K}_k(\mathbf{y}_k - \bar{\mathbf{y}}_k)\\ \hat{\mathbf{P}}_k &= \bar{\mathbf{P}}_k - \mathbf{K}_k\bar{\mathbf{P}}_k\mathbf{K}_k^T\end{aligned} \tag{7}$$

This part of the algorithm, as can be seen, is governed by the equations (4) and (5) with usage of the Jacobi matrices evaluated for the best available estimate, which is the a priori estimate.

3.2 Filter Deficiencies

For application of the EKF for the state estimation it is necessary to symbolically evaluate the Jacobi matrices which came out by approximating nonlinearities in the state-space model. There is a more serious limitation of the EKF which appear when the functions in the state and measurement equation are non-differentiable or even worse unknown (a black-box system). Therefore it is suitable to use other approximation techniques which don't require the differentiation. With these estimators deal the next section.

4 Discrete Difference Filters

As mentioned in the preceding section, usage of the Taylor approximation for linearization of nonlinear relations in state evolution or/and measurement equation (1) should be the main limitation of the EKF. A derivative-free interpolation like the one described in next section is therefore desired. The algorithms presented in [6] underline the proceeding sections and cover detailed derivation of presented methods.

4.1 First-Order Discrete Difference Filter

This filter uses the first two terms from Stirling's interpolation formula and the Cholesky factorization of the covariance matrices. Therefore let me at first introduce these factorizations of individual covariance matrices:

$$\mathbf{R} = \mathbf{S}_w \mathbf{S}_w^T \qquad \mathbf{Q} = \mathbf{S}_v \mathbf{S}_v^T \qquad \hat{\mathbf{P}} = \hat{\mathbf{S}}_x \hat{\mathbf{S}}_x^T \qquad \overline{\mathbf{P}} = \overline{\mathbf{S}}_x \overline{\mathbf{S}}_x^T \tag{8}$$

where the matrices $\mathbf{S}_w$ and $\mathbf{S}_v$ can be directly computed from known $\mathbf{R}$ and $\mathbf{Q}$. The whole algorithm works with the factored covariance matrices, therefore also the resulting a priori and a posteriori covariance will be in factored form.

A Priori Update
The a priori state estimate is the same as for the EKF, i.e.

$$\overline{\mathbf{x}}_{k+1} = \mathbf{f}(\hat{\mathbf{x}}_k, \mathbf{u}_k, \overline{\mathbf{v}}_k) \tag{9}$$

The corresponding factored covariance matrix is:

$$\overline{\mathbf{S}}_{x,k+1} = \begin{bmatrix} \mathbf{S}_{x\hat{x},k} & \mathbf{S}_{xv,k} \end{bmatrix} \tag{10}$$

A posteriori update
The a priori measurement estimate is in accordance with that in the EKF, i.e.

$$\overline{\mathbf{y}}_k = \mathbf{g}(\overline{\mathbf{x}}_k, \overline{\mathbf{w}}_k) \tag{11}$$

The covariance matrix of the measurement estimate again in the factored form $\mathbf{P}_{y,k} = \mathbf{S}_{y,k} \mathbf{S}_{y,k}^T$ can be expressed as

$$\mathbf{S}_{y,k} = \begin{bmatrix} \mathbf{S}_{y\overline{x},k} & \mathbf{S}_{yw,k} \end{bmatrix} \tag{12}$$

For the a posteriori state estimate it is necessary the Kalman gain evaluation which with accordance to (5) requires the cross-covariance $\mathbf{P}_{xy}$ given by:

$$\mathbf{P}_{xy,k} = \overline{\mathbf{S}}_{x,k}\mathbf{S}^T_{y\bar{x},k} \tag{13}$$

By applying this to (5):

$$\mathbf{K}_k = \overline{\mathbf{S}}_{x,k}\mathbf{S}^T_{y\bar{x},k}\left[\mathbf{S}_{y,k}\mathbf{S}^T_{y,k}\right]^{-1} \tag{14}$$

The a posteriori state estimate is then given by the equation (6), i.e.:

$$\hat{\mathbf{x}}_k = \overline{\mathbf{x}}_k + \mathbf{K}_k(\mathbf{y}_k - \overline{\mathbf{y}}_k) \tag{15}$$

and the appropriate factored covariance

$$\hat{\mathbf{S}}_k = \left[\overline{\mathbf{S}}_{x,k} - \mathbf{K}_k\mathbf{S}_{yx,k} \quad \mathbf{K}_k\mathbf{S}_{yw,k}\right] \tag{16}$$

It is noteworthy that the matrices $\mathbf{S}_{y,k}$, $\overline{\mathbf{S}}_{x,k}$ and $\hat{\mathbf{S}}_k$ are rectangular and for their later usage by computing the covariance matrices it is necessary to transform them to a quadratic form using the Householder triangularization [5].

4.2 Second-Order Discrete Difference Filter

Utilization of the second-order Stirling's interpolation formula for the state estimation also known as the DD2 filter evaluates its covariance estimates in factored form in the same way as the DD1 filter. Therefore the factorization given by the equation (8) will be used here.

A Priori Update

The equation for computing the a priori state estimate is more complex than in the DD1 for the reason of including the higher-order terms of the interpolation formula.

$$\begin{aligned}\overline{\mathbf{x}}_{k+1} &= \frac{h^2 - n_x - n_v}{h^2}\mathbf{f}(\hat{\mathbf{x}}_k, \mathbf{u}_k, \overline{\mathbf{v}}_k) + \\ &+ \frac{1}{2h^2}\sum_{i=1}^{n_x}\mathbf{f}(\hat{\mathbf{x}}_k + h\hat{\mathbf{s}}_{x,i}, \mathbf{u}_k, \overline{\mathbf{v}}_k) + \mathbf{f}(\hat{\mathbf{x}}_k - h\hat{\mathbf{s}}_{x,i}, \mathbf{u}_k, \overline{\mathbf{v}}_k) + \\ &+ \frac{1}{2h^2}\sum_{j=1}^{n_v}\mathbf{f}(\hat{\mathbf{x}}_k, \mathbf{u}_k, \overline{\mathbf{v}}_k + h\mathbf{s}_{v,j}) + \mathbf{f}(\hat{\mathbf{x}}_k, \mathbf{u}_k, \overline{\mathbf{v}}_k - h\mathbf{s}_{v,j})\end{aligned} \tag{17}$$

The symbols stands for dimensions of the state and process noise respectively.

The corresponding Cholesky factor of a priori covariance is:

$$\overline{\mathbf{S}}_{x,k+1} = \left[\mathbf{S}_{x\hat{x},k} \quad \mathbf{S}_{xv,k} \quad \mathbf{S}^{(2)}_{x\hat{x},k} \quad \mathbf{S}^{(2)}_{xv,k}\right] \tag{18}$$

A Posteriori Update

The equation for computing the a priori measurement estimate is given by:

$$\overline{\mathbf{y}}_k = \frac{h^2 - n_x - n_w}{h^2}\mathbf{g}(\overline{\mathbf{x}}_k, \overline{\mathbf{w}}_k) + \frac{1}{2h^2}\sum_{i=1}^{n_x}\left[\mathbf{g}(\overline{\mathbf{x}}_k + h\overline{\mathbf{s}}_{x,i}, \overline{\mathbf{w}}_k) + \right.$$
$$\left. + \mathbf{g}(\overline{\mathbf{x}}_k - h\overline{\mathbf{s}}_{x,i}, \overline{\mathbf{w}}_k)\right] + \frac{1}{2h^2}\sum_{j=1}^{n_w}\left[\mathbf{g}(\overline{\mathbf{x}}_k, \overline{\mathbf{w}}_k + h\mathbf{s}_{w,i}) + \mathbf{g}(\overline{\mathbf{x}}_k, \overline{\mathbf{w}}_k - h\mathbf{s}_{w,i})\right] \quad (19)$$

where n_w represents the dimension of the process noise.

The Cholesky factor of corresponding covariance matrix is then:

$$\mathbf{S}_{y,k} = \left[\mathbf{S}_{y\overline{x},k} \quad \mathbf{S}_{yw,k} \quad \mathbf{S}^{(2)}_{y\overline{x},k} \quad \mathbf{S}^{(2)}_{yw,k}\right] \quad (20)$$

The equations used for the Kalman gain and the a posteriori state update evaluation are the same as in DD1 filter.

However, the equation for the appropriate Cholesky factor of the covariance matrix is different:

$$\hat{\mathbf{S}}_k = \left[\overline{\mathbf{S}}_{x,k} - \mathbf{K}_k\mathbf{S}_{yx,k} \quad \mathbf{K}_k\mathbf{S}_{yw,k} \quad \mathbf{K}_k\mathbf{S}^{(2)}_{yx,k} \quad \mathbf{K}_k\mathbf{S}^{(2)}_{yw,k}\right] \quad (21)$$

As in the DD1 filter, the matrices $\mathbf{S}_{y,k}$, $\overline{\mathbf{S}}_{x,k}$ and $\hat{\mathbf{S}}_k$ are rectangular and they need to be transformed using the Householder triangularization.

4.3 Discrete Difference Filter Algorithm

For lucidity Table 1 shows the whole algorithm for both the discrete difference filters.

Table 1. DD1 and DD2 algorithm

	DD1	DD2
Initialization	$\overline{\mathbf{x}}_0, \overline{\mathbf{P}}_0 \Rightarrow \overline{\mathbf{S}}_{x,0}$	
A priori measurement estimate $\overline{\mathbf{y}}_k$	(11)	(19)
Covariance matrix of the a priori estimate $\mathbf{S}_{y,k}$	(12)	(20)
A priori cross-covariance matrix $\mathbf{P}_{xy,k}$	(13)	
Kalman gain $\mathbf{K}_k$	(14)	
A posteriori state estimate $\hat{\mathbf{x}}_k$	(15)	
Covariance matrix of the a posteriori estimate $\hat{\mathbf{S}}_{x,k}$	(16)	(21)
A priori state estimate $\overline{\mathbf{x}}_{k+1}$	(9)	(17)
Covariance matrix of the a priori state estimate $\overline{\mathbf{S}}_{x,k+1}$	(10)	(18)

5 Matlab/Simulink Model

The concept of model in Simulink environment illustrates Fig. 1. It shows a universal model with all three filters which are simply editable according to given dynamic system.

Measurement equations along with state equations were created as a user defined library blocks consisting from Embedded Matlab function block where the equations themselves were written. This made the whole model much user-friendlier and simply editable.

Every filter, represented by gray subsystem block, process information from measured quantities and inputs to produce its estimate and covariance matrix. The estimates from each filter are graphically compared with other filters estimates and the values computed via the mathematical model.

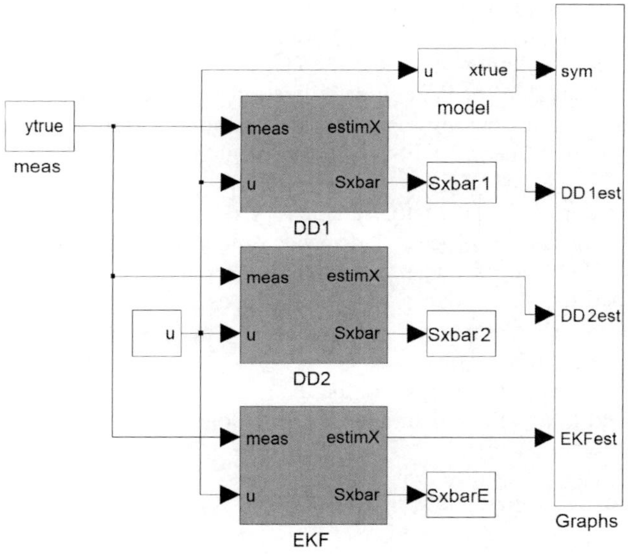

Fig. 1. Simulink model

6 Filters Utilization and Results Comparison

Correct application of preceding filter algorithms for processing responses of vehicle system is conditioned by formulation of the state equations describing vehicle dynamics. Furthermore, it is necessary to define which quantities are state values, which represent system inputs and which system outputs. Also, it must be determined which of the measured quantities will be used for the algorithm implementation. All this information together with measurement and results comparison on real test manoeuvre is beyond the range of this text and are detailed described in [1].

7 Conclusion

A theoretical base for estimation of nonlinear state-space models was presented with emphasis placed on derivative-free algorithms and their practicability for the vehicle parameters identification.

For filter performance examination a single-wheel vehicle model was used. So called Magic formula provided nonlinearity in state equations and therefore the conventional linear Kalman filter was inapplicable. The results showed that the estimates acquired from DD1 filter are approximately identical to the one obtained from DD2 filter certainly as a result of unsatisfactory information from measurements. Too short total time of the testing manoeuvre (approx. 60 seconds) and too small sampling rate (10 Hz) caused this identicalness.

The EKF was computationally too difficult because of the Jacobi matrices evaluation and the solver finally refuses to symbolically evaluate these entries. The Magic formula complexity fully shows the limitation of this filter. Its comparison with derivative-free filters DD1 and DD2 could be found in [7].

For software implementation the MATLAB/SIMULINK was used. This choice of software is suitable for later filter utilization because the vehicle dynamics can be modelled in another way, using SimMechanics or Adams. System modelled this way can subsequently be easily implemented into already created Simulink model.

The most important factor for satisfactory function of mentioned filters is the exact knowledge in the process and measurement noise covariance matrices. Especially the DD2 filter is very sensitive to inaccurate noise identification inflicting divergence of the filter. Also a good initial guess of the state and corresponding covariance matrix are important for the convergence speed. When the information about the process or/and measurement noise is not well known, then an adaptive filter is used. Filters of this type are the main area of our future work together with their application for vehicle parameters estimation.

References

[1] Laurinec, M., Porteš, P., Blaťák, O.: Discrete-difference filter in vehicle dynamics analysis. In: Recent Advances in Mechatronics, pp. 19–24. Springer, Heidelberg (2009) ISBN 978-3-642-05021-3

[2] Kalman, R.E.: A new approach to linear filtering and prediction problems. Transaction of the ASME - Journal of basic engineering, 35–45 (1960)

[3] Havlena, V., Štecha, J.: Modern control theory. Scriptum ČVUT (2000), ISBN 80-01-02095-9

[4] Julier, S., Uhlmann, K.J.: A general method for approximating nonlinear transformations of probability distributions. Robotics Research Group, University of Oxford (1996)

[5] Grewal, M., Andrews, A.: Kalman Filtering: Theory and Practice Using MATLAB, 2nd edn. John Wiley and Sons, Chichester (2001)

[6] Nørgaard, M., Poulsen, N.K., Ravn, O.: Advances in Derivative-Free State Estimation for Nonlinear Systems, Revised Edition. IMM-Technical Report-1998-15 (2004)

[7] Laurinec, M.: Extended and Derivative Free Kalman Filter. In: Advances in Automotive Engineering, vol. II, pp. 135–279. Tribun EU, Brno (2008), ISBN: 978-80-7399-497-6

[8] Vlk, F.: Motor vehicle dynamics. VLK, Brno (2000), ISBN 80-238-5273-6

[9] Scheit, T.S.: A Finite-Difference Method for Linearization in Nonlinear Estimation Algorithms. Automatica 33(11), 2053–2058 (1997), ISSN 0005-1098

[10] Froberg Carl, E.: Introduction to Numerical Analysis, 433 p. Addison-Wesley, Reading (1970), ASIN B000NKJ5LC

[11] Moler, C.: Numerical Computing with MATLAB Interpolation (April 23, 2011), `http://www.mathworks.com/moler/interp.pdf`

[12] CORRYS-DATRON: GPS sensors-MicroSAT (May 10, 2007), `http://www.corrsys-atron.com/gps_sensors.htm`

[13] ISO/WD 3888-2, Passenger cars Test track for a severe lane-change manoeuvre, Part 2: Obstacle avoidance (1999)

[14] Web source (April 23 , 2011), `http://en.wikipedia.org/wiki/Taylor_series`

Active Vibration Control Using DEAP Actuator

A. Nagy

Széchenyi István University, Egyetem tér 1.,
Győr, H-9026, Hungary

Abstract. The dielectric electro active polymers actuators (DEAP) have several advantages over the conventional actuators, for example low cost, small weight and silent working. Because of the properties of the polymers, the precise control of this actuator type requires some special considerations. In this paper a dynamic model of a DEAP actuator was created. Additionally a closed loop control of a DEAP actuator in a vibration attenuator application was demonstrated.

1 Introduction

In recent years the interest on the dielectric electro active polymers has been increasing. The main cause of that is the expectedly low producing cost of the DEAP actuators, the very small weight and the silent operation that could be very important in lot of application. In the other hand the DEAP actuators these days could not develop enough force or stroke which is required. DEAP actuator suppliers present only static characteristics that are not sufficient for designing dynamic systems. The main goal of this paper is to develop a simple dynamic model, as well as to demonstrate its availability.

2 Properties of the DEAP Foil

In our test experiments the active material of the actuator was silicon foil produced by the company Danfoss Polypower. From point of view of controller design the foil have some relevant properties which should be taken into consideration.

2.1 Quadratic Characteristics

The DEAP actuator is essentially a capacitor with compliant plates and whose dielectric is made from soft material. It is well known that the extent of the force is proportional to the square of the applied voltage.[1]

2.2 Polarity Independent Stroke

The DEAP actuator we used is a spring roll actuator (Fig. 1). This type of actuator consists of a DEAP foil rolled around a coil spring. In idle state the foil is preloaded by the spring. In order to ensure stroke in both direction according to the polarity of the applied voltage the voltage has to be biased.[1]

3 DEAP Actuator Model

The mechanical model of a spring roll DEAP actuator is shown in the Fig. 1. The spring constant c_m, damping coefficient b and lumped mass m characterize the mechanical part.

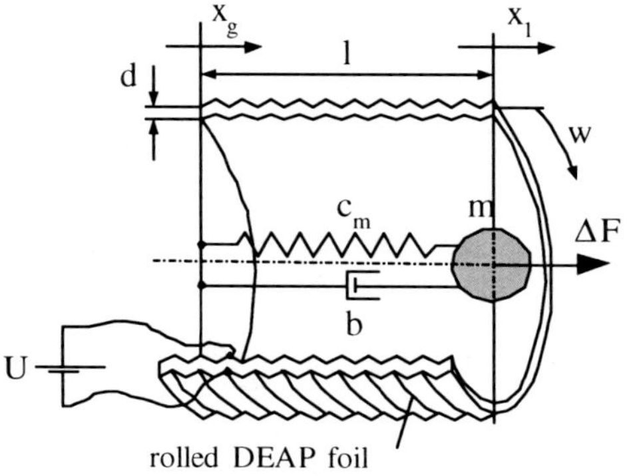

Fig. 1. Main parts of the actuator

Assuming that the volume $V=d\cdot l\cdot w$ of the soft DEAP foil is constant the Maxwell forces inside the foil can be calculated by derivation of the energy function of a capacitor (1).

$$W = \frac{1}{2}\cdot\varepsilon\frac{A}{d}\cdot U^2 = \frac{1}{2}\cdot\varepsilon\cdot\frac{w^2\cdot l^2}{V}\cdot U^2 \tag{1}$$

Denoting U^2 by P, the linearized force is

$$F = \frac{\partial W}{\partial l} = \varepsilon\frac{w^2}{V}P\cdot l \cong \varepsilon\frac{w^2}{V}\left(l_0\cdot P_0 + l_0\cdot\Delta P + P_0\cdot\Delta l\right) \tag{2}$$

The deviation of the force at P_0, l_0 is

$$\Delta F = F - \varepsilon \frac{w^2}{V} l_0 \cdot P_0 \cong \varepsilon \frac{w^2}{V} \left(l_0 \cdot \Delta P + P_0 \cdot \Delta l \right) = a \cdot \Delta P + c_e \cdot \Delta l \quad (3)$$

Inserting equation (3) into the mechanical model we get the equation of motion of the actuator and using $\Delta l = x_1 - x_g$

$$m \cdot \ddot{x}_1 + (c_m - c_e) \cdot x_1 + b \cdot \dot{x}_1 = a \cdot \Delta P + (c_m - c_e) \cdot x_g + b \cdot \dot{x}_g \quad (4)$$

Applying this result one can create the transfer functions of the actuator (Fig.2). To eliminate the nonlinear characteristics of the actuator, the model is extended by a square root stage.

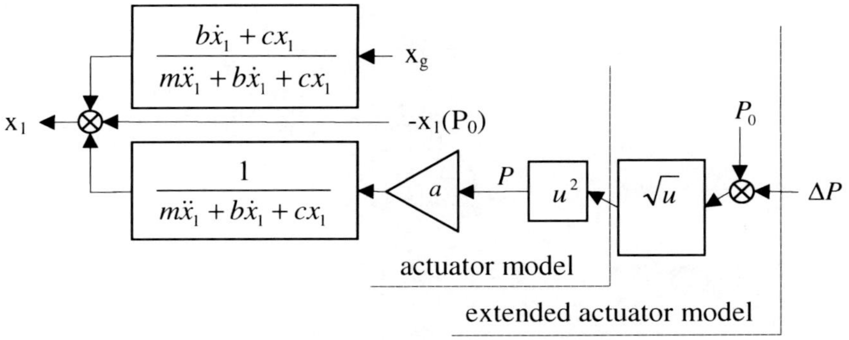

Fig. 2. Transfer functions of the actuator

4 Parameter Identification of the Transfer Function

To identify the parameters of the model we used the set-up shown in Fig. 3. As excitation a sweep signal from 0 to 160 Hz was applied on the input of the high voltage amplifier. Displacement of the moving part of the actuator was measured by a laser distance sensor. The measured data and the Bode diagram of the identified model are shown in Fig. 4. The matching between the curves was quite good.

The actuator proved to be a second order system with natural frequency of 80 Hz and damping ratio of 0.088.

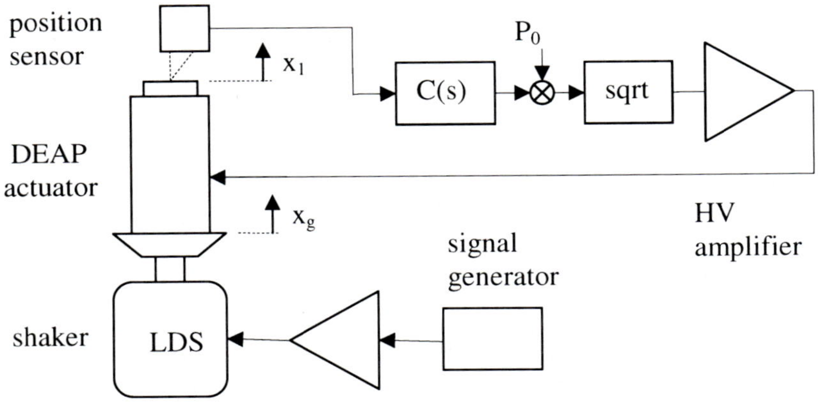

Fig. 3. Set-up of the active vibration control

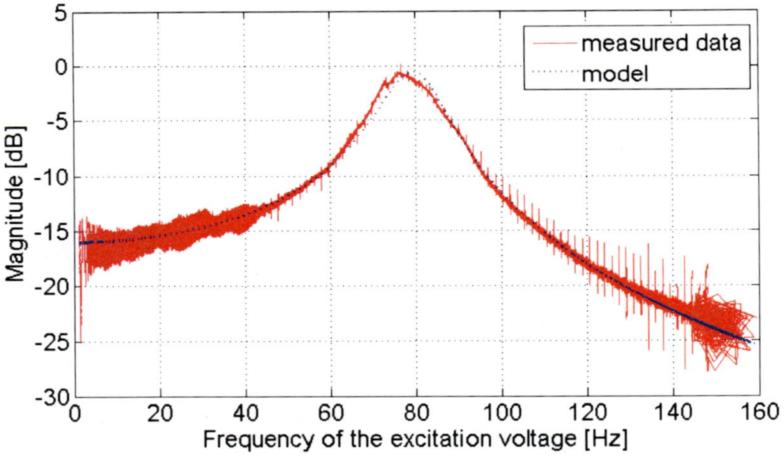

Fig. 4. Bode diagram of the actuator

5 Test Result of the Vibration Control

As shown in the Bode diagram the actuator function has a pair of complex poles. This pair can be cancelled by two zeros of the compensator. As a disturbance the actuator housing was moved by a shaker. The measured vibration and the current of the high voltage amplifier are shown on Fig. 5.

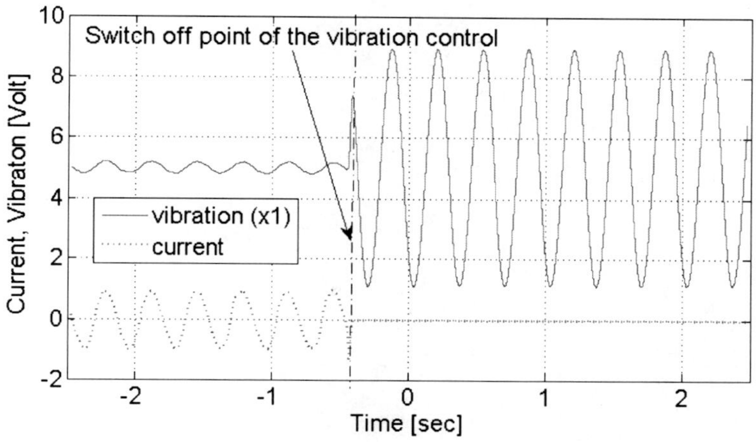

Fig. 5. Test result of the vibration control

6 Conclusions

This paper has presented dynamic modeling of DEAP actuator. In order to identify the parameters of the model measurements were carried out. Applied model agreed well with the measured data. As an example, an. active vibration control system designed on the basis of this model proved the availability of DEAP actuators.

Acknowledgement

Our research was supported by the OMFB -01007/2009 project of the Hungarian Ministry of Education.

Reference

[1] Kim, K.J., Tadokoro, S. (eds.): Electroactive Polymers for Robotic Applications. Springer, London (2007)

Contact Controller for Walking Gait Generation

V. Ondroušek* and J. Krejsa

Brno Technical University, Technická 2,
Brno, 616 69, Czech Republic

Abstract. This contribution is focused on the lowest layer of the control unit, which can be used for walking gait generation of the four legged robot in the rough terrain. The main aim is to design the feedback contact controller. This controller ensures optimal distribution of reacting forces in the landing points of the chosen legs using movements of the robot's body or changes the configuration of one leg to achieve the best position of the landing points in the unknown environment. Used approach enables to connect this base controller to another one in the exactly defined manner to achieve more complicated and sophisticated behaviors of the robot. The set of such composite controllers represents the lowest layer of the whole control unit based on the Discrete Event Dynamic Systems paradigm (DEDS). The dynamic simulation of the four legged robot is used to verify the proposed solution.

1 Introduction

The main issue of this article is to design the feedback contact controller, which can produce elemental behaviors of the four legged robot. This controller is a part of the complex architecture of the control unit, which ensures stable walking of the robot in the rough terrain. The architecture is based on utilization of the DEDS paradigm, and can be divided into three layers [1]. The lowest layer consists of the set of basic feedback controllers that ensure the elementary behaviors of the robot. The coordination mechanism, which operates with the formal descriptions of the basic and composite controllers, represents the middle layer. This layer predicts which instances of controllers are allowed to be activated in the specific configuration of the robot. These predictions are used by the highest layer, which can be implemented using any technique of reinforcement learning domain, e.g. Q-learning algorithm.

* Corresponding author.

2 Used Approach

The lowest layer of the architecture consists of a small set of the basic feedback controllers. These controllers are connected to the sensors and actuators and produce elementary behaviors of the robot, e.g. changing yaw angle of the body, changing configuration of one leg, etc. Basic controller can be defined by its potential surface. Each point of this surface represents a configuration of the robot. Activation of the controller means looking for the global extreme of the surface, which is an iteration process. Configuration of the robot is changing between each two steps of this process. The global extreme of the function can be found using gradient method, because of the strictly increasing character of the surface.

The members of the set of basic controllers, $C = \{\Phi_1, \Phi_2, \ldots, \Phi_N\}$, are designed in the general manner. Particular behaviors of the robot are formed due to connecting IO resources and the general design of the controller together. Lets have a contact controller Φ_f. The goal of this controller is the optimal distribution of reacting forces in the landing points of the legs using body movements. Legs of the robot, formally denoted as $0,1,2,3$ are used as an input or output resources. Position and orientation of the robot's body x, y, φ are used as output resources. The configuration of the leg 1 will be changed by activating controller instance, $\phi_{f\,\underline{1}}^{\overline{1,2,3}}$, to achieve a stable stance on the legs 1,2,3. Similarly, the translatory movements of the robot's body in the x a y directions will be performed by activating the controller instance, $\phi_{f\,\underline{x,y}}^{\overline{1,2,3}}$.

3 Implementation

The contact controller must be able to operate the robot in a rough terrain. The robot of interest is the four legged robot EQ3-KT produced by Lynxmotion Company. This robot has 3 DOF on each leg. This robot was placed on the inclined plane for the purpose of the controller design, see fig. 1. The superscripts A, B, C denote landing points of the legs, where the reacting forces have to be figured out. The angle α (γ) denotes angle between the plane Ω and the x (y) axis of the basic system of coordinates, $S_1 \equiv 0xyz$.

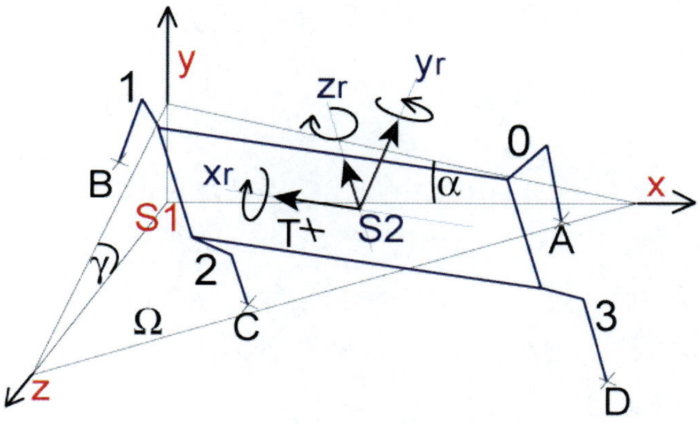

Fig. 1. Scheme of the robot on the inclined plain

All acting forces are shown on the figure 2, where F_G denotes gravity force affecting the robot in the COG point, $T = [T_X, T_Y, T_Z]$. The force $\vec{F}_N^A = \left(F_{NX}^A, F_{NY}^A, F_{NZ}^A\right)$ denotes normal component of the reaction force acting in the landing point of the leg 0 in the perpendicular direction to the plane Ω. Finally the force $\vec{F}_T^A = \left(F_{TX}^A, F_{TY}^A, F_{TZ}^A\right)$ denotes tangent component of the reaction force lying in the plane Ω.

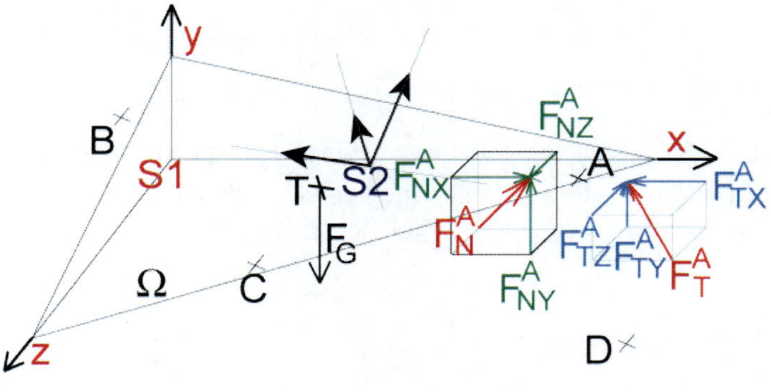

Fig. 2. Forces acting in the landing point A

Following power equations $(1)(2)(3)$ and torque equations $(4)(5)(6)$ can be used for description of the statically stable stance of the robot on the inclined plane.

$$F_{NX}^{A} - F_{TX}^{A} + F_{NX}^{B} - F_{TX}^{B} + F_{NX}^{C} - F_{TX}^{C} = 0 \tag{1}$$

$$F_{NY}^{A} + F_{TY}^{A} + F_{NY}^{B} + F_{TY}^{B} + F_{NY}^{C} + F_{TY}^{C} - F_{G} = 0 \tag{2}$$

$$F_{NZ}^{A} - F_{TZ}^{A} + F_{NZ}^{B} - F_{TZ}^{B} + F_{NZ}^{C} - F_{TZ}^{C} = 0 \tag{3}$$

$$\begin{aligned} &-F_{NY}^{A} A_Z - F_{TY}^{A} A_Z + F_{NZ}^{A} A_Y - F_{TZ}^{A} A_Y - F_{NY}^{B} B_Z - F_{TY}^{B} B_Z + F_{NZ}^{B} B_Y \\ &-F_{TZ}^{B} B_Y - F_{NY}^{C} C_Z - F_{TY}^{C} C_Z + F_{NZ}^{C} C_Y - F_{TZ}^{C} C_Y + F_G T_Z = 0 \end{aligned} \tag{4}$$

$$\begin{aligned} &+F_{NX}^{A} A_Z - F_{TX}^{A} A_Z - F_{NZ}^{A} A_X + F_{TZ}^{A} A_X + F_{NX}^{B} B_Z - F_{TX}^{B} B_Z \\ &-F_{NZ}^{B} B_X + F_{TZ}^{B} B_X + F_{NX}^{C} C_Z - F_{TX}^{C} C_Z - F_{NZ}^{C} C_X + F_{TZ}^{C} C_X = 0 \end{aligned} \tag{5}$$

$$\begin{aligned} &-F_{NX}^{A} A_Y + F_{TX}^{A} A_Y + F_{NY}^{A} A_X + F_{TY}^{A} A_X - F_{NX}^{B} B_Y + F_{TX}^{B} B_Y + F_{NY}^{B} B_X \\ &+F_{TY}^{B} B_X - F_{NX}^{C} C_Y + F_{TX}^{C} C_Y + F_{NY}^{C} C_X + F_{TY}^{C} C_X - F_G T_X = 0 \end{aligned} \tag{6}$$

The components of the normal and tangent forces can be find out based on the known geometry:

$$F_{NX}^{A} = F_{N}^{A} \sin\alpha, \quad F_{NY}^{A} = F_{N}^{A} \cos\alpha, \quad F_{NZ}^{A} = F_{N}^{A} \sin\chi \qquad (7)(8)(9)$$

$$F_{NX}^{B} = F_{N}^{B} \sin\alpha \quad F_{NY}^{B} = F_{N}^{B} \cos\alpha \quad F_{NZ}^{B} = F_{N}^{B} \sin\chi \qquad (10)(11)(12)$$

$$F_{NX}^{C} = F_{N}^{C} \sin\alpha \quad F_{NY}^{C} = F_{N}^{C} \cos\alpha \quad F_{NZ}^{C} = F_{N}^{C} \sin\chi \qquad (13)(14)(15)$$

$$F_{TX}^{A} = F_{T}^{A} \cos\alpha \quad F_{TY}^{A} = F_{T}^{A} \sin\alpha \quad F_{TZ}^{A} = F_{T}^{A} \cos\chi \qquad (16)(17)(18)$$

$$F_{TX}^{B} = F_{T}^{B} \cos\alpha \quad F_{TY}^{B} = F_{T}^{B} \sin\alpha \quad F_{TZ}^{B} = F_{T}^{B} \cos\chi \qquad (19)(20)(21)$$

$$F_{TX}^{C} = F_{T}^{C} \cos\alpha \quad F_{TY}^{C} = F_{T}^{C} \sin\alpha \quad F_{TZ}^{C} = F_{T}^{C} \cos\chi \qquad (22)(23)(24)$$

The reaction forces, $F = \left(F^A, F^B, F^C\right)$, in dependence on the robot's COG position, can by calculated using equations (1) to (24). The potential surface of the controller must have only one local extreme, which should be the global extreme as well. Thus the following equation for the potential surface of the contact controller design can be used:

$$\Phi_f = \sigma(F) = \sqrt{\frac{1}{N}\sum_{i=1}^{N}\left(F_i - \overline{F}\right)^2} \quad , N = 3, \tag{25}$$

where Φ_f denotes the contact controller, $\sigma(F)$ denotes standard deviation of reaction forces in the legs, that are connected to the specific instance of the controller as the input resources. Resulting potential surfaces of different instances of this contact controller are shown in the following chapter.

The design of the contact controller, described above, has to be verified. To avoid a damage of the robot, we have decided to use the software simulation for this purpose. First of all, it was necessary to find suitable tools that meet our requirements, e.g. high accuracy, incorporation of existing programming code of the controllers design, 3D graphical representation of simulation, involving static and dynamic effects from acting forces, etc. We have decided to build our own solution based on combination of Open Dynamics engine (ODE) and OpenGL, [3]. The engine is used to create virtual world with colliding bodies and acting forces. The OpenGL library is used for 3D graphical representation of the simulation. The Python is used as the main programming language for implementing whole solution. The Xml ODE format is used to describe simulated world with all its bodies, geometries, joints, acting forces and many others. The prepared model of four legged robot EQ3-KT was placed to the simulated world to verify the abilities and suitability of implemented contact controller [4].

4 Achievements

Resulting potential surface for the instance $\phi_{f\,\underline{x,y}}^{0,2,3}$ of the contact controller is shown on fig. 3. Calculation of this surface is performed for the initial configuration 1 of the robot, where all of the angles of joints are set to its mean value, i.e. the manipulability, M , of the robot in this position is maximal, [2]. Another potential surface of the same instance of implemented contact controller is shown on fig. 4. In this case, the initial configuration 2 of the robot is less favorable, because all of the hip joints angles are changed, see fig 5a.

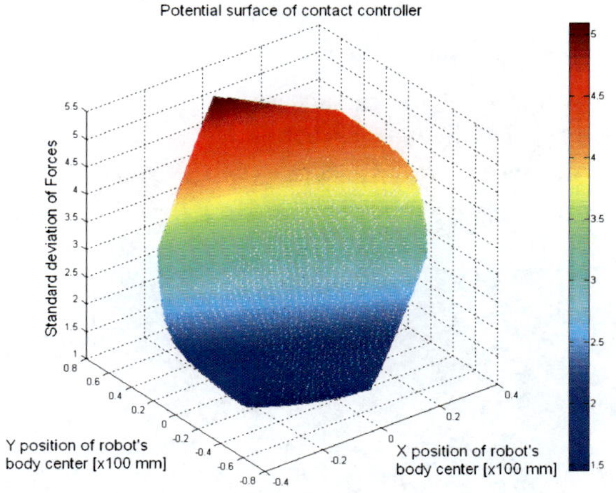

Fig. 3. Potential surface of $\phi_{f\,\underline{x,y}}^{0,2,3}$ for initial configuration 1

It is obvious, that there is only one global minimum representing the goal of an activation of the contact controller, in both cases. The goal can be reached using the gradient algorithm with regard to strictly decreasing character of the surface. Such a gradient algorithm has been used in the following verification experiment.

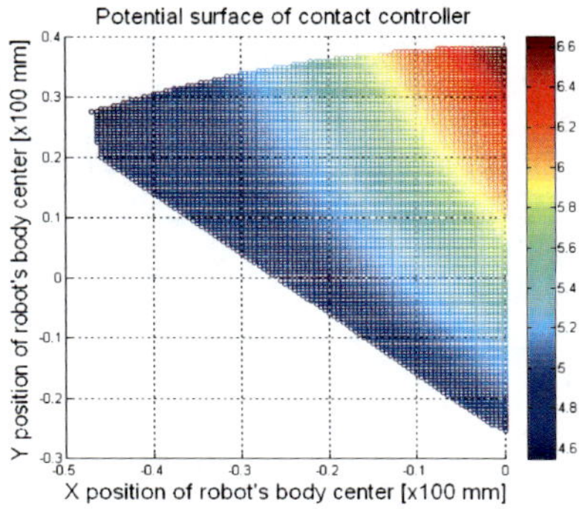

Fig. 4. Potential surface of $\phi_{f\,\underline{x,y}}^{0,2,3}$ for initial configuration 2

Two main different simulating experiments were performed to verify proposed design of the contact controller. The robot model was placed on the inclined plane in the first case. The position and yaw angle of the robot's body x, y, φ were used as output resources. Various combinations of legs were used as input resources. For example fig. 5 shows the process of activating the instance $\phi_{f\,\underline{x,y}}^{0,2,3}$, i.e. translatory movements of the robot's body from initial configuration to the new one with optimal distribution of reacting forces in the landing points of the legs $0, 2, 3$. Another set of various instances of the contact controller was simulated as well. The robot always reached the statically stable target position.

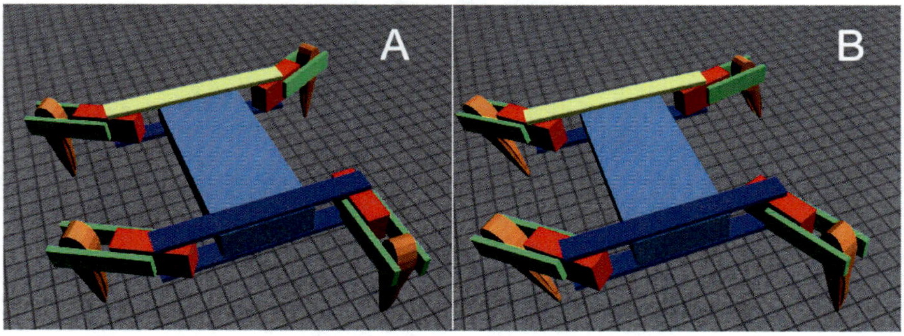

Fig. 5. Inclined plane (a) initial configuration 2 (b) final configuration

The planar surface with unknown static obstacles was used in the second set of experiments. Various instances of the contact controller with only one leg as an output resource were used. The activation of such instance changes configuration of one leg including the landing point of this leg. Configuration of the other legs as well as position and orientation of the robot's body are not changed during the activation process. The goal of such instance is to achieve the best configuration of the robot for translatory or rotary movements of the body, which will be performed in the next step of walking gait generation algorithm. The process of operating of the contact controller, $\phi_{f\underline{2}}^{\overline{0,2,3}}$, is shown on fig 6. The final configuration of the leg, see fig. 6d, can not be figured out prior to activation of the controller, because the robot has no information about a surface or obstacles and the output resource is equipped with one sensor only, which is able to detect contact between the leg and the surface or with the obstacle respectively. Thus the operating of the contact controller is an iterative process and the landing point of the leg (see the red crosses) is changed in each step of this process.

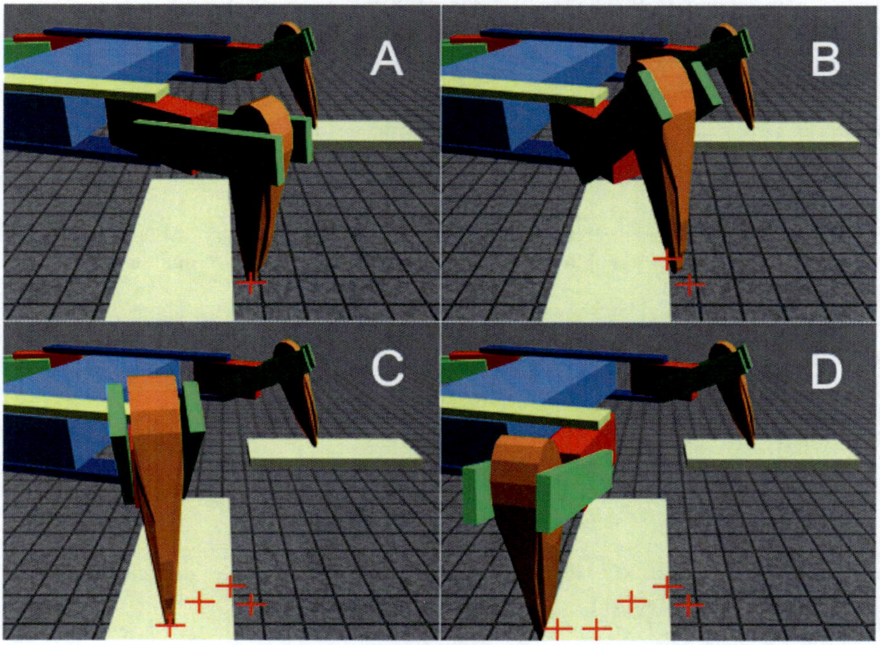

Fig. 6. Controller $\phi_{f\underline{2}}^{\overline{0,2,3}}$, planar surface with unknown static obstacles.

5 Conclusion

This contribution is focused on the design of feedback contact controller, which can be used for the four legged robot walking in the rough terrain. Each controller is designed in the general manner in the used approach. Particular behaviors of the robot are formed due to connecting IO resources and the general design of the controller together. The main goal of this controller is to achieve a statically stable stance with optimal distribution of reacting forces in the landing points of the chosen legs. This goal is reached using translatory or rotary movements of the robot's body or changing the configuration of one leg as well. The analytical analysis of the problem is given in the chapter 3. Various resulting potential surfaces of different instances of the designed contact controller are shown in the chapter 4. Each surface has a strictly decreasing character with one local minimum only, which is global minimum as well. This extreme represents the target configuration of the robot within an activation process of the controller. The suitability of the whole approach has been shown using the complex dynamic simulation of the robot acting on the inclined plane and planar terrain with unknown static obstacles as well. The accomplished tests has proven the ability to reach optimal distribution of reacting forces in the landing points of selected legs from various initial configurations of the robot. Future work will be focused on the design of composite controllers, which are able to produce more complex behaviors of the robot.

Acknowledgement

Published results were acquired using the subsidization of the Ministry of Education, Youth and Sports of the Czech Republic, research plan MSM 0021630518 "Simulation modelling of mechatronic systems".

References

[1] Huber, M.: A Hybrid Architecture for Adaptive Robot Control. MIT Press, Amherst (2000)

[2] MacDonald, W.S.: Legged Locomotion over Irregular Terrain Using Control Basis Approach. MIT Press, Amherst (1996)

[3] Ondroušek, V.: The solution of 3D indoor simulation of mobile robots using ODE. In: Recent Advances in Mechatronics 2008-2009., pp. 215–220. Springer, Berlin (2009), ISBN 978-3-642-05021-3

[4] Ondroušek, V., Věchet, S., Krejsa, J.: Design of the Control Basis for Walking Gait Generation Based on DEDS. In: EPE-PEMC 2010, Skopje, Republic of Macedonia, pp. 415–421 (2010), ISBN 978-1-4244-7854-5

Design and Application of Multi-software Platform for Solving of Mechanical Multi-body System Problems

A. Sapietová*, M. Sága, P. Novák,
R. Bednár, and J. Dižo

University of Žilina, Univerzitná 1,
Žilina, 010 26, Slovakia

Abstract. In this paper is presented the design and application of multi-software platform for solving of mechanical multi-body system problems. Built ADAMS working interface and open architecture of MATLAB programming language enable share common data during parallel run of simulations. This was used for design and implementation of evaluation algorithm and optimization of technical equipment parameters in terms of their mechanical properties. The goal is to present the maping process of the working space of the chosen manipulating equipment and aplication of the optimising approach to find its geometry parameters. Multi-software technique will be applied for computational realisation i.e. Adams and Matlab.

1 Introduction

The problem during the production machine synthesis solving is to find the suitable construction and the corresponding suitable parameters of the mechanism (i.e. materials, dimensions of body shapes and cross-sections, algorithms for the movement controlling, etc.). The parameters must fulfill some functional and technological requirements. These requirements may relate only to the position of the mechanism elements. In this case we call it the geometrical synthesis. In the other cases we must deal with the velocity and acceleration requirements. The methods solving these problems are the scope of the kinematics synthesis.

The dynamical requirements (i.e. balancing, reaction forces reduction, motion irregularity reduction, required motion properties, etc.) are solved by the dynamical synthesis [1].

* Corresponding author.

The area of the mechanisms synthesis has been primarily defined as a relationship between the input and the output (1:1). However, this area is more complicated for the production machine mechanisms and it covers the whole mechanics and the control area. The solutions can be characterized as multiple inputs – outputs (M:N). Therefore it is needed to extend the traditional mechanic techniques with the techniques focused on the general design theory [1].

2 Files Preparation for the Kinematics Synthesis of the Solid System Virtual Prototype

The solution objective is to design and implement the algorithm of the evaluation of the operating gear parameters from the mechanical properties view [2]. The approach is based on the solution of the fixed solid system (FSS) virtual prototypes by means of the computer simulation, using the kinematics and the optimization [6]. There is tendency to eliminate the time-consuming analyses so that the program interconnection of the ADAMS and MATLAB systems will be designed and implemented. For the inverse kinematics problem solving the virtual prototyp (VP) manipulator robot has been chosen (fig. 1). The solution contains the following processes:

- Virtual prototype preparation in the ADAMS/View environment,
- modification of the simulation,
- creation of an "*.acf" file,
- data export into an "*.adm" file,
- creation of programs in MATLAB environment,
- inverted kinematics problem,
  - the creation of the point of interest working positions map,
  - the movement trajectory specification for the point of the interest,
  - the selection of the points from the working positions map that copy the prescribed
  - trajectory in the best way,
  - the optimization of the solid system element lengths.

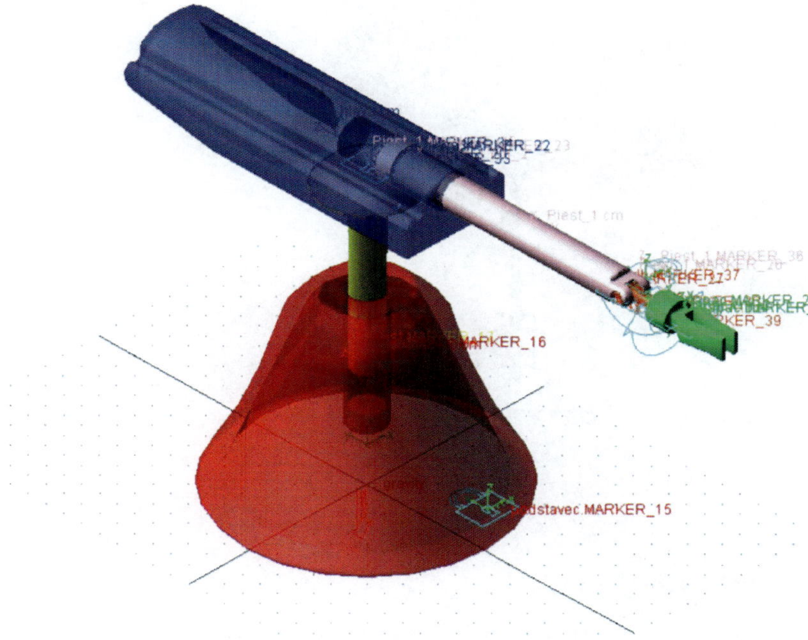

Fig. 1. Virtual prototyp manipulator robot in environment ADAMS/View

2.1 Virtual Prototype Preparation in the ADAMS/View Environment

The particular elements of the manipulator robot were designed in the ADAMS/View environment. Kinematic schema of the manipulating equipment has four movement degrees of freedom (fig. 2). Quantity "v" is speed of motion and quantity "ω" is speed of angle.

During the creation of virtual prototypes the models in the ADAMS environment are saved in the various file formats according to their purpose. The manipulator robot has been created in the binary file format (model.bin). This database contains the information about the working space configuration, all properties of one or more models and the analyses and the simulations results too. The binary format allows the quick writing and loading of the data and it can be transferred between computers with the different operating systems but it does not allow their reading and editing.

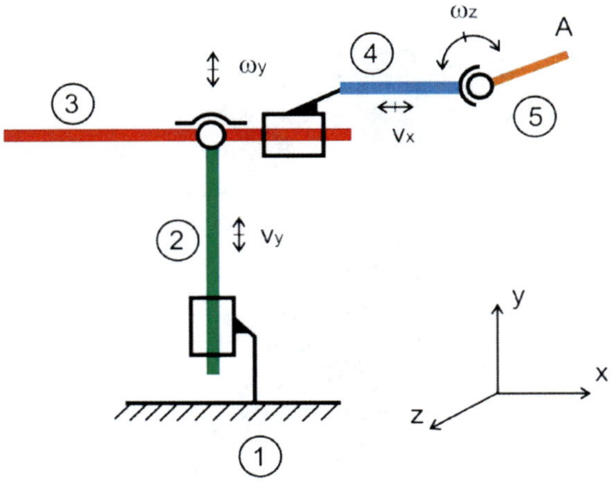

Fig. 2. Kinematic schema manipulating equipments

2.2 Modification of the Simulation

In the modification of the simulation, the prescribed process of the movements for all working cycles is defined. Modification of the simulation is created so that motion was directed gradually. In the particular joints the following movements are prescribed (Fig. 2, Fig. 3):

MOTION_1 ["-1*time"] % displacement PART_2 - v_y

MOTION_2 [-200] % resistance of the PART_2 in the given position

MOTION_3 ["1d *(time-200)"] % rotation of the PART_3- ω_y

MOTION_4 [180d] % resistance of the PART_3 in the given position

MOTION_5 ["1*(time-380)"] % displacement PART_4 - v_x

MOTION_6 [100] % resistance of the PART_4 in the given position

MOTION_7 ["-1d*(time-480.0)"] % rotation of the PART_4- ω_z

When the solid moves and it needs to stop afterwards, the zero movement must be activated for it in the time of the stop. Therefore more movements (MOTION_1, MOTION_2) are included into the JOINT (fig. 3).

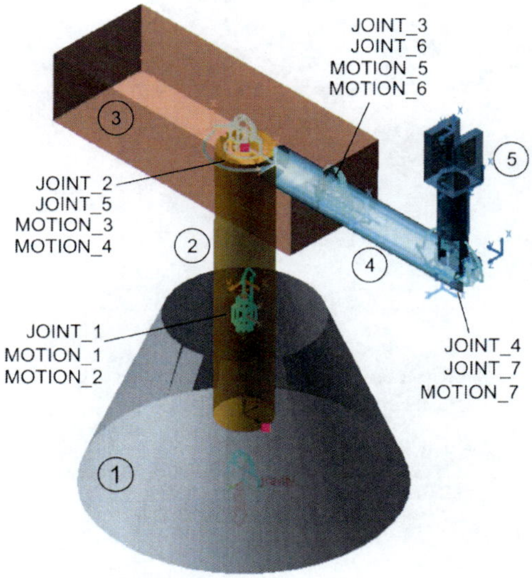

Fig. 3. Joints and motions of the manipulating equipment

Functionality is the virtual prototipe (VP) verified by the simulation in MSC. ADAMS/View environment and the kinematics parameters are calculated. Diagrams on figures 4 to 6 illustrative trajectory centre of gravity particular parts.

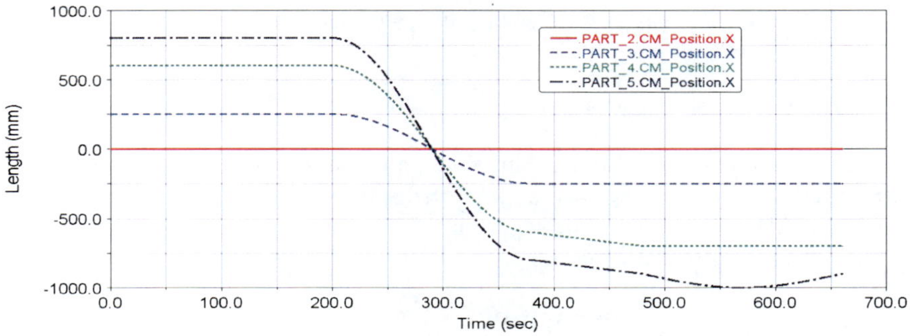

Fig. 4. Trajectory centre of gravity particular parts – position x

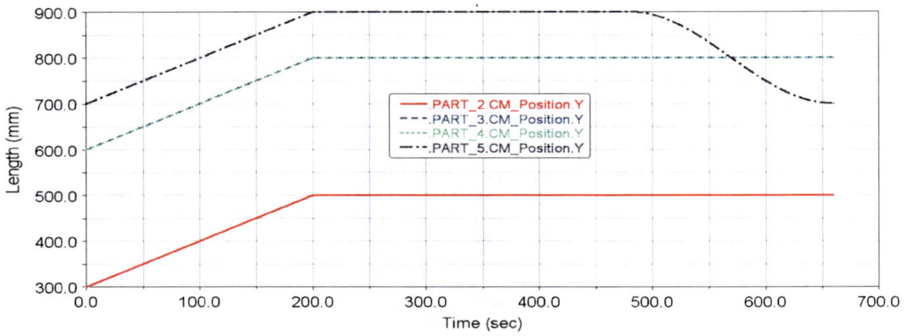

Fig. 5. Trajectory centre of gravity particular parts – position y

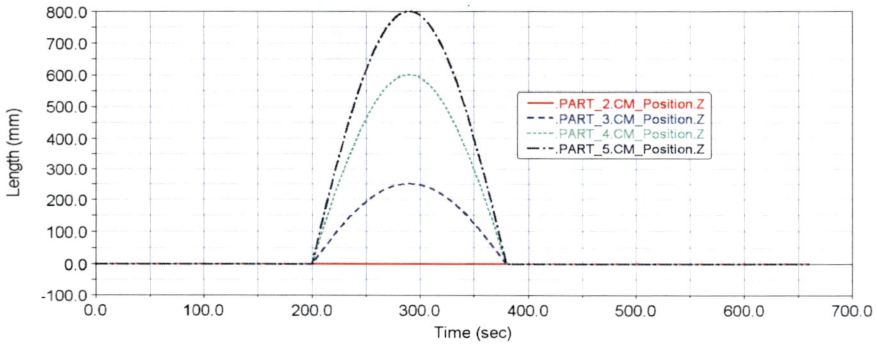

Fig. 6. Trajectory centre of gravity particular parts – position z

2.3 Creation of an "*.acf" File

Values for simulation modification instruction are stored into the new created file with "*.acf" suffix. This file can be imported into ADAMS environment with the aim to run ADAMS/Solver with the prescribed movement simulation modification and the consequent system solution

2.4 Data Export into an "*.adm" File

Model in the "*.adm" format is the record in its most saving form for sending to the ADAMS/Solver environment. It contains the data only for one model; it does not store the environment configuration. The advantage is the ASCII format in which the "*.adm" file can be comfortably read, edited and transferred between computers with the different operating systems [5]. This file can be again imported into ADAMS environment in order to run the ADAMS/Solver.

2.5 Creation of Programs in MATLAB Environment

Programs created in the MATLAB environment control the computation approach. These programs are stored in the following files:

- ***Optim_GENERAL.m*** – this file contains optimization methods with the call of ***dona.m***,
- ***dona.m*** – the file for computation of the objective function and for the solution of the point positions with the minimal distance from the prescribed trajectory. This program calls the subprogram ***runAdams.m***,
- ***runAdams.m*** – it runs the ADAMS system and creates the result file ***model.res***,
- ***opt_uhl.m*** – it optimizes the output parameters, e.g. angels of rotation in order to obtain the continuous movement.

2.6 Inverted Kinematics Problem

The inverted kinematics problem solves the suitable input kinematics parameters multibody of the systems (MBS) for obtaining of the described movement [2]. The graphical diagram of the particular program relations is shown on the figure 7. Assuming the stepping motor realistic usage with step of 5 degrees, all working positions of the manipulator robot endpoint "A" are mapped (fig. 8). From the obtained working position map the coordinates of those points have been chosen which correspond with the prescribed trajectory in the best way and which meet the condition of the continuous movement MBS at the same time.

The input kinematics parameters ensuring the given element prescribed movement MBS are the solution results. The chosen results of the rotation angle values a posunutí are shown in the table 1. Their number depends on the solution step size. Parameter „r" is minimal distance of the solved point from trajectory.

Element lengths have been changed from the original $L_2 = 500$ mm and $L_5 = 200$ mm to the optimal ones by the coordinate parametrising of the markers. The optimal values of the manipulator robot element lengths are: L2 = 515,92 mm, L5 = 191,85 mm.The genetic algorithm method has been used. The objective function value was a numerical zero. In order to obtain the continuous movement the optimal length values has been used in the program ***opt_uhl.m***.

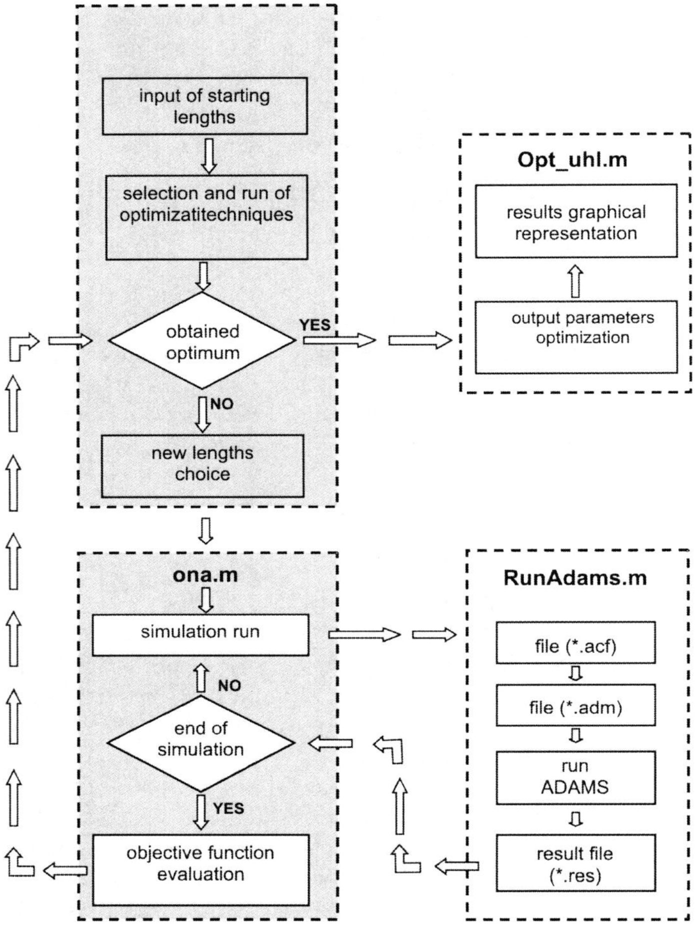

Fig. 7. Graphical representation of the inverted kinematics problem solution.

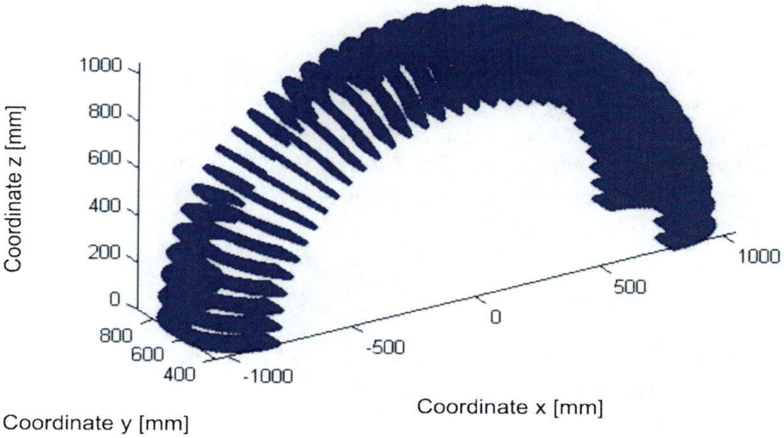

Fig. 8. Point "A" – map of working positions

Table 1. Evaluation of monitored values in the MBS synthesis.

displacement PART_2 [mm]	angle PART_3 [deg.]	displacement PART_4 [mm]	angle PART_5 [deg.]	position x [mm]	position y [mm]	position z [mm]	min. distance [mm]
60	2,5	50	80	1022,0935	549,9583	44,6256	0,0792
80	7,5	30	80	1014,3148	569,9583	133,5371	17,8553
100	10	30	80	1007,5246	589,9583	177,6538	8,9280
10	15	15	45	987,9135	610,0000	264,7106	8,9294
30	17,5	15	45	975,4267	630,0000	307,5509	17,8540
50	22,5	15	45	944,9101	650,0000	391,3946	0,2367
70	27,5	15	45	907,2021	670,0000	472,2595	17,8540
90	30	15	45	885,7390	690,0000	511,3817	8,9294
0	35	35	10	838,0477	710,0417	586,8073	8,9294

3 Conclusion

The goal of the paper was to suggest and apply the approach for the synthesis solution and the consequent manipulation device analysis using the interconnection of ADAMS and MATLAB programs. From the experiences obtained during the problem solution, the next goals can be formulated; to modify the program in order to shorten the computation times and to use the whole range of possibilities offered by the ADAMS software.

Acknowledgement

The part of the results of this work has been supported by VEGA grant No. 1/0463/08 and the KEGA grant No. 3/5028/07.

References

[1] Valášek, M.: Synthesis of mechanisms of production machines. In: 22 th Conference with International Participation, Computational Mechanics 2006, pp. 29–38. TYPOS-Digital Print, Plzeň (2006)
[2] Arnold, M., Schiehlen, W.: Simulation Techniques for Applied Dynamics: CISM Courses and Lectures, vol. 507, p. 313. Springer, Wien, New York (2008)
[3] De Jallón, J.G., Bayo, E.: Kinematic and Dynamic Simulation of Multi-body Systems. The Real-Time Challenge. Springer, New-York (1994)
[4] Segľa, Š., Ciupitu, L., Reich, S.: Optimization of a spring balancing mechanism for parallelogram robor mechanisms. Journal mechanisms and Manipulators 5(2), 43–48 (2006)
[5] Vaško, M., Sága, M., Handrik, M.: Porovnávacia štúdia vhodnosti vy-braných numerických postupov používaných v IKPA. Acta Mechanica Slovaca 12(3), 413–422 (2008)
[6] Dekýš, V., Sága, M., Žmindák, M.: Some aspects of structural optimiza-tion by finite element method. In: Proc. of the international scientific conference: Innovation and utility in the Visegrad fours. Agricultural and mechanical engineering, Nyiregyhaza, pp. 605–610 (2005)

The Map Merging Approach for Multi-robot Monocular SLAM

D. Shvarts[1] and M. Tamre[2]

[1] Tallinn University of Technology, Jarvekulla tee 75, Kohtla -Jarve, 30322, Estonia
[2] Tallinn University of Technology, Ehitajate tee 5, Tallinn, 19086, Estonia

Abstract. In carrying out the movement in an unfamiliar area, a mobile robot must solve a number of important tasks. He should be able to measure, build a three-dimensional map and to determine its position on the map. With access to the map, the robot is able to plan its route. However, such an effective navigation is limited to the boundaries of the area.

If it is necessary to move beyond a well-known area, the robot needs more information. Having access to maps built by other robots that work in adjacent areas, the robot receives the missing information.

In this paper the known methods of collaborative maps created by different robots are presented. Methods for obtaining three-dimensional maps of autonomous robots based on the flow of video data, as well as methods for combining maps of adjacent areas are investigated. A modified method of connecting the maps created by different robots is proposed.

1 Introduction

The term navigation is commonly understood as the technology of computing the optimal route of the mobile robot. In the different applications of navigation the area of effective planning is limited to the available map of the environment. To construct this map SLAM technology is widely used. The robot builds a map and localizes itself on it in the real time. However, in many practical applications, there is a need to make movement in the adjacent area about which the robot has no prior information. In such situations, the robot is unable to navigate effectively. To increase the accuracy of the maps created by the construction of maps of large areas a group of robots is used. An example of such application is shown in Figure (1).

The Robot number 1 is operating in the R1 area. Using the SLAM technology, the robot builds a map of this area. When you receive an order to move into the region of R8, in front of the robot arises the problem of calculating the optimal route in a given area. Not having a global map, the robot comes across the problem of the further development of the map. In this case, we cannot undertake a navigation at all. In order to navigate, the robot should get maps of the missing areas. The problem is to combine these maps with the map of the area R1 creates by the first robot. In this paper, methods for measuring 3-dimensional coordinates of points in world frame are investigated, an option of the map merging created by different robots which move in the neighboring areas is suggested.

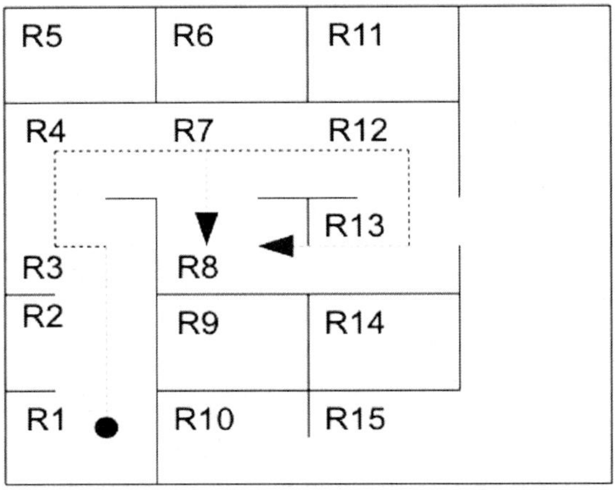

Fig. 1. An example application using the union maps, created by various autonomous robots.

2 Overview

The Problem of co - SLAM consists of 2 parts:

- individual monocular SLAM
- maps merging

Technology of individual SLAM is described in detail in [1], [2]. We provide more details only about the methods of measuring of 3D coordinates of points.

2.1 Methods for Constructing 3D Coordinates

Nowadays, there is a great interest in the methods of construction of 3D coordinates with monocular video system. Such methods are used in individual SLAM.

The problem of finding the three-dimensional coordinates of objects is the inability to measure it only in one image. The only way to obtain 3D coordinates of the environment is to use in the calculations a series of successive frames of the same point. Several ways of measuring the 3D coordinates of points are known and used to construct its three-dimensional model. The method described in [1] is based on time-lapse evolution of the probability density. Evolution of finding the coordinates of a landmark is shown in figure 2 [2].

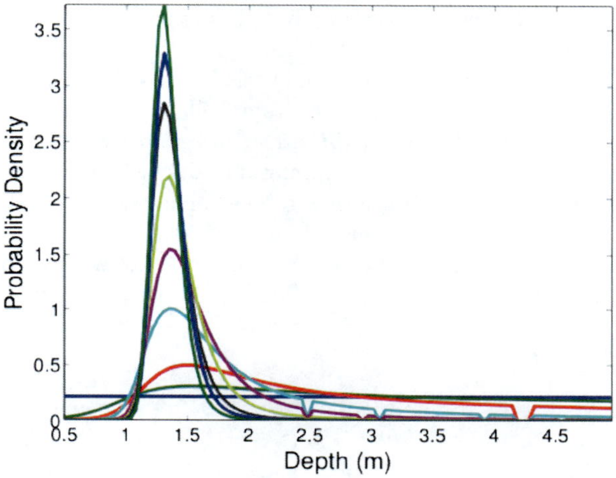

Fig. 2. Frame-by-frame evolution of the probability density over feature depth represented by particle set. 100 equally-weighted particles are initially spread evenly along the range 0.5m to 5.0m; with each subsequent image measurement the distribution becomes more closely Gaussian. [2]

The disadvantage of this method is the restrictions on the likely position of a point on the beam from the optical center of the camera (in this case, five meters). This is because only the points located at a short distance from the optical center of the camera can be used with respect to its movement. In such case a useful parallax is created. As is known, the point located at a considerable distance from a camera is useful in determining the rotation of the camera.

The second method, described in [2] uses the property of the camera to measure the angle between the optical axis and a point lying 3D world. This method uses a sequence of images created by a moving camera. The disadvantages of the method are volumetric calculations associated with an unusual presentation of 3D point. Each point is represented by 6 - dimensional vector, in contrast to the standard representation in Euclidean XYZ coordinates. The key to this method is inverse depth representation of a distance from the optical center of the camera to the observation point. An advantage of this method is that the accuracy of the determined depth of the point is high enough, the coordinates of a point can be determined and with a slight parallax.

2.2 Map-Alignment Problem

In [3] two ways are suggested to solve the problem of combining maps created by different robots. The problem lies in finding the transformation matrix between two coordinate systems of mobile robots. The first approach suggests finding matching points in the two systems. The desired transformation matrix is the one that produces the greatest possible number of correspondences between points. An alternative approach represents measuring the missing data for the calculation. It is proposed to use the computational capabilities of robots, as well as additional equipment installed on them to measure both the distance between them and their relative orientation. Due to noise in this measurement, the estimated transformation may be inaccurate, which may adversely affect the ultimate result of the merged maps. This method assumes map overlapping and direct communication between robots. But in many practical applications, robots cannot see each other, hence the direct measurement of the distance and bearing would not be possible. In this case we need to impose additional conditions that could link the two coordinate systems of robots. As suggested in [3], this step is called hypothesis generator step.

2.3 Problem Formulation

In this section, we discuss the problem of map merging. Our method allows us to avoid direct contact between the two robots. This means that we cannot measure the distance and bearing directly by the devices installed on the robot. In addition, we want to get rid of the expensive laser measuring systems, thus simplifying the design of the robot and reducing its cost. To solve this problem, we propose to introduce a third, independent system of coordinates. If we know the orientation and distance of mobile systems, relative to a third fixed system, the problem of calculating the relative position of robots becomes trivial. As the fixed system the coordinate system of a stereo rig is taken. For simplicity of presentation of mathematical expressions, we consider the case of two robots in 2D space. These expressions can be easily extended to situations with a large number of robots operating in 3D space.

2.4 Metrological Estimation of the Distance between Two Robots

Consider two coordinate systems $\{C_1\}$and $\{C_2\}$, associated with cameras mounted on two mobile robots, and coordinate system of a stereo rig $Z_c O_c X_c$. Let's call it a landmark. Position of the landmark is determined with respect to each camera in a fixed time. Coordinates of each moving camera $\{C_1\}$ and $\{C_2\}$ are given in the image plane of the landmark "M" by their own transition vector ρ_M^{C1} and ρ_M^{C2} with the appropriate variance σ_M^{C1} and σ_M^{C2} . Transition vector that connects the landmark and the camera and defined relative to the camera, is estimated by ρ_{C1}^{M} and ρ_{C2}^{M} respectively, with variance σ_{C1}^{M} and σ_{C2}^{M}. Since estimates of the transition vectors ρ_{M}^{C1} and ρ_{C1}^{M} are independent, a more accurate estimate of the transition vector ρ_1 between the camera and the landmark is calculated as a weighted average between the two above estimates of measurements.

$$\rho_1 = \sigma_\rho^2 \left(\frac{\rho_M^{C1}}{\left(\rho_M^{C1}\right)^2} + \frac{\rho_{C1}^{M}}{\left(\rho_{C1}^{M}\right)^2} \right), \ \sigma_\rho^2 = \frac{1}{\left(\rho_M^{C1}\right)^2} + \frac{1}{\left(\rho_{C1}^{M}\right)^2} \quad (1)$$

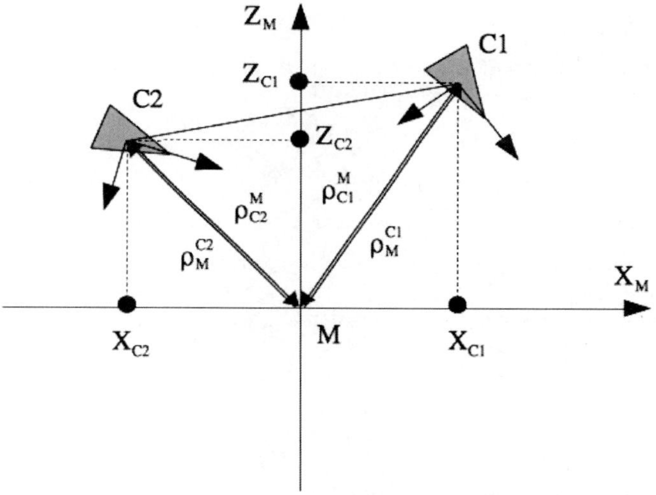

Fig. 3. The position of the coordinate system of two cameras in the stereo rig frame.

The resulting vector of measurements is:

$$Z = \begin{bmatrix} \rho_1 \\ \rho_2 \end{bmatrix} + \begin{bmatrix} \eta_{\rho 1} \\ \eta_{\rho 2} \end{bmatrix} = Z(t) + \eta \quad (2)$$

where η - white zero-mean Gaussian noise process.

Knowing ρ_1 and ρ_2 we can find the coordinates of cameras in the stereo camera frame. Then the estimate of a distance between the points C_1 and C_2 is defined:

$$\rho = C_1 C_2 = \sqrt{(X_{C1} - X_{C2})^2 + (Z_{C1} - Z_{C2})^2} \tag{3}$$

2.5 Estimating the Relative Orientation of Two Robots

In calculating the relative orientation of the cameras mounted on mobile robots, fixed coordinate system of a stereo rig is used as a linking element. The orientation of coordinate system of each moving camera relative to coordinate system of the fixed stereo rig is determined. However, in this case, the angles that are necessary for calculating are defined in both as coordinate system of cameras and in the landmark frame. The angle φ_1 is the angle between the optical axis of the camera and the transition vector ρ_1 with is defined in a coordinate system of the stereo rig.

$$\varphi_1 = a\tan\left(\frac{X_{Ci}}{Z_{Ci}}\right) \tag{4}$$

Angle β_1 is defined in the coordinate system of the camera. The Angle ψ_j indicates the orientation of the camera with respect to the coordinate system of the stereo rig and can be calculated from the expression:

$$\psi_{i,j} = 90 - \varphi_{i,j} - \beta_{i,j} \tag{5}$$

For the case, where $X_{ci} < 0$

$$\psi_{i,j} = 90 + \varphi_{i,j} - \beta_{i,j} \tag{6}$$

Figure (5) shows estimated orientation of coordinate system of each camera in coordinate frame of the stereo rig. The mutual orientation of the camera may be estimated from the following expression:

$$\Theta_i^j = \Theta_{Ci}^M - \Theta_{Cj}^M \tag{7}$$

where $\Theta_{Ci,j}^M$ is orientation of the camera in the coordinate system of stereo rig.

$$\Theta_{Ci,j}^M = 180 + \psi_{i,j} \tag{8}$$

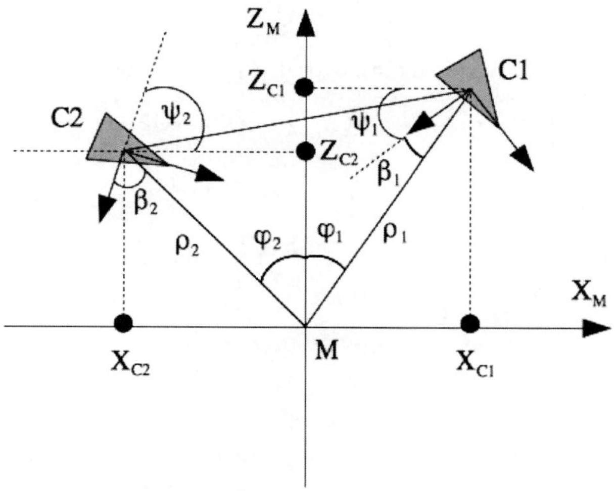

Fig. 4. Estimation of a bearing of mobile cameras in the coordinate frame of a stereo rig

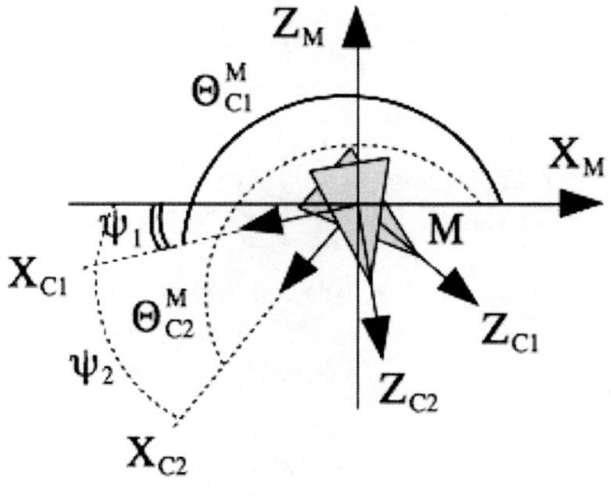

Fig. 5. Orientation of coordinate frame of mobile cameras in the stereo rig coordinate frame.

2.6 Building a Global Map

The purpose of this chapter is the transformation of the state vector X_2^{G2} defined in a coordinate system $\{G_2\}$in the state vector defined in $\{G_1\}$ (Fig. 6).

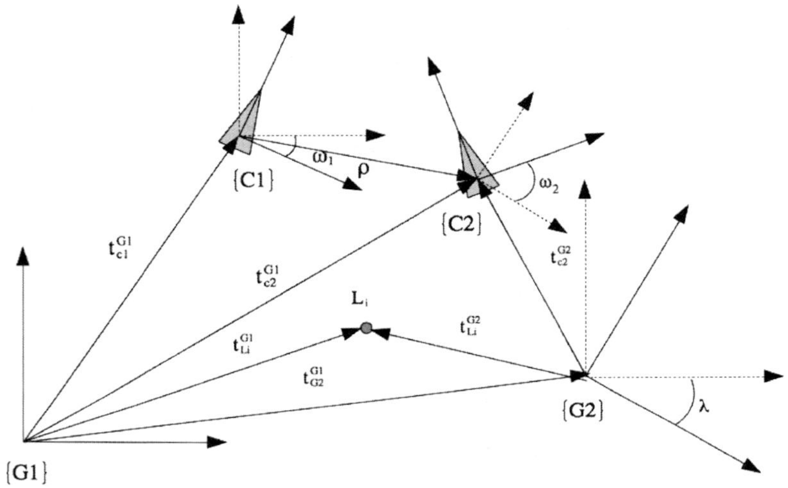

Fig. 6. The relative orientation of the global coordinate system $\{G_1\}$and $\{G_2\}$

State vector X_i^{Gi} is expressed by the following function:

$$X_i^{Gi} = f(X_{Ci}^{Gi}, x_{L1}^{Gi} ..., x_{Ln}^{Gi}) \tag{9}$$

where $X_{Ci}^{Gi} = \begin{bmatrix} x_i & y_i & \omega_i \end{bmatrix}^T$ - position of a camera and $x_{L_i}^{Gi} = \begin{bmatrix} x_{L_i} & y_{L_i} \end{bmatrix}$ - position of landmarks.

Knowing the transition vector t_{G2}^{G1} and the rotation matrix ${}^{G1}_{G2}T(\lambda)$ we can express the state vector X_2^{G2} defined in a coordinate system $\{G_1\}$as a function of X_1^{G1}, X_2^{G2} and Z:

$$X_2^{G1} = f(X_1^{G1}, X_2^{G2}, Z) \tag{10}$$

Rotation matrix can be easily calculated, as a matrix product:

$${}^{G1}_{G2}T(\lambda) = {}^{G1}_{C1}T\left(\omega_1\right){}^{C1}_{C2}T\left(\theta\right){}^{G2}_{C2}T\left(\omega_2\right) \tag{11}$$

and the transformation vector (the radius – vector) is calculated (see Figure 6) as follows:

$$t_{G2}^{G1} = t_{C1}^{G1} + {}_{C1}^{G1}T(\omega_1)\rho - {}_{G2}^{G1}T(\lambda)t_{C2}^{G2} \quad (12)$$

Camera position and landmarks, which are represented in the coordinate system, $\{G_2\}$ can be represented in $\{G_1\}$ as follows:

$$t_{C2}^{G1} = t_{C1}^{G1} + {}_{C1}^{G1}T(\omega_1)\rho \quad (13)$$

and position of any points as

$$t_{Li}^{G1} = t_{G2}^{G1} + {}_{G2}^{G1}T(\lambda)t_{Li}^{G2} \quad (14)$$

Consequently, the transformation of the state vector X_2^{G2} into a state vector X_2^{G1} performed according to expressions (13) and (14).

The accuracy of this transformation depends on the accuracy of the estimates of state vectors in two coordinate systems, as well as on the estimation of relative orientation of the cameras and the distance between them.

3 Conclusion

In this paper, an effective method to estimate the metrological distance between mobile cameras and bearing is proposed. This method will allow merging the maps created by different autonomous robots in a global three-dimensional map. The materials proposed in this paper are the basis for practical research in Co-SLAM.

References

[1] Davison, A.J.: Real-Time Simultaneous Localisation and Mapping with a Single Camera. In: Ninth IEEE International Conference on Computer Vision, p. 1403 (2003)
[2] Civera, J., Davison, A.J.: Inverse Depth Parametrization for Monocular SLAM. IEEE Transactions on Robotics 5, 932 (2008)
[3] Zhou, X.S., Roumeliotis, S.I.: Multi-robot SLAM with Unknown Initial Correspondence: The Robot Rendezvous Case. In: IEEE/RSJ International Conference on Intelligent Robots and Systems, p. 1785 (2006)

Optimization of Stator Channels Positions Using Neural Network Approximators

M. Sikora* and Z. Ančík

Institute of Solid Mechanics, Mechatronics and Biomechanics,
Faculty of Mechanical Engineering, Brno University of Technology
Technická 2, Brno, 616 69, Czech Republic

Abstract. This paper deals with an improvement of synchronous generator cooling. Finding of the best positions of stator radial channels is a way to cooling improvement. An optimization method for finding of the best channels positions is presented here. The main goal of optimization is to achieve uniform temperatures field of winding along slot length, without overheated locations. The Computational Fluid Dynamic (CFD) is used for estimating of stator temperatures. Input parameters for CFD models are computed by optimization algorithm which is able to predict stator temperatures behavior. This algorithm is based on neural networks approximators. There is a feedback between CFD models and optimization algorithm.

1 Introduction

At present, a high power-dimensions ratio is typical for many electric machines. It is therefore necessary to ensure a good removal of losses heat from compact machine interior. A using of stator radial channels is one of ways how to improve cooling of synchronous electric generators. So, an air which flow from air gap into these channels removes heat from stator packet (convective heat transfer) and cools the stator, as shown in Fig.1.

A uniform distribution of radial channels positions along stator length is used in majority of synchronous machines. This distribution is able to bring sufficient results for this application [1] but probably not the best results. There can be overheated locations, see Fig. 5. Hence, we should try to find optimal channels positions in stator in order to ensure uniform temperatures distribution in stator, without overheated locations. This optimization can extend possibility of machine overload, improve lifespan and reduce ohmic resistance of winding.

* Corresponding author.

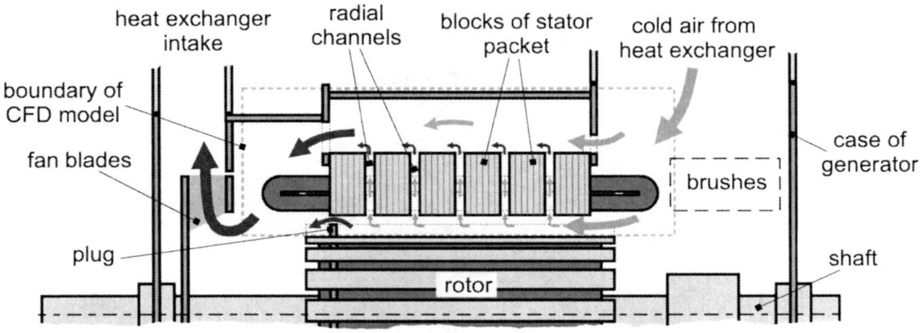

Fig. 1. Scheme of synchronous generator air cooling.

2 Used Approach

Our aim is optimization of generator cooling system. The optimization is focused on position of radial channels in stator. Therefore, we need to investigate and evaluate behavior of many thermo-mechanical and fluid parameters in optimized machine. This can be realized by experiment with real machine, by flow networks [2], thermal networks [3], [4] or by CFD modeling [5], [6], [7]. However, the way of experiment would be very expensive and difficult. That is why we decide to use the computational modeling and Computational Fluid Dynamic (CFD) methods.

We should also mention that we will consider invariable quantity of radial channels and also invariable width of these channels. The optimization will be focused only on finding of radial channels positions (respectively length of blocks of stator packet between radial channels).

2.1 Optimization Approach

There are several different approaches which can be used for channels positions optimization. The approach that we have chosen can be briefly described as follows:

- This approach uses knowledge of some physical phenomena in this real system.
- Is based on detailed assessment of previous CFD results, not only on resulted temperature field assessment.
- We can create approximating relations by the help of resulted CFD data and these relations can serve for prediction of physical values in various modifications of channel positions.
- We can use the predicted physical values and knowledge of physical phenomena in this real system. Therefore, we can calculate expected temperatures of stator without many CFD simulations and mesh modifications.

2.2 Physical Values Prediction

The second optimization approach is based on evaluation of physical values which affect resulted temperatures in stator. This approach is similar to approach presented in [8] and [9]. We have not to calculate these physical values by physical relations (it is often possible only by CFD), but we can evaluate these values by the help of approximation in imported CFD data. We should only find the variables which decisively affect the resulted (approximated) physical values. The idea of this concept is presented in Fig. 2. There are several methods for values approximation. We decide for approximation by neural network. It should serve very well for prediction of values which we need for calculation of stator temperatures and others variables. The scheme of considered optimization process is in Fig. 3.

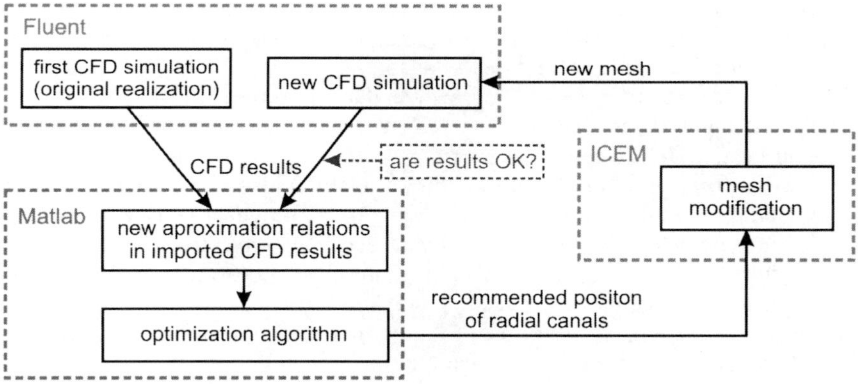

Fig. 2. Diagram of used optimization approach (second approach) and relations among separate steps.

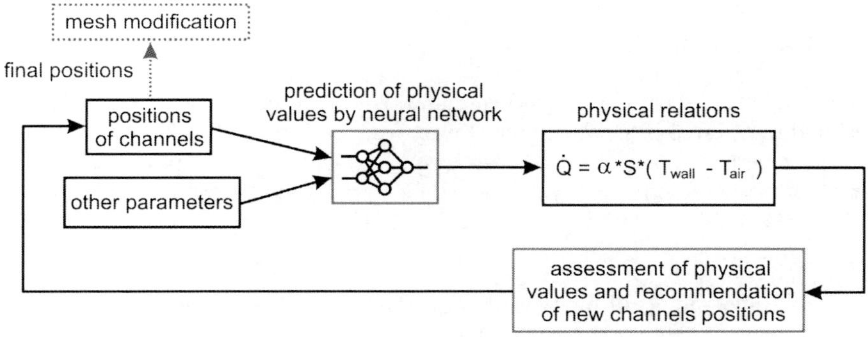

Fig. 3. Scheme of optimization process.

3 Implementation

3.1 Mesh and CFD Model

The CFD model is important for evaluation of temperatures field in original realization and in modified arrangements of stator radial channels. If we consider some simplifications in the machine, we will be able to use a rotational symmetry of generator in CFD model. These used simplifications relate mainly to rotor geometry and its influence on air flow in rotor-stator gap. Because we are limited by the computational performance, we will not model whole rotor pole, but we create only its simplified equivalent. We use Ansys-ICEM mesher for mesh creating and upgrading. For solution was used CFD software Fluent.

3.2 Reworded Goal of Optimization

A reaching of uniform temperatures in stator is our original goal of channels positions optimization. The stator winding and its impregnation and isolation of wires are the most temperature-dangerous locations in stator. Therefore, we should try to ensure uniform temperature along the winding without dangerous overheated locations.

A comparing of heat, that is generated in each block of stator packet (losses heat) and heat, which is getting away from each block, constitutes a way to solution of this optimization problem. In addition, it should happen by constant winding temperatures in each block - this is our goal. The generated heat has to be same as the heat, which is getting away by air cooling (convective heat transfer). It is every fulfilled in real machine in stable state. But in our optimization task we have to ensure balance of these heats. So, we can reword our goal: We should ensure the balance of these heats in all stator blocks at uniform temperatures of blocks (temperatures of winding in each block).

We have to also consider that the winding temperatures in each block will be same at the end of optimization - this is main goal of this task. So, there should not be any conductive heat transfers among stator blocks. This makes easier this task. But we have to consider heat flows into end winding outside stator packet.

3.3 Generated Heat and Heat Transferred by Convection

We should specify parameters and variables, which influence the generated losses heat and transferred heat, which is got away from each stator block.

The generated losses heat is very easy to calculate and depends mainly on length of block (determined by positions of channels) and specific heat losses.

The heat which is getting away (convective transfer) depends on these variables and parameters:

- Temperatures of block walls T_{wall}. These temperatures are predictable related to winding temperatures.
- Temperatures of cooling air near to block walls T_{air}.
- Area of block walls A. Depends only on block length (area of radial channel walls is constant).
- Heat transfer coefficients α.

The heat which is getting away from the block can be expressed as follows:

$$\dot{Q} = \alpha \cdot A \cdot \left(T_{wall} - T_{air}\right) \tag{1}$$

It is difficult to calculate variables α , T_{wall} , T_{air} in relation (1) by physical relations. Therefore, we try to predict these physical values by neural network approximators and by resulted values of these variables in previous CFD simulations.

3.4 Prediction of Physical Values

The detailed scheme of values predictions is presented in Fig. 4 and is implemented in Matlab. We used the neural networks as approximators for analyse of resulted CFD values. The neural network was trained by using of the appropriate CFD resulting data. At beginning, we started only with one file of results - CFD results of original stator realization with uniform channel distribution. It is evident that informative value of this initial data file is relatively poor and first values will not be probably predicted too accurately. But when we have further CFD results for neural network training, the approximators will be able to bring better and more accurate predictions.

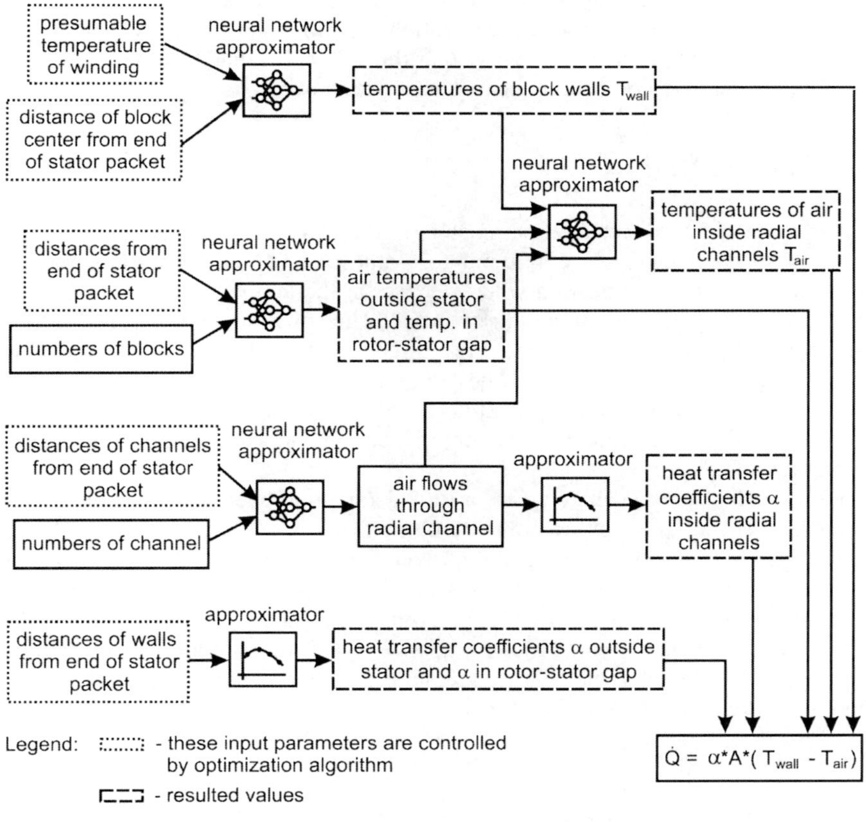

Fig. 4. Scheme of values prediction which is implemented in Matlab.

3.5 Optimization Algorithm

Our aim is balancing of generated heat and heat which is getting away from the block by convection. So, we can try to ensure this aim by optimization algorithm. This algorithm should minimize the differences between generated and transported heats; the test function can be expressed as follows:

$$K = \sum_{i=1}^{n} (\dot{Q}_{gen\ i} - \dot{Q}_{away\ i})^2 \tag{2}$$

Where n is number of blocks of stator packet. These differences depend on channel positions and also on wall temperatures which are in predictable relation with winding temperatures. These winding temperatures have to be the same in each block - this is main goal of this task. Therefore, the optimization algorithm should be able to change the channel positions and also the supposed collective temperature of winding which predictable influences temperatures of blocks walls.

So, there are n-1 positions of channels in n blocks and one collective temperature of winding. It constitutes n inputs parameters together and optimization algorithm have to balance heats in n stator blocks. So, this task must have only one exact solution. We used simply optimization algorithm which assess and reward responses (2) of input changes. This algorithm is implemented in Matlab and applies values predicted by neural network approximators.

4 Obtained Results

We obtain a plenty of resulted parameters at the end of each CFD simulation. A lot of these parameters serve as input for values predictions in Matlab. We can assess a benefit of optimization when we look at the temperatures fields in plane of slot symmetry. These temperatures fields of original stator realization and of first two modifications are presented in Fig. 5, 6 and 7. The maximal temperature of winding lowered after second mesh modification. The decrease of maximal temperature is approximately 4 °C. But, there is not significant difference between average winding temperature of original realization and average temperature after second modification.

We can also see that there is not a satisfactory improvement of winding temperatures distribution when we compare figure 5 and 6. There are overheated locations in both figures. Insufficient improvement of first modification results is probably related with poor quantity of CFD data which we could use for training of arpproximators. We had only CFD data from original stator realization at that time. But we can use two data files for approximators training after first modification. So, the predicted values during second run of optimization algorithm will be more accurate as a consequence of this.

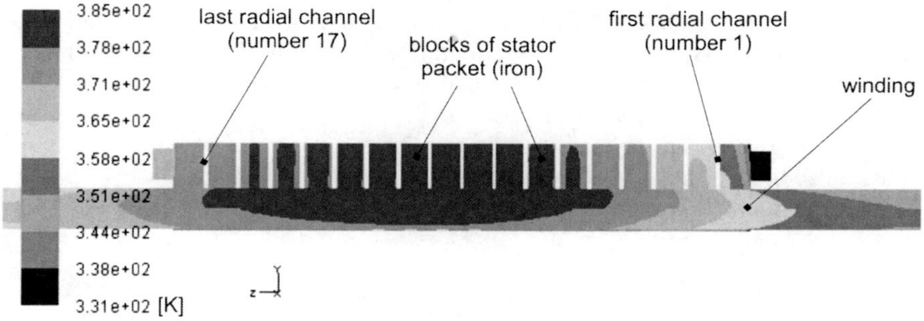

Fig. 5. Contours of temperature in symmetry plane of stator slot - original stator realization.

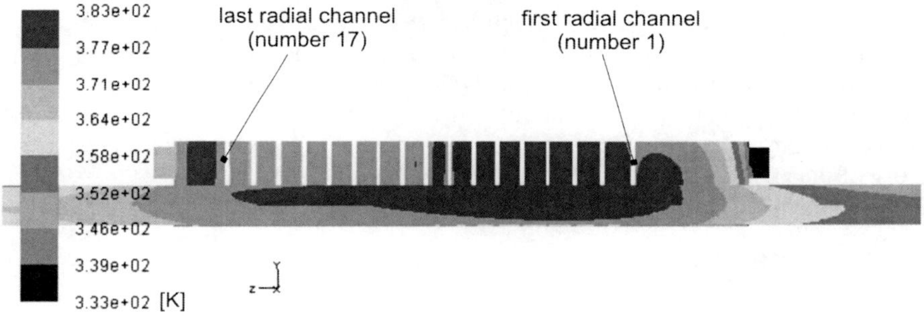

Fig. 6. Contours of temperature in symmetry plane of stator slot - after first mesh modification.

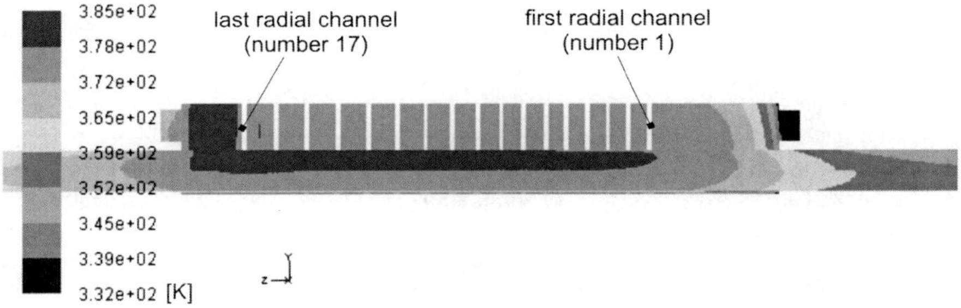

Fig. 7. Contours of temperature in symmetry plane of stator slot - after second mesh modification.

We have to also consider that we will not be able to create approximating relations with two inputs when the inputs are in linear relation. For example, channels numbers and channels positions in original stator realization are in linear relation (uniform distribution of channels positions). We had to use only one-input approximation for first optimization run in this case. We can approximate these two inputs as soon as we obtain data from first modification simulation. It is evident that this two-input approximation will be more accurate.

5 Conclusions

The optimization of radial channels positions constitutes one of ways to improvement of generator cooling. In addition, it is very inexpensive way because we have to change only quantity of laminations in blocks of stator packet. The mentioned method of optimization is very effective; we obtained almost uniform temperatures along winding very early, after two mesh modification. So this method is relatively time-undemanding. Results which were obtained after second stator modification brought decrease of maximal winding temperature (4 °C) against original stator realization. This is not a big decrease, but it surely improves possibility of machine overloading and winding lifespan.

Acknowledgment

This work presented in this paper has been supported by project Introduction of Problem Based Learning to Mechanical Engineering Curricula (CZ.1.07/2.2.00/07.0406) and research project FSI-S-10-15.

References

[1] Traxler-Samek, G., Zickermann, R., Schwery, A.: Cooling Airflow, Losses, and Temperatures in Large Air-Cooled Synchronous Machines. IEEE Transactions on Industrial Electronics 57(1) (2010), ISSN: 0278-0046

[2] Shiyuan, C.: Network analyses of ventilation system for large hydrogenerator. In: Proc. 5th ICEMS, August 18–20, vol. 1, pp. 137–140 (2001)

[3] Rajagopal, M.S., Seetharamu, K.N., Aswathnarayana, P.A.: Transient thermal analysis of induction motors. IEEE Trans. Energy Convers. 13(1), 62–69 (1998)

[4] See, Y., Hahn, S., Kauh, S.: Thermal analysis of induction motor with forced cooling channels. IEEE Trans. Magn. 36(4), 1398–1402 (2000)

[5] Depraz, R., Zickermann, R., Schwery, A., Avellan, F.: CFD validation and air cooling design methodology for large hydro generator. In: Proc.17th ICEM, pp. 1–6 (2006)

[6] Vlach, R.: Computational and Experimental Modeling of DC Machine Cooling by Heat Pipe. In: Proceedings of the 6th IASME/WSEAS international conference on heat transfer, thermal engineering and environment (HTE 2008), PTS I and II – New aspects of heat transfer, thermal engineering and environment, Rhodes, GREECE (2008), ISBN: 978-960-6766-97-8

[7] Mugglestone, J., Pickering, S.J., Lampard, D.: Effect of geometric changes on the flow and heat transfer in the end region of a TEFC induction motor. In: Proc. 9th IEEE Int. Conf. Elect. Mach. Drives, Air-cooled Large Turbine Generator with Multiple-pitched Ventilation Ducts, Canterbury, U.K., September 1999, pp. 40–44 (1999)
[8] Vlach, R., Grepl, R., Krejci, P.: Control of stator winding slot cooling by water using prediction of heating. In: 2007 IEEE International Conference on Mechatronics, Kumamoto, JAPAN, pp. 467–471 (2007)
[9] Vlach, R., Grepl, R., Krejci, P.: Predictor for control of stator winding water cooling of synchronous machine. In: Recent Advances in Mechatronics, pp. 190–194 (2007), ISBN: 978-3-540-73955-5

Identification of EMA Dynamic Model

Martyna Ulinowicz* and Janusz Narkiewicz

Department of Automation and Aeronautical Systems,
Warsaw University of Technology, Warsaw, 00-665, Poland
martyna.ulinowicz@meil.pw.edu

Abstract. The electromechanical actuator (EMA) model is presented with the methods for identifying its design parameters. The actuator is part of the system for flap deployment on the commercial transport airplane. The differential equations with the feedback control describe behaviour of the actuator deflection. For parameter identification the Maximum Likelihood (ML) method with two minimization algorithms: linearized Gauss-Newton and Levenberg-Marquardt is applied. Both approaches were effective, while tested on hypothetical data.

1 Introduction

The concept of More Electrical Aircraft MEA, recently attracts more interest due to trend for "greening" aviation operations. In aircraft control, the electric system will replace the hydraulic one used nowadays, resulting in saving weight and operational costs [1].

In the project NEFS - New Track integrated Electrical Single Flap Drive System, funded by EC under 6 FP the concepts are investigated for replacing conventional hydraulic drive system used for deploying wing flaps of large transport aircraft by individual, distributed electrical drive actuators integrated into each flap track beams [5].

The system performance is tested at the laboratory rig, but the system failures and their influence on aircraft is investigated by simulations. To make the integrated simulations of system and aircraft reliable, the actuator simulation model should reflect behaviour of a real device in a correct way. The required accuracy level of simulation model may be achieved by identification of the system model parameters using data from laboratory tests.

In this paper parameters of the nonlinear actuator model are identified using ML method.

2 EMA Model

In the project two concepts of drive systems are considered: a high torque/low speed and a geared low torque/ high speed. The actuator models of both systems are similar.

* Corresponding author.

The EMA system considered in this paper contains two drive stations 1 and 2 connected to aircraft flap. Each drive station contains a ball-screw driving the flap connected to common DC motor by U-joint and a gearbox. The motor is controlled by performing assumed function of flap deflection.

The actuator is described by two equations, electrical equation

$$L_A \frac{dI_A(t)}{dt} = -R_A I_A(t) - k_s \Omega(t) + U_A(t) \tag{1}$$

and mechanical equation:

$$J \frac{d\Omega(t)}{dt} = M_n - \left(M_{B1} + M_{TPER1} + M_{TPER2}\right), \tag{2}$$

where:

Ω - motor shaft angular velocity, R_A - armature resistance U_A - armature voltage, I_A - armature current, k_s - motor magnetic flux coefficient, J - inertia moment of rotating parts reduced to shaft [3].

The torque generated by the motor is calculated as:

$$M_n = k_n I_A, \tag{3}$$

where:

k_n - motor constant.

The M_{TPER1} and M_{TPER2} are the external torques from station 1 and 2.

Each station contains bevel gear, U-joint and ball-screw.

The M_B is the primary brake torque applied to the motor modeled as:

$$M_B = M_{B\max} - 0{,}5 M_{Bmax} \left(1 - e^{-K_B\left(\frac{\Omega}{\Omega_{0,5T_{\max}}}\right)}\right), \tag{4}$$

where:

K_B - electromagnetic brake constant coefficient, $M_{B\max}$ - maximal torque value (for $\Omega = 0$), $\Omega_{0,5M_{B\max}}$ -angular velocity value for half a maximal torque $M_{B\max}$.

In bevel gear the motor angular velocity Ω is reduced to Ω_{TP} with reduction ratio i_{TP}:

$$\Omega_{TP}(t) = \frac{\Omega}{i_{TP}} \tag{5}$$

External torque acting on the motor shaft is calculated as the sum of external torque from the U-joint M_{UER}, torque resulting from the losses in bevel-gear

M_{TP} and secondary brake torque M_{B2},which is equal to zero when the brake is not released:

$$M_{TPER} = \frac{M_{UER}}{i_{TP}} + M_{TP} + M_{B2} \tag{6}$$

The bevel gear moment due to mechanical energy loses is composed of viscous damping represented by B_{TP} coefficient and Coulomb friction C_{TP} [3]:

$$M_{TP} = B_{TP}\Omega_{TP}(t) + C_{TP}sign(\Omega_{TP}(t)) \tag{7}$$

The screw is connected to the gear-box by U-joint where the angular rotation is changed :

$$\Omega_U(t) = g_U\Omega_{TP} \tag{8}$$

where:

$$g_U = \left(\frac{\cos\beta}{1 - \sin^2\beta\cos^2\theta_{TP}} \right), \tag{9}$$

$$\theta_{TP} = \theta / i_{TP} , \tag{10}$$

β -angle between the input and output shaft axis,
θ - motor shaft angle of rotation.

The additional moment M_U due energy loses in U-joint is assumed in the same form as for the gear box:

$$M_U = B_U\Omega_U(t) + C_U sign(\Omega_U(t)) , \tag{11}$$

so the U-joint output shaft moment takes the form:

$$M_{UER} = \frac{M_{TSER}}{\cos\beta} + M_U , \tag{12}$$

In the ball screw the shaft rotation is transferred into nut translation:

$$\frac{dy(t)}{dt} = r \cdot tg\gamma \cdot \Omega_U(t) \tag{13}$$

where:
r -screw rolling radius
γ -nominal lead angle

External loads F_{ER} from the flap act on the ball-screw. The moment on the screw resulting from external loads is calculated as :

$$M_{ER} = -r \cdot F_{ER1} \sin \gamma \cos \gamma , \tag{14}$$

The output moment from the ball screw is composed from moment resulting from external loads M_{ER} and additional torque loses in the ball screw M_{TS} modeled analogically to the damping in bevel-gear and U-joint:

$$M_{TSER} = M_{TS} + M_{ER} , \tag{15}$$

where:

$$M_{TS} = B_{TS} \Omega_U (t) + C_{TS} sign(\Omega_U (t)) . \tag{16}$$

2.1 Control

The input signal for the system is the required flap deflection angle φ_r (Fig. 1).

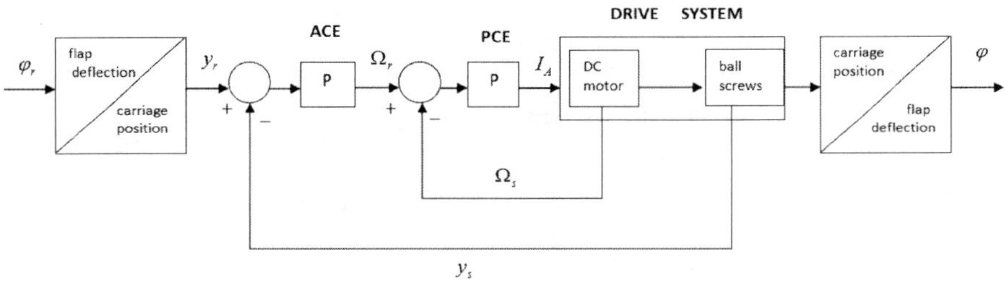

Fig. 1. Electromechanical actuator control system

This signal is transformed into required carriage position y_r which is compared with the position measured by the sensor y_s and changed proportionally into angular velocity of the screw Ω_r command in ACE module using proportional regulator with p_{ACE} coefficient. The control signal I_A for the motor is calculated in PCE module, so the control equation stands as:

$$I_A = \frac{p_{PCE}}{k_n} \left(p_{ACE} (y_r - y_s) - \Omega_s \right) \tag{17}$$

where:

$$y_s = 0.5 \cdot (y_1 + y_2) , \tag{18}$$

y_1, y_2 - carriage position of station 1 and 2 respectively,

Ω_s -measured angular velocity,

p_{PCE} -a proportional control coefficient.

Appropriate value of the signal should be contained in the boundaries follow from DC motor performance:

$$M_{\min} \le p_{PCE}\left(\Omega_r - \Omega_s\right) \le M_{\max} \tag{19}$$

As a result actuator is described by the state equations:

$$\begin{cases} \dot{\Omega} = \dfrac{1}{J}\left[M_n - \left(M_{B1} + M_{TPER1} + M_{TPER2}\right)\right] \\ \dot{y}_1 = r \cdot tg\gamma \cdot \Omega_{U1} \\ \dot{y}_2 = r \cdot tg\gamma \cdot \Omega_{U2} \end{cases}, \tag{20}$$

thus $x = \begin{bmatrix}\Omega & y_1 & y_2\end{bmatrix}^T$ is vector of the model states.

The actuator model is nonlinear with respect to states and is formulated in time domain.

3 Identification Methods

The aim of the EMA system identification is to calculate the actual values of mathematical model parameters.

The output vector of the system is formulated as:

$$\begin{aligned} y = [\Omega \quad y_1 \quad y_2 \quad I_A \quad \Omega_{TP1} \quad \Omega_{TP2} \quad \Omega_{U1} \quad \Omega_{U2} \quad M_{ER1} \quad M_{ER2} \cdots \\ M_{TSER1} \quad M_{TSER2} \quad M_{UER1} \quad M_{UER2} \quad M_{TPER1} \quad M_{TPER2} \quad p_{ACE} \quad p_{PCE}]^T \end{aligned}. \tag{21}$$

The observation parameters z are:

$$\begin{aligned} z = [\Omega_s \quad y_{1s} \quad y_{2s} \quad I_{As} \quad \Omega_{TP1s} \quad \Omega_{TP2s} \quad \Omega_{U1s} \quad \Omega_{U2s} \quad M_{ER1s} \quad M_{ER2s} \cdots \\ M_{TSER1s} \quad M_{TSER2s} \quad M_{UER1s} \quad M_{UER2s} \quad M_{TPER1s} \quad M_{TPER2s} \quad p_{ACE} \quad p_{PCE}]^T \end{aligned}, \tag{22}$$

where: I_{As} denotes the current passed to the DC motor, Ω_s -angular velocity on the motor output shaft, Ω_{TP1s}, Ω_{TP2s} - angular velocities on the bevel-gears, Ω_{U1s}, Ω_{U2s} -angular velocities on the U-joints. The external torques acting on ball-screws of two stations are M_{ER1s} and M_{ER2s}, while torques before the ball-screws states as M_{TSER1s} and M_{TSER2s}. The torques M_{UER1s} and M_{UER2s} are those before the U-joints, M_{TPER1s}, M_{TPER2s} are moments before the bevel-gears.

The system is observable. The proportional regulators coefficients p_{ACE} and p_{PCE} are taken from regulator adjustment on the test rig.

The estimated parameters are combined in column vector Θ:

$$\begin{gathered}\Theta = [J \quad B_{TS1} \quad C_{TS1} \quad B_{U1} \quad C_{U1} \quad B_{TP1} \quad C_{TP1} \quad B_{TS2} \\ C_{TS2} \quad B_{U2} \quad C_{U2} \quad B_{TP2} \quad C_{TP2} \quad k_n \quad p_{ACE} \quad p_{PCE}]^T\end{gathered} \tag{23}$$

Considering that in U-joint the β angle reaches maximally 5 degrees, the values of both g_U depending of angle θ_{TS} is close to 1, thus it is assumed that $g_{U1} = g_{U2} = 1$. Parameters β_1, β_2, γ, i_{TP1}, i_{TP2}, r are known from the system design.

3.1 Identification Algorithm

The Maximum Likelihood (ML) method in time domain with two alternative minimisation methods: linearized Gauss-Newton (ML_GN) and Levenberg-Marquard (ML_LM) was chosen for identification in this research.

In both cases, ML estimates are obtained by minimization of the cost function $J(\Theta, R)$, which in case of considered EMA system is assumed as [1, 4]:

$$J(\Theta) = \det(R), \tag{24}$$

where covariance matrix R is defined in the following form:

$$R = \frac{1}{N_{data}} \sum_{k=1}^{N_{data}} [z(t_k) - y(t_k)] \cdot [z(t_k) - y(t_k)]^T \tag{25}$$

The ML_GN algorithm with relaxation strategy contain the following steps [2, 4]:

1) Assuming initial values of parameters $\Theta(t_0)$ and states $x(t_0)$.

2) Computation of gradients of cost function with respect to parameters that are identified and gathering them in the matrix G:

$$G_i = \frac{\partial J}{\partial \Theta} \tag{26}$$

3) Information matrix (Hessian) F evaluation.

4) Solving the following equation with respect to $\Delta\Theta_i$:

$$F_i \cdot \Delta\Theta_i = -G_i \, , \tag{27}$$

5) Updating parameters:

$$\Theta_{i+1} = \Theta_i + \Delta\Theta_i \tag{28}$$

6) Checking convergence or maximum iteration number limit

$$\left| \frac{J_i - J_{i-1}}{J_i} \right| < 10^{-5} \tag{29}$$

The same algorithm but with Levenberg-Marquard (ML_LM) method of optimization used for cost function minimization may be applied in the fallowing way [2], [4]:

1) Fallow steps 2-5 from the ML_GN algorithm and compute the cost function $J(\Theta) = \det(R)$, which will be further considered as $J_{i-1}(\Theta)$.

2) Solve the equation with respect to $\Delta\Theta_i$:

$$(F + \lambda I) \cdot \Delta\Theta_i = -G \tag{30}$$

for Levenberg-Marquard (LM) parameter $\lambda = \lambda^{i-1}$ and $\lambda = \lambda^{i-1}/\nu$, where ν is a reduction factor $(\nu > 1)$ and I -the identity matrix.

The LM parameter λ enables to control the update search direction. If $\lambda \to \infty$ the algorithm is reaching steepest- descent variant but while $\lambda \to 0$ it becomes closer to the Gauss-Newton.

3) Update parameters for each of above solution $\Delta\Theta_i\left(\lambda^{i-1}\right)$ and $\Delta\Theta_i\left(\lambda^{i-1}/\nu\right)$.

4) Compute respective cost functions $J_i\left(\Delta\Theta\left(\lambda^{i-1}\right)\right)$ and $J_i\left(\Delta\Theta\left(\lambda^{i-1}/\nu\right)\right)$

5) Compare two above cost functions: $J_i\left(\Delta\Theta_i\left(\lambda^{i-1}\right)\right)$ and $J_i\left(\Delta\Theta_i\left(\lambda^{i-1}/\nu\right)\right)$ with the one from previous iteration and choose those which corresponds to the greatest reduction by reaching the smallest value.

6) Select parameters update corresponding to the cost function chosen in previous step.

7) Update parameters vector (Eq. 28).

8) Check convergence or maximum iteration number limit (Eq. 29).

Both identification procedures were implemented in Matlab software environment and applied for the system parameters estimation.

4 Results

Instead of laboratory test data, there were used perturbed by random signal simulation ones to make sure that the algorithms are working correctly and the calculations derived by them are reliable. The results tests of both algorithms are presented on the Figure 2. On each diagram, the measured output signal is compared to output signals of the model with identified parameters using ML_GN algorithm and ML_LM one.

The results indicates that implementation of both methods leads to accurate parameters and recreates behaviour of EMA system properly.

Currently, the laboratory test are in progress and EMA system will be identified as soon as the data will be available.

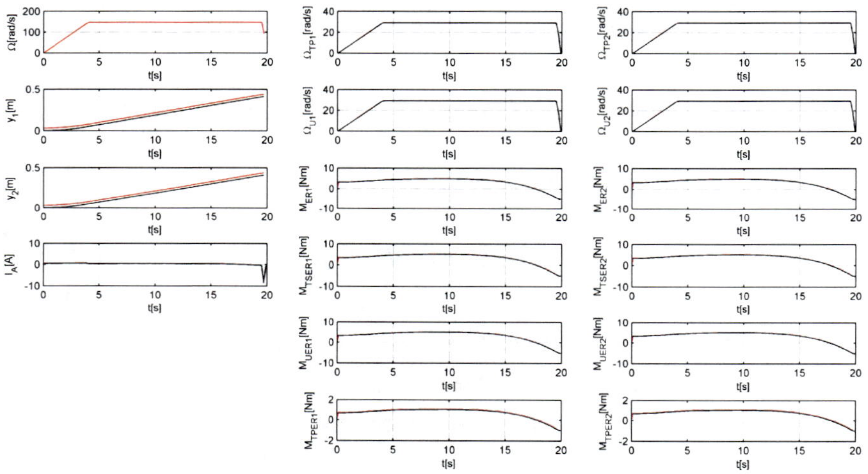

Fig. 2. Output signals comparison (measured, identified with ML_GN,ML_LM).

5 Conclusions

The dynamical model describing system behavior in form of differential equations is developed and ML method with two alternative optimization algorithms (linearized GN and L-M) is implemented in Matlab software environment. As a result the system model parameters are successfully estimated using test data. After laboratory test completion, the experiment data will be used for the system parameters identification and more reliable model for simulation will be obtained.

Acknowledgments

The paper was prepared under EC funded 6 FP project NEFS - New Track integrated Electrical Single Flap Drive System, Contract No 030789, coordinated by EADS IW.

References

[1] Andrzejczak, M.: Methods applied to aircraft identification. In: III Int. Conf. "Intelligence, Integration, Reliability", papers, Kijev (2010)

[2] Bonnas, J.F.: Numerical optimization: Theoretical and practical aspects. Springer, Berlin (2006)

[3] Esfandiari, R.S., Bei, L.: Modeling and analysis of dynamic systems. CRC Press/Taylor& Francis Group, Boca Raton (2010)

[4] Jategaonkar, R.V.: Flight vehicle system identification, A Time Domain Methodology. AIAA, Reston, Virginia (2006)

[5] New Track integrated Electrical Single Flap Drive System (NEFS) 6-th European Union Framework Programme research project, http://www.nefs.eu/

Introduction to DiaSter – Intelligent System for Diagnostics and Automatic Control Support

M. Syfert, P. Wnuk, and J.M. Kościelny

Institute of Automatic Control and Robotics,
Warsaw University of Technology, ul. Św. A. Boboli 8,
Warsaw, 02-525, Poland

Abstract. The paper presents general view of the DiaSter system implementing advanced methods of modeling, diagnostics and supervisory control for industrial processes. The scope of the tasks realized in the system as well as the system software platform were characterized, in particular: the software structure, central archival and configuration databases, the way of data exchange in the system and the modules of modeling and calculations.

1 Introduction

In recent years there have been significant developments in techniques for modeling [3, 5] advanced control [1, 11] and process diagnostic [2, 4, 6, 10]. Modern computer systems enable the application of complex computational algorithms, developed on the basis of recent research in computing, automatic, diagnostics and knowledge engineering. They use artificial intelligence techniques such as artificial networks, fuzzy logic, rough sets, evolutionary algorithms and methods for knowledge discovery from databases [9].

Development of methods outlined above in conjunction with the rapid progress of computer technology (computing power of new generation processors, memory capacity, speed of data transmission in LAN networks and field bus, the Internet growth) leads to a new generation of control systems. It is characterized by the introduction of advanced software for modeling, control and diagnosis processes. This software is a special software modules, which are part of the automation system, or expert systems integrated with automation systems.

Such software package is DIASTER system. It is developed by a research team composed of specialists from the Warsaw University of Technology, Silesian University of Technology, Rzeszów University of Technology and University of Zielona Gora and supported by polish grant: An intelligent system for diagnosis and control of industrial processes support "DIASTER". It is brand new, functionally and software extended version of AMandD system, developed in Institute of Automatic Control and Robotics, Warsaw University of Technology [7, 8]. The system is dedicated for use in the energy industry, chemical, pharmaceutical, metallurgical, food and many other.

The system is world-wide unique solution. It includes implementation of a wide range of the latest algorithms in the field of intelligent computation, used to system modeling, supervisory control, optimization, fault detection and isolation. Thanks to its open architecture, connections to virtually any automation system are possible and easy to implement. The position of DiaSter system in industrial process management tasks is presented in Fig. 1.

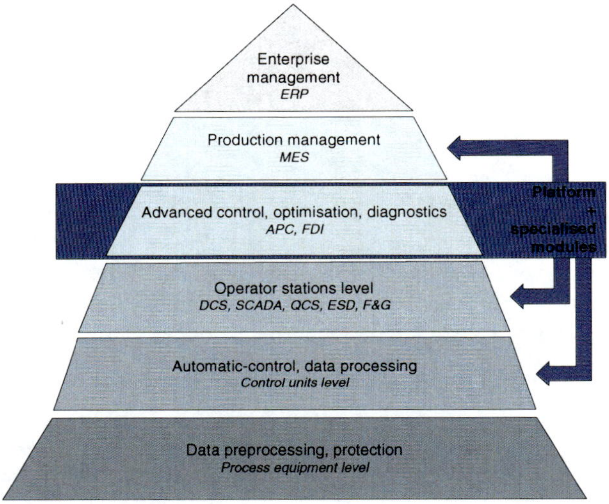

Fig. 1. The position of DiaSter system in respect to the hierarchy of tasks (and system classes) connected with production process

DiaSter system allows to realize following functions: process variables processing. simulation and modeling, virtual sensors and analyzers, process simulators, fault detection, fault diagnosis, monitoring of degradation degree of technological equipment, support of process operators decisions, knowledge discovery in databases, advanced control and optimization, superior tuning and adaptation of control loops.

2 Software Platform

DiaSter system consists of **a software platform** and **several specialized packages** cooperating with its use. The software platform is a core of a system. Its main components are: archival data processing and model identification module, module of on-line processing of system variables, visualization module, central configuration and archival databases and communication server. They are described in the consecutive sections.

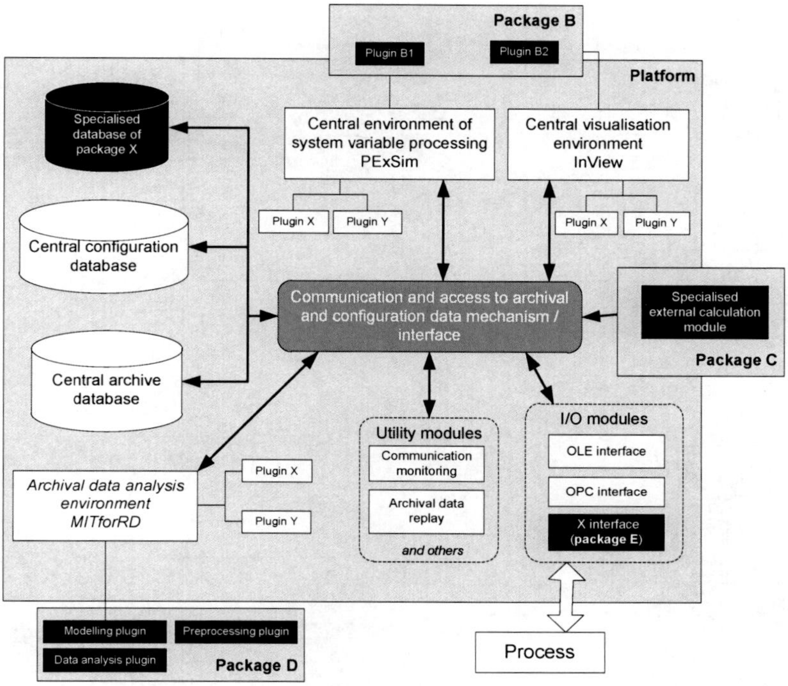

Fig. 2. General structure of DiaSter system. Black blocks symbolize specialized packages realized as different system components cooperating with the platform

There are also available several utility module. There are not responsible for main information processing but are very useful during system configuration and tests.

Advanced system functionality is delivered, developed and realized by specialized packages, called user packages. They are realized as independent software modules cooperating with other system components or in the form of plug-ins of software platform modules (modeling, processing and visualization).

3 Central Archival and Configuration Databases

In order to make possible the cooperation of all system components the following elements were worked out: common information model, central configuration environment and central archival database.

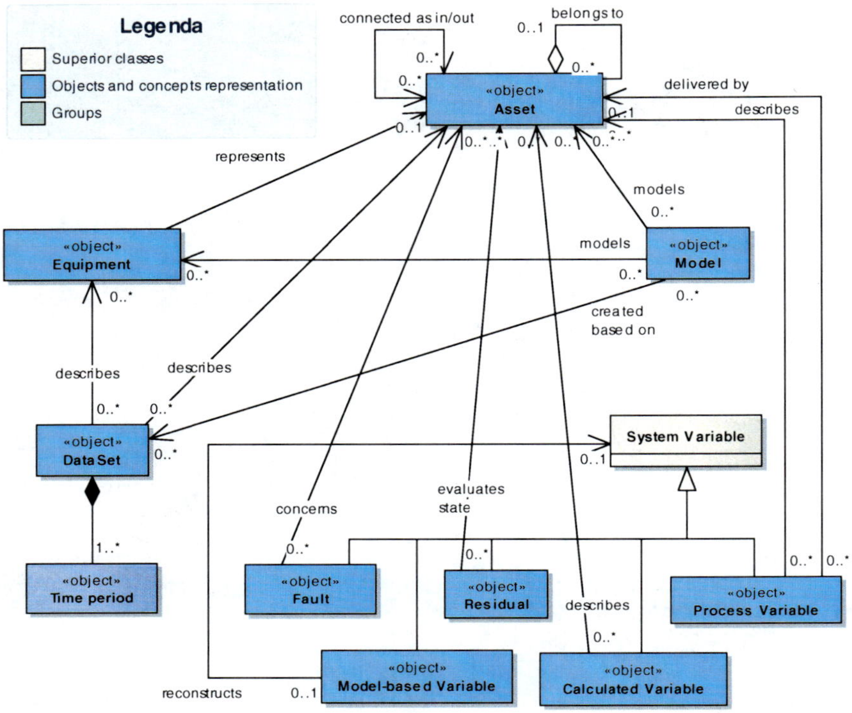

Fig. 3. Main objects and relations of the information model

The **information model** specifies the definition of the information being exchanged between the system components (Fig. 2). It constitutes the basis of the data exchange in respect to configuration as well as processed system variables. First of all, the information model defines:

- the elements connected with process / installation description, e.g., division into subsystems and particular components including the specification of logical as well as physical units, specification of process variables and their groups,
- the logical components of processing algorithms connected with realized by the system tasks, e.g., models, residuals, calculated and simulated variables, faults,
- the relation between the above elements, e.g., the relation "is a part of" and "is connected as input / output" defined between process subsystems and components or the relation "controls the behavior of" between residuals and monitored process components.

The information model also defines the types of system variables and the types of their values processed by the platform. It constitutes the framework of process and processed variables description. It is designed in such a way to be extendable by the particular user packages, or a group of packages.

The **central configuration environment** is a repository of configuration data for all system modules. It consists of:

- Central configuration database.
- The set of configuration interfaces.
- The interface to user-defined databases.

The **central archival database** is capable to store any system variables defined in the system configuration. It can store the variables of build-in data types (process variables, residuals, faults, etc.) as well as user-defined data types. Also any type of variable values can be stored. There build-in as well as user-defined value types are handled.

4 Data Exchange

DiaSter system is fully distributed software system with possibility to work on many PC-s connected with local area TCP/IP compliant network. Communication subsystem is a native solution, working with specially designed protocol independent on any external software. Communication library delivers simple API to system modules programmers. Thanks to this library the connection of an external program to DiaSter system is reduced to loading library and setting IP address of communication server. To send a message only one function call is needed.

More sophisticated communication mechanisms are provided by some modules working on-line. Additional communication interfaces build on the basis of CORBA standard are used to access the properties and methods of each function block embedded in the on-line calculation module (PExSim), and to remotely control the process of simulation / calculations realized by this module. Similar mechanism is used in visualization. It allows to transmit complex and infrequently changing information and direct access to system objects.

5 Modeling Module

For the identification purposes the **MITforRD (Model Identification Tool for Diagnosis and Reconstruction) module** is designed in DiaSter system. It allows to create models without the knowledge of the analytical form of the relationship between modeled variables. MITforRD module allows to identify both static and dynamic models of different types starting from well-known linear transmittances to neural network or fuzzy logic based models. The identification is carried out off-line using measurement data from central system archives. The models implemented in MITforRD module belongs to the group of partial parametric models of the process variables (time series). Models are obtained in a semi-automatic identification process.

One of the main assumptions for DiaSter system was to make it flexible and expandable. Therefore, the main program MITforRD Model Builder does not implement a possibility of model estimation of any type. All model types are implemented as plug-ins. Similar mechanism is provided also in other DiaSter modules. All plug-ins are based on dynamic link libraries (dll's).

MITforRD module may employ a wide range of evolutionary algorithms in order to explore the structure of linear or fuzzy models. Such algorithms, besides many advantages, also have one serious disadvantage - usually are very time consuming. To reduce computation time MITforRD provides distributed computing environmenyouyout. It can be used to utilize free computation power of the classic office PCs running on Windows and connected to the local area network.

6 On-Line Calculation Module

The main element of DiaSter platform for on-line use is an **calculation module called PExSim**. It is dedicated for advanced system variables processing. The processing algorithms are written and stored in the form of configurable function block diagrams. The primary task for PExSim module is to process the information circulating in the system in a way defined by the user. From this point of view PExSim module can be treated as specialized programming language. An algorithm of information processing is defined graphically by creating so-called processing paths. Each path consists of a set of interconnected function blocks, which carries out various tasks on signals (see Fig. 4).

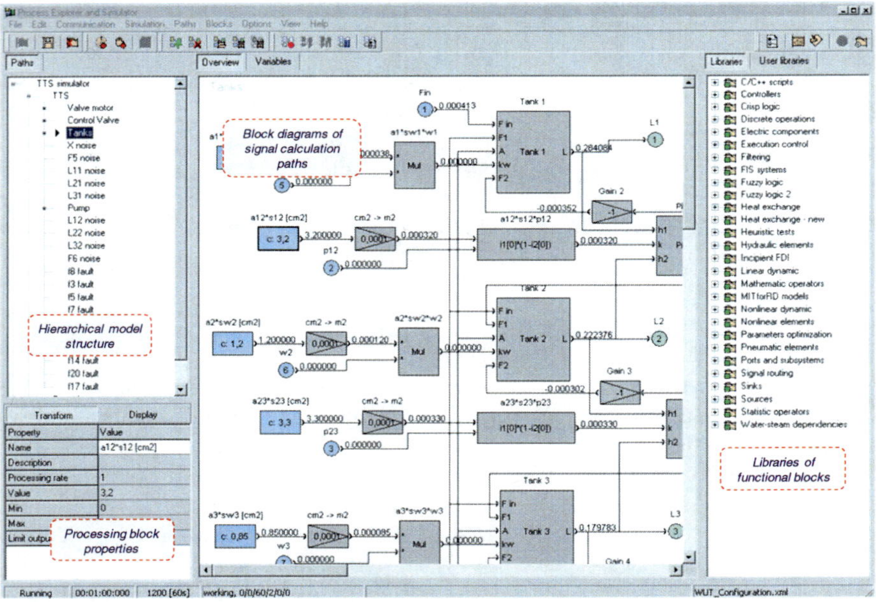

Fig. 4. Graphical programming of desired calculations in a form of processing paths

Function blocks are provided as PExSim plugins, and are grouped into thematic libraries, e.g.: surces, sinks, statistical operations etc. Each function block has the set of parameters defining the way of signal processing by this block, and stored by the platform. Various block inputs and outputs can transmit data of various types (e.g.: floating numbers, fuzzy values, vectors etc.) including user data types. PExSim can run as stand-alone tool (simulator mode), or as a module working in a distributed system (multi-module mode). Multi-module mode allows to exchange the data with other DiaSter modules.

7 Visualization Module

The platform delivers also a visualization module (graphic user interface) called PExSim. It can be used to realized advanced operator interfaces. It is organized in a similar way as tools for configuring process mimics in supervisory, control and data acquisition systems (SCADA) or decentralized control systems (DCS). The set of synoptic screens organized in hierarchical structure is prepared during the configuration stage for particular application. Dedicated displays visualizing the values of system variables in particular way are placed on that mimics.

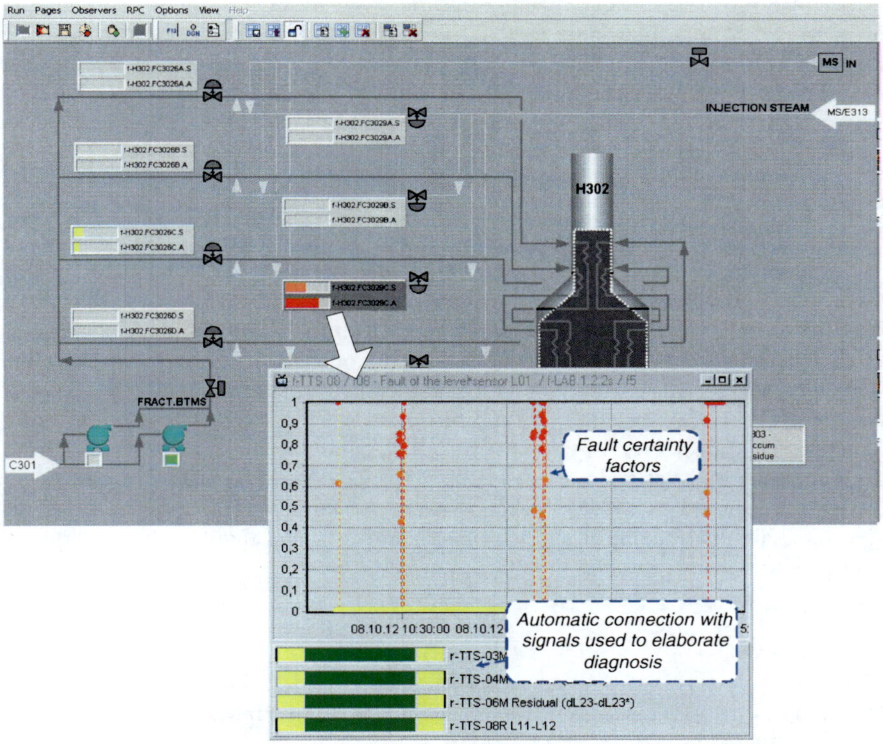

Fig. 5. Exemplary process operator interface dedicated to deliver diagnostics information about process state. Dedicated display is used to automatically present the information that was used to generate the diagnosis – the set of useful residuals in respect to particular fault is retrieved automatically from the diagnostics relation configuration stored in diagnostic package private database

The displays are realized as visualization module plug-ins. Such approach enables to elaborate specialized displays designed by the users. Such displays can even visualize process variables of user types (unknown by the platform itself), e.g. visualization of fuzzy signals or dedicated diagnostic messages. The only limitation is that the plug-in creator must have the knowledge about the processed signals structure.

8 Summary

Presented system was created as a result of development grant that finished in 2009. Currently test of the system are conducted. Their aim is to remove bags and finally prepare the system to implementation, commercial ones as well as research-development.

The most important part of the system is its software platform and the possibility to easy extend its functionality. The tasks that can be realized by the system can be also fulfilled, however usually in limited scope, by the class of commercial SCADA, DCS or similar systems that are available on the market. However, due to specialized system structure, there is a possibility to implement, apply and test modern and innovative techniques in the field of monitoring, advanced supervisory control, modeling and diagnostics of industrial processes which application in classical control and monitoring systems is difficult or even impossible.

The system structure was designed taking into account wide range of its possible fields of application. The use of the software platform and elaborated packages is planned not only in commercial applications but also in research-development and didactic tasks. The development of new system packages is conducted all the time. In respect to commercial application it is planned to used the system to realized the set of simulators of power generation units of conventional power stations. It is planned to use that simulators in the process of training the operators and other power station technical stuff. In the field of research and development the application of a system as monitoring system of a gas network is currently realized. The didactic use of the system is realized all the time. The system is used during the laboratories and project of courses related with the problems of process modeling (physical as well as based on process data and parametric models), diagnostics, automatic-control and applications of artificial intelligence methods in automatic-control.

References

[1] Brdyś, M.A., Tatjewski, P.: Iterative Algorithms for Multilayer Optimizing Control. Imperial College Press, London (2005)

[2] Chen, J., Patton, R.: Robust model based fault diagnosis for dynamic systems. Kluwer Akademic Publishers, Boston (1999)

[3] Cholewa, W., Kiciński, J.: Technical diagnostics. Inverse diagnostic models (in Polish). Wydawnictwo Politechniki Śląskiej, Gliwice (1997)

[4] Gertler, J.: Fault Detection and Diagnosis in Engineering Systems. Marcel Dekker, Inc., New York (1998)
[5] Janiszowski, K.: Identification of parametric models in examplesc (in Polish). Akademicka Oficyna Wydawnicza EXIT, Warszawa (2002)
[6] Korbicz, J., Kościelny, J.M., Kowalczuk, Z., Cholewa, W. (eds.): Fault Diagnosis. Models, artificial intelligence, application. Springer, Heidelberg (2004)
[7] Kościelny, J.M., Syfert, M., Wnuk, P.: Advanced monitoring and diagnostic system 'AMandD'. Problemy Eksploatacji 2(61), 169–179 (2006)
[8] Kościelny, J., Syfert, M., Wnuk, P.: Advanced monitoring and diagnostic system 'AMandD'. In: Safeprocess 2006 – A Proceedings Volume from the 6th IFAC Symposium on Fault Detection, Supervision and Safety of Technical Processes, Beijing, P.R. China, August 29-September 1, vol. 1, pp. 635–640. Elsevier, Amsterdam (2006)
[9] Moczulski, W.: Technical diagnostics. The methods of knowledge acquiring (in Polish). Wydawnictwo Politechniki Sląskiej, Gliwice (2002)
[10] Patton, R., Frank, P., Clark, R. (eds.): Issues of fault diagnosis for dynamic systems. Springer, Heidelberg (2000)
[11] Syfert, M., Rzepiejewski, P., Wnuk, P., Kościelny, J.M.: Current diagnostics of the evaporation station. In: 16th IFAC World Congress, Praga, lipca, pp. 4–8 (2005)
[12] Tatjewski, P.: Advanced Control of Industrial Process. Structures and Algorithms. Springer, Heidelberg (2007)

Simulation of Vehicle Transport Duty Cycle with Using of Hydrostatic Units Control Algorithm

P. Zavadinka

Institute of Solid Mechanics, Mechatronics and Biomechanics
Faculty of Mechanical Engineering, Brno University of Technology
Technicka 2896/2, 616 69 Brno, Czech Republic

Abstract. Presented paper described some information about results of four wheel drive vehicle model simulation in transport mode with using of control algorithm for hydrostatic units (pump and motor). The aim is to get information about possible reduction of fuel consumption with proposed control algorithm in selected working conditions.

1 Introduction

Emission rules have significant impact on whole vehicle market. They bring new challenges for transmission, working, controlling and vehicle management functions [2], [3]. In case of new concepts it is appropriate to create complex dynamic model of vehicle [5], [7]. Every type of mobile working machine (MWM) has a characteristic duty cycle [4]. The duty cycle of mobile working machine must be defined at the beginning of hydrostatic transmission control design [6].

The basic duty cycle can be divided to vehicle transportation to workstation - Mode A and vehicle work on workstation - Mode B. A telescopic loader can be selected for instance with following modes: vehicle transportation to workstation (uphill and downhill driving) and work on workstation (stuff loading and unloading) as it is shown in Fig. 1 as Mode A and Mode B.

A regulation of pump and motor is typical for telescopic loaders, when the pump and motor displacement is dependent on the combustion engine speed [9].

2 Concept of Driving Modes Control

Control of driving modes is method how to control the powertrain modes of mobile working machine [7]. With hydrostatic transmission (HST) it is possible to change driving mode according to driver or safety demands. It means that the driver can change MWM driving mode by pressing correspondent button in dependence on current situation.

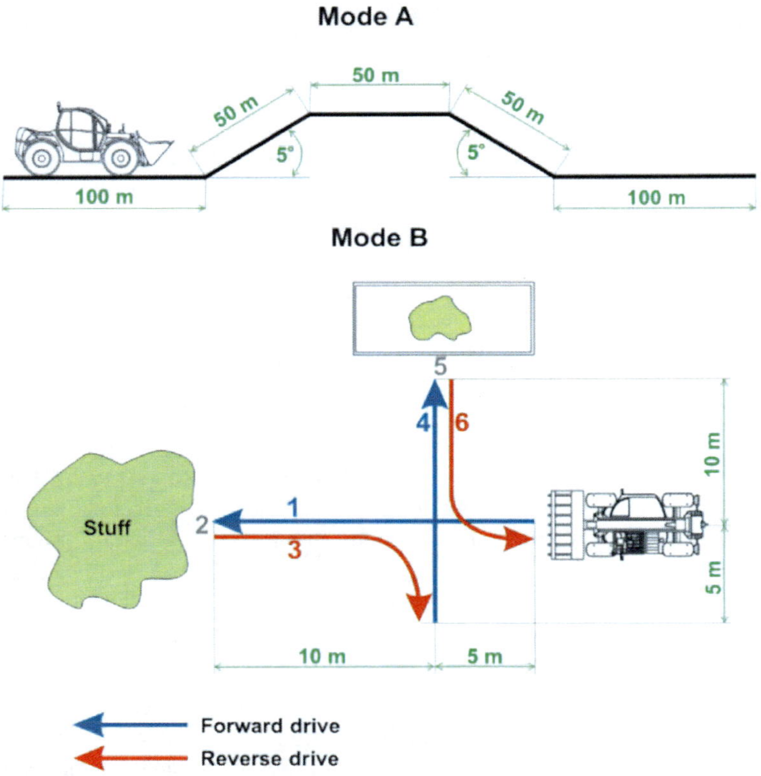

Fig. 1. Selected duty cycle with Mode A and Mode B 1 and 4 - forward drive, 3 and 6 - reverse drive, 2 - loading, 5 - unloading

A regulation of pump and motor is typical for telescopic loaders, when the pump and motor displacement is dependent on the combustion engine speed [9]. The primary regulation (regulation of pump) is active by lower engine speed. The secondary regulation (regulation of motor) is active by the higher engine speed. The primary and secondary regulation is not active together. During engine speed increasing the pump displacement is increasing too (primary regulation). When all the options of primary regulation are ran out, the secondary regulation starts to be involved, that it is mean that the displacement of motor is reduced. By decreasing of the engine speed the hydrostatic transmission control acts analogues.

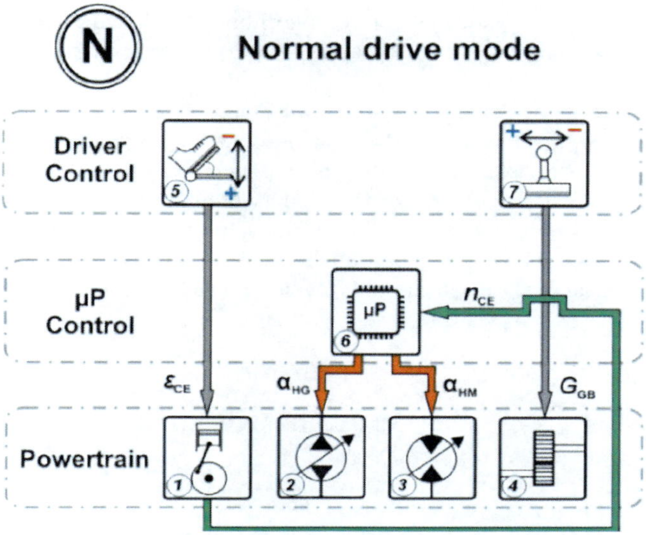

Fig. 2. Normal drive mode with Combustion engine 1, Pump 2, Motor 3, Gearbox 4, Accelerator pedal control 5, Microprocessor 6 and Joystick control 7

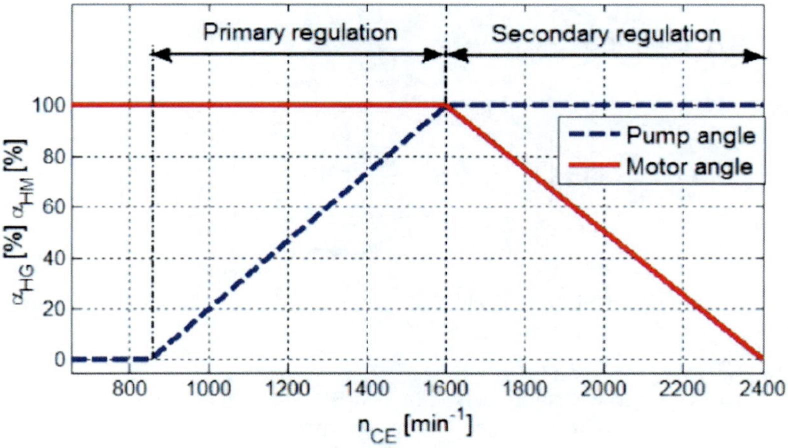

Fig. 3. Normal drive mode, Pump angle α_{HG} and Motor angle α_{HM} vs. Combustion engine speed.

As an examples, sophisticated control algorithms developed by tractor producer Fendt (Vario transmission) or John Deere (AutoPower transmission) are described in [1].

Normal drive mode principle is shown in Figure 2. The speed is controlled by accelerator pedal; the speed gear shift is controlled by joystick by this mode. The swashplate and bent axis angles are linear dependent on the engine speed (as shown in Figure 3). The beginning of secondary regulation is in 1600 min^{-1}, where the maximal engine torque is available. This setting of hydrostatic transmission control allows secure machine operation. There is no focus on operation costs. The working mode can be used in the duty cycle or machine transportation to work-station.

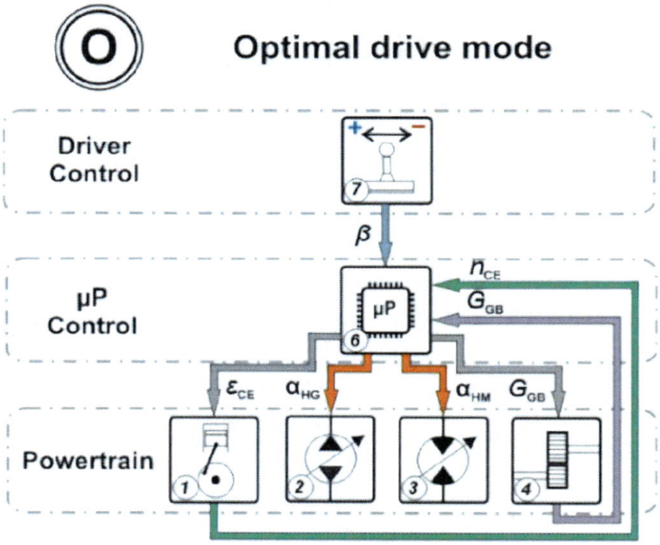

Fig. 4. Optimal drive mode with Combustion engine 1, Pump 2, Motor 3, Gearbox 4, Microprocessor 6 and Joystick control 7

Optimal drive mode principle is shown in Figure 4. This driving mode allows vehicle control only with joystick. The pump, motor, engine and gearbox are controlled in dependence on driver demand. The swash plate and bent axis angles are linear dependent on the engine speed (as shown in Figure 5). The beginning of secondary regulation is in 1100 min^{-1}. This value was adjusted by simulations. This setting of hydrostatic transmission control allows decrease of fuel consumption. It is optimal mode regarding operation costs. The working mode can be used in the duty cycle (the automatic gear shift is blocked) or machine transportation to workstation.

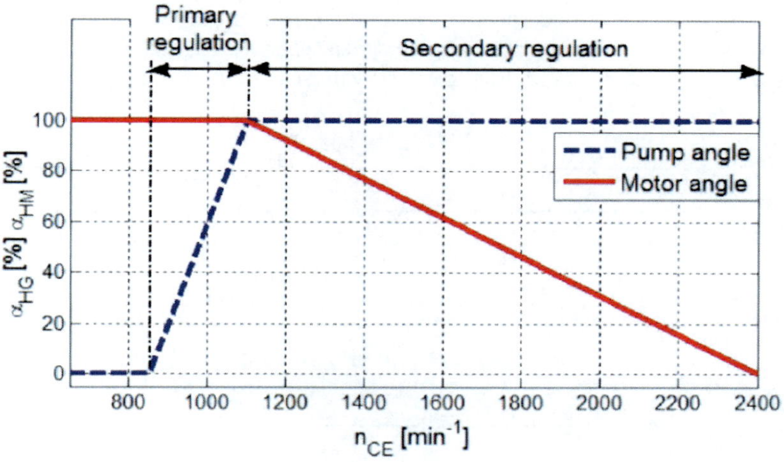

Fig. 5. Optimal drive mode, Pump angle α_{HG} and Motor angle α_{HM} vs. Combustion engine speed

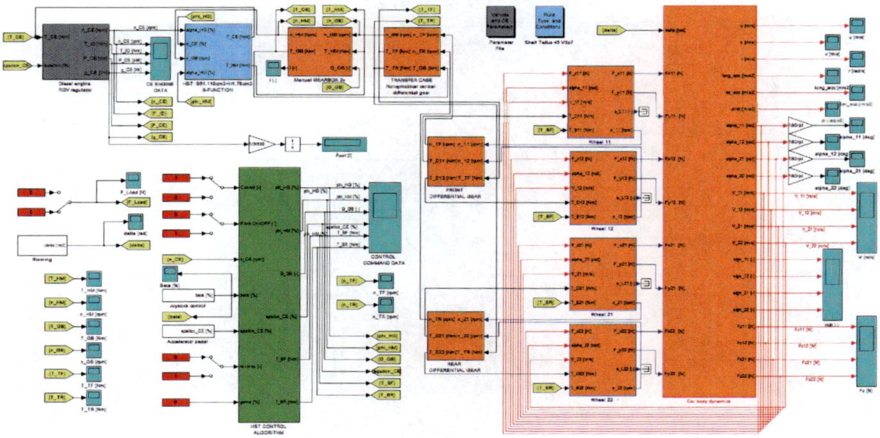

Fig. 6. Model of vehicle built in Matlab-Simulink

3 Simulation of Transport Duty Cycle (Mode A)

The vehicle model described in [7], [8] and shown on Fig. 6 in subsystems structure was tested for various driver demands and various speeds in mode A. The model consists from dynamic model of vehicle and control algorithms of hydrostatic transmission. The road surface was assumed to be a dry field road. The mobile working machine was unloaded. During the testing of simulation model by normal drive mode a change of speed gear in 4th second was simulated. The optimal drive mode shifted speed gears automatically in dependence on combustion engine speed. The demanded engine speed was set up by accelerator pedal (in optimal drive mode by joystick) to demand values according to Tab. 1.

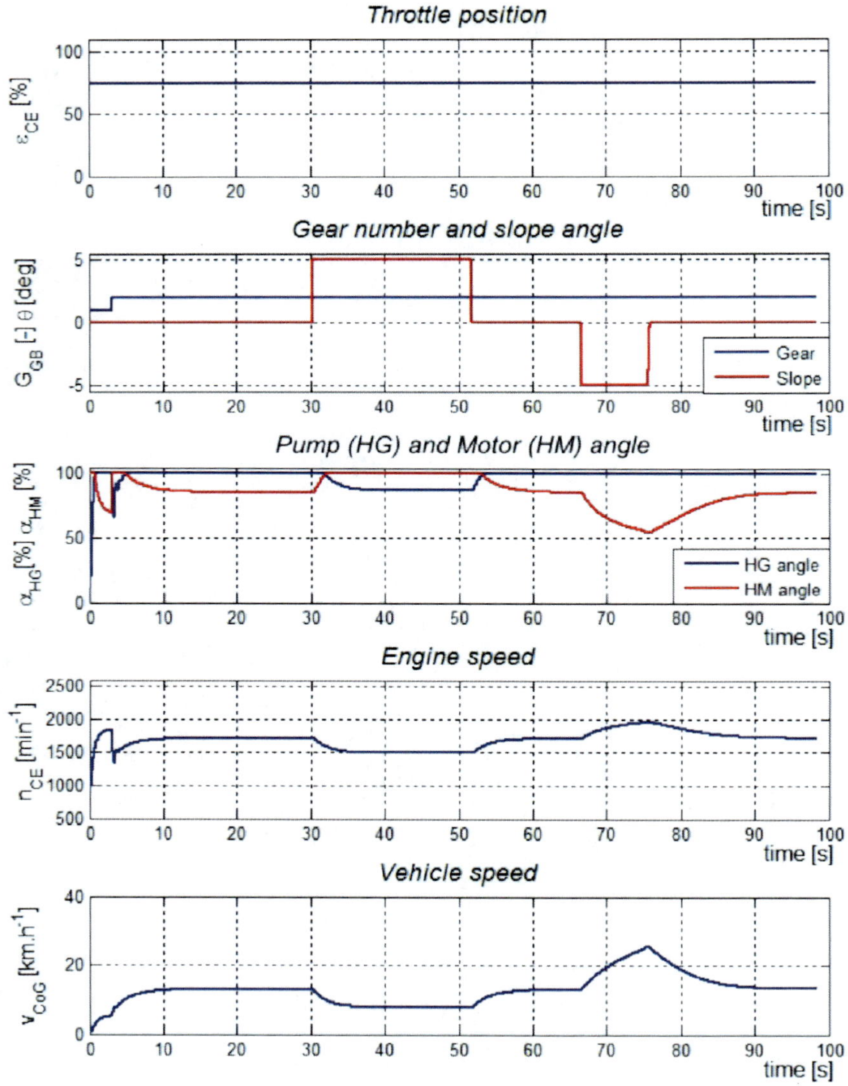

Fig. 7. Mode A simulation, Optimal drive mode - field path

Fig. 7 represents one simulation example by selected vehicle parameters during Mode A. In this simulation the optimal drive mode of powertrain control is active (vehicle/engine speed is set up by joystick).

The Tab. 1 represents comparison between normal drive mode and optimal drive mode during Mode A simulation. The drive mode (setting of hydrostatic transmission control) can significantly influence vehicle performance and operating costs. On the basis of intuitive premises and accessible information of telescopic loader's from manufacturers, obtained results can be considered as realistic. The simulation results must be considered carefully at the beginning of simulation. According to Tab. 1 it can be deduced, that for higher engine speed (vehicle speed) the optimal drive is more suitable in duty cycle - Mode A.

Table 1. The control algorithm comparison for test range in Mode A

OPTIMAL vs. NORMAL DRIVE MODE		
Demanded CE speed [min^{-1}]	Fuel consumption [%]	Range time [%]
1500	+16.4	+41.9
1800	-13.7	-26.4
2000	-5.4	-21.6
2200	-13.1	-16.4

4 Conclusions

In this paper some information and simulation result of vehicle powertrain model with proposed control algorithm of hydrostatic pump and motor were presented. The results show strong non-linear behavior of model, which decelerates simulation mainly at the beginning.

All performed simulations show acceptable results however in some cases (start, gear shift) it is necessary to consider physical and mathematical complexity of model's structure. During the simulation, it is important to know relationships and principles behind the user interface of these blocks, otherwise the simulation results can be misinterpreted.

We can assume that the reduction of fuel consumption can be done by suitable modification of pump and motor control characteristics. The hydrostatic transmission setting in the optimal drive is more suitable. For selected application and testing cycles the using of primary regulation is more suitable for lower engine speed and smaller speed range.

In the future, it is necessary to continue with developing of new blocks of drive and control and also with debugging of the initial conditions and parameter settings. This work provides good base for further model development of mobile working machines.

Acknowledgement

This paper has been supported by research project FSI-S-11-15 "Design, testing and implementation of control algorithms with use of nonlinear models of mechatronics systems".

References

[1] Bauer, F., Sedlak, P., Smerda, T.: Traktory. Prori Press, Praha (2006)

[2] Carter, D.E., Alleyne, A.G.: Earthmoving vehicle powertrain controller design and evaluation. In: Proceedings of the American Control Conference, 2004, June 30 -July 2, vol. 5, pp. 4455–4460 (2004)

[3] Grepl, R.: Real-Time Control Prototyping in MATLAB/Simulink: Review of tools for research and education in mechatronics. In: IEEE International Conference on Mechatronics ICM 2011, Istanbul, April 13-15 (2011)

[4] Grepl, R., Vejlupek, J., Lambersky, V., Jasansky, M., Vadlejch, F., Coupek, P.: Development of 4WS/4WD Experimental Vehicle: platform for research and education in mechatronics. In: IEEE International Conference on Mechatronics ICM 2011, Istanbul, April 13-15 (2011)

[5] Kiencke, U., Nielsen, L.: Automotive Control Systems for Engine, Driveline and Vehicle, 2nd edn., vol. XVIII, 512 p (2005), ISBN 978-3-540-23139-4

[6] Kriššák, P., Kučík, P.: Computer aided measurement of hydrostatic transmission characteristics. In: Hydraulika i Pneumatyka, vol. 4, pp. 22–25, INDEKS 37726 (2005), ISSN 1505-3954

[7] Zavadinka, P.: Modeling and Simulation of Mobile Working Machine Powertrain. M.S. Brno University of Technology (2009)

[8] Zavadinka, P., Kriššák, P.: Modeling and simulation of mobile working machine powertrain. In: Technical Computing Prague 2009, Praha, Czech republic, November 19, p. 118 (2009), ISBN 978-80-7080-733-0

[9] Zavadinka, P., Kriššák, P.: Modeling and simulation of diesel engine for mobile working machine powertrain. In: Hydraulics and pneumatics 2009, Wroclaw, Poland, October 7-9, pp. 339–348 (2009), ISBN 978-83-87982-34-8

[10] Zhang, R., Alleyne, A.G., Carter, D.E.: Robust gain scheduling control of an earthmoving vehicle powertrain. In: Proceedings of the American Control Conference, June 2003, vol. 6(4-6), pp. 4969–4974 (2003)

Part IV

Robotics

Development of Experimental Mobile Robot with Hybrid Undercarriage

P. Čoupek, D. Škaroupka, J. Pálková, M. Krejčiřík, J. Konvičný, and J. Radoš

Faculty of Mechanical Engineering, Brno University of Technology, Brno, Czech Republic

Abstract. This paper deals with the project of experimental mobile robot with the undercarriage which combines both legged and wheeled locomotion. The robot can drive on wheels on flat terrain and walk on irregular terrain, e.g. on stairs. The novel topology with seven servo drives and four independent wheel drives is described. The application of laser sintering technology is also discussed.

1 Introduction

This paper deals with development process of experimental mobile robot with hybrid undercarriage. The idea of nontraditional way of movement can be found in many publications.

The Roller-Walker with four legs and passive wheels is presented in [1]. Every leg has 3DOF + 1DOF for the flipping wheel. Maximum speed of the robot is 0.8 m/s. Climbing ability is limited to 10° because of passive wheels. Height of the robot is 25cm and weight 24kg.

The mobile robotic platform AZIMUT [2] is based on the idea of independent leg-track-wheel articulations. Movement can be realized by legs, wheels or tracks. Locomotion is provided by 12 motors for 12 DOF. The robot is able to achieve speed of 0.35 m/s to the maximum elevation 28° and speed of the movement over flat terrain is 1.2 m/s. Weight of the robot is 63kg with carrying capacity 10.4kg. Highly modular design needs to have distributed embedded systems to control every single part of the body, which predetermines the robot for experimental purposes.

Another example of the robot with hybrid undercarriage is the PAW robot [3], which combines wheeled and leg movement. PAW was large weight (over 20kg). The robot uses positioning wheels for the braking and turning. The main control hardware of the robot is PC/104 with QNX 6.1 operation system.

The interesting conception of Walk'n roll [4] robot is also using combination of wheeled legs. In this case, the robot can drive on the flat terrain and for roughed terrain it uses front pair of the legs with braked wheels to attract the back pair of wheeled legs. Movement of this robot is effective, but it cannot pass the step obstacle.

Experimental robots with hybrid undercarriage are well suited to reach high efficiency of locomotion and high level of ability in overcoming obstacles such as a step (stair) or steep terrain.

The role of the wheels can be used not only for increasing efficiency of locomotion, but also for overcoming obstacles [4]. Independent leg articulation [2] supplemented by wheel can be effective solution for experimental robot suited for research and education in mechatronics.

The developed novel construction of the hybrid undercarriage described in this paper was designed to fulfill following requirements: speed 8 km/h, 2.2 m/s; climbing ability 30°; step obstacle - ability to go upstairs (maximum height of a step is 10 cm). After many preliminary conceptual variants (Fig. 1) the resulting robot equipped with seven servo drives and four independent wheel motors was built.

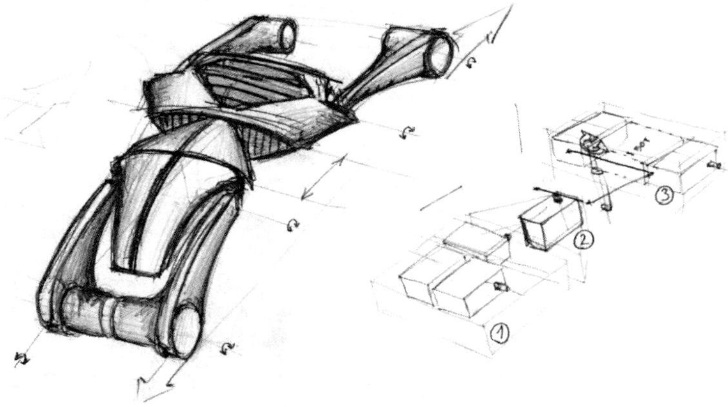

Fig. 1. Design of preliminary conception

2 MBS Simulation: The Support for the Design

During the development of several variants of the topology and geometry of the robot, the multi body system simulation was extensively used. The complete model of robot, servodrives and environment was built in Matlab/SimMechanics [7] environment. Fig. 2 shows the example of the kinematical analysis of one free leg movement possibilities. Three rigid legs are fit to the ground by spherical kinematic links (thus cannot slip on the ground). Free leg is excited by a random force and can freely move. Movement restrictions are implemented in rotary links (penalty method).

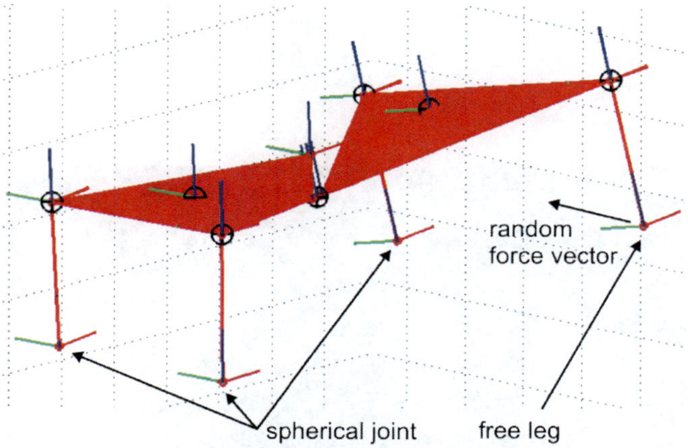

Fig. 2. Visualization of MBS SimMechanics simulation

The important part of MBS simulation is the model of unilateral constraints implemented according to [8, 9, 10, 14].

After the optimal variant selection, the fast computational kinematics was derived based on homogeneous transformations [11, 13]. The simulation model in SimMechanics was used as a reference for the computation verification.

3 Construction and Production Technology

The primary layout of the construction was made by an approximate calculations and rough sketches. After optimization in SimMechanics the accurate data for 3D modeling process in the Rhinoceros 3D and SolidWorks was obtained.

The construction of all basic parts is designed as a tin shell. This conception provides both important features: the high stiffness as well as low weight. Production technology for main parts was the SLS (Selective Laser Sintering process). The material used for final production (Fig. 3) is polyamine PA 2200. Sintered material reached about 90% of strength of base material.

Driving of the robot is providing by legs equipped by driven wheels. Torque is on driven wheal distributed by toothed belt from brushed electric DC motor (size 280) equipped transmission (gear 5:1).

Walking ability is enable by four legs, each leg has one DOF (degree of freedom), and central join which disposing of two DOF. Walking stability is guaranteed by balancer with 1 DOF. All this parts are driven by modeling servo actuators Hitec HSR-5990TG. Torque of this servo motor is 24 [kg.cm] in 6V, weight 68g. Generally robot is equipped by 7 peaces of servo motor and four peaces of brushed electric motors. The control of all servos during the testing phase was achieved through rapid control prototyping [6] hardware dSPACE. The final controller inspired by [5] was based on ARM 9 microcontrollers (particularly NXP LPC 2326).

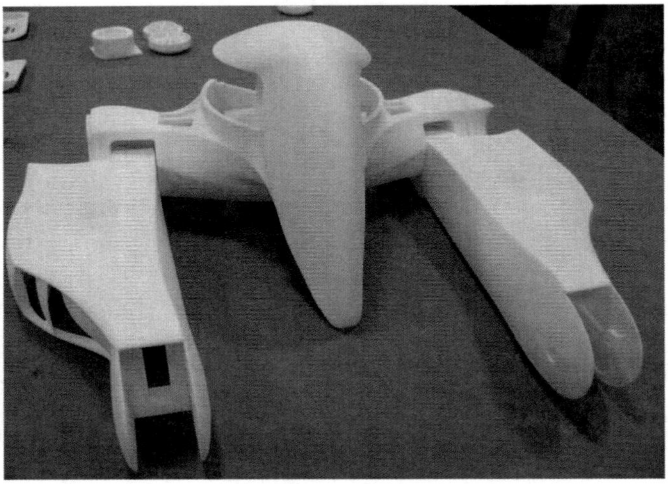

Fig. 3. Robot components made by Rapid prototyping technology SLS process

4 Design

The used morphology follows the inspiration of cuttlefish's body. Organically shaped surfaces are joining together in the dynamic edges splitting the robot's body into flatter faces lightweighing the body mass. Central part of the body (Fig.6) is naturally respecting the united functional form and the balancer stands out above the main body and create the main dominant element. The morphology of the undercarriage deals with strong functional requirements, which affect mainly wheeled legs. Necessity of inner cooling components of the leg brings a new possibilities for using other type of shape solution. For this parts are used a new approach of shape solution based on unification of organic shaping and theme inspiration of air intake channels of jet engines (Fig.5).

One of the main contributions of the described project is the innovative approach to the design process. The philosophy of the industrial design says that aesthetics of the product is non-separable to its functionality. We considers as necessary to keep this philosophy, even in the case of developing an experimental robot. We created outside form of the robot as external skeleton with smooth shaping of the robot's body associates a hybrid kind of animal (Fig. 4, 5).

Fig. 4. Visualization of final design of experimental mobile robot

Fig. 5. Final design of experimental mobile robot with hybrid undercarriage

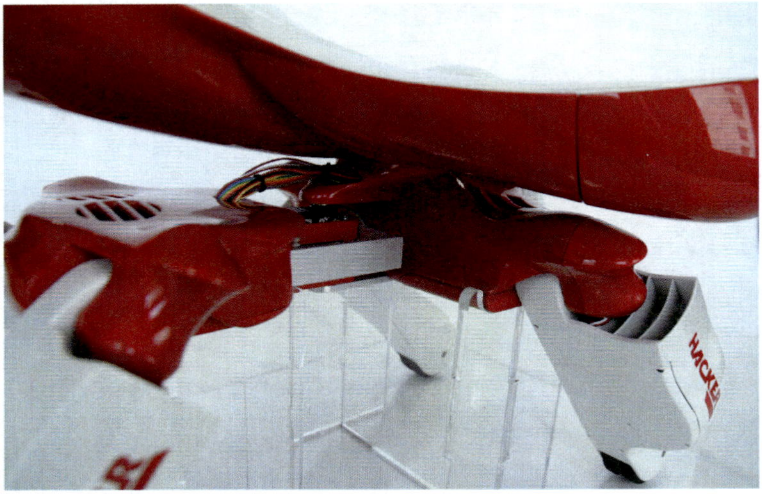

Fig. 6. Solution of central part of the body

5 Conclusion

The new mobile robot presented in this paper is equipped with the hybrid walking/wheeled undercarriage, which combines advantages of both types of locomotion. Therefore, the robot can easily move itself through obstacles. The independent leg articulation enables experiments with algorithms for walking and movement in general. The novelty of our conception is in the unique composition of seven servo drives and four independent wheel drives. The possible application areas include the exploration, rescue, pyrotechnic and entertainment.

Acknowledgment

This research was supported by project FSI-J-11-47 / 1436.

References

[1] Hirose, S., Takeuchi, H.: Study on Roller-Walk (Basic CHaracteristics and its Control). In: International Conference on Robotics and Automation, Minneapolis, Minnesota, April 1996, pp. 3265–3270 (1996)

[2] Michaud, F., Létoumeau, D., Arsenault, M., Bergeron, Y., Cadrin, R., Gagnon, F., Legault, M.A., Millette, M., Paré, J.F., Tremblay, M.C., Lepage, P., Morin, Y., Bisson, J., Caron, S.: AZIMUT, a Leg-Track-Wheel Robot. In: lntl Conference on lntelligent Robots ana Systems, Las Vegas, Nevada, October 2003, pp. 2553–2558 (2003)

[3] Trentini, M., Smith, J.A., Sharf, I.: Intelligent Mobility for Dynamic Behaviours of PAW. In: A Hybrid Wheeled-Leg Robot Robot Design and Control Algorithm Development for Wheeled Mobility Behaviours, Defence R&D Canada – Suffield, Technical Memorandum, DRDC Suffield TM 2005-203, pp. 1–16 (2005)
[4] Adachi, H., Koyachi, N.: Development of a Leg-Wheel Hybrid Mobile Robot and Its Step-Passing Algorithm. In: International Conference on Intelligent Robots and Systems Mad, Hawaii, USA, October 29 - November 3, pp. 728–733 (2001)
[5] Vlachy, D., Zezula, P., Grepl, R.: Control unit architecture for biped robot. In: International Conference on Mechatronics, Warsaw Univ. Technol, Warsaw (2007)
[6] Grepl, R.: Real-Time Control Prototyping in MATLAB/Simulink: review of tools for research and education in mechatronics. In: IEEE International Conference on Mechatronics ICM 2011, Istanbul, April 13-15 (2011)
[7] Wood, G.D.: Simulating Mechanical Systems in Simulink with SimMechanics. Tech. rep., The MathWorks, Inc., 3 Apple Hill Drive, Natick, MA, USA (2003), http://www.mathworks.com
[8] Grepl, R., Vlach, R., Krejci, P.: Modelling of unilateral constraints for virtual prototyping in SimMechanics. In: IEEE International Conference on Mechatronics, Kumamoto, JAPAN, May 8-10 (2007)
[9] Bottasso, C.L., Trainelli, L.: Implementation of effective procedures for unilateral contact modeling in multibody dynamics. Mechanics Research Communications 28, 233–246 (2001)
[10] Grepl, R.: Simulation of unilateral constraint in MBS software SimMechanics. In: International Conference on Mechatronics, Warsaw Univ. Technol., Warsaw (2007)
[11] Grepl, R.: Extended kinematics for control of quadruped robot. In: International Conference on Mechatronics, Warsaw Univ. Technol., Warsaw (2007)
[12] Zezula, P., Vlachy, D., Grepl, R.: Simulation modeling, optimalization and stabilisation of biped robot. In: International Conference on Mechatronics, Warsaw Univ. Technol., Warsaw (2007)
[13] Mechatronics – selected problems, Brno University of Technology (2008) (in Czech), ISBN 978-80-214-3804-0
[14] Faik, W.: Modeling of Impact Dynamics: A Literature Survey. In: 2000 International ADAMS User Conference (2000)

Dynamic and Interactive Path Planning and Collision Avoidance for an Industrial Robot Using Artificial Potential Field Based Method

A. Csiszar[1,*], M. Drust[2], T. Dietz[2], A. Verl[3], and C. Brisan[1]

[1] Department of Mechanisms, Precision Mechanics and Mechatronics, Technical University of Cluj Napoca, Cluj Napoca, 400114, Romania
[2] Fraunhofer Institute for Manufacturing Engineering and Automation (IPA), Stuttgart, 70569, Germany
[3] Institute for Control Engineering of Machine Tools and Manufacturing Units (ISW), University of Stuttgart, Stuttgart, 70174, Germany

Abstract. In this paper the dynamic path planning with collision avoidance of an industrial robot is addressed. The Artificial Potential Field (APF) method is used and, as a contribution, it has been expanded with obstacle-charges having different geometrical forms in order to positively influence the generated trajectory. Also an analysis method is presented that ensures that no violation of a specified safety envelope can occur around the obstacles. Implementation of the theoretical consideration and experimental validation using industrial equipment is shown.

1 Introduction

All industrial systems nowadays use safety measures to assure that nobody (operator, worker, etc.) can be harmed by any industrial equipment. Today's practice, in most cases, is to prevent access to the robot's workspace or to stop the robots when external sensors detect unpredicted changes in the occupancy of the workspace [1]. Path planning algorithms with collision avoidance extend the applicability of robots in unpredictable situations (including also physical human–robot interaction).

Local and global path planning algorithms can be distinguished. Global path planners, also called offline path planners (A* algorithm [2], genetic algorithm [3], etc.) require the knowledge of the workspace prior to planning. Obstacle movements are not considered.

Local path planning algorithms on the other hand are online, real-time capable. The fact that the motions (position changes) of the obstacles are handled makes this type of path planning suitable for online collision avoidance. Vector Field Histogram (VFH) [4] and its derived methods (VFH+ [5], VFH* [6]) are local

* Corresponding author.

path planning algorithms, producing near optimal global paths. These algorithms, widely used in mobile robotics deal with sensor uncertainty, while also taking into account the shape of the robot while moving through gaps, between obstacles. Dynamic Window approach [7] takes into account the dynamic capabilities of the robot while dealing with unforeseen scenarios.

Artificial Potential Field (APF) [8] based methods realize the path planning by finding a physical analogy. Most used analogy is modeling the obstacles and target as electrostatic charges interacting with each other, creating a potential field, with a global minimum in the target point. The method allows a flexible weighting of different objectives for the path generation. Main disadvantage of the method is the deadlock positions, impeding reaching the target. Most papers consider only point charges when modeling an obstacle. This is difficult and/or disadvantageous when dealing with obstacles of line shape, wall type or in the case when not all coordinates of the obstacles are known.

In this paper the commonly used point charges for APF solutions are completed with charges of other geometry (line, plane) in order to better influence the generated trajectory. The aim of the article is to experimentally validate these contributions. To allow interactivity and to demonstrate the dynamic capabilities of the approach a joystick is used for robot target point definition, this also helps dealing with deadlock positions.

2 Artificial Potential Field Based Path Planning

The artificial forces path planner is based on a virtual electric field approach. The obstacle-charges in the workspace of the robot exert a repulsive effect on the robot while the desired target-charge position has an attractive effect on the robots TCP (Tool Center Point) position. In a first step the strength of the fields of obstacles and target on the points of interest on the robot have to be computed. For this field computation a number of charge types have been proposed to reflect various shapes of obstacles and suitable evasion strategies for collision avoidance. The simplest case is the point charge obstacle. The field vector $\vec{E}_D$ of this charge type can be derived from the Maxwell equations using the divergence theorem [9] as:

$$\vec{E}_D = \frac{f_Q}{\left|\vec{r}_{OP}\right|^2} \frac{\vec{r}_{OP}}{\left|\vec{r}_{OP}\right|}, \tag{1}$$

where the factor $f_Q = \frac{Q_{ES-P}}{4\pi\varepsilon}$ collects the obstacle charge Q_{ES-P} and the product $4\pi\varepsilon$ of the electric permittivity ε and a constant occurring in the denominator of the equation to respect real-world physics and $\vec{r}_{OP}$ is the relative displacement vector pointing from the point charge to the point of interest.

This obstacle is preferably used for objects that are relatively small and that do not possess a preferred direction for evasion.

For an infinitively long line charge the field vector $\vec{E}_L$ in the point of interest is computed using the equation:

$$\vec{E}_L = \frac{f_\lambda}{|\vec{r}_{LP}|} \frac{\vec{r}_{LP}}{|\vec{r}_{LP}|}; \tag{2}$$

where the factor $f_\lambda = \frac{\lambda_{ES}}{2\pi\varepsilon}$ collects the obstacle line charge λ and the product $2\pi\varepsilon$ of the electric permittivity ε and a constant occurring in the denominator of the equation from real-world physics and $\vec{r}_{LP}$ denotes the distance between line and point of interest.

Line charge fields are used for bar-shaped obstacles that can be best avoided in a plane perpendicular to their length. A typical application is the collision mitigation with humans working inside the workspace of the robot. The robot is usually not allowed to move above the head of humans due to the risk resulting from the potential energy stored in robot joints and handled parts.

Infinite charged planes create a uniform electric field $\vec{E}_P$ with constant field strength. The field of plane charges can be used to create a constant drift in the workspace of the robot.

There are other desirable obstacle characteristics whose physical equivalent cannot be easily recognized. First of all, a common application is to limit the workspace of the robot by introducing boundaries. Since the TCP movement should be affected by these boundaries only if it is in the close vicinity of these, it is desirable to have a kind of field charge with a field strength that is decreasing with the distance from the field charge:

$$\vec{E}_P = \frac{f_\sigma}{|\vec{r}_{DP}|^2} \frac{\vec{r}_{DP}}{|\vec{r}_{DP}|}, \tag{3}$$

where $\vec{r}_{DP}$ is the distance of point of interest and plane and $f_\sigma = \frac{\sigma_{ES}}{\varepsilon}$ is a factor collecting the area charge σ of the plane and electric permittivity. The physical analogy of this type of field is a point charge moving in the specified plane with minimal distance to the point of interest.

In particular for trajectory planning it is desired that the attractive force of the target point does not decrease with the distance from the target point, i.e. the field of a point charge multiplied with the factor $|\vec{r}_{OP}|^2$.

The field vectors of the different obstacles and targets can be superimposed at the point of interest. A resulting force $\vec{F}_{res}$ is computed by multiplying the resulting field vector $\vec{E}_{res}$ with the virtual electrostatic charge of the point of interest Q_{ES-TCP} which corresponds to a gain controlling the speed of the trajectory.

Additional forces $\vec{F}_M$, e.g. virtual forces with mechanical analogies, can be added at this point.

$$\vec{F}_{res} = \vec{E}_{res} Q_{ES-TCP} + \vec{F}_M = \left(\sum \vec{E}_D + \sum \vec{E}_L + \sum \vec{E}_P\right) Q_{ES-TCP} + \vec{F}_M \tag{4}$$

For path planning and robot control a method to compute a position offset from the force vector needs to be applied.

For static path planning a unit step in the direction of the resulting force vector is applied, i.e. the force vector is normalized and multiplied with the desired step-width of the path-planner. This way a collision free path (having no velocity component) can be obtained

Trajectory (velocity) planning is possible with the above described approach. In this case, an offset vector $\vec{\Delta}_P$ for the point of interest is computed in each time step from the resulting force.

In the simplest case, the point of interest is modeled as an object of mass zero moving in a viscous liquid with a damping force proportional to its speed. In this case, using forward Euler integration, the speed and the offset vector are directly proportional to the resulting force:

$$\vec{v}_S = \frac{\vec{F}_{res}}{d}; \qquad \vec{\Delta}_P = \vec{v}_S \Delta t_i; \tag{5}$$

In order to reflect the limited bandwidth of the robot's actuators a non-zero mass can be assigned to the point of interest.

In some particular situations these equations fail to produce a path leading to the target point. This is the case if either the electric field vector becomes zero or if an oscillating behavior occurs in the vicinity of a zero field vector point due to the discretization. In these situations, to avoid deadlock, artificially introduced potentials help to guide the robot to the right path, in combination with the user joystick input.

To ensure a collision-free path a check that the superimposed field of all obstacles results in a repulsive field is imposed. For this purpose a required safety envelope is defined for each obstacle. The safety envelope is defined as smooth, closed surface. The superimposed field of all obstacles is computed on the surface envelope and projected on the normal vector of the safety envelope for each obstacle. This results in a measure for the repulsive effect of the obstacle for the defined safety margin. By finding the minimum of this measure on the safety envelope the worst case approach direction can be found and analyzed. This procedure is demonstrated below on the example of a point charge obstacle with a spherical safety envelope with radius r.

The superimposed field of all charges is given by

$$\vec{E}_{Tot}(x,y,z) = \sum_i \vec{E}_i(x,y,z), \tag{6}$$

where $\vec{E}_i$ is the field of the i-th charge as stated in equations (1) to (3). For the obstacle charge under analysis $\vec{E}_{Tot}$ is computed on the surface of a sphere around the obstacle by introducing spherical coordinates r, ϑ and φ, where is r defined by the required diameter for the safety envelope.

$$(x \quad y \quad z)^T = [x_c + r\sin(\vartheta)\cos(\varphi), \quad y_c + r\sin(\vartheta)\sin(\varphi), \quad z_c + r\cos(\vartheta)]^T \tag{7}$$

By computing the scalar product of the field and the normal unit vector of the sphere the repulsive component of the field is obtained:

$$E_{\text{Rep}}(\vartheta,\varphi) = \vec{E}_{Tot}^T(\vartheta,\varphi)\cdot[\sin(\vartheta)\cos(\varphi), \quad \sin(\vartheta)\sin(\varphi), \quad \cos(\vartheta)]^T; \tag{8}$$

The function $E_{\text{Rep}}(\vartheta,\varphi)$ describes the repulsive effect of the field along the safety envelope. By searching for the minimum of the unconstrained, multivariable function the worst-case approach scenario and the related repulsive effect can be determined. Optimization methods can be used for finding the minimum. A descent starting value is found by sampling the function with a narrow grid and using the location of the found minimum.

3 Implementation and Experimental Validation

The experimental setup combines industrial standard hardware and software components with presented algorithm for path planning and individual interaction with a robot system. It focuses on the interoperability and the portability in order to achieve comparable validation results of the implementation.

The experimental setup consists of KUKA KR500 industrial robot with 6 dof; KUKA robot controller KR C2; PC running Matlab/Simulink; Bridge-PC for visualization/communication handling; RTX real-time system; Sick safety laser scanner S3000; . Fig. 1a shows an overview of the system architecture.

The controlled robot is an industrial KUKA robot KR500 with six degrees of freedom. The motion control is realized with the KUKA KR C2 controller. For the external motion control the KUKA RobotSensorInterface (RSI) is used. A PC (Bridge PC) is used to collect and broadcast data and information. The running WIN32 software acts as bridge and receives the data of path planning module. The data is then transmitted to the motion controller of the robot. By using a kinematic model of the robot this software component can be used to alternatively visualize the motions in the CAD framework CADability. This offers the possibility to test the proposed algorithms without the robot hardware and root controller.

The path planning is implemented as a model using Matlab/Simulink and incorporates a subset of modular, configurable subsystems. The model tasks as high performance tasks run on a second PC, which is connected to the Bridge PC via TCP/IP.

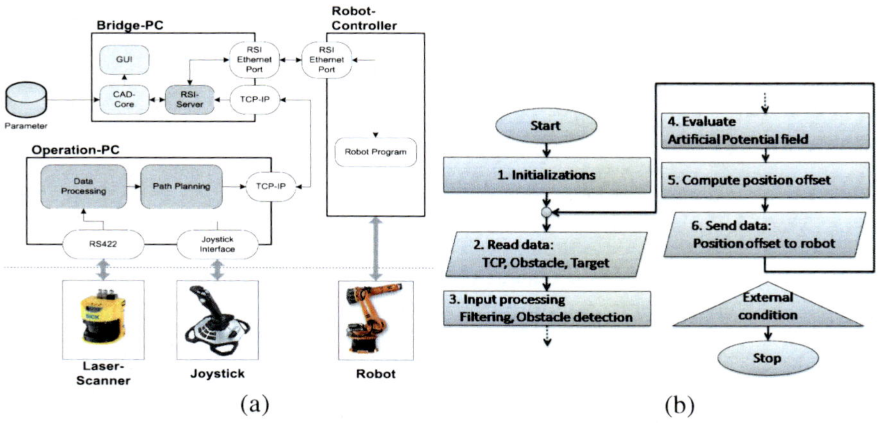

Fig. 1. (a) Schematic drawing of the overall system architecture. (b) flowchart of path planning algorithm

A serial communication was set up to gather sensor information from a Sick S3000 safety laser scanner. The range of the used safety laser scanner is larger than the workspace section of the robot. The laser scanner observes the area around the robot in two dimensions (in this case XY). Currently the height of an obstacle is considered to be infinite large and thus the measured contour is mapped into 3D space. This approximation can cause uncertainties particularly when an obstacle is unknown or the shape is not consistent, like a human that moves his upper part of the body or his arms. In order to acquire 3D information of the environment 3D laser scanners can be used. But those are typically more expensive. Cost-effective alternatives are nowadays the combination of 2D lasers canners with a turning device, camera systems based on fixed triangulation stations (stereo camera system) or finally measurement devices that examine structured light patterns. Due to the high lot size the Microsoft Kinect sensor device which uses infrared grids for depth determination is an alternative (but not safety certified) especially in service robotics.

The option of code generation of Matlab and Simulink allows use of the models functionality on a wide range of target platforms including real-time operating systems. Although the Simulink model is running in quasi real-time (RT Simulink block) for the presented architecture the use of the real-time system RTX is planned and the compatibility of the implementation and code generation has been already verified.

As already mentioned the path planning algorithm, described above in theory, is implemented as Simulink function blocks. Considering the inputs (current obstacle positions and type, current TCP position, current target position) it makes the necessary computations in order to move the robot on a collision free path. At each cycle the path planning equations are reevaluated, positions (TCP, obstacles and target) are updated and the new TCP motion (position offset) is generated by the path planning algorithm. This implementation does not consider the volume of the robot for obstacle avoidance. Future work is planned to extend the method with this aspect. The steps of the algorithm, as presented in fig. 1b, are as follows:

1. At the beginning of the simulation variables Q_{ES-P}, λ_{ES}, σ_{ES}, d, Δt_i, etc. and the communication are initialized.

2a. Obstacles in the workspace of the KUKA robot are detected by the Sick safety laser scanner. The measured values, obtained from the sensor are in polar coordinates. Obstacle detected is assumed to be line type (described by equation (2), resembling a human body)

2b. The target position is considered to be a point charge whose field strength does not decline with distance, attracting the point charge which is representing the TCP. The position of the target charge is moved by the operator, by the means of a joystick.

2c. The current TCP position (in Cartesian coordinates) is also needed as input for the path planning. This is obtained from the robot controller.

3a. A median filter to average measurement errors applied to the polar measurements is the least computationally demanding; a one dimensional median filter is applied over a window of 5 elements.

3b. The obstacle is detected based on the filtered laser scanner data. A mean filter is applied for measurements inside the workspace to find the center of the obstacle. This simplification only allows the detection of one symmetric obstacle in the workspace. A sensor system allowing detection of multiple obstacles is currently under development.

3c. Coordinates are transformed to the robot base coordinate system, using standard transformation matrices. Parameters of this transformation are the relative coordinates between the robot and the sensor.

4. The path planning block operates based on equations (1) - (6). Having as input the relative coordinates of the obstacle(s) depending on the type of the obstacle $\vec{r}_{OP}$ in equation (1), $\vec{r}_{LP}$ in equation (2) and/or $\vec{r}_{DP}$ in equation (3) the potential field of the obstacles are calculated independently based on the parameter f_Q equation (1), f_L equation (2) and/or f_D equation (3). The superposition of the effects is calculated based on equation (4). In this implementation the parameter $\vec{F}_M$ is considered 0 and will be used in future implementations. The parameter Q_{ES-TCP} (electrostatic charge assigned to the TCP) is considered to be 1. This way equation (4) becomes:

$$\vec{F}_{res} = \vec{E}_{res} Q_{ES-TCP} = \left(\sum \vec{E}_D + \sum \vec{E}_L + \sum \vec{E}_P\right) Q_{ES-TCP} \,. \quad (9)$$

5. The correlation between the output value of the path planning block (cyclical position offsets, $\vec{\Delta}_P$) and the electrostatic force $\vec{F}_{res}$ acting upon the unitary charge Q_{ES-TCP} representing the TCP is implemented as shown in equation (5). Since data transmitted to the robot are cyclical relative position offsets $\vec{\Delta}_P$ this command mode has a similar behavior as a velocity command mode.

6. Position offsets $\vec{\Delta}_P$ are sent to the robot controller and the algorithm restarts cyclically from step 2.

For the purpose of validating the above described theoretical considerations and implementation the following scenario is considered:

The intended robot path, a straight line, is obstructed with a line obstacle (as described in equation (2), having $Q_{ES-TCP} = 1e^{-7}C$) and a point obstacle (as described in equation (3), having $\lambda = 1e^{-6}C$). The resulting robot path with either the line obstacle or both the line and the point obstacle are recorded. The line obstacle is chosen since only the X and Y coordinates of the obstacle can be measured by the laser scanner and this obstacle shape could e.g. represent a human in the robot workspace. The point obstacle (virtual obstacle) is added to demonstrate the influence of different obstacle types on the path. The velocity of the robot is calculated based on equation (5), with parameter $d = 1Ns/m$,and $\Delta t_i = 12ms$. The path planning algorithm modifies the intended path, in order to avoid collision.

Experimental results are presented in fig. 2. The robot moves in a smooth shape to avoid the line obstacle and continues in an almost straight line to the target if only the line obstacle is used. If, additionally, a point obstacle is introduced the robot shows another smooth avoidance maneuver towards the end of the path. As expected, this avoidance movement is no longer executed in the plane perpendicular to the line obstacle. Using the analysis method presented in section 2, based on equation (8), using the descent gradient minimization, the repulsiveness of the situation with both obstacles is analyzed. This is minimization is done prior to the path planning, and is not yet real time capable. With a safety radius of $r = 100mm$ the minimum repulsiveness for the line obstacle is $E_{\text{Rep}}(\vartheta, \varphi) = 20N/C$ while around the point charge the minimum repulsiveness is $E_{\text{Rep}}(\vartheta, \varphi) = 9e^4 N/C$. The large difference between the two values aligns well with the different behavior of the different potential fields in close vicinity of two different charge types. Since both repulsiveness values are positive no violation of the safety envelope can occur.

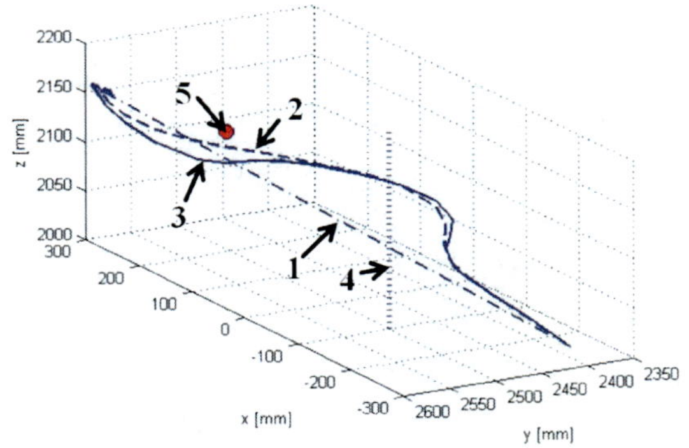

Fig. 2. Path of the robot for different obstacles showing the intended path (1), the recorded path (2) with one line obstacle (4) and the recorded path (3) with one line obstacle (4) and an additional point obstacle (5).

4 Conclusions and Outlook

The proposed approach for fast path and trajectory planning for industrial robots appears promising for pragmatic obstacle avoidance. The path planning approach is also applicable with dynamically moving obstacles. The proposed set of obstacle models allows the handling of common interaction scenarios occurring during robot operation. The safety envelope analysis guarantees that no collision can occur in case of static obstacles. Paired with a surveillance mechanism to monitor the separation of robot and obstacles also in the case of hardware failure [11] the approach could be used for dynamic path generation in human-robot interaction. The proposed tools and methods for implementation show promising results and allow the expansion to real-time systems. Further research is planned to extend the safety envelope analysis also for mobile obstacles. Also it is planned that the volume of the robot will be included in the obstacle avoidance.

Acknowledgement

This research was partially supported by the DAAD (German Academic Exchange Service) and partially funded by Empire Dynamic Structures, Vancouver, Canada.

References

[1] ISO 10218-1:2006: Robots for industrial environments – Safety requirements – Part 1: Robot (2006)
[2] Yao, J., Lin, C., Xie, X., Wang, A.J., Hung, C.-C.: Path Planning for Virtual Human Motion Using Improved A* Star Algorithm. In: Proc. 7th Int. Conf. on Information Technology: New Generations, pp. 1154–1158 (2010)
[3] Gao, M., Xu, J., Tian, J., Wu, H.: Path Planning for Mobile Robot Based on Chaos Genetic Algorithm. In: Proc. 4th Int. Conf. on Natural Computation, vol. 4, pp. 409–413 (2008)
[4] Borenstein, J., Koren, Y.: The vector field histogram-fast obstacle avoidance for mobile robots. IEEE Transactions on Robotics and Automation 7(3), 278–288 (1991)
[5] Ulrich, J.B.: VFH+: reliable obstacle avoidance for fast mobile robots. In: Proc. IEEE Int. Conf. Robotics and Automation, vol. 2, p. 1577 (1998)
[6] Ulrich, J.B.: VFH*: local obstacle avoidance with look-ahead verification. In: Proc. IEEE Int. Conf. on Robotics and Automation, vol. 3 (2000)
[7] Seder, M., Petrovic, I.: Dynamic window based approach to mobile robot motion control in the presence of moving obstacles. In: Proc. IEEE Int. Conf. on Robotics and Automation, pp. 1986–1991 (2007)
[8] Lee, M.C., Park, M.G.: Artificial potential field based path planning for mobile robots using a virtual obstacle concept. In: Proc. Int. Conf. on Advanced Intelligent Mechatronics, vol. 2, pp. 735–740 (2003)
[9] Alonso, M., Finn, E.: Physics, Workingham. Addison-Wesley, Reading (1992)
[10] Arias-Castro, E., Donoho, D.L.: Does median filtering truly preserve edges better than linear filtering? Annals of Statistics 37(3), 1172–2009 (2009)
[11] Dietz, T., Pott, A., Verl, A.: Simulation of the Stopping Behavior of Industrial Robots. In: New Trends in Mechanism Science, vol. 5, Part 6, pp. 369–376 (2010)

The Design of Human-Machine Interface Module of Presentation Robot Advee

J. Krejsa[1,*] and V. Ondrousek[2]

[1] Institute of Thermomechanics AS CR v.v.i. Brno branch, Technicka 2, Brno, 616 69, Czech Republic
[2] Brno University of Technology, Faculty of Mechanical Engineering, Technicka 2, Brno, 616 69, Czech Republic

Abstract. This paper presents the approach used in the design of the highest software layer of autonomous mobile robot Advee used commercially for presentation purposes. The requirements for the layer are discussed together with available means. The overall structure of the module is shown and key features are described in detail. The module serves successfully when exploited to variable environment represented by people with computer literacy of great variation.

1 Introduction

With mobile robots appearing nowadays outside research laboratories more than ever, human-robot interface (HRI) related issues attract great attention. Most of the research is focused on improving certain means of communication, especially the voice dialog (sound used both as inputs – natural language understanding and outputs – robot voice) and utilization of computer vision (human face recognition, face related higher features recognition, gesture recognition, etc.).

Various means of communication can be combined to increase the level of interaction. In (Park, 2009) the speech recognition is combined with user localization using two processed microphones signals to obtain the location of the source, that can be further utilized in the response of the robot. Face recognition is accompanied with gesture recognition in (Chang, 2009) bringing wider variety of possible inputs to the robot. Work of (Perzanowski, 2001) is an example of multimodal HRI, combining speech and gesture recognition with external input from PDA.

This paper is focused on HRI of presentation robot Advee, whose purpose is to serve as an autonomous mobile source of information, transferable to people of different computer literacy. Such requirement brings the necessity of combining all robot means to get the redundancy in communication channels, so the less computer literate people can still get the message while more literate user is not repelled.

* Corresponding author.

2 Module Requirements and Robot Means

In order to design the HRI module properly the requirements must be stated first. The overall goal is to create the interface that is friendly and universal (can be used by variety of users), while the underlying engine is robust, modular and flexible. The requirements are listed below in order of importance.

- redundancy: information should be transferred to user in all possible ways in parallel (e.g. both visual and sound)
- robustness: HRI module must be capable of operation even when minor hardware failure occur (e.g. camera fails)
- flexibility: simple addition of new feature, sensor, etc
- parametrization:overall behavior can be easily changed, certain features can be disabled/enabled upon request
- adaptability: HRI should adapt to the user abilities, e.g. longer timeouts for slower users, etc.
- modularity: several programmers can work independently on the project, particular modules can be tested independently

Based on the requirements, what are the means of the robot? We can divide the means into two basic categories: inputs and outputs, where inputs represent the information from the user to the robot and outputs represent robot reaction. Some of the means are bidirectional, e.g. the touch screen acts both as input and output. The overview of Advee's means is shown together with its physical location on Fig. 1.

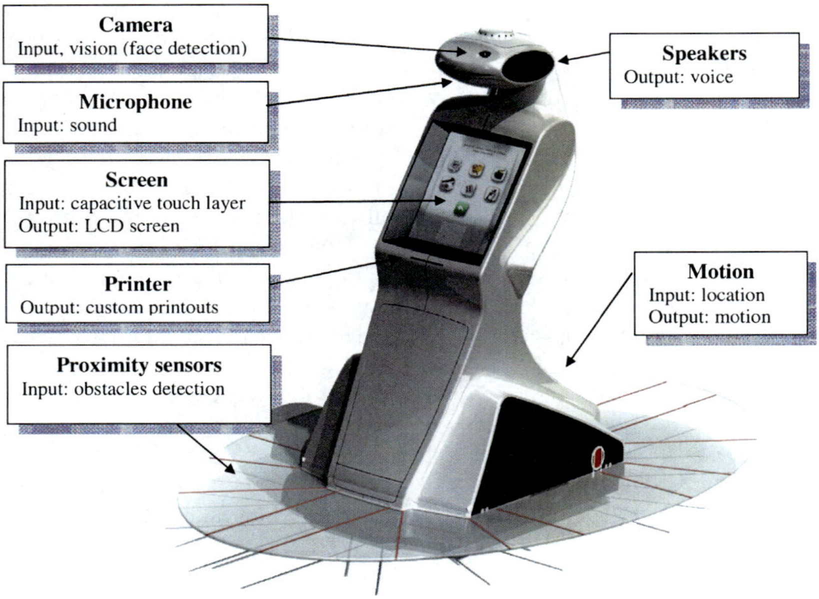

Fig. 1. Overview of Advee's means

3 HRI Structure and Behavior

The key idea in designing the HRI module is to separate the interaction into independent blocks, that are sequentially activated by main HRI engine. Only single block is active at the time. The overall behavior of the robot can then be defined in a single routine controlling blocks activation.

Each block uses different means of the robot, however in most cases the majority of the means are used. To enable access to the means, the particular means are encapsulated into so called sources. Programmatically the blocks and sources are represented by certain classes. Each block is inherited from *BaseBlock* class and each source is inherited from *BaseSource* class. *BaseBlock* contains links to all the sources instances, therefore any subsequent block inherited can access all the sources. As the sources can serve to single block only, the mechanism that assigns and releases sources is implemented. Sources are assigned prior to block activation and released once the block is finished.

Currently implemented sources and blocks are shown on Fig. 2. (left), with an example of *Map* as active block having access to all the sources. The sources directly depend on the means of the robot from previous chapter. As for the blocks, the key ones are:

- *Motion* –active when the robot is moving and not interacting with people. Motion can end for a number of reasons: people are detected (combination of proximity sensors data, face detection of the camera images, etc), somebody touches the screen, etc.
- *Selector* – block that serves as the menu allowing the users to select further actions of the robot.
- *Catcher* – block that verifies whether there is a person the robot can talk to. The block serves as decision maker for uncertain situations (low confidence in face detection, etc)
- Custom blocks – the names are selfexplanatory: *Map* shows interactive map of the surroundings, *Video* runs the video, *Games* enables users to play games, *Print* handles printing, etc.

Whole behavior of the robot is determined by the sequence of blocks. The sequence depends partially on user and partially on behavior definition. An example of the signal flow is shown in Fig. 3. The flow can be arbitrarily modified, certain features (blocks or block sequences) omitted or added.

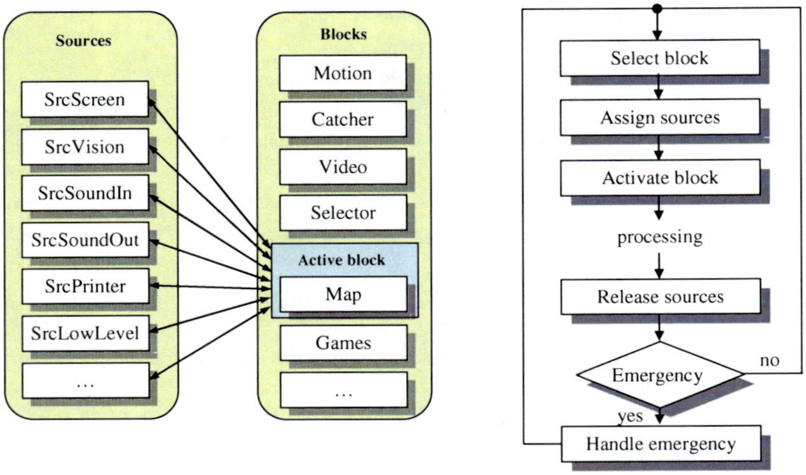

Fig. 2. Resources management and HRI structure

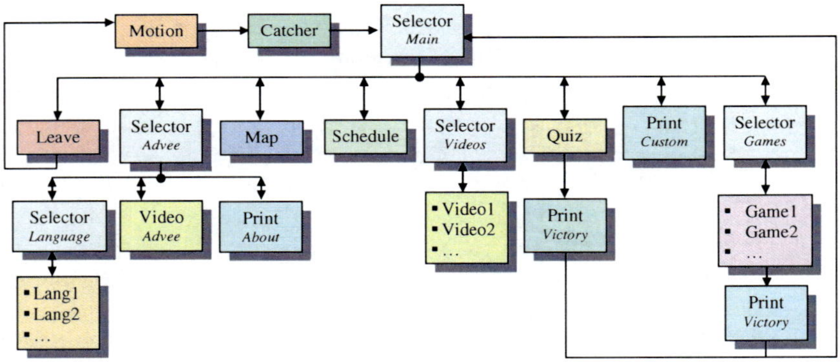

Fig. 3. Example of behavior definition

4 Verification and Conclusions

The HRI was initially verified using released version of HRI together with simulator of the lower layers of robot. During the initial tests the HRI was presented mainly to students and basic settings were tuned, such as timeouts for particular blocks, understandability of the graphics, etc. Once the basic setup was done the tests were performed using real robot on a number of people from various backgrounds and data in the form of questionnaires were collected afterwards together with the camera recordings.

Based on the collected data the HRI was further modified, with the emphasis on redundancy of given information, e.g. all voice outputs (talk of the robot) are accompanied with the text on the screen of the same meaning. Further tests lead to minor modifications in rephrasing the speech output and changing the voice modulation for further clarification. During the use of the robot all the user inputs are logged and collected data are used for further analysis and improvements.

In order to get a measure of success of the HRI operation, a set of following indicators was proposed:

ATI - average time of interaction with single user [sec]: interval between the start of the interaction and its end (user left the robot). Longer ATI means single user spent more time with the robot, it is only indirect indicator of how well the interaction went.

CSR - user catching success rate [-]: rate between successful catch of user (interaction starts) and failure (user walks away). Lower CSR means that people are afraid of the robot, do not know how to start an interaction, etc.

TIP – interaction timeout percentage [%]: percentage of interactions ended with timeout in arbitrary block other than main menu. Higher percentage means that user left the robot during interaction.

PPE – person per event [-]: total number of interactions divided by total number of users. If the PPE is higher than 1 then some users used the robot repeatedly.

BPU – blocks per user [-]: number of blocks activated per user in single interaction. Higher BPU means that user exploited more options during interaction.

The values of the indicators obtained during the operation in real environment with people coming from all the backgrounds are shown in Tab. 1. High variations in ATI and BPU values are caused by different offer of the HRI on different events. CSR value of about 70% is considered successful, the PPE value is higher than initially expected. The PPE value is essentially influenced by the rewards the robot can provide for given event. With no reward the value of PPE quickly goes to about 1.05.

Table 1. Values of HRI quality indicators

indicator	ATI	CSR	TIP	PPE	BPU
value	75±22	0.69	14.2	1.20	6.1±4.5

HRI of robot Advee is now fully operational and used in commercial application of the robot with approximately 200 hours of operation in interaction mode. HRI can serve both young and senior users, as illustrated on Fig. 4. The results so far are satisfactory and prove the proposed concept to be viable in real world application.

Fig. 4. Test runs in shopping center with wide range of users

Acknowledgement

The research was funded by the AS CR under the research plan AV0Z20760514 and by Operational Programme Education for Competitiveness, project number CZ.1.07/ 2.3.00/09.0162. Authors would like to thank Petr Schreiber from Bender Robotics for his work on HRI coding.

References

[1] Perzanowski, D., Schultz, A.C., et al.: Building a multimodal human-robot interface. IEEE Intelligent Systems 16(1), 16–21 (2001)
[2] Chang, M., Chou, J.: A friendly and intelligent human-robot interface system based on human face and hand gesture. In: AIM 2009, pp. 1856–1861 (2009)
[3] Park, K., Lee, S.J., et al.: Human-robot interface using robust speech recognition and user localization based on noise separation device. In: ROMAN 2009, pp. 328–333 (2009)

Force-Torque Control Methodology for Industrial Robots Applied on Finishing Operations

T. Kubela*, A. Pochyly, V. Singule, and L. Flekal

Institute of Production Machines, Systems and Robotics, Faculty of Mechanical Engineering, Brno University of Technology, Technicka 2, Brno, 616 69, Czech Republic

Abstract. The paper deals with the force-torque control methodology for industrial robots that can be more or less applied on a variety of industrial applications such as grinding, drilling, automatic assembly or other applications where the key aspect is to control the physical contact between a tool and a workpiece mounted on the robot gripper or vice versa. In other words, contact forces and respective moments need to be controlled in real-time. The experimental set-up is based mainly on a KUKA robot (KR3, KR 16), FTC sensor (SCHUNK FTC-050), external PLC system (Embedded PC Beckhoff) for main control structures and other devices. Main practical results are concerned with determining the contact between a workpiece, grasped in the robot gripper, and the surface while maintaining a constant force during robot motion.

1 Introduction

Industrial robots have become standard equipment for flexible manufacturing systems. It is mainly due to the flexibility and universality of a 6 DOF industrial robot that are integrated into manufacturing systems fulfilling various tasks and based on statistics of the International Federation of Robotics (www.ifr.org), the number of robots used in practice is still rising. However, industrial robot programming is still an annoying task and expert knowledge is expected in most cases. It is the main reason that the robots are used in fixed installations performing periodic and repeatable tasks (without changes) and programmed either online or off-line by an expert. Another limitation is concerned with a very rare interaction with the worker; the worker and the robot are usually separated through mechanical fences or other safety systems. Finally, such robot utilization concept is considered to be profitable rather only with medium to large lot sizes.

Nevertheless, new demands on robot programming and applicability are emerging nowadays. At first, there is a need to automate applications that are nonstandard for robots. Such applications require the robot end-effector (tool) to be in

* Corresponding author.

direct contact with the environment. It means that the contact forces an moments need to be controlled to ensure a proper task performance and right results. This field belongs to the finishing operations [1, 2] such as grinding, polishing, deburring etc. Despite many research projects and results were presented and published, it has not reached a proper industrial usage so far.

The second demand is concerned with a new robot programming technique that is based on human robot interaction and cooperation. This technique along with a new generation of industrial robots [3] allows the robot to be flexibly relocated within the shop floor, the tasks can be frequently and easily changed and robots can be instructed online: by demonstration [4] of a non-expert or supported by off-line techniques using CAD objects data.

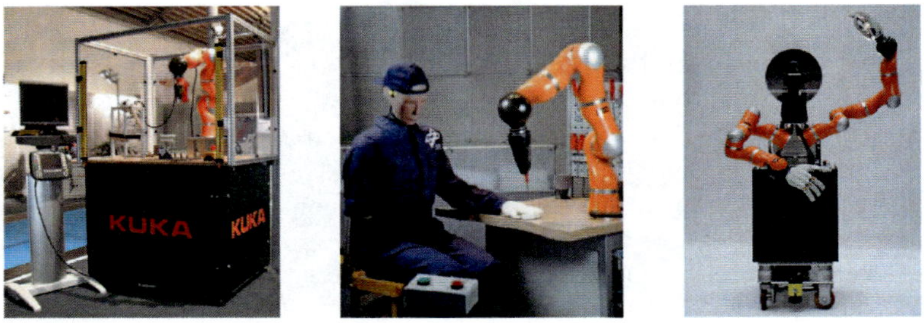

The SME worker's third hand | Safe human-robot collaboration | Mobile dual arm/hand manipulation

Fig. 1. Ongoing R&D projects with the KUKA Light-Weight Arm [3]

There is also a frequent robot interaction with the worker (even the force interaction) and concerning the safety issue, the robot workspace is fully shared with the worker. Finally, such concept is expected to be sustainable and profitable even for small lot sizes; typical for SMEs. A typical example of this next generation robot is the KUKA Light-Weight Arm (Fig. 1).

To implement a robotic cell concerned with finishing operations or human robot interaction, there is required a proper force-torque feedback based on a force-torque sensor [5]. Various control architectures can be applied [6, 7, 8]. Theoretically, the best way would be a force controller based on the complete model of the robot. Concerning this type of controller, there is a problem that the outputs from such controller are motor torque (force) what is difficult to implement on the actual range of industrial robots since they have had more or less a closed architecture rarely allowing the direct control of the motor torque. However, in this work we applied a cascade force controller (Fig. 2) regardless of the fact that it does not take into account the whole model of the robot. Depending on the "closed architecture" of standard robot controllers (e.g. KUKA KR C2/C3), the cascade force controller allows rather only the tool position to be controlled.

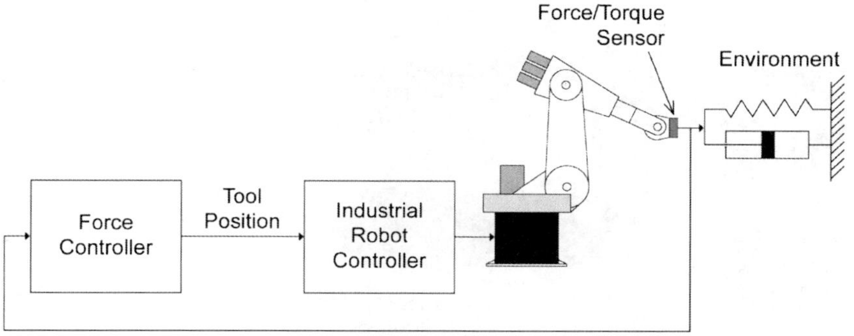

Fig. 2. Cascade force controller conception for industrial robot

2 Motivation and Real Industrial Demands

This project of force-torque control of industrial robots is motivated by real industrial demands and it is focused on particular applications [9]. In collaboration with our industrial partner we received various objects for testing and currently we are dealing mainly with components intended for finishing operations - especially grinding applications.

To answer the needs, the project have been initiated (we successfully asked for funding - Ministry of Industry and Trade of the Czech Republic) and we establish a research platform based on academic-industry partnership.

Despite many models for technology transfer in practice have been described in the literature, few successful R&D projects fulfilled the industrial (customer) requirements concerned mainly with advanced robotic applications such as force-torque robot control, bin-picking based on 3D vision, continuous image analysis of various surfaces in high speed etc.

In general, one extreme approach starts from a researcher doing basic research that might lead to a successful industrial application in the near future. Another approach concerns with specific market needs and a search for a researcher able to cope with the problem. Many models also claim that a successful technology transfer in practice needs to be based on close collaboration between researchers and practitioners.

In our case, the following model (Figure 3 on the right) has been used in this project. The model is partly based on some ideas of a seven-step technology transfer model [10] concerning with identifying industry needs, formulating a research agenda, formulating a solution and comprehensive validation based on small lab experiments, pilot projects leading to release the full solution step by step.

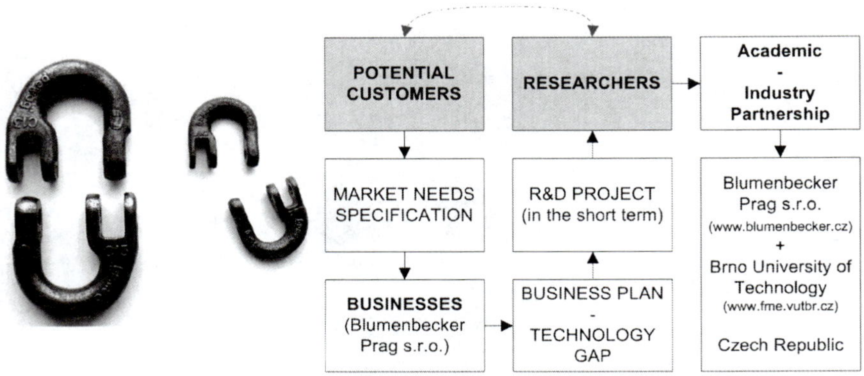

Fig. 3. A component example (on the left) and technology transfer model used in this project (on the right)

3 Design of the End Effector

Basic principle of object (Fig. 3 on the left) grasping is presented on the following figure (Fig. 4). A 2-finger parallel gripper equipped with angled prismatic inserts (parts 1 and 2, Fig. 4) is used to grasp the component. As an additional supporting element a pneumatic cylinder is used to hold the component using another prismatic element (part 3, Fig. 4).

The main problem of such grasping is a fact that in the pressing process of fingers with angled prismatic inserts, a partial displacement of the grasped object in the direction outwards of the fingers may occur what is caused by the respective force of the gripper. This force is needed to be minimized by designing a suitable tilt angle of fingers, angle of prismatic elements and using a material with an adequate friction coefficient. However, details concerned with this procedure including the action of force on the grasping element are not covered in this contribution.

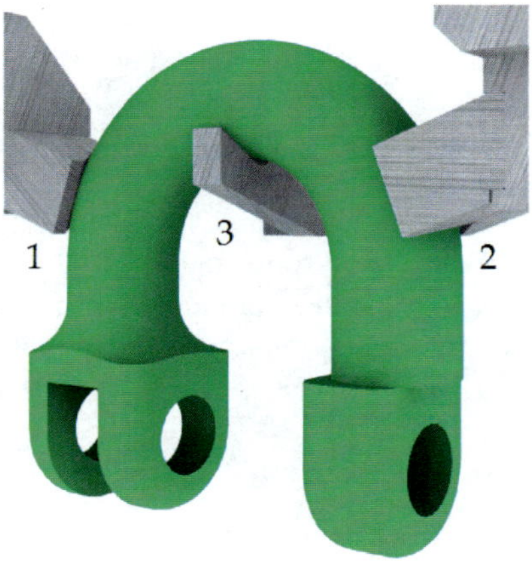

Fig. 4. Basic principle of object grasping

Various alternatives of the end effector have been designed taken into account either a particular (single) object size and shape or even objects within the whole production range (Fig. 5) with a big variety of object sizes and shapes.

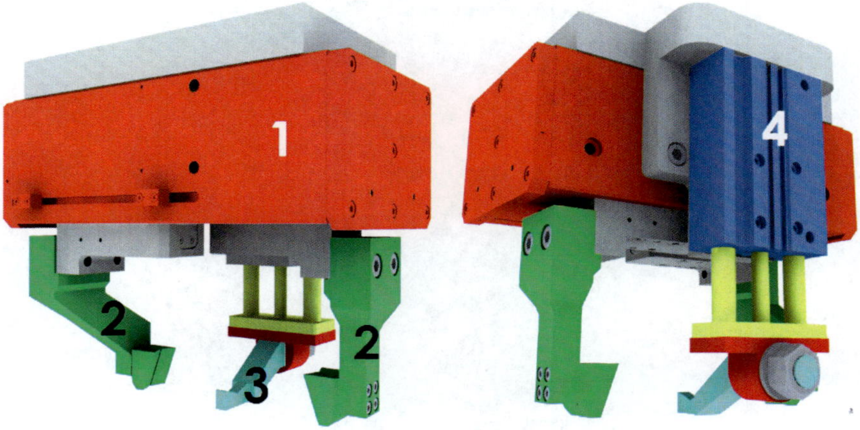

Fig. 5. A model of the end effector (for the whole production range)

The end effector is based on the 2-finger parallel gripper SCHUNK PFH (part 1, Fig. 5) that was chosen for its relative high stroke and other parameters. Further, there is a compact pneumatic cylinder (part 4, Fig. 5) with guiding for the prismatic element (part 3, Fig. 5). However, for the experimental testing we finally modified the model (Fig. 5) for a more limited production range in order to reach a more compact end effector.

4 Control Methodology

The hardware structure of the robotic cell (Fig. 6) is mainly composed of the following components. A PLC system (Embedded PC Beckhoff) is used to read data from the force sensor (SCHUNK FTC-050) via serial connection RS232. In the PLC additional data processing is performed in order to send control deviations to the robot controller via DeviceNet interface.

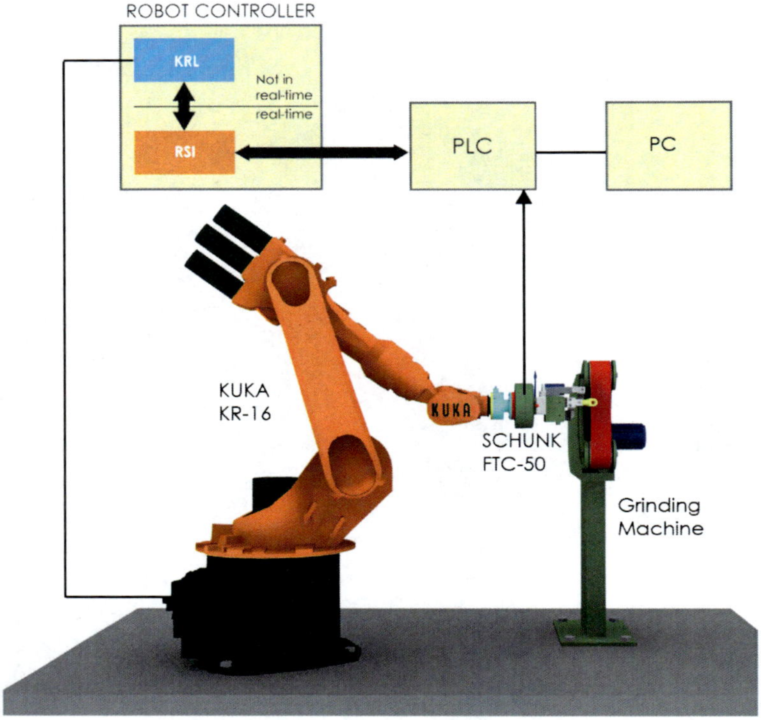

Fig. 6. Main structure of the robotic cell with a general control structure

Concerned with the KUKA robot controller, KUKA RSI (Robot Sensor Interface, v.2.3) software package has been used to fulfill the condition of real-time data processing. Based on this software package, RSI objects (e.g. object for path correction), that run in the defined 12ms cycle, are used to adjust the robot trajectories on the basis of control deviations.

The control system (Fig. 7) is based on two PID controllers (Reg I and Reg II) that are applied either on the beginning of the contact or when moving along with a workpiece to maintain a constant contact force. In other words, the first PID controller (Reg I) is used for greater deviations (errors) whereas the second controller (Reg II) is used only when smaller deviations occurs. It is done by means of the Switch Condition block (Fig. 7) where the compensated force (f_{cont}) and desired force (f_{des}) is to be compared. Desired position (x_{des}) of the robot is set to the Position Controller.

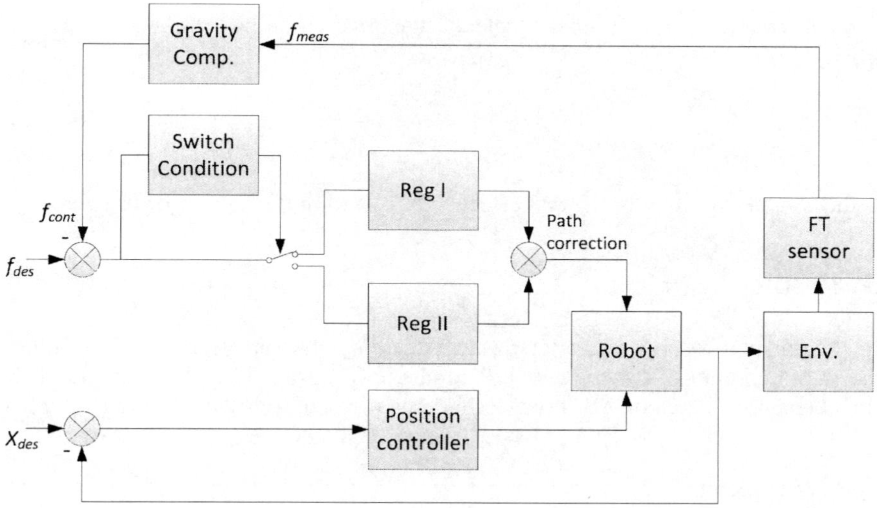

Fig. 7. Main block diagram of control system

The gravity compensation is characterized by the following equations. Contact force between the robot and the environment is based on measuring and prediction of gravity forces (torques):

$$\vec{F}^{contact} = \vec{F}^{meas} - \vec{F}^{pred}$$
$$\vec{M}^{contact} = \vec{M}^{meas} - \vec{M}^{pred} \tag{1}$$

Load compensation is based on the identification of the values in relation to following equations:

$$\vec{F}^{pred} = m\vec{g}_{sensor}$$
$$\vec{M}^{pred} = \vec{c}_{sensor} \times m\vec{g}_{sensor} \tag{2}$$

where

$$\vec{g}_{sensor} = R_{base}^{sensor}(\theta) \cdot \vec{g}_{base} \tag{3}$$

is the direction of gravitational acceleration expressed in the FT sensor frame and where

$$\vec{g}_{base} = [0 \quad 0 \quad -9.81]^T \tag{4}$$

is the direction of gravitational acceleration expressed in the robot base frame and finally where

$$\vec{c}_{sensor} = \begin{bmatrix} x_c & y_c & z_c \end{bmatrix}^T \tag{5}$$

is the center of gravity of the end effector expressed in the FT sensor frame.

5 Results

In this chapter we present pilot results regarding the force-torque robot control based on the control structure described above (Fig. 6 and 7). An example of the time behavior of the contact force (Fz) is shown on the following figure (Fig. 8).

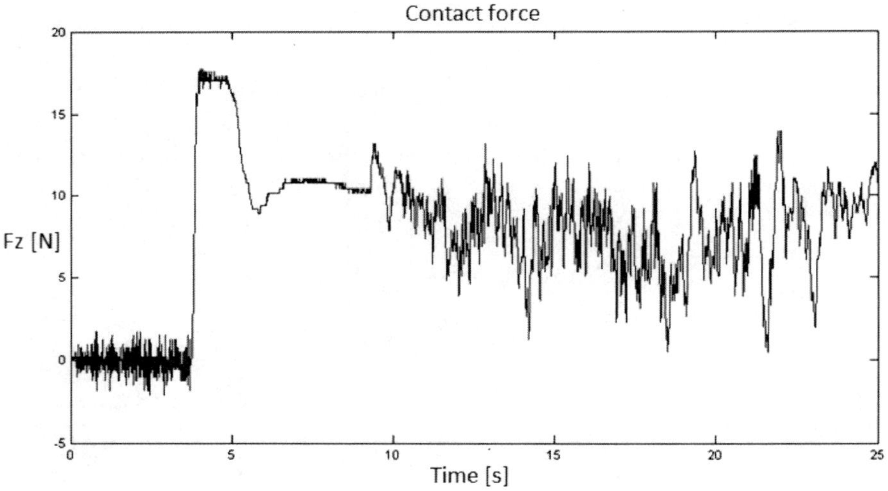

Fig. 8. Contact force during the robot motion along with the surface

The desired contact force to maintain while the robot is in contact with the surface is set to 10 N. The time interval between about 6 and 9 seconds can be determined as maintaining the contact force one a single robot position whereas the rest of the graph represents the motion of the robot trying to maintain the contact force given in advance. It is apparent that the robot alternates in gaining and losing the contact and the process need to be further optimized and this will be the main part of our future work.

6 Conclusion

This R&D project is focused on the industrial robot force control applicable in various applications though in this case it is concerned with particular objects supplied by industrial partner. Pilot results were presented using the control architecture described above where a desired contact force with object surfaces can be controlled allowing the robot to follow object profiles (e.g. for grinding applications).

The advantage of the control architecture is concerned with the independence of the robot controller. All control algorithms and data processing is performed by the external PLC system (Embedded PC Beckhoff). The only condition for KUKA robot controller is the necessity of RSI (Robot Sensor Interface) software package to be installed in order to guarantee the real-time operation.

Acknowledgement

This work has been supported and funded by the Ministry of Industry and Trade of the Czech Republic within the industrial research and development program TIP (2009); project number FR-TI1/169: Adaptive Force-Torque Control of Robots for Industrial Applications.

References

[1] Bogue, R.: Finishing robots: a review of technologies and applications. Industrial Robot 36(1), 6–12 (2009)

[2] Pires, J.N., Ramming, J., Rauch, S., Araujo, R.: Force/torque sensing applied to industrial robotic deburring. Sensor Review 22(3), 232–241 (2002)

[3] Bischoff, R.: From research to products: The development of the KUKA Light-Weight Robot. In: Proceedings of the 40th International Symposium on Robotics (ISR 2009), Barcelona, Spain (2009)

[4] Pires, J.N., Veiga, G., Araujo, R.: Programming-by-demonstration in the coworker scenario for SMEs. Industrial Robot 36(1), 73–83 (2009)

[5] Perry, D.: Optimize your robot's performance by selecting the right force/torque sensor system. Industrial Robot 29(5), 395–398 (2002)

[6] Bigras, P., Lambert, M., Perron, C.: New formulation for an industrial robot force controller: Real-time implementation on a KUKA robot. In: Proceedings of the IEEE International Conference on Systems, Man and Cybernetics (SMC 2007), Montreal, QC, Canada, pp. 2794–2799 (2007)

[7] Blomdell, A., et al.: Extending an Industrial Robot Controller: Implementation and Applications of a Fast Open Sensor Interface. IEEE Robotics & Automation Magazine 12(3), 85–94 (2005)

[8] Caccavale, F., Natale, C., Siciliano, B., Villani, L.: Integration for the next generation: embedding force control into industrial robots. IEEE Robotics & Automation Magazine 12(3), 53–64 (2005)

[9] Pochyly, A., Kubela, T., Singule, V.: Force-torque control of industrial robot for industrial applications. In: Proceedings of the Engineering Mechanics 2010 - Book of Extended Abstracts, Prague, Czech Republic, pp. 105–106 (2010)

[10] Gorschek, T., Garre, P., Larsson, S., Wohlin, C.: A Model for Technology Transfer in Practice. IEEE Software 23(6), 88–95 (2006)

Tactical and Autonomous UGV Navigation

J. Mazal*, P. Stodola, and I. Mokrá

University of Defence, Kounicova 65,
Brno, 66210, Czech Republic

Abstract. The paper is focused on the task of autonomous navigation of a ground vehicle in the general environment under the tactical conditions. This problematic consists of a wide set of operational, tactical and technical problems, generally containing the sort of multi-criteria decision tasks and in many cases converge to a battle process optimization. This article is dedicated to one of the thorniest problems of contemporary asymmetric battlefield, namely math solution of a real-time tactical maneuver and its automatic execution. The article presents the basic goals and approaches to this problem. The basic principles and algorithms of autonomous motion were implemented into an experimental robot which is used primarily for verification and demonstration purposes.

1 Introduction

The unmanned systems have recently become very popular. There are many ways to use them in various fields of human activities. In recent years the development of these systems has moved significantly forward particularly in the field of automation of control. The role of operator of semi-autonomous and autonomous unmanned means has moved from the management and decision making sphere to the sphere of supervision and situation overview. Due to the rapid development of military technologies and increasing requirements to combat information systems, the technology level enables more advanced capabilities mediating through the new components of these systems. As the latest upgrade, we could consider for instance decision support system for tactical commander, whose integral part could contain a real time maneuver management of friendly units or components (UGVs) on the battlefield combined with fully automatic execution of commander's intent. Theoretical approach linked with that aspect can be decomposed into a series of problems intervening field of multi-criteria decision making process, game theory, graph theory, and other areas of mathematics.

* Corresponding author.

2 Primary Level of Tactical Path Planning

The fundamental solution is based on a series of procedures and weighted integration of several separate layers, where all phases converge to a construction of maneuver graph of tactical environment. Next, the graph is processed by set of algorithms, mostly issued from the Roy-Warshall principle to find the most appropriate way to the target location considering tactical/environmental conditions. These solutions come from a particular math modeling of the tactical environment, which (in our case) is described by a 3D array (set of multiple 2D arrays) [1]. Integrating process dealing with many layers can be relatively complex and therefore the initial solution is treated only in approximate scenario, where input factors (layers) were derived from the following effects:

- The influence of altitude
- The effect of vegetation (forest)
- The influence of water
- The impact of communications
- The impact of terrain obstacles
- The effect of visibility of threatening element

Setting the input criteria of solution (decision) is conceived in part of the integration model of quantification of input characteristics, whose description would significantly exceed the framework of the article. But in the general overview it is set of models of the multi dimension functions, and their task is to properly incorporate the influence of external conditions and characteristics to the numeric form of the set of pragmatical coefficients. Which are then applied with mathematical methods and transformations, leading to the final model construction, in which to find the optimal solution is already trivial. Therefore, as an example may serve the following function (1) defining the simplified calculation of pragmatical (linked to a particular position) coefficient in the context of the visibility by endangering element and close contact options. General model takes algorithmic nature, since under different conditions are applied to different calculations, nevertheless main part f_{Ou} is demonstrated as follows:

$$f_{Ou} = \frac{(1-0{,}002n_3)\left(0{,}7e^{-\left(\frac{n_3-200}{100}\right)^2}\right)+\frac{(1-0{,}002n_3)}{\frac{n_3}{1000}+20}}{n_1+1} \times \frac{\tan^{-}\frac{n_2}{n_3}}{2} \tag{1}$$

n1 - shortest distance to the nearest vegetation
n2 - altitude difference of the threatening element and the friendly element
n3 - distance between threatening element and the friendly element

In the outcome of experiments, there is an apparent desire of motion, which is situated close to a communications and intent of avoidance of visible areas as is illustrated on following fig. 1. Threatening visibility laid out had a decisive impact on final criteria displacement of the model. Next, as displayed the shadow map, the total sum of weights of the minimum path to each point of the map under the above conditions.

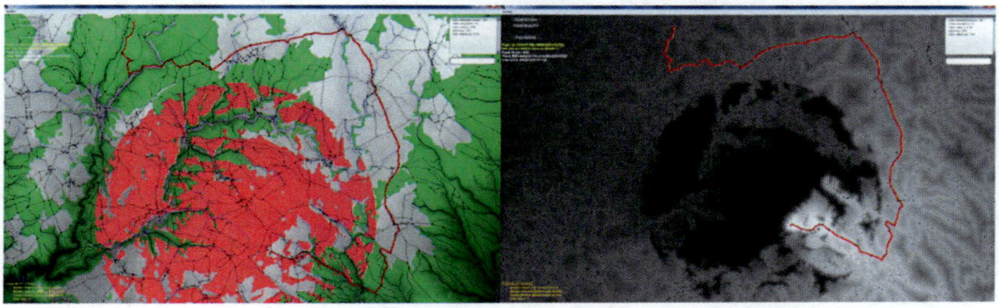

Fig. 1. Tactical path optimization and layout of maneuver optimality weighting

3 Autonomous Navigation and 2D/3D Experiments

The vehicle autonomous navigation and obstacle avoidance in the 3D space is a relatively complex task [2]. The input parameters of the algorithm are the current vehicle position and orientation in the space and the target (desired) position we want to achieve, what is actually iterative parts of the curve (solved in previous part) obtained by multi-criteria decision modelling on global area of operation. The algorithm runs continuously while the vehicle is moving in the steps defined by the time intervals up to achieving the final/partial coordinates. The algorithm consists of four key processes and several partial activities [3] as follows:

- Localization process
- Virtual environment reconstruction process
- Optimal path searching process
- Motion control process

The results and practical experiments were carried out in a hall of about 17 × 29 m. basic task was the optimized movement of vehicle through the experimental environment with 24 obstacles as is shown on fig. 2. Figure on the upper-left side shows the general configuration of the environment, in the next ones gradual development of vehicle motion. Green colour shows the route run through, and the red shows the optimum path from the current vehicle position to the target position.

The major problem in the real environment is the localization of the robot, especially the precise determination of its orientation. Any small inaccuracy in determining the angle of orientation is substantially reflected in the reconstruction of the obstacles. Also small errors appeared in determining the orientation were

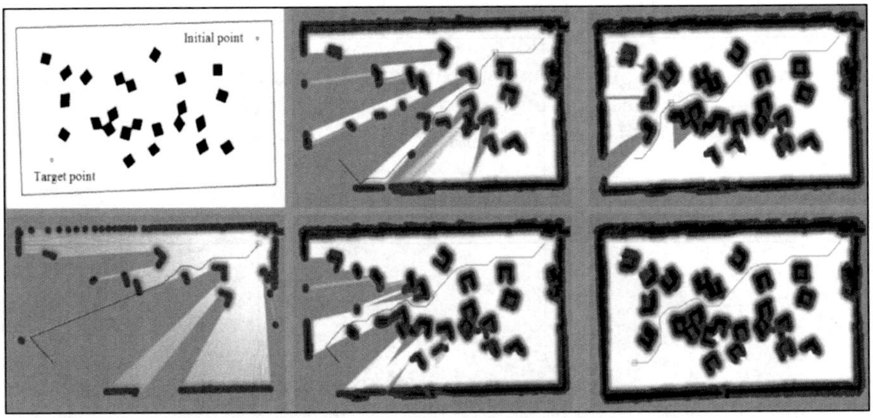

Fig. 2. Real tests of autonomous motion of experimental vehicle

caused by an incorrect synchronization of the time of position and orientation calculation of the vehicle by means of model and time of environment mapping by the laser scanner. Nevertheless, as the results show, this error does not have a significant impact on the overall function of the autonomous motion. Next, two-dimensional limitation in experiments was primarily due to an available laser scanner that is able to scan the environment only in one plane. The algorithm can be implemented even in a fully three-dimensional space with only some modifications of certain parts. There exists the solution (modification of pan and tilt) of utilization of contemporary laser radar (2D) to obtain 3D results, but it is slow for real-time experiments. Therefore, it generates a set of points, which represent obstacles in the area. On this set we can apply advanced analyses, so that the result is a cost map of the environment respecting the laws of physics and vehicle design features, this map will be, as in the previous case, used as a gate into the optimal path-finding algorithm.

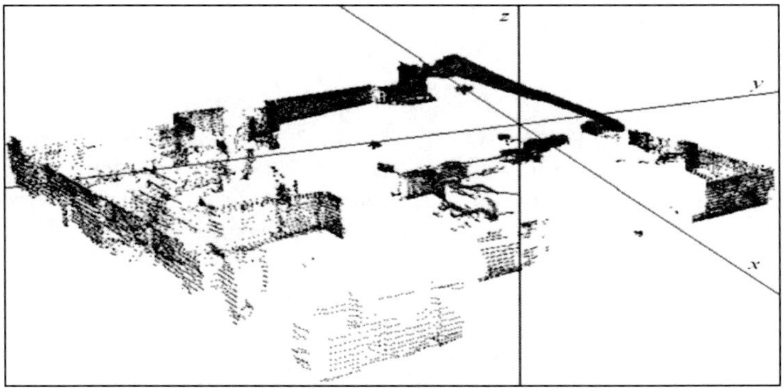

Fig. 3. Sample of 3D environment reconstruction

4 Conclusion

The capabilities of this presented solution demonstrate one possible approach to some problems, especially in the area of tactical decision support, optimal tactical manoeuvre and autonomous navigation. It is nnecessary to understand the fact that by that time this is sufficiently solved only algorithmic and mathematical side of a general problem. But the most important correct setting of tactical criteria (responding to real impacts) would be a task for additional experiments and statistical evaluation. However, current solutions should comply with the basic requirements and the advanced solution is intended to be further developed.

References

[1] Stodola, P., Mazal, J., Rybanský, M.: Dead Reckoning Navigation for Autonomous Unmanned Ground Vehicles in a Real Terrain. In: Proceedings of the Joint 9th Asia-Pacific ISTVS Conference, Sapporo (2010)

[2] Buehler, M., Iagnemma, K., Singh, S.: The DARPA Urban Challenge: Autonomous Vehicles in City Traffic. In: Springer Tracts in Advanced Robotics, vol. 56. Springer, Heidelberg (2009)

[3] Stodola, P., Mazal, J.: Optimal Path-finding Algorithm for Autonomous Unmanned Ground Vehicles. In: International Unmanned Vehicles Workshop, Turkish Air Force Academy, Istanbul (2010)

DELTA - Robot with Parallel Kinematics

M. Opl, M. Holub, J. Pavlík, F. Bradáč, P. Blecha, J. Kozubík, and J. Coufal

Brno University of Technology, Technická 2896/2
Brno, 616 69, Czech Republic

Abstract. A complex solution of a design of the DELTA type robot, i.e. the robot with parallel kinematics, is described in this paper. The project ensued from cooperation with the company Dyger, s.r.o., with the objective to develop a robotic device suitable for high-speed handling of small objects and applicable in many industrial configurations. The task was not only to design the mechanical part of the robot that would work together with the provided hardware and software, but also to develop our own algorithms for calculations of the target positions, singular positions of the mechanism etc.

1 Introduction

This project has been motivated by a requirement from industry. Our department received a proposal for possible cooperation in development of the robot with parallel kinematics – DELTA type robot. The company Dyger, s.r.o., is an exclusive representative of the company Beckhoff and has at its disposal the recently-released data library for control of robotic appliances with parallel kinematics. Hence, we attempted to make use of these new possibilities and to develop an affordably priced DELTA robot which could be further offered for solutions of specific industrial applications.

As the supplied software is not an open source software and it is not possible to simply verify if the mathematical description of the mechanism are valid, we proceeded to develop our own algorithms for calculations of position, speed, acceleration and, first of all, for calculation of the mechanism's working space and equally important singular positions. Our objective was also to verify the mathematical apparatus used in the supplied software [1].

2 Kinematic Analysis of DELTA Robot

The kinematic configuration of the DELTA robot presents one of the more simple types of parallel structures. This is achieved through implementation of three parallelograms, ensuring mutual parallelism of the stationary base and the moving platform [2].

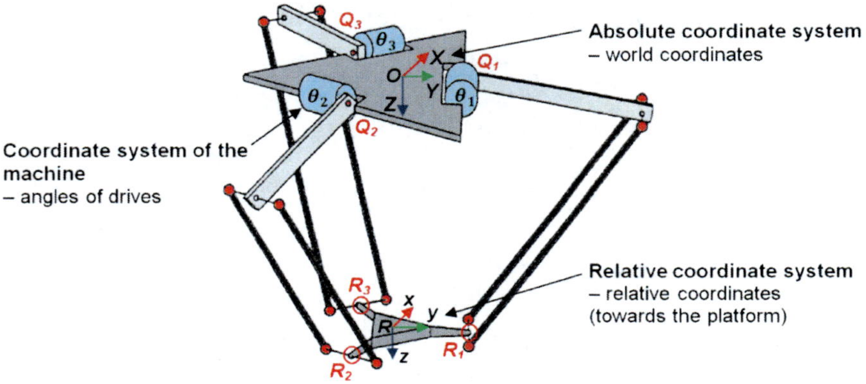

Fig. 1. Kinematic scheme of DELTA robot [5]

Based on an analysis of the individual kinematic links and their degrees of freedom (DOF) it is possible to calculate the total number of DOF of the mechanism with the equation (1).

$$F = b(n - g - 1) + \sum_{i=1}^{g} f_i - f_{id} + s \qquad (1)$$

Where b – number of DOF in space, n – number of elements, g – number of joints, f_i – DOF of i-th element, f_{id} – number of identical DOF, s – number of passive joints.

After substitution, the total number of DOF equals three. These DOF represent only translational motions of the platform, hence the absence of rotary motions ensures the desired parallelism.

Another complex parameter affecting applicability of a robot in a specific industrial configuration is the size and shape of the working space. However, the working space is an unknown quantity, ensuing from the construction and therefore also from the individual partial parameters of the mechanism. To delimit the working space, it is first necessary to define these parameters. It is mainly the geometric parameters (lengths of the arms, positions of joints etc.) shown in Fig. 2.

To determine the working space size parameters, we have used a problem from indirect kinematics, following from the geometric approach to the whole solution. The principle is displayed in Fig. 3. This method was chosen because it is simple, easy to understand and the calculations can be realized with high efficiency.

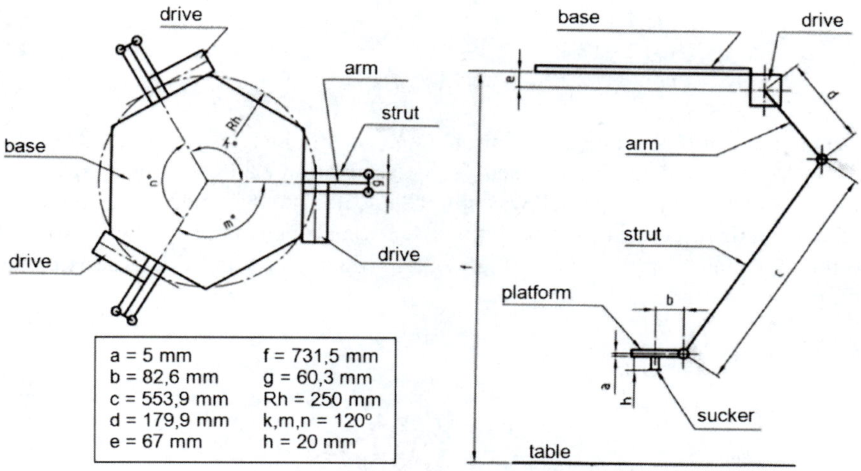

Fig. 2. DELTA robot parameters definitions

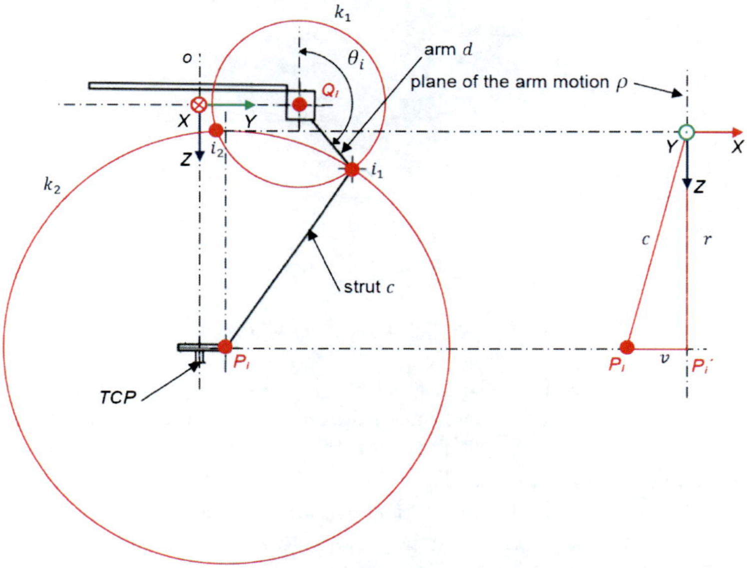

Fig. 3. Geometric approach to solution

$$\boldsymbol{A} = \boldsymbol{i}_1 \tag{2}$$

$$\boldsymbol{B} = [-1{,}0] \tag{3}$$

$$\boldsymbol{\theta}_i = arccos((\boldsymbol{A} \cdot \boldsymbol{B})/(|\boldsymbol{A}| \cdot |\boldsymbol{B}|)) \tag{4}$$

The results of this geometric solution are the desired angles of the arms Θ_i. The calculated angles were further used for determination of the working space. A numerical incremental approach was applied to calculate the angles Θ_i on the basis of the predefined criteria in a predefined space and their relevance was assessed (solutions i_1 and i_2 are shown in Fig. 3). The final shape and dimensions of the working space can be seen in Fig. 4.

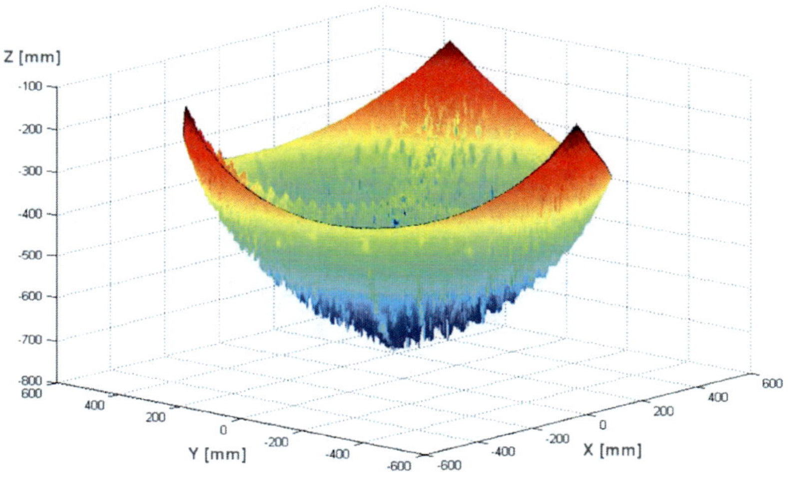

Fig. 4. Working space of DELTA robot

This step has served for definition of the size and shape of the DELTA robot working space from the parameters according to Fig. 2. Specific requirements on position optimization of the TCP (TOOL CENTER POINT) can be achieved through changes of the individual parameters. Regarding the concept of parallel mechanisms it is necessary to take into account the possibility of existence of singular positions, which, when reached, cause loss of power transmission or loss of some DOF. From the aspect of practical use of the device, these positions are undesirable. For this reason it is necessary to examine all relevant singular positions of the mechanism so that restricting criteria of the positioning could be set. In our case, the geometric approach was also applied. The result is displayed in Fig. 5 right. [3]

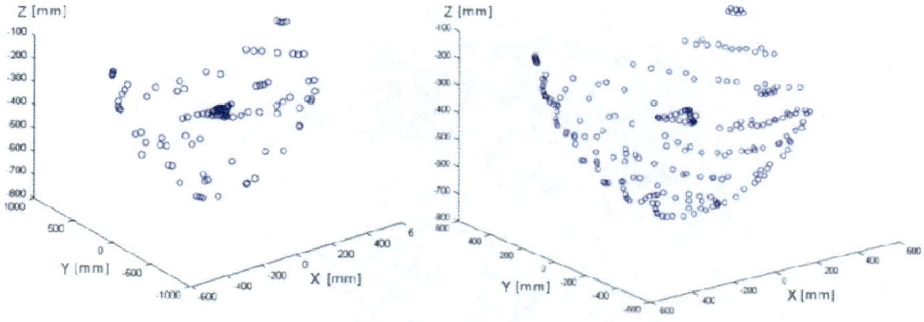

Fig. 5. Singular positions – Jacobian verification on the left

To verify our presumptions and correctness of the calculation algorithm, we have performed a check calculation with a Jacobi determinant according to López [4]. The obtained results have verified sufficient accuracy of the approach used in our work for this specific type of kinematics.

3 Design Solution

The DELTA type robots with the parallel mechanism are characterized with high dynamics of the end effector (TCP). This comes together with the requirements on minimum weight of the moving components, sufficiently sized drives including transmissions, stiffness of the individual components and mainly minimum backlash in the joint couplings.

The selected type of drive (a servo motor with a planetary transmission) with the maximum output torsion moment of 90 Nm provides a sufficient potential for development of a highly dynamic device. In accordance with this criterion, the components and their material, directly affecting the overall TCP dynamics, were suitably selected. For most of the components, aluminium alloy was chosen as the construction material with suitable mechanical properties. The struts were made of a composite material on the basis of carbon fibres due to its more suitable weight/stiffness ratio.

For the purpose of experimental tests focused on accuracy and durability of joint couplings, several variants of the solution were proposed. They are built of components readily available on the market. Fig. 7 right shows a classic design solution employing a ball joint to realize the joint couplings. A drawback of this solution is the need to exert retention force; in this case it is realized by springs between the individual struts.

The second variant of solution is realized with the components produced by the IGUS company (Fig. 7 left). The joint is made in the form of two mutually perpendicular rotary couplings ensuring sufficient DOF. An advantage of this solution is that mutual preloading of the rods is not necessary. A drawback lies in higher complexity due to higher number of components needed to realize this link and lower resistance to possible accidents [6].

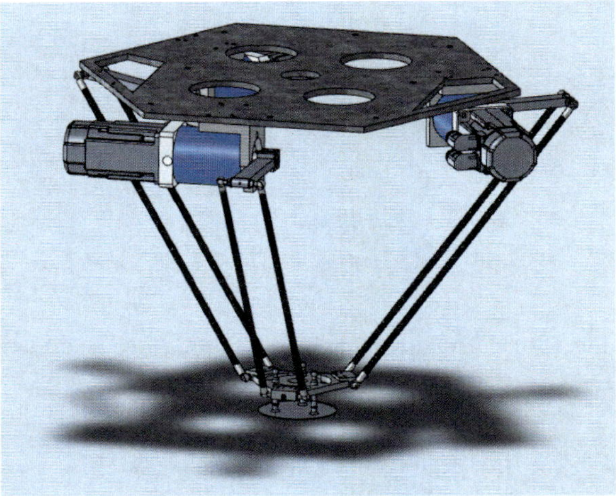

Fig. 6. 3D-model of DELTA robot

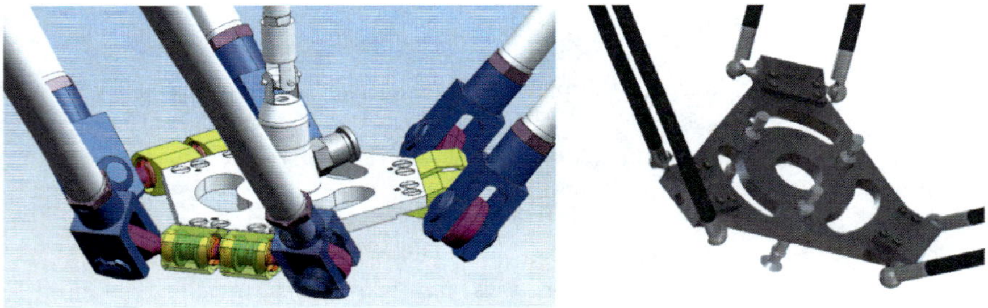

Fig. 7. Constructional variants of the joint couplings

4 Conclusion

Based on the described proposals, the DELTA robot prototype has been built that is currently being subject to intensive experimental tests focused on improvement of the accuracy and dynamics of the mechanisms and also on durability of the individual components.

These experiments have sufficiently verified the control system functions including the Beckhoff company library, as well as the theoretical knowledge regarding the working space size and singular positions of the DELTA robot.

The obtained results were presented in an exposition of the Dyger company at the AMPER international fair trade which took place in Brno in spring 2011.

Another task within the project, which is still being worked on, will be rotation of the end effector. This additional construction component of the DELTA robot seems to be essential for most of the industrial applications. Also, a suitable user interface for control of the robot functions will be supplemented in cooperation with the Dyger company. The objective is to design a marketable product which could be offered for applications in industry (see Fig. 8).

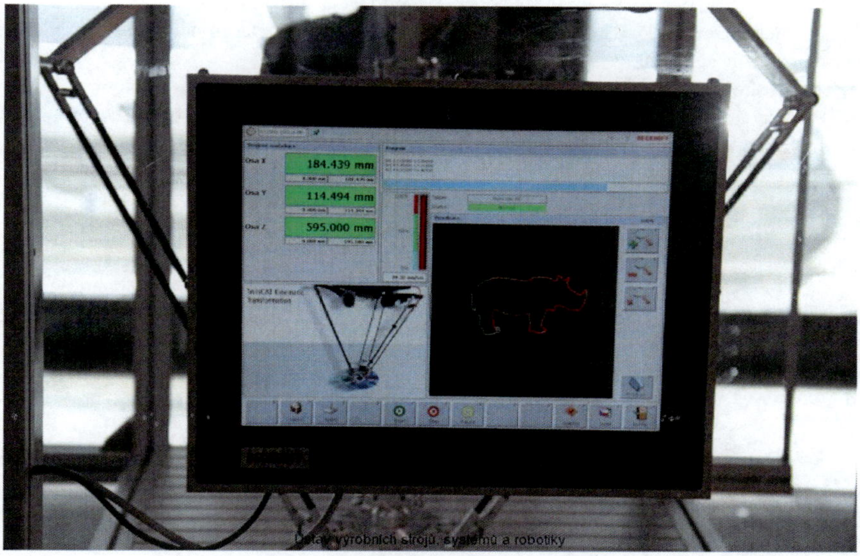

Fig. 8. DELTA robot control system touch panel

Acknowledgemet

This work is supported by Brno University of Technology, Faculty of Mechanical Engineering project NO. FSI-S-10-27, CZ.1.07/2.3.00/09.0162 and by company DYGER, s.r.o.

References

[1] Brezina, T., Brezina, L.: Controller Design of the Stewart Platform Linear Actuator. In: 8th International Conference on Mechatronics, Luhacovice, Czech Republic, pp. 341–346 (2009)

[2] Briot, S., Arakelian, V., Glazunov, V.: Design and analysis of the properties of the DELTA inverse robot (2008), http://www.irccyn.ec-nantes.fr/~briot/PubliConf/Delta%20Inverse.pdf

[3] Laribi, M.A., Romdhane, L., Zeghloul, S.: Analysis and dimensional synthesis of the DELTA robot for a prescribed workspace (2006), http://www.sciencedirect.com/science?_ob=MImg&_imagekey=B6V46-4M33VTM-1-17&_cdi=5750&_user=640830&_pii=S0094114X0600142X&_origin=&_coverDate=07%2F31%2F2007&_sk=999579992&view=c&wchp=dGLbVlWzSkWb&md5=5324c88af9e2fcdf01484204ffc21a90&ie=/sdarticle.pdf

[4] López, M., Castillo, E., García, G., Bashir, A.: Delta robot: inverse, direct, and intermediate Jacobians, In: IMechE (2006), http://robotics.caltech.edu/~jwb/courses/ME115/handouts/DeltaRobotInverseDirectIntermediateJacobians.pdf
[5] Neugebauer, R., Wittstock, V., Drossel, W.: Werkzeugmaschinen-Mechatronik, Arbeitsblätter. Chemnitz: TU-Chemnitz (2010)
[6] Traslosheros, A., Roberti, F.: Servo control visual para tareas de seguimiento dinámico tridimensional, mediante la utilización de una cámara en un robot Delta (2008), http://oa.upm.es/3651/1/INVE_MEM_2008_56343.pdf

Motion Planning of Autonomous Mobile Robot in Highly Populated Dynamic Environment

S. Vechet and V. Ondrousek

Brno University of Technology, Faculty of Mechanical Engineering, Technicka 2, Brno, 616 69, Czech Rep.
vechet.s@fme.vutbr.cz

Abstract. The number of applications where the autonomous mobile robot is placed into a highly populated area increases in recent years. Presented paper describes a finite state machine based method used to successful navigation through environment highly populated by a number of dynamic obstacles. Paper includes the test results obtained by the real robot in a shopping mall representing the attractive example of such an environment.

1 Introduction

The motion planning in highly populated environment is a critical issue in applications where the common obstacles are made by moving people. There are many methods and approaches to plan a motion of an autonomous robot through such environment. We present a method to navigate the robot in case where the robot is surrounded by people while taking into consideration the safety restrictions. The safety restrictions are based on industrial standards for moving vehicles in public areas.

The motion planning is a part of navigation problem and it can be divided into a two main groups. First is called the global motion planning (path planning) in which the robot tries to plan a path from starting point to the goal. Usually the robot has information about the nature of environment: static/dynamic obstacles, free spaces, landmarks etc.

Second group is called the local motion planning, in which the robots try to generate a sequence of a few motion steps based on local information gathered from sensors, e.g. ultrasonic sensors, laser range finders, camera, etc.

The overall review of motion planning techniques is shown in [1]. A number of methods solve the global path planning problem in populated urban environments using C-Space and Voronoi diagram [2].

Also the classical approach using the artificial potential field is very successful. This approach is based on an attractive potential fields around the goal and repulsive potential fields around obstacles [3].

The local path planning problem is a part of navigation tasks in which the robot movement depends only on local information about the environment. Such information is usually represented by proximity sensor readings, e.g. ultrasonic, infrared or laser. Based on this information the robot must create a local control strategy for the movement. The local planning problem can be solved by neural networks [4] or fuzzy reactive control [5].

The practical applications of motion planning in highly populated environment are the motion in multi-room environment [6] or shopping mall [7].

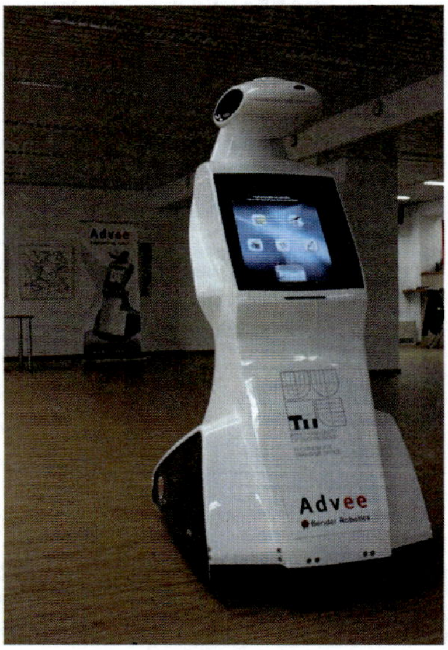

Fig. 1. The prototype of autonomous mobile robot Advee

In this paper we present a method based on finite state-machine theory to generate a local control sequence to reach the global goal. This method is currently used in autonomous robot Advee (see figure 1.). Advee has a common platform based on Ackerman steering principle as it is described in [8][9]. The paper is organized as follows: The Mealy state-machine is briefly described in chapter 2. The dynamic environment is presented in chapter 3. The practical implementation of finite state-machine is described in chapter 4. The real-world experiments are shown in chapter 5.

2 Mealy Machine

Mealy machine is a finite state-machine, where outputs are based on both inputs and current state. The main advantage of Mealy machine (in comparison with Moore machine) is that the output can change even when there are changes in inputs only and no change in the state occurs. In Moore machine the outputs change only when current state is changed.

The Mealy machine is defined as a set of six parameters $\left(S, S_0, \Sigma, \Lambda, T, G\right)$, where:

- S set of possible states
- S_0 initial state
- Σ input alphabet $\left(x_0, \ldots, x_n\right)$; x_i is possible input value
- Λ output alphabet $\left(z_0, \ldots, z_n\right)$; z is possible output value
- T transition function; $T : S \times \Sigma \rightarrow S$
- G transition function; $G : S \times \Sigma \rightarrow \Lambda$

Such a finite state-machine can be used for intelligent movement in unknown environment. The sensor measurements are mapped to the input alphabet Σ. Output alphabet Λ is afterwards used for robot motion planning.

3 Problem Definition

The main issue in motion planning is a planning in dynamic environment. Such an environment usually occurs in highly populated public places e.g. shopping malls.

Presented method for motion planning was designed mainly for the motion in congress centers and shopping malls where the place to maneuver is of limited size and there is a lot of people moving or standing around the robot.

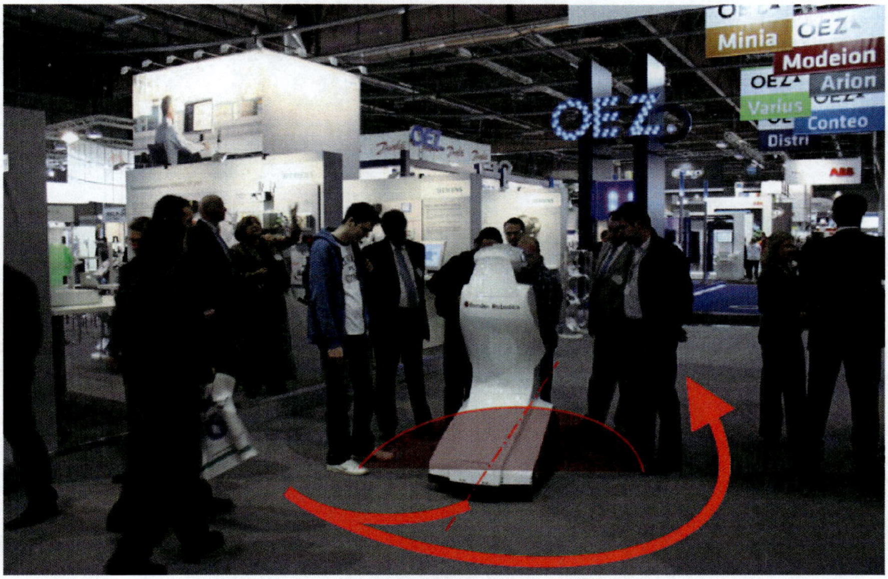

Fig. 2. Representation of a possible motion through people standing around robot

The method to generate the local motion plan was designed with respect to keep the safety of the people moving around. On the other hand, the method has to be able to generate motion plan according to a global goal to satisfied the need of traversing as most of given place as possible. This need is based on main purpose of the robot, which is promotion. So the robot should meet many people on different places during its path. These two needs (safety and long distance traveling) are in opposite e.g. the highly safe robot will stop right after detecting any human around, therefore the totally traveled distance during its operational time its very short. The typical example of environment is shown on figure 2.

The overall view of the designed motion planner framework is on figure 3.

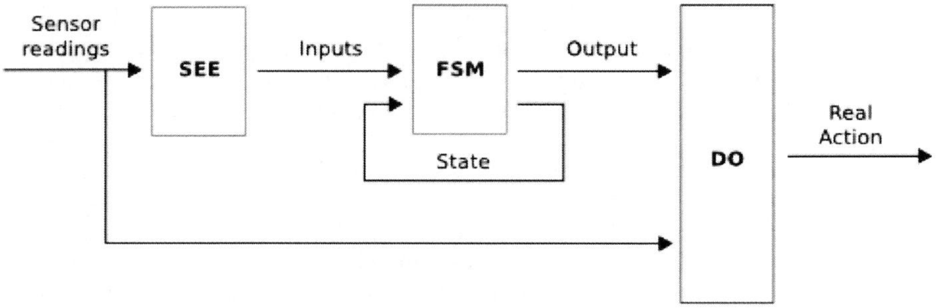

Fig. 3. Motion planner framework

The sensor reading are transformed to the input alphabet Σ in (block SEE) and then used in state-machine (FSM). The output from FSM is transformed (block named DO) with respect to the original sensor readings to a real action. In our case it is a translation and rotation velocity which is directly used to control the robots engines.

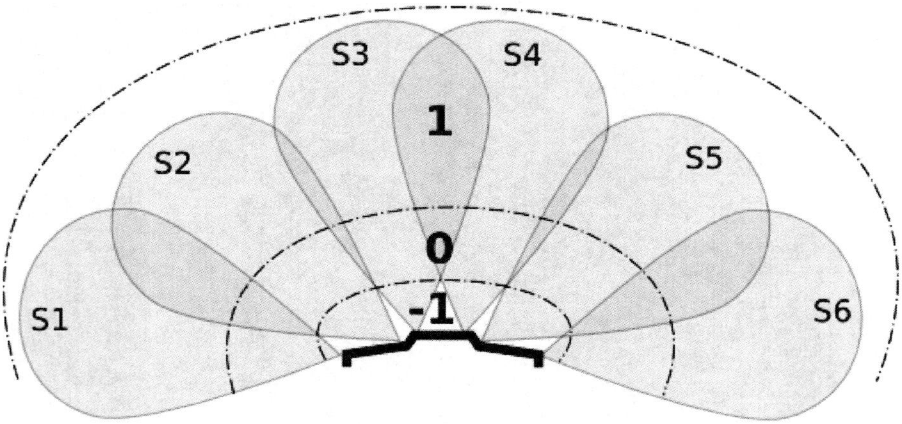

Fig. 4. Detection areas of sensors placed on the robot

Figure 4 shows the sensors location and orientation on the robot (top view). Each sensor has its detection range divided into the three discrete areas, which are used in FSM as inputs. Due to the fact that the input alphabet Σ of state-machine is discrete, the sensor inputs needs to be discretized. The discretization is accomplished via the block SEE. The FSM is then used to make a decision what action should the robot perform. After the action is chosen the real action (real velocities of the robot) are calculated (block DO, see figure 3) also in dependence on real sensors measurements.

4 State-Machine Description

The finite state-machine used in robot Advee is defined, with respect to description above, as follows:

- Possible states $S=\{s_0,s_1,s_2,s_3\}$; where $s_0=Forward$, $s_1=Backward$, $s_2=Pause$, $s_3=Stop$
- Initial state $S_0=s_2=Pause$
- Input alphabet $\Sigma=\{-1,0,1\}$
- Λ output alphabet $\Delta=\{z_0,\ldots,z_8\}$; where $z_0=Follow\_a\_goal$, $z_1=Turn\_right\_sharply$, $z_2=Turn\_left\_sharply$, $z_3=Turn\_right$, $z_4=Turn\_left$, $z_5=Follow\_a\_corridor$, $z_6=Backward$, $z_7=Back\_right$, $z_8=Back\_left$,
- Transition functions T and G are joined into one transition function $T:S\times\Sigma\rightarrow S\times\Lambda$.

The transition function is defined as a transition table. For implementation details see table 1 where the part of the used transition table is shown. Presented part of the transition function is used for forward navigation through the environment full of dynamic obstacles. The transition function cannot be shown in its complete form due to the fact, that the full transition table has 59049 rows. On the other hand only 86 combinations have the real meaning, however the number of rows is significantly reduced as it is still too high for the complete list.

Table 1. The transition function for Mealy state-machine

Inputs						State	New State	Output
S1	S2	S3	S4	S5	S6			
1	1	1	1	1	1	s_0	s_0	Follow a goal
-	0	1	1	1	1	s_0	s_0	Turn right: sharply
1	1	1	1	0	-	s_0	s_0	Turn left: sharply
0	1	1	1	1	1	s_0	s_0	Turn right
1	1	1	1	1	0	s_0	s_0	Turn left
-	1	1	1	1	-	s_0	s_0	Follow a corridor
1	1	-	-	-	-	s_0	s_1	Backward
-	-	-	-	1	1	s_0	s_1	Backward
1	-	-	-	-	-	s_0	s_1	Backward
-	-	-	-	-	1	s_0	s_1	Backward

5 Real-World Application

The autonomous robot Advee is used in shopping malls and congress centers. Currently (April 2011), it has about six month of full operation behind. The typical environment to deal with is shown on figure 5. In such highly populated environment where the robot Advee is surrounded by people, the development of usable motion planner is very complicated. In such cases the robot usually changes the current state to $Pause$ so it stands and waits for the moment when the people make some space.

It is obvious that in real experiments it is impossible to reach the same configuration of environment which contains a moving people. Due to this fact we created the average wandering criterion $AWDR$ for comparison of different motion planners.

Fig. 5. Typical situations where the robot Advee is surrounded by people

The criterion is bases on traveled distances in forward and backward directions and on number of times that the planner chooses to change the movement from forward to back. The equation of used criterion is shown below:

$$AWDR = \frac{\sum_{i=0}^{n}\left(1-\frac{1}{s_b^i+1}\right)\cdot\frac{d_b^i}{d_f^i}\cdot\left(d_f^i+d_b^i\right)}{\sum_{i=0}^{n}\left(d_f^i+d_b^i\right)} \qquad (1)$$

Where n is number of performed runs of the planner (note, that in practical experiments, one run takes about 4 hours of usage), i is the actual attempt, s_b^i is the number of changes in directions from forward to back in one run, d_b^i and d_f^i are traveled distances in meters in backward and forward directions.

The typical path traveled during the one attempt is shown on figure 6. There are some problematic parts for the planner, typically represented by many changes in forward/backward directions. The robot was surrounded by people in those cases and tried to avoid them. The proposed criterion is used to eliminate the motion planners which attract to never-ending forward/backward movement in case of the robot is surrounded by people.

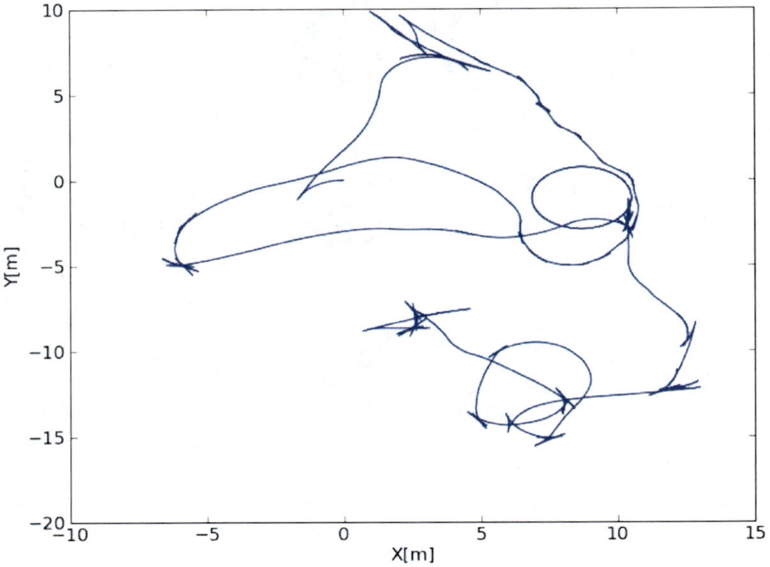

Fig. 6. Path traveled during testing phase.

A motion planner based on Bayesian filters (very similar to fuzzy reactive control [5] was used for comparison with the described FSM planner. We also use a user remote control as a third planner for comparison. Table 2 shows the comparison of these tree motion planners.

Table 2. Comparison of different motion planners

	FSM	Bayesian	User
Average WDR [-]	0.353	0.662	0.193

The FSM planner is almost twice as efficient as the Bayesian based planner, however, as we expected the efficiency of the motion planned manually on-site is the best one.

6 Conclusions

In this paper was presented the method for motion planning of autonomous mobile robot in highly populated environment based on Mealy finite-state machine. This method was successfully tested in both simulation and real environment. The efficiency defined via $AWDR$ criterion and the successful application of presented motion planning subsystem in real-world is the motivation for future work in developing more sophisticated solution coming up from finite-state machine.

Acknowledgement

Published results were acquired with the support of the Ministry of Education, Youth and Sports of the Czech Republic, research plan MSM 0021630518 "Simulation modeling of mechatronic systems".

References

[1] Calisi, D.: Motion planning, reactive methods, and learning techniques for mobile robot navigation contents (2007)
[2] Lidoris, G., Rohrmüller, F., Wollherr, D., Buss, M.: The autonomous city explorer (ace) project – mobile robot navigation in highly populated urban environments
[3] Leng, Q.C.C., Huang, Y.: A motion planning method for omnidirectional mobile robot based on the anisotropic characteristics. International Journal of Advanced Robotic Systems (2008)
[4] Yang, S.X., Meng, M.Q.-H.: Real-time collision-free motion planning of a mobile robot using a neural dynamics-based approach. IEEE Transactions on Neural Networks 14(6), 1541–1552 (2003)
[5] Xu, W.L., Tso, S.K., Fung, Y.H.: Fuzzy reactive control of a mobile robot incorporating a real/virtual target switching strategy. Robotics and Autonomous Systems 23(3), 171–186 (1998)
[6] Thrun, S., Buecken, A.: Integrating grid-based and topological maps for mobile robot navigation. In: Proceedings of the AAAI Thirteenth National Conference on Artificial Intelligence (1996)
[7] Shiomi, M. Takayuki, K., Glas, D.F., Satake, S., Ishiguro, H., Hagita, N.: Field trial of networked social robots in a shopping mall
[8] Hrbáček, J., Ripel, T., Krejsa, J.: Ackermann mobile robot chassis with independent rear wheel drives. In: Proceedigns of EPE-PEMC 2010 conference, Republic of Macedonia, pp. T5/46-T5/51 (2010), ISBN 978-1-4244-7854-5, IEEE cat num: CFP1034A-DVD
[9] Ripel, T., Hrbáček, J., Krejsa, J.: Design of the Frame for Autonomou Mobile Robot with Ackerman Platform. In: Proceedings Engineering Mechanics, pp. 515–518 (2011)

Multi-purpose Mobile Robot Platform Development

J. Vejlupek and V. Lamberský

Brno University of Technology, Faculty of Mechanical Engineering,
Technická 2896/2, Brno, 616 69, Czech Republic
b.j.vejlupek@seznam.cz

Abstract. This paper deals with the design of a four wheel drive mobile robot platform. It describes complex design process including the chassis, hardware, control and power electronics, low level software and communication with high-level control. Robot is intended to be used for testing of various control algorithms including DC motor position, speed, momentum; traction control; and further for the higher level algorithms such as: obstacle detection and avoidance and path planning. Robot can be used for research and development, and also it presents great educational platform for mechatronic engineering. Main asset for education is the complex design including kinematics, dynamics, embedded computing, electronics design, and also closes connection to automotive industry. And as a practical task, it increases students motivation.

1 Introduction

There are many different chassis options for wheeled mobile robot platforms: differential chassis, two-wheeled, three-wheeled, chassis with omnidirectional wheels and probably the most common type: car-like with four wheels, where the steering is done by one axle – the Ackerman chassis [1], [12]. For our goals, we needed chassis with all four wheels driven and steerable, which is not very common. There is a solution from [10], but this one is too big and heavy, which could be a safety issue. Vehicles with AWS are produced by several car makers; there are various motivations such as vehicle stability and maneuverability. Platform is intended to be used for verification of model simulation [2], [11], algorithm development and testing. Intentions are to develop and test parking algorithms such as [3], algorithms for traction control and vehicle road-stability [4], [6], and also steering [5]. Our goal is to build a small scaled model. The mechanical construction is described in Section 2, the drive system and the steering in Section 3, the electronics design in Section 4, and low level firmware including the ackerman steering for AWS is in Section 5.

2 Mechanical Construction of the Robot

Mechanical construction (Fig. 1.) is based on the key requirements such as "All Wheel Drive" (AWD), "All wheel steered" (AWS) capability and compact size: robot is about 80 cm long and 45 cm wide. Chassis frame is aluminum custom made; wheel suspension and a part of power transmission is based on 1:5 RC buggy parts available from hobby shops. RC buggy front wheel assembly is used for both axles. Requirements for the chassis came from the idea to have a car-like platform to test various algorithms for traction control (ABS, DSTC), and driving comfort (parking assistant, cruise control, etc.) In some cases the small size has proven to be a complication as it significantly increases the requirements for mechanical accuracy and precision. Additional protective frame was made from plastic plumbing tubes to increase safety of the vehicle and the operator.

Fig. 1. Mechanical construction

3 Drive and Steering System

Based on the initial project requirements for AWD and AWS capability, the platform is equipped with independent drives for each wheel consisting of a brushed DC motor (35W 24V) with planetary gear head (14:1) and belt transmission to wheel and model hobby servo for steering the wheel. This allows rapid acceleration and even when using all four wheels the platform is capable of slippage (tested on dry linoleum) as well as skid while turning at full speed.

As a power source, there are in total three battery packs: 28.8V LiFe (8x A123) for the DC motors, 6V NiMH pack for the servos, and another 6V NiMH pack for the on-board controllers. Reasons for this independent power sources are the high power drains: each servo can drain up to 3 Amps at peak which may lead to voltage drop on the battery and may cause a brown-out reset on microcontrollers. Power converter from 30V to 6V at 12A for servos and another one for 5V at 1A for control electronics are under consideration for further improvements of the platform.

4 Electronics Design

Electronics design is one of the key parts in the whole design, final schematics is shown in fig. 2: On board control is realized using two microcontrollers dsPIC33fj128mc804, each for one axle driving two h-bridges and two RC servos. First one, designated as *Master* is driving the front axle and second one designated as *Slave* is driving the rear axle. They are connected together via CAN bus (automotive standard); this is an important feature of the design as it allows more devices to be connected easily: i.e. additional unit for secondary sensory array data acquisition and processing. Another option is to use more powerful platform to run the driving algorithms such as MPC555 or cRIO. In our case, MPC555 has been successfully used for ABS and simple ESP.

Communication with operator (PC with joystick, running MATLAB Simulink model) is realized via wireless UART and converted to RS232.

H-bridge is based on full bridge N-channel FET driver ISL83204A allowing up to 60V power source, and IRFB4115 MOSFET allowing currents up to 104A. Part of the h-bride is a current transducer LEM LTS-25NP which provides analogue voltage output. Each h-bridge is driven with one PWM signal, and two other digital lines: direction and disable.

Each *servo interface board* works as a power and signal router. RC servo require PWM signal with certain frequency (50 Hz); position is determined by the build-in controller from the pulse width (0.9-2.1ms). Slight disadvantage of using RC servos is no position feedback. Power for the servos is provided form separate NiMH 6V 2Ah battery pack.

Main sensory data are drive currents, positions and speeds (quadrature incremental encoder - QEI), which are measured on each axle control unit, and tri-axial acceleration, tri-axial angular speed and tri-axial compass - these are embedded in ADIS16405, which is linked to the "master" unit via SPI.

Position is acquired from QEI by counting the edges with respect to direction, and speed is directly derived from time that passes between two or more edges (number of edges varies with speed to maintain certain level of accuracy.)

Secondary sensory array includes drive temperature monitoring: measuring the stator temperature of the drive, IR-range sensors: measuring the distance to obstacle in front and behind the vehicle to avoid collision. Future plans are to add battery power monitor to ensure the safety of the batteries (protection from deep discharge) and of the high level mission.

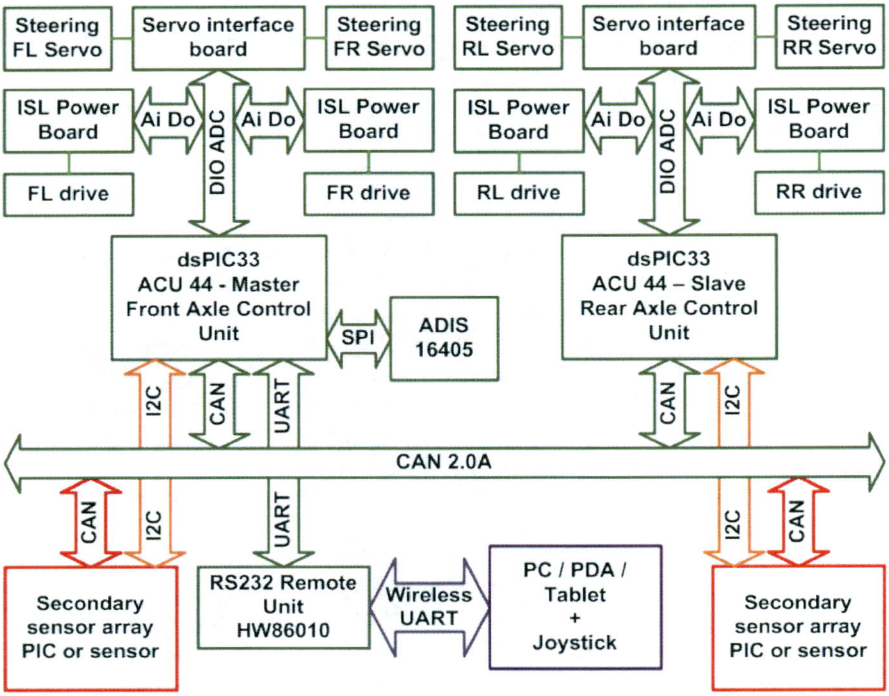

Fig. 2. On-board electronic scheme

5 Low Level Software Architecture

5.1 On-Board Communication

On-board electronics is designed as a distributed system; therefore there are several independent units with various protocols: Communication between axle control units is implemented using CAN bus. This provides flexibility and extensibility of the control system i.e.: additional (more powerful) control unit(s) could be added. Communication with human operator is realized using wireless UART currently running at 115200 bps. Custom communication packet with CRC has been designed to transfer control data to the platform and sensory data back to the PC [13].

Master (front) unit is receiving commands from operator through UART; received data are unpacked and checked for validity. Main message includes four 12-bit values: Motor command (speed / momentum), front steering angle, rear steering angle, and buttons, though it is possible to assign a different meaning to them. These values are then transmitted via CAN bus to the *slave* (rear) unit. *Slave* unit provides via CAN bus *Master* unit with its sensory data: motor positions (or speeds), currents, temperatures (or general disable). Future plan is to improve the UART communication to transmit more type of messages.

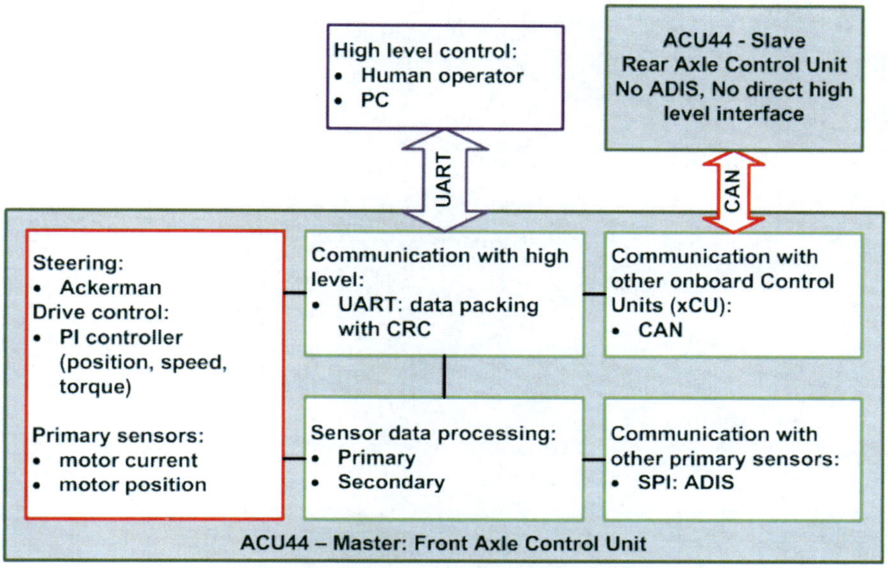

Fig. 3. Firmware scheme: control loops inside ACU44

5.2 Ackerman Steering for AWS Vehicle

Steering angle derived from Ackerman algorithm is computed on both units separately to distribute the computation power. In case, the position or speed regulator is required, Ackerman has to be used as well. For the purposes of traction algorithms testing, current (torque) regulator has proven to be sufficient and easier to use as it does not require the Ackerman algorithm (besides the steering angles). This approach has been also proven by [6].

One of the vehicle stability algorithms uses directly the information about torque (current) to detect the slippage of the wheels. Similar approach is shown in [7].

As each wheel is steered independently, we do not need to design steering trapezium, but we still need to compute steering angles for each wheel. These are shown in figure 4 and relevant equations are 1.1-4.4, where 4.1-4.4 are the actual angles. Angles α and β are given by operator as a steering angles for each axle, or they could be determine by a drive-comfort algorithm i.e.: they could be a function of the vehicle speed. In case that both steering angles are zero all the steering angles are set to zero and no calculation is executed.

$$R_\alpha = l \cdot \frac{\sin\left(\pi/2 - \beta\right)}{\sin\left(\alpha + \beta\right)} \tag{1.1}$$

$$R_\beta = l \cdot \frac{\sin\left(\pi/2 - \alpha\right)}{\sin\left(\alpha + \beta\right)} \tag{1.2}$$

$$H_{\alpha} = R_{\alpha} \cdot \sin\alpha \tag{2.1}$$

$$H_{\beta} = R_{\beta} \cdot \sin\beta \tag{2.2}$$

$$R = \frac{\cos\alpha}{R_{\alpha}} = \frac{\cos\beta}{R_{\beta}} \tag{3}$$

$$\alpha_L = \tan^{-1}\left(\frac{H_{\alpha}}{R - d/2}\right) \tag{4.1}$$

$$\alpha_R = \tan^{-1}\left(\frac{H_{\alpha}}{R + d/2}\right) \tag{4.2}$$

$$\beta_L = \tan^{-1}\left(\frac{H_{\beta}}{R - d/2}\right) \tag{4.3}$$

$$\beta_R = \tan^{-1}\left(\frac{H_{\beta}}{R + d/2}\right) \tag{4.4}$$

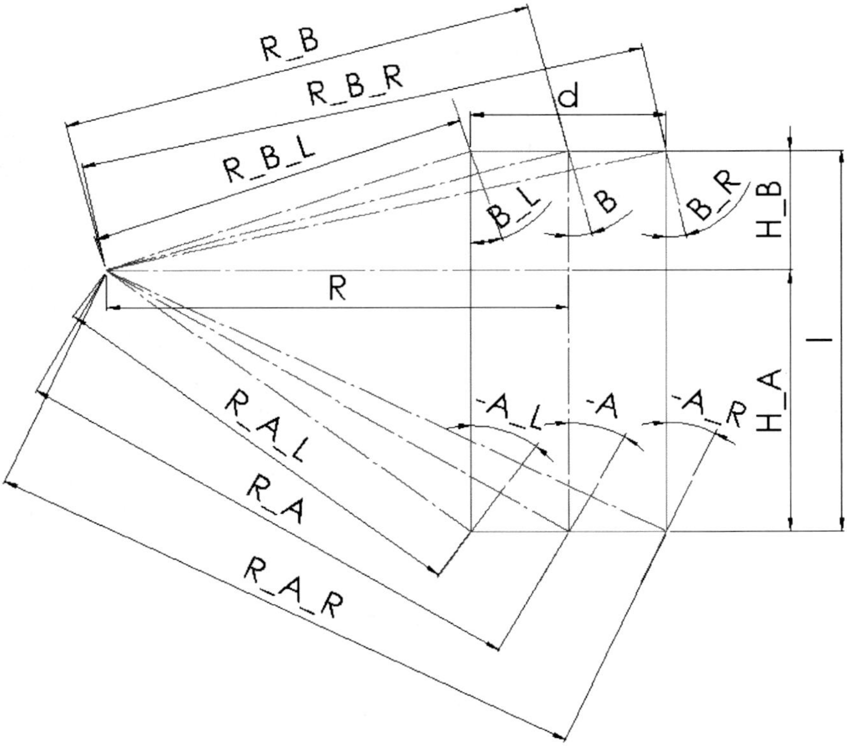

Fig. 4. Ackerman AWS steering (A denotes α ; B: β ; R and L: Right and Left)

5.3 Rapid Code Generation

Low level programming is mostly done with rapid code generation [8] using Kerhuel Toolbox [9] and Real Time Workshop for MATLAB Simulink. This provides C code generated from Simulink, which can be further modified or expanded, and then it is compiled into binary code by C30 compiler from Microchip.

6 Conclusion

Design process of an experimental vehicle is presented in this paper; it has been approached as a mechatronical problem and the result is a working platform ready to be used in education and also research and development for experiments. At the moment, several student projects have already been finished and others are on going.

Acknowledgment

The work presented in this paper has been supported by research project FSI-S-11-15 "Design, testing and implementation of control algorithms with use of nonlinear models of mechatronics systems".

References

[1] Hrbáček, J., Ripel, T., Krejsa, J.: Ackermann mobile robot chassis with independent rear wheel drives. In: 14th International Conference on Power Electronics and Motion Control (EPE/PEMC), September 6-8 pp. T5-46-T5-51(2010)

[2] Yang, M., Gao, X.H., Wang, H., Wang, C.: Simulation analysis of multi-axle vehicle braking stability based on all-wheel active steering technology. In: 2nd International Conference on Industrial Mechatronics and Automation (ICIMA), May 30-31, vol. 1, pp. 366–369 (2010)

[3] Gupta, A., Divekar, R., Agrawal, M.: Autonomous parallel parking system for Ackerman steering four wheelers. In: IEEE International Conference on Computational Intelligence and Computing Research (ICCIC), December 28-29, pp. 1–6 (2010)

[4] Hakima, A., Ameli, S.: Improvement of vehicle handling by an integrated control system of four wheel steering and ESP with fuzzy logic approach. In: 2nd International Conference on Mechanical and Electrical Technology (ICMET), September 10-12, pp. 738–744 (2010)

[5] Kim, Y.-C., Min, K.-D., Yun, K.-H., Byun, Y.-S., Mok, J.-K.: Steering control for lateral guidance of an all wheel steered vehicle. In: International Conference on Control, Automation and Systems ICCAS 2008, October 14-17, pp. 24–28 (2008)

[6] Hallowell, S.J., Ray, L.R.: All-wheel driving using independent torque control of each wheel. In: Proceedings of the American Control Conference, June 4-6, vol. 3, pp. 2590–2595 (2003)

[7] Zhu, Z., Yuan, K., Zou, W., Hu, H.: Current-based wheel slip detection of all-wheel driving vehicle. In: International Conference on Information and Automation ICIA 2009, June 22-24, pp. 495–499 (2009)

[8] Grepl, R.: Real-Time Control Prototyping in MATLAB/Simulink: review of tools for research and education in mechatronics. In: IEEE International Conference on Mechatronics ICM 2011, Istanbul, April 13-15 (2011)
[9] Kerhuel, L.: Simulink block set embedded target for microchip devices (2010), http://www.kerhuel.eu/wiki/Simulink_-_Embedded_Target_for_PIC
[10] Superdroid 4WD 4WS mobile robot platform build for DSpace inc., http://www.superdroidrobots.com/shop/custom.aspx?recid=2
[11] Portes, P., Laurinec, M., Blatak, O.: Discrete-difference filter in vehicle dynamics analysis. In: 8th International Conference on Mechatronics, pp. 19–24 (2009)
[12] Kotzian, J., Konecny, J., Prokop, H., Lippa, T., Kuruc, M.: Autonomous explorative mobile robot navigation and construction. In: 9th Roedunet International Conference (RoEduNet), pp. 49–54 (2010)
[13] Vlachy, D., Zezula, P., Grepl, R.: Control unit architecture for biped robot. In: International Conference on Mechatronics 2007, Recent Advances In Mechatronics, Warsaw, pp. 6–10 (2007)

Part V

Mechatronic Systems

Integration of Risk Management into the Machinery Design Process

P. Blecha*, R. Blecha, and F. Bradáč

Brno University of Technology, Faculty of Mechanical Engineering,
Technická 2896/2, Brno, 616 69, Czech Republic

Abstract. Risk management is currently a key component of any design process. The designers are required to identify all hazards within the whole life cycle of the machine and, if necessary, to take appropriate measures in order to reduce the risk of these hazards to an acceptable level. Furthermore, the new machinery directive 2006/42/EC requires this process of assessment and minimization of risk to be documented, which is an essential prerequisite of placing the product on the market or into operation. As people generally tend to underestimate the risks, it is necessary to pay adequate attention also to elaboration of the instructions for use of the designed machine, and besides, to take into consideration any reasonably foreseeable misuse. This paper presents a concept of application and documentation of the technical risk management process during the design of machinery, with an emphasis on its transparency.

1 Introduction

Due to its varied applications and permanent use in industry, machinery is associated with high number of occupational accidents causing high social costs connected with incapacity of employees to work. This situation is reflected in the ever-increasing strictness of the requirements on machinery safety. The new directive 2006/42/EC toughens up the requirements on performance and documentation of the risk assessment process for all phases of the life cycle of machinery prior to issue of declaration of conformity. According to this directive, machine designers must identify all hazards that could occur during the whole life cycle of the machine, estimate the risks of these hazards and take preventive measures for reduction

* Corresponding author.

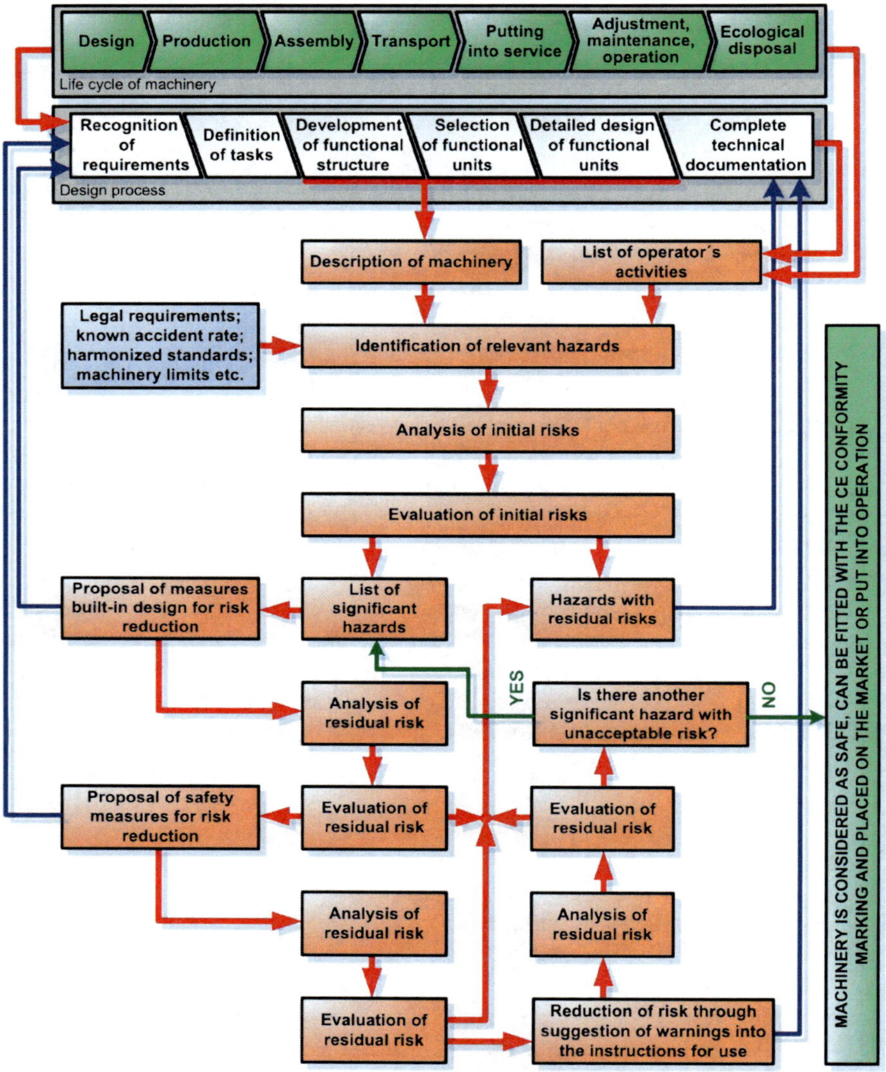

Fig. 1. The process of risk assessment and risk reduction according to 2006/42/EC

of risk in hazards reaching unacceptable size of risk. Moreover, they should consider not only the intended use of the machine, but also any reasonably foreseeable misuse. For the identified hazards with unacceptable size of risk, the designer must propose measures for their reduction in the following order of priorities: constructional measures, safety measures, information for use of the machine. The identified hazards with acceptable size of risk (so called residual risk) must be included in the instructions for use of machinery in order to fulfill the requirement of the directive on making the customer aware of these risks. The described process of hazard identification and gradual reduction of risk for assurance of machinery safety, as shown by the diagram in Fig. 1, can be described as "risk management". In general sense, however, risk management is perceived as a tool for management of financial risks in the first place, where risk is a contrariety to an opportunity (or wasted opportunity). For this reason, in this work we use the term "technical risk management" in order to stress the different philosophy of risk management. The financial aspect of risk reduction can be taken into consideration only when selecting the form of fulfillment of the machinery directive requirements, if, however, the current state of the art and technical requirements allow alternative methods of risk reduction.

According to the available statistical data [1], approximately 80% of all occupational accidents are due to incorrect or insufficient estimation of risk. This fact itself presents a significant hazard with unacceptable risk in the process of machinery design. Application of a methodological procedure supporting the design of safe machinery is a possibility how to preventively reduce this risk.

The concept of safe design and construction of machinery has been integrated into a number of methods of scientific design [2], [3], [4], [5]; however, these methods can be successfully applied only by highly qualified specialists [6]. Most of the small and medium-size enterprises do not have enough personnel with appropriate qualification for application of the scientific design methods. For this reason it was necessary, on the basis of the demand from industry, to develop such a methodological procedure of technical risk management that would be transparent and easily applicable.

2 System Approach to Risk Management

The following subchapters describe the presented system approach to technical risk management in machinery based on the proven techniques supporting hazard identification and risk estimation [7].

2.1 Strategic Risk Management

The policy of risk perception must be clarified on the level of top management that has to set up a team responsible for realization of the risk assessment process and to specify its competences. This team must determine the rules for risk estimation with the use of the risk estimation graph (Fig. 2). It means that categories of the individual risk components must be defined: S – seriousness of possible health damage (S0 - irrelevant injury; S1 - slight injury; S2 - serious injury; S3 - death), A - frequency and length of threat (A1 - seldom to occasional; A2 - frequent to continuous); E - possibility of avoiding the hazard (E1 - possible; E2 - possible at certain conditions; E3 - hardly possible), W - probability of hazardous event occurrence (W1 - low; W2 - medium; W3 - high). Besides, accurate explanation of these categories in word must be attached (e.g. W3 means a hazardous event occurrence more than once per shift).

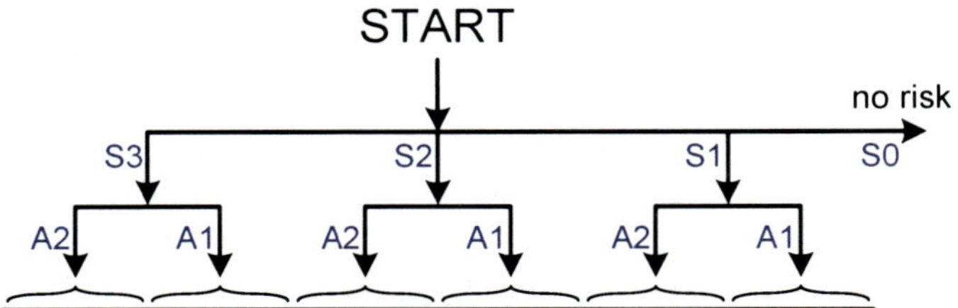

E3	E2	E1	E3	E2	E1	E3	E2	E1	E3	E2	E1	E3	E2	E1	E3	E2	E1	
16	15	14	13	12	11	10	9	8	7	6	5	4	3	2	1	0	0	W1
17	16	15	14	13	12	11	10	9	8	7	6	5	4	3	2	1	0	W2
18	17	16	15	14	13	12	11	10	9	8	7	6	5	4	3	2	1	W3

0 - 4 acceptable risk

5 - 6 risk acceptable with review

5 - 18 unacceptable risk

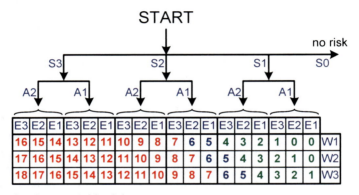

Fig. 2. Example of risk estimation graph

Furthermore, on the basis of the risk perception policy in the enterprise, the team must set the risk acceptance criteria. The risk estimation graph shows an example of acceptability of the individual sizes of risk indicated by different colours.

2.2 System Analysis

At this stage, within the „Planning of risk analysis and risk assessment", it is first necessary to gather information on the current state of the art of science and technology, solutions of safety risks in comparable products, information on previous occurrences of risks (i.e. accident occurrence) in similar equipment etc. In the next step „System analysis of machinery", the team elaborates a comprehensive description of the machinery including a written section and a block diagram on suitable level of resolution, which shows all relevant interactions between its individual elements (subsystems). Fig. 3 shows an example of a block diagram for automatic tool changer. Here it is important to pay attention to sufficiently detailed description of all interaction between individual elements of the block diagram as it helps the team members to obtain a thorough idea of the construction and functions of the machinery, which is essential for successful identification of all hazards, carried out for all elements depicted in the block diagram. For example, in the analysis of a servomotor from Fig. 3, interaction T indicates the hazard of material ejection or winding up. Interaction Th signalizes the hazard of burn injury, E warns of the possible electrical injury and I draws attention to the hazards connected to functional safety.

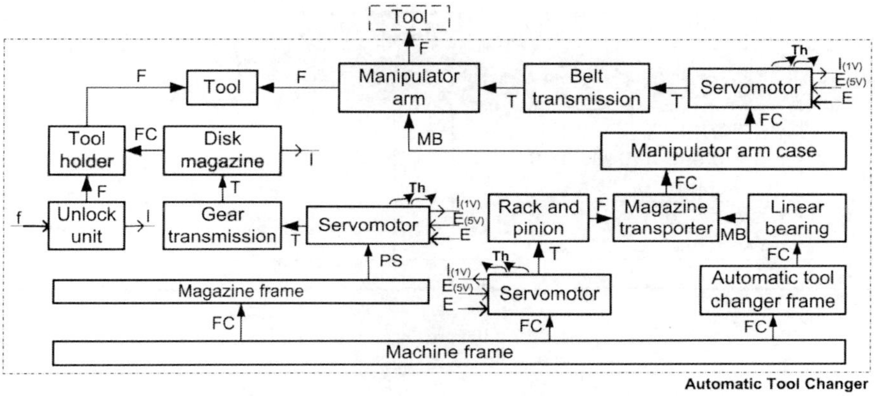

E – electrical energy, **F** - force, **f** - fluid, **FC** – fixed connection, **I** - information, **T** - torque, **MB** – movable bearing, **Th** – thermal energy

Fig. 3. Example of block diagram

Moreover, it is necessary to list and describe activities of all the professions involved in the individual stages of the machinery life cycle. Equally important is determination of the limit values of machinery. Limit values represent the limit possibilities of the machine or staff and include: limits of use (e.g. various operational states of the machine), spatial limits (e.g. range of movements) and time limits (e.g. expected service life of the machine). Identification of relevant hazards in machinery stems from the previous two steps and is performed in accordance with the harmonised EN standard EN ISO 12100 (replacing standards EN ISO 12100-1, EN ISO 12100-2, EN ISO 14121-1 from June 2011) for the whole life cycle of the machine. Using the previously elaborated block diagram and a description of the activities of personnel, the team determines, in accordance with the EN ISO 12100, all relevant hazards associated with the machinery and analyzes the dangerous zones within the machinery.

2.3 Risk Analysis

At this stage, the initial risks of the previously identified relevant hazards are conveniently estimated with the use of the risk estimation graph (Fig. 2). Based on the size of the initial risk and risk acceptance criteria, these identified relevant hazards are classified into two groups: those with significant risks requiring reduction through appropriate measures, and hazards with residual risks which the producer must report to the machinery user. The initial estimation of risks is carried out for the case when no preventive risk-reducing measures are applied. Fig. 4 presents an example of documentation of this stage of the risk assessment process.

List of identified hazards		**Initial risk estimation**					**Type of hazard**
No.	**Hazard description**	**S**	**A**	**E**	**W**	**Size of init. risk**	
1.1-1	Hazard of crushing during manipulation with the machine or its parts.	S3	A1	E3	W2	**14**	**significant**
4.1-1	Hazard provoked by noise from the belt gear.	S1	A2	E1	W1	**2**	residual
etc.							

Fig. 4. List of identifed hazards – documentation example

LIST OF SIGNIFICANT HAZARDS						Machine:			
Date:		Responsible:				Model:			
		Requirements					**Technical solutions**		
No. of hazard	TYPE OF HAZARD	Relevant articles of 2006/42/EC	Relevant articles of EN ISO 12100	Articles of repealed EN ISO 12100-1	Articles of repealed EN ISO 12100-2	DESCRIPTION OF HAZARDOUS SITUATIONS	Relevant articles of corresponding safety standard B or C type	Relevant articles of corresponding safety standard B or C type	Relevant articles of corresponding safety standard B or C type
Hazards, hazardous situations and hazardous events									
1	Mechanical hazards caused by:								
1.1	Crushing hazard	1.3.7 1.3.8 1.3.9	6.2.1 6.2.2 6.2.3	4.2.1					
1.2	Shearing hazard	1.3.7 1.3.8 1.3.9	6.2.1 6.2.2 6.2.3	4.2.1 4.2.2	4.2.1 5.2.1 5.3.2.6				
etc.									

Fig. 5. List of significant hazards – documentation example

In significant hazards it is then necessary to prove in what way the producer fulfilled the requirements of the machinery directive 2006/42/EC and of the harmonized standard EN ISO 12100. For this purpose, it is possible to use the form shown in Fig. 5. For each type of significant hazard, the requirements on risk reduction ensuing from the directive 2006/42 EC and the standard EN ISO 12100 are listed in the form of references to the corresponding articles of these documents. According to the type of the machine assessed and to the identified hazards, the descriptions of the hazardous situations are filled in, together with the references to the articles of safety directives specifying the requirements on the corresponding technical realization of risk reduction in the corresponding hazard.

If the hazards arise at more places within the machinery and during different operational statuses of the machinery, it is necessary to perform the estimation of risk for all places of their occurrence and for all the operational statuses needed, in separate forms. The estimated initial size of risk and its individual components (S, A, E, W – see Fig. 2) are then entered also into the risk estimation form (Fig. 6), which documents the process of risk reduction in a very transparent way.

2.4 Risk-Reducing Measures

This step is performed in accordance with the requirements of EN ISO 12100 (iteration method of risk reduction). First, the measures built in design are proposed, followed by safeguard measures reducing the risks; finally, information for users is provided. Each proposal of a measure must include a detailed written description of the risk reduction method. In Fig. 6, the hazard of crushing during workpiece clamping with a chuck in a lathe has been chosen as an example well illustrating this step. After every proposal of a risk-reducing measure, the residual risk must be estimated. Each of the estimated risks must be evaluated and classified as acceptable or unacceptable. It is necessary to bear in mind that total elimination of risk is not possible. If the size of risk is not acceptable, further risk-reducing measures must be proposed.

The hazards, whose risk can not be totally eliminated through construction and safeguard measures, must be listed in information on residual risks. The descriptions of measures in the third iteration step of the risk reduction called "information for use" are then copied into the instructions of use of the machinery.

2.5 Evaluation of Overall Safety of the Machinery

This phase of the technical risk management has not been included in any legal regulation so far. It is a task for the strategic risk management where the management of an enterprise must decide, whether the overall size of residual risks associated with the machine after fulfilment of the legal requirements on the safety of the machine is acceptable.

Brno UT FME	RISK ESTIMATION FORM responsible: (name)		machine: lathe date: 10.12.2008
No. of hazard	ID	**Hazard identification** according to EN ISO 14121-1: 1. Mechanical hazards	
1.1	**3**	**Crushing**	
Phase of life cycle: *use*		Danger zone: *operating area*	
Exposed persons: *operator*		Operational state: *machining and adjustment*	
Description of hazardous event:	*Crushing during workpiece clamping with a chuck. Danger of crushing of fingers or parts of hands.*		
Initial risk	Seriousness of possible health damage:	*S2 - serious injury*	Size of risk **12**
	Frequency and length of threat:	*A2 - frequent to continuous*	
	Possibility of avoiding the hazard:	*E3 - hardly possible*	
	Probability of hazardous event occurrence:	*W3 - high*	
STEP 1: Measures built in design (according to EN ISO 12100-1)			
Description of measure:	*The design of the machine respects the ergonomical principles reducing the possibility of inserting fingers or parts of hands between the moving parts.*		
Reduced risk after measure	Seriousness of possible health damage:	*S2 - serious injury*	Size of risk **10**
	Frequency and length of threat:	*A2 - frequent to continuous*	
	Possibility of avoiding the hazard:	*E2 - possible at certain conditions*	
	Probability of hazardous event occurrence:	*W2 - medium*	
STEP 2: Safeguards and additional protective measures (according to EN ISO 12100-1)			
Description of measure:	*It is possible to activate only one dangerous movement at a time in a clearly arranged space. Workpiece clamping with a safe speed of 2m/min. Use of enabling switch. Dangerous movements of the clamping device can be activated only by a deliberate action of the operator, i.e. pressing down the control pedal requiring permanent actuation.*		
Reduced risk after measure	Seriousness of possible health damage:	*S1 - slight injury*	Size of risk **4**
	Frequency and length of threat:	*A2 - frequent to continuous*	
	Possibility of avoiding the hazard:	*E2 - possible at certain conditions*	
	Probability of hazardous event occurrence:	*W2 - medium*	
STEP 3: Information for use (according to EN ISO 12100-1)			
Description of measure:	*Warning in the instruction manual: "It is not possible to totally eliminate the hazards associated with workpiece clamping by technical means; therefore, behave in such a way so as to avoid injury of fingers or parts of hands!"*		
Residual risk	Seriousness of possible health damage:	*S1 - slight injury*	Size of risk **3**
	Frequency and length of threat:	*A2 - frequent to continuous*	
	Possibility of avoiding the hazard:	*E2 - possible at certain conditions*	
	Probability of hazardous event occurrence:	*W1 - little*	
VALIDATION:	**Measures are sufficient:** (name)		date:

Fig. 6. Example of risk estimation form

The overall safety of the machinery may be assessed, for example, as a comparison of the sum of all initial risks and the sum of all residual risks, mainly for automatic and adjustment regime of the machinery, and for handling activities within the machinery (Fig. 7).

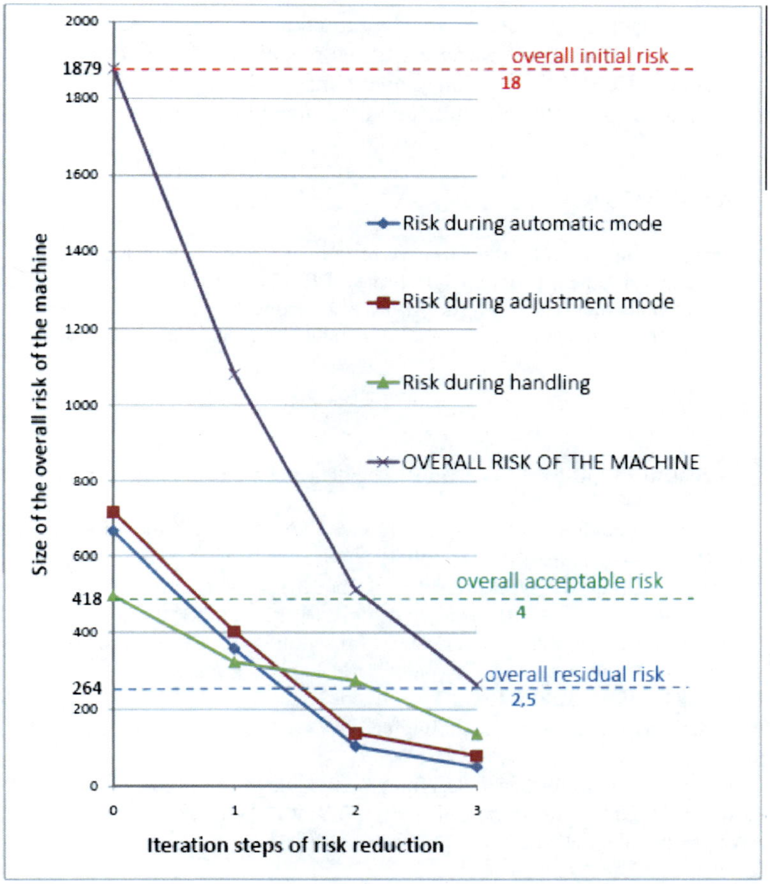

Fig. 7. Assessment of the overall safety of the machinery

3 Conclusion

The presented system approach to technical risk management objectifies to a great extent identification of hazards and estimation of risks. At the same time, it supports realization of risk analysis for all phases of life cycle of a machine. Working with a block diagram showing interactions between individual subsystems and elements of a machine, this approach increases transparency of the whole risk assessment process and helps to estimate the risks correctly and thoroughly. The possibility of overlooking a hazard or underestimating a risk is minimized. Even the persons not directly involved in the machinery development can perform risk evaluation on highly professional level as they are guided through the whole risk assessment process by the described procedure. The documentation obtained throughout the technical risk management process enables the manufacturer to prove that the machinery matches the requirements of the Directive 2006/42/EC and that adequate effort to ensure safety of the machine has been made.

The presented technical risk management approach has been several times successfully applied in industry on multifunctional machine tools.

Acknowledgements

This research was supported by the Ministry of Industry and Trade of the Czech Republic (Research and Development Program TIP, project FR-TI3/780 "Support of development of fully satisfying machines") and by the Ministry of Education, Youth and Sports of the Czech Republic (project 1M0507 "Research of Production techniques and technologies").

References

[1] Czech Statistical Office, Occupational accident occurrence in the Czech Republicin 2007, 57 p (2008)
[2] Krause, F.-L., Kimura, F., Kjellberg, T., Lu, S.C.-Y.: Product Modelling. In: Annals of the CIRP, vol. 42(2), pp. 695–706 (1993)
[3] Bernard, A., Hasan, R.: Working situation model for safety integration during design phase. In: Annals of the CIRP, vol. 51(1), pp. 119–122 (2002)
[4] Marek, J.: Management of risk at design of machining centres. In: Proceedings of Machine Tools, Automation and Robotics in Mechanical Engineering, ČVUT Praha, pp. 91–98 (2004), ISBN 80-903421-2-4
[5] Eder, W.E., Hosnedl, S.: Design Engineering - A Manual for Enhanced Creativity (2007)
[6] Marek, J.: How to go further in designing methodology of machine tools? In: Proceedings of AEDS 2006 Workshop, University of West Bohemia, pp. 81–88 (2006)
[7] FAA System Safety Handbook, December 30 (2000)

The Problems of Continuous Data Transfer between the PC User Interface and the PCI Card Control System

P. Bureš

VÚTS a.s, U Jezu 525/4,
Liberec, 461 19, the Czech Republic

Abstract. The problematic of a continuous transfer of large data volume between the user interface and the control system covers the control system as a whole, from hardware design, via communication choice up to defining the principals of data transfer control. The aim is a hardware design from the available products for industrial application and a determination of the principals of data transmission.

1 Introduction

Currently, stand alone programmable logic controllers (PLC) are normally used in connection with touch screens implementing „Human Machine Interface" (HMI) in industrial automation. In case of a requirement for processing a large volume of data in the PLC in the order of tens of megabytes, a memory limitation of the PLC data of tens of kilobytes will occur. This limitation of the PLC memory size can be solved by a sequential transfer of data between the PLC and HMI which enables an extension with the FLASH memory at present. HMI screens, just like PLC, are programmed by means of development environment delivered from a HMI producer, which places restrictions on the possibilities of HMI devices given by the means of development environment. Another issue of data transfer concerns communication interface whose administration is implemented in the HMI device and it cannot be intervened in that administration, thereby it cannot be guaranteed continuous data transfer from the HMI into the PLC whose program cycle is in units of milliseconds.

In order the automation system can have conditions for the continuous data transmission, it is necessary to eliminate the above written constraints. One of many manufacturers' variants is the PLC controller in the form of an extending PCI card to an industrial personal computer (IPC). The PLC on a PCI card is a completely independent system that provides an IPC access to the PLC memory via the PCI.

2 PCI Bus

The PCI bus is a standard parallel interface for communication of PC components between each other, as shown in Figure 1 below.

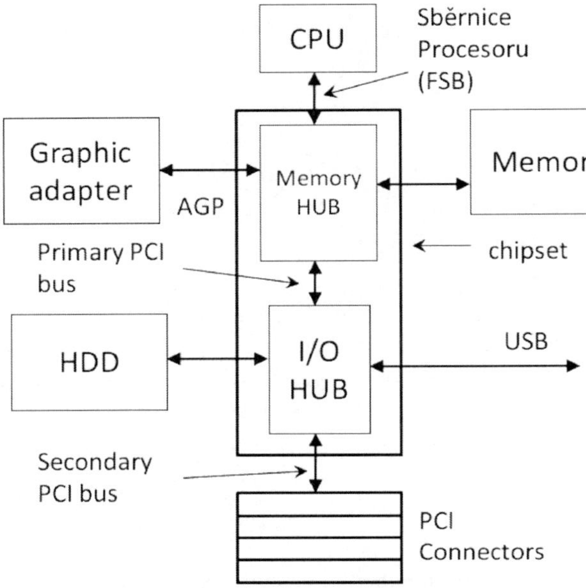

Fig. 1. PC architecture with PCI bus

The PCI specification includes several variants according to the level of logic signals, the bus data part width and the frequency of clock pulses. The basic frequency of clock pulses is 33 MHz and the bus data part width is 32 bits. Table 1 shows the maximum theoretical data flow for different variants of the PCI bus.

Table 1. Theoretical maximum data flow of PCI bus

Clock Frequency [MHz]	Data part width [bit]	Maximal data flow [MBps]
33	32	132
33	64	264
66	32	264
66	64	532
133	32	532
133	64	1066

The actual speed of communication of the application with the device is lower for several reasons however. One of them is the number of running applications when the processor handles particular applications in parts. This reduction of the PCI extent of utilization lowers the ability of BUS-MASTERING on the PCI device when the device itself will take communication control on the PCI bus and the processor is free for other applications. Another speed reduction is to share the bus data part for data and address when an address for write /read is sent at first in the bus and data subsequently. This reduction in speed reduces BURST mode which is a data block transfer when the starting address is transferred, followed by a continuous block of data.

3 Communication Control (Management)

The PCI bus is controlled by the IPC system; therefore, it is not possible, according to the preceding description of the PCI, to control the communication of the entire bus. Within the HMI application, however, we can propose the principle of communication control with regard to the minimization of the bus load. When reading more visualization elements in the HMI application in the IPC from a single register in the PLC, the PCI is more loaded with communication, that communication increase can be reduced through virtualization. The virtual register in the application repeatedly updates its value from the PLC register, the visualization elements then read the value stored in the IPC memory. In case of writing a value from the visualization into the PLC, a record in the virtual register will occur at first and this will carry out an entry into the PLC as per Fig. 2.

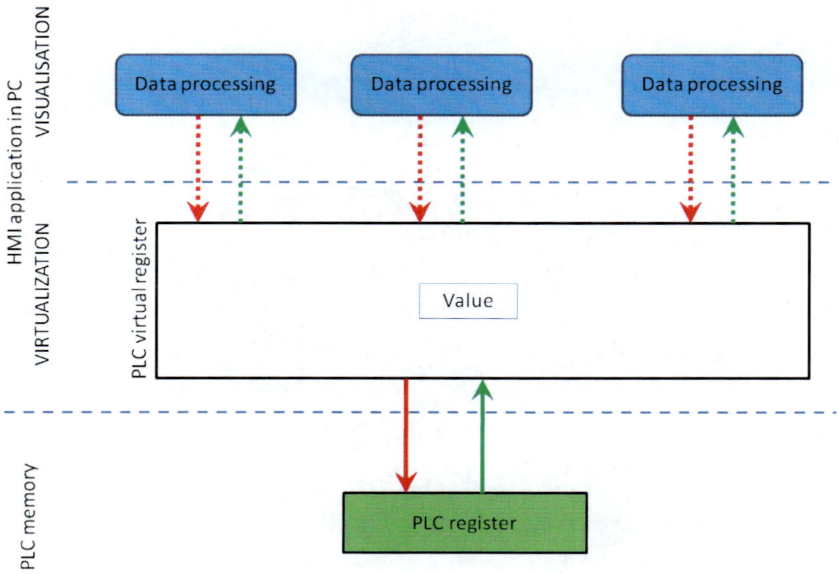

Fig. 2. Principle of communication PC - PLC

According to the importance of the given register in the process of control, cycle time can be then changed in which value update from the PLC occurs, and thus the priority of data transfer can be defined.

This reduction of communication load through the PCI interface will release the bus for the transfer of large blocks of data that can replace a continuous transfer.

4 Transfer of Large Blocks of Data

Memory size limitation in the PLC results in using the transfer of the parts of the required data from the IPC into the PLC in small blocks in such a way so that the memory for the PLC program calculations should remain always free.

A block transfer of data will reduce the PCI bus load and data transfer will be provided in a sufficient advance. But it is also necessary to ensure homogeneity of data, thus to secure not to write data from the IPC to the PLC in an area where there is just reading values by the PLC program. We will ensure it by dividing the data block in the PLC into 2 parts which are switched each other between reading mode by the PLC program and recording from the IPC, the principle is given in Fig. 3.

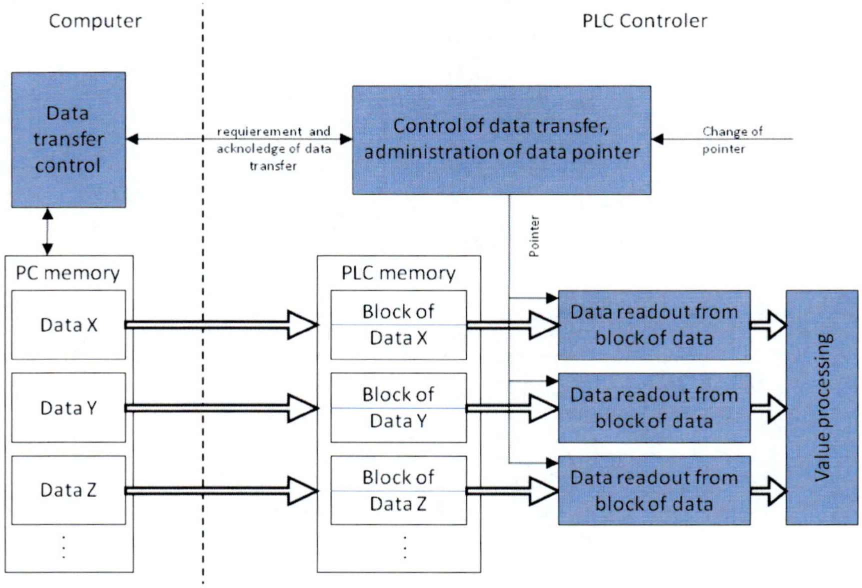

Fig. 3. Data block transfer principle

5 Conclusions

The defined principles can be generally applied to any PC and PCI controllers. A control system of the industrial PC and PLC of YASKAWA' s MP2100 controller has been created and an application of the HMI interface has been elaborated. Testing was carried out on the application. To transmit a single value it was measured a data flow of 5 KB/s; when transferring a data block in a size of 800 Byte, the data flow was 929 KB/s, which offers a great acceleration of communication for substituting the continuous data transfer per individual values with block transfer.

References

[1] http://computer.howstuffworks.com/pci7.htm
[2] http://www.pcguide.com/ref/mbsys/buses/types/pci.htm

The Control System of a Cam Grinding Machine in Connection with Yaskawa's Electronic Cams

V. Crhák and P. Jirásko

VÚTS, a.s., (Plc), U Jezu 525/4,
Liberec 1, 46119, Czech Republic

Abstract. The paper deals with the problems concerning the cam production by grinding technology in relation to the CNC machine control system. Cam production is a specific field in both the theoretical preparation of production data and the production technology on the grinding machine itself. In the final phase, those aspects of the technical preparation of production are realized in the control system of grinding machine. The technological and system requirements predestinate cam production machines as single-purpose, which presents a unique design and a unique CNC control system. The control system is proposed on a software and hardware (SW and HW) platform of components from Yaskawa, a Japanese company. Furthermore, there are outlined connections of the CNC control system with electronic cams that are built on the same HW and SW platforms.

1 Introduction

Contemporary control systems of working machines are very sophisticated systems which are generally prepared for the control of many interpolating axes with a possibility of computing and controlling the correction tool paths, inclusive many supporting programs and cycles. That complexity often goes at the expense of the required simplicity due to special operations and ever repeating activities. If we deal with the problems of grinding, it is obvious that a machine of a vertical milling machining center type must be prepared for broad applicability and the machine producer cannot be limited by a certain type of machine part. The applicability of such a center must be ensured by the *CNC* control system. The system must have the possibilities of extending the linear and rotary interpolating production *NC* axes, it must meet many requirements on the processing of production data and the methods of creating NC programs (data from *CAD/CAM* systems), it must be ready for the usability of a number of cutting and non-cutting production technologies (drilling, lathe turning, milling, application of measuring probes, etc.). The application of those systems is demanding and expensive.

Cam production with a final grinding technology is a specific problem. It is a unique production machine on which only cams are ground productively. The special technology of cam grinding also determines the conception of a *CNC* control system, production data processing and communication with operators. Of course, the systems of well-known producers (*Siemens, Heidenhain, Fanuc*) may be applied as *CNC control means*, but they require additional SW amendments or special versions of their control systems (*Heidenhain-Atek*). The research institute VÚTS, a.s, (Plc), in Liberec, has been involved in the calculations of cam mechanisms and cam production for decades. In connection with that field of conventional cams, we are now engaged in the applications of electronic cams based on *Yaskawa´s* components and cam control systems which we implement on our machines (*CamEl* weaving machine). At the same time, we deal with the possibilities of using *Yaskawa's SW* and *HW* on single-purpose machines for cam manufacture. As grinding the cam active surfaces is more demanded from the technological point of view than their milling, the further text will depict the technology of grinding solely.

2 Defining the Scope of Application of Yaskawa's Control Systems in VÚTS, a.s., (Plc)

Almost every today's processing machine is controlled by its control system based on the *PLC* automate. The *PLC*, with its inputs, outputs and communication, controls the machine operation in a technological period in time flow (*scan*). Through necessary communication with the operator by means of the *HMI* (*Human Machine Interface* – touch screen), machine modes are defined, such as *JOG* mode (manual control) and *AUTO* (machine production run). Our field of applications of *Yaskawa's* control systems are machines in which an emphasis is placed on *electronic cams*. From the viewpoint of mechanical engineering applications of the drives of mechanism working links, it is meant by the term of *electronic cam* such use of servomotor (as a servo (active) force link) which is alternative to the drives that are possible combinations of cam and linkage mechanisms driven by conventional asynchronous motors. Furthermore, by an *electronic cam* it is meant the own (proper) drive (servomotor and frequency inverter) inclusive *HW* and *SW* of drive control (controller). The *Yaskawa HW* and *SW* system will be depicted hereinafter; we only declare here that the choice of the system was determined by the requirements on the high dynamics and positional accuracy of electronic cams in demanding applications.

2.1 VÚTS's, a.s., (Plc), Single-Purpose Processing Machines

There are drives and controls of the CamEl air-jet and water-jet weaving machine (leno weave of technical fabrics). The problems of weaving machine control includes the high-performance electronic cams of the main drive (beat-up and shedding mechanisms) as to Fig.1 and the drive of a conventional double cam mechanism of the water insertion pump as to Fig. 2.

Fig. 1. Main drive (SGMGH-75D)

Fig. 2. Water pump drive (SGMGV-30D)

2.2 VÚTS's, a.s., (Plc), Working Single-Purpose Machines

As it was indicated, within the *Yaskawa* system, it is possible to program comprehensively extensive machine control systems. Thanks to the quality of drives and experience with the dynamics and positional accuracy of electronic cams, we have developed within a research project a common one-axis control system of electronic cam which can be extended according to the number of required axes. The difference of the electronic cam system and the common *CNC* machining system will be explained hereinafter. The first test application of the *CNC* system was tested on the *Strausak* grinding machine frame according to Fig. 3. The results proved the ability of the *Yaskawa* system to control the process of radial cam grinding effectively. The results of shape accuracy (+/- 0,01 mm) and surface roughness (Ra 0,3 - 0,5) according to protocols exceeded the expectations. The grinding tool was a conventional disc that was dressed (trued) during grinding. Based on the experience with the prototype of the control system, it is currently being developed a new grinding machine of radial cams of middle and small dimensions in concept at which an emphasis will be placed on the shape accuracy of the cam active surfaces or contours.

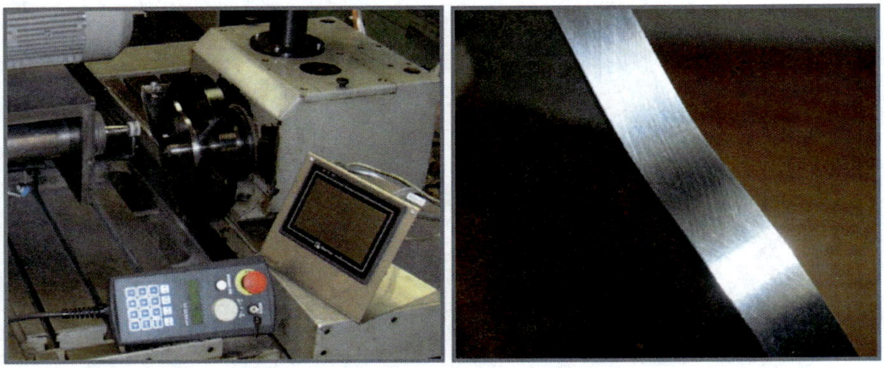

Fig. 3. A Strausak testing machine and the ground surface of a radial cam

2.3 The Conception of Cam Grinding Machines

With regard to the usability of the *Yaskawa* system on the cam grinding machine, we define cam types and interpolating *NC* axes resulting from them. The issues of cam mechanisms are systematically described in [1]. As a *cam* mechanism we will designate a *mechanism* with one degree of freedom consisting of at least one cam connected with other links with one common kinematic pair minimally. As a *cam* we indicate a mechanism link which implies with the movement of its active surface the motion of the driven link through the common kinematic pair and the cam is usually a mechanism driving link then. The basic cam types are *radial, axial*, and *globoid*. In Fig. 4 and Fig. 5, there are schemes of production coordinates (interpolating axes) for *radial* and *axial* cams. Axial cams are further divided as per the motion of the driven link. This is either sliding motion (*cylindrical* cams) or rotational motion (*axial cams with rocker*). The basic

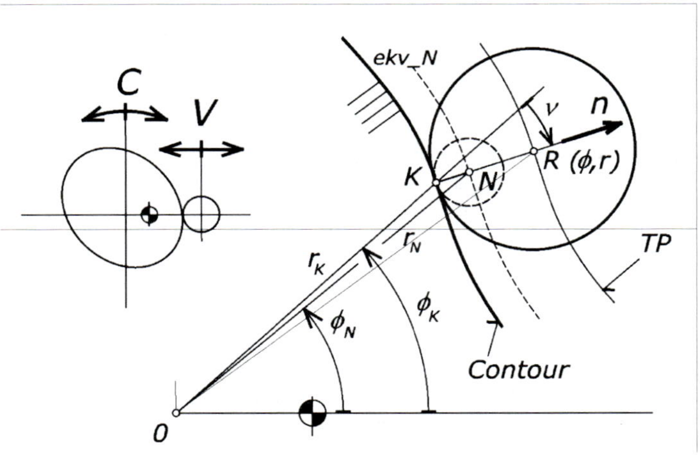

Fig. 4. Interpolating axes (radial cam)

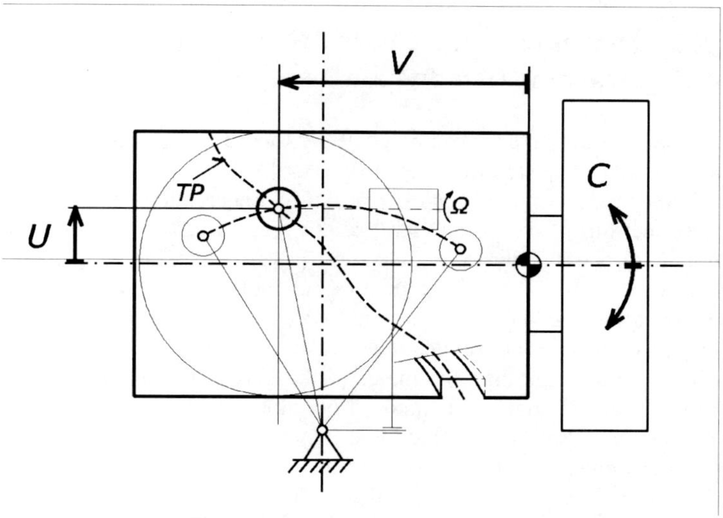

Fig. 5. Interpolating axes (axial cam)

NC axis is always the rotational motion of a cam workpiece. By turning „C“ rotational axis by 90° (at a radial cam grinding machine) in such a way that the rotational axis is parallel to „V“ axis, we can achieve the NC configuration of axes for the production of cylindrical cams. It results from the Figs. that radial, cylindrical and globoid cams require an interpolation motion of both NC axes; axial cams with rocker require an interpolation of three axes. The additional positional tool axis „Z“ is not shown.

In the test as follows, we will focus on the production of radial cams because axial and globoid cams principally use only the theoretical profile coordinates (roller center path). On the contrary, the control system of the radial cam grinding machine must process equidistance data to the theoretical profile or the cam contour because during grinding a material removal occurs according to the surpluses (allowances) for one machining cycle and the grinding disc is being reduced due to its diamond dressing. The active surface of axial and globoid cams is a common undevelopable surface which must be produced by the tool diameter identical to the mechanism roller. This is achieved by the eccentricity of the planetary motion of the grinding tool. The eccentricity is ensured by the controlled NC axis which has a constant value during the working cycle however (it is not interpolating). The production data are always the same then. From this point of view, the grinding machine of globoid cams is the simplest both from the point of view of production data computation and the point of view of the CNC control system. Paradoxically, globoid cams are the most expensive element of all the cam mechanisms.

3 SW Requirements on the Control System of the Radial Cam Grinding Machine (Production Data Processing)

Production data are the polar coordinates of the theoretical profile (angle and vector) or cam contour. The step of production coordinates depends on the interpolation type between „*C*“ and „*V*“ axes. In case of linear interpolation, we did not notice any reflections of the ground surface at a step of 0,2° of „*C*“ axis. The data are continually processed into equidistances according to the tool diameter and surplus (allowance) size. It can be solved in the cycle or by recalculating the data out of the working cycle. The chosen method of recalculating in the cycle requires the knowledge of the normal angle. It is the simplest way to deliver the normal angle to the system as another parameter to the given contour point because the computation of normal angle is a standard operation of the computation software of cam kinematic analysis by which the production coordinates are prepared. Thus, the system must save the production data into external data registers with a sufficient reserve. The program possibilities of data processing are given in the following paragraph.

4 Yaskawa's Hardware

There are two drives of „*C*“ and „*V*“ interpolating axes, positioning „*Z*“ axis and a frequency inverter of the grinding spindle drive in the structure of machine.

In case of the conception of the „*V*“ linear axis with a ball screw linked directly with the servomotor and the „*C*“ rotational axis with a backlash free rotary reducer is easy to choose the servomotor type. With regard to shifts (slides) during grinding and the used servomotors of the *SGMGV* series with a 20bit encoder, a drive with a sufficiently suitable moment characteristics can be chosen from that series.

The most important part of the system is a *MP2300* series controller. It is a top controller which can be configured with a number of additional cards.

5 Yaskawa's Software (Development Environment)

The mightiest means are the software possibilities of the controller and the *MPE720v6* program development environment. In principle, the development environment is divided into two parts which work under the same field of external variables.

The first field is the *PLC* automate with two types of scans (*High-Speed Scan* and *Low-Speed Scan* in the shortest time of 1 ms). The way of programming is carried out on the basis of the very user friendly *Ladder Diagram.* The hierarchy system of the local and global variables, mathematical and logic operators, control structures, relay logics, three levels of nested subprograms, the possibilities of user functions and their calling, etc., that everything is in the direction of a *PLC* programmer and in such a way it makes a very strong tool from the *MPE720* development environment. In this field, all variants of the user functions of *electronic cams* are programmed.

The second part is the field of continuous motion programming. The way of programming is solved on the basis of an instruction set in a well-arranged text editor. In this field, the continuous motions of interpolating axes based on the positional information and shifts (slides) are programmed.

6 The Structure of the Control System of Electronic Cam

The electronic cam realizes the motion exciting function on the servomotor shaft. The problems of the application of electronic cams are included in the work [2]. We only state that the electronic cam control system is programmed on the basis of a *state one-axis scheme* according to Fig. 6 and it is implemented in the *PLC* field.

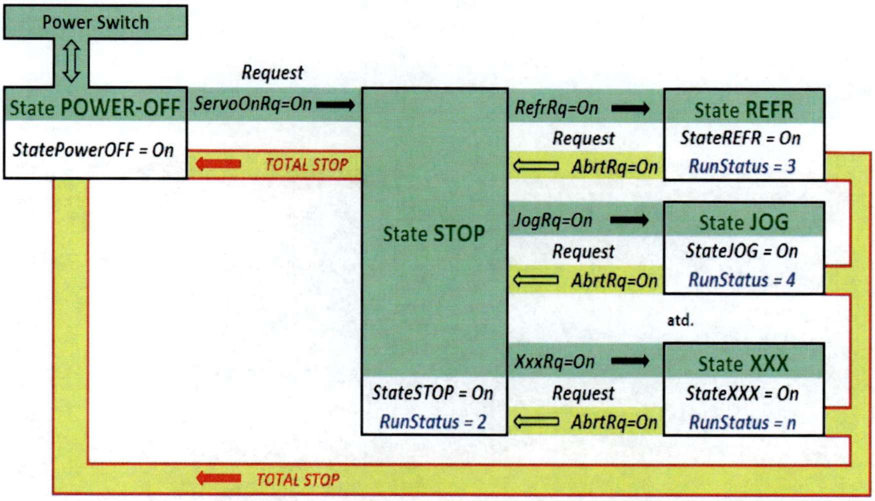

Fig. 6. Machine state control system scheme

7 The Control System Structure of Radial Cam Grinding Machine

The control system structure of cam grinding machine is identical with the scheme of the state machine according to Chapter 6. The one-axis system is only extended to the three-axis system in which two axes are interpolating. The continuous control according to production data and axial interpolation takes place in *Motion*, in the *CNC* state then. The *CAM* state is suppressed.

In the text editor of *Motion*, grinding technology is programmed as per the input information from the *HMI*. As to the size of the removals (reductions) of particular grinding cycles and the changes of grinding disc diameters due to dressing, the source data of cam contour are corrected in the nested cycles *For/Next*. In the internal nested cycle, polar coordinates of the instantaneous equidistance in the specific working cycle are calculated on the basis of an instantaneous state of the grinding disc diameter and the surplus (allowance) size.

8 Communication

Communication is arranged by a system function by a defined communication protocol. It is that fact important that by pressing the button on the *HMI*, a bit variable will be defined which represents the global communication variable concerning the appropriate (corresponding) machine state. Buttons on the *HMI* are associated (matched) or divided into groups (screens) which correspond to the four basic function modes *INITIAL*, *BASIC*, *MANUAL* and *AUTO*. A demonstration of *AUTO* function mode on the *Weintek* touch screen is shown in Fig. 7. The program core of the state machine in the *PLC* is independent on the tree structure of *HMI*.

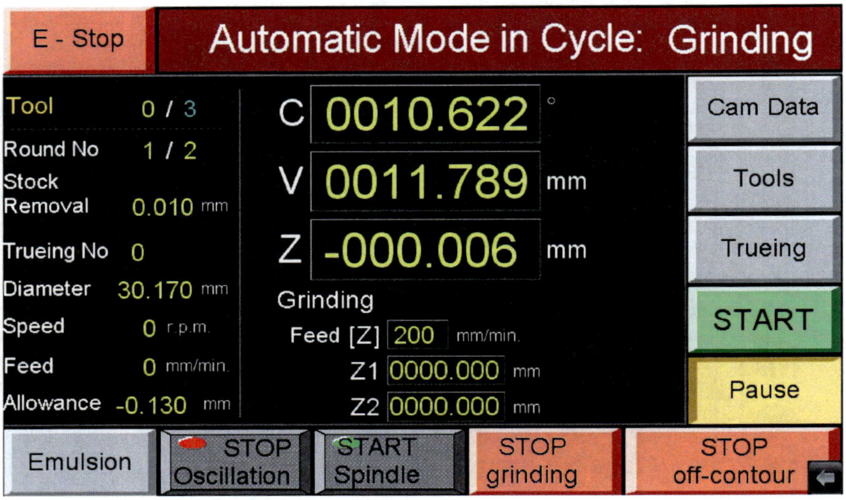

Fig. 7. AUTO mode

9 Proposal of a New Conception of the Control System of the Radial Cam Grinding Machine and Its Continuity (Relationship) with the Functions of Electronic Cams

The new conception will take two directions. In the first direction, it will come fully to the application of „object“ function modules of user functions. The function blocks with their inputs and outputs will be autonomous units which will represent machine states. The machine states for particular axes will not be structured in the *Ladder diagram* by nested subprograms. The structures will concern the individual axes in *MANUAL* mode only.

In the second direction, it will be a replacement of *Motion* by the functions of electronic cams for individual *NC* axes. The motions of the axes will be not bound by interpolation, but they will be bound by a common virtual axis with the use of the feed forwards constraints of the velocity and moment of particular axes. This conception will enable data correction as a reaction to systematical technology errors [3] and deviations due to the dynamics of the inertia forces of individual axes inclusive the external force technology load. This direction is in its beginning, but we have great experience in the control of the positional accuracy of electronic cams with the use of all *SW* and *HW* means of the *Yaskawa* system.

10 Conclusion

The configuration of axes (rotational, sliding) and the number of axes of the control system of cam grinding machine is a very simple case. In the case of axial and globoid cams where the accuracy of the active surface is attained *by the eccentricity* of the planetary motion of the grinding tool *in principal*, the control *CNC* system is simpler in comparison to the radial cam grinding machine system. Due to the open structure of the control system as a whole (*PLC*, *Motion*, *HMI* communication) it can be stated that the *SW* and *HW* components of *Yaskawa* meet the requirements for the control systems of grinding machines of all types of cams. At the same time, it is possible to solve the shape accuracy of cam contour in the control system of radial cam grinding machine. These problems are dealt with in [3].

References

[1] Koloc, Z., Václavík, M.: Cam Mechanisms. Elsevier, Amsterdam (1993)

[2] Jirásko, P.: Methodology of electronic cam applications in drives of working links of mechanisms of processing machines. Dissertation TU Liberec CZ (2010)

[3] Crhák, V., Jirásko, P.: The Problematic of the Production Accuracy of Radial Cams. In: Conference ECMS 2011(2011)

Kiwiki: Open-Source Portal for Education in Mechatronics

J. Dudak[1], G. Gaspar[2], D. Maga[3], and S. Pavlikova[4]

[1] European Polytechnical Institute Ltd, Osvobozeni 699, 686 04 Kunovice, Czech republic
[2] Faculty of Materials Science and Technology, Paulinska 16, 917 24 Trnava, Slovakia
[3] CTU Faculty of Electrical Engineering, Technicka 2, 166 27 Prague, Czech Republic
[4] Faculty of Mechatronics, Pri Parku 18, 911 50 Trencin, Slovakia

Abstract. With arise of modern interdisciplinary subjects as mechatronics; a huge demand for new technologies in education appears. This leads to a broad spread of modern technologies both in education (eg. distance laboratories [1, 2]) and educational materials [3-5]. A systematic pressure is applied to universities due to insufficient availability of relevant educational literature and materials. Since many years ago, this has been a question of the budget (because of the price of printed material, which usually becomes out of date after 2-4 years), but today it is a question of technology. The times when the base medium of materials has been changed from paper to digital are here today. The today's technology offers a better approach – interactive, open and free knowledge based form. One of the offered technologies is the one created by authors – portal *kiwiki*, specialized in mechatronics. It is available at *kiwiki.info* web sites.

1 Introduction

The *Kiwiki* educational portal for (and not only for) mechatronics has been built 2 years ago. It has been realized within the Department of Computer Sciences at Faculty of Mechatronics in Trencin, and its realization has been based on following:

- demands of the local legislative,
- self-test of plagiarism,
- free and on-line access to students final works.

The local Slovak legislative changes according to international standards and statutes the universities to publish the students' final works. This goes for every level of university education – beginning with bachelor works, through diploma works to PhD theses. The access to these works in public or university libraries is strictly limited to the number of copies or distance of the interested person. Another important task is the question of plagiarism. It is very complicated, usually based on co-incidence of several factors, to reveal the plagiarism. These can easily be achieved by on-line system; with possibilities of public watch and supervision, where the projects are published in their earliest phase of writing and everyone can follow the progress of the process. In case of *Kiwiki* portal even everyone can edit the work, since system allow this activity, nevertheless the history of changes, including the editors registration data are recorded in the system.

2 Motivation to "Get In"

As mentioned above, the portal *Kiwiki* is able to publish the students' final works. So the students are able to commend themselves with their first public work. This can be a base reference when applying for a job after finishing the study. The students are also able to read, criticize and classify the works of theirs schoolmates. In the opposite way – their schoolmates are able to criticize their work and in fact they contribute to the higher level of the work.

One of the most important features of *Kiwiki* portal is to publish the study materials created and published for students at the faculty. The open-access policy will allow changing the contents even after publishing. The access to carry out these changes has no restrictions. The portal *Kiwiki* can be, and already is, used by lecturers to create their textbooks or scripts. Since the publishing tools have their prescribed structure, the results are very professional and may be highly interactive.

The third party interested in the portal is the faculty itself. Besides the satisfied students and lecturers, the open-source policy leads to continual increase of published works quality and gives an un-negligible stamp of quality.

3 Ideas of Free Teaching System

There are professional systems (Learning Management System – LMS) for supporting the electronic education. A part of them is distributed as open-source – together with theirs source code. Nevertheless, the expression "free education system" has a different meaning here. In case of *Kiwiki* the "freedom" is bounded with free access to presented information. The building of this system has been inspired by basics freedoms defined by Richard Stallman at 1986, especially freedom nr. 3: "The freedom to distribute copies of your modified versions to others. By doing this you can give the whole community a chance to benefit from your changes. Access to the source code is a precondition for this" [6].

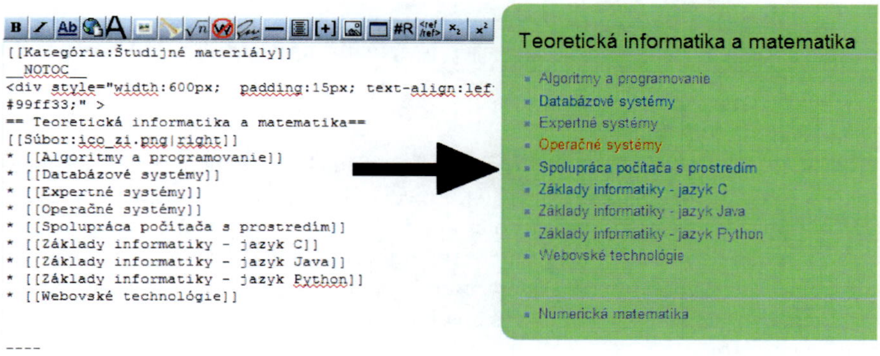

Fig. 1. Mediawiki environment and the result (in original language mutation)

The suitable platform to realize these ideas is the *mediawiki* (www.mediawiki.org) environment (Figure 1). The base of the systems (without additional modules) is offered to designers. The system build on this technology has the following properties (in contrast to LMS systems):

a) Open System

One of the most important web-space with free information access is the Wikipedia. The web of Wikipedia is also based on *mediawiki* technology. Data stored at *mediawiki* based systems (as wikipedia.org, wikibooks.org, commons.wikimedia.org, etc.) are fundamentally free accessible for public. The accompanying effect to this way published materials is that the contents in indexed by internet searching engines (*google*, *bing*, *yahoo*,...) and can be suggested to potential students (or employers), resp. other partners. The general opinion on education process quality, level of graduates or lecturers can be built according to these, which can be an important step to find a partner for different level of project teams or student/lecturer exchanges.

b) Two-Way Data Flow

The contents of the materials can be changed by everybody, both student and lecturer. Of course, the damages based on dab motivated person can easily be caused – however these practices are eliminated by system, which records the history of changes including the user data and the belonging timestamps. The possibility of editing the presented materials has also the following consequences on belonging user:

1. The user/student is fully responsible for changes made,
2. Positive motivation – the user is actually a co-author of the presented materials

c) A Simple Feedback
The part of the systems core is the discussion module. This module is working as the discuss room for a concrete *Kiwiki* project, what in terms of *mediawiki* means one wiki-page. The information not suitable for direct publishing in the project web-page can be released there within the frames of free discussion. The remarks can be seen and read by the system users with no limits. According to practical experience of the authors, the discussion is mostly focused on tendency or routing of the main project.

The basic difference if compared to logics of LMS systems, where the communication is typically focused on one lecturer – one student, is the way of communication [7]. The comments and ideas of one user can be seen and responded by everybody else.

Table 1. Basic data for accessing the *Kiwiki* web page.

Period	Nr. of visits	Nr. of views	Average visit time in minutes
January, 2011	9481	19804	6:38
Fabruary, 2011	8203	18720	2:00
March, 2011	13283	33566	2:43
April, 2011	12803	26857	2:44
Total	43770	98947	3:26

4 Additional Services

There is a print module build into the *Kiwiki*. The user can create the high quality outputs in the camera ready form in pdf standards. The module is also able to merge the various numbers of *Kiwiki* pages to one pdf output with synchronized format of the documents. Also the chapters, table of contents, numbering, etc. are generated automatically.

There also exists a module for creating the mathematical expressions. The module is called *mediawiki-math* and its usage is similar to *tex* standards. The evaluation of *Kiwiki* pages based on users' opinion can be built in the systems. This will increase the amount of trust to the actually read information.

special page

Popular pages

Showing below **50** results starting with **# 1.**

View (previous 50) (next 50) (20 | 50 | 100 | 250 | 500).

1. Home (29256 visits)
2. Materials for lectures and exercises (8714 visits)
3. Undergraduate work (4428 visits)
4. Offer student works for 2010/11 FE (4051 visits)
5. Theses (3576 visits)
6. Expert Systems (3541 visits)
7. Measures to improve the quality of the car companies (2824 visits)
8. Draft set of measures to improve the quality of the automobile in society (2541 visits)
9. The introduction of quality management system in a company with major activities in prc
10. Algorithms and Programming (2297 visits)
11. Coursework (2260 visits)
12. Simulation and modeling (1993 visits)
13. Hella Slovakia Signal Company - Lighting Ltd. (1940 visits)
14. Configurable microprocessor systems (1816 visits)
15. Foundations of Computer Science - Java (1779 visits)
16. Kolokviálna test (1761 visits)
17. Offer student essays (1761 visits)
18. Theoretical basis of assessment of quality parts (1586 visits)
19. Proposal to introduce a quality management system in the company IORC (1447 visits)
20. Capacity of the production process (1386 visits)

Fig. 2. Screenshot of page-counter service results (translated into English)

5 Conclusions

The *Kiwiki* system (Figure 3) has been on-line since the beginning of 2010. After more than one year of its lifetime the authors realized a huge interest of users to get in. More than 160 users have registered in the system as *Kiwiki* redactors. Recently it reached 2646 pages, 1716 files (means figures and pictures in the pages) and 542 projects with contents.

Also the accesses to *Kiwiki* web pages are supporting the opinion of the authors on this successful project. For instance, during April 2011 the *Kiwiki* registered more than 14500 visitors with 30600 page views. More information about visiting the web page can be seen in Table 1.

Fig. 3. View on Kiwiki.info main page (in original language mutation)

Projects published on *Kiwiki* have been used to create a printed materials for subject Database systems at Faculty of Mechatronics. The content of the text book is licensed by Creative Commons BY-SA, which guarantees the freedom of propagation, copy or distribution with obligation to quote the original source (http://kiwiki.info).

It is clear to the authors of the project that the successful implementation of *Kiwiki* very strongly depends on the size of the domestic organization. This project is suitable for small or medium size organizations or teams. One of the most important features of the created system which has not been mentioned in the paper is the support of the project teams. The free access and the presented logic of it implementation is a very good tool for team-built documents, when the members of the team have different tasks to solve but the results should be synchronized with knowledge of each team member.

References

[1] Maga, D., Sitar, J., Bauer, P.: Automatic Control, Design and Results of Distance Power Electric Laboratories. In: Recent Advances in Mechatronics 2008-2009, pp. 281–286. Springer, Heidelberg (2009), ISBN 978-3-642-05021-3

[2] Bauer, P., Dudak, J., Maga, D., Hajek, V.: Distance Practical Education in Power Electronics. International Journal of Engineering Education 23(6), 1210–1218 (2007), ISSN 0949-149X

[3] Fabo, P., Siroka, A., Maga, D., Siroky, P.: Interactive Framework for Modeling and Analysis. Przeglad Elektrotechniczny 83(11), 79–80 (2097), ISSN 0033-2097

[4] Bauer, P., Leuchter, J., Stekly, V.: Simulation and Animation of Power Electronics in Modern Education. In: 4th WSEAS AEE, pp. 48–52. WSEAS Press, Praha (2005), ISBN 960-8457-13-0
[5] Manas, P.: Experience with Cad Systems Teaching at Army Academy. In: WSCG 1997, pp. 656–659 (1997)
[6] GNU operating system. The Free Software Definition. Free Software Foundation, Inc. (2010), `http://www.gnu.org/philosophy/free-sw.html` (cited April 25, 2011)
[7] Brezina, T., Hadas, Z., Singule, V., Blecha, P.: European Projects for Support of Mechatronic Fields at Brno University of Technology. In: 13th Int. Symposium on Mechatronics, Mechatronika 2010, pp. 109–111. TnUAD Trencin, Trencin (2010), ISBN 978-808075461-7

Increased Performance of a Hybrid Optimizer for Simulation Based Controller Parameterization

R. Neugebauer, K. Hipp*, A. Hellmich, and H. Schlegel

Chemnitz University of Technology, Faculty of Mechanical Engineering,
Institute for Machine Tools and Production Processes,
Reichenhainer Str. 70, 09126 Chemnitz, Germany
wzm@mb.tu-chemnitz.de

Abstract. The controller parameterization is often carried out by applying basic empirical formulas within an integrated automatic design. Hence, the determined settings are often insufficient verified by the resulting system behavior. In this paper an approach for the controller parameterization by using methods of simulation based optimization is presented. This enables the user to define specific restrictions e.g. the complementary sensitivity function to influence the dynamic behavior of the control loop. A main criterion for practical offline as well as controller internal optimization methods is the execution time, which can be reduced by applying a hybrid optimization strategy. Thus, the paper presents a performance comparison between the straight global Particle-Swarm-Optimization (PSO) algorithm and the combination of the global PSO with the local optimization algorithm of Nelder-Mead (NM) to a hybrid optimizer (HO) based on examples.

1 Introduction

In the field of operations research a large number of methods were developed to support decision-making processes. It has been proven, that there is a wide field of application. In this paper a brief introduction using these methods for mechatronic controller parameterization is given with the goal of increased speed using a hybrid optimizer. In section 2 the basics of simulation optimization as well as the used optimization algorithms are stated. Subsequently in section 3 the application for controller parameterization is briefly introduced. The structure and functionality of the hybrid optimizer is subject of section 4. A performance evaluation of the hybrid optimizer is done in section 5. The paper closes with a comparison and conclusions given in section 6.

* Corresponding author.

2 Simulation Based Controller Parameterization

Generally, simulation based optimization is a methodology of searching for the global extremum of an objective function by the coupling of a simulator with an optimizer [1]. It results a cyclic sequence between the optimizer and the simulator (figure 1).

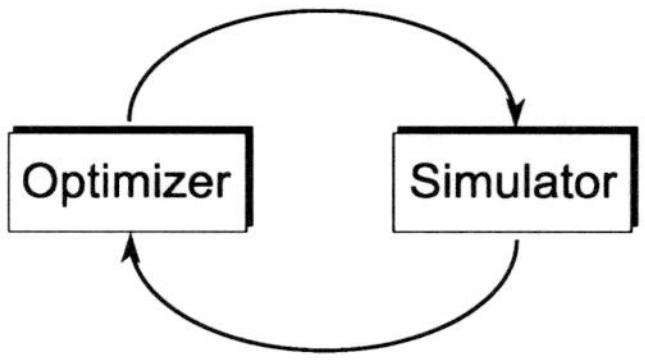

Fig. 1. Cyclic sequence

The optimizer determines a possible solution and passes it to the simulator for evaluation. According to the result of the simulator the optimizer calculates a possible better solution. The core of the simulator is a model of the entire system which is examined.

Therefore, an optimization problem (eq. 1) has to be solved [1].

$$F(\theta) \rightarrow \min_{\theta \in \Theta}(F(\theta)) \tag{1}$$

F(), called fitness function [2], is a real-valued function which represents the evaluation of the actual solution. In general, the implementation of constraints is realized by using punishment values. If a constraint is violated, a punishment value is added to the evaluation of the actual solution. Therewith it is depreciated and avoided by the optimizer. The evaluation of a solution is calculated in accordance to equation 2.

$$F(x_n) = \text{Main\_Criterion}\,(x_n) + \sum_j \text{Punishment\_Constraint}_j(x_n) \tag{2}$$

Optimization techniques are divided into global and local algorithms [3]. The objective of global optimization is to find the global extremum over the entire function space. In contrast, local methods start from a defined point in the search space and try to determine a better solution. According to [4] simulation based optimization could be used to adjust controller parameters considering definable constraints.

It exits a large number of optimization algorithms for different application fields. Hereafter the PSO and the NM algorithm are described.

2.1 Particle-Swarm-Optimization

PSO is a common heuristic technique [5], which is based on the simulation of the movement of herds or swarms. An individual of a swarm is called particle. The trajectory of each particle depends on the movement of the other individuals of the swarm and random influences. The advantages of the algorithm are among other things its simple structure, no need for gradient information and its performance. The position of every particle i in the k -th step is described by the vector x_i^k . The position of each particle in the $(k+1)$-th step is update according to equations 4 and 5.

$$x_i^{k+1} = x_i^k + \Delta x_i^{k+1} \tag{3}$$

$$\Delta x_i^{k+1} = \omega \Delta x_i^k + c_1 r_{1,i}^k (x_i^{best,k} - x_i^k) + c_2 r_{2,i}^k (x_{swarm}^{best,k} - x_i^k) \tag{4}$$

Where c_1, c_2 and ω are positive constants, $r_{1,i}^k$ and $r_{2,i}^k$ are two random values in the range [0, 1]. The term $x_i^{best,k}$ represents the best previous position of particle i till step k and $x_{swarm}^{best,k}$ as the best known position among all particles in the population. Therefore, $x_i^{best,k}$ is called "simple nostalgia" because the individual tends to return to the place that most satisfied it in the past. The term $x_{swarm}^{best,k}$ realizes the publicized knowledge, which also every individual tends to [6].

2.2 Nelder-Mead

The NM algorithm (or simplex method), which was originally presented in [7], uses a geometric structure, the simplex, with $n+1$ points in the search space with the dimension n , e.g. for $n = 2$ the simplex is a triangle.

At the beginning the simplex is constructed around a committed start point. The edges of the simplex are called vertex and have to be arranged equidistant from each other. The basic principle of the algorithm is the modification of the simplex towards the extremum. In general this is achieved by replacing the worst vertex by a better one using four functions: reflect(), expand(), contract() and shrink(). A detailed description of the algorithm can be found in [8].

3 Optimization Problem

Assuming the stated closed loop system structure [9], the system behavior is described with a transfer function G_S (eq. 5).

$$G_S(s) = K * \frac{1 + b_1 s + b_2 s^2 + ... + b_m s^m}{1 + a_1 s + a_2 s^2 + ... + a_n s^n} \tag{5}$$

It is supposed to use a PID controller G_R in the additive structure (eq. 6).

$$G_R(s) = K_R + \frac{K_I}{s} + K_D s \tag{6}$$

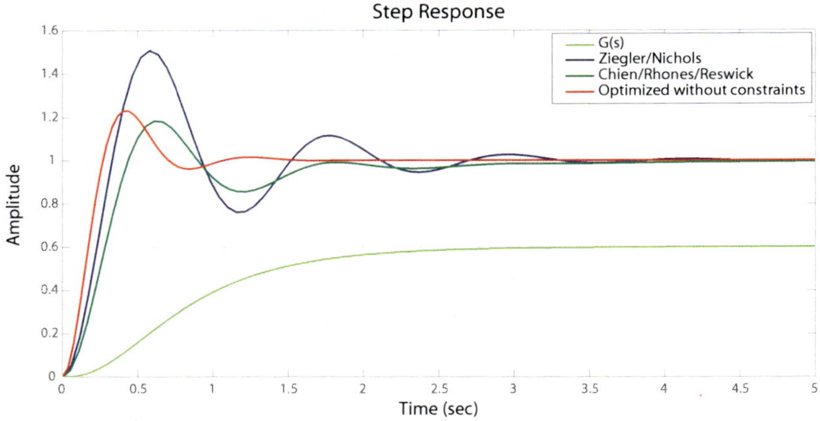

Fig. 2. Comparison of parameter settings

In figure 2 possible attainable transition functions of a PT3 plant (K=0.6, a_1= 0.92, a_2= 0.234, a_3=0.018) with a PID controller are shown. The main optimization criterion is the control area. No constraints were defined. The results of the optimization process are $K_R = 12.327$, $K_I = 17.936$ and $K_D = 2.07$.

To reduce the overshoot of the system the complementary sensitivity function (CSF) T(s) could be used [10].

The mathematical structure is:

$$T(s) = \frac{G_R(s) * G_S(s)}{1 + G_R(s) * G_S(s)}. \tag{7}$$

It allows an evaluation of the influence of changes in the command signal. By specifying the CSF it is possible to affect the dynamic of the control loop. Therefore, by setting

$$T(s) = \frac{1.1}{0.1s + 1} \tag{8}$$

the permissible amplification is limited. The new results are $K_R =$ 12.327, K_I = 17.936 and K_D = 2.07. As expected the overshoot is reduced (figure 3) while the rise time increases.

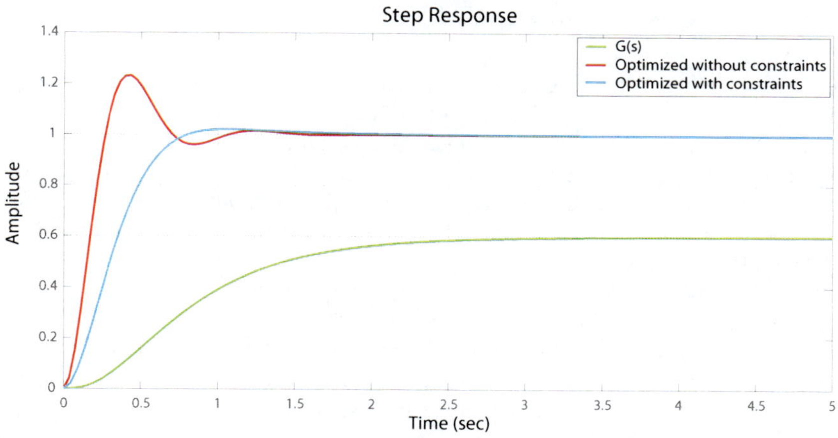

Fig. 3. Comparison of parameter settings

4 Hybrid Optimizer

In this paper the HO is a combination of the global PSO and the local NM algorithm with the objective of better performance in comparison to a standalone optimization approach.

The operation of the algorithm is the following: First the PSO is performed with a small number of calculations and then terminated. Hence, the PSO is only used for global exploration of the search space. Subsequently three instances from NM algorithm start from the best, the 3rd best and the 5th best point examined by the PSO realizing a local search. The solution of the optimization is the best result of the three NM instances (figure 4).

The reason behind starting the local search from different points is the robustness against local extremes. It has been investigated, that if only one instance is used, the hit rate of the HO to find the extremum is reduced [11].

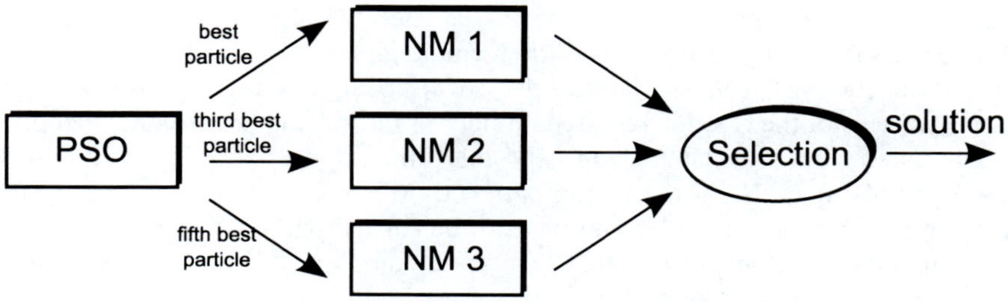

Fig. 4. Structure of hybrid optimizer

5 Performance Comparison

The performance tests have been carried out using a self developed modular optimization application written in C# using VisualStudio 2010 supporting different optimization algorithms. The simulation models were implemented in MATLAB®.

Four different transfer functions (table 1) were utilized to compare the required number of calculations to determine the global extremum with a defined tolerance to the best known solution. For every transfer function the optimization was performed thirty times.

As the results in table 1 show the HO only requires 6% to 69% invocations of the simulator in contrast to the standalone PSO. This reduces the execution time of an optimization run tremendous. Furthermore the HO was always able to detect the global extremum. The reason for the high number of necessary calculations for system three can be justified with the complex shape of the search space. The gradients around the global extremum are very high and therefore the location of the extremum is very small.

Table 1. Overview test functions.

Controlled System	Transfer function	Calculations	
		PSO	Hybrid
1 (PT3)	$K=1,\ a_1 = 2,\ a_2 = 2,\ a_3 = 1$	1737	342
2 (PT3)	$K=1,\ a_1 = 3.1,\ a_2 = 2.3,\ a_3 = 0.2$	2007	459
3 (PT3)	$K=1,\ a_1 = 2,\ a_2 = 2,\ a_3 = 3$	23459	1464
4 (PT2)	$K=1,\ a_1 = 3,\ a_2 = 2$	542	374

6 Conclusion

The combination of the PSO and NM to a hybrid optimizer increased the performance dramatically in comparison to the standalone PSO algorithm. The advantage of the HO is the switch from the well performing global optimization technique of the PSO to the NM, which is more effective in local exploration. Even with the three instances of the NM the required number of the calculations is still smaller.

This is essential to enable online and real time applications. But furthermore investigations in adjusting the tuning parameter of the algorithms concerning the problem of controller parameterization must be carried out. Moreover even different optimization techniques e.g. the Newton's method, genetic algorithms and different combinations to a hybrid optimizer must be investigated. Furthermore it is conceivable to use the methodology of simulation based optimization for tuning more complex systems like a controller cascade with filters.

Acknowledgement

Funded by the European Union (European Social Fund) and the Free State of Saxony.

References

[1] Köchel, P.: Simulation Optimisation: Approaches, Examples and Experiences, TU Chemnitz (2009), ISSN 0947-5125
[2] Carson, Y., Maria, A.: Simulation Optimization: Methods and Applications. In: Proceedings of the 1997 Winter Simulation Conference, pp. 118–126 (1997)
[3] Tekin, E., Sabuncuoglu, I.: Simulation optimization: A comprehensive review on theory and applications. IIE - Transactions 36(11), 1067 (2004)
[4] Neugebauer, R., Hipp, K., Hofmann, S., Schlegel, H.: Application of simulation based optimization methods for the controller parameterization considering definable constraints. In: Mechatronik 2011, pp. 247–252 (2011)
[5] Eberhart, R., Kennedy, J.: A new optimizer using particle swarm theory. In: Proceedings of the Sixth International Symposium MHS 1995, pp. 39–43 (1995)
[6] Eberhart, R., Kennedy, J.: Particle Swarm Optimization. In: IEEE International Conference on Neural Networks Proceedings, pp. 1942–1948 (1995)
[7] Nelder, J.A., Mead, R.: A Simplex Method for Function Minimization. The Computer Journal 7(4), 308–313 (1965)
[8] Schwefel, H.: Evolution and Optimum Seeking. Wiley VCH, Chichester (1995), ISBN 0471571482
[9] Lunze, J.: Regelungstechnik 1: Systemtheoretische Grundlagen, Analyse und Entwurf einschleifiger Regelungen, 8th edn. Springer, Berlin (2010), ISBN 9783642138072
[10] Aström, K., Hägglund, T.: Advanced PID Control. In: ISA – The Instrumentation, Systems and Automation Society (2006), ISBN 1556179421
[11] Hipp, K.: Entwurf, Implementierung und Test eines Software-Werkzeuges zur Bestimmung optimaler und robuster Regler mittels Verfahren der simulationsbasierten Optimierung. Diploma Thesis, Chemnitz (2010)

Parameter Identification of Civil Structure by Genetic Algorithm

L. Houfek[1], P. Krejci[1], and Z. Kolarova[2]

[1] Institute of Solid Mechanics, Mechatronics and Biomechanics, BUT, Faculty of Mechanical Engineering, Technicka 2896/2, 61669 Brno, Czech Republic
houfek@fme.vutbr.cz, krejci.p@fme.vutbr.cz
[2] Institute of Building Structures, BUT, Faculty of Civil Engineering, Veveri 331/95, 60200, Brno, Czech Republic
kolarova.z@fce.vutbr.cz

Abstract. This paper deals with the identification of building construction simulation model parameters. We carried out experiments determining construction behavior when actuated by external force applied to real small-scale model. The measured frequency characteristics are used to identify parameters of the simulation model created in the ANSYS APDL environment using the Genetic algorithm. The resulting (tuned) simulation model in the FEM environment will be used to simulate and investigate behavior of buildings using various types of external actuating effects.

1 Introduction

Noise and vibrations are dynamic powers resulting from phenomena including, for example, movement of vehicles, activity of various machines and tools, rotation movements of machinery, operation of varied building equipment, but also normal activities related to living and working in the building.

Nowadays, the aforementioned noise sources are being installed in buildings where they used to be rare. A good example would be residential rooms. Not long ago, machines were scarce in residential buildings, but ventilation and air conditioning, elevators, heating systems and other household equipment, such as washing machines and kitchen fume hoods, are quickly becoming a regular feature.

Mechanical vibration of buildings and its negative impact on work (life) quality are nearly impossible to avoid. However, incorrect or imperfect solving of the issue during the designing stage may result in expensive repairs, not necessarily resulting in the required remedy.

For the aforementioned reasons, understanding of the causes of vibrations is one of the basic requirements when working with mechanical oscillations, be it computer modeling, practical prevention of these phenomena or protection against their undesired effects on a real construction.

Since it is practically impossible to examine impacts of individual sources of vibration on real objects, we plan to apply simulation modeling to investigate a whole range of possible problematic situations.

2 Real Experimental Model

Within the frame of this project, the Faculty of Civil Engineering at the Brno University of Technology built an experimental building model for measuring of expansion of vibrations through constructions.

The experimental model (see Fig. 1) represents a real model for computer simulations, but it serves for numerous other experiments as well.

Fig. 1. Real experimental model

The described experimental object is a section of a building construction in 1:5 scale.

The model is an object made of bricks with a 100 mm thick ferroconcrete base construction and a 60 mm thick ferroconcrete ceiling construction. Walls are made of full bricks laid on VC mortar.

Precise dimensional characteristics of the model were adapted to our ability to lift the experimental model by available technology, thus expanding its applicability for subsequent measuring. Reinforcing of ferroconcrete sections was also adapted to construction lifting.

See Fig. 2 for a schematic layout and cross-section of the model. Fig. 3 documents building of the experimental object.

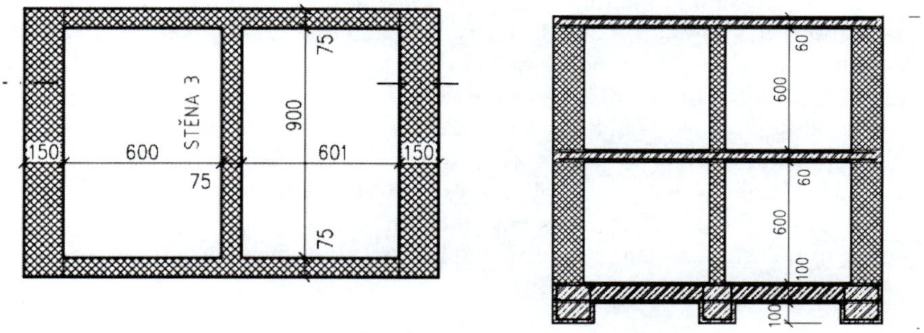

Fig. 2. Layout and cross-section of the experimental model

Fig. 3. Building of the experimental model

To facilitate measuring for subsequent simulations, the model was laid on the foundation ferroconcrete plate without any spring-cushioning.

3 Finite Element Model (FEM)

As we have already said, the aim of the project is to assemble the FEM computer simulation model fully describing real behavior of the experimental model. The computer model will be used to carry out simulations that model loading with various types of technological equipment (e.g. elevator movement in the shaft) and we will investigate expansion of vibrations through the construction and impact on the noise spectrum inside the building.

The computer model was based on the geometry of the real model described above. We first created a 2D model for initial outlining (see Fig. 4). We created the model using the parametric macro for the ANSYS computation software. The macro contains information on the geometry of the experimental object, assigning of material, creation of the finite element mesh, defining of limit conditions, and it also defines the solution method and type of required analysis results. Material properties can only be estimated within certain limits. This is why we identified them. The purpose of identification was to find such material characteristics of the simulation model that would guarantee identical behavior of the experimental and FEM simulation model for selected behavior characteristics. We had to identify four parameters (E_1, E_2, ρ_1 and ρ_2) representing the modulus of elasticity and the density of the two used materials with the aim to find the identical first eign frequency of the simulation and experimental models (see the table below).

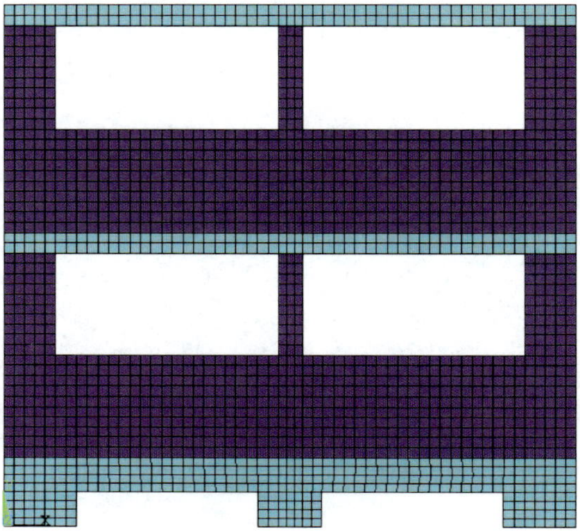

Fig. 4. 2D FEM model

While defining individual materials, the FEM model was simplified compared to the real model. The geometrical model does not include all partial elements in detail - bricks, joints and reinforcements, but it is divided to larger modules to limit the model complexity. The partial modules are assigned one of the two material types.

The computer model assumes solid anchoring of the model in the foundation. We used 8-node SHELL93 elements for model discretization.

4 Experimental Modal Analysis

With regard to the required tuning of the computer model we had to determine dynamic properties of the experimental model. To acquire these values, we used the experimental modal analysis method. The analysis has produced the frequency spectrum, eign shapes for the given spectrum and modal damping for each shape.

The experimental modal analysis method is a very effective method for acquiring of the structural analysis of the given object. It is divide into two parts. In the first stage, „field“ data are gathered, then analyzed and the aforementioned parameters are acquired. We used the Pulse Labshop measuring system, which contains an experimental modal analysis model.

We installed 139 measuring points on the experimental object. We measured responses to vibrations produced by a hammer for all combinations of measuring points and the frequency response functions were calculated from these responses. Data measured in this way were used to establish eign frequencies and eign shapes for the experimental object.

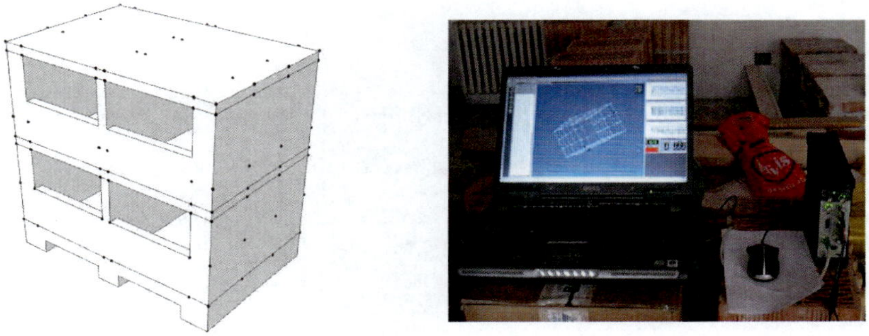

Fig. 5. EMA network on the real experimental object

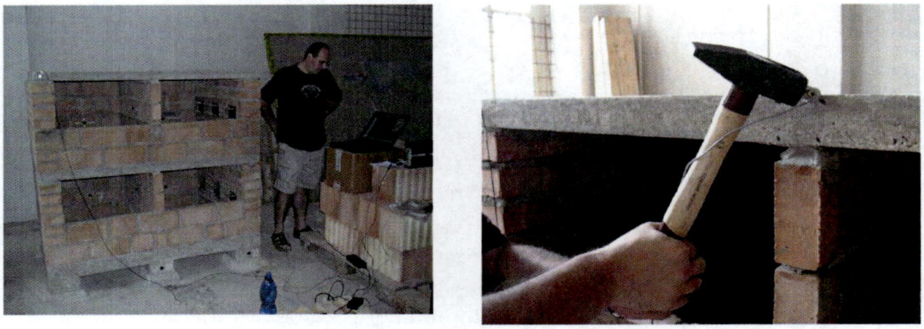

Fig. 6. Measuring of experimental modal analysis

The experimental modal analysis has established the first eign frequency of the experimental model of 25.56Hz, used as a characteristic value for the identification of the FEM model parameters.

5 Identification of the Model Parameters Using the Genetic Algorithm

We applied the Genetic algorithm [2] implemented in the MATLAB environment as the „genetic algorithm and direct search“ toolbox to identify the FEM model parameters. This identification method required the data flow shown on Fig. 7. We used the equation (1) as a cost function of identification problem.

$$err = abs(\frac{f_{sim}^2 - f_r^2}{f_r^2}) \tag{1}$$

Where

f_{sim} is the FEM model first eign frequency

f_{sr} is the first eign frequency measured on the experimental model

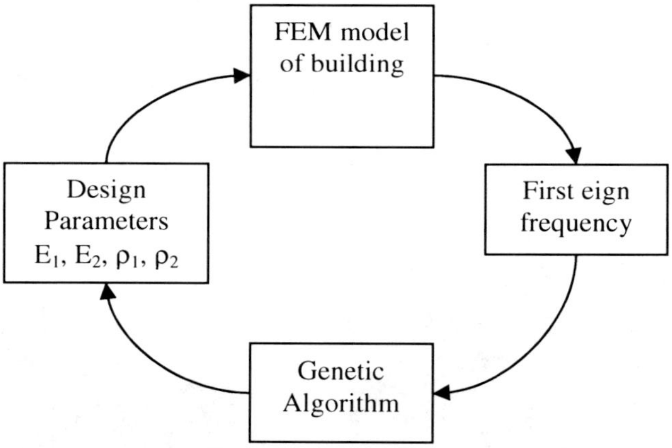

Fig. 7. Data flow

The identification process results in the model parameters shown in the Tab. 1. With these parameters, the FEM model has shown identity at the first eign frequency with the value measured on the real experimental model.

Table 1. Final parameters of FEM model

Ferroconcrete		Brick	
E_1 [GPa]	ρ_1 [kg/m^3]	E_2[GPa]	ρ_2 [kg/m^3]
214.498	2205	141.458	1584

6 Conclusion

The article deals with the identification of parameters of the 2D simulation model of a construction (building) created in the ANSYS FEM environment. We used the genetic algorithm implemented in the Matlab environment and the characteristics measured on the real experimental model for identification. The identification results show high similarity in the behavior of the simulation and experimental models. In the future, we plan expanding of the 2D simulation model to a 3D model, which will give a truer picture of the real construction.

Acknowledgement

Published results were acquired using the subsidization of the Ministry of Education, Youth and Sports of the Czech Republic, research plan MSM 0021630518 "Simulation modeling of mechatronic systems" and project FSI-S-11-15 "Design, testing and implementation of control algorithms with use of nonlinear models of mechatronics systems".

References

[1] Ansys, Inc., Theory Reference, Release 10, Southpointe, 275 Technology Drive, Canonsburg, PA 15317
[2] Goldberg, D.: Genetic Algorithms in Searching, Optimisation and Machine Learning. Addison-Wesley, Reading (1989)
[3] Zaveri, K.: Modal Analysis of Large Structures - Multiple Exciter Systems. Brüel&Kjar, Denmark (1985)
[4] Ewins, D.J.: Modal Testing - Theory, Practice and Application. Research Studies Press Ltd., England (2000)

Modeling and H^{∞} Control of Centrifugal Pump with Pipeline

T. Brezina*, J. Kovar, T. Hejc, and A. Andrs

Faculty of Mechanical Engineering-Brno University of Technology/
Institute of Automation and Computer Science, Brno, Czech Republic

Abstract. The present paper deals with the control of system consisting of induction motor driving the centrifugal pump with pipeline at outlet. For purpose of control is proposed an attitude to develop a robust controller based on H^{∞} method and its use with and without estimator of pipeline states. At the end of the article are presented numerical solutions of this model. The whole system is created and numerically solved in Matlab/Simulink.

1 Introduction

Control of water supply systems in nowadays is current topic. The pump is a fundamental part of such systems; therefore to govern those systems is necessary to develop a control of pump. Such systems are nonlinear with many unpredictable interventions, thereby controller must be "robust" – that means must be able to govern not only in the neighborhood of working point of system. For purpose of controller design is advantageous to develop a mathematical model of system.

The model of pump can be reached by several ways [1, 2]. To create a mathematical model of pump is advantageous to split whole system into several interdependent subsystems. Model of centrifugal pump is therefore divided into two parts - induction motor part and hydraulic part. To obtain a mathematical description of induction motor can be used many methods. One way is to use field coordinate system description [3, 4, 5, 16, 17], which method is used in this model. Hydraulic part of pump is considered as a source of pressure head and load torque [1, 2].

Usually the mathematical model of flow in pipeline contains partial differential equations, which has no closed-form solution. Therefore are solved by numerical methods. Another method is to use a description by hydro-electrical analogy [6, 7, 8].

Control of pump in water supply systems can be obtained by many methods. Most spread solution is to use PID (or PSD) regulator (which is usually cheap to realize but not robust), or regulator designed by pole-placement method [8]. Another approach is to use a robust controller based on H^{∞} method [9, 10, 11].

* Corresponding author.

This article shows an approach to design a robust regulator based on H^{∞} method with and without estimator to control a centrifugal pump with pipeline at outlet.

2 Model of System

System is composed of induction motor driving pump (centrifugal) with pipeline at outlet (Fig. 1a). Whole system can be described by set of partial differential equations. Solution of those equations can be obtained by numerical methods. But if we consider this system as mechatronic system [15, 18], it is beneficial to divide this system into several subsystems (Fig.1b) [1]. Therefore we have to solve (numerically) three subsystems (for inductance motor, hydraulic part of centrifugal pump and for pipeline) of differential equations.

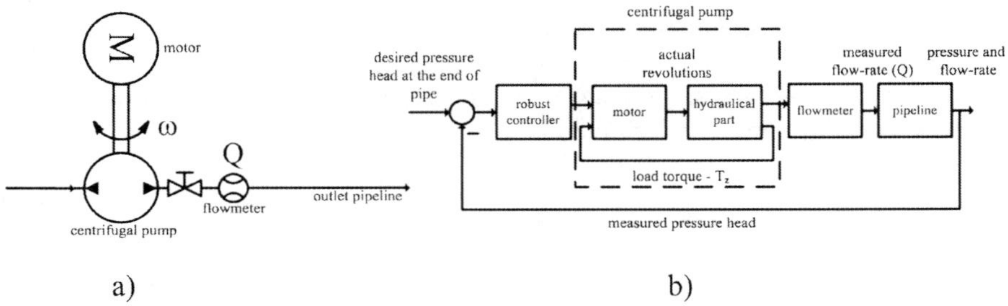

Fig. 1. Scheme of system (a) and block diagram of system without estimator (b)

2.1 Model of Centrifugal Pump

Centrifugal pump is divided, as we mentioned before, into two parts. First one is mathematical model of induction motor; second one is a model of hydraulic part on pump (Fig. 1b).

The mathematical model of induction motor is defined at field coordinate system, which is described at [9, 10, 11, 16]. Motor itself has speed regulator which is based on well known vector control method (Fig. 2) [4].

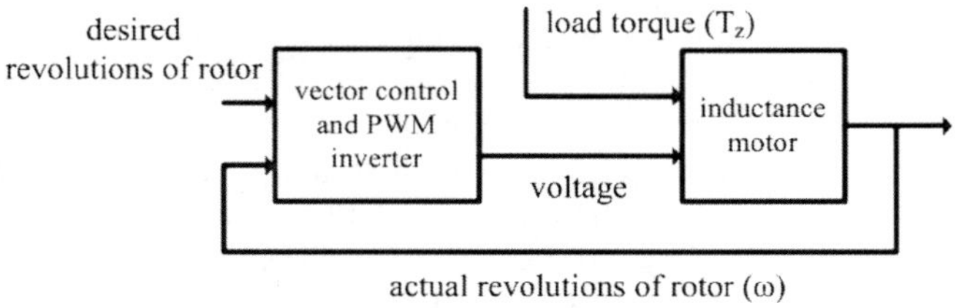

Fig. 2. Block diagram of motor with speed regulator

Hydraulic part of pump is considered (Fig. 3) as a source of pressure head and load torque depending on angular velocity of rotor and flow-rate in pump [1, 2].

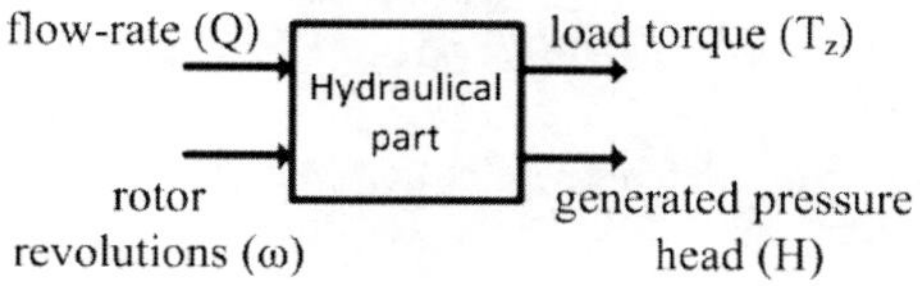

Fig. 3. Block diagram of hydraulic part

Equations describing these facts are (more described at [1, 2]):

$$T_Z = -a_{t2}Q^2 + a_{t1}\omega Q + a_{t0}\omega^2 \tag{1}$$

where T_z is load torque caused by flow through hydraulic part of pump and parameters a_{t2}, a_{t1}, a_{t0} are depending on physical properties of pump defined as [1, 2]:

$$a_{t2} = \rho\left(\frac{r_2\cot(\beta_2)}{A_2} - \frac{r_1\cot(\beta_1)}{A_1}\right) \tag{2}$$

$$a_{t1} = \rho\left(r_2^2 - r_1^2\right) \tag{3}$$

$$a_{t0} = K_n \tag{4}$$

where r_2, r_1 , A_2, A_1 are diameters and areas of inlet and outlet of impeller, ρ is a density of transmitted medium, and parameters β_2, β_1 are angles depending on geometry of impeller blades.

Pressure head generated by pump can be described as [1, 2]:

$$H = -a_{h2}Q^2 + a_{h1}\omega Q + a_{h0}\omega^2 \tag{5}$$

where a_{h2}, a_{h1}, a_{h0} are defined as:

$$a_{h2} = K_s + K_f \tag{6}$$

$$a_{h1} = \sigma_s\left(\frac{r_2\cot(\beta_2)}{gA_2} - \frac{r_1\cot(\beta_1)}{gA_1}\right) - K_sK_d^2 \tag{7}$$

$$a_{h0} = 2K_sK_d - \sigma_s\left(\frac{r_2^2}{g} - \frac{r_1^2}{g}\right) \tag{8}$$

where σ_s is slip factor and constant parameters K_s, K_d, K_f (depending on friction etc.) are described in [1].

Slip factor can be defined as [12]:

$$\sigma_s = 1 - \frac{\sqrt{\cos \beta_2}}{Z^{0.7}} \tag{9}$$

where Z is a number of impeller blades (called Weisner formula).

2.2 Model of Pipeline

Flow of fluid in pipeline is commonly described by partial differential equations, which are usually solved numerically. For purpose of governing it is sufficient to use a hydro-electrical analogy, which is based on this assumption [6, 7, 8, 14]:

$$\begin{aligned} Q &\approx I \\ p &\approx U \end{aligned} \tag{10}$$

where Q is flow-rate, p is pressure (or pressure head), I is current and U is voltage. These equations represent quantitative similarity between qualitatively different systems.

Flow in pipeline can be described by two ordinary differential equations. To achieve a numerical stability of calculation is necessary to divide pipeline into several segment (usually same) and therefore flow in pipeline can be described by (considered is one dimensional flow) as:

$$\begin{aligned} \dot{p}_{j+1} &= \frac{1}{C_H}(Q_{j-1} - Q_j) \\ \dot{Q}_{j+1} &= \frac{1}{L_H}\left(p_{j+1} - R_H Q_{j+1} - p_{j+1}\right) \end{aligned} \tag{11}$$

where p is pressure, Q is flow-rate if pipeline and C_H, L_H, R_H are hydraulic impedances and index i means the order of each segments [7, 8]. Pipeline is in model represented by state space model (A, B, C, D), where state vector is represented as:

$$x = \left[p_1, Q_1, p_2, Q_2, \ldots\ldots, p_n, Q_n\right]^T \tag{12}$$

where n is number of segments and must satisfy a condition of stability (meaning numerical stability) [8]:

$$\frac{lf}{c_S} < n < 10\frac{lf}{c_S} \tag{13}$$

where c_s is speed of sound in transmitted medium, l length of pipeline and f denotes a frequency of pressure change in medium. In this model is number of segments 320.

3 H^{∞} Controller

Controlled variable is pressure head at the output of pipeline. Change of this pressure head is due to change of rotor revolutions (with delay depending on the speed of sound in medium and length of pipeline).

Robust controller based on H^{∞} method is proposed by well known method described at [9, 10, 11, 13]. Due the complementary sensitivity of system (major part is an induction motor) was chosen those weighed functions:

$$W_2 = \frac{40(s+200)}{s} \tag{14}$$

$$W_2 = 0.4 \tag{15}$$

and robust regulator is:

$$K = W_2 K_{\infty} W_1 \tag{16}$$

where K_{∞} is controller gains.

3.1 Estimator

Response time of pressure head at the end of pipeline on change of rotor revolutions depends on length of pipeline and speed of sound in transmitted medium. This response delay cause undesirable overshoot of pressure head at the end of pipeline. For this reason is necessary to use an estimator of pipeline states (Fig. 4) (is necessary to estimate pressure head at the end of pipeline with time advance).

Estimated system (pipeline) has (in this case) 320 states and estimator with this number of states cannot be proposed by ordinary methods. Because this estimator was modeled as shorter pipeline (14m) – response of pressure head at the end of pipeline on change of rotor revolutions is therefore faster.

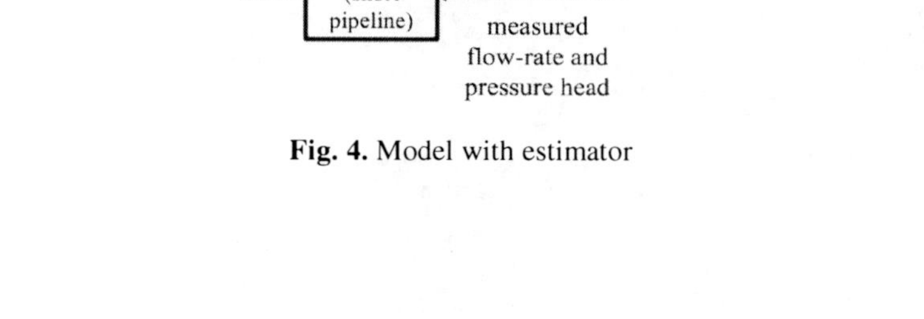

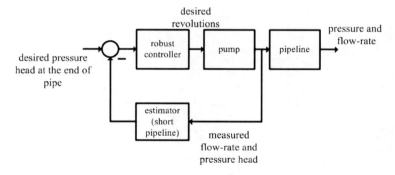

Fig. 4. Model with estimator

4 Results

The model was numerically solved with these parameters (Tab. 1):

Table 1. Table of model parameters

Rotor resistance	0.38 Ω	Stator resistance	0.21 Ω
Stator inductance	$0.31*10^{-3}$ H	Rotor inductance	$0.64*10^{-3}$ H
Magnetizing inductance	$4.1*10^{-3}$ H	Number of poles	6
Length of pipeline	800m	Rated power of motor	1.25kW
Desired value of pressure head at the end of pipeline	10m	Transmitted medium	Water
Flow-rate at required state	0.18 m^3/m	Considered type of flow in pipeline	Laminar

The examples of step responses of the controlling system with and without estimator are shown at Fig. 5 and Fig. 6. An advantage of robust controller is that we can regulate a motor from stop state to required state (approximately 240 rotor revolutions per minute). Disadvantage of this type of regulation is that the target state is achieved after approximately 2 seconds.

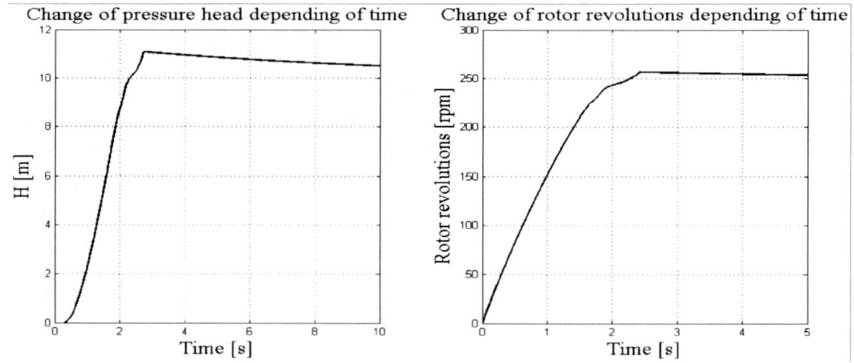

Fig. 5. Numerical results of model without estimator. The left graph shows pressure head at the end of pipeline (desired value is 10m), right shows rotor revolutions

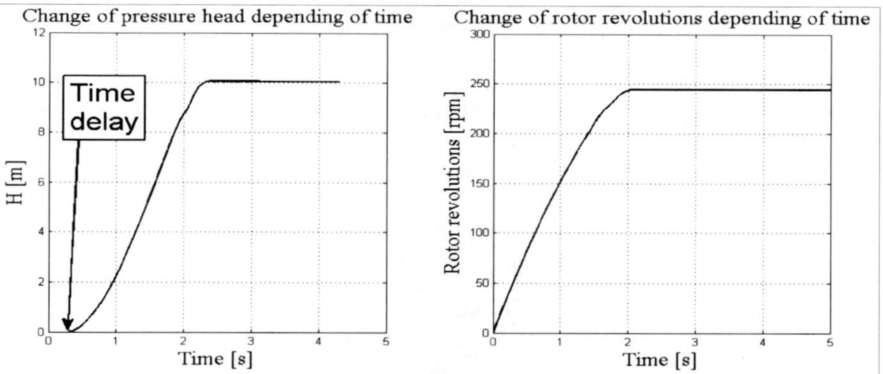

Fig. 6. Numerical results of model with estimator. The left graph shows pressure head at the end of pipeline (desired value is 10m), right shows rotor revolutions

5 Conclusions

In this paper a control based on H^{∞} method for centrifugal pump with pipeline on output was developed. Model of whole system was created and numerically solved in Matlab/Simulink. Parameters of simulation are shown in Tab. 1.

Numerical results of control are shown at Fig 5. and 6. These results indicate that if the length pipeline is 800m than the attitude of control without estimator leads to overshoot of pressure head at the end of pipeline. Therefore is appropriate to use an estimator. The restriction of this control is that time for which the system achieved the desired value of pressure head at the end of pipeline is relatively long. This approach of control can be used for next improvement.

Acknowledgements

This work is supported from research plan MSM 0021630518 Simulation modeling of mechatronic systems, FSI-J-11-39 and FSI-S-11-23.

References

[1] Kallesøe, C.S.: Fault Detection and Isolation in Centrifugal Pumps, Ph.D. dissertation, Department of Control Engineering, Aalborg University (March 2005)
[2] Kallesøe, C.S., Cocquempot, V., Izadi-Zamanabadi, R.: Model Based Fault Detection in a Centrifugal Pump Application. IEEE Transactions on Control Systems Technology 14(2), 204–215 (2006)
[3] Krause, P.C., Wasynczuk, O., Sudhoff, S.D.: Analysis of Electric Machinery and Drive Systems, 2nd edn. Wiley Interscience, IEEE Press (2002)
[4] Ozpineci, B., Tolbert, L.M.: Simulink implementation of induction machine model - a modular approach. In: IEEE International Electric Machines and Drives Conference, IEMDC 2003, June 1-4, vol. 2, pp. 728–734 (2003)

[5] Mezyk, A., Trawinsky, T.: Modeling of drive system with vector controlled induction machine coupled with elastic mechanical system. In: Recent Advances in Mechatronics, pp. 248–252. Springer, Heidelberg (2007)
[6] Tao, W.Q., Ti, H.C.: Transient analysis of gas pipeline network. Chemical Engineering Journal 69(1), 47–52 (1998)
[7] Kovar, J., Brezina, T.: Model of pipeline with pump for predictive control. In: Annals of DAAAM for 2010 & Proceedings of the 21st International DAAAM Symposium "Intelligent Manufacturing & Automation: Focus on Interdisciplinary Solutions", pp. 1363–1364 (2010)
[8] Brezina, T., Kovar, J., Hejc, T.: Modeling and Control of System with Pump and Pipeline by Pole Placement Method (in press, 2011)
[9] El-Zobaidi, H., Leigh, J.R.: Robust control of an induction motor: the H^{inf} comprime factors approach. In: International Conference on Control, Control 1994, vol. 2, pp. 1538–1541 (March 1994)
[10] Chen, B.-S., Chang, Y.-T.: Fuzzy State-Space Modeling and Robust Observer-Based Control Design for Nonlinear Partial Differential Systems. IEEE Transactions on Fuzzy Systems 17(5), 1025–1043 (2009)
[11] Doyle, J.C., Glover, K., Khargonekar, P.P., Francis, B.A.: State-space solutions to standard H_2 and H_∞ control problems. IEEE Transactions on Automatic Control 34(8), 831–847 (1989)
[12] Frenk, A.: A slip factor calculation in centrifugal impellers based on linear cascade data (2005), on-line source,
`http://ae-www.technion.ac.il/~jetlab/6thsmp/frenk.pdf`
[cited on 4.5.2011]
[13] Kwakernaak, H.: Robust control and H[infinity]-optimization–Tutorial paper. Automatica 29(2), 255–273 (1993)
[14] Nicolet, C.: Hydroacoustic modelling and numerical simulation of unsteady operation of hydroelectric systems (2007), on-line source,
`http://library.epfl.ch/theses/?nr=3751` (cited on 31.3.2011)
[15] Hadas, Z., Vechet, S., Singule, V., Ondrusek, C.: Development of Energy Harvesting Sources for Remote Applications as Mechatronic systems. In: Proceedings of EPE-PEMC 2010 - 14th International Power Electronics and Motion Control Conference, Ohrid, Macedonia, pp. 13–19 (2010)
[16] Maga, D., Hartansky, R., Manas, P.: The Examples of Numerical Solutions in the Field of Military Technology. Advances in Military Technology 1(2), 53–66 (2006)
[17] Fabo, P., Pavlikova, S.: Gsim - Software for Simulation in Electronics. In: Proceedings of EPE-PEMC 2010 - 14th International Power Electronics and Motion Control Conference, pp. S414-S417 (2010)
[18] Fabo, P., Siroka, A., Maga, D., Siroky, P.: Interactive Framework for Modeling and Analysis. Przeglad Elektrotechniczny 83(11), 79–80 (2007)

Optimization of a Heat Radiation Intensity on a Mould Surface in the Car Industry

J. Mlýnek[1,*] and R. Srb[2]

[1] Department of Mathematics and Didactics of Mathematics,
Faculty of Sciences, Humanities and Education – Technical University of Liberec,
Studentská 2, Liberec 1, 461 17, Czech Republic
[2] Institute of Mechatronics and Computer Engineering, Faculty of Mechatronics,
Informatics and Interdisciplinary Studies – Technical University of Liberec,
Studentská 2, Liberec 1, 461 17, Czech Republic

Abstract. This article is focused on the process of the heat radiation intensity optimization on an aluminium mould surface intended for the production of artificial leather in the car industry (e.g. the artificial leather on a car dashboard). The inside of the mould surface is sprinkled with a special powder and its outside is heated by infra heaters located above the mould, up to a temperature of 250°C. It is necessary to carry out the configuration optimization of the infra heater locations above the mould in such way that the heat radiation intensity on the mould surface is approximately the same. The procedure of configuration optimization of the infra heater locations by use of a genetic algorithm is described in this paper. Experimental measured values for the heat radiation intensity in the surroundings of an infra heater are used for the calculation procedures.

1 Introduction

This article is focused on the technical problem of aluminium moulds warming for the production of artificial leather in the car industry. Let us think about a mould weighing approximately 300kg. The mould is warmed using infra heaters of the same capacity, located over the mould at the distance of between 5 and 30cm form mould circumference. The inside of the mould is sprinkled with a special powder and is subsequently warmed to a temperature of 250°C. It is necessary to ensure the same heater radiation intensity (within the given tolerance) on the whole circumference and the same material structure and color of the artificial leather produced will thereby be obtained. Infra heaters have a tubular form and their length is usually between 15 and 30cm. The heater is equipped with a mirror located above radiating tube, which reflects back heat radiation in the adjusted direction (see Figure 1).

* Corresponding author.

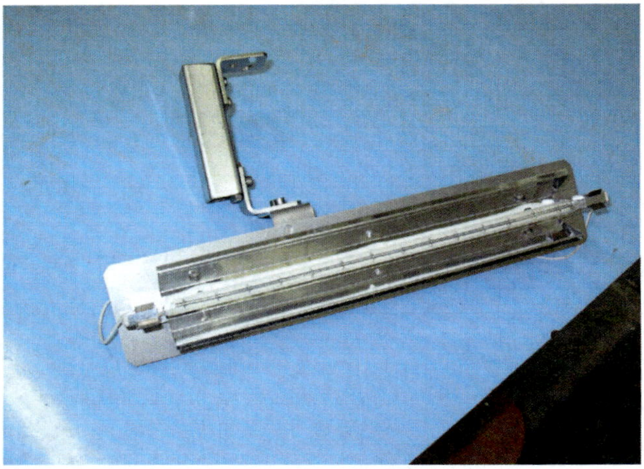

Fig. 1. Philips infra heater with a 1000W capacity

Hence, we can't use features of point heat radiation. We also don't know heater distribution function from the heater manufacturer. Consequently, we measured experimental values for the heat radiation intensity in the surrounding of an infra heater.

The setting of the heater locations in production was adjusted upon the basis of the experience of technicians and the ensuring of the same radiation intensity on the whole mould circumference was achieved by way of successive refinements. However, the adjusting of the heater location and fixation of the heaters (usually from 50 to 150 depending up mould size) in construction in this way is labour-intensive and time-consuming.

The surface of mould is described by its elementary surfaces. A producer uses moulds of different sizes and moulds are often very rugged. We keep track of the following types of possible collisions when locating heaters: one heater doesn't radiate on a second heater more than the given limit, one heater has sufficient distance from second heater, a heater has sufficient distance from the mould surface and is over the surface. The technical problem of optimization is rather complicated, we used a genetic algorithm as the method of reaching a solution.

A model of the radiated mould by infra heaters was proposed for this reason and will be described in more details in the next chapter.

2 Model of Heat Radiation on the Mould Circumference

We will describe a model of heat radiation on the mould surface by use of infra heaters. We will assume the representation of the heaters and mould in a 3-dimensional Euclidean space E_3 with a coordinate system $(O; x_1, x_2, x_3)$.

The mould circumference is given by the elementary surfaces p_j, where $1 \leq j \leq N$, i.e. N elementary surfaces are defined in all of them. $\bigcup_{1 \leq j \leq N} p_j = P$, is true, where P denotes the total circumference of the mould and $\operatorname{int} p_i \cap \operatorname{int} p_j = \emptyset$ for $i \neq j$. Every elementary surface is represented by the following parameters:

- its centre of gravity $T_j = \left(x_1^{T_j}, x_2^{T_j}, x_3^{T_j} \right)$,
- the outer normal vector $v_j = \left(x_1^{v_j}, x_2^{v_j}, x_3^{v_j} \right)$ in point T_j,
- the area of elementary surface s_j [m^2].

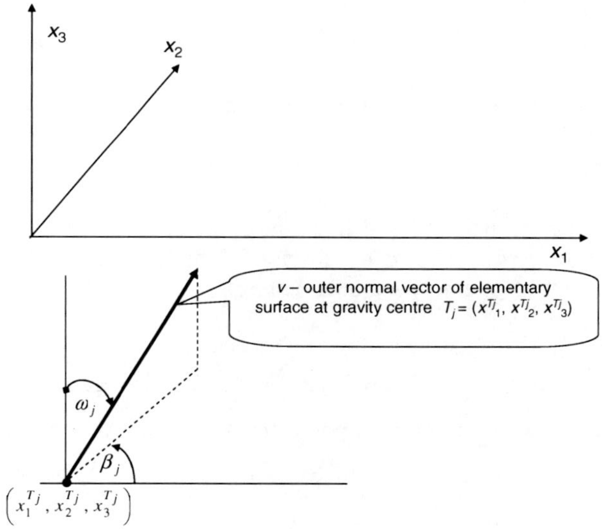

Fig. 2. Parameters determining the elementary surface p_j.

We suppose that the outer normal vector v_j at the point T_j has a unit length. It is then possible to unambiguously enter the vector v_j through angles β_j and ω_j, where positively oriented angle $0 \leq \beta_j < 2\pi$ determines the size angle of the positive part of axis x_1 and the vertical projection of the outer normal vector to a plane given by axes x_1 and x_2 (ground plane). The angle ω_j is determined by the outer normal vector v_j and positive part of axis x_3 (see Fig. 2), where $0 \le \omega_j \le \pi/2$. Every elementary surface p_j is defined then by 6 parameters:

$$p_j : \left(x_1^{T_j}, x_2^{T_j}, x_3^{T_j}, \beta_j, \omega_j, s_j \right), \quad 1 \le j \le N . \tag{1}$$

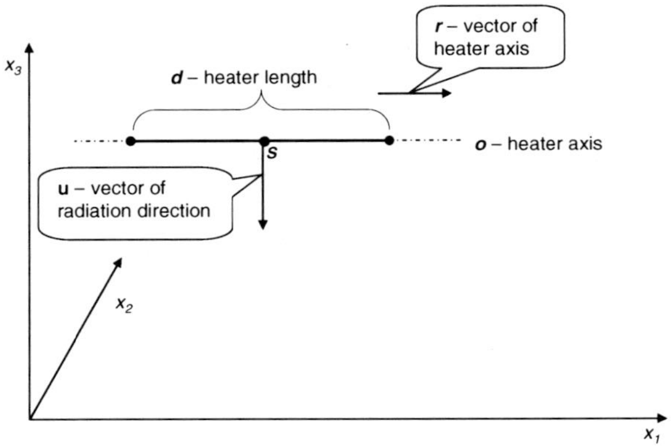

Fig. 3. Schematic representation of the heater.

We assume that all heaters used have the same capacity and are of the same type. Every heater is represented by abscissa d[m] in length. The location of a heater is described by the following parameters:

- coordinates of the heater centre $S = [\, x_1^S, x_2^S, x_3^S \,]$,
- radiation direction vector $u = \left(x_1^u, x_2^u, x_3^u \right)$, we assume a unit vector length u and we assume component x_3^u is negative, i.e. the heater radiates "down", then the coordinate x_3^u of vector u is explicitly allocated,
- the vector of the heater axis $r = \left(x_1^r, x_2^r, x_3^r \right)$: we assume that the length of the vertical projection of vector r to a plane given by the axes x_1 and x_2 (ground plane) is 1. This projection and positive part of axis x_1 define angle φ, where $0 \le \varphi < \pi$, then $x_1^r = \cos\varphi$, $x_2^r = \sin\varphi$, component x_3^r is explicitly defined (vector u and r are orthogonal).

The location of every heater Z is described by the following 6 parameters:

$$\text{Z:} \left(x_1^S, x_2^S, x_3^S, x_1^u, x_2^u, \varphi \right). \tag{2}$$

An infra heater is schematically demonstrated in Fig. 3. The location of M heaters is described by $6M$ parameters.

3 Radiation Intensity Calculation

In this chapter we will describe a procedure for radiation intensity calculation on a particular elementary surface of a mould for a given configuration of heaters. The parameters of the elementary surfaces are entered.

We denote L_j a set for all the heaters radiating on the j-th elementary surface for the defined location of the heaters, $1 \le j \le N$. Furthermore, we will denote I_{jl} [W/m²] radiation intensity of l-th heater on the j-th elementary surface. The total radiation intensity I_j on the j-th elementary surface is defined by the relation (in more detail e.g. in [3])

$$I_j = \sum_{l \in L_j} I_{jl} \ . \tag{3}$$

We denote an I_{opt} recommended radiation intensity on the surface of the mould by the producer. We determine the difference F_j of the radiation intensity on the j-th elementary surface I_j ($1 \le j \le N$) from optimal intensity I_{opt} upon the basis of the relation

$$F_j = \left| I_j - I_{opt} \right| \tag{4}$$

and the average difference of radiation intensity F given the relation

$$F = \frac{\sum_{j=1}^{N} F_j s_j}{\sum_{j=1}^{N} s_j} \ . \tag{5}$$

We recall that s_j denotes the area of the elementary surface p_j.

We will use experimental measured values for the heat radiation intensity in the surroundings of an infra heater to determine values I_{jl} in relation (3). We will use the linear interpolation of a function of 5 variables (corresponding to the first 5 parameters of the elementary surface in relation (1), in more detail e.g. in [1], page 148).

4 Optimization of Heater Locations by the Use of a Genetic Algorithm

We note that the terms and relations used in this chapter are described in more detail e.g. in [2]. We will proceed with the subsequent method of optimization of heater locations by the genetic algorithm method. We suppose that M heaters is used for the heat radiation of a mould surface. The location of every heater is defined by 6 real parameters according to relation (2). Then $6M$ parameters are necessary to the definition of the locations of all the heaters. One chromosome will represent one individual (one possible location of heaters), particular genes of the chromosome will represent the determining parameters of heater locations. A population will contain Q individuals. Continuous generated individuals will be saved in the matrix $\boldsymbol{A}_{Q \times 6M}$. Every row of matrix $\boldsymbol{A}$ represents one individual. Our aim is to find such individual y, that radiation intensity on the mould surface approaches the value I_{opt} recommended by the producer, i.e. we seek individual $y_{\min} \in B$ satisfactory condition

$$F(y_{\min}) = \min_{y \in B} F(y), \tag{6}$$

where $B \subset E_{6M}$ is searched space and function F is defined by relation (5).

Schematic description of a genetic algorithm:
1/ the creation of a specimen and a initial population of individuals,
2/ the evaluation of the all individuals, the execution of all individuals sorted according to their fitness F
3/ **while** a condition of termination isn't fulfilled **do**
  if operation crossover is randomly chosen **then**
    randomly select a pair of parents,
    execute of crossover
  else
    randomly select an individual,
    execute of mutation
  end if,
  integration and evaluation (fitness) of new calculated individuals ,
  sorting of all individuals in accordance with evaluations ,
  storage only first Q individuals with the best evaluation for subsequent calculation
 end while,
4/ output of the best individual found – 1. row of matrix $\boldsymbol{A}$.

The setting of the starting individual (specimen) is chosen in such a way that all the centres of heaters create nodes in a regular rectangular network and lie over the mould, in parallel plane with the plane defined by axes x_1 and x_2. We will consequently generate next Q -1 individuals by random modifications of genes values.

The following functions and operations are used in the schematic description of a genetic algorithm.

Evaluation function (*fitness*) F for individuals y is defined by relation (5).

Testing procedure against collisions: one heater doesn't radiate on a second heater more than the given limit, one heater has sufficient distance from the second heater, a heater has sufficient distance from mould surface and is over the surface. If a collision exists, individual y is penalized and consequently expelled from population.

Individual selection to crossover or to mutation is implemented on the principle *fitness-proportionate selection*. Let us consider a population with Q individuals (i.e. $y_1, y_2, \ldots, y_Q$). We will specify

$$G = \sum_{j=1}^{Q} \frac{1}{F(y_j)}, \tag{7}$$

where F is the evaluation function defined by relation (5). We suppose that the radiation intensity of heaters on the surface of the mould isn't wholly uniform when the location of heaters are represented by individual y_j, i.e. $F(y_j) \neq 0$ for $j = 1, 2, \ldots, Q$. The probability of selection of individual y_j is defined by relation

$$p(y_j) = \frac{1}{F(y_j) \cdot G}. \tag{8}$$

We generate a random number $r \in <0, 1>$ in every selection of an individual from the population. We will select an individual y_i if and only if the following holds true

$$\sum_{j=1}^{i-1} p(y_j) < r \leq \sum_{j=1}^{i} p(y_j) \text{ for } i = 1, 2, \ldots, Q, \tag{9}$$

where we put $\sum_{j=1}^{i-1} p(y_j) = 0$ for $i = 1$.

During *operation crossover* we do only *one point crossover* and modify variants of crossover. Two children $(c_1, \ldots, c_{6M})$ and $(d_1, \ldots, d_{6M})$ originate from parents $(a_1, \ldots, a_{6M})$ and $(b_1, \ldots, b_{6M})$. We denote $m = |a_i - b_i|$. If $a_i \geq b_i$ then $a_i = a_i + m, b_i = b_i - m$ else $a_i = a_i - m, b_i = b_i + m$ for $i = 1, 2, \ldots, Q$.

During the operation of a mutation of an individual $(a_1, \ldots, a_{6M})$ we select randomly $\delta \in <-\varepsilon, +\varepsilon>$, where $\varepsilon > 0$. We generate randomly natural number $p \in <1, 6M>$ and we obtain two new individuals $(c_1, \ldots, c_{6M})$ and $(d_1, \ldots, d_{6M})$ under the rule $c_i = d_i = a_i$ for $i = 1, \ldots, 6M$ and $i \neq p$; $c_p = a_p + \delta$, $d_p = d_p - \delta$.

5 Practical Example of Radiation Intensity Optimization

We will describe the results of the radiation intensity optimisation calculation on the mould surface for a given aluminium mould in this chapter. A software application was programmed in the Matlab language. 20 heaters were used for the mould radiation. The parameters of all the heaters were the following: producer Philips, capacity 1600W, length 15cm, width 4cm.

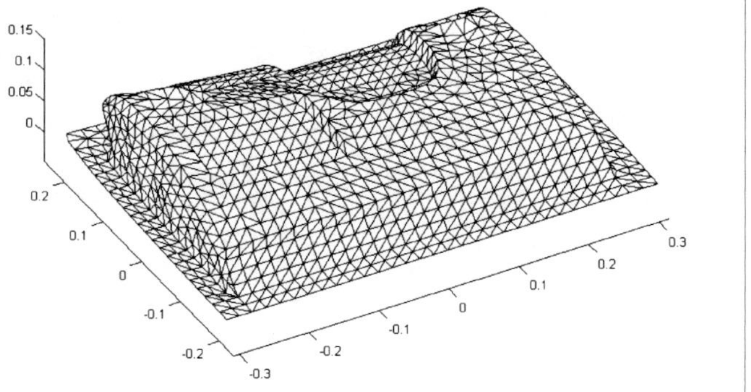

Fig. 4. Aluminium mould

The calculation was accomplished for an aluminium mould displayed in Fig. 4. The mould is described by 2187 triangular elements. The size of the mould is 0,6x0,4x0,12[m^3]. The recommended radiation intensity by a producer is $I_{opt} = 44$ [kW/m^2].

We will choose a parallel plane ρ with the plane given by the axes x_1 , x_2. The plane ρ is at distance of 10 cm over the maximum of values x_3^{Tj} of the all the centres of gravity T_j of the elementary surfaces. Firstly, we will construct a specimen y_1 in keeping with genetic algorithm described in previous chapter. We will create a regular rectangular network in the plane ρ and we will locate centres of heaters to network nodes. Vectors of heater axes will be parallel with axis x_1. The average difference of radiation intensity of specimen y_1 is $F(y_1) = 24{,}74$ [kW/m^2]. Consequently we will construct an initial population of individuals by modifying parameters of the specimen. The population will contains 30 individuals (Q = 30). Now we will apply the genetic algorithm described in the previous chapter to get the new individuals. The finding of individual $y_{\min}$ defined by relation (6) isn't realistic in practice. But we obtain the optimized individual $y_{loc\,\min}$ (local minimum), the determination of the individual $y_{loc\,\min}$ and the size of value $F(y_{loc\,\min})$ dependent on the number of genetic algorithm iterations executed (see Fig. 5).

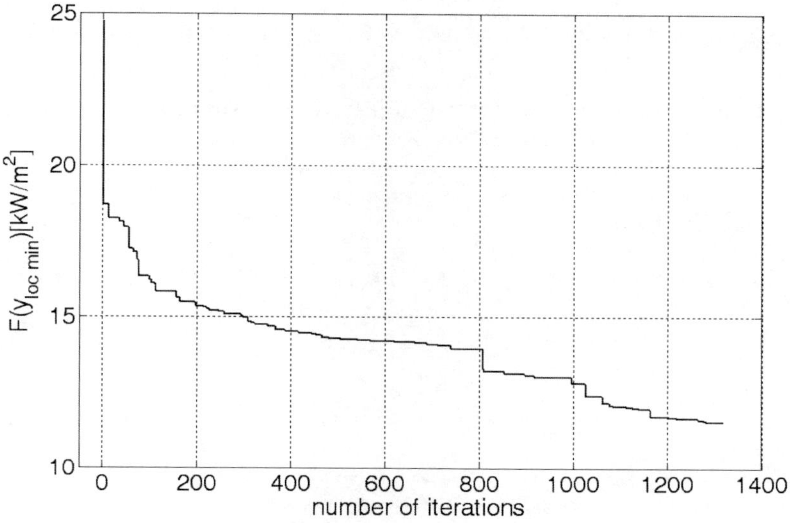

Fig. 5. Dependence of value $F(y_{locmim})$ on the number of iterations

We see value $F(y_{loc\min})$ of the best individual found is decreasing in dependence of genetic algorithm iterations number. We finished the calculation after 1318 iterations. Heater locations parameters (corresponding to relation (2)) of the finding individual $y_{loc\min}$ are as follows:

(-0.3029; -0.2137; 0.1162; 0.0133; 0.4185; 0.4256)
(-0.1581; -0.2226; 0.2084; 0.0007; -0.3399; 2.0005)
(0.0032; -0.2199; 0.1355; -0.2228; 0.0888; 0.5117)
(0.1027; -0.2194; 0.2049; 0.2182; 0.0109; 2.0339)
(0.2617; -0.2120; 0.2073; 0.0173; -0.3274; 0.5337)
(-0.3116; 0.0141; 0.2113; -0.3010; 0.0423; 2.4640)
(-0.1447; -0.0625; 0.2060; 0.0843; -0.0310; 0.2742)
(0.0087; -0.0637; 0.2053; 0.4695; -0.2895; 0.7380)
(0.1532; -0.0727; 0.2046; 0.0358; -0.3338; 1.5319)
(0.3058; -0.0713; 0.2091; 0.0142; -0.1271; 2.1253)
(-0.3004; 0.0789; 0.2094; -0.3923; 0.2563; 1.8799)
(-0.1481; 0.0776; 0.2027; 0.0453; -0.0429; 0.6662)
(-0.0069; 0.0866; 0.2031; -0.0232; 0.3357; 2.9576)
(0.1593; 0.0863; 0.2100; -0.1178; 0.0868; 1.6483)
(0.3208; 0.0803; 0.2077; 0.2078; 0.0148; 1.3284)
(-0.2995; 0.2275; 0.2051; -0.0387; 0.0986; 1.8883)
(-0.1410; 0.2365; 0.2102; 0.0219; 0.2943; 0.4439)
(-0.0044; 0.2245; 0.1517; 0.0167; 0.4032; 1.6617)
(0.1547; 0.2290; 0.2073; -0.2672; 0.1590; 0.6964)
(0.3083; 0.2282; 0.1290; 0.1671; 0.0357; 0.7507).

The average difference of radiation intensity of individual $y_{loc\,\min}$ is $F(y_{loc\,\min}) = 11{,}21$ [kW/m^2]. We recall that $F(y_1) = 24{,}74$ (where y_1 is specimen) and heater locations of y_1 were relatively good.

The example described is illustrative. It is obvious it will be suitable to increase the number of heaters and then accomplish optimization.

Acknowledgement

This work was supported by MPO project No. FR-TI1/266.

References

[1] Antia, H.M.: Numerical Methods for Scientists and Engineers. Birkhäuser, Basel (2002)
[2] Affenzeller, M., Winkler, S., Wagner, S., Beham, A.: Genetic Algorithms and Genetic Programming. Chapman and Hall/CRC, Boca Raton (2009)
[3] Linhard IV, J.H., Linhard V, J.H.: A Heat Transfer Textbook, `http://web.mit.edu/lienhard/www/ahtt.html`

Machine-Tool Frame Deformations and Their Mechatronic Monitoring during Machine Operation

L.W. Novotný* and J. Marek

TOSHULIN, a.s., Wolkerova 845,
Hulín, 768 24, Czech Republic

Abstract. Based on experience of TOSHULIN company, the authors have elaborated studies and descriptions of possibilities how to measure motions of frames at machining centres with great loading capacity (approximately tens of tons); these motions result from heating or from mechanic loading. This paper summarizes the problem substance and it shows the solution way, stated knowledge and commercial use.

1 Introduction

We know nothing about thermal deformations of the machine frame (i. e. bed, columns, cross rail, etc.), about their size and especially about their shape, until measuring and simulation – computational modelling – are performed [2]. We only know their outside reflection which was measured e. g. at the machine specified for many machines [1]. $\Delta H'$ was measured and it is expected that $\Delta H'_t = \Delta H_t$. Deformation $\Delta H''_t$ is not known and it can have various appearance as it is shown in the figure. There are some possible ways available how to perform measuring.

Measuring possibilities:

- optical measuring,
- piezoelectric measuring,
- tensometric measuring,
- capacity measuring, etc.

Requirements put on measuring and measuring technique:

- possible in the technical aspect
- verifiable
- affordably priced (maybe also e. g. an option),
- feasible in the assembly aspect by the existing staff.

* Corresponding author.

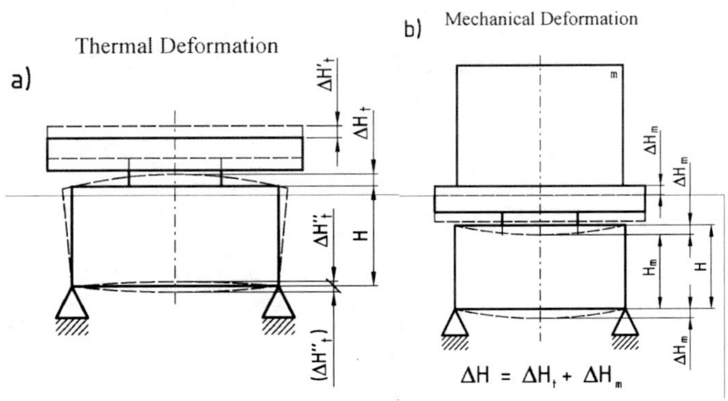

Fig. 1. Example – behaviour of machine frame deformations during operation [1]

Having analysed the measuring methods, tensometric measuring and measuring by proximity switches were selected to be feasible ones. Measuring consists in deformation sensing by means of a sensor which shall be enough sensitive, precise and it shall be in the sufficient distance for a heat source. Deformation transfer occurs by a so-called deformation carrier (rod), which has considerably bigger rigidity compared with the measuring sensor and thermal dilatation is smaller than cast iron dilatation. Measuring by proximity switches uses sensors which are located on a suitably selected place regarding to the measured point.

The realization process is described here:

- negotiation and approval by responsible company's bodies,
- selection of the machine where the pilot project has been performed,
- issue of the internal order p54,
- contract conclusion with subdeliverers and purchase of necessary sensors, etc.,
- designing work (mechanical design and electrical design),
- production and assembly,
- technical experiment itself,
- evaluation and implementation in internal processes and commercial application.

2 Mechanical Deformation Issue

Mechanical bed deformations ΔH_m caused by the influence of a workpiece, of cutting force, etc. are shown in Fig. 2. Fig. 3 shows the FEM analysis of the bed designated as 1030/0001A04 [1]. This analysis confirms that the bed behaves as a thick beam. The total deformation is $\Delta H = \Delta H_t + \Delta H_m$.

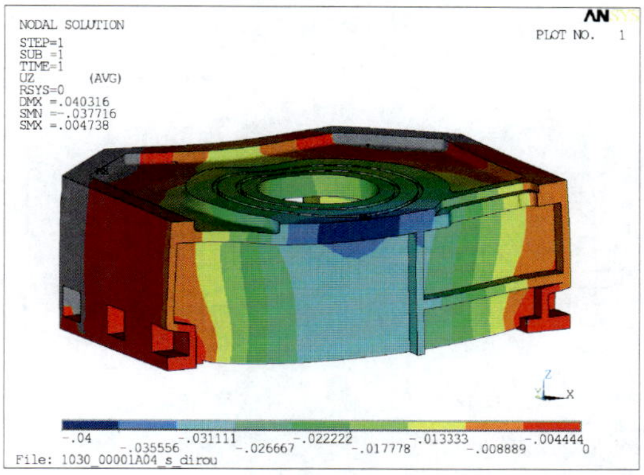

Fig. 2. Example – machine frame deformations during operation determined by FEM [1]

3 Design of the Measuring Sensor Location

Fig. 3 shows the possible location of sensors considering the machine as well as considering its foundation.

Variant No. 1 – it senses thermal deformations well; however we do not know anything about the deformation direction ΔH''t, which makes this measuring indeterminate. The same thing is valid for the directions X and Y (which is valid for all variants). It is assumed that mechanical deformations are not sensed – motions of the upper surface and of the lower surface are identical ones (Fig. 3).

Variant No. 2 – it senses thermal deformations as well as mechanical deformations and the particular deformations can be differentiated. However, this variant is the more expensive one.

Variant No. 3 – it senses deformations; however, it is not possible to differentiate, whether these deformations are thermal or mechanical ones. The change is shown (measured) only.

In order to obtain in-process deformation check, in all cases it would be necessary to input the new initial measuring state, after a workpiece or the pallet is loaded.

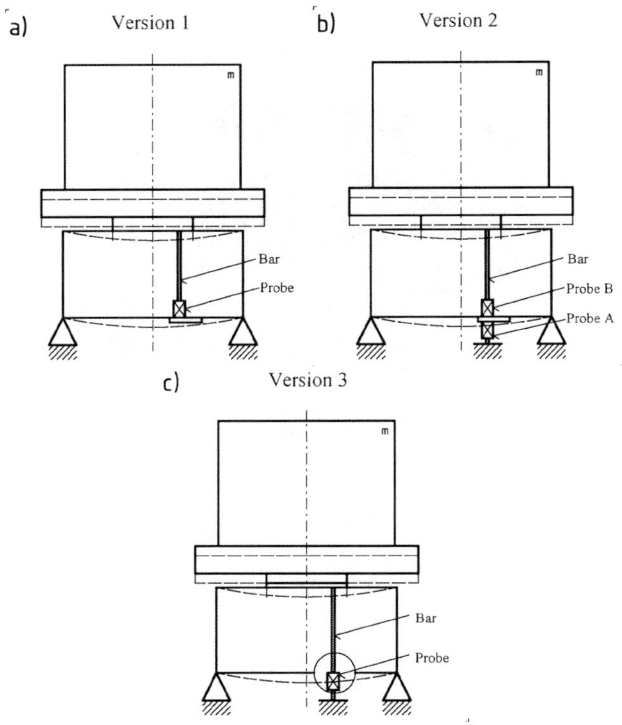

Fig. 3. Three variants of deformation measuring and location of sensors [1]

4 Risks and Risk Analysis of Measuring (Experiment)

A number of risks is connected with measuring. The part written below mentions some items of risk analysis, which was performed on the study and on preparation of the particular experiment. Possible risks at measuring and realization are:

- Insufficient difference between the thermal dilatation coefficient of the frame and the thermal dilatation coefficient of the rod (deformation carrier) causes that the measuring will be not precise and sensitive enough.
- Tendency to measuring of vibrations – high sensitivity of the tensometric sensor will result in sensing the frame vibrations – distortion of the measuring.
- Insufficient sensitivity of the measuring sensor will result in the fact that deformations will not be measurable.
- Some of the above-mentioned presumptions is not true.

5 Production, Assembly and Experiment Itself

This measuring is determined especially for machines at very demanding customers. It is necessary to set the new initial sensor state every time, when the loading situation is changed, most of all at the workpiece exchange – this is a software matter.

The measured deviation is entered to the PLC analog input, the PLC program performs evaluation and the resulting value is added to the current position value of the Z-axis. This occurs using the control system option "thermal compensation".

The price for subdeliveries to the measuring equipment is approximately 200000 Kč. The most problematical rod, which is also the most demanding one in the manufacturing aspect, seemed to be the rod made of carbon composite, invar, etc. However, this issue was solved successfully with deliverers. The other necessary components can be manufactured as a standard at TOSHULIN by its own means [1].

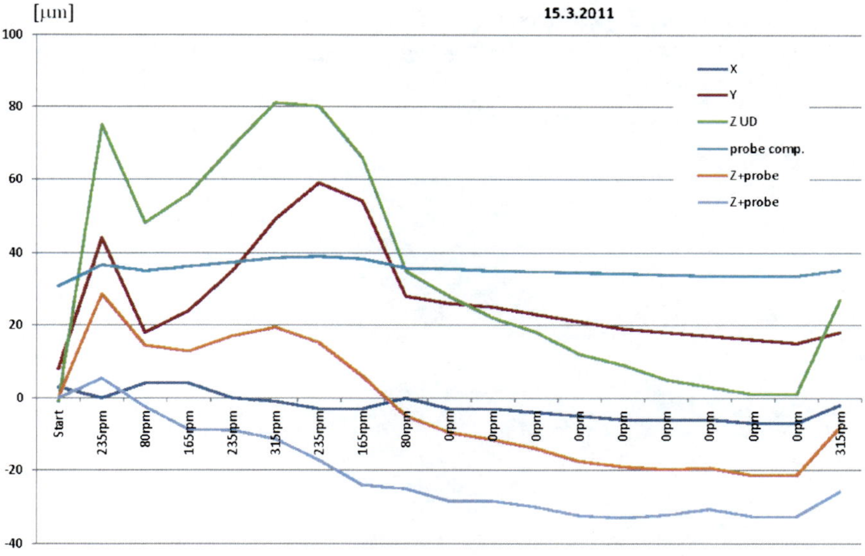

Fig. 4. Dependences of machine-tool deformations, measured experimentally [1]

The designed sensor directly includes tensometric elements and elements for measuring by proximity switches in the other case, calibration is performed in PLC; therefore, there are no special demands put on training of assembly technicians. The assembly process is described in the separate assembly instructions.

The enclosed figure shows a graph of machine thermal deformations in a time period in dependence on machine loading. The figure shows clearly that machine frame deformations have an important influence on machine manufacturing accuracy.

6 Conclusion

The designed solution helped to solve the issue of measuring the thermal bed deformations in the Z-axis (X- axis, Y-axis) direction which was very often discussed on a great number of places. The measuring signal output to the control system serves e. g. directly to compensate deformations or to check, whether the maximum deformation is not exceeded, etc. The advantage of this measuring way is that this is an in-process one, i. e. it has the direct relation to the machining process. The solution itself, i. e. integration to the machine tool is a part of TOSHULIN company's know-how.

References

[1] Internal documents and test records by TOSHULIN, a.s
[2] Marek, J., et al.: Design of CNC machine tools, MM publ., s.r.o (2010)

Dynamic Model of Piston Rings for Virtual Engine

P. Novotný*, V. Píštěk, and L. Drápal

Brno University of Technology, Technická 2896/2,
Brno, 616 69, Czech Republic

Abstract. Piston rings play important roles in the lubricant characteristic of reciprocating engines with the consequences on engine wear and a vast amount of lubricating oil consumption. Piston ring dynamics is a very complex problem and it is solved numerically using the virtual engine. This paper shows a work describing a simulation algorithm of the piston ring dynamics.

1 Introduction

Piston rings of reciprocating engines have several functions, mainly

- to seal the clearance between the piston and cylinder in order to retain gas pressure and to minimize blow-by,
- to ensure adequate lubricant applied on the cylinder surface to sustain high thrust and gas force loads at high speed while maintaining oil consumption at acceptable levels and
- to keep piston temperatures relatively low while transferring heat to cylinder walls and coolant.

There are two types of piston rings: compression rings and oil rings. Automobile reciprocating engines normally use three rings, two compression rings and one oil ring.

This work describes a simulation method of piston ring dynamics. The algorithm includes the piston ring motions and blow-by/blowback flow models. The results are presented on Diesel tractor engine.

2 Piston Ring Forces

The motion of piston ring is caused by forces and moments including:

- the static ring tension and torsion from the installation of a piston ring in the cylinder liner,

* Corresponding author.

- the inertia forces and moment related to component mass and engine speed and
- surface forces.

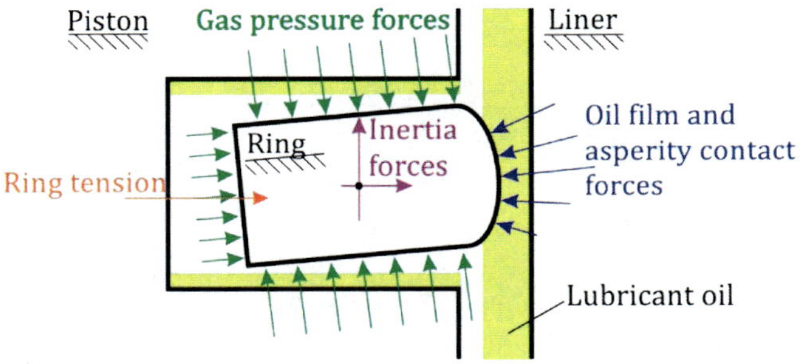

Fig. 1. Description of forces acting on the ring

2.1 Hydrodynamic Lubrication Model

When a gap is fully lubricated the hydrodynamic lubrication can be considered. The problem is described by Reynolds equation, which indicates the relationship between the pressure and the film shape as a function of viscosity and velocity. The Reynolds equation can be simplified by assumptions of constant viscosity. For the analysis of lubrication in a piston ring pack, the oil film thickness between the piston ring and the cylinder wall is much smaller than the piston ring radius. It can be assumed as an infinite flat bearing. The average Reynolds equation can be simplified to an ordinary differential equation [3]

$$\frac{\partial}{\partial x}\left(\frac{\rho h^3}{12\eta}\frac{\partial \bar{p}}{\partial x}\right)=\frac{U}{2}\frac{\partial(\rho \bar{h}_T)}{\partial x}+\frac{\partial(\rho \bar{h}_T)}{\partial t} \quad . \tag{1}$$

The h symbol presents oil film thickness, x is coordinate, t is time, p is circumferential averaged pressure, ρ is oil density, η is oil viscosity and U is relative velocity.

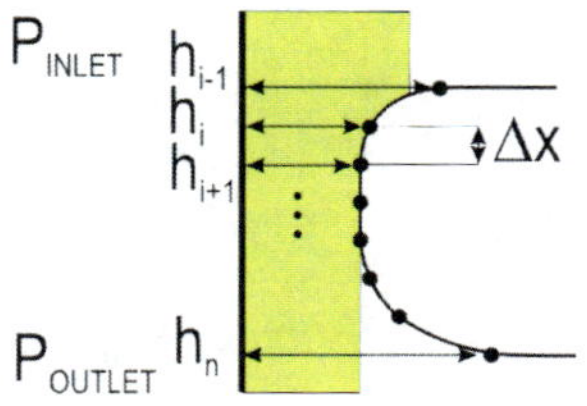

Fig. 2. Computational grid for a numerical solution of 2D Reynolds equation

The Reynolds equation is solved numerically using Finite Difference Method. Figure 2 presents computational grid for numerical solution of 2D Reynolds equation. Numerical solution details can be found in [1, 2, 3].

Viscous friction force exerting on bearing area are calculated as

$$F_{th} = \int_{x_1}^{x_2} \left(-\frac{\mu U}{h} + \frac{h}{2} \frac{\partial \overline{p}}{\partial x} \right) * \pi d * dx \quad . \tag{2}$$

Oil film thickness changes during the engine cycle. This change can be derived from the piston ring motions, oil film evaporation, etc. It affects the piston ring lubricant condition, especially the inlet condition. Fig. 3 illustrates the film thickness on the engine liner wall. For the first and the second rings, the film inlet position is the first contact point between the oil film surface and ring face profile. For the film exit position, the oil film thickness was assumed equal to the minimum clearance between the ring surface and the engine liner wall. The oil ring is supposed to be always soaked in oil. Thus, the fully flooded condition is applied to this ring.

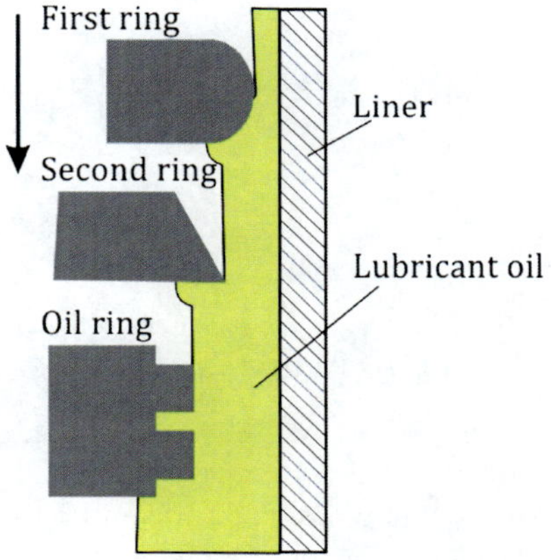

Fig. 3. Oil film thickness interferences for all piston rings

2.2 Boundary Lubrication Model

The pure boundary lubrication according to Greenwood and Tripp [4] is used when oil supply is insufficient. The nominal pressure can be calculated as

$$P_c = KE'F(h/\sigma) \quad , \tag{3}$$

where

$$K = \left(8\pi \frac{\sqrt{2}}{15} \right) (N\beta\sigma)^2 \left(\sqrt{\frac{\sigma}{\beta}} \right) \tag{4}$$

and

$$E' = \frac{2E_1E_2}{E_1(1-\upsilon_1^2)+E_2(1-\upsilon_2^2)}. \quad (5)$$

E_1 and E_2 denotes Young's modulus of the liner and piston rings respectively, υ_1 and υ_2 are poison numbers, σ is composite RMS roughness, β is radius at asperity summit and N is number of asperities per unit area.

The boundary lubrication forces are included only if

$$h/\sigma < w\ , \quad (6)$$

where w parameter is usually equal to 3.

2.3 Gas Blow-By/Blow-Back Flow Model

Gas pressure in piston–cylinder system is divided into several groups. The gas flow through piston ring gap is assumed to be the isentropic orifice flow. The flow between two volumes can be solved using

$$\dot{m} = A_E\rho_1\sqrt{\kappa RT_1} \qquad \text{for} \qquad \left(\frac{p_1}{p_0}\right) \le \left(\frac{2}{\kappa+1}\right)^{\frac{\kappa}{\kappa-1}}, \quad (7)$$

or

$$\dot{m} = A_E\sqrt{\rho_0 p_0}\sqrt{\frac{2\kappa}{\kappa-1}\left(\left(\frac{p_1}{p_0}\right)^{\frac{2}{\kappa}} - \left(\frac{p_1}{p_0}\right)^{\frac{\kappa+1}{\kappa}}\right)} \ \text{for} \ \left(\frac{p_1}{p_0}\right) > \left(\frac{2}{\kappa+1}\right)^{\frac{\kappa}{\kappa-1}}. \quad (8)$$

The symbol $\dot{m}$ in equations (7) and (8) is mass flow, A_E is efficient area R is gas constant, T is temperature, p_0 and p_1 are pressures in first volume and second volume respectively, κ polytrophic exponent.

Figure 4 presents the piston assembly flow model scheme.

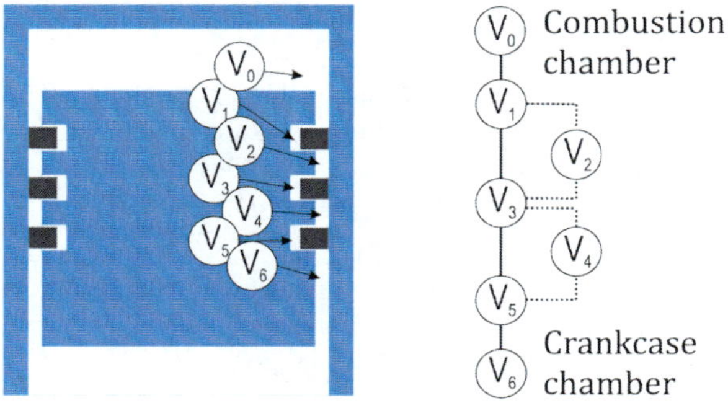

Fig. 4. Blow-by/Blow-back piston assembly flow model scheme

3 Conclusions

The new simulation algorithm of piston ring dynamics has been developed. The virtual modeling can be applied for different combustion engines and it enables to compare different piston ring parameters before complicated measurements on prototype engines.

Figure 5 presents calculated minimal oil film thickness for Diesel in-line four cylinder tractor engine with cylinder bore 105 mm.

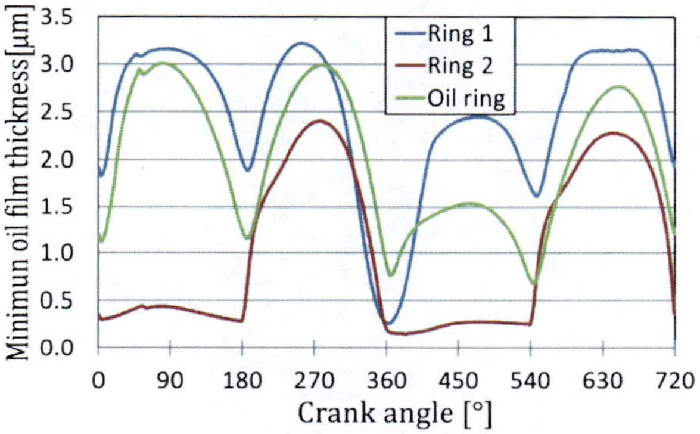

Fig. 5. Calculated oil film of Diesel in-line four cylinder tractor engine with cylinder bore 105 mm and three piston ring configuration

Acknowledgements

The above activities have been supported by the grant provided by the GAČR (Grant Agency of the Czech Republic) reg. No. 101/09/1225, named "Interaction of elastic structures through thin layers of viscoelastic fluid". The authors would like to thank GAČR for the rendered assistance.

References

[1] Novotný, P., Píštěk, V.: New Efficient Methods for Powertrain Vibration Analysis. Proceedings of the Institution of Mechanical Engineers, Part D. Journal of Automobile Engineering (2010) ISSN 0954-4070

[2] Novotný, P.: Virtual Engine – A Tool for Powertrain Development. Brno University of Technology, Brno (2009), Inaugural Dissertation

[3] Wannatong, K., et al.: Simulation Algorithm for Piston Ring Dynamics. Simulation Modelling Practice and Theory (2008)

[4] Greenwood, J.A., Tripp, J.H.: The Contact of two Nominally Flat Rough Surfaces. In: Proc. Instn. Mech. Engrs (1970)

A Cost-Effective Approach to Hardware-in-the-Loop Simulation

M.M. Pedersen[1,3,*], M.R. Hansen[2], and M. Ballebye[3]

[1] Aalborg University, Dept. of Mechanical and Manufacturin Engineering, DK-9220, Aalborg, Denmark
mmp@me.aau.dk
[2] University of Agder, Department of Engineering, Grimstad, Norway
[3] Højbjerg Maskinfabrik A/S, Oddervej 200, DK-8720, Denmark

Abstract. This paper presents an approach for developing cost effective hardware-in-the-loop (HIL) simulation platforms for the use in controller software test and development. The approach is aimed at the many smaller manufacturers of e.g. mobile hydraulic machinery, which often do not have very advanced testing facilities at their disposal. A case study is presented where a HIL simulation platform is developed for the controller of a truck mounted loader crane. The total expenses in hardware and software is less than 10.000$.

1 Introduction

The safety demands for machinery are ever increasing, as described in e.g. the new Machinery Directive [1] or EN13849 [2]. More and more electronic safety and control systems are therefore implemented in order to avoid potential danger, both from conventional use, but also from foreseeable misuse.

As an example from the loader crane industry, the new revision of EN12999 [3] requires the stability of the truck to be monitored and secured via the crane safety system, in order to avoid overturning of the truck. In the case of using the loader crane for personal lifts, the safety requirements are further increased due to the obvious risk to the person(s) being lifted.

The safety system for an advanced loader crane as shown in Fig. 1 may include in the order of 50 IOs and up to some 20 CAN modules connected to a main controller. The trend is towards more distributed IO systems communicating via CANbus networks, but also using sensors with integrated CANbus interface.

The complexity of the controller software increases in the same rate as the safety demands, and is further increased by a high level of product configurability. It thus becomes increasingly difficult to perform end-to-end tests for such systems, using a breakout board with switches, diodes and potentiometers.

* Corresponding author.

In other industries with complex safety systems, e.g. the automotive, aerospace or offshore industries, hardware-in-the-loop simulation have long been the standard platform for testing controller software. In the crane industry and other similar industries, controller software testing have typically been performed using the real machine, though this is neither safe nor efficient.

Fig. 1. Truck with loader crane used for personal lift.

Earlier HIL simulation (HILS) platforms typically considered a single component or minor isolated systems for testing controller software with limited visual output. However, advances in computational power have propelled a rapid evolution of HILS platforms which now considers entire systems, such as heavy construction machinery or offshore structures and offer rich 3D animation output.

A HILS system for a hydraulic excavator with both audio and visual simulation output integrated in the operator cabin of a real excavator, was presented by Elton et al. [4]. The system included a remote hydraulic load emulator for a variable displacement pump, connected via Internet. The simulation are thus feed accurate (non-modeled) information from the pump.

Ghabcheloo and Hyvnen [5] discusses development of a motion control system for a hydraulic wheel loader in a HILS platform called GIMsim. It consists of 5 PCs communicating via LAN to accommodate animation, dynamics, tire contact, IO + CAN and simulation control. The simulation is developed in Matlab/Simulink, compiled and transfered to an xPCtarget for hard real-time operational speed.

Recently, Trydal [6] presented a very advanced proprietary HILS platform from National Oilwell Varco. It concerns an offshore drill-rig including the entire drilling process equipment and displays a 3D animation in a dome in order to completely immerse the operators. Instead of physical IOs, each controller are connected to a similar controller which simulates the IO functionality.

The general trend is to devélop the simulation in Matlab/Simulink, use several PCs connected via LAN and use hardware IO connectivity from dSpace, National Instruments or others. Simulation input are given from real operator interfaces and results are presented using more and more realistic 3D animation. The animation software is generally developed specifically for the considered system using OpenGL.

In this paper, a cost effective approach to HIL simulation is demonstrated for a non time-critical application using only standard high-level engineering software. It should thus be possible for it to be implemented and maintained by engineers, without in-depth software development background.

2 Considered System

We consider a truck with an advanced folding loader crane attached to it, Fig. 1. The crane is equipped with a so-called flyjib, and has 6 degrees of freedom. In the tool point, a variety of accessories can be mounted, such as a rotator and grab, a personal lift basket, and/or a winch.

The crane is mounted with a stabilizer system containing 4 extensible booms each with a stabilizer leg, and an additional fixed front stabilizer leg. In total we have 19 hydraulically actuated DOFs controlled by 12 directional valves and several shift valves.

The safety and control system, Fig. 3, of such a crane primarily covers a number of capacity limitation, stability monitoring, collision avoidance and workspace limiting features, but also several secondary safety features, such as smart speed reduction and selective stopping of load-increasing motion. All safety features are supported by visual and audible warning signals.

The electronic hardware consists of a CAN network with a main controller, 3 I/O modules (flyjib, winch and stabilizers), a radio remote control, a slew sensor, 2 dual axis inclinometers, and up to 12 electro-hydraulically actuated directional control valves. Inputs and outputs are thus distributed to the corresponding function via the CAN network. The critical components are redundant, e.g. pressure transducers and inclinometers. Also the I/O modules and controller consist of two separate processors, each of which monitors the operation of the other, in order to achieve a high level of safety.

3 HIL Simulation

The developed HIL simulator consists of three subsections; 1) a simulation PC, 2) a hardware-gateway PC and 3) the real controller hardware, as shown in Fig. 2.

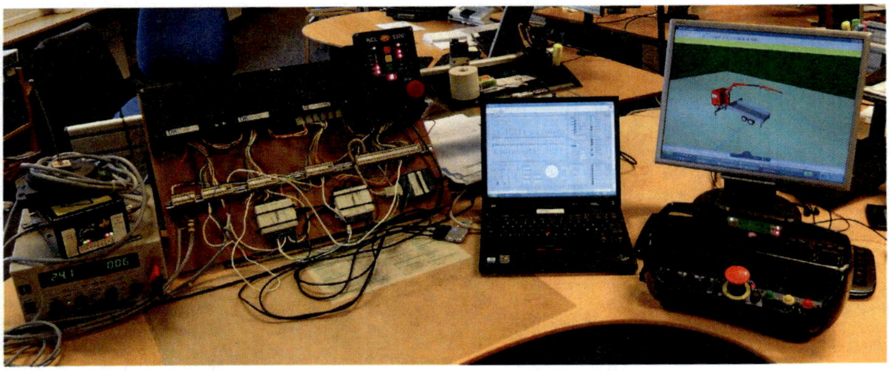

Fig. 2. Physical setup, from left: 1) controller hardware, 2) Gateway PC, 3) Simulation PC and operator interface.

The simulation is developed in Matlab using the Simulink 3D animation toolbox (which doesn't actually require Simulink) for visualization. The Matlab and Simulink programming environments are well known for their simple syntax and high productivity.

The hardware-gateway is developed in LabView using low-cost USB I/O and CAN devices from National Instruments for low level communication with the controller hardware. The link between the simulation running in Matlab and the hardware controlled by LabView is achieved over the local area network (LAN), using the IP/UDP protocol.

3.1 Co-simulation via LAN/UDP

LAN communication using the UDP protocol is an effective approach to achieve co-simulation with two different programs, e.g. Matlab and LabView or others. The programs can run on either two different PCs connected by LAN, or on a single PC sending data to itself - between programs. Using several PCs gives a performance advantage though and enables additional monitoring possibilities and distributed location of equipment.

The UDP protocol is typically used for streaming media over the Internet and other applications where it is preferable to drop packets rather than wait for them. In many applications, this is not a problem, if the same data are transmitted repeatedly. Sending data with UDP requires specification of the receiver IP address and port number, receiving requires only specification of the local port number and the size of the packet in bytes.

In this case study, the UDP protocol is used to communicate between the simulation running in Matlab on the Simulation PC and the Hardware Gateway PC running LabView, see Fig. 3. An UDP packet are sent both ways at every timestep in the simulation, all necessary data are collected and transferred in these packets.

The simulation model is thereby updated with the valve flow command sent from the controller to the virtual valves, and the state of all outputs from the controller and modules, e.g. to activate shift valves or perform engine control. In the other direction, the controller hardware is updated with the state of the virtual sensors in the simulation, e.g. hydraulic pressure, truck inclination, slew angle and estimated flow from the simulated valves, etc.

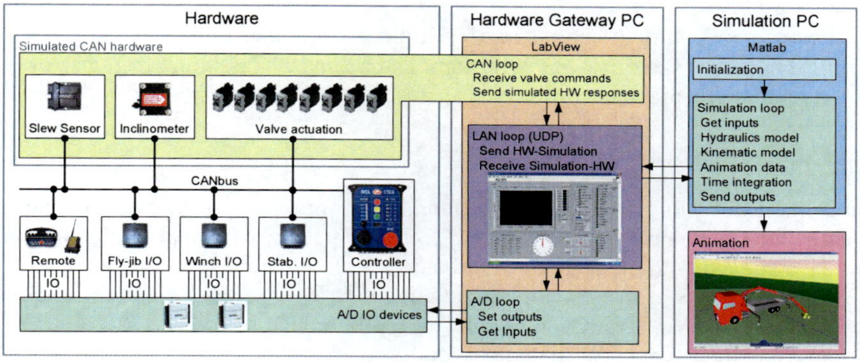

Fig. 3. Overview of the HIL simulation.

3.2 Hardware Gateway

Using a separate PC with LabView to facilitate communication between the controller hardware and the simulation has several advantages. Firstly, the price is low compared to dedicated systems. Secondly, the system is very flexible and scalable. The gateway PC can furthermore be used to control and monitor values in the HIL simulation while it runs. Additional filtering or noise can also be added to the signals here.

As shown in Fig. 3 the LabView program is divided in three separate loops; a LAN loop, CAN loop and A/D loop. Each loop covers communication between the gateway and LAN network, CAN network or analog/digital hardware. Separate loops in LabView runs in parallel for efficient execution and at different rates of 200Hz, 50Hz and 20Hz respectively. The hardware gateway additionally simulates a number of CAN devices, i.e. sensors and valve actuation.

3.3 Simulating CANopen Devices

The communication profile of most CAN devices such as sensors and valves is relatively simple and these can therefore be simulated quite easily. For a thorough overview of CAN and CANopen, see Pfeiffer et al. [7]. A first approach is to let LabView send CAN frames corresponding to the necessary transmit PDO(s) of the devices to the controller.

If needed, it is also possible to implement a limited object dictionary and SDO server in the LabView program containing the entries the controller (or other devices, e.g. a service terminal) will inquire on. Typically, to simulate e.g. a generic CAN I/O module, it will only be necessary to include in the order of 10 entries in the object dictionary.

The SDO protocol is straightforward for the expedited transfers (up to 4bytes), however if segmented or block transfers are required, this will take some more work, see CiA [8].

For the application at hand, strict real-time performance is not critical. Most of the CAN devices sample at 10-20Hz only and a small delay in the response of the simulation is acceptable.

3.4 Simulation Model

Since the simulation model must run in real time (or close to) in order to communicate with the controller, it is essential to limit the computational burden to what can be achieved with the available hardware.

Most dynamics of the system is omitted, which is acceptable since the controller mainly considers steady state values of the input. This is necessary in order to achieve near real-time performance in Matlab.

Referring to Fig. 3, the simulation is initiated by loading the crane parameters, such geometric and inertial properties, joint states and limits, hydraulic system properties, etc.. The main simulation loop is then started which is executed as fast as possible (around 200Hz) and updates all simulation parameters and controls the animation.

All required data is retrieved from the UDP packet; flow command to the valves, state of the outputs on the controller and modules, and necessary simulation control parameters which are set in the Hardware Gateway PC, i.e. mass of payload, mass of truck, crane setup, accessories mounted, etc.

Based on the flow commands and output states, the reference velocity of all DOFs are then calculated. Since several outputs controls shift valves, in order to use the directional control valves for multiple functions, the state of these leads to different modes of operation; i.e. crane mode, stabilizer mode, winch mode, etc.

Then a complete kinematic analysis is performed in order to determine position, orientation and velocity of all parts and the resulting tool point motion. The kinematic parameters are needed in order to drive the animation.

A simple kinetic analysis is also conducted, which determines the needed parameters; crane lifting moment, truck inclination due to crane load, hydraulic pressures in the main cylinder, flyjib cylinder and winch. Additionally, the ground projection of the total Center of Mass (COM), are calculated and displayed in the animation, in order to visualize the stability margin.

The only dynamics included is for a suspended payload, which is modeled as a 2 DOF pendulum. This gives the user the impression of operating a dynamic system and thus experience the consequences of abrupt motion.

The state of all virtual sensors are then determined. These include digital sensors, monitoring whether e.g. stabilizers are fully deployed, the jib is above horizontal, the extension system is fully retracted, etc. Several analog sensors are also simulated, e.g. pressure transducers for measuring pressure in selected cylinders, crane slew angle, truck inclination and directional control estimated valve flow, etc.

Time integration of the calculated velocities for all DOF is then performed using the trapezoidal rule and the animation is updated. All the simulated devices are modeled as ideal, i.e. they respond instantaneously and with with perfect precision. In the Hardware Gateway, however, noise is added to selected signals, in order to test the controller filtering capabilities.

The necessary values are then assembled in a UDP packet and feed back to the controller via the Hardware gateway, and the simulation loop starts over.

3.5 3D Animation

Using the Simulink 3D Animation Toolbox, a Virtual Reality (VR) world can be manipulated from Matlab by translating, rotating or scaling the 3D parts at each time-step, to make it appear that the crane is moving. The VR world is defined using the VRML markup language, i.e. the relation between the parts and the surroundings.

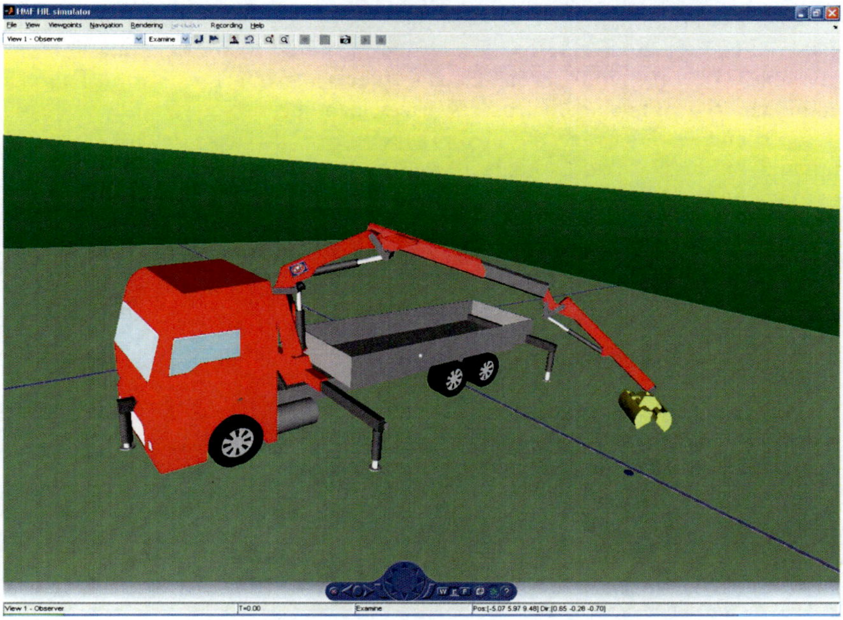

Fig. 4. Screenshot from the 3D animation.

Fig. 4 shows a screenshot from the implemented 3D animation. Some CAD programs can export parts to VRML format, otherwise, simple geometries, e.g. boxes and cylinders, can be directly defined in the VRML language.

Using the Simulink 3D Animation Toolbox makes it simple to create an advanced 3D environment for simulation output. Its flexible and scalable and based on a simple markup language, unlike the typical custom-made OpenGL applications used for HILS animation. Output from the animation can either be displayed directly or recorded for later use.

4 Conclusion

The HIL simulator described in this paper have proven to posses the necessary level of detail to test all the features of the controller concerned. Indeed, its possible to create much more precise testing conditions, compared to using the real machine or a board with switches and potentiometers.

Especially when testing dangerous conditions, such as truck overturning or faults in the control system, simulation is preferred, since it is both expensive and time consuming to perform such test using the real machine.

The HIL simulator currently only comprise ad-hoc testing capability, however, it could be extended to included fully automated end-to-end testing of the controller, which will lead to additional savings on the testing expenses.

As a replacement for ad-hoc testing of controller software using the real machine or a breakout board, HIL simulation is a very effective tool. With the additional control over the testing conditions and increased monitoring possibilities, it is possible to create more repeatable and precise testing conditions. The effort required and risk of testing controller software can thus be significantly reduced.

It's difficult to determine the payback period on the investment of such HIL simulator, but a qualified estimate is 1-2 years. This is due to saved engineer-hours for testing, when comparing to testing using the real machine.

References

[1] EU Machinery Directive, 2006/42/EG European Union (2006)
[2] EN ISO 13849-1:2006 Safety of machinery - Safety-related parts of control systems - Part 1: General priciples for design. CEN (2006)
[3] FprEN12999:2010 (Final Draft) Cranes - Loader cranes. CEN (2010)
[4] Elton, M.D., Enes, A.R., Book, W.J.: A Virtual Reality Operator Interface Station with Hydraulic Hardware-in-the-Loop Simulation for Prototyping Excavator Control Systems. In: IEEE/ASME International Conference on Advanced Intelligent Mechatronics, Singapore, July 14-17 (2009)
[5] Ghabcheloo, R., Hyvnen, M.: Modeling and motion control of an articulated-frame-steering hydraulic mobile machine. In: 17th Mediterranean Conference on Control and Automation, Thessaloniki, Greece (2009)
[6] Trydal, S.: HIL Simulator. Hardware In Loop Simulator med fokus på integrerte prosesser. (Norwegian) NODE seminar, Grimstad, Norway (September 29, 2010), `http://www.nodeproject.no/index.php?articleid=393`
[7] Pfeiffer, O., Ayre, A., Keydel, C.: Embedded Networking with CAN and CANopen Copperhill Media, MA, USA (2003)
[8] CAN in Automation (CiA), e.V. CANopen, Application Layer and Communication Profile. CiA 301 (2002)

Stability of Magnetorheological Effect during Long Term Operation

J. Roupec* and I. Mazůrek

Institute of Machine and Industrial Design,
Brno University of Technology,
Technická 2896/2; 616 69, Brno; CZ
jroupec@hotmail.com

Abstract. The article describes a determination of magnetorheological (MR) effect stability during long term operation. The experiments were carried out in a custom design rheometer, which is built in a mechanical pulsator. The rheometer design allows measurement of the rheological properties of MR fluid and its exposure to a long-term loading simultaneously, without any manipulation of the measured sample. The experiments were performed under conditions which correspond to real conditions in an automotive MR damper. The results of long term operation have shown that during the durability test the yield stress increases up to five times in the fluid off-state due to *in-use-thickening (IUT)*. The same increase was identified in the on-state. These results disprove the hypothesis that a measurable decrease of MR effect arises during the long term operation due to the degradation of magnetic properties.

1 Introduction

1.1 MR Fluid

Current magnetorheological (MR) fluids are suspensions formed by a carrying fluid and ferromagnetic particles, most commonly powdered iron. Upon the application of an external magnetic field (Fig.1) the MR fluids can change their state from fluid to a semi-solid or plastic state, and back, in several milliseconds. This property is known as the MR effect.

* Corresponding author.

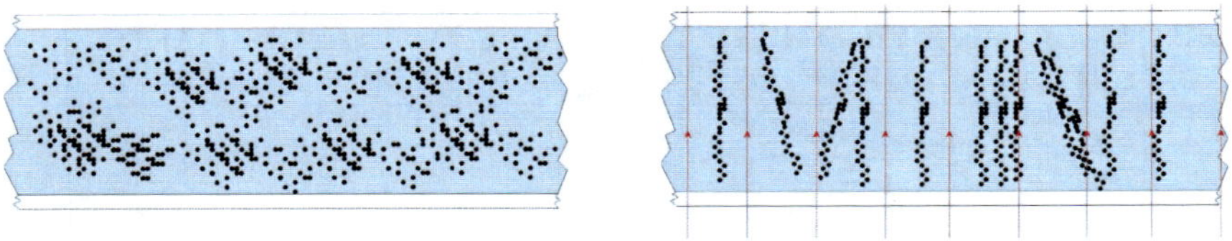

Fig. 1. MR fluid behavior (a) without application of a magnetic field; (b) with application of a magnetic field [1]

This property can be appropriately utilized for regulation of linear and rotary motion. The widest commercial application of these fluids is in mechatronic damping elements–MR dampers. MR dampers are frequently used in car axle suspension systems, suspension of driver seats in goods vehicles, vibration damping during seismic activity or damping of cable bridge vibrations caused by wind and rain.

For rotary or linear MR devices the resultant effects of the magnetic field can be described simply by the Bingham model which is represented by the following equation:

$$\tau = \tau_y(H) + \eta \cdot \dot{\gamma}, \tag{1}$$

where τ_y is a yield stress component which depends on the intensity of the magnetic field and acts in the direction of fluid flow, η is the dynamic viscosity of MR fluid in off-state and $\dot{\gamma}$ is the shear rate.

1.2 In-Use-Thickening

Resistance force of an MR fluid increases with the number of loading cycles during long-term operation in a linear MR device, see [2,3]. This phenomenon is called *in-use-thickening (IUT)*. Carlson et al. advocate a hypothesis that the IUT is most likely caused by oxidation of the iron micro particles. The oxidized layer is both hard and brittle [4]. Nanoparticles of oxides are separated from this oxidized layer due to particle interaction during the MR fluid flow. These nanoparticles cause a change of MR fluid consistency. MR fluid thickens. Contrary to the claims by Carlson, the thicker consistency (Fig.2) is characterized by an increase in initial tension (yield stress) [5] and not in the MR fluid viscosity. The measured viscosity of samples in Fig. 2 is identical.

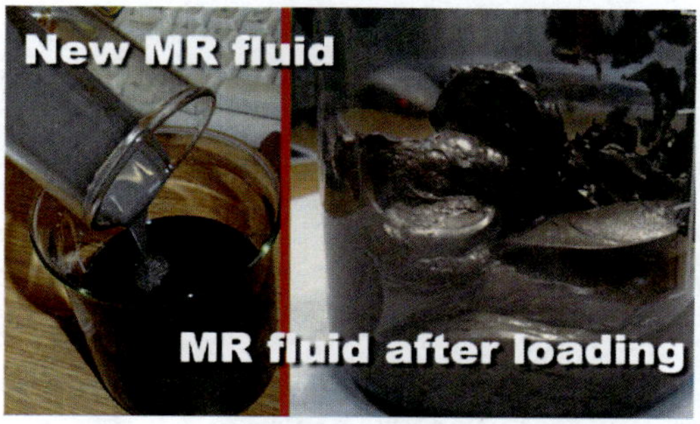

Fig. 2. New MR fluid (left), after the durability test in 2010 (right)

1.3 MR Effect Stability

MR effect stability is dependent on stability of magnetic properties of MR fluid iron particles. The magnetic field H causes a magnetic induction B in iron particles which is directly proportional to relative permeability μ_r and magnetic constant μ_0, see equation (2):

$$B = \mu_0 \cdot \mu_r \cdot H \tag{2}$$

Cheng's [6] experimental results show (Tab. 1) that oxidized iron particles have poorer magnetic properties than non-oxidized particles. The degradation of magnetic properties is evident after only three days of air drying. In the first line of the table the non-oxidized Fe particles are shown as *CI particles*, and oxidized particles as *B'*. The coercive field value has doubled and magnetic remanence even tripled during the experiment. These changes can significantly influence the behavior of the fluid outside the MR valve active area. The value of magnetic saturation of particles has a direct influence on the MR effect. This value has decreased from 186.6 to 175.6emu.g^{-1} at the same magnetic field intensity, i.e. by about 5,9%.

Table 1. Magnetic properties of conventional and composite CI particles [6]

Sample	M_s/(emu·g^{-1})	$\Delta M_s/M_s$	H_c/Oe	$\Delta H_c/H_c$	M_r/(emu·g^{-1})	$\Delta M_r/M_r$
CI particles	186.6		4.07		0.1871	
B′	175.6	−5.9%	9.12	+124.1%	0.5670	+203.0%

To date, the decrease of magnetic saturation during long term operation has not been discussed in literature and, consequently, neither has the decrease of MR effect. In estimating of the MR effect stability it is necessary to take into consideration *IUT* because both phenomena can act simultaneously. *IUT* increases the yield stress, while the expected decrease of MR effect decreases the yield stress.

The significant decrease of MR effect together with IUT would significantly limit the ability to dynamically control the MR device during its operation.

2 Methods

2.1 Experimental Configuration

A unique custom design of slit-flow rheometer (Fig.3) was used for durability tests. The rheometer design allows measurement of the rheological properties of MR fluid (yield stress, viscosity) and its exposure to a long-term loading simultaneously, without any manipulation of the measured sample. Operating conditions during loading (temperature, shear stress and shear rate) are similar to those of a real automotive MR damper. The entire loading cycle is permanently monitored and the measured quantities are recorded for the whole duration of loading. The rheometer design and operation parameters are described in [7]. The measurement chain and calculation methodology of flow and viscosity curves are discussed in [8].

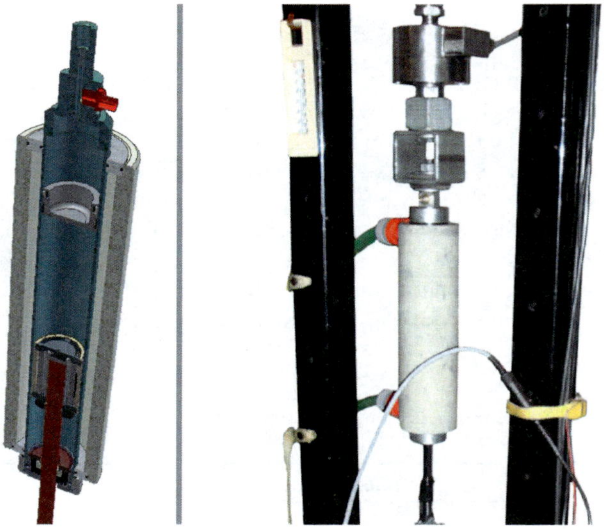

Fig. 3. (a) cross section of rheometer; (b) fixture of rheometric unit in pulsator

2.2 Determination of MR Effect Stability

The MR effect evaluation during long term loading can be carried out after the IUT quantification. IUT appears in both, the on-state and the off-state, while MR effect is pertinent to the on-state only. Therefore, IUT dependence on LDE parameter *(Lifetime Dissipated Energy)* [2,3] was evaluated in the off-state (fig. 4) [5].

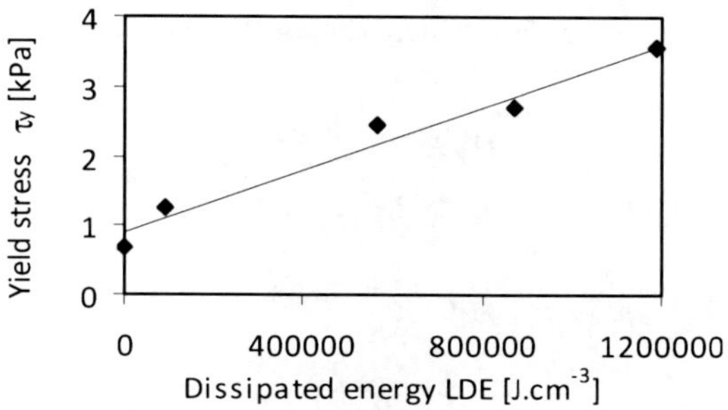

Fig. 4. Yield stress during the durability test [5]

For the entire duration of loading, the force at constant revolutions of the pulsator and constant magnetic field intensity inside the active zone was monitored. This force was then converted into shear stress. The rheometer is equipped with effective water cooling to maintain a steady temperature. In spite of this, the temperature oscillated in a range from 54 to 65°C during the loading. The temperature has an influence on the MR fluid viscosity but not on the shear stress, see Fig.5 [5]. If we accept the simplified Bingham model of MR fluid behavior in Equation (1), then it is sufficient to only correct the second term of the equation by the temperature influence. This term denotes the viscosity component of MR fluid shear stress.

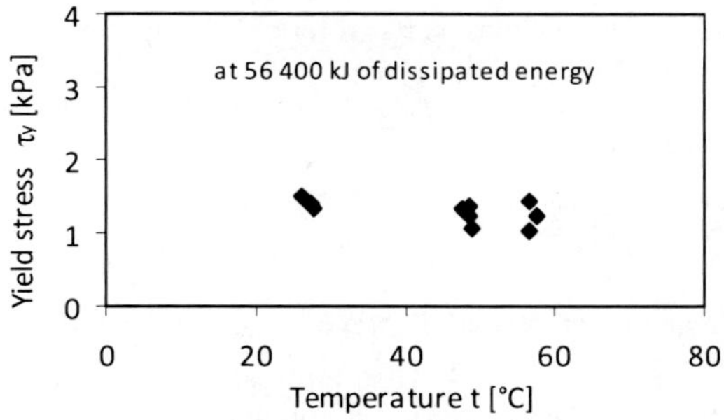

Fig. 5. Yield stress dependence on temperature [5]

Correction of the temperature influence K on total shear stress is calculated by modifying Equation (1):

$$\tau = \tau_y(H) + \eta \cdot \dot{\gamma} + K \tag{3}$$

Correction K can be determined as the difference of the viscosity components at the reference temperature and at the measured temperature:

$$K = \tau_{visco-ref} - k \cdot \tau_{visco-ref} = \tau_{visco-ref} \cdot (1 - k), \tag{4}$$

where k is the correction factor which shows the ratio between viscosity at the reference temperature and at the measured temperature:

$$k = \frac{\eta(t_{ref})}{\eta(t_{meas})} = \frac{0{,}7838 \cdot e^{-0{,}0265 \cdot t_{meas}}}{0{,}7838 \cdot e^{-0{,}0265 \cdot t_{ref}}} = e^{-0{,}0265 \cdot (t_{meas} - t_{ref})}, \tag{5}$$

where t is the temperature and $\eta(t)$ is the measured viscosity dependence on temperature [5]:

$$\eta(t) = 0{,}7838 \cdot e^{-0{,}0265 \cdot t} \tag{6}$$

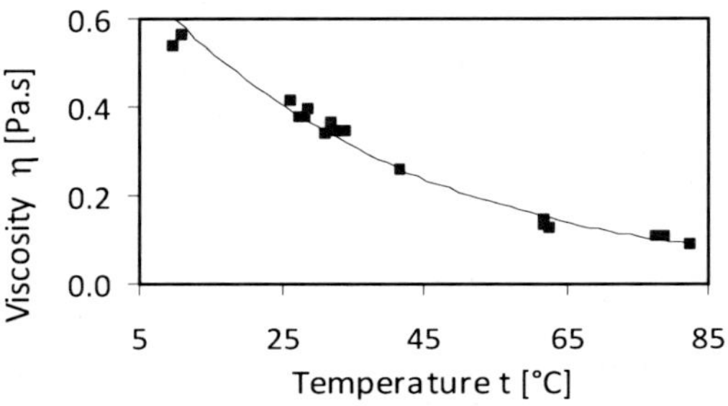

Fig. 6. Viscosity dependence on temperature [5]

Temperature 60°C was selected as the reference temperature for the experiment because it lies in the center of the measured temperature interval. The viscosity component for this temperature can be calculated from the measured temperature dependence (Fig.6):

$$\tau_{visco} = \eta(t_{ref}) \cdot \dot{\gamma} = 0{,}7838 \cdot e^{-0{,}0265 \cdot t_{ref}} \cdot \dot{\gamma} = 5{,}75 kPa, \tag{7}$$

where $\dot{\gamma}$ is the shear rate at maximal force. The shear rate at constant revolutions of pulsator has a value of 35 956 s^{-1}. This approach can be applied to the entire set of measured data because the viscosity did not change during the loading [5]. Evolution of the shear stress without IUT influence can be obtained by subtracting the yield stress in the off-state from the total corrected shear stress.

3 Conclusions

Figure 7 shows MR effect evolution without IUT influence over the entire duration of loading. Regression equation shows moderate decrease of MR effect:

$$\tau(LDE) = -1{,}06 \cdot 10^{-6} \cdot LDE + 40{,}436 \tag{8}$$

This decrease confirms the hypothesis of MR effect degradation caused by oxidation. However, this decrease is so small, that, with respect to the total forces during the control of an MR device, it can be ignored. IUT influence is much more important because it increases the total shear stress and also manifests itself in the off-state. It is important to note that MR fluid was operated in an air-free rheometric unit without free access of air.

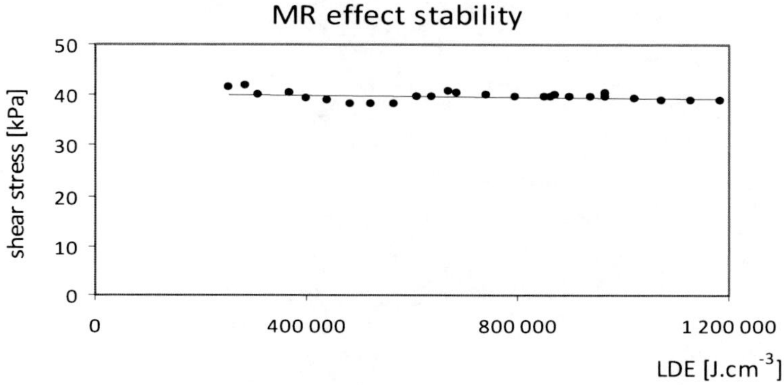

Fig. 7. Evolution of shear stress without IUT influence during the entire loading

References

[1] Arne, L.: Norwegian Defence Research Establishment, pp. 14–15 (2008)
[2] Carlson, J.D.: J. of Intelligent Material Systems and Structures 13, 431 (2002)
[3] Carlson, J.D.: Int. J. of Vehicle Design 33, 207 (2003)
[4] Ulicny, J.C., et al.: Material Science and Engineering 443, 16 (2006)
[5] Roupec, J., et al.: In: 17. Int. Conf. Engineering Mechanical Svratka, p. 519 (2011)
[6] Cheng, H.B.: Acta Physico-Chimica Sinica 24, 1869 (2008)
[7] Roupec, J., Mazůrek, I.: FSI Junior konference, p. 120 (2009)
[8] Roupec, J., et al.: Recent Advances in Mechatronics, p. 115 (2009)

Simulation Modelling and Control of Mechatronic Systems with Flexible Parts

T. Brezina, J. Vetiska, Z. Hadas, and L. Brezina

Faculty of Mechanical Engineering, Brno University of Technology,
Technicka 2896/2, 616 69 Brno, Czech Republic

Abstract. This paper deals with a simulation modelling of mechatronic systems with flexible parts. The presented approach can be used for a development of the mechatronic system which contains flexible parts. Deformations of these flexible parts affect behaviour of the whole mechatronic system. The flexible parts are included in the most of engineering applications and during a development cycle the behaviour of such parts is usually assumed as the behaviour of rigid parts and spring elements. The presented simulation modelling of the mechatronic system includes the behaviour of a multi-body system with the flexible parts using co-simulation techniques and it can be useful for a control design and a better prediction of the mechatronic system behaviour especially in systems where a deformation of flexible parts is significant for a correct operation of the system.

1 Introduction

This paper provides a solution for a simulation modelling of a mechatronic system in a co-simulation technique with only one simulation model of the developed mechatronic system. It includes analysis like finite element methods, multi-body systems, models of electro-mechanical conversion, actuators, control systems, signal processing etc.

A modern mechatronic system is developed in accordance with a mechatronic approach [1] and this approach guarantees required properties of the developed mechatronic system. The mechatronic approach for the development of modern mechatronic systems use simulation modelling for design of individual systems of new mechatronic products [2]. The mechatronic system usually consists of a precise mechanical system, actuators, sensors and control system. The special software is used for the traditional development of individual subsystems and the simulation modelling of the whole system goes several development cycles [3] where the results of individual models from different software are connected with correct feedback of the system.

This paper presents example of the simulation modelling of the mechatronic system under a co-simulation technique, where the complex model is solved together with respects feedback of individual mechatronic subsystems. The simulation modelling of such system [4] is solved in the co-simulation of software MSC.ADAMS and MATLAB.

2 Mechatronic Systems with Flexible Parts

The mentioned approach is very useful for development of the modern mechatronic system. However, the behaviour of several mechatronic systems, especially in machine tools [5], robotics [6], applications of precision mechanics [7], etc., is affected by deformation of several components – flexible parts. Oftentimes this deformation is ignored and the mechatronic system is developed as the multi-body system [8] with rigid parts and ideal spring elements. In case that a frequency of operating loads is close to Eigen modes of the flexible part, the deformation of such part can be significant and the correct operation of the mechatronic system can be affected due to a resonance operation of the flexible part.

The traditional solution of this problem with the deformation or oscillation of the flexible part is substituted the flexible part for a simple resonance system (mass-stiffness-damper). This resonance system can be used for multi-body analysis instead of the flexible part. A disadvantage of this solution is only 1 degree of freedom (DOF) of the substituted resonance system.

The second solution is using of a FEA model of the flexible part in the multi-body system. This solution makes demands on computing time. However the flexible body in FEA software can be reduced for several DOF by a Graig-Bampton method [9]. This method is included in program ANSYS for a finite element analysis. An "ANSYS-ADAMS Interface" is included in ANSYS solver and the flexible body can be reduced using finite element methods for the flexible part with several DOF. This reduced part can be used in ADAMS/Flex toolkit and this flexible part is included with other rigid parts of the mechatronic system in the MSC.ADAMS multi-body model with respect for a flexible behaviour of a reduced DOF.

3 Graig-Bampton Method

The "ANSYS-ADAMS Interface" and ADAMS/Flex toolkit for modelling of flexible bodies is generally based on the Craig – Bampton method [9] which is used for reducing DOF of the FEA model, thus making the model less computationally demanding.

The idea of this method is to separate physical DOF of the system into boundary DOF and interior DOF. Then there are defined two shapes of modes. These are constraint modes which are obtained by prescribing unit displacement to each of boundary DOF and consequently the rest of boundary DOF are fixed and fixed boundary normal modes are obtained by fixing the boundary DOF and evaluated the Eigen solution. The number of fixed boundary modes is defined by the user.

The relation between physical DOF and modal coordinates of the constraint and fixed-boundary normal modes is then described as

$$\mathbf{u}=\begin{Bmatrix}\mathbf{u}_b\\ \mathbf{u}_i\end{Bmatrix}=\begin{bmatrix}\mathbf{I} & \mathbf{0}\\ \mathbf{\Phi}_{ic} & \mathbf{\Phi}_{in}\end{bmatrix}\begin{Bmatrix}\mathbf{q}_c\\ \mathbf{q}_n\end{Bmatrix}, \tag{1}$$

with transformation matrix $\mathbf{\Phi}=\begin{bmatrix}\mathbf{I} & \mathbf{0}\\ \mathbf{\Phi}_{ic} & \mathbf{\Phi}_{in}\end{bmatrix}$, where $\mathbf{u}_b$ is the boundary DOF, $\mathbf{u}_i$ is the interior DOF, $\mathbf{\Phi}_{ic}$ represents physical displacements of the interior DOF in the constraint modes, $\mathbf{\Phi}_{in}$ represents physical displacements of the interior DOF in the normal modes, $\mathbf{q}_c$ represents the modal coordinates of the constraint modes and $\mathbf{q}_n$ represents the modal coordinates of the fixed-boundary normal modes.

A general undamped free vibrating system is defined as

$$\mathbf{M}\ddot{\mathbf{u}}+\mathbf{K}\mathbf{u}=\mathbf{0}\,. \tag{2}$$

The system may be then transformed by substituting (1) to (2) and multiplying the equation by $\mathbf{\Phi}^T$ as

$$\mathbf{\Phi}^T\mathbf{M}\mathbf{\Phi}\begin{Bmatrix}\ddot{\mathbf{q}}_c\\ \ddot{\mathbf{q}}_n\end{Bmatrix}+\mathbf{\Phi}^T\mathbf{K}\mathbf{\Phi}\begin{Bmatrix}\mathbf{q}_c\\ \mathbf{q}_n\end{Bmatrix}=\mathbf{0}, \tag{3}$$

where **M**, **K** are matrices of mass and stiffness. The transformed system may be with generalized mass and stiffness matrices

$$\hat{\mathbf{M}}=\mathbf{\Phi}^T\mathbf{M}\mathbf{\Phi}=\begin{bmatrix}\mathbf{I} & \mathbf{0}\\ \mathbf{\Phi}_{ic} & \mathbf{\Phi}_{in}\end{bmatrix}^T\begin{bmatrix}\mathbf{M}_{bb} & \mathbf{M}_{bi}\\ \mathbf{M}_{ib} & \mathbf{M}_{ii}\end{bmatrix}\begin{bmatrix}\mathbf{I} & \mathbf{0}\\ \mathbf{\Phi}_{ic} & \mathbf{\Phi}_{in}\end{bmatrix}=\begin{bmatrix}\hat{\mathbf{M}}_{cc} & \hat{\mathbf{M}}_{nc}\\ \hat{\mathbf{M}}_{cn} & \hat{\mathbf{M}}_{nn}\end{bmatrix} \text{ and}$$

$$\hat{\mathbf{K}}=\mathbf{\Phi}^T\mathbf{K}\mathbf{\Phi}=\begin{bmatrix}\mathbf{I} & \mathbf{0}\\ \mathbf{\Phi}_{ic} & \mathbf{\Phi}_{in}\end{bmatrix}^T\begin{bmatrix}\mathbf{K}_{bb} & \mathbf{K}_{bi}\\ \mathbf{K}_{ib} & \mathbf{K}_{ii}\end{bmatrix}\begin{bmatrix}\mathbf{I} & \mathbf{0}\\ \mathbf{\Phi}_{ic} & \mathbf{\Phi}_{in}\end{bmatrix}=\begin{bmatrix}\hat{\mathbf{K}}_{cc} & 0\\ 0 & \hat{\mathbf{K}}_{nn}\end{bmatrix}$$

written as

$$\begin{bmatrix}\hat{\mathbf{M}}_{cc} & \hat{\mathbf{M}}_{nc}\\ \hat{\mathbf{M}}_{cn} & \hat{\mathbf{M}}_{nn}\end{bmatrix}\begin{Bmatrix}\ddot{\mathbf{q}}_c\\ \ddot{\mathbf{q}}_n\end{Bmatrix}+\begin{bmatrix}\hat{\mathbf{K}}_{cc} & 0\\ 0 & \hat{\mathbf{K}}_{nn}\end{bmatrix}\begin{Bmatrix}\mathbf{q}_c\\ \mathbf{q}_n\end{Bmatrix}=\mathbf{0}, \tag{4}$$

where indexes *i, b, n* and *c* describes internal DOF, boundary DOF, normal mode and constraint mode.

4 Example of Co-simulation ADAMS – SIMULINK

The simple model of the mechatronic system with flexible part was chosen from presenting of this co-simulation technique. The whole mechatronic system, Fig. 1, consists of the multi-body system with the flexible part in ADAMS, actuating mechatronic/adaptronic system (electromagnet model) and control system in MATLAB.

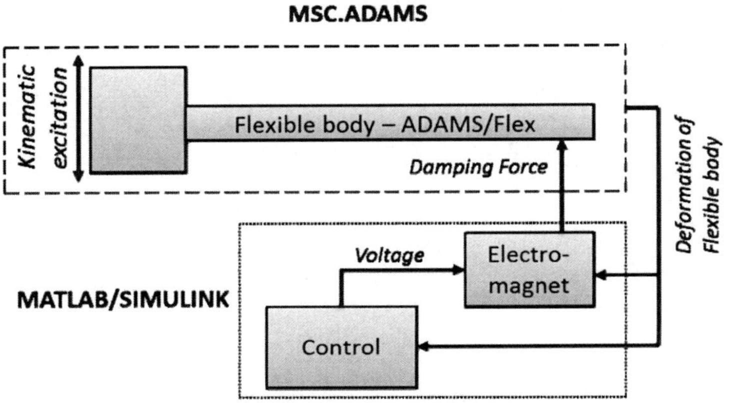

Fig. 1. Matlab/Simulink – Adams co–simulation model

The multi-body system in ADAMS environment consists of a rigid ground, flexible part from ADAMS/Flex toolkit, prismatic joint of the flexible part, motion with a kinematic excitation (vibrations) and an external damping force, which is controlled from SIMULINK model.

The mechanical ADAMS model contains a model of the external excitation and the system behaviour is affected by the damping force, which is calculated in the mechatronic model in SIMULINK on the base of the deformation of the flexible part. The whole mechatronic system is solved together in the co-simulation [10].

A simple steel cantilever with a massive base is chosen for presenting of the co-simulation technique of the mechatronic system with flexible parts. This cantilever can be modelled as mechanical spring with known stiffness and damping. This resonance system has 1 DOF but the real cantilever has several Eigen modes which can affect the system behaviour. The steel cantilever was analysed in ANSYS and transformed for ADAMS.

5 FEA Model of Flexible Part in ANSYS

The FEA model of the chosen flexible part was analysed in ANSYS and 4 Eigen modes of the steel cantilever are shown in Fig. 2. The frequencies of the Eigen modes are shown and compared in Tab. 1. This model can be used for

"ANSYS-ADAMS Interface" and the flexible part is transformed in a modal neutral file (*.MNF) that contains the flexibility information for the component. This file is written in the format required by ADAMS/Flex toolkit on the base on Graig-Bampton method [9].

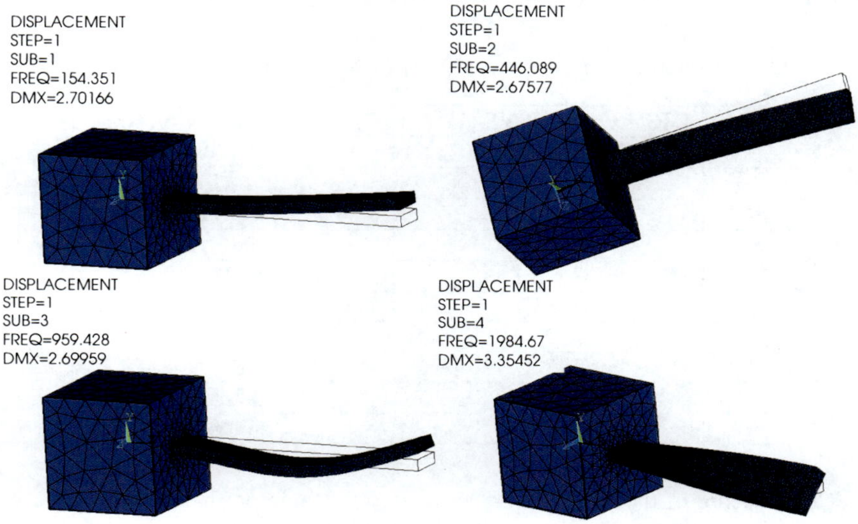

Fig. 2. Modal analysis of flexible cantilever in ANSYS – 4 Eigen modes

6 Flexible Part in ADAMS

The FEA model from ANSYS is used for the Graig-Bampton reduction of DOF and interface points of the flexible part are chosen. The interface points are very important for the multi-body model in ADAMS because the multi-body system can applied loads, measures and joint connected only in the interface points of the flexible part. The modal neutral file was created and transformed from ADAMS to the ADAMS/Flex toolkit by "ANSYS-ADAMS Interface".

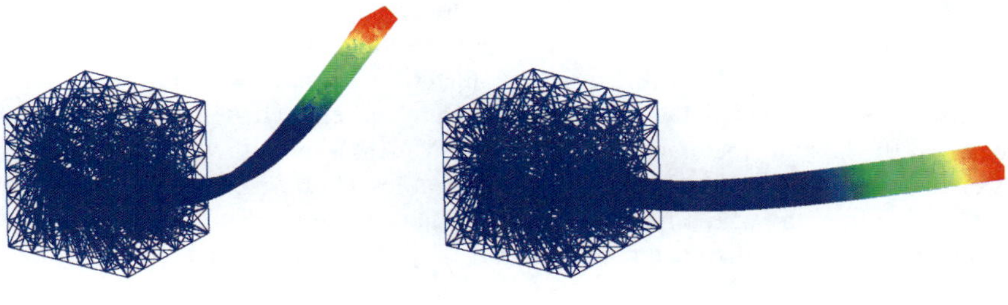

Fig. 3. The first and second Eigen mode of reduced flexible part in ADAMS

A multi-body model with the flexible part is created in ADAMS environment with its toolkit ADAMS/Flex and the first and second Eigen mode of the cantilever are shown in Fig. 3. The other ADAMS flexible part modes are shown in Tab. 1. This reduced part has 6 DOF as the free body and additionally 12 Eigen modes, which are contained in the modal neutral file. The 10 Eigen modes of the model in ANSYS and the reduced model in ADAMS are compared in Tab. 1.

Table 1. Eigen Modes of FEA model in ANSYS and reduced model in ADAMS

MODE	FREQ – ANSYS [Hz]	FREQ – ADAMS [Hz]
1	154.35	154.35
2	446.09	446.14
3	959.43	959.72
4	1984.7	1986.1
5	2613.1	2619.3
6	2653.1	2657.6
7	5108.2	5126.5
8	5352.3	5899
9	5979.9	6760
10	6644.9	6713

The reducing of the flexible part by the Graig-Bampton method provides very good agreement of 6 Eigen modes. The 7th mode is near the FEA model and next modes cannot be used for the multi-body analysis with this flexible part.

In the ADAMS model the base of the flexible part is connected by the prismatic joint to the ground body and the rest of the flexible part can oscillated in a relevant Eigen mode. The base of the body is actuated by mechanical vibrations through the prismatic joint as was mentioned.

7 SIMULINK Control System of Model in ADAMS

The oscillation amplitude of the interface point on the cantilever free end is controlled [11] via an external damping force produced by a simple electromagnet model which is modelled in Matlab/Simulink [12]. The Simulink also provides the control design [13] of the magnetic damping force.

Input and output state variables of the multi-body model in ADAMS are defined in ADAMS/Control toolkit and the connection of the co-simulation ADAMS and Simulink is provided by an s-function which is created in ADAMS/Control toolkit on the base of the state variables. The ADAMS model (*adams_sub*) of the system is excited by vibrations and state variables (*Displacement_Y* and *Velocity_Y*) of the free end of the flexible cantilever is observed in Simulink model. The measurement of the displacement and velocity is used for the control model

(*Control*) of the damping magnetic force produced by the electromagnet model (*System*). The system model can be replaced by a corresponded model of adaptronic or mechatronic system). The computed force in Simulink model is linked back to the ADAMS model to provide the damping of the free end. Fig. 4 presents the co-simulation model of the mechatronic system example.

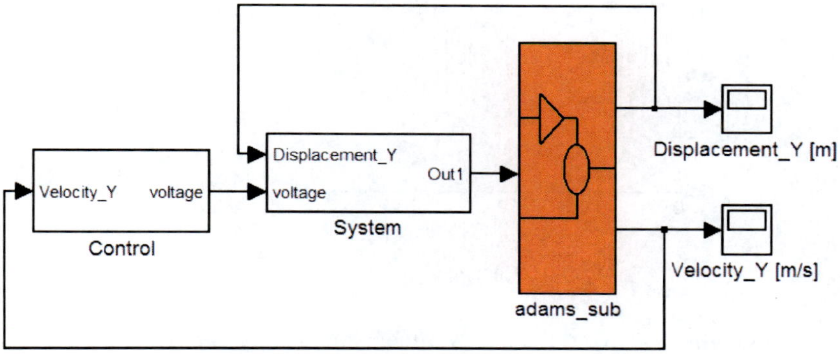

Fig. 4. Matlab/Simulink – Adams co–simulation model

The magnetic force is applied only in an attracting direction, thus the control system of the electromagnet is restricted to a simple switching of an electromagnet voltage. The control system of the electromagnet switching is based on the actual velocity of the free end measured from ADAMS model. The desired velocity of the free end is 0 m/s; it means behaviour of the flexible part in ADAMS without vibrations of the cantilever end.

8 Simulation Results

The ADAMS multi-body system with the flexible part is excited by sinusoidal movement with the amplitude 0.05 mm with the resonance frequency 154 Hz. The movement amplitude of the free end oscillations is shown in Fig. 5 and for the system without external damping force is 1.2 mm. A material damping of the flexible part is defined in ADAMS/Flex toolkit. The SIMULINK control system with the model of the electromagnet, Fig. 4, starts absorb the oscillation of the cantilever free end in the time 0.15 second and the oscillation of the free end of the flexible cantilever in ADAMS is reduced.

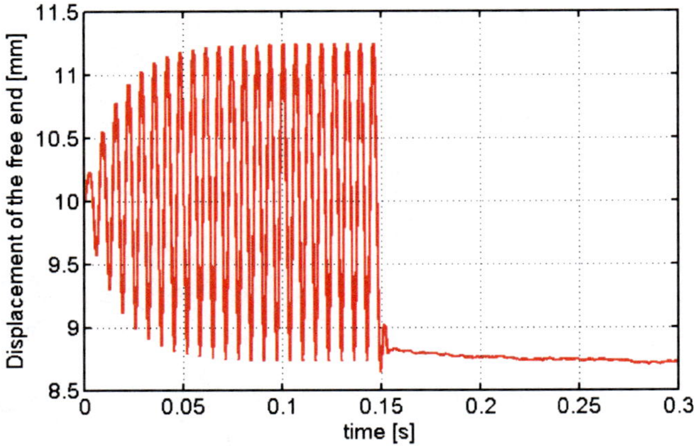

Fig. 5. Displacement of free end

The oscillations were consequently almost completely eliminated by the operation of the electromagnet. The waveform of the electromagnetic damping force is shown in Fig. 6. This system can operate with the maximal damping force approximately 200 N with the corresponding current around 0.5 A for an applied driving voltage 48 V.

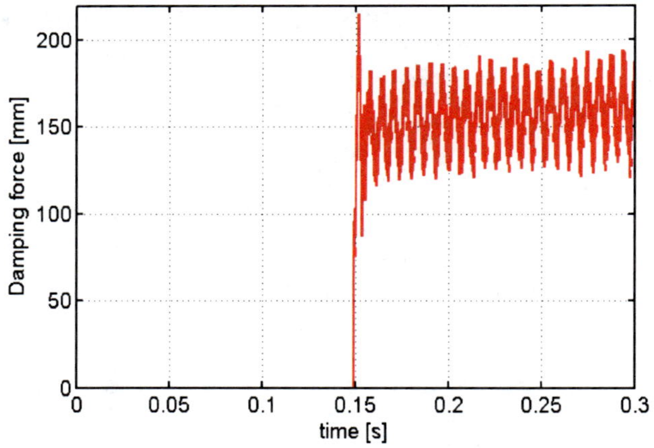

Fig. 6. Damping force produced by electromagnet

9 Conclusions

The simulation modelling with the co-simulation technique was illustratively presented on the simple multi-body system with the flexible cantilever, which is excited by vibrations and affected by the magnetic damping force. The model of the electromagnet is switched on the base of the control system which processes values of the displacement and velocity of the flexible cantilever.

The presented methodology of the mechatronic system co-simulation with the flexible parts can be very useful for the development [1] and optimization [14] of mechatronic systems in machine tools, robotics applications, precise engineering etc.

Acknowledgment

Published results were acquired using the subsidization of the Ministry of Education, Youth and Sports of the Czech Republic (research plan MSM 0021630518 "Simulation modelling of mechatronics systems") and University projects FSI-J-11-39 and FSI-S-11-23.

References

[1] VDI 2206, Design methodology for mechatronic systems, Association of German Engineers, 118 pages (2004)

[2] Hadas, Z., Singule, V., Vechet, S., Ondrusek, C.: Development of energy harvesting sources for remote applications as mechatronic systems. In: Proc. of EPE-PEMC 2010 - 14th Int. Power Electronics and Motion Control Conference, Ohrid, Macedonia, pp. 13–19 (2010)

[3] Fabo, P., Siroka, A., Maga, D., Siroky, P.: Interactive Framework for Modeling and Analysis. Przeglad Elektrotechniczny 83(11), 79–80 (2007)

[4] Fabo, P., Pavlikova, S.: Gsim - Software for Simulation in Electronics. In: Proc. of EPE-PEMC 2010 - 14th Int. Power Electronics and Motion Control Conference, pp. 414–417 (2010)

[5] Reinhart, G., Weissenberger, M.: Multibody simulation of machine tools as mechatronic systems for optimization of motion dynamics in the design process. In: Advanced Intelligent Mechatronics, IEEE/ASME Int. Conference, pp. 605–610 (1999)

[6] Yongxian, L., Chunxia, Z., Zhao, J.: Research on Co-simulation of Rigid-flexible Coupling System of Parallel Robot. In: IEEE Int. Conference on Industrial Technology, ICIT 2009, pp. 10–13 (2009)

[7] Hadas, Z., Ondrusek, C., Singule, V.: Power sensitivity of vibration energy harvester. Microsystem Technologies 16(5), 691–702 (2010)

[8] Maga, D., Hartansky, R., Manas, P.: The Examples of Numerical Solutions in the Field of Military Technology. Advances in Military Technology 1(2), 53–66 (2006)

[9] Craig, R.R., Bambton, M.C.C.: Coupling of substructures for dynamics analyses. AIAA Journal 6(7), 1313–1319 (1968)

[10] Merlo, A., et al.: Active Vibration Control of Precision Machine Tools through a Modular Adaptronic Device. In: ICMC 2010, Mechatronics for Machine Tool, Sustainable Production for Resource Efficiency and EcoMobility, pp. 639–654 (2010)

[11] Zhu, D.L., Qin, J.Y., Zhang, Y., Zhang, H., Xia, M.M.: Research on Co-simulation Using ADAMS and MATLAB for Active Vibration Isolation System. In: Proc. of IEEE Computer Society, the 2010 Int. Conference on Intelligent Computation Technology and Automation, Washington, DC, USA, vol. 2, pp. 1126–1129 (2010)
[12] Turek, M., Brezina, T.: State Controller of Active Magnetic Bearing. In: Recent Advances in Mechatronics, pp. 92–96. Springer, Heidelberg (2007)
[13] Brezina, T., Brezina, L.: Controller design of the Stewart platform linear actuator. In: Recent Advances in Mechatronics 2008-2009, Springer, pp. 341–346. Springer, Heidelberg (2009)
[14] Hadas, Z., Ondrusek, C., Kurfurst, J.: Optimization of Vibration Power Generator Parameters Using Self-Organizing Migrating Algorithm. In: Recent Advances in Mechatronics 2008-2009, pp. 245–250. Springer, Heidelberg (2009)

Orthotic Robot as a Mechatronic System

J. Wierciak*, D. Jasińska-Choromańska, and K. Szykiedans

Warsaw University of Technology, Faculty of Mechatronics,
Institute of Mechatronics and Photonics, sw. A. Boboli 8
Warszawa, 02-525, Poland

Abstract. Mechatronics is at present a dominating concept of designing machines and devices of various kinds. High speeds of data acquisition and processing achieved by means of microprocessor controllers used in such systems, as well as high dynamics of the actuators applied, make it possible to employ mechatronic devices in quite new fields. This refers to the medical domain as well, and wearable robots are one of the most spectacular examples. They are person-oriented robots that can be defined as devices worn by human operators, whether to enhance the function of a limb or to replace it completely. The authors work on an orthotic robot for disabled people suffering from paresis of lower limbs. In the paper, there is proposed a functional structure of such device as well as review of similar systems under development around the world.

1 Introduction

At the Department of Design of Precision Devices, Warsaw University of Technology, there have been begun works on a device for verticalization and aiding the gait of persons suffering from paresis of lower limbs. The designed device can be numbered among so-called „wearable robots" [10] – robots that are a kind of a wear for the user. Their task is to replace or aid human limbs. Dependently on the function realized, these robots are classified into one of the following three groups [10]:

a) exoskeletons – strengthening the force of human muscles beyond their natural abilities,
b) orthotic robots – restoring lost or weakened functions of human limbs,
c) prosthetic robots – replacing an amputated limb.

* Corresponding author.

Numerous design similarities exist between the three kinds of robots listed above, what allows the engineers to make use of experiences of designers originating from various centers, who have been working on all sorts of solutions.

2 Structure of the Device

The designed device (Fig. 1) can be numbered among the orthotic robots. Its main function is to enforce the movement of the lower limbs of the user in a manner similar to a regular human gait. Additionally, the system is to provide the user with the ability of taking some obstacles, including going up and down the regular stairs, as well as sitting on standard furniture and getting up from it. It has been assumed that the device will realize movements of the user's leg according to determined programs. The other elements of the gait, i.e. unloading the leg being shifted and keeping one's balance is left to the user equipped with additional orthopedic devices such as crutches.

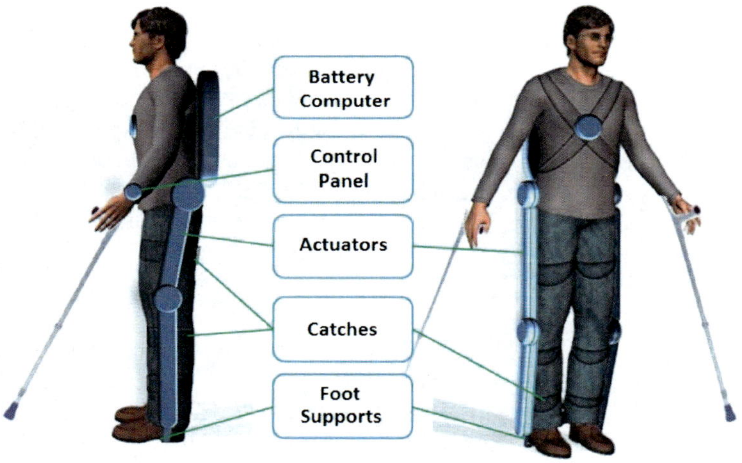

Fig. 1. The designed system

Devices aiding human movements, as robots, are typical examples of mechatronic systems [1, 2, 8], whose structure is presented in a form of a general schematic, as in Fig. 2. In the next section solutions of selected robots aiding human gait are reviewed, with reference to particular elements of their structure.

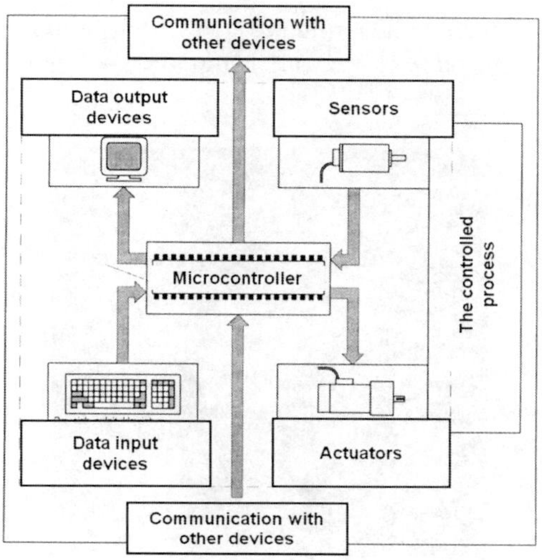

Fig. 2. A general schematic of a mechatronic device [5]

3 Analysis of the System with Regard to Its Main Function

3.1 The Controlled Process

In the case of the designed device, according to the accepted assumptions, the controlled process is putting the paralyzed lower limbs of the user into motion that imitates a regular human gait or it is enforcing movements corresponding to taking obstacles, standing up or sitting down.

3.2 Actuators

Actuators of the device are used to move the lower limbs of the user. According to the commonly accepted simplifications [10], each leg has seven degrees of freedom: the hip joint and the ankle joint - each 3 degrees, and the knee joint - 1 degree. In the case of systems aiding the gait, the number of the degrees of freedom is usually limited to 3 degrees for each limb. The driven mechanical members of these robots are attached to the user's limbs by means of strips that provide a possibility of small relative displacements of the legs with respect to the rigid structure of the robot.

In the case of Japanese exoskeletons *HAL-5 (Hybrid Assistive Leg)* [9, 12], designed both for enhancing the capability of human strength as well as for rehabilitation purposes (Fig. 3), the user's legs are driven by DC motors coupled with harmonic drives, located exactly in the axis of the articulation, and ensuring bending/extension of the hip and the knee. The *HAL* device, unlike many other exoskeletons, transmits the load to the ground using the human skeletal system, and its only task is to increase torques in the hip, the knee and the ankle joints.

Fig. 3. HAL-5 exoskeleton [12]

The *RoboKnee* exoskeleton is a relatively simple device for supporting the knee joint while ascending the stairs and squatting [11]. The device operates as a follow-up system, using a special spring-loaded linear actuator (*Series Elastic Actuator*) (Fig. 4a) attached to a knee fixator, just below the hip, at the calf (Fig. 4b). The task of the device is to provide an additional torque to the knee at the least physical effort of the user. The *RoboKnee* control system employs force sensors measuring a reaction of the ground in a vertical direction and sensors locating the center of gravity within the fibular plane. This information is used in a positive feedback system that is to increase the torque at the knee. However, it is unable to measure neither the lateral forces nor the angle of the force with respect to the ground. The bending angle of the knee is determined by the position of the actuator, which is measured by a linear incremental sensor.

The *Sarcos* exoskeleton [7] supports the muscular systems of the whole body of a user (Fig. 5). It is supplied by a portable energy source. It employs hydraulic rotary actuators located exactly at the driven joints of the device. For the control purposes, the *Sarcos* senses the force of the interactions between the user and the robot, what is meant to recognize the user's movements and adapt to them the movements of the exoskeleton ("get out of the way" control). The foot of the device communicates with the exoskeleton by a rigid metal plate containing force sensors, what prevents it from bending.

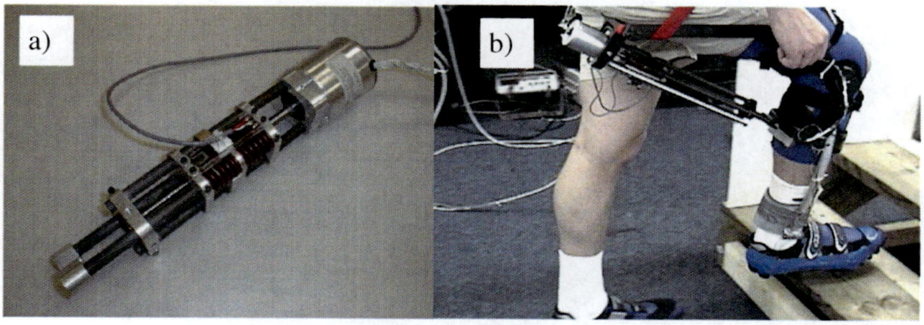

Fig. 4. a) Spring-loaded linear actuator, b) application example of the RoboKnee exoskeleton [7]

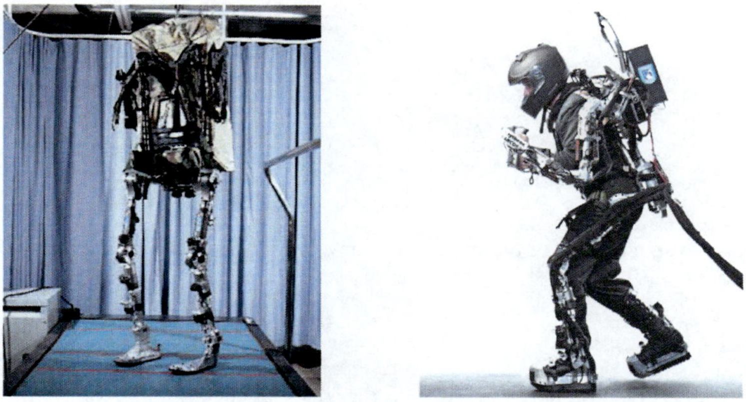

Fig. 5. Sarcos exoskeleton and a user aided with the device [3]

Currently, in terms of preparation for a widespread use, the most advanced exoskeleton of lower limbs designed exclusively for the disabled is a *ReWalk* device designed at Argo Medical Technologies Company [9]. The inventors of the device have shunned publishing their solutions, so far. However, analysis of numerous video footages promoting the achievements of the company provides some information. The robot is equipped with electric drives. Several possible solutions of systems that drive the robot joints are presented in a related patent [6] - (Fig. 6).

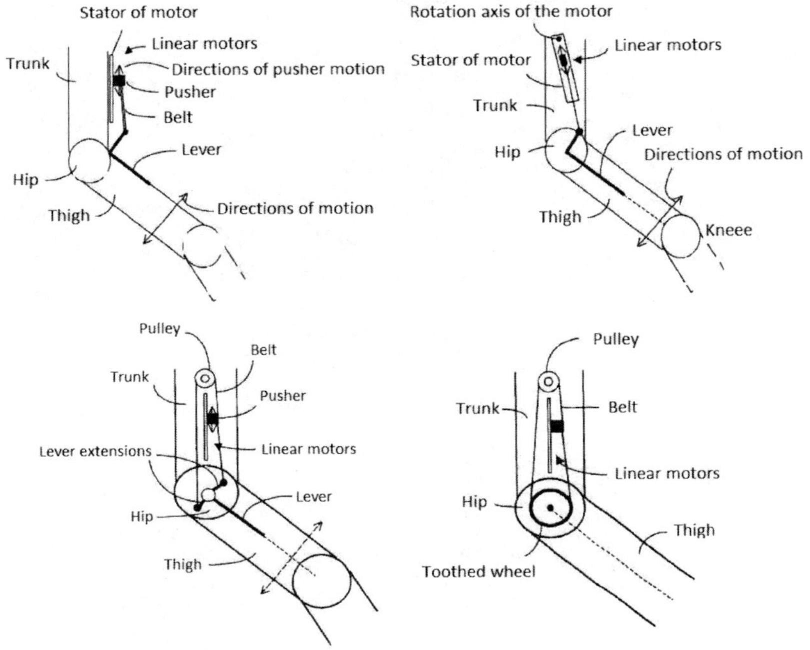

Fig. 6. Design variants of the drive systems for moving the legs, according to the patent for the ReWalk robot [6]

3.3 Measuring Systems

In the case of exoskeletons controlled on the basis of the signals from the human nervous system, their sensory systems are significantly developed, both in terms of hardware and software. For example, the control system of the *HAL* robot uses *EMG* electrodes placed on the skin surface below the hip, over the knee, and additionally force sensors measuring reaction of the ground. A gyroscope and an accelerometer mounted in a knapsack are used to determine an instantaneous position of the user's trunk. The sensors collaborate with two control systems: one based on *EMG* signals and the other based on patterns of gait. The first of them recognizes the user's intentions, the second controls the regulation units [9].

In the designed device, due to the accepted concept of a program realization of the movements of the limbs, the measurement systems will be used to verify correctness of the motions being realized, and in particular to verify their execution (completion). Much greater role will be played by transducers generating signals to the safety system and the system of balance keeping.

3.4 Input Devices

In the first place, devices for inputting information into an orthotic robot, are used to:

- input data necessary to start the system,
- select a specific operation program and its parameters,
- initiate movements.

In the *ReWalk* device [13], a basic input device for a user is a special control panel (remote control) attached to the wrist, equipped with buttons for selecting appropriate operation mode (Fig. 7).

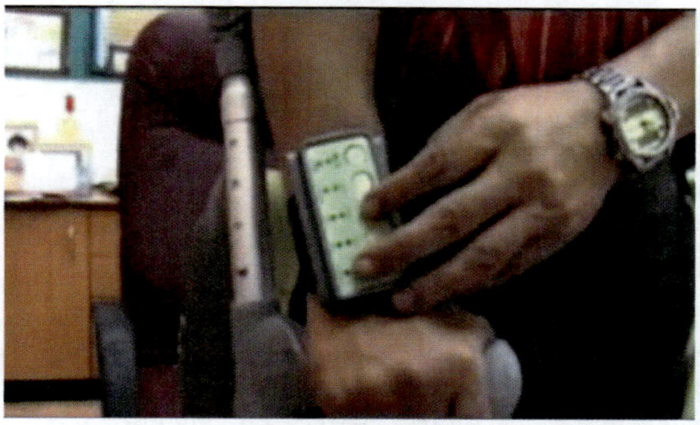

Fig. 7. Control panel of the *ReWalk* device [13]

3.5 Output Devices

The task of the devices to output information is: notifying the user about all the relevant facts that might affect the functioning of the system. In the leg motion system information will include such things as:

- possible faults detected in the system by the diagnostic systems,
- current level of auxiliary energy in a portable power source.

A poll taken among the disabled does not give an unequivocal answer to the question about the expected manner of informing the users. Therefore, it can be planned to adapt this manner to individual needs, i.e. to design and use interchangeable video and audio devices or devices of other kind.

4 Safety System

The device for aiding the gait is an advanced mechatronic system whose use is related with many hazards for the user, and to some extent also to its environment. In addition to performing the main functions of motor system must take an action to eliminate hazards or minimize the consequences of their occurrence. It has been assumed that the superior in a relation to limb movements the device function should be to ensure users safety both in the course of his movement and at rest. This assumption leads to the conclusion that the control system of the designed device should be constructed in such a way that, depending on the situation in which user is located, receives the appropriate objective of the action:

- Launching, implementation and the supervision of motion functions, where the conditions of the moving patient safety to accomplished.
- User health and life saving, in case of emergency.

Mechatronic systems, which independently alter the purpose of the action as a result of changing environment state are known as self-optimizing systems [4, 5]. The construction of such systems is one of the leading contemporary development trends of mechatronics [3].

5 Structure of the Control System

Assumptions for designing device as self-optimizing system concept will lead to an extended functional diagram of the control system (Fig. 8) having a hierarchical structure, in which decisions concerning an instantaneous objective of the operation of the device are made at the highest level of management. The decision-making system must be equipped with a knowledge base containing criteria

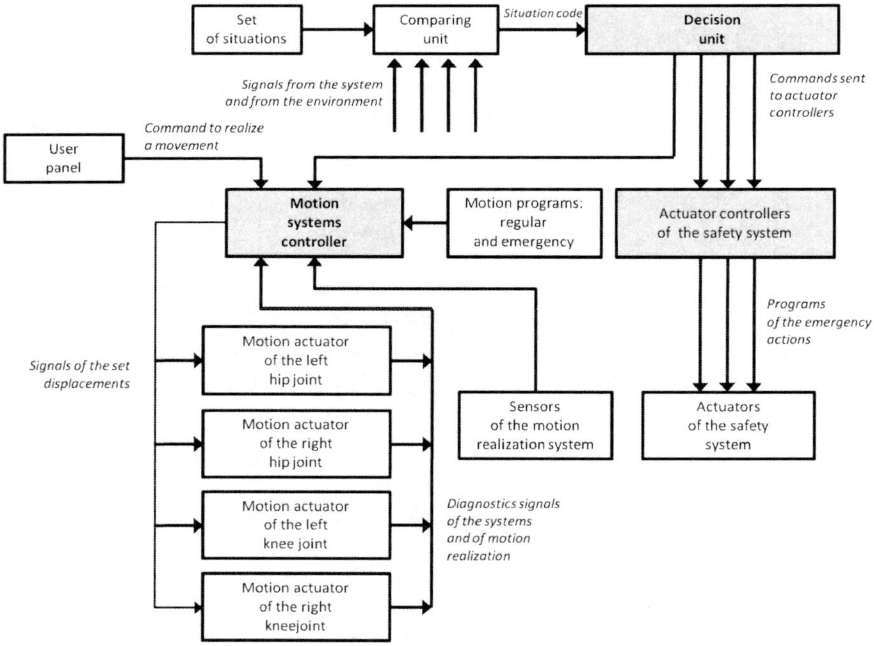

Fig. 8. Schematic of the control system of the device for verticalization and adding the motion

for assessing a set of the values of the signals characterizing the state of the device. Depending on the outcome of the assessment of the state of the device, of its user and of the environment, the system is to send, to the controllers of specific actuators, commands concerning actions to be undertaken. One of these controllers will be a unit synchronizing operation of the systems realizing motion functions of the robot. In situations where the criteria for assessing a current state of the user and the device will allow specific motion actions to be safely realized, the controller will have a right to receive and realize the program commands of the patient. Realization of these commands will consist in synchronized performance of movements by particular actuators coupled with the user's legs. On the other hand, when the analysis of the signals characterizing the state of the device and the user indicates a dangerous situation, the decision-making system is to send to the controllers commands for particular systems to take actions aimed to eliminate the threat or minimize its effects.

6 Summary

The presented considerations determine directions of further works on the device for aiding the gait of the disabled suffering from paresis of lower limbs. The basic design works are focused on the system realizing movements of the user's legs. Tasks of the control system are enhanced with a safety function. In order to realize this function, actuators driving the user's limbs, as well as additional systems that are a subject of distinct works will be employed.

The presented works have been realized within the "**ECO-Mobility**" project No UDA-POIG.01.03.01-14-154/09-00, financed by the European Union.

References

[1] Bishop, R. (ed.): Mechatronic Systems, Sensors and Actuators. Fundamentals and Modeling. CRC Press, Boca Raton (2008)

[2] de Silva, C.W. (ed.): Mechatronic Systems: Devices, Design, Control, Operation and Monitoring. CRC Press, Boca Raton (2007)

[3] Gausmeier, J., Rammig, F., Schafer, W.: Self-optimizing Mechatronic Systems. Design the Future, HNI-Verlagsschriftenreihe, Band 223 (2008)

[4] Gawrysiak, M.: Analiza systemowa urządzenia mechatronicznego, Politechnika Białostocka. Rozprawy naukowe nr 103. WPB. Białystok 2003 (in Polish)

[5] Gawrysiak, M.: Mechatronika i projektowanie mechatroniczne, Politechnika Białostocka. Rozprawy Naukowe nr 44. Białystok (1997) (in Polish)

[6] Goffer, A.: Gait-locomotor apparatus, Patent specification EP 1260201B1 (December 10, 2008)

[7] Guizzo, E., Goldstein, H.: The rise of the body bots. IEEE Spectrum 42(10), 50–56 (2005)

[8] Isermann, R.: Mechatronic Systems – Fundamentals. Springer, Heidelberg (2005)

[9] Kawamoto, H., Lee, S., Kanbe, S., Sankai, Y.: Power Assist Method for HAL-3 using EMG-based Feedback Controller. In: Proc. IEEE International Conference on "Man and Cybernetics Systems", pp. 1648–1653 (2003)
[10] Pons, J.L.: Wearable Robots: Biomechatronic Exoskeletons. John Wiley & Sons, Chichester (2008)
[11] Pratt Jr., E., Krupp, B.T., Morse, C.J., Collins, S.H.: The RoboKnee: An exoskeleton for enhancing strength and endurance during walking. In: Proceeding of the IEEE International Conference on Robotics and Automation, New Orleans, pp. 2430–2435 (2004)
[12] Sankai, Y.: Leading Edge of Cybernics: Robot Suit HAL. In: SICE-ICASE International Joint Conference (2006)
[13] ARGO Medical Technologies, Restoring Upright Mobility, Rediscovering Inclusion, http://www.argomedtec.com

Part VI

Biomedical Engineering

Biomechanical Monitoring in Hip Joint Prosthesis

M. Abounassif, B. Hučko, and Ľ. Magdolen

Faculty of Mechanical Engineering,
Slovak University of Technology in Bratislava,
Nam.Slobody 17, Bratislava, 00421, Slovakia

Abstract. This paper discusses the construction of a monitoring unit for patients that have their hip joints replaced with a prosthesis, the objective is to monitor the stress applied on the hip joint where it alerts the patient when the levels of the load reaches critical levels thus, prolonging the life span of the prosthesis. The step reached for now is the construction of a functional telemetric system connected to three strain gauges that are to supply us with needed data for monitoring the loads, these data are processed according to simulations in FEM environments, and also from derived equations describing the relation between strain, normal stress and bending moments respectively. The telemetry is supplied with energy by induction, the antenna serves as an energy source as well as a receiver of RF signal to be processed. The next step will be clinical testing and later on is to be reduced in size to fit the prosthesis.

1 Introduction

The problem of measurement in vivo has been in the center of interest in the biomechanical community.

The reason to measure the acting forces and moments in vivo is the decreasing age and increasing activity levels of patients. This not only requires a prolongation of implant life-time by reducing wear but may also necessitate a rising of the loads applied in fatigue tests above the current level as given by ISO standards. Even the contact forces measured in elderly patients often exceed the current ISO limits.

Last known innovations in this part, were by Bergmann, Graichen and Rohlmann[1].The constructed a multi-channel measurement chip which was inserted into the hip joint prosthesis which surgically replaced the hip joint, the telemetry is based on the concept shown in Fig.1.[1]

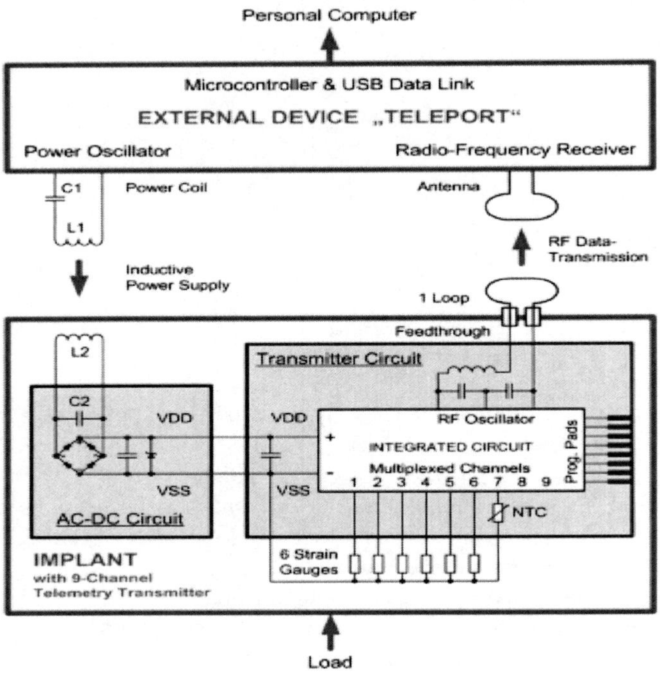

Fig. 1. General concept of Bergmann's telemetry [1]

The telemetric system is consists of external responsible for measuring and gathering data and then sending it and internal components that register and store the data, shown in the Fig. 2.

The external telemetry system consists of the following components: power generator, twin RF receiver, oscilloscope, S-VHS video cassette recorder, video mixer, title generator, personal computer, video LC display, VGA LC display, S-VHS cameras. All components except the cameras are built into two 19A "flight case" racks for easy transportation and mobile measurements.

The internal components serve to measure the loads and to send them to external unit for analysis; using a twin RF-receiver demodulates the signals for computation and storage.[1]

This concept is for clinical applications and measurements only, that is why our concept and approach to the problem is from the another point of view, where we want to construct a monitoring system that gives the patient alerts of overload and wearing rate.[1,2]

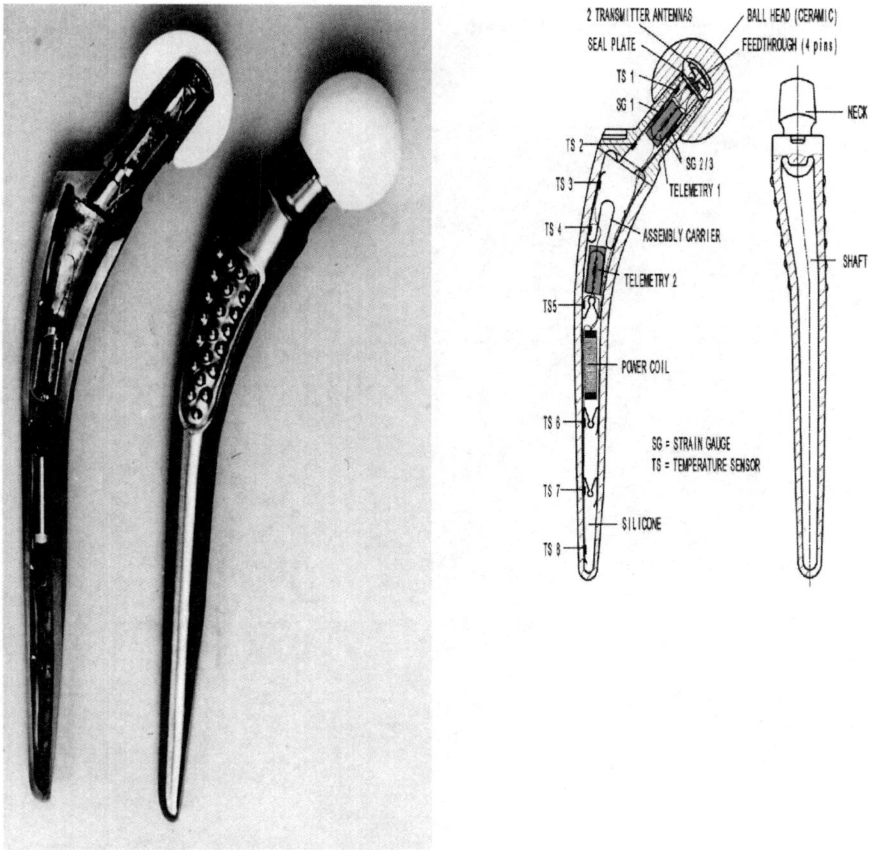

Fig. 2. Internal compomenets of Bergmann's telemetry [1]

2 Our Concept and Approach

As we mentioned before our concept is to construct a monitoring system for the patient that he can use anywhere, its function is to alert the patient that his action in real-time is affecting the prosthesis in a negative manner, this occurs when its overload, this would cause the prosthesis to wear and thus decreasing its life-time.

Theoretical basis:

The unit will be measuring strain occurring in the prosthesis, and according to the following deduction, its calculates the respective load of each strain from normal stress to bending moments around predetermined axes.

The data received is the processed in MATLAB[4], according to results gained from simulation in MEP environment, as well as using equations that describe the relation between strain, and normal stress and bending moment, since the three strain gauges will be mounted on three points separated by angle 120 show in figure below:

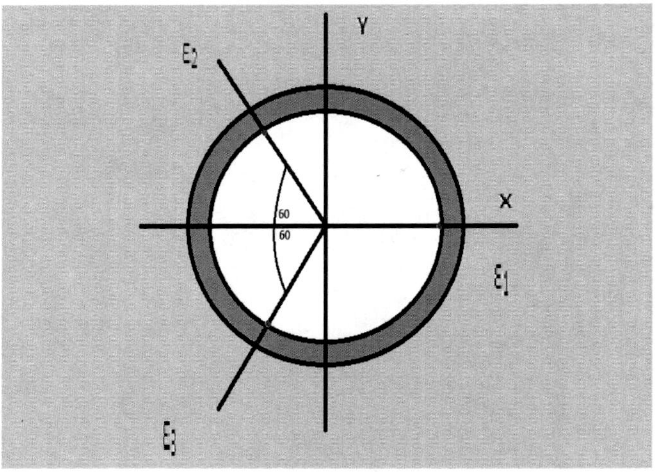

Fig. 3. The positions of the strain gauges around the internal walls of the prosthesis

We know that the strains:

$$\varepsilon_1 = \varepsilon_1^{Mox} + \varepsilon_1^{Moy} + \varepsilon_1^{N} \tag{1}$$

$$\varepsilon_2 = \varepsilon_2^{Mox} + \varepsilon_2^{Moy} + \varepsilon_2^{N} \tag{2}$$

$$\varepsilon_3 = \varepsilon_3^{Mox} - \varepsilon_3^{Moy} + \varepsilon_3^{N} \tag{3}$$

and we know that:

$$\varepsilon_x = \frac{1}{E}\left[\sigma_x - \mu\left(\sigma_y + \sigma_z\right)\right] \tag{4}$$

$$\varepsilon_y = \frac{1}{E}\left[\sigma_y - \mu(\sigma_x + \sigma_z)\right] \tag{5}$$

$$\varepsilon_z = \frac{1}{E}\left[\sigma_z - \mu\left(\sigma_y + \sigma_x\right)\right] \tag{6}$$

$$\gamma_x = \frac{\tau_x}{G} \tag{7}$$

$$\gamma_y = \frac{\tau_y}{G} \tag{8}$$

$$\gamma_z = \frac{\tau_z}{G} \tag{9}$$

and that

$$\sigma = -\frac{M_o}{I}.y \tag{10}$$

where M_o is the bending moment

and $\frac{y}{I}$ is the quadratic moment of the cross-section.[3]

Since our goal is to get the values of the bending moments and axial force from measured strain we deduce that:

$$\varepsilon_1 = \frac{1}{\pi E}\left(\frac{N}{r_0^2 - r_1^2} - \frac{4(M_{oy}.r_1 cos\propto + M_{ox}.r_1 sin\propto)}{r_0^4 - r_1^4}\right) \quad (11)$$

$$\varepsilon_2 = \frac{-1}{2\pi E}\left(\frac{2N}{r_0^2 - r_1^2} - \frac{4.r_1(M_{ox}.\sin(60-\propto) + M_{oy}.\cos(60-\propto)}{r_0^4 - r_1^4}\right) \quad (12)$$

$$\varepsilon_3 = \frac{-1}{2\pi E}\left(\frac{2N}{r_0^2 - r_1^2} - \frac{4.r_1(-M_{ox}.\sin(60-\propto) + M_{oy}.\cos(60-\propto)}{r_0^4 - r_1^4}\right) \quad (13)$$

As the telemetric system constructed by Bergmann, our concept is also divided into internal components and external, but with several differences.

Internal components:

The internal components are the three strain gauges connected to a chip where we have the resistor bridges as well as AD7797 shown in the fig, which serves for calibration and amplification, also it has a mux for multiplexing the signals. This chip in turn is connected to the TMS37157 from Texas Instruments. Which role is to send the data to the receiving unit via RF at a frequency of 134.2 kHz.[5]

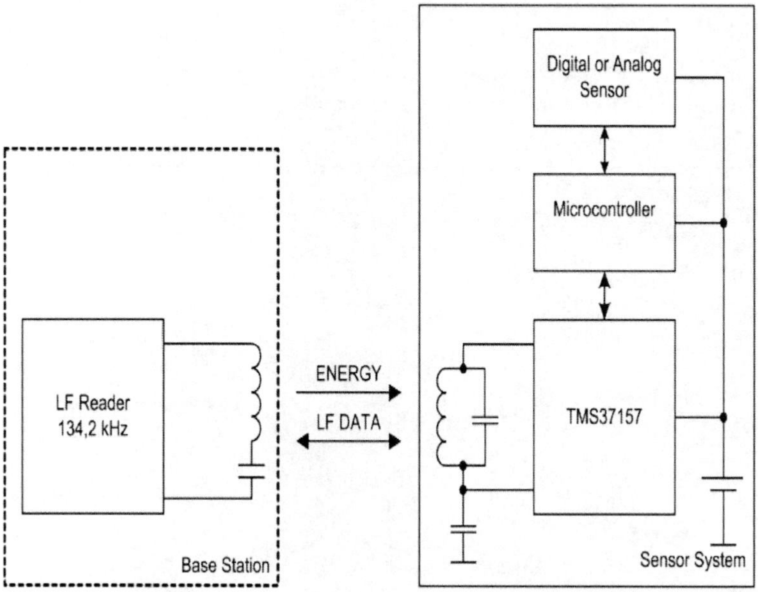

Fig. 4. General concept of our telemtry regarding is function

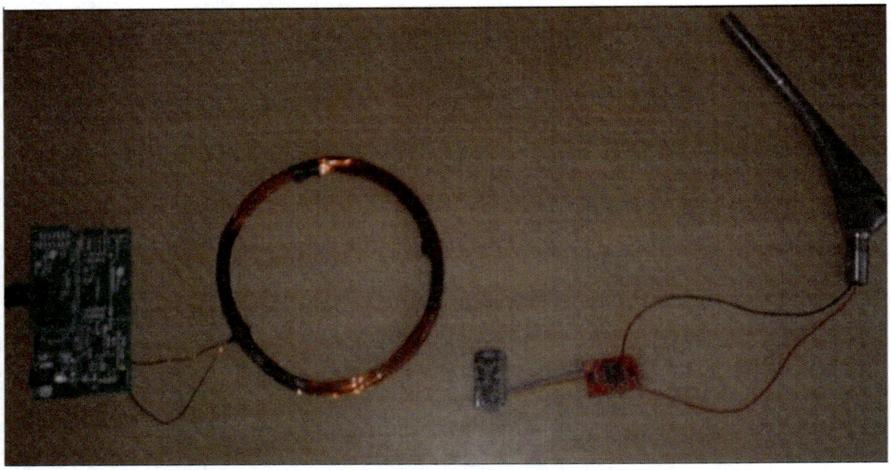

Fig. 5. Our constructed telemetry measuring on prosthesis

The external components are the receiving unit connected to coil antenna that serves to power up the whole telemetric system as well as a receiver, where the data is sent to the PC via USB port to be processed for evaluation.

3 Discussion

Our constructed telemetry was tested in vitro, by applying harmonic loading of value around 35 MPa, for a period of 20 seconds.

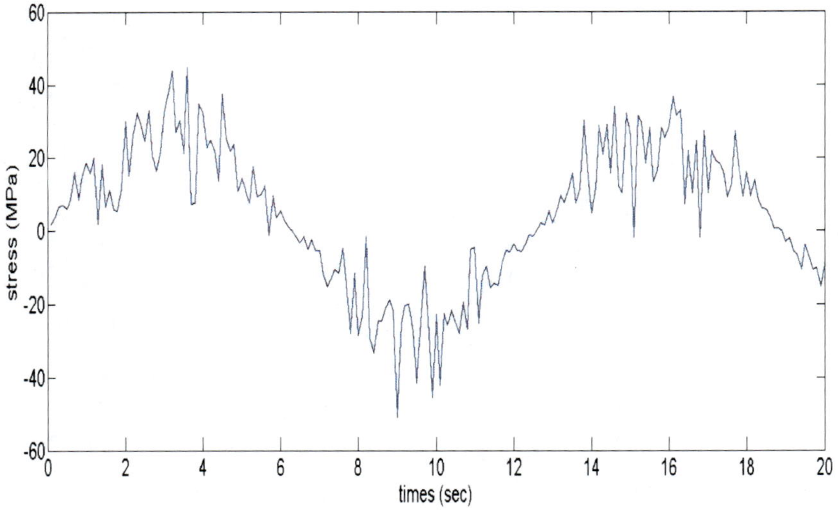

Fig. 6. Obtained results from in vitro testing.

After constructing a functional telemetry, and testing it, the next stage is to minaturising it using micro technologies, to make it fit the prosthesis and after that its the stage where we are to deal with paperwork to get permission to clinically test this system, on living being in our case probably on pigs and later on humans.

Acknowledgement

The support of the Slovak Ministry of Education grant VEGA 1/0571/11 is gratefully acknowledged.

References

[1] Bergmann, G., Graichen, F., Siraky, J., Jendrzynski, H., Rohlmann, A.: Multichannel strain gauge telemetry for orthopaedic implants. Journal of Biomechanics (1988)
[2] Bergmann, G., Graichen, F., Rohlmann, A.: Hip joint forces during walking and running, measured in two patients. Journal of Biomechanics (1993)
[3] Timoshenko, S.P.: Theory of elasticity. Mcgraw-Hill Book Company, New York (1951)
[4] MATLAB user's manual
[5] Texas Instruments data sheet

Systolic Time Intervals Detection and Analysis of Polyphysiographic Signals

Anna Strasz[1], Wiktor Niewiadomski[1,2], Małgorzata Skupińska[3], Anna Gąsiorowska[1,4], Dorota Laskowska[1], Rafał Leonarcik[1], and Gerard Cybulski[1,3]

[1] Department of Applied Physiology, Mossakowski Medical Research Centre of PAS, 5 Pawińskiego St., Warsaw, 02-106, Poland
[2] Department of Experimental and Clinical Physiology, Medical University of Warsaw, 26/28 Krakowskie Przedmieście St., Warsaw, 00-927, Poland
[3] Institute of Metrology and Biomedical Engineering Department of Mechatronics, Warsaw University of Technology, 8 St A. Bobola St., Warsaw, 02-525, Poland
[4] Laboratory of Preclinical Studies in Neurodegenerative Diseases, Nencki Institute of Experimental Biology, 3 Pasteur St, Warsaw, 02-093, Poland

Abstract. The newly developed computer program allows the continuous measurement of the following systolic time intervals (STI): *PEP* – preejection period, *Q-S2* - time between trough of Q wave and aortic valve closure, *Q-D* - time between trough of Q wave and dicrotic notch, *S2-D* - time between aortic valve closure and dicrotic notch, as well as QQ interval. The program was tested on the recordings obtained in 30 young, healthy subjects remaining in supine position who performed twice two-minute isometric handgrip (HG). First HG was followed by four-minute rest, second HG by two-minute occlusion of the working arm. Since there were distinct oscillation in the time course of *Q-S2* intervals and the time course of *Q-D* intervals was relatively smooth, we suggested that *S2* and *D* might reflect different events which is in contrary to common notation.

1 Introduction

The careful analyses of systolic time intervals (STI) may provide information on heart muscle contractility and sympathetic activity (SNS) [1] – [8]. Te aim of our study was to verify reliability of STI as indices of SNS activity. Determination of STI requires combined analysis of several biological signals. The following signals were used in polyphysiographic analysis: electrocardiogram (ECG), carotid pulse, and phonocardiogram. We performed 'manual' analysis of some chosen intervals which are believed reflect changes in sympathetic activity induced by experiment. We decided to replace cumbersome procedure with automated detection of key events, what in turn allowed determination of consecutive systolic time intervals in the whole period of observation.

2 Method

2.1 The Detection of Characteristic Points

Our newly developed computer program, prepared in MATLAB environment, allows the continuous automatic (and also manual) measurement of the systolic time intervals basing on the polyphysiographic signals. Some characteristic points on the analyzed curves were detected in the following order:

Q_i - trough of Q wave (ECG), local minimum preceding peak of R_i wave (Fig.1.),
S_i - trough of S wave (ECG), local minimum following peak of R_i wave (Fig.2.),
E_i – beginning of blood ejection from left ventricle (carotid pulse), local minimum of carotid pulse curve immediately following S_i (Fig.3.),
M_i – local maximum of carotid pulse curve immediately following E_i (Fig.4.),
D_i – dicrotic notch (carotid pulse), local minimum occurring during predetermined period (250 ms) after M_i (Fig.5.),
$S2_i$ – beginning of second heart tone caused by aortic valve closure (phonocardiogram), maximum peak of the first positive deflection of the S2 complex preceding dicrotic notch (Fig.6.).

Figures 1-6 illustrate the method of the characteristic point detection.

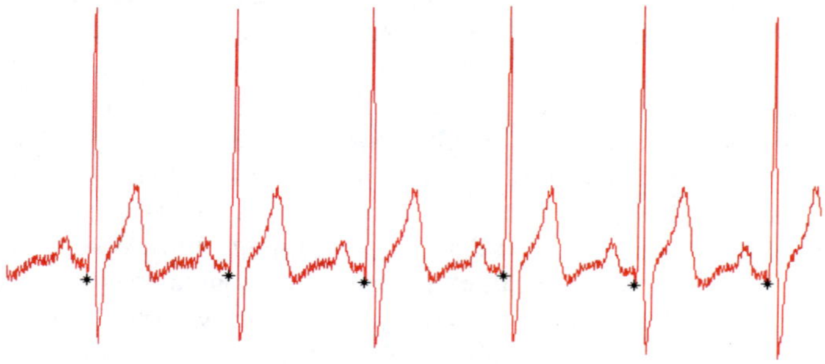

Fig. 1. *Q* - trough of Q wave; ECG.

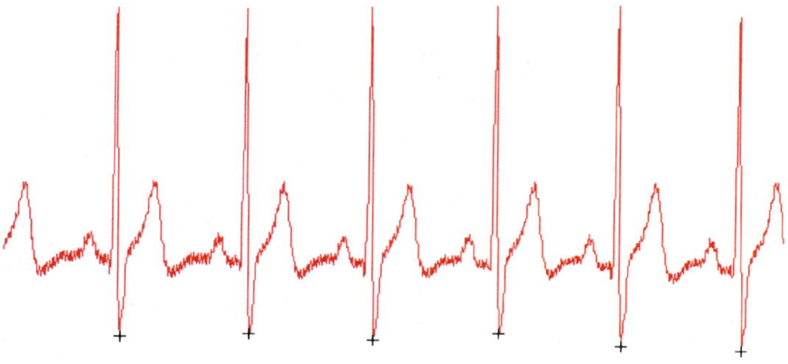

Fig. 2. *S* - trough of S wave; ECG.

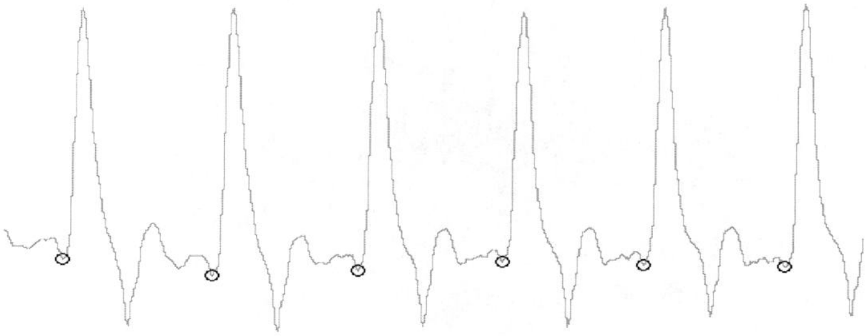

Fig. 3. *E* - beginning of blood ejection from left ventricle; carotid pulse.

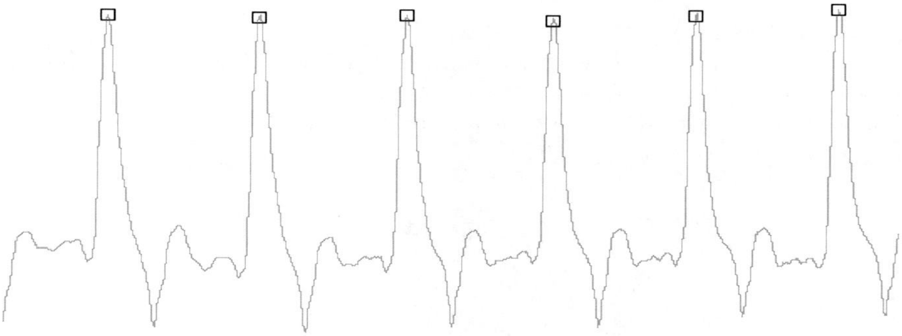

Fig. 4. *M* – local maximum in carotid pulse curve immediately following *E*.

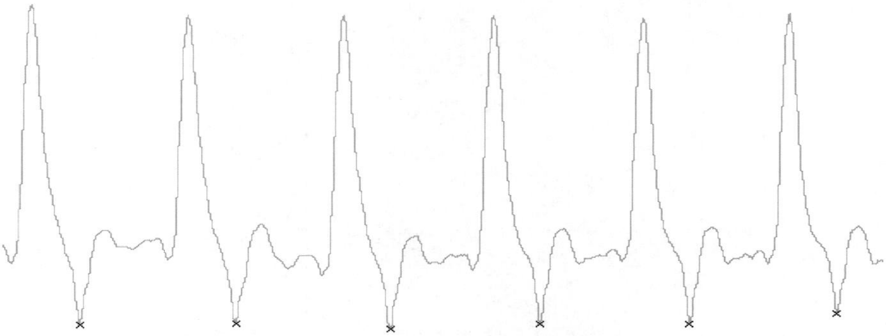

Fig. 5. *D* – dicrotic notch; carotid pulse.

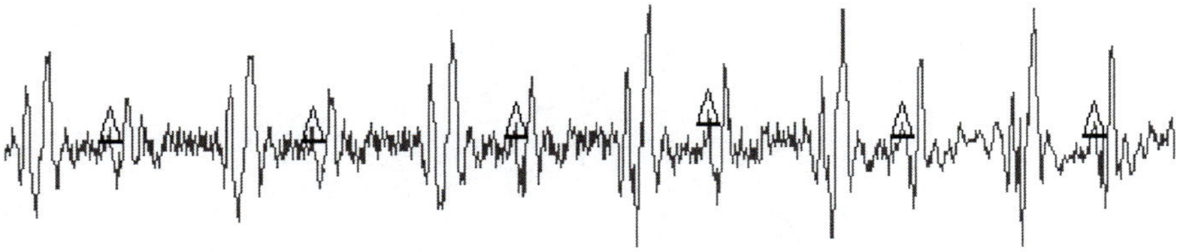

Fig. 6. Phonocardiogram: *S2* - beginning of second ton caused by aortic valve closure

2.2 Determination of Time Intervals between Chosen Events

The method of time intervals determination was presented in (Fig.7.). Time of event X_i appearance is denoted as $t(X_i)$. The following time intervals have been determined:

1. pre-ejection period: $PEP_i = t(E_i) - t(Q_i)$
2. time between trough of Q wave and aortic valve closure: $Q\text{-}S2_i = t(S2_i) - t(Q_i)$
3. time between trough of Q wave and dicrotic notch: $Q\text{-}D_i = t(D_i) - t(Q_i)$
4. time between aortic valve closure and dicrotic notch: $S2\text{-}D_i = t(D_i) - t(S2_i)$
5. QQ_i interval: $QQ_i = t(Q_{i+1}) - t(Q_i)$

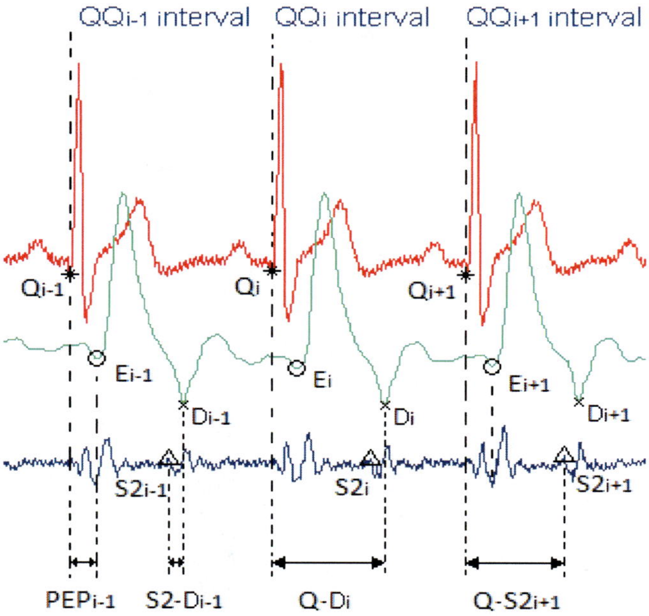

Fig. 7. Traces, events, time intervals. Traces: from the top: ECG, carotid pulse, phonocardiogram. Events: *Q* (trough of Q wave) – '*'; *E* (beginning of blood ejection from left ventricular) – 'o'; *D* (dicrotic notch) – 'x'; *S2* (beginning of second ton caused by aortic valve closure) – 'Δ'. Time intervals: *PEP* (preejection time period); *Q-S2* (interval between *Q* and *S2*); *Q-D* (interval between *Q* and *D*); *S2-D* (interval between event *S2* and *D*); *QQ* (interval between consecutive *Q*).

2.3 Experiment and Case Presentation

Thirty young, healthy subjects participated in the study. When subjects remained supine, they performed two-minute isometric handgrip (HG) at 30% of the maximal voluntary contraction. The first HG was followed by four-minute rest. After second HG, two-minute occlusion of the working arm was applied, followed by two-minute rest. The study`was undertaken to verify reliability of some indices of SNS activity, particularly PEP and PEP/LVET.

Case A (Fig. 8)
In this case *Q-S2* curve are followed faithfully by changes of *Q-D* curve. The constant time lag between D_i (dicrotic notch) and $S2_i$ (aortic valve closure) may be interpreted as transition time from aortic valve to carotid artery; thus D_i and $S2_i$ are the same event observed in different places. Accordingly, the constancy of time lag from D_i to $S2_i$ (ca. 30 ms) is evident in time course of *S2–D* curve. This curve remains almost flat through the experiment. Basing on the continuous time courses of selected time intervals it may be concluded that: 1/ the distinct oscillations of *QQ* seems to not influence strongly these intervals, 2/ shortening of *QQ* during handgrip is paralleled by slight decrease of *Q-D* and *Q-S2*, 3/ during occlusion, when *QQ* intervals return to baseline values also *Q-D* and *Q-S2* return to the baseline values, i.e. *Q-D* and *Q-S2* do not depend on arterial pressure which remains elevated.

Case B (Fig. 9)
It is evident that *Q-D* oscillate only slightly, the oscillations of *Q-S2* are much greater. This is also evident in *S2-D* interval (from 10 ms to 70 ms). Assuming that the detection of *S2* and *D* is correct it is rather impossible that the *S2* and *D* reflect the same event. In 1984 O'Rourke et al. [9] stated that usually the beginning of the diastolic wave, termed dicrotic notch, occurs immediately after aortic valve closure. Although these two events are sometimes believed to be synonymous. However they may stem from different processes. Closure of aortic valve is predominantly a cardiac phenomenon whereas dicrotic wave is predominantly a vascular one. Is it possible that this explanation applies also to our results; thus *S2* reflects the cardiac event – valve closure, whereas *D* reflects beginning of the diastolic wave. If such hypothesis holds true next question arises: why *Q-D* interval is much more stable than *Q-S2*. Case B demonstrates also the constancy of *PEP* which changes only slightly during handgrip despite considerable shortening of *QQ*.

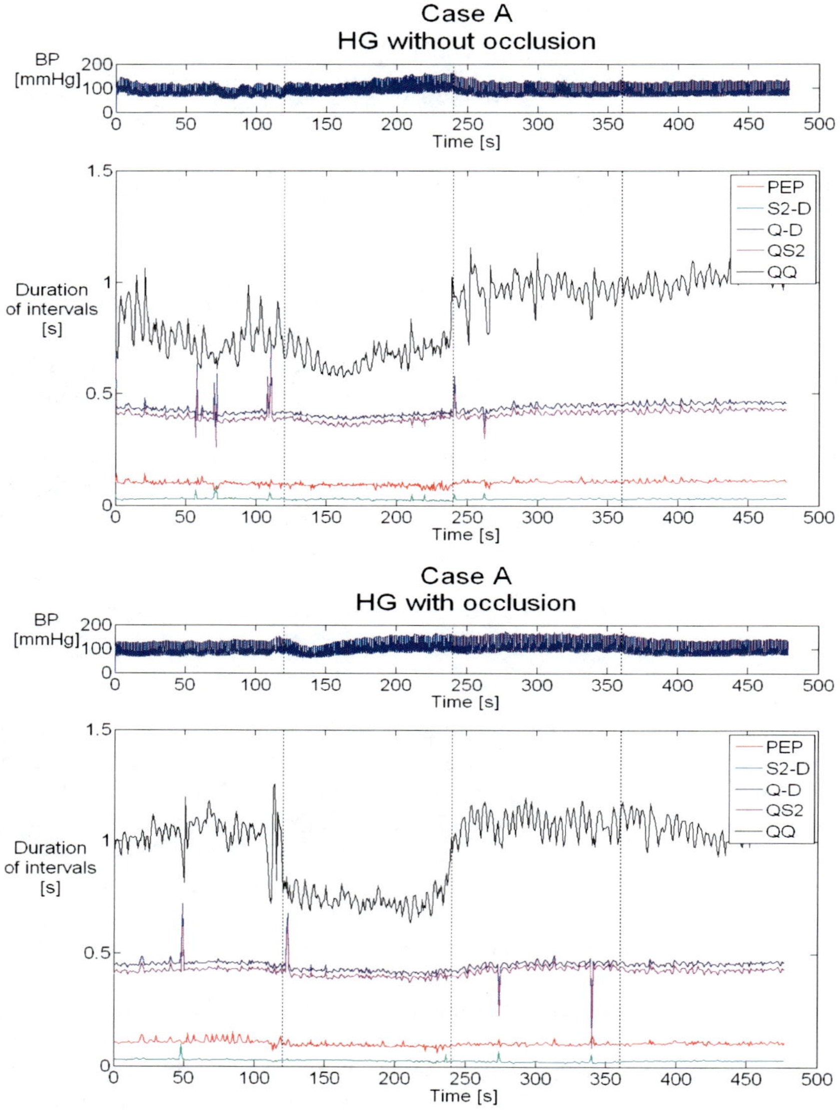

Fig. 8. Presentation of time courses of chosen time intervals for case A. S2–D curve remains almost flat through the experiment.

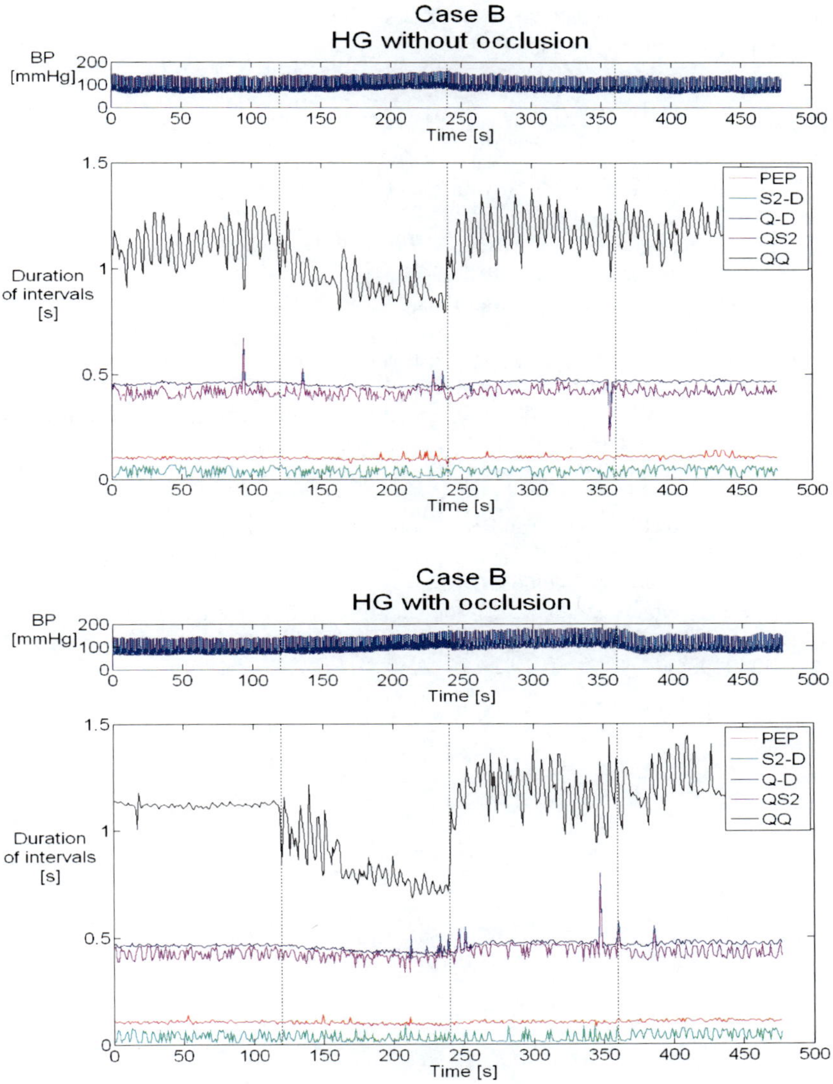

Fig. 9. Presentation of time courses of chosen time intervals for case B. Q-D curve oscillates only slightly, the oscillations of Q-S2 curve are much greater.

3 Conclusion

The automated beat-to-beat event detection and time intervals calculation allows to present continuous time course of systolic time intervals. Such presentation demonstrates oscillatory pattern of intervals changes, interdependence or lack of thereof between time intervals.

References

[1] Ahmed, S.S., Levinson, G.E., Schwartz, C.J., Etringer, P.O.: Circulation 46, 559–571 (1972)
[2] Cacioppo, J.T., Berntson, G.G., Binkley, P.F., Quigley, K.S., Uchino, B.N., Fieldstone, A.: Psychophysiology 31, 586–598 (1994)
[3] Schächinger, H., Weinbacher, M., Kiss, A., Ritz, R., Langewitz, W.: Psychosom Med. 63, 788–796 (2001)
[4] Brownley, K.A., Hinderliter, A.L., West, S.G., Girdler, S.S., Sherwood, A., Light, K.C.: Med. Sci. Sports Exerc. 35, 978–986 (2003)
[5] Goedhart, A.D., Willemsen, G., Houtveen, J.H., Boomsma, D.I., De Geus, E.J.: Psychophysiology 45, 1086–1090 (2008)
[6] Meijer, J.H., Boesveldt, S., Elbertse, E., Berendse, H.W.: Physiol. Meas. 29, S383–S391 (2008)
[7] Licht, C.M., Vreeburg, S.A., van Reedt Dortland, A.K., Giltay, E.J., Hoogendijk, W.J., DeRijk, R.H., Vogelzangs, N., Zitman, F.G., de Geus, E.J., Penninx, B.W.: Clin. Endocrinol. Metab. 95, 2458–2466 (2010)
[8] Hinnant, J.B., Elmore-Staton, L., El-Sheikh, M.: Dev. Psychobiol. 53, 59–68 (2011)
[9] O'Rourke, M.F., Yaginuma, T., Avolio, A.P.: Annals of Biomedical Engineering 12, 119–134 (1984)

Transcutaneous Blood Capnometry Sensor Head Based on a Back-Side Contacted ISFET

M.A. Ekwińska[1], B. Jaroszewicz[1], K. Domański[1], P. Grabiec[1], M. Zaborowski[1], D. Tomaszewski[1], T. Pałko[2], J. Przytulski[2], W. Łukasik[2], M. Dawgul[3], and D. Pijanowska[3]

[1] Institute of Electron Technology, Al. Lotników 32/46, Warsaw, 02-668, Poland
ekwinska@ite.waw.pl
[2] Warsaw University of Technology, Faculty of Mechatronics, Św. A. Boboli 8, Warsaw, 02-525, Poland
[3] Nałęcz Institute of Biocybernetics and Biomedical Engineering Polish Academy of Sciences, Trojdena 4, Warsaw, 02 – 109, Poland

Abstract. Authors present the construction of an electrochemical sensor head for non-invasive measurements of CO_2 tension in arterial blood. The main element of the capnometric sensor under development is a Back Side Contacted Ion Selective Field Effect Transistor. The transcutaneous sensor head consists of a plastic housing, in which the ISFET chip, an Ag/Cl reference electrode, a thermistor, a heating element, and a membrane that is permeable by gases are mounted. The design and fabrication technology of the BSC ISFET, as well as some preliminary experimental results obtained using the designed capnometric sensor head, are presented in this paper.

1 Introduction

Capnometry is a measurement technique used to determine the content of carbon dioxide (CO_2) in respiratory gases or in blood. It allows evaluating the efficiency of lung ventilation and the ability of cardiocirculatory and pulmonary system to eliminate CO_2 from tissues. The capnometric examinations of the blood gases are very important for the diagnosis and treatment of cardiopulmonary disorders, and allow the control of the course of treatment or circulatory and pulmonary support of severely ill patients [1].

The CO_2 content in blood can be expressed as a CO_2 tension named also as partial pressure in pressure units (kPa or mm Hg) or as a volumetric concentration value (vol %). There are some methods for the measurements of CO_2 in blood. The most classical and popular is the electrochemical potentiometric method introduced by Severinghaus and Bradley (1958). The method is based on the Stow-Severinghaus concept. In this concept, a glass pH electrode and reference electrode are placed in a bicarbonate solution. The measuring chamber (space where electrodes submerged in solution are located) are separated from the examined area of blood by a hydrophobic membrane. The voltage between the electrodes is a function of pCO_2 in blood (where pCO_2 is carbon dioxide partial pressure in blood which expresses the pressure of the dissolved carbon dioxide in the blood plasma). Nowadays, this concept is mainly used to design the blood sensors intended for in vitro method of measurement of pCO_2 in a stable temperature, usually, of 37°C.

This concept, however, can be used for in vivo non invasive method. In this case the partial pressure of CO_2 is measured transcutaneously using sensor that is fixed to skin surface. Such a sensor contains a heating system. The skin is heated to a temperature of 39-45°C for arterialisation of the capillary vessels in the skin. The heating system promotes better CO_2 diffusion through the skin to the sensor. The main disadvantages of this type of sensors are high impedance of an electrical source (voltage between electrodes), sensitivity to electromagnetic interferences, a high electrical noise level and also difficulty to design a miniaturized glass pH electrode.

The purpose of this article is to present a new design of a non invasive blood capnometric sensor having none of the above-mentioned disadvantages.

2 The Idea of Transcutaneous Capnometric Sensor of CO_2 Partial Pressure in Arterial Blood

The proposed idea of a blood capnometric sensor is based on a modified Severinghaus-Stow concept. In this idea, instead of a glass pH electrode, the semiconductor Back Side Contacted Ion Selective Field Effect Transistor (BSC ISFET) structure is used. The design of the transcutaneous sensor head for CO_2 measurements in blood is shown in Fig.1.

CO_2 exhaled from human skin penetrates through the sensor head membrane. The membrane material should assure gas diffusion (CO_2) and isolation from water and other liquids (hydrophobic properties). Material such as teflon, polypropylene, polytetrafluoroethylene provides not only high CO_2 permeability but also provides high chemical resistance, satisfactory mechanical properties and stability during measurement. Then, CO_2 reacts with the bicarbonate solution located in the sensor head. As a result of this chemical reaction pH of the solution changes. This small change of the pH can be measured with the use of a BSC ISFET.

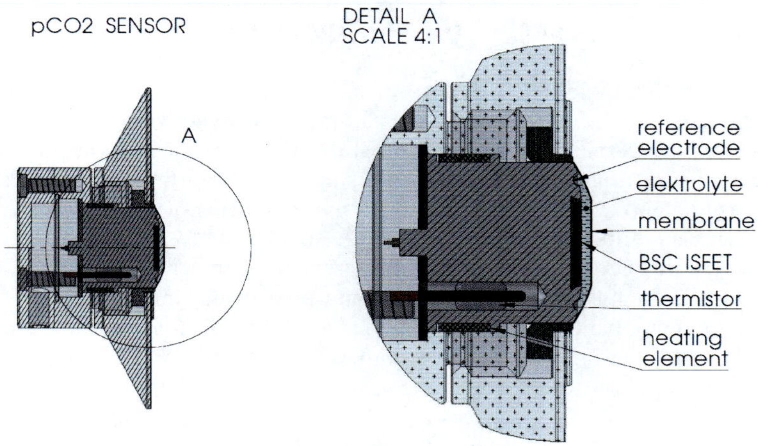

Fig. 1. Cross-section of a transcutaneous sensor head for CO2 measurements in the blood

The equilibrium in the electrochemical system is:

$$CO_2+H_2O <-> H_2CO_3 <=> H^+ + HCO_3^-$$

The electrolyte pH is changed depending on the quantity of CO_2 diffusing through the membrane. The relationship between the variables is:

$$pH=A-\log pCO_2,$$

where A is constant.

It was experimentally proved that sensor response stability to concentration of HCO_3^- ion requires at least 20 mmols of sodium bicarbonate in water in measuring chamber (the area between ISFET and reference electrode).

It is obvious in our experiment that CO_2 determination is normally carried out at the temperature of human blood (i.e. 37 ºC).

Increase of the human skin temperature from 39 to 45ºC increases the amount of CO_2 that penetrates through skin. This process causes the change of the pH value.

The sensor head contains also a resistive heater and a termistor for controlling temperature between 39 and 45°C. The heater was applied also for arterializations of capillary vessels in the skin. At higher temperature gas penetration through membrane increases. The sensor head is fixed to the surface of the skin by double–sided adhesive tape. pCO_2 in blood is determined by means of signal processing of measured pH of fluid (water solution) in which CO_2 is diffused.

3 ISFET – The Heart of Blood Capnometric Sensor

An Ion Sensitive Field Effect Transistor is a semiconductor-type electrochemical sensor. It converts the physicochemical parameter characterizing the tested sample into an electric signal. The ISFETs are still one of the most promising sensors. There are a large number of applications using ISFETs starting from simple in-lab pH measurements, to much more complex measurements of different and specific ions having an influence on water quality [2, 3].

After the invention of ISFETs with SiO_2 sensing membrane [4] and having possibility to determine human electrolytes and urea in dialysate [5], ISFETs have been applied in the bio-medical laboratories. In bio-medical and chemical applications the gate sensing material determines the ISFET microsensor quality. To obtain device pH sensitivity in the range from 45 to 50 mV/pH and a high output signal stability a stacked gate sensing membrane composed of silicon oxide (SiO_2) and silicon nitride (Si_3N_4) has been commonly used. Nowadays, a single hafnium oxide (HfO_2) layer, deposited directly on a Si gate area, has been widely used to improve device pH sensitivity [6, 7].

Currently, the requirements for bio-medical laboratories are focused on the minimization of the examined sample volume. A significant progress in the development of miniaturized systems for chemical analysis towards Micro Total Analysis System (μTas) and lab-on-a-chip is presently observed.

In this paper an innovative BSC ISFET (Fig.2) has been implemented in a new transcutaneous pCO_2 sensor head (Fig.1).

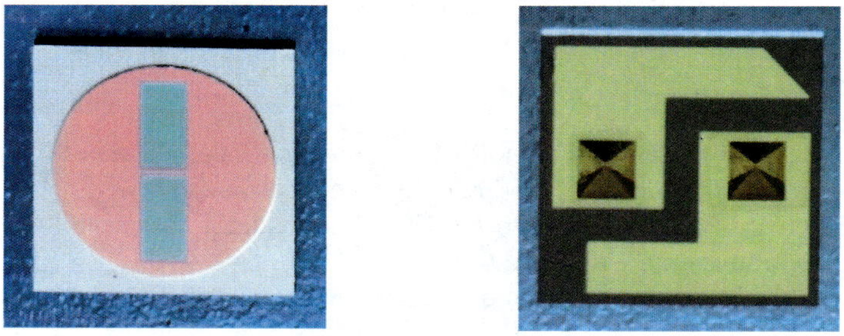

Fig. 2. BSC ISFET structure a) front side, c) back side

In this concept (see Fig.3), the Source/Drain (S/D) Au-metalized contacts are placed at the passive (back) side of the Si-chip that is completely separated from the front ion detection side. At the front side of the chip a gate stacked sensing membrane is exposed to a sensed medium. It comprises 2 layers: the silicon dioxide and silicon nitride (both 65 nm thick).

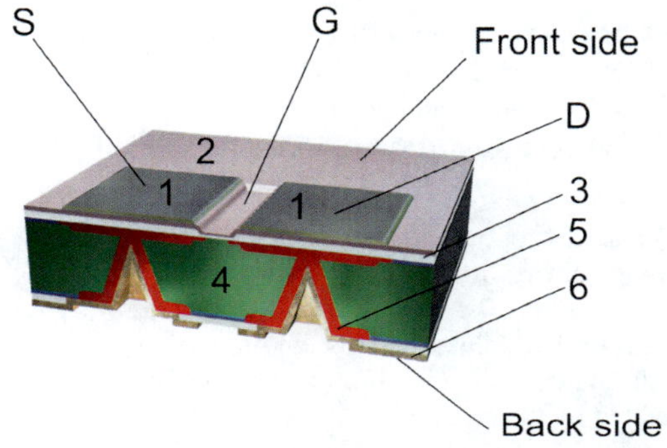

Fig. 3. BSC ISFET – cross section: 1 – Si_3N_4 layer; 2 – SiO_2 layer; 3 – p+ channel stopper; 4 – p- type substrate; 5 – n+ diffusion path, 6 – Cr-Au electrode; S – source, G – gate; D – drain, dimensions are not in the same scale

Silicon nitride is a well known material in CMOS microelectronics, especially for its long term stability and high resistance to hydration. An appropriate drain current value and device pH sensitivity are achieved by setting the W/L ratio to 14/600 μm. The BSC ISFET concept and manufacturing technology are widely presented in our other papers [8, 9]. The BSC ISFETs has been produced at ITE since 2005. To approach the main idea of the pCO_2 sensor heads as well as to fulfill such requirements as: maximum 20nl capacity of measuring cell and minimalization of silicon surface in the measuring cell our standard BSC ISFET chip was specially adapted (Fig.3). Both, a decreased chip size (from 4.9×4.9 mm^2 to 3×3 mm^2) and a 100 μm lowered front side chip surface out of 2.7 mm circle (with S/D areas) were done to fulfill sensor head design requirement and its water resistance. In order to achieve described above shape and requirements an additional photolithography step and wet etching of silicon dioxide followed by deep plasma etching of silicon were made.

4 Experimental Results of BSC ISFET Parameters Study

Experiments were held in laboratory conditions at a temperature of 20ºC ± 1ºC. During measurements a control system with a floating bridge was used [10], where 1.5 V was applied as the drain-source voltage, and 100 μA was applied as the drain current. 15 buffer solutions (0.005 M NaH_2PO_4, 0.1 M NaCl with additives of HCl or NaOH to adjust the desired pH value) of various pH were prepared. pH values of prepared solutions were: 2.010; 2.500; 2.970; 3.499; 4.009; 4.509; 5.013; 4.498; 6.003; 6.999; 8.015; 8.998; 9.996; 10.999; 11.996. Measurements were made every 10 minutes starting from pH=5.013 and changing solutions firstly in the direction of decreasing pH (to pH 2.010), then in the direction of increasing pH (to pH 11.999), and finally back to the starting point. The flow rate was 2ml/min and was forced by a peristaltic pump.

Using such obtained experimental results it was possible to determine the sensitivity of the tested BSC ISFET structure as well as hysteresis of BSC ISFET. Results of measurements for three chosen BSC ISFETs are compared in Table 1. on Fig.4 sensitivity for one choosen BSC ISFET is presented.

Table 1. Sensitivity and drift of 3 measured BSC ISFETs, S (mV/pH) – average sensitivity in the investigated range of pH; S_a (mV/pH) – average sensitivity for 2 – 7 range of pH; S_b (mV/pH) – average sensitivity for 7 – 12 range of pH, H – hysteresis, ΔU_{GS} (mV) for pH=7; R^2 – coefficient of determination (It is the proportion of variability in a data set that is accounted for by the statistical model. It provides a measure of how well future outcomes are likely to be predicted by the model)

	1	R^2	2	R^2	3	R^2
S	49.6	0.998	48.9	0.998	49.4	0.998
S_b	50.0	0.993	49.8	0.991	50.2	0.993
S_a	50.8	0.993	49.8	0.993	50.5	0.993
H	25.2		26.5		24.6	

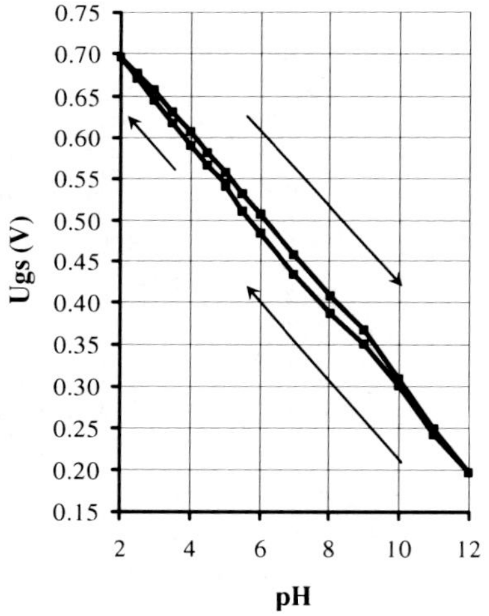

Fig. 4. Example results of BSC ISFET sensitivity – result for one of structures under investigation, U_{gs} – voltage gate – source

The average value of sensitivity in the investigated range of pH is 49.3 ± 0.29 mV/pH.

Next, the tested BSC ISFET structures were treated with a buffer solution of known acidity (pH = 5.013) for 20 days. As in the previous case, the measurements were made every 10 minutes and the flow rate forced by peristaltic pump was 2ml/min.

Average value of the drift is 0.0363 pH/10h ± 0.0004 pH/10h.

5 Experimental Results for Transcutaneous Blood Capnometric Sensor

The experiments were carried out in water before and after saturation of 100% CO_2 gas, which flowed through the examined fluid (water solution). During this process the output voltage of the CO_2 sensor decreased more than 100 mV. This means that the acidity of water solution increased more than two pH units because the sensor sensitivity was about 50 mV/pH. (Fig.5).

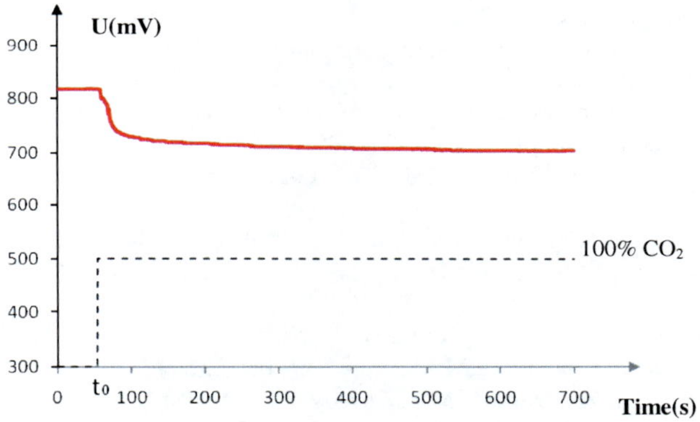

Fig. 5. pCO_2 sensor output voltage before and during 100% CO_2 gas flowing through the water, t_0 – beginning of CO_2 flowing, sensor head response time is less than 1 min.

6 Summary

The fabrication technology of BSC ISFET developed in ITE was successfully modified to adapt the ISFET chips to the designed capnometric sensor head. The sensitivity and long term stability of manufactured chips fulfilled the requirements of this application. The preliminary measurement of the designed sensor head proved its sensitivity to high concentrations of carbon dioxide dissolved in water and reasonably short response time (less than 1 min). We expect that the monitoring of pCO_2 in arterial blood will be possible by measurement of the output signal of the ISFET – based sensor. The output signal reflects changes of the ISFET threshold voltage, which is directly correlated with ion concentration in solution remaining in contact with the ISFET gate region. The ion concentration in turn is established by penetration of CO_2 molecules into the sensor gel longer through the patient skin and the sensor membrane. Hence the sensor output signal is induced by the CO_2 concentration in arterial blood.

Acknowledgements

This work was partially supported by Polish Ministry of Science and Higher Education within the project POIG 01.03.01-00-014/08-00 MNS DIAG co-financed by the European Union.

References

[1] Pałko, T.: Frontiers Med. Biol. Engnng. 10(3), 185–198b (2000)
[2] Filipkowski, A., Brzózka, Z., Wróblewski, W., Opalski, L., Ogrodzki, J., Kobus, A., Jaroszewicz, B., Moscetta, P., Sanfilippo, L., Alabashi, R., Temple-Boyer, P., Humenyuk, I.: Proceed. of the 19th Int. Conf. Informatics for Environmental Protection Enviro. Info., pp. 49–56 (2005)
[3] Zaborowski, M., Jaroszewicz, B., Tomaszewski, D., Prokaryn, P., Malinowska, E., Grygołowicz-Pawlak, E., Grabiec, P.: Proceedings of the 14th International Conference Mixed Design of Integrated Circuits and Systems (2007)
[4] Bergveld, P.: IEE Transactions of Biomedical Engineering BME-17, 70–71 (1970)
[5] Dawgul, M., Pijanowska, D., Jaroszewicz, B., Kruk, J., Grabiec, P., Torbicz, W.: Przegląd Elektrotechniczny 5, 27–32 (2008)
[6] Lai, C.-S., Lu, T.-F., Yang, C.-M., Lin, Y.-C., Pijanowska, D.G., Jaroszewicz, B.: Sensors and Actuators B 143, 494–499 (2010)
[7] Jaroszewicz, B., Tomaszewski, D., Grabiec, P., Wzorek, M., Yang, C.-M., Lai, C.-S., Pijanowska, D., Dawgul, M., Torbicz, W.: Proceedings of the Electron Technology Conference ELTE (2007)
[8] Jaroszewicz, B., Grabiec, P., Koszur, J., Kociubiński, A., Brzóska, Z.: Proceedings of 9th International Conference Mixdes, pp. 139–141 (2002)
[9] Jaroszewicz, B., Zaborowski, M., Tomaszewski, D., Taff, J., Panas, A., Skwara, K., Malinowski, A., Grabiec, P.: Proceedings of the Electron Technology Conference ELTE (2007)
[10] Krzyśków, A., Pijanowska, D., Kruk, J.: Controller of the operation point of the chemical sensor of the ion sensitive field effect transistor ISFET type, Polish Patent No. 178242 (1996)

The User Interfaces of the Polish Artificial Heart Monitoring System

B. Fajdek, M. Syfert, and P. Wnuk

Warsaw University of Technology, Faculty of Mechatronics,
Automation and Robotics Institute,
ul. Św. A. Boboli 8, Warsaw, 02-525, Poland

Abstract. The paper gives an overview of the remote artificial heart monitoring system, however, the main emphasis is placed on the presentation of available user interfaces that are used by the maintenance staff and doctors. At the beginning, the short description of the whole system structure together with presentation of its main modules is given. Then, there are presented two main user interfaces. The first called virtual console is dedicated for on-line monitoring of currently assembled devices. The second one is a web-based application used mainly for life-cycle management, maintenance and archival data analysis. Finally, the future development of the system is discussed.

1 Introduction

Nowadays a number of patients that are waiting for heart transplantation is still growing. Moreover, there is sufficient number of donors. One of the possible treatment method in these cases is mechanical heart supporting. Also, many patients which require long-time support involve the use of mechanical supporting devices outside the hospital [1]. It is very important to provide doctors and other staff essential remote systems, which can monitor physiological-related information as well as device-related information. Nowadays there is a growing number of systems for that purpose[2][3][4]. Mainly these are the systems dedicated to a particular, specific solutions / devices.

This paper presents the short description of the software solutions which have been applied in the system of remote monitoring, control and management of the Polish heart supporting system (POLCAS) [5]. A short summary of the functionalities implemented in the central monitoring system is presented. The described system has distributed structure. Its main components are: the embedded software system that runs on a computer built into the control unit of the artificial heart, a group of native communication programs created in C++ for Windows/Linux, and the central monitoring system based on the MySQL database and web-based application created with use of PHP, Javascript and AJAX technologies. The software modules which are located and implemented in the control unit of the artificial heart and working in real-time. They are responsible for collecting and buffering the current values of the parameters and measurements describing the state of device and the patient to which they are attached. The main part of the system is located on the central server of monitoring system. It is responsible for data storage, life-cycle management, maintenance and archival data analysis. It can be also used for on-line device monitoring, however, the data are read from the database in that mode. The important aspect of such kind of monitoring system is to provide flexible and intuitive tools to access to stored and processed data. The target customers of the system are technical and medical staff. For this reason, the proposed system should be adopted to fit their needs. There are two main user access interfaces to the system: the virtual console and web-based monitoring system. First, general system structure is described. Then, in consecutive sections, the main user interfaces are presented.

2 Core Modules of the System

The designed system consists of several interacting components. General structure of the system is shown in Fig. 1. One can distinguish the following elements of the system:

I. CMS² Server

a. **Connection module** - is responsible for communication between the POLPDU control units (its telemetric modules) and the central monitoring server. It receives: device-related information, physiological-related information and sends control commands to the POLPDU control units using the dedicated exchange protocol. It uses two main channels of communication: through Internet and GSM networks.

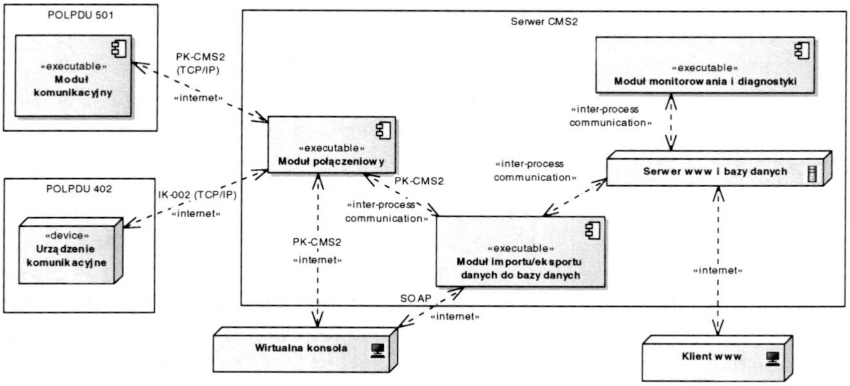

Fig. 1. The structure of the CMS2 system

b. **Import/export data module** – the main task that this module perform is storing / retrieving data to / from MySQL database.
c. **GUI module** – the set of PHP modules that are responsible for the server-side of web-based graphical interfaces.
d. **Monitoring and diagnostic module** – this is specialized module that performs advanced diagnostics of the POLPDU control units using measurements data stored in the database.

II. POLPDU device

a. **Communication module –** is responsible for communication between the connection module and POLPDU control unit. It sends device-related and physiological-related information to the central monitoring center. It is also responsible for handling orders received from the central monitoring center. It operates in the real-time mode.

III. User access interfaces

a. **Virtual console** – it is an independent software written in C++. It can communicate with the POLPDU control unit using two modes: direct connection and indirect connection through the connection module. Also using virtual console it is possible to retrieve archival data from database.
b. **Web-based user interface –** web-based (client side) module that allows management of the central monitoring system (e.g. user management, permission management, control units configuration management). It also allows online monitoring device-related information and analysis of archival data. It is written with use of JavaScript, HTML, Smarty, CSS and Ajax technologies.

3 Virtual Console

The virtual console is a dedicated software which allows remote monitoring and supervision of the POLCAS system. The application is intended to be used by medical and technical when the particular unit is in active work (connected to the patient), as well as during passive work (without the patient – service mode). The software can be also used for training medical personnel. For this reason, the graphical interface of the application has been adopted to the real look of the POLPDU control unit panel. The application allows:

- remote connection to the POLPDU control unit that is registered in the central database system,
- verification of the user trying to connect remotely to a specified unit (through login and password),
- to show basic control unit identification data with the type of connection mode,

- distinction of the working modes of the application according to the user permission (different permissions for medical and technical staff),
- support for the basic modes of device operation:
  - o BiVAD mode – supporting two artificial heart chambersm
  - o LVAD mode – supporting only left ventricle,
  - o RVAD mode – supporting only right ventricle,
  - o TAH mode – the device work without the ECG synchronization,
  - o SYNCHRO ECG – the device works with ECG synchronization,
- to show the main device-related information (the pressure in the tanks: hypertension, vacuum, the major power status, battery charge status, etc.),
- to visualize, so-called, fast-changing parameters (e.g., the control pressure in the left chamber, the control pressure in the right chamber, the ECG signal),
- to visualize the archival data that are stored in the central database of the monitoring system.

The graphical interface of the application is shown in Fig. 2.

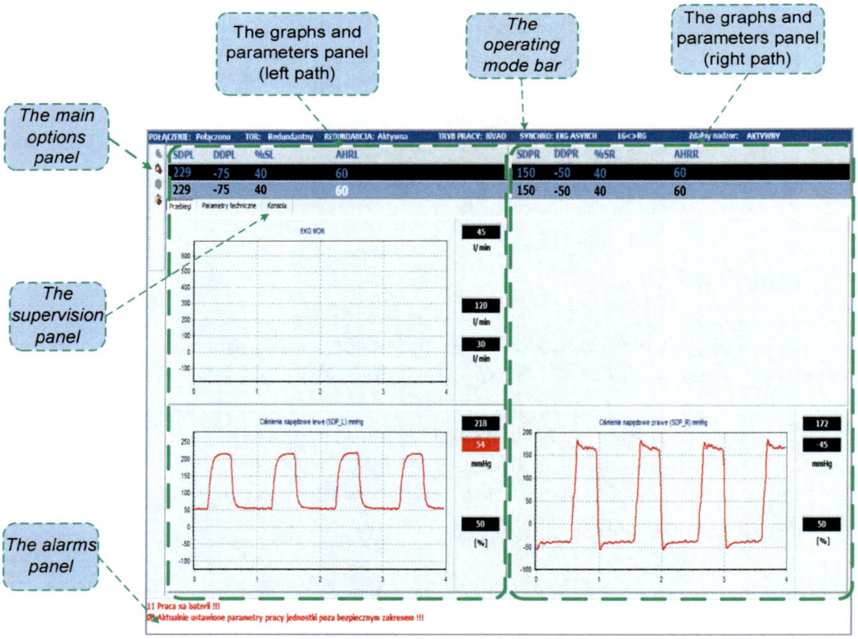

Fig. 2. The graphical interface of the virtual console application.

The main application window consist of:

- main options toolbar – includes basic application options that allows: connection to the POLPDU control unit, terminate connection, configuration of the basic connection parameters,
- graphs and parameters panel - the left path – includes basic parameters of the left ventricle (the left pneumatic path) such as pumping/suction pressure, left systolic percent, etc., and waveforms of left ventricle pressure and ECG signal,
- graphs and parameters panel - the right path – includes basic parameters of the right ventricle (the right pneumatic path) such as pumping/suction pressure, right systolic percent etc., and waveforms of right ventricle pressure,
- operating mode bar – the bar indicates the current operation mode of the control unit and the status of the connection,
- alarm panel – displays the queue of alarms raised by the control unit.

Depending on the operating mode of the control unit only the relevant parameters and graphs are displayed. Therefore, it is possible to modify only the parameters that are appropriate for a given operating mode. The appearance of the left pneumatic path parameters panel is shown in Fig. 3.

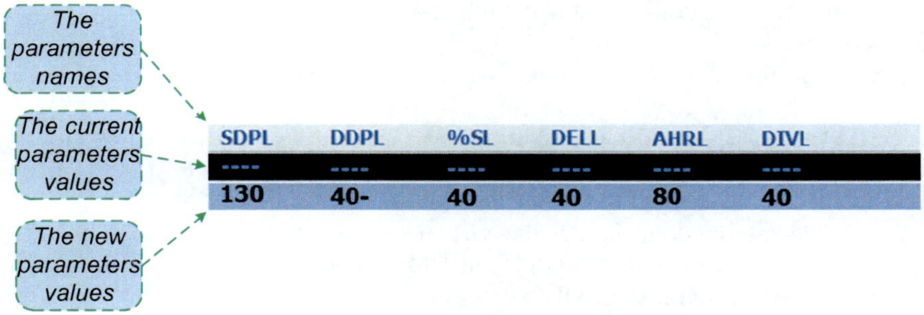

Fig. 3. The left pneumatic path parameter panel

In comparison to the actual view of the real POLPDU control unit an additional bar of parameters was added to allow applying new parameter values. Sending new parameter settings must be preceded by the user confirmation of the changes and internal verification.

The application has been also equipped with the supervision module (Fig. 4).

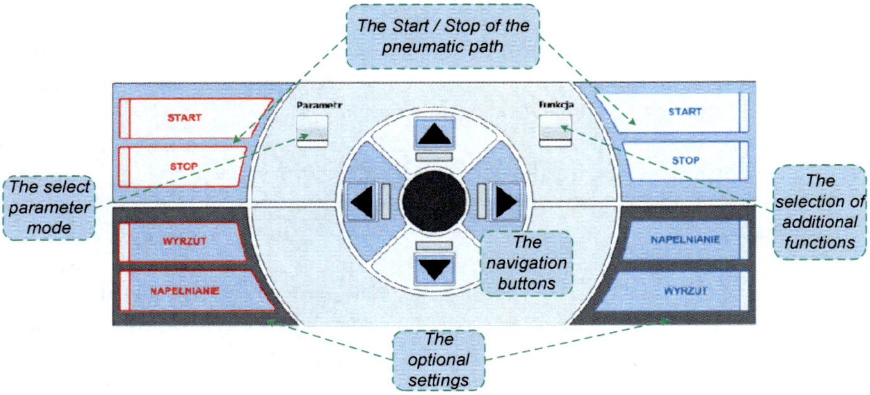

Fig. 4. The graphical interface of the supervision module of the virtual console

The graphical interface of the control console has been adopted to the real look of the POLPDU-402 control console. Such approach allows quickly familiarize with the interface, especially by the medical staff. The module allows to change the basic parameters of the POLPDU control unit remotely (e.g., the discharge/suction pressure for the left/right path, the operating frequency of the left/right path, etc.).

The application can operate in the two different modes:

- The direct mode: the virtual console application communicates directly with the POLPDU control unit using the GPRS network. It is the emergency mode, used mostly in cases of the central server failure
- The intermediate mode: the virtual console application communicates with the connection module on CMS[2] central server (the server communicates with the POLPDU control units). Is is the default operation mode.

3.1 User Interfaces of the Monitoring Center

The user interfaces of the central server allows user to fulfill the task from three main groups:

- management of configuration data,
- manage and analyze archival data,
- on-line device monitoring.

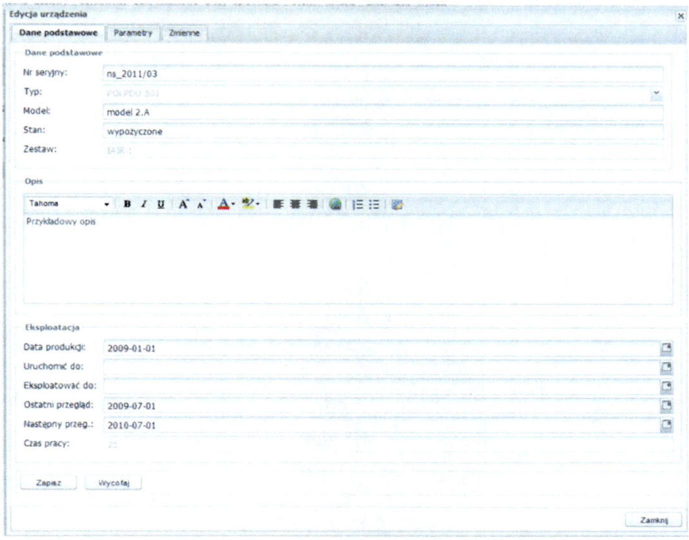

Fig. 5. An example of the graphical web interface: the artificial heart sets configuration management

In the field of devices configuration management the center allows to perform the necessary operations connected with the lifecycle management of the devices that are the part of artificial heart assemblies (the sets to which the patients are connected), such as:

- management of individual devices (defining according to the patterns, identifying the set of measurement variables and parameters, setting the operational parameters),
- assembly management (creating, deleting, etc.),
- recording service activities that are performed at the level of devices and assemblies,
- analyzing the history of use of particular devices in particular assemblies.

In the successive versions of the system there will be added the following functionality: the possibility to define new pattern devices (new products), management of patients and doctors, tracking the parts supply chain, selling and borrowing devices including the database of clients, advanced maintenance operations and advanced reporting.

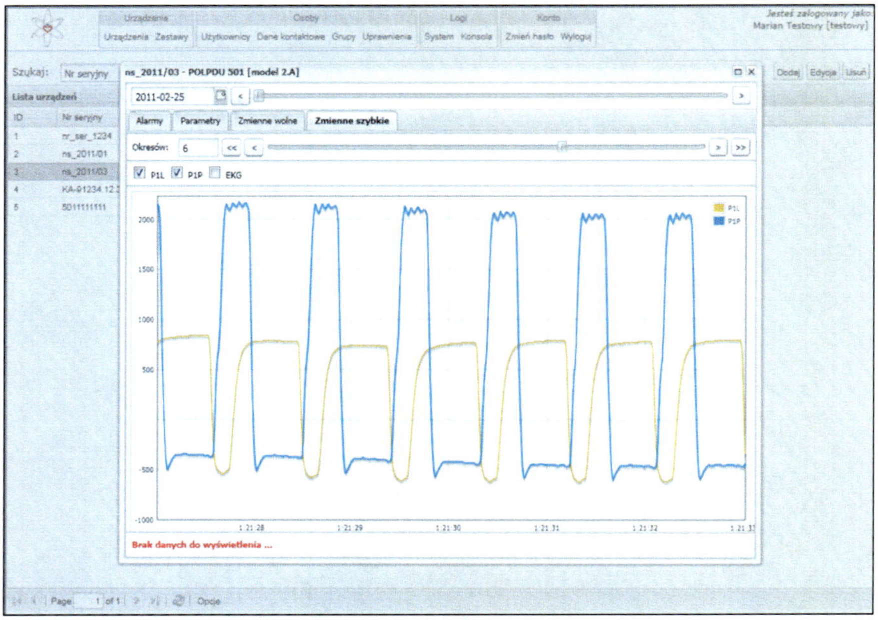

Fig. 6. The preview of the current state of the artificial heart control unit in web browser

The second group of tasks allows the user to monitor the behavior of the devices working in assemblies by analyzing the archival data collected during its operation as well as reported alarms. Through the specialized interface the user has access to short-term history that is automatically created for each working device. The time horizon of the history is defined by the system administrator. It is normally set to 1 month. The user interface is equipped with dedicated selection mechanism that allows for quick search of the time periods where the data from particular device are available. There is also a possibility to create long-term (in fact eternal) archives. For the particular device the user can point out the time window from which the long-term archive will be created. The data stored in such archives can be also easily searched with use of implemented dedicated search engine.

In addition, the Web interface of the CMS^2 system allows on-line data monitoring for particular device or assembly. This part of the interface is realized analogously to virtual console, but its operation is different. This module displays the data stored in the central database, it does not use direct communication with devices.

All the data are presented in a graphic form (automatically or manually updates charts), and can be exported for use in external systems.

The superior in respect to the above mentioned basic tasks of the system is the user management (including the division into the different roles and groups) and the advanced permission management system in which permissions are allocated at the level of devices, users, groups, roles and individual activities.

This part of the system is implemented in accordance with the Web-desktop technique, which is based mainly on JavaScript an Ajax technology. The server part was implemented in PHP language.

4 Conclusion

Reliable continuous remote monitoring systems for patients with an artificial heart implant which are discharged from the hospital after suitable stabilization period is gaining practical meaning. The paper presents short description of the designed remote monitoring system for the Polish heart supporting system. The main core modules were shortly presented. The system has been tested during the measurement series that was carried out in the Foundation of Cardiac Surgery Development in Zabrze and with the data generation emulators installed on remote sites for simulation studies. The system showed acceptable functionality and reliability. The main emphasis in this Section was placed to present the user access interfaces to the system. The description of the main features that provides the virtual console application and the web-based monitoring system were presented. Further works are in progress to complete the technical and clinical integration tests.

Acknowledgements

This work was supported by the National Centre for Research and Development (NCBiR) under the Project "Development of metrology, information and telecommunications technologies for the prosthetic heart" as part of multiannual "Polish Artificial Heart" Program (task 2.1 – Developing a system of automatic control and supervision of work for extracorporeal cardiac prosthesis).

References

[1] Zielinski, T., Browarek, A., Zembala, M., Sadowski, J., Zakliczynski, M., Przybylowski, P., Roguski, K., Kosakowska, A.B., Korewicki, J., POLKARD HF investigators: Risk Stratification of Patients With Severe Heart Failure Awaiting Heart Transplantation, Prospective National Registry POLKARD HF, Transplantation

[2] Choi, J., Park, J.W., Chung, J., Min, B.G.: An inteligent remote monitoring system for artificial heart. IEEE Trans. Inf. Technol. Biomed. 9(4), 564–573 (2005)

[3] Bingchuan, Y., Herbert, J.: Web-based real-time remote monitoring for pervasive healthcare. In: 2011 IEEE International Conference on Pervasive Computing and Communications Workshops (PERCOM Workshops), March 21-25, pp. 625–629 (2011)

[4] Wu, S., Jiang, P., Yang, C., Li, H., Bai, Y.: The development of a tele-monitoring system for physiological parameters based on the B/Smodel. Computers in Biology and Medicine 40(11-12), 883–888 (2010)

[5] Kustosz, R., Gawlikowski, M., Darłak, M., Kapis, A.: Pneumatyczny system wspomagania serca POLCAS do zastosowań długoterminowych. Śląskie Warsztaty Biotechnologii i Bioinżynierii Medycznej – BIO-TECH-MED. Silesia 2005, Cardiac Surgery Development Foundation, Zabrze (2005)

The Latest Developments in the Construction of an Electrogustometer

A. Grzanka*, T. Kamiński, and J. Frączek

Warsaw University of Technology, Institute of Electronic Systems,
ul. Nowowiejska 15/19, 00-665 Warszawa, Poland

Abstract. This paper presents a new concept for the gustatory sense examination. The gustatory sense examination methods used so far made use of gustometry and electrogustometry. All the former examination methods for the gustatory sense threshold were subjective by nature. After delivering a stimulus the examined person had to decide whether he or she senses the taste and to signal this to the examiner. In order to eliminate the subjective influence of the patient on the outcome of the measurement, we have used the gustatory evoked potentials. To trigger and register the gustatory evoked potentials, we have designed and developed a special system described in this paper.

1 Introduction

Gustometry is a diagnostic technique determining the sensitivity of the taste. The examination is very important in the diagnostics and prophylaxis of many disease and physiological entities, e.g. in gustatory stimuli sensing disorders, the so called dysgeusia, mental diseases, neurological diseases, diabetes, sclerosis multiplex or the effects of substances harmful for the gustatory threshold. The reasons for the gustatory disorders may be of various kinds, from pathologies localized in the oral cavity, through gustatory duct /tract damages, to nutritional deficiencies and metabolic disorders associated with systemic diseases. In each case of gustatory disorder, detailed diagnostics is essential, since it may be the first symptom of many serious diseases [1, 2, 3, 4, 5]. In the gustatory sense research there are used two methods employing chemical or electric stimulation. Classical gustometry based on stimulating the gustatory sense with chemical solutions is very hard in application and hardly reproducible[6, 7]. Among the various methods of gustometry, electrogustometry deserves particular attention. It is a method of examining the taste with the use of small electric stimulations applied to the surface of the tongue. This method was for the first time proposed by Bent Krarup in 1958 [8]. The Bent Krarup device is equipped with two stainless electrodes: the active one – the anode (flat, oval, 5 mm in diameter), applied to the tongue and the neutral one (rectangular, 4 cm x 5 cm) – placed on the wrist, on a piece of fabric damped with

* Corresponding author.

sodium chloride. The active electrode was flat. The tests showed that the oval electrodes of approx. 5 mm in diameter are the best. The electrode was placed on a short holder to examine the tip of the tongue, or on a long holder to examine the back of the tongue. There was also a possibility to place the two electrodes on a double holder for simultaneous examination of the both sides of the tongue. The button in the handle switched the electric stimulation on or off.

2 The Construction of the System for Recording Gustatory Evoked Potentials (GEPs)

All the former examination methods for the gustatory sense threshold were subjective by nature. After delivering a stimulus the examined person has to decide whether he or she senses the taste and to signal this to the examiner. Applying advanced measurement algorithms and computing methods we may decrease the subjective influence of the examined subjects on the results of the measurement. A development of an objective gustatory sense measurement method seems however a much better solution. A concept has been developed to construct a system enabling registration and analysis of the GEPs – Gustatory Evoked Potentials. The system comprises the following three interacting and cooperating modules (subsystems):

- The electrogustometer
- The GEP recorder – a device for gathering GEP signal
- A PC with suitable software – the "GEP Reader"

The electrogustometer is designed to stimulate the gustatory sense, the GEP recorder module is used to register and pre-process the signal, the PC with software ("the GEP Reader") enables recording, visualizing and digital processing of the GEPs.

As opposed to the earlier applied solutions in this study there were used a current of rectangular waveform and a bipolar application of the stimulus on the tongue. It consists in both electrodes being simultaneously positioned on the tongue, thus the stimulus is delivered locally. The stimulation is conducted with impulse current of rectangular waveform.

The electrogustometer is constructed from z 6 components (modules): the ATMEGA microcontroller, a current source controlled by microcontroller, a 6 – button membrane keypad, an LCD, power management circuit and an optional USB communication module. The power system supplies 3 voltages from 5 V and ±15 V. The USB module enables communication of the electrogustometer with the PC and cooperation with specially designed software. The electrogustometer block diagram is presented in fig.1.

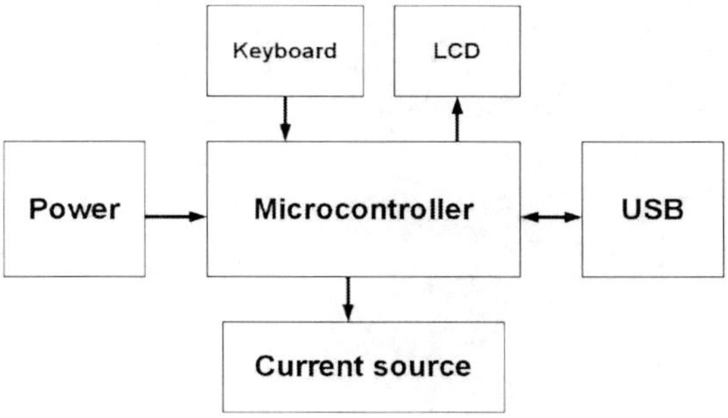

Fig. 1. The electrogustometer block diagram

The electrogustometer is made in a portable version. The device is supplied with five AA batteries (or NiMH or NiCd batteries), it has a 4 row display, a keypad and three connectors. These allow for connecting the measuring electrode, the GEP recorder (the trigger) and the patient's button. The button is used if the electrogustometer works as an independent device in the subjective measurements mode. The patient's button enables the patient to signal the change from 'sensing' to 'not sensing'. The trigger connector enables synchronizing the operation with the GEP recorder. The device makes use of an electrode of our own design. The construction of the electrodes enables taking measurements with the patient's mouth closed. This has great influence on the repeatability of the GEP examinations and on the patient's comfort.

The information on the device status (operation, standby) and on the preset parameters is displayed on a 4 row LCD. The stimulation parameters: the current intensity, the stimulation duration, the frequency of the stimulation waveform, the number of single stimulations are all to be set by means of a 6 – button membrane keypad. Using the keys one may launch or stop the measurement sequence. The electrogustometer may operate in three modes:

- Operation in the GEPs registration mode
- Independent operation enabling 'standard' electrogustometric examination
- Operation with the PC – the electrogustometer is controlled from the PC level and the examination results are stored in a database

In the GEPs registration mode, the electrogustometer is the master device and then it launches the measurement sequence. Before commencing with the measurement, it is necessary to preset the stimulation signal parameters and the number of stimulations in the electrogustometer. The device allows presetting 1 up to 200 stimulations. Before each stimulation, the device sends a synchronizing TTL signal enabling the synchronization of the GEP recorder. This signal is synchronized by the GEP recorder. Thus it is possible later on to establish the mean value of the registered single GEPs, to process and analyze them.

The second module of the system is a complete device for gathering and transmitting in real time the gustatory evoked potentials (GEP) signal from the human brain to the PC. The device consists of the application, control and transmission sections. You can see the schematic diagram of the device below:

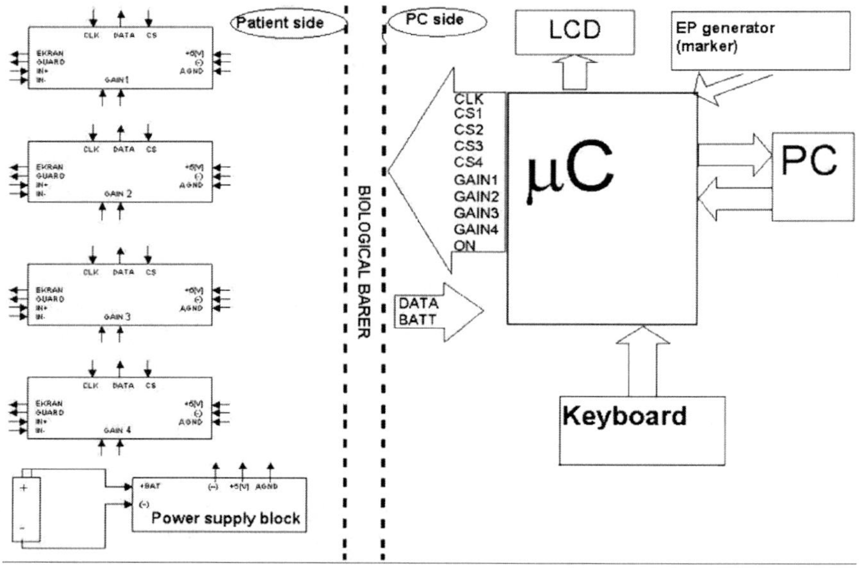

Fig. 2. Schematic diagram of the GEP recorder

On the left side there are four biological amplifiers and a power supply block with a battery. On the right side there are a microcontroller, a keyboard, an alphanumeric LCD screen and a transmitter which communicates with the PC. Between the left and the right sides there is a biological barrier separating the patient from the mains power supply. The following paragraphs describe the details of each element of the diagram in figure 2. Every part on the "patient side" is battery powered, but on the "PC side" it is powered from USB port.

Power block supplies four biological amplifiers with positive voltage +2.5 V, negative voltage -2.5 V and ground. It takes energy from a single 9 V battery or six AA accumulators. Additionally it alerts when the supply voltage drops below the limit value, which means that battery is discharged.

Four biological amplifiers, each biological amplifier is a single channel of the device. Their task is to prepare a biological signal, such as an EP to be received by an analog-to-digital converter. Each amplifier, after receiving the signal from the electrodes on the patient's head, amplifies it 2000, 20000 or 50000 times and filters it with band pass frequency response. Frequencies from 1[Hz] to 30[Hz] are amplified but others are attenuated. It also clears the signal from the 50[Hz] noise by means of a notch filter which is based on the Wien-Robinson bridge. The amplified and filtered signal is fed to the ADC which has 12-bit resolution. The table below shows how maximal amplitude and resolution depends on the gain of the amplifier.

Table 1. How maximal amplitude and resolution depend on the gain.

Gain [V/V]	Maximal amplitude [μV]	Resolution [μV]
50 000	50	0,049
20 000	125	0,122
2 000	1250	1,222

Biological barrier, its function is to separate the patient from the mains and prevent from an electric shock. The biological barer is made of two types of isolators: transoptors (for slow signals) and digital isolators based on transformers (for fast signals).

The microcontroller collects data from the analog-to-digital converters (ADC), sets the gain of each channel, transmits the samples to the computer via an FTDI chip, controls the LCD screen and the keyboard and receives the marker from the EP generator. The keyboard allows the user to turn on and off the biological amplifiers, set gain of each channel, start and stop the measurements and set the sampling rate. The alphanumeric LCD screen is used to communicate the user with the system, for example to inform about the current working mode of the device such as gain in each channel.

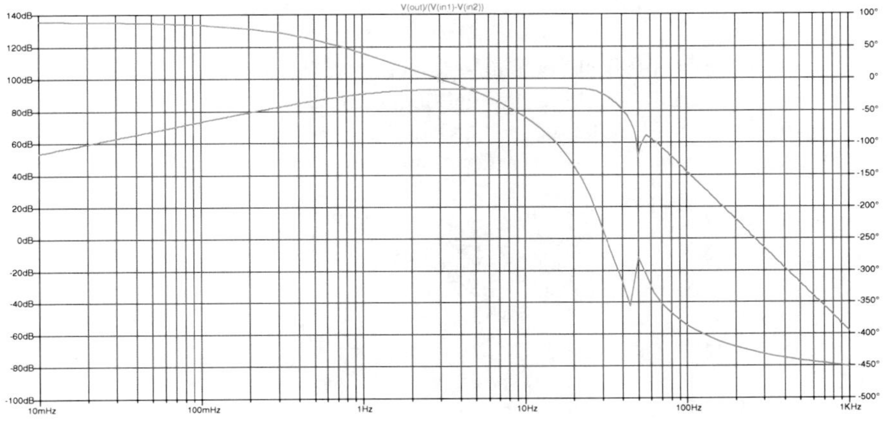

Fig. 3. Frequency response of the device in logarithmic scale of the gain

It supplies the device with power (only the "PC side"), receives data from the device, averages samples and presents the processed data on the computer screen. The transmission P2 protocol between the device and the PC is shown below, in table 2.

Table 2. Single frame transmitted to the PC

Byte	1	2	3	4	5	6	7
	10100101	1011010	00000010	Packet Counter	MSB Ch1	LSB Ch1	MSB Ch2
	Synchronization		Version				

8	9	10	11	12	13	14	15	16	17
LSB Ch2	MSB Ch3	LSB Ch3	MSB Ch4	LSB Ch4	MSB Ch5	LSB Ch5	MSB Ch6	LSB Ch6	EP Marker

After the bytes from all the channels have been transmitted, the EP marker becomes the last byte of the frame. It comes from the EP generator. When the EP generator is stimulating the patient's tongue, the EP marker is sent to the microcontroller of the GEP device and the GEP device sends the EP marker to the PC in the last byte of the frame. Thanks to that the computer program is informed when the single stimulation has begun.

The third element of the system is the PC with appropriate software designed specially to control the GEP registering and processing system. The application has been developed by Igor Podobiński as part of a student project. Mr Podobiński made use of the C# programming language for .NET 4.0 and Windows Presentation Foundation Interface. As mentioned before, the data from the GEP recorder is transmitted to the PC by means of a USB interface, with the use of the P2 protocol. The GEP data from the device is kept in the GEPData container. In general, this container consist of three main fields:

- Raw data – this is all the data received from the device. (for six channels)
- Type of data – could be normal data or a marker (which is set when the next sequences to be meant start)
- Mean data – these are averaged sequences. (for six channels)

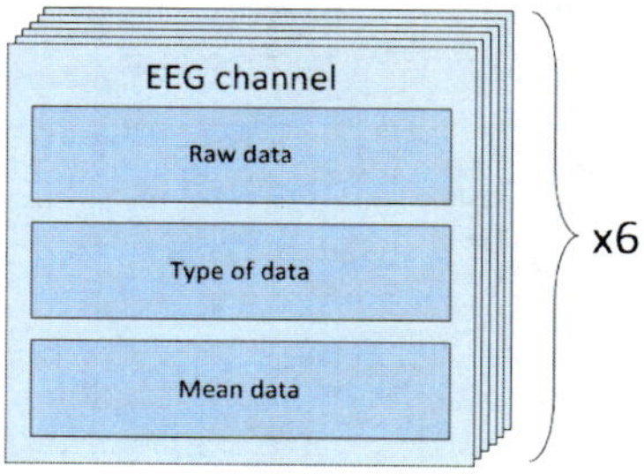

Fig. 4. GEPdata container

The operation cycle of the application may be divided into two stages: Get data from the device and process the received data and visualize it. The reception of the data from the GEP recorder is shown in the diagram flowchart fig. 5. The application reads the serial port buffer to get the data from the device using the P2 protocol. If there is any data then the programme puts this into the GEP container (GEPData). Having received the data, the programme can draw charts of raw data, mean data or one sequence (as necessary). Raw data can be drawn in whole or from markers (synch points). Sequences of data can be displayed as averaged separately in each channel.

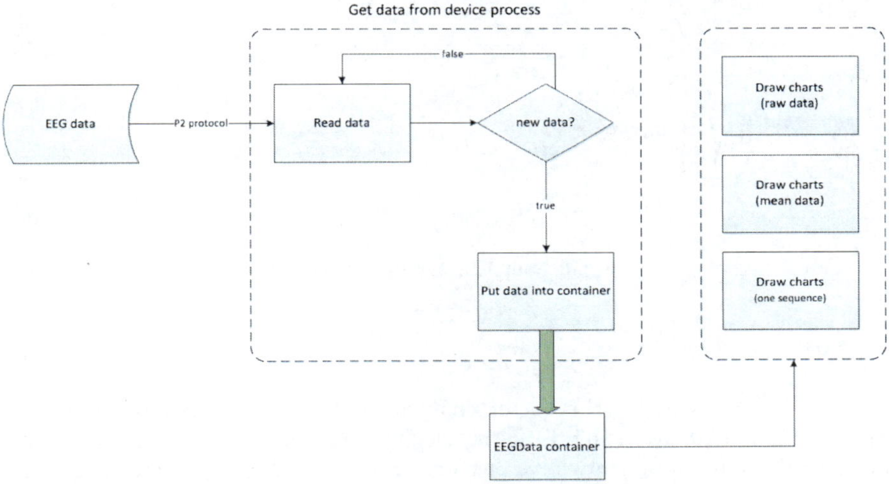

Fig. 5. Flowchart

On the screen there can also be presented waveforms with raw data as well as waveforms of averaged data. It is possible to set chart axes by specifying the minimum and maximum values for the Y (amplitude) and the X (time) axes. The raw GEPs as well as averaged GEPs can be saved into a file and then retrieved. This enables processing and analyzing the registered GEPs in other programmes. The GEPRecorder application is under constant development, we are planning to extend it with an advanced digital signal processing module. A sample programme window is presented in fig. 6.

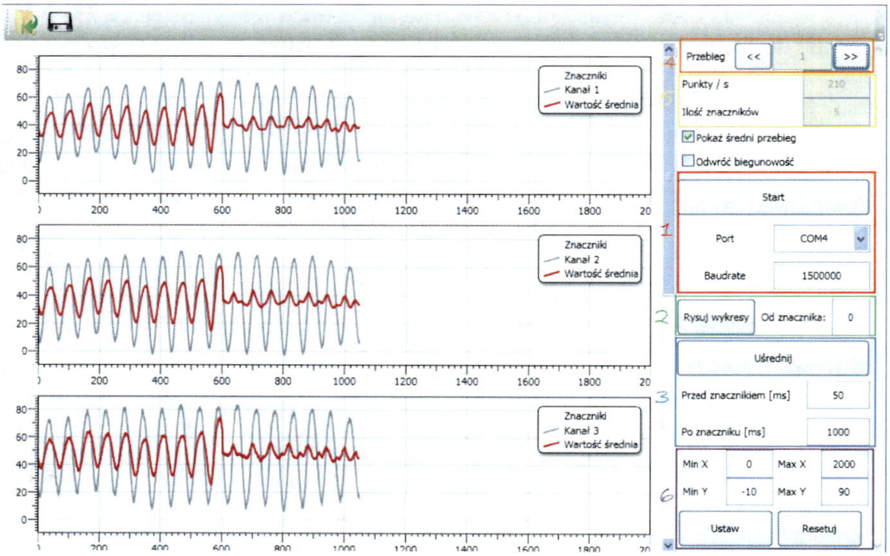

Fig. 6. Sample programme window

3 Summary

The system presented in this paper enables examining the gustatory sense by means of two methods: using the current techniques and electrogustometric methods as well as using the innovative gustatory evoked potentials GEP technique. At present we are conducting patient examinations making use of the gustatory evoked potentials technique. The results should enable the development of effective gustatory sense examination methods.

References

[1] Kamiński, T., Grzanka, A.: A Concept of Objective Electrogustometry. Biomedical Engineering, IASTED, Innsbruck (2010)

[2] Knychalska-Karwan, Z.: Jezyk: fizjologia i patologia, WUJ, Kraków (2000)

[3] Ryglewicz, D.: Zaburzenia smaku w chorobach neurologicznych. Terapia 1(146), 30–33 (2004)

[4] Balczewska, E., Nowak, A.: Zaburzenia smakowe – dysgeusia. Nowa Stomatologia 1-2, 12–16 (2000)

[5] Sienkiewicz-Jarosz responsem. i in., Taste responses in patients with Parkinson's disease. Journal Neurology Neurosurgery Psychiatry 76, 40–46 (2005)

[6] Konopka, W., Dobosz, P., Kochanowicz, J.: Zaburzenia smaku w otolaryngologii. Otolaryngologia 2(4), 145–149 (2003)

[7] Dobosz, P., Konopka, W., Grzanka, A.: Dwubiegunowa elektrogustometria pradowo-impulsowa. Prace Naukowe Politechniki Warszawskiej Elektronika, z.157 (2006)

[8] Krarup, B.: Electrogustometry: a method for clinical taste examinations. Acta Oto-laryng 49, 294 (1958)

[9] Prutchi, D., Norris, M.: Design and development of medical and electronic instrumentation. John Wiley & Sons, Chichester (2005)

[10] Augustyniak, P.: Przetwarzanie sygnałów elektrodiagnostycznych, AGH Uczelnianie Wydawnictwa Naukowo – Dydaktyczne, Kraków (2001)

In Silico Simulator as a Tool for Designing of Insulin Pump Control Algorithm

H.J. Hawłas* and K. Lewenstain

Warsaw University of Technology, Faculty of Mechatronics,
Institute of Metrology and Biomedical Engineering,
8 Sw. A. Boboli Str., 02-525 Warsaw, Poland

Abstract. The Diabetes Mellitous treatment is becoming more important issue due to significantly increasing number of affected persons. There is a need for new methods that will facilitate treatment and improve life quality of diabetic patients. An automatic insulin pump control is one of possible solutions. This paper concerns in silico test as a tool for investigation for such control. The most important facts about in silico trials and in silico simulator are discussed. Shown differences of in silico subject response to meal / insulin scenario indicates need for individualized insulin pump control algorithm. Benefits of premeal insulin delivery and inter-individual metabolism differences prove that concept of intelligent control algorithm that "learns" typical day of patient would be a great medical tool.

1 Introduction

Type 1 Diabetes Mellitous (T1DM) in simple terms is a disease in which the body is not able to independently regulate blood glucose levels due to lack of insulin. Insulin is a hormone that lowers blood sugar levels. Diabetes is a chronic disease with no cure, and if left untreated leads to severe complications and even death. According to The World Health Organization Fact Sheet [1] Diabetes is becoming an increasingly important issue for the health of people around the world.

Key facts about diabetes are:

- More than 220 million people worldwide have diabetes.
- In 2004, approximately 3.4 million people died from the effects of high blood sugar.
- Over 80% of deaths in people with diabetes occur in low and middle income countries.
- WHO predicts that the number of deaths due to diabetes will double between 2005 and 2030.

---

* Corresponding author.

This paper focus on technology that will facilitate and improve the life quality of patients with T1DM. Ongoing studies aim at developing an algorithm to allow the automatic control of a personal insulin pump based on signals received from Continuous Glucose Monitoring device (CGM). Such a system would enable a device that could be perceived as an artificial pancreas, that controls level of glucose in patient's blood. Preliminary requirements for such system were stated in previous paper [2].

2 Use of Medical Tests

The project of every complex device must be subjected to verification confirming the correctness of its operation. Developing of any device and its verification is quite simple if the mechanisms occurring in a controlled phenomenon are fully known, and the process is stable, repetitious and random parameters are negligible. For medical devices supporting the living organisms it becomes complicated issue, because the processes are not completely repetitious, they include a number of input variables, and some of them are random. Verification of such a device is even more important because health or even life depends on its operation.

Facing the problem is not trivial because of the complexity of the processes occurring in the body and the strong influence of interindividual characteristics of each subject. There is significant influence of random variables as meals (quantity, duration, time) and the body demands for a glucose (physical activity, agitation). It is very important because the consequence of exceeding the acceptable range of blood glucose in either direction, is life threatening.

For these reasons, a purely theoretical concept of control does not allow to cover all data, therefore, an algorithm and its verification requires testing.

Regarding stated subject there are two main possibilities of conducting research in biomedical engineering. Due to the high risk, preliminary even advanced experiments can not be conducted on humans.

The tests may be performed by the traditional in vivo - in a living organisms. In this case, the object of research is likely to be dogs with surgically removed pancreas. Such tests require appropriate licensing, provision of medical care to the animals during the tests and their proper preparation to the tests. In summary, such studies would have been lengthy and costly. The results would have be available within several years, and to some extent inaccurate due to interspecific differences.

Research may also be implemented as in silico, or using a computer model that reflects the function of living organism. In short words, to conduct such studies we need a computer with an appropriate software.

In silico research is fast, cheap and the most important do not pose risks to living creatures. Testing the device in extreme conditions, often leading to death of the virtual subject, does no ethical doubt. The results of such studies are available immediately. Furthermore subjects can be re-examined to verify how a patient reacts to even minor changes in treatment. In silico subject is always in the same "initial condition" regardless of their medical history and well-being. In addition, the test is free from errors or inaccuracies caused by human factors such as measuring inaccuracy or a dose omission.

In silico simulations enable cheap and fast rejection of treatments that are doomed to failure, or to be ineffective.

Computer test performs many experiments covering variation occurring in the whole population rather than carrying out tests on animals on a small scale (several individuals).

The main problem associated with in silico research is creation of the appropriate computer model adequate to the planned research. This is mainly due to the prediction of parameters describing the behavior of the object and then designate the values of these parameters.

Despite many advantages it should be noted that data from in silico simulations will be burdened with some error, which arises from the model approximation. However, even in the in vivo studies carried on animals are subject to error associated with interspecies differences. Sometimes interspecies differences may have dramatic impact on test results. An example of this might be the case of the first test of penicillin antibiotics. If penicillin was tested on guinea pigs instead of mice, it is unlikely it that would wait its implementation. Surely further research would be abandoned because of the severe toxicity of penicillin for guinea pigs.

Therefore, a model based on human body responses, to the given stimulation, can be regarded as a very reliable or comparable to in vivo tests on animals. Also in silico tests can be repeated without prejudice to the animals.

3 Simulation Model

A typical computer model, used for instance in the construction of machines is determined by averaging the parameters from a study of a whole population. As the result of the simulation we achieve response from the model that is the most likely value of the entire population. So we conduct a single simulation, and model answer with single averaged response.

The in silico computer model to determine the body's response to some type of stimulation must be implemented in different way. We can not get the response that is averaged but one that takes into account the specific characteristics of the test subject. In this case, it is required to obtain a separate response for each subject, i.e. we perform n simulations for all n individuals on the basis of which the model was created. This approach takes into account the results of intra-individual characteristics of each virtual patient. This allows us to check whether the results are correct for all subjects, and not only that statistically averaged result is correct. This is crucial for testing the body's response to insulin. It allows to pick the features that has a significant impact on the results. It also increases the reliability of tests, and hence subsequent patient safety. Taking into account the specific response for each individual provides a better fitment of blood glucose control algorithm.

In conclusion, in silico studies provide a reliable preclinical alternative to animal tests and enables them to be carried out in a less time.

Model used for this studies was developed by a team of Boris Kovatchev et al. from the University of Virginia, USA [3]. Simulator is a software package that runs under Matlab/Simulink platform.

The simulator is designed in two versions. In the version with a limited number of subjects to 30, which can be used locally by interested researchers, and the full 300 subjects version in which tests are performed only by simulator developer's site. Results from the full version of the simulator can be used interchangeably with animal testing as approved by the U.S. Food and Drug Administration in 2008.The simulator has a graphical interface to configure and run tests.

The model was created specially to enable research into methods of glucose control with closed feedback loop. The model represents a whole system which consists of: a blood glucose sensor, insulin pump and controller that calculates a dose of insulin delivered by an insulin pump (Fig.1). If we switch off controller we get standard open-loop circuit, that enables to perform typical treatment tests. Simulator, due to the fact that its main purpose is for testing of control systems takes into account the specificities of real system.

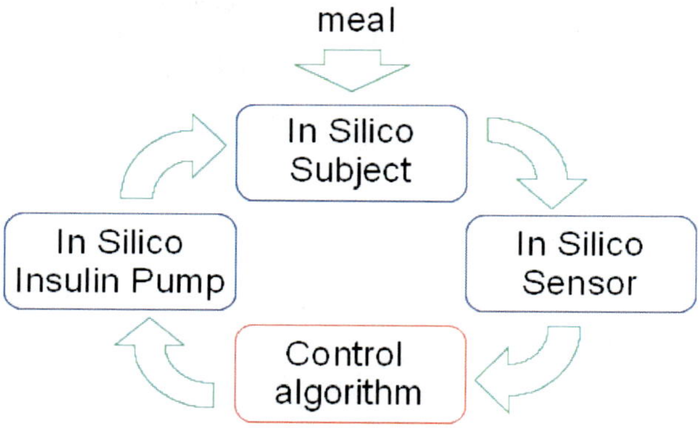

Fig. 1. Schematic diagram of in silico simulator.

In Silico Subject

Simulation environment is based on glucose-insulin meal model presented by Chiara Dalla Man et al. [4]. The model of the in silico subject is multi-compartment one. It assumes that insulin controls glucose uptake and internal glucose production. Briefly, these compartments represent the relevant glucose-insulin organs and the main relationships between them. These compartments are: the stomach, plasma, intestine, liver, kidney, brain, erythrocytes or peripheral tissues that utilize glucose.

Data for patients were determined on the basis of studies conducted on a population of 300 subjects in which 100 were adults, 100 were children and 100 were adolescent. This model uses approximately 30 parameters to determine the behaviour of each of 300 virtual patients. These parameters were determined on the basis of a triple tracer meal studies [5].

In Silico Glucose Sensor

Blood glucose measurement system modelled in the simulator takes into account parameters such as delays associated with subcutaneous glucose measurement, measurement uncertainty and measurement noise. Model reproduce the actual distribution of measurement glucose sensor errors. The software allows to select one of three commercial Continuous Glucose Monitoring (CGM) systems [3]. These are: Freestyle Navigator ™ (Abbott Diabetes Care, Alameda, CA)[7], Guardian RT (Medtronic, Northridge, CA)[8], and Dexcom ™ STS ™, 7-day sensor (Dexcom, Inc.., San Diego, CA)[9].

In Silico Insulin Pump

Since the insulin pump software takes into account parameters such as delay of delivery of insulin subcutaneously (s.c.) and the discretisation of insulin on the amount and the timing [3]. The data correspond to the actual two devices: OmniPod Insulin Management System (Insulet Corp., Bedford, MA) [10] and the Deltec Cozmo ® (Smiths Medical MD, Inc.., St. Paul, MN) [11].

4 Method and Results

Testing scenario is written in a ASCII text file. The controller block is defined and connected in the Simulink. The input to the model is a list containing information about meals. In the case of tests in open-loop list also includes data on the delivery of insulin. Once the simulator is in closed-loop mode the controller instructs the insulin pump with dose parameters.

Data obtained from the model are the traces of blood glucose and a statistical data to help assess the quality of glycaemic control.

The ultimate result of research is to create an algorithm for insulin dose calculation based on information received from the CGM system. The algorithm primarily aims to avoid hypoglycaemia and hyperglycaemia. It is planned that the control algorithm will be "intelligent" and in the course of the work of the entire system it will be tuning its control parameters. Algorithm will recognize the typical behaviour and food habits of the subject as the hours and amount of ingested food and the demand for glucose and insulin.

Before the direct examination of the control algorithm, in silico subjects as well as in vivo subject can be checked in order to obtain anthropometric data. In silico patients may also be subjected to tests such as glucose tolerance test, which allow to infer additional parameters to initialize the control algorithm.

The intelligent controller can be taught the best insulin dosage on the basis of the subsequent iteration and a treatment quality evaluating system.

It is planned to create a universal general algorithm for selecting the dose, which will then be tuned to a specific patient based on test results. Such idea appears to be a reasoned direction of study because of the very large differences between the behaviour of individuals.

Presented on Fig.2. blood glucose trace recorded by in silico CGM after a meal containing 75g of CHO and lasting 5 minutes with a dose of 5U of insulin (bolus) through insulin pump at the same time.

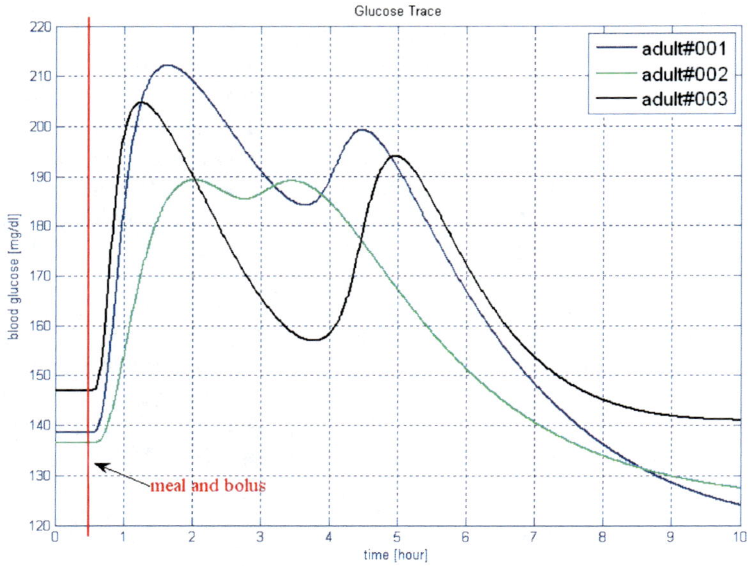

Fig. 2. Glucose traces for adult subjects. (75g CHO and 5U at the same time)

Meals and insulin are given 30 minutes after the start of the measurements. Traces are shown for three different adult subjects. Please note the different nature of the body's response to the meal and boulus. There are three clinically different glycaemic regions suggested by [12]. These bands are (i) hypoglycaemia with range BG<70 mg/dl, (ii) hyperglycaemia with range BG>180 mg/dl, and (iii) range between (i) and (ii) known as a target range. Percentage of time spent in certain BG ranges may be used as a measure to indicate differences between subjects.

Subject adult# 001 is in the range of hyperglycaemia for 29% of the time while the adult#002 is less sensitive and the percentage of time spent in hyperglycaemia is 14%. Percentage of time spent in BG>180 mg/dl for adult#003 subject is equal to 15% of the total testing time but it affects large fluctuations. In this case glucose trace is growing rapidly than to fall quickly and then re-crossing of BG=180 mg/dl again.

Even greater inter-individual differences exist between subject in different age groups, which is shown in Fig.3. All three patients were given the test meal and insulin dose exactly as in the previous case (Fig.2).

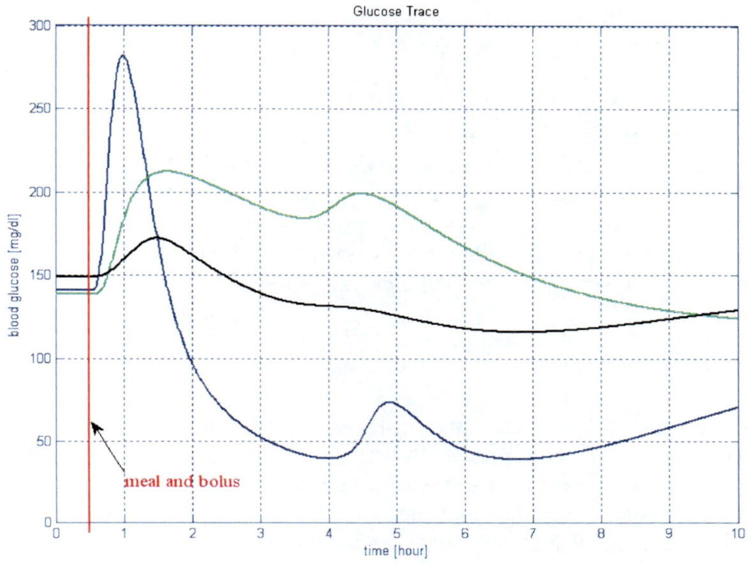

Fig. 3. Glucose traces for subject in different age.(75g CHO and 5U at the same time)

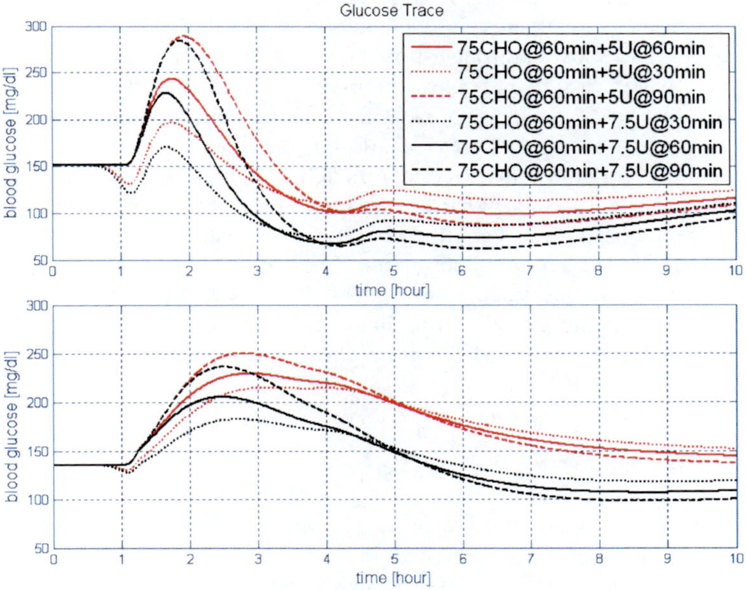

Fig. 4. Influence of insulin dose amount and timing to the BG trace.

Glucose trace for the child subject is the most rapid in the first stage of inducing hyperglycaemia 8%, and then going to hypoglycaemia for the 70% of the time. In the case of adolescent subject glucose trace is the most subdued and is located in the accepted limits of 70mg/dl <BG <180 mg/dl. An adult subject is the same as the adult # 001 on Fig.2.

Fig.4 shows the effect of time and amount of insulin delivery to the glucose trace. For all tests, patients received 75g of CHO in the 60th minute, and a typical dose of 5U or increased 7.5U of insulin, in different timing. Both the dose before the meal and increase in its value lowers postprandial glucose. However, the dose escalation is not justified by medical and causes too low blood glucose.

5 Conclusion

In silico testing allows quicker and easier medical research. In addition, it allows perform repetitive tests under small changes of parameters and experiments that would have been dangerous to subjects in vivo.

Presented results show the need for individualized of control algorithm for each patient individually. We anticipate that after the creation of a universal algorithm for whole population, each control algorithm will require fine fitting to subjects biological parameters. In addition, we expect that more accurate control may be obtained, if we use several universal algorithms types different for some population subgroups. Such subgroups may be achieved by dividing the population by age or biometric characteristics e.g. weight, obesity, Oral Glucose Tolerance Test.

The results also show significant time effect of drug dispensing in the advance of the meal, which in turn indicates importance of timing in the glucose regulation. This also proves that the approach in which the regulator learns the patient's typical day and in advance can provide a dose is the right direction of study.

Also, the current study, in silico significantly facilitate the subsequent implementation of the real system. One can imagine the actual implementation of the following insulin delivery system in closed loop. The patient initially uses an insulin pump and CGM in an open loop mode. Implementation of the system beyond its core functions at this stage is to collect data about the subject. These data would be used to create a model of an in silico subject on the basis of which in a safe manner will, be fine-tuned control algorithm designed for work in a closed loop.

References

[1] WHO Fact sheet N°312, WHO Word Wide Web, `http://www.who.int/mediacentre/factsheets/fs312/en/`

[2] Hawłas, H.J., Lewenstein, K.: The Design of an Insulin Pump – Preliminary Requirement. In: Recent Advances in Mechatronics, Part 4, pp. 329–334 (2010)

[3] Kovatchev, B., Breton, M., Man, C.D., Cobelli, C.: In Silico Preclinical Trials A Proof of Concept in Closed-Loop Control of Type 1 Diabetes. Journal of Diabetes Science and Technology 3(1), 44–55 (2009)

[4] Man, C.D., Rizza, R., Cobelli, C., Fellow, IEEE: Meal Simulation Model of the Glucose-Insulin System. IEEE Transactions on Biomedical Engineering 54(10), 1740–1749 (2007)
[5] UVa T1DM simulator user guide
[6] Frequently Asked Questions about the UVa Type1 Metabolic Simulator
[7] FreeStyle Navigator, Abbott Word Wide Web, `http://www.freestylenavigator.com/index.htm`
[8] Guardian® RT Continuous Glucose Monitoring System, Medtronic Word Wide Web, `http://wwwp.medtronic.com/newsroom/content/1168605107804.pdf`
[9] 7-day sensor, Dexcom Word Wide Web, Navigate to menu The SEVEN Difference, `http://www.dexcom.com`
[10] OmniPod Insulin Management System, Insulet Corp. Word Wide Web, `http://www.myomnipod.com/`
[11] Deltec Cozmo®, Smiths Medical MD Word Wide Web, `http://www.smiths-medical.com/landing-pages/promotions/md/coz-home.html`
[12] Clarke, W., Kovatchev, B.: Statistical Tools to Analyze Continuous Glucose Monitor Data. Diabetes Technology & Therapeutics 11(suppl. 1), 45–54 (2009)

A Pressure Based Volume Sensing Method

S. Mohammadi* and M. HajiHeydari

Islamic Azad University (south Tehran branch), Ahang highway,
Tehran, 1767969473, Iran

Abstract. Volumes of vessels are determined by various methods. Ultrasonic and light based methods, Thermal approaches in fluid sensing and so on. More over Floaters are used widely. These methods are unstable and erroneous for unbalanced static (e.g. tilted status) and dynamic vessel situations. In this article we present an approach for detection of volume in the vessels. This approach use pressurizing the vessel and the principle that fluids (or solids like granular substances or powders) aren't compressible. This approach provides a stable method for volume detection of vessels (fuel tank of automotive vehicles, a silo etc.). This approach can precisely determine volume of the vessel in unstable static and dynamic situations.

1 Introduction

Volume of fluid or solid in the vessels (usually based on level of it) is sensed in engineering and art for a long time by now. More over precise sensing of volume in vessels in motion and dynamic statuses (e.g. fuel tank of vehicles, or a silo when is filling) are more important. By now many methods for sensing of Volume has introduced, from very basic ones [1] to more complicated. One of this methods that has saved his usability by now, is floater based level (and then volume) sensing. These floaters stay on liquid surface and its motions change to mechanical motions by mechanical linkages, or to electrical signals by Electrical Components for use.

In many situations (e.g. fuel tanks) because of exposure of electrical instruments of transduction to corrosive media, many efforts have made for isolation of electrical and mechanical parts of these transducers [3-10]. Beside, floater based volume sensing methods has mechanical mobile parts that has special wearing, accordingly by now some methods has introduced. Methods like a resistive hot tape vertically placed in the tank [11], use of a sonic conductive rod vertically placed in the tank [12], Use of ultrasonic distance measuring techniques [13-15], Use of light and mirror [16], capacitive sensors [17-18], and approaches based on pressure [19-23]. All of these methods have appropriate and special use of themselves. But in dynamic statuses we need some tricks to eliminate noises.

---

* Corresponding author.

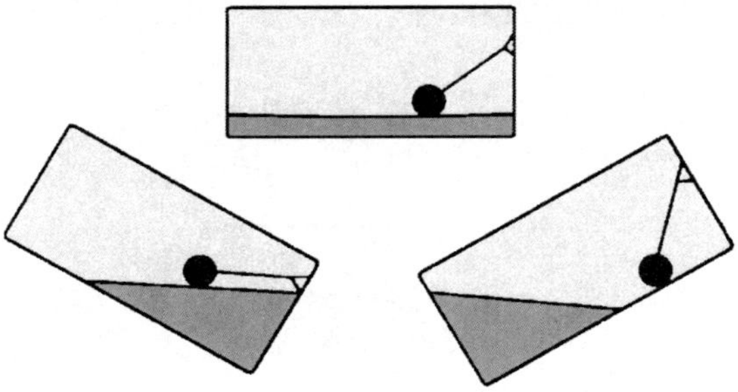

Fig. 1. Floater imprecision

One of approaches is Integration on the time [24]. Another approach is take account for fuel usage rate, which can be precisely determined even in motion, in calculations [25]. Here we present a pressure based method. This method is a stable method, independent of static or dynamic status of vessel. Sensors that use this method don't contact the fluid, and can mount easily and out of the vessel.

2 Pressure Based Volume Sensing

Two methods mentioned in references [21-22] have used Manometric approach. One of the problems of these methods is change of pressure in its pressure tube based on solution of trapped air in the manometric column over the time, or for turbulent motions of fuel in semiempty vessel and penetration of air into the tube. Venting manometric column on refueling to equalizing pressure of the tube with the atmosphere on refueling may resolve this problem. In another invention [23] by periodic generation of pressure and use of a bubbler approach for determining of pressure on bottom of tank, level and then volume has sensed. In another effort to use of pressure for control of liquid volume, a pressure based valve designed for water volume controlling in a clothes washer machine [19].

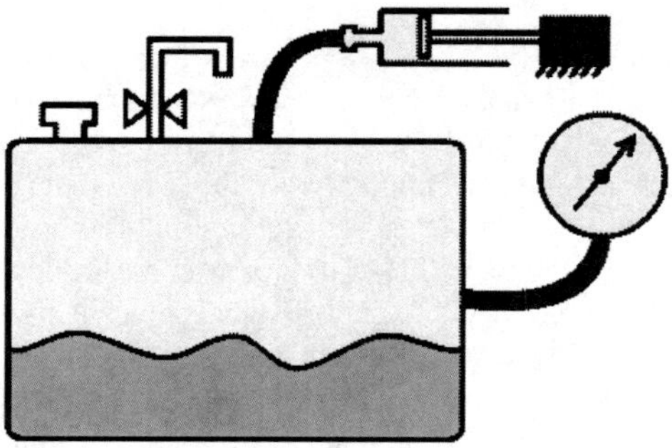

Fig. 2. Typical schematic of approach

In this method the principle of incompressibility of liquid in opposite of gas has used. Here we use this principle and propose a method for sensing of volume. We blow a certain volume of air into the vessel that its air vents are temporarily closed, and based on increasing of pressure in the vessel, determine the volume of gas(compressible substance) and then volume of liquid in it.

3 Method

Process that periodically repeats is: First we close air vents. Then blow a certain and predefined volume of air (or another suitable gas, based on fluid in the vessel), based on this over mass, an over pressure will produces.

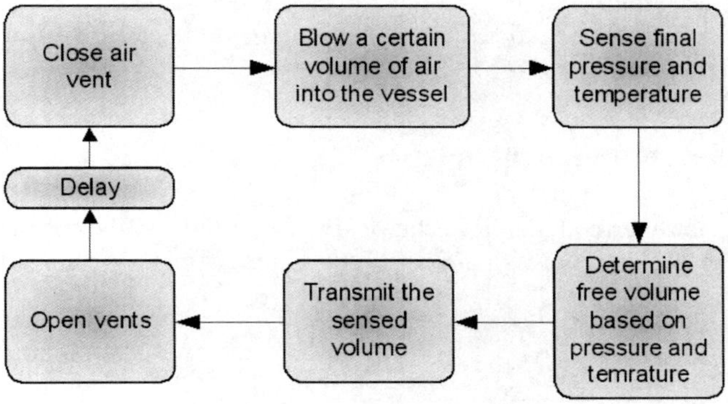

Fig. 3. Process block diagram

By measuring of this pressure we can determine the volume of gas (and then volume of liquid) of a dcetermined vessel. One of key advantages of this method is that it's vessel shape independent:

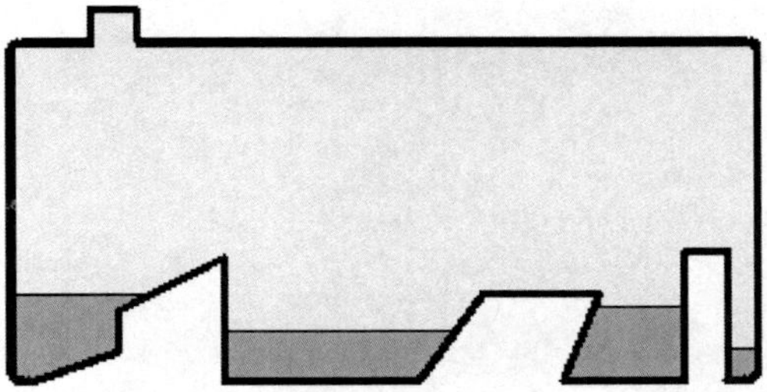

Fig. 4. Irregular vessel shape

Now we have two ways for implementation of "Determine free volume based on pressure and temperature" box in the block diagram. Based on our usage we can use each of the ways. First and simple and limited method is obtain a few tables of free volume based on pressure and temperature. In this method, we may be take account seasons and... Another method is to use of thermodynamics tables, diagrams and equations.

4 Specification Tables

In first mentioned method. We should take account season Or/And weather Or/And humidity and so on. But in some applications, for example in non-ideal Gases, may it be appropriate method. Tables can obtained in a test stage and record them on a micro controller or peripheral replaceable EEPROM. Then the MCU can use those data, and based on pressure and temperature estimate the free volume of the vessel.

5 Thermodynamic Approach

In this approach thermodynamics relations will be used. At the beginning we assume that the gas is a pure gas (a gas with same chemical composition in each phase of it).

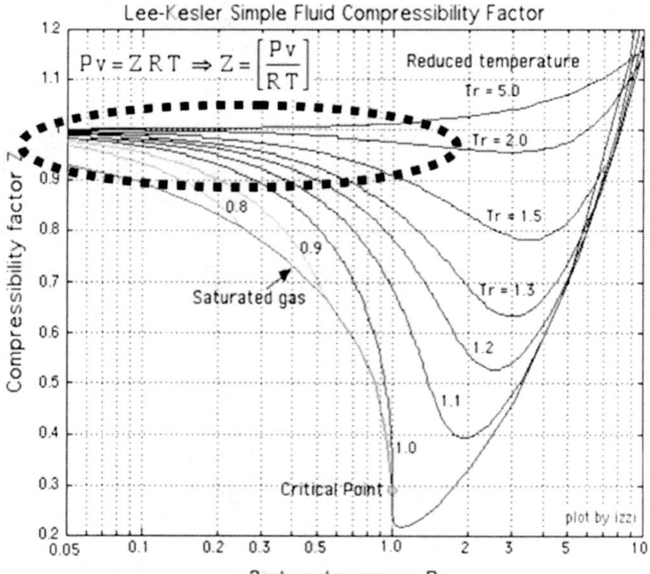

Fig. 5. Ideal gas with simple thermodynamics relationships zone

But it can be ideal or non-ideal. If it's a non-ideal gas we should use thermodynamic diagrams and tables. Same approach as described in the last section can use to implement this approach. Now let's assume the gas or vapour is ideal (really for gases that are almost ideals, those with z-factor depicted below in Lee-Kesler diagram). Then the process of compression of the vapour will be an adiabatic process, If vessel is isolated or, we compress the vapour rapidly, and read temperature and pressure as quick as possible. Nowadays, there are pressure transmitters with a few milliseconds response time. There are rapid temperature sensors and temperature sensing methods too [26-27]. Then for an ideal gas we have:

$$Pv = nRT \tag{1}$$

Accordingly:

$$V_f = \frac{T_f}{T_i}\frac{P_i}{P_f}v_i \tag{2}$$

And from definition:

$$\begin{cases} \Delta v = v_i - v_f \\ \Delta T = T_i - T_f \end{cases} \tag{3}$$

By substitution and a few mathematical operations can be seen:

$$V_f = \frac{(T_i - \Delta T)P_i}{T_i P_f - (T_i - \Delta T)P_i}\Delta V \tag{4}$$

This equation is independent of v_i, Δv is the volume of the cylinder (if pressurizing done by a cylinder and a piston) and temperatures can acquire by temperature transmitters. If temperature can considered constant, The equation become very simple:

$$V_f = \frac{P_i}{\Delta P}\Delta V \tag{4}$$

Now let's study the process. Base on the block diagram, process starts with closing of air vents. Then we pressurize the gas or vapor of the vessel. At the first and end of the pressurizing temperature and pressure should be record. Note that pressure and temperature transmitters are rapid response time. Now by substitution these data in (5.4) equation vf will optain. vf is the volume of vessel occupied by gas or vapor, what we wants.

6 Conclusion

The In this article we propose an approach for volume detection. This method is independent of static or dynamic status of vessel. To use this method practically, our volume transducer apparatus may comprise one or more (based on whole volume of vessel) pressure transducers by different ranges of pressure detection. In the other hand, wall strength of vessels, like fuel tank of automotive vehicles that has made by thin metal films, should be studied, because periodical blowing of air into the vessels exert fatigue stresses on walls. Another point is that pressure transducers used in this method, for vessels in dynamic status (e.g. fuel tank of vehicles) should not having massive moving parts(e.g. LVDTs), because its inertia may interfere with data acquisition.

References

[1] Davies, H.R., Industrial Research Corporation: Fuel-level indicator, U.S. Pat. 1497291 (1924)
[2] Dollar, F.W.: Fuel level indicator, U.S. Pat. 2708897 (1955)
[3] Nishida, K., Shibata, H., Tsuchiya Co.: Fuel level detector for automobile fuel tank, U.S. Pat. 4627283 (1985)
[4] Herford, M.L., Walbore Corporation: Fuel level sensor, U.S. Pat. 6571626 B1 (2003)
[5] Brozozowski, M.A., Gilmour, D.A., TI Group Automotive System: Sealed fuel level sensor, U.S. Pat. 6868724 B2 (2005)
[6] Davila, G., Delphi Technologies: Fuel level sensor, U.S. Pat. 7062966 B1 (2006)
[7] Newman, T.R., Washeleski, J., Nartron Corporation: Fuel level sensor, U.S. Pat. 7093485 B2 (2006)
[8] Bremmer, L.M., Enge, T., Schmidt, R.W., Teets, M.R., Chastang, R., Chrysler LLC: U.S. Pat. 7392700 B2 (2008)
[9] Roth Robert, A., Aikman Steven, W., Delphi Technologies Inc.: Light sensing contactless fuel level sensor, E. Pat. 2023098 A1 (2008)
[10] Ferreira, D.M., Johansen, M.R., Ti Group Automotive Systems: Fuel level measurement device, U.S. Pat. 7409860 B2 (2008)
[11] Uriu, E., Nishida, K., Tabata, M., Matsushita Electric industrial Co.: Thermal fuel level detector, U.S. Pat. 5022263 (1991)
[12] Duckart, A., Hauler, P., Zabler, E., Robert Bosch GmbH: Tank level meter, U.S. Pat. 4896535 (1990)
[13] Crayton, J.W., Boston, T.A., Betts, D.A., Caterpillar Inc.: Ultrasonic fuel level sensing device, U.S. Pat. 5319973 (1994)
[14] Hale, D.M., Isle, N.M., Borza, D.P., DaimlerChrysler Corporation: Fuel level sensor, U.S. Pat. 7062967 B2 (2006)
[15] Reimer, L.B., Murphy, G.P., SSI Technologies Inc.: Immersed Fuel level sensor, U.S. Pat. 7542870 B2 (2009)
[16] Gilmour, D.A., Forgue, J.R., TI Group Automotive systems: Optical fuel level sensor, U.S. Pat. 6668645 B1 (2003)
[17] Hauzeray, S., Intertechnique: Gauge for measuring fuel level in a tank, and a system for measuring the weight of fuel in the tank, U.S. Pat 7010985 B2 (2006)

[18] Da Silva, J.R.V., Indebrás Indústrial Electomecãnica Brasileira Ltda: Apparatus for measuring and indicating the level and/or volume of a liquid stored in a container, U.S. Pat. 7441455 B2 (2008)
[19] Knoop, D.E., Panstian, J.K., Whirlpool Corporation: Automatic water level control system for an automatic washer, U.S. Pat. 4835991 (1989)
[20] Benjey, R.P., Eaton Corporation: Solid state fuel level sensing, U.S. Pat. 6282953 B1 (2001)
[21] Krahn, H., Gűnther, A., Weindle, W.: Measuring system for liquid volumes and liquid levels on any type, U.S. Pat. 5802910 (1998)
[22] Meagher, J.T., Motorola Inc.: Fuel level sensor, U.S. Pat. 6907780 B1 (2005)
[23] Casey, G., Kavlico Corporation: Fuel tank module control system, U.S. Pat. 7251997 B1 (2007)
[24] Koboyashi, H., Kita, T., Nissan Motor Company: Fuel gauge for an automotive vehicle, U.S. Pat. 4470296 (1984)
[25] Davison, L.E., Melville, T.R., Ford Global Technologies Inc.: Fuel level monitor, U.S. Pat. 6397668 B1 (2002)
[26] Heichal, Y., Chandra, S., Bordatchev, E.: A fast-response thin film thermocouple to measure rapid surface temperature changes. Experimental Thermal and Fluid Science 30(2), 153–159 (2005)
[27] Howard, S.L., Chang, L.-M., Kooker, D.E.: Thermocouple sensor for rapid temperature measurements during ignition and early phase combustion of packed propellant beds. Review of Scientific Instruments 66(8), 4259–4266 (1995)

Design of Electric Drive for Total Artificial Heart with Mechatronic Approach

R. Huzlik[1,*], S. Fialova[2], and F. Pochyly[2]

[1] Brno University of Technology, Faculty of Electrical Engineering and Communication, Technicka 10, Brno, 616 00, Czech Republic
[2] Brno University of Technology, Faculty of Mechanical Engineering, Technicka 2 Brno, 612 00, Czech Republic

Abstract. This paper describes mechatronic approach in design of drive for total artificial heart. Total artificial heart (TAH) is one way, how can help the people with cardiovascular diseases. TAH can substitute real human heart for some times. In this paper human heart is briefly described and some variant of drive and approach of design are discussed as well.

1 Introduction

In this day, one of the most deathly diseases are cardiovascular diseases. One of way how can be healed is transplantation of heart. There are two trends in transplantation – transplantation real human heart by other people or using total artificial heart (TAH). Real heart transplantation has some disadvantages. One of these disadvantages is the absence of enough people for donor and incompatibility of heart based on blood type. TAH can be used as a bridge to real heart.

2 Human Heart and TAH

As this work deals with design of a drive for a blood pump, it will be good to say some information about a real heart. The heart weight is about only 300 grams, achieves to pump over a five liters of blood every minute in the rest conditions. A heart rhythmic activity consists of two phases, which repeats periodically. The heart expels blood in a systole phase, which changes with diastole phase when blood flows contrariwise inside. The diastole is more considerable for a normal heart function. Only in this phase of a heart activity blood can flow freely thru wreath arteries, which supply own heart with nutrients and oxygen. Systolic normal blood pressure is from 90 up to 150 mmHg in resting state and diastolic blood pressure from 60 up to 80 mmHg. The heart power grows significantly at a physical work. Heart in these conditions must pump over 20 up to 35 liters of blood per

[*] Corresponding author.

minute, so as body were supplied with needed amount of oxygen and nutrients. Of course this achieves healthy heart only.

An acute heart failure can happen during complex cardiovascular operations often, then immediate mechanical assistance is necessary. The device performing this function is sometimes called at special literature as ventricular assist device (VAD). An alone substitution of heart leads to two trends. Either it is a substitution with a biological heart from other person or it is an artificial heart system, which is called at literature most often as total artificial heart (TAH), providing the same function. In 1967 the South African surgeon Christian Barnard did the first heart transplantation in the world. It has taken did take years when transplantation of the heart became a routine operation, which saves thousands people before certain death. But this solution has one basic limitation and it is the fact, that donors never will be enough. That is why we have to take way, which will not need any biological heart. Even before we describe synchronous machine developed at our department, we acquaint with blood flow in the human body. The blood circulation has two separate circuits, which are sorted consecutively. Each of them belongs to one heart ventricle. A small (pulmonary) circuit is driven by right heart ventricle, big circuit is driven by left ventricle. A volume of blood pumped over small and big circuit is the same. But the small and big circuits are different in pressure. The pressure in big circuit is 4 up to 5 times higher than in pulmonary circuit and therefore also power of left heart, which works at conditions of greater pressure, is 4 up to 5 times higher in comparison with power of right heart

The small circuit starts in right atrium, while upper and inferior hollow vein leads. The blood comes from right atrium to right ventricle thru tricuspid valve. Systole forces pull out blood after fulfilling right ventricle thru pulmonary valve into pulmonary artery. In lungs blood gives off waste gasses and picks up oxygen. Through pulmonary veins the oxygen rich blood returns to the left atrium.

The big circuit starts in the left atrium, from here blood flows thru mitral valve into the left ventricle. The blood forces out into the aorta thru aortic valve with the help of a left ventricle muscle. The oxygenated blood leads from aorta to single parts of body. After that blood gives off a part of oxygen and picks up a certain amount of carbon dioxide, returns back to the heart in upper and inferior hollow vein.

On the base of [1], we can set basic demands for TAH construction:

- Size limitation - maximal diameter 85mm, maximal length 95 mm. This value is based on size of women's and child's thorax.
- Weight limitation – natural heart weight 300 – 400g.
- Maximal losses should be lower than 20W. This value is given by heat tolerance limit for human body, because maximal temperature in human body is 42°C.
- Average pumping capacity 6 l/min with pressure 100mmHg (13,33 kPa).

3 Mechatronic Approach

TAH is complex mechatronic system with many parts. Three basic parts can be defined:

- Pump (mechanical part)
- Motor (electro-mechanical part)
- Inverter and control (electronic part)

In this system, every part influences behavior of another part and has effect to maximal efficiency of this system. For this reason mechatronic approach for design of TAH drive is chosen.

All part of this system is developed as part of complex system in which all parts are mutually influenced. This is main idea of mechatronic approach [2]. Complex simulation of this system is one of the main way of this design.

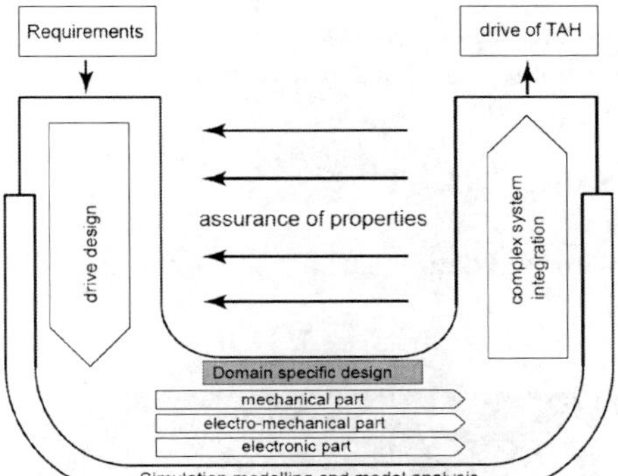

Fig. 1. Macro-cycle of TAH drive of development

When some version of the system is developed, we can use several repeating macro cycle, when for example we can change parameters of final system or mathematical model of system. Using of repeating macro cycle leading to upgrade of the system (and to an innovation) and degree of product maturity.

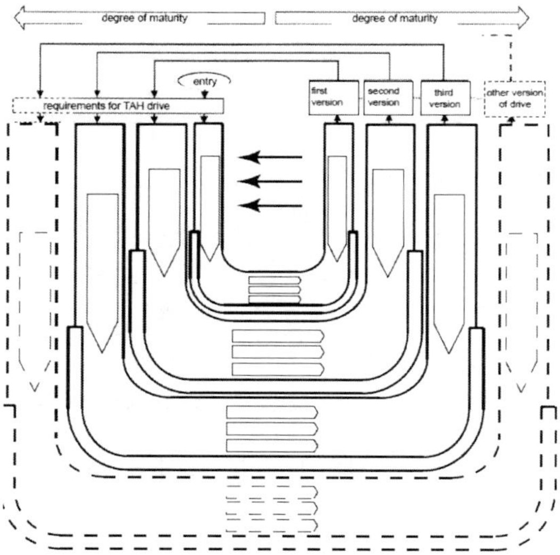

Fig. 2. Several repeating macro-cycle of TAH development [2]

4 First Variants of TAH Drive

Axial BLDC motor with magnetic bearings was choosen as the first conception of TAH drive [6](Fig.3).

Fig. 3. First variant of TAH with axial motor and magnetic bearing.

Magnetic bearing was removed in second variant, because it was needed to decrease losses of whole system [4] (Fig.4 and Fig.5.).

Fig. 4. Part of a second variant of TAH with axial motor.

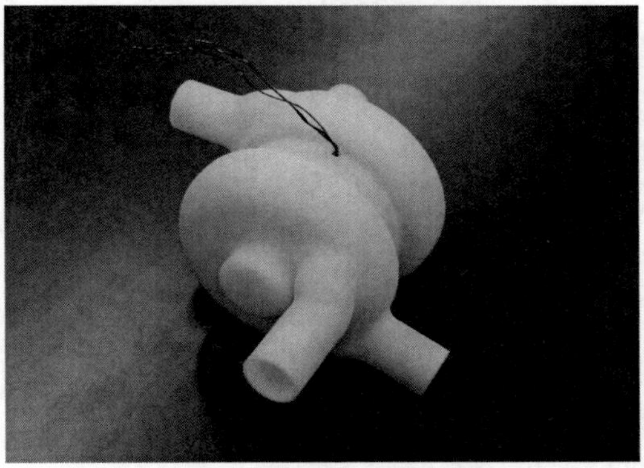

Fig. 5. Second variant of TAH with axial motor.

Total artificial heart consists from:

- slot-less axial flux synchronous motor with two rotors
- two turbine runner
- inverter

The new double turbine blood pump solution consists of a fix part of pump stator with two inlets port of the pump and two outlets for the left and the right ventricle. In the middle part of the stator is slot/less disc winding of the permanent magnet axial flux synchronous motor, which is without iron core. The rotor discs of the motor are integrated into the turbine rotor and this system is equipped with two active permanent magnet discs, whose magnetic flux crosses thru the stator winding.

The pump for this TAH was design in Kaplan Department of Hydraulic Machines, Institute of Power Engineering, Brno University of Technology.

Electronics of small volume and maximum efficiency is needed for motor control. In the first and second variant of drive RC model's inverter was used. This inverter is used as sensor-less BLDC controller based on switching on zero-crossing detection. This method is used for switching of BLDC on high speed (greater than 1000 rpm).

5 Design of Axial Motor

Back EMF line to line voltage can be calculated on the base of equation (1)[9]

$$U_i = 4 \cdot N_1 \cdot k_{w1} \cdot \alpha \cdot B_g \cdot \pi \cdot \left(D_{out}^2 / 4 - D_{in}^2 / 4 \right) \cdot n \tag{1}$$

where N_1 is number of turns in one phase, k_{w1} is winding factor, α effective pole arc coefficient, B_g is maximal value of magnetic flux density in windings, D_{out} is outer diametr, D_{in} is inner diametr and n is rotational speed.

Torque of motor can be calculated on the base of equation (2)[9]

$$M = 2 \cdot N_1 \cdot k_{w1} \cdot \alpha \cdot B_g \cdot \left(D_{out}^2 / 4 - D_{in}^2 / 4 \right) \cdot I_a \tag{2}$$

where I_a is stator current.

In order to optimize the performance of the axial flux machine the ratio of the inner and outer diameter of the machine should be chosen carefully since this ratio has as well as the air-gap flux density have significant effect on the machine characteristics.

The machine total outer diameter D_0 is given by

$$D_t = D_0 + 2W_{cu} \tag{3}$$

where W_{Cu} is the protrusion of the winding in the radial direction both in the outer and to the axis of the machine.

This dimension must be checked with respect to dimensions according to the pump body.

$$W_{cu} = \frac{D_i - \sqrt{\left(D_i^2 - 2AD_g\right) / K_{cu} J_s}}{2} \tag{4}$$

where J_s is the current density and K_{Cu} is the copper fill factor.

Outer surface diameter can be calculated from sizing equations

$$D_0 = \sqrt[3]{\frac{P_R}{m \cdot \frac{\pi}{2} \cdot K_e \cdot K_i \cdot K_p \cdot \eta \cdot B_g \cdot A \cdot \frac{f}{p} \cdot \lambda\left(1 - \lambda^2\right) \frac{1 + \lambda}{2}}} \tag{5}$$

where K_Φ is electrical loading ratio, m is number of machine phases, K_e is EMF factor, K_i is current waveform factor, K_P is electrical power waveform factor, η is machine efficiency, A is total electrical loading, f is frequency, p is number of machine pole pairs, λ is diameter ratio.

The axial length of the machine L_e is given by

$$L_e = L_s + 2L_r + 2g \tag{6}$$

where L_S is axial length of the stator, L_r is axial length of the rotor and g is the air gap length.

The axial length of rotor L_r becomes

$$L_r = L_{cr} + L_{PM} \tag{7}$$

The axial length of rotor disc core L_{cr} is

$$L_{cr} = \frac{B_u \pi D_0 (1+\lambda)}{B_{cr} 8p} \tag{8}$$

where B_{cr} is the flux density in the rotor disc core and B_u is the attainable flux density on the surface of the PM.

The length of permanent magnet LPM can be calculated from equation 9

$$L_{PM} = \frac{\mu_r B_g}{B_r - \frac{K_f}{K_d} B_u}(g + W_{cu}) \tag{9}$$

where B_r is the residual flux density of the PM material, K_d is the leakage flux factor and $K_f = B_{gpk}/B_g$ is the peak value corrected factor of air-gap flux density in radial direction.

Combining these equations the axial length L_e becomes

$$L_e = \frac{\pi D_0 (1+\lambda)}{4p}\frac{B_u}{B_{cr}} + (2W_{cu} + 2g) \times \left(1 + \frac{\mu_r B_g}{B_r - \frac{K_f}{K_d} B_u}\right) \tag{10}$$

On the base of these value was designed motor, which can get minimum supposed value of torque and rotation speed.

This design can be confirmed via software with finite element method.

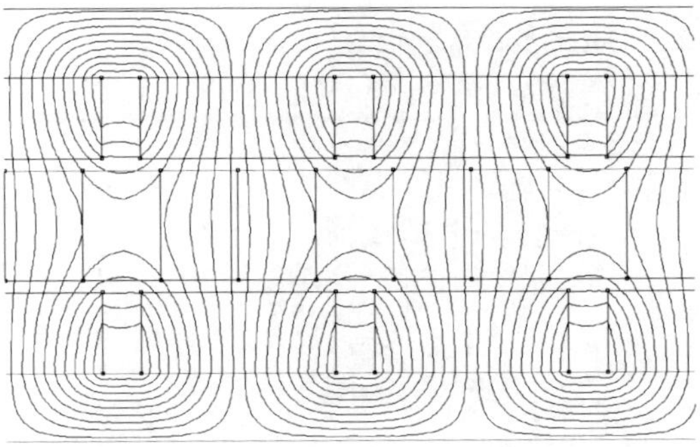

Fig. 6. Distribution of flux lines in TAH motor.

6 Design of Control System

For developing of control system CompactRIO and program LabVIEW was choosen. Compact RIO is platform for measuring and controlling from National Instruments. CompactRIO includes FPGA in chassis and real-time operating system in controller.

Beaucase we expected lower speed than 1000 RPM, zero crossing method for switching has problem. In the design, we set Hall sensor for detection of switching point.

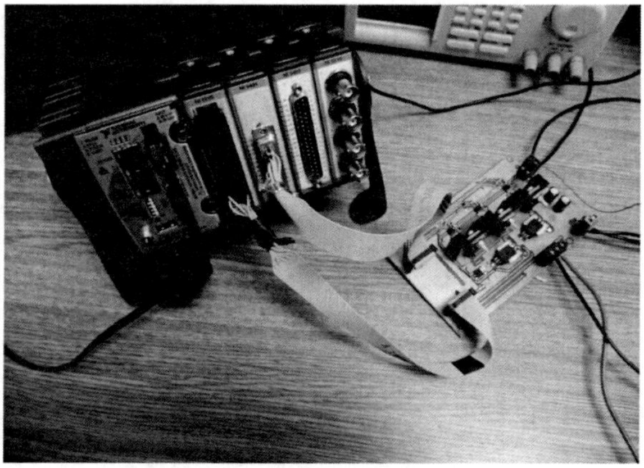

Fig. 7. CompactRIO with controlling board for TAH.

In FPGA running controlling loop comprising:

- PWM generation,
- switching of transistor,
- reading of measured value,
- running of safety part.

In controller running control loop with PID regulator

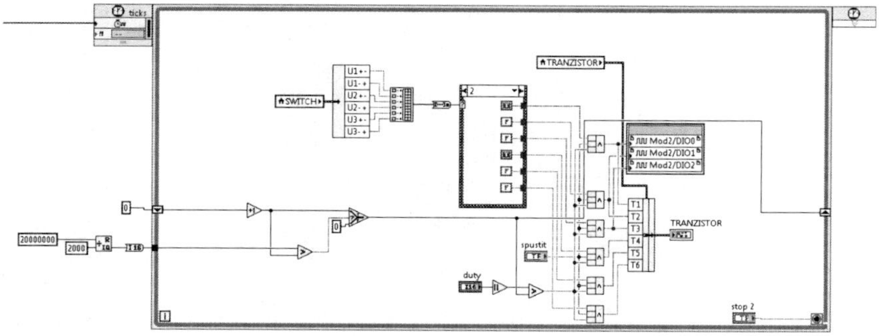

Fig. 8. Switching loop in FPGA.

7 Conclusion

TAH is one way how people with cardiovascular diseases can survive to transplantation on near real heart. TAH is complex mechatronic system including pump, motor and control system. Mechatronic approach was used for the design of such complex system.

Acknowledgment

This paper was supported by Ministry of Education, Youth and Sports of the Czech Republic research grant MSM 0021630518 "Simulation modeling of mechatronic systems". and GACR 102/09/1875.

References

[1] Lessmann, M., Finocchiaro, T., Steinseifer, U., Schmitz Rode, T., Hameyer, K.: Concepts and designs of life support systems. Science, Measurement & Technology, IET 2(6), 499–505 (2008)
[2] Hadas, Z., Singule, V., Vechet, S., Ondrusek, C.: Development of energy harvesting sources for remote applications as mechatronic systems. In: 2010 14th International Power Electronics and Motion Control Conference (EPE/PEMC), September 6-8 (2010)
[3] Lapčík, J., Huzlík, R.: Sealless Industrial Pump With Permanent Magnet Slotless Disc Motor And Magnetic Bearings. Prace Naukowe Instytutu Maszin, Napedów I Pomiarów Elektrycznych Politechniki Wroclawskiej 1(62), S.209–860 (2008) Issn: 1733-0718
[4] Lapčík, J., Huzlík, R.: Electric Drives for Special Types of Pumps: A review. In: Recent Advances In Mechatronics 2008- 2009, pp. s.293–s.297. Springer, Berlin (2009) ISBN: 978-3-642-05021-3
[5] Huzlík, R., Lapčík, J.: PM synchronous disc motor for total artificial heart. In: ISEM 2009 - XVII. International Symposium on Electric Machinery, pp. s.85–s.88. ČVUT, Praha (2009)
[6] Lapčík, J., Láníček, T., Ondrůšek, Č.: Slotless Synchronous Motor For Blood Pump Drive. In Simulation Modelling Of Mechatronic Systems I. Brno University of Technology, Faculty of Mechanical Engineering: 2005, s. 120 (s.) (2005) ISBN: 80-214-3144- X
[7] Gieras, J.F., Wang, R., Kamper, M.J.: Axial Flux Permanent Magnet Brushless Machines, 2nd edn., p. 364 s. Springer, Heidelberg (2008)
[8] Aydin, M., Huang, S., Lipo, T.A.: Design and electromagnetic field analysis of non-slotted and slotted TORUS type axial flux surface mounted machines. In: IEEE International Conference on Electrical machines and Drives, Boston, pp. 645–651 (2001)
[9] Huang, S., Aydin, M., Lipo, T.A.: A Direct Approach To Electrical Machine Performance Evaluation: Torque Density Assessment And Sizing Optimization. In: 15th International Conference on Electrical Machines, ICEM 2002, Belgium (2002)

Apparatus System for Detection of Cardiac Insufficiency

M. Jamroży and K. Lewenstein

University of Technology in Warsaw, Mechatronics Faculty,
ul. Andrzeja Boboli 8, Warszawa, 02-525, Poland

Abstract. Measurement and medical signal analysis is still a big challenge for engineers. Despite the technological development and increase in processing speed, we still have problems with designing the appropriate measurement devices and proceeding with patients' examination. It is very important to ensure the patient's safety and to observe the regulations specifying the use of medical devices. Heart auscultation is conducted during almost each appointment with the doctor. On its basis it is possible to diagnose many cardiological diseases, including cardiac muscle insufficiency. This article discusses problems, which arise while preparing measuring track for simultaneous measurement and acquisition of the digital acoustic data, vibrations of heart registered on the patient's skin surface and ECG signals. We present system solutions applied during constructing measurement station, as well as author's computer software – interface allowing easy and effective service of all the constituent devices. We attach particular importance to the issue of simultaneous registration of sampling signals in the case of devices with different clockwork generators. The aim of the following study is a description of the station which can be used for the data acquisition. After further processing, the data will be used for automatic support of the cardiac muscle insufficiency diagnosis and also for enabling early detection of this disease. This problem, which seems simple to solve in case of detecting severe abnormalities, becomes complicated when we speak of patients with slight disorders.

1 Introduction

A routine physical examination, which is usually accompanied by a medical interview in nearly every visit at the doctor's is heart auscultation. On this basis a doctor can draw many conclusions about the patient's health status. In particular, it can help in diagnosing any cardiological problems, such as arrhythmia or valvular status. Assessment of myocardial tissue and its lesions during heart auscultation is impossible due to the limitations of human hearing.

Rebuilt muscle tissue changes its mechanical parameters causing as a consequence the entire organ dysfunction. During operation, it generates pressure waves on the outside of the chest, which in case of far-reaching pathology fundamentally differ from those whose source is a healthy muscle. It is a signal containing mechanoscopic information about the heart operation parameters [1, 2].

Basically, this signal is evaluated subjectively by the doctor using studio headphones. It can also be measured with the use of accelerometer, a microphone with a fixed membrane or an auscultation funnel. Then signals are digitally processed and loaded into the computer so that they can be transformed to obtain maximum information on the status of the patient. Measured simultaneously with the above-described ECG signal synchronizes it and also provides additional information on the electrophysiological state of the tissue [3].

The aim of the following work is a description of a station which can be used for the data acquisition. After further processing the data will be used for automatic support of the cardiac muscle insufficiency diagnosis and also for enabling early detection of this disease. This problem seems to be relatively simple to solve in case of detecting severe abnormalities, but becomes complicated when we speak of patients with minor disorders.

In preliminary studies ,conducted previously, we recorded one bipolar ECG off-take (Eindhoven I or II) and alternatively signal registered by microphone or accelerometer [4]. Measurements were taken of both of these signals synchronically in patients with heart failure and healthy patients. The ECG signal was designed to indicate the phase of the cycle of depolarization and repolarization of the muscle tissue in order to allow finding states corresponding with movements of the heart wall cavities recorded as vibrations. Despite repeated attempts, we found no characteristic features of amplitude courses for both groups of patients (healthy and ill). On the other hand, the course spectral analysis was found to be strongly dependent on heart rate, that is why defining a simple relationship between the spectrum and cardiac insufficiency, has proved to be impossible. Because of the fact that the measurement station used during the initial experiments for data acquisition had a limited capacity - only two channels, it was decided in the first place to design and construct the apparatus complying with anticipated requirements in excess.

2 Assumptions of the Measurement Station

Block diagram of the proposed measurement station is shown in figure 1. It includes: ECG apparatus, a group of sensors (microphone and accelerometer), preamplifier sensor and signal amplifier, transmitters, data acquisition card (containing AC converters), and finally a computer with specialized visualization software.

Our aim is to create software for assisting diagnostics, processing and analyzing signals and for assessing the state of the myocardium.

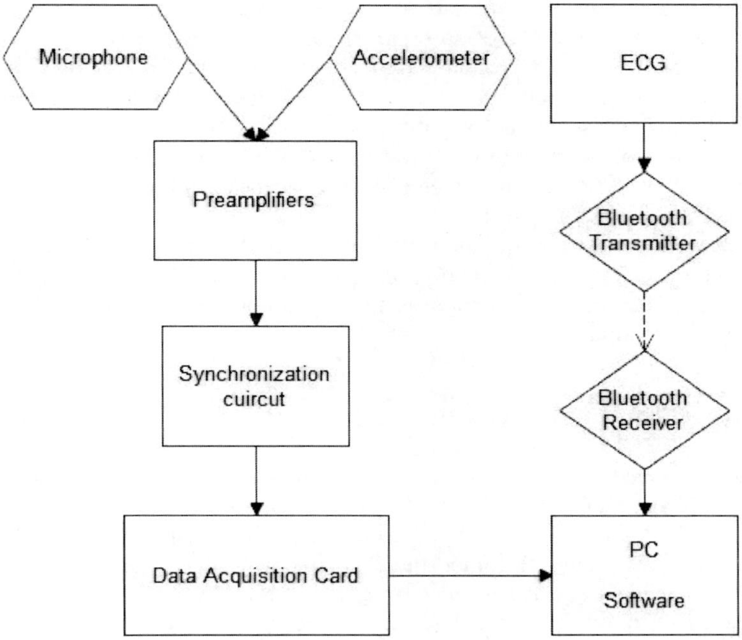

Fig. 1. Block diagram of measurement system.

Each measurement station having direct contact with the patient must first meet certain safety requirements. In Poland, there are detailed rules determining the conditions for the applicability of the devices, including the allowable apparatus parameters for leakage currents, voltages, protective barriers, etc. In this case, self-construction of the equipment such as in figure 1, which would receive certificates allowing its use, is at this stage of research pointless, and it is much better to use commercially available ready-made solutions with minor adaptations.

After consulting cardiologist we decided that the ECG apparatus should make measurements of Eindhoven's or Goldberger's limb off-takes and at least one Wilson's off-take. We are not able to diagnose cardiac insufficiency on the basis of these signals, however they are a very good source of information on other diseases that are relevant to the occurrence of the insufficiency.

According to the assumptions, the measurement station should enable measuring the vibration signals via accelerometers and microphone recording infrasound. The requirement for the selection of recorded low frequency range is caused by the observation of heart activity in patients with heart insufficiency examined by echocardiograph.

Microphone, in its measuring part, has to be equipped with an auscultation funnel of a specific shape, selected on the basis of subjective experience of a doctor. To this end a slotted line should be provided with high quality headphones.

We need an accelerometer of an adequate sensitivity and bandwidth that allows measurement of the vibrations of an organ situated inside the chest and recorded from the surface.

The entire slotted line is designed to make the acquisition of all signals simultaneously. This is necessary because of the clinical significance of timing between particular phases of heart contractions. Applied filters must have a linear phase characteristic, in order not to introduce distortion of the amplitude's spectral components of different frequencies [5].

The software should be equipped with user-friendly interface, allowing efficient service and simple record and reading of the recorded signals. The whole measuring apparatus should be light and small to allow easy transport and testing patients in different places.

3 Hardware Implementation

As it was mentioned above, due to legal requirements related to patient's safety, we used ECG apparatus available on the market. We applied FARUM device, model E600GC, additionally equipped with Bluetooth transmission module [6]. The apparatus software has been modified in such a way so as to allow, in addition to digital ECG measurement data recording, the wireless transmission to a computer. It helps us to avoid the problems associated with patient protection against any potential electric shock when using wired connections.

As we assumed, the auscultation funnel was chosen subjectively, and its shape has been used to implement cover plate connected to the Bruel & Kjaer measurement microphone ,type 4144 with a diameter of one inch [7]. As the second complementary transmitter we selected Brüel & Kjaer accelerometer, model 4507B [8]. The accelerometer was fitted to the surface of the skin using double-sided adhesive tape.

Signals from both transducers are led to a specially crafted input preamplifier consisting of two slotted lines. The former is a microphone preamplifier, and the latter – system for supplying and receiving the signal from the accelerometer interface IEPE (Integrated Electronic Piezo Electric as defined in IEEE 1451.4). Microphone preamplifier works with a headphone amplifier with adjustable gain, which are connected to the headphones from Sennheiser HD280Pro [9]. The choice was dictated by their acoustic parameters and assemble abilities during transport.

Analogue signals from the preamplifier are carried via concentric conductors to the system synchronizing them temporarily. Data acquisition is done with the data acquisition card from National Instruments 6008USB [10]. Sampling frequency of all signals, both ECG on the apparatus, as well as those coming from the preamplifier, is 800 Hz, and the ADC converter resolution is 12 bits. Figure 2 presents a model of measurement station, realized according to the above mentioned description.

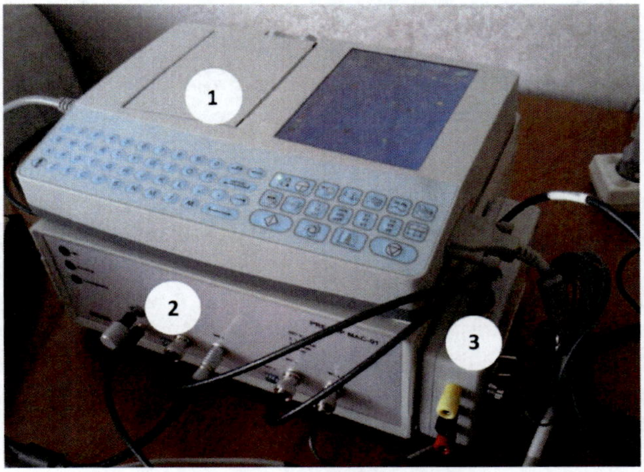

Fig. 2. Measurement devices (1 – ECG apparatus, 2 – Preamplifier, 3 – Data acquisition card with synchronization circuit)

4 Communication and Data Storage

As we mentioned before, the ECG apparatus software has been modified for data transmission via Bluetooth. By pairing the apparatus with computer and creating a virtual COM port for the device, data can be transmitted from the apparatus after sending the command "s ". The apparatus gives the recorded data from I and II Eindhoven off-takes and one Wilson's off-take, that you can choose from the menu. Each channel is encoded by two bytes. At the beginning of the reading frame is assigned a counter that allows to control transmission errors. The last byte from the frame is the least significant byte of CRC-16 CCITT calculated from the previous 7 bytes.

ECG data are sent every 2 seconds, so that between successive sessions there are time intervals. Intervals allow to send control commands to the apparatus, and for example stop the transmission. Software that lets to make the transmission and its service has been prepared by the author himself in NI LabView environment. Patient's examination record is possible not until entering his PESEL number (national personal identification number) or - in case of foreigners – appropriate identification data.

The program searches the list of available devices, paired by Bluetooth in Windows, and after finding the ECG (searching the physical address) establishes a connection. At the same time signal reception is activated via card USB6008. Despite the fact that the sampling frequency for both- the ECG apparatus and measuring card is 800 Hz, the fact of using two different clock generators in the ECG apparatus and the measuring card prevents a full synchronization. It should be noted that the use of algorithms such as NTP, which allow time synchronization with different clocks is impossible due to the limited capacity of the apparatus memory. After receiving and decoding the frame is checked whether it is a subsequent one and, if necessary, corrects the error by issuing the appropriate flag. Such a flag prevents from saving incorrect data. After entering the patient's data, and synchronizing it you can move on to the recording, which lasts 30 seconds. If an error occurs during transmission, a message appears informing you of this fact, and the record will be broken. The information contains the description of the steps you should take to make a correct record.

The number of transmission errors depends on the level of disturbance in the measurement point. During the test with the use of devices that generate electromagnetic interference, which was carried out in Warsaw, the average number of errors occurring during an hour broadcast was 5. In case of transmission outside the town, we didn't register any error in 5 hours. Signals observed on the screen and saved on the disc are not filtered, so we can determine the level of disturbance reaching ,for example, ECG electrodes, or a microphone, which allows to minimize their impact. Figure 3 shows a window from the measurement program with an exemplary record of the signals.

5 Analysis of the Simultaneity of Signals

During the presentation of assumptions and the description of the hardware devices, the issue of synchronicity of differently recorded ECG and acoustic signals was raised. To meet this requirement, the signals from the preamplifier are sent via concentric conductors to a special subsystem (with the microcontroller Atmega16) synchronizing and making acquisitions. This system, after activating, sends rectangular pulse whose rising edge is used for synchronization. It has two switches allowing the card to disconnect the measuring signal from the microphone, and add a microcontroller signal instead. At the same time a signal is sent from the controller to 3 sockets for connecting ECG electrodes. Triggering pulse by the microcontroller causes the appearance of a rectangular signal at each slotted line. The program, after pressing "synchronization button", waits for rising edge and set a time shift in transmission.

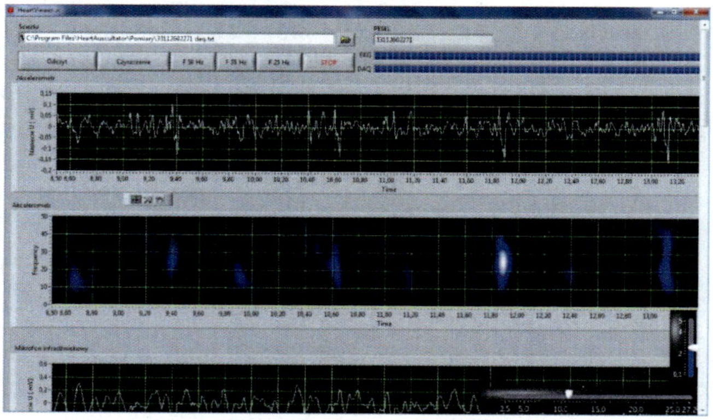

Fig. 3. Measurement application with acquired mechanoscopic signals

Bluetooth transmission doesn't allow determining the initial moment of passing data or controlling the moment of start downloading data to the buffer store. While measuring USB6008 card acquisition begins almost immediately (average delay of about 20 ms), the establishment of a Bluetooth transmission can take up to 3 seconds.

ECG and measurement card signals possess additional sample counters, which are equalized after synchronization, what makes controlling the moment of start possible. At the same sampling frequency this kind of synchronization would be sufficient. However, since the accuracy of the sampling frequency of 800 Hz in two sampling systems is different, after receiving a million samples, the signals are shifted toward each other. Time needed for sending a million samples is approximately 20 minutes and 50 seconds.

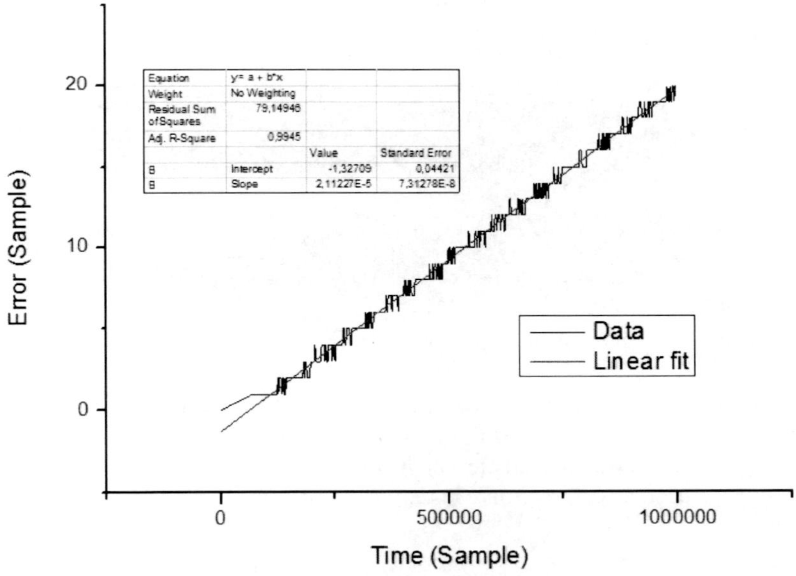

Equation	y= a + b*x		
Weight	No Weighting		
Residual Sum of Squares	79,14946		
Adj. R-Square	0,9945		
		Value	Standard Error
B	Intercept	-1,32709	0,04421
B	Slope	2,11227E-5	7,31278E-8

Fig. 4. Synchronization error increase during measurement

In order to determine the difference in clock frequency and define the correction resulting from this difference, we conducted several series of measurements. For this purpose we transmitted a rectangular pulse of a 1 Hz frequency and examined the number of samples of rising edge between the two channels. Figure 4 shows the displacement of the edge measured by the number of samples in relation to the number of samples with the entry of approximately one million samples. "Adj. R-Square" is statistical measure of how well a regression line approximates real data points.

After a series of 10 examinations we established that displacement after a million samples is 20 ± 2 samples. After making correction consisting in incrementation of the counter of the proper passage after every 50 000 samples, inaccuracy after 20 minutes was± 2. When converted to time equivalent it is about 2.5 ms and meets strict requirements for synchronicity as time distance between particular waves of ECG are one rank larger. Their range is from 20 to 400 ms and depend on the patient's heart rate. In order to maintain maximal non-synchronic error below the acceptable level of 2.5 ms, we entered into the program a special procedure for synchronization, which activates every 20 minutes.

6 Conclusions

The paper above presents hardware implementation of the ECG apparatus for acquisition of heart mechanoscopic signals registered with the use of an infrasound microphone, accelerometer and ECG apparatus. We show software visualizing data and also discussed the problem of signal synchronization. Further examinations will concern processing the obtained data in such a way so as to define a set of features allowing detecting insufficiency of the cardiac muscle.

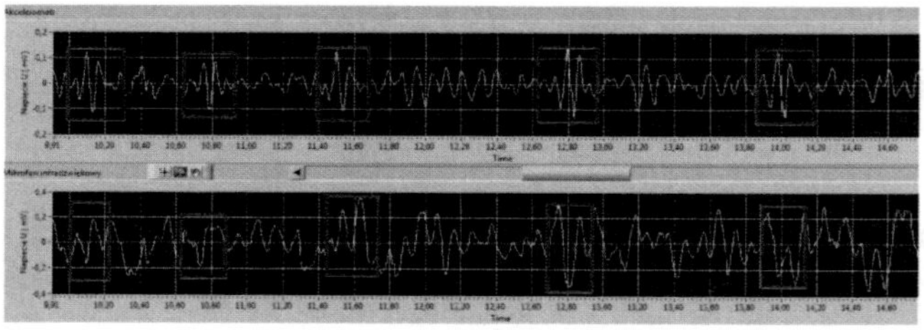

Fig. 5. Signals from the healthy patient's heart acquired by accelerometer and microphone

Figure 5 presents the complete record of signals originating from the healthy patient and figure 6 shows the result of the appropriate data registration in the case of patients with a very advanced cardiac insufficiency. In the record of the healthy patient left ventricular contraction causes the cyclical increase in the amplitude of the signals.

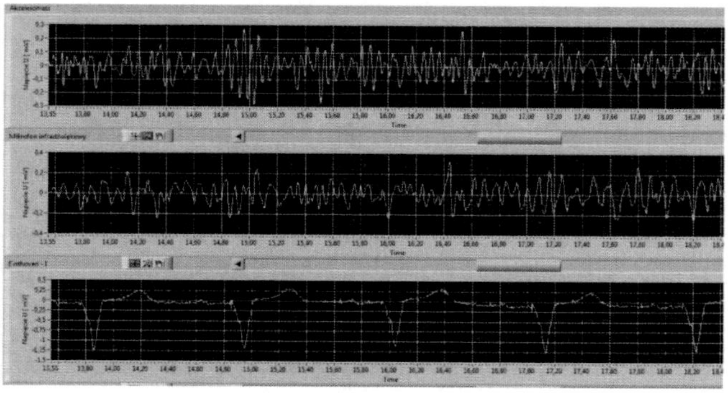

Fig. 6. Signals from the patient with very advanced cardiac insufficiency

Figure 6 is additionally enriched with ECG mileage showing lack of correlation between signal's amplitude and contraction of the left ventricle. The difference in this two examples are very big so it's easy to distinguish this two patients. Unfortunately for those cases where heart disease is poorly advanced signals are nearly identical to those in healthy individuals, and the doctor is unable to discern a significant difference. We think it's worth trying to find this difference by computer analysis of signals, which we would like to apply to the data collected in near future.

References

[1] Colucci, W., Brunwald, E.: Atlas of cardiac insufficiency (2005)
[2] Leyko, T., Ranachowski, Z.: Analysis of cardiac acustic emission –Implementation in cardiology – General Practice Nr 1/2007
[3] Gracia, T.B., Holtz Neil, E.: EKG: Art of interpretation (2007)
[4] Jamroży, M., Leyko, T., Lewenstein, K.: Early detection of the cardiac insufficiency. In: Recent Advences In Mechatronics 2008-2009, pp. 407–413. Springer, Heidelberg (2009)
[5] Zieliński, T.: Digital Signal processing – From theory to applications (2005)
[6] Manual - Farum E600GC
[7] Datasheet of microphone Bruel&Kjaer typu 4144
[8] Datasheet of accelerometer Bruel&Kjaer typu 4507B
[9] Datasheet of headphones Sennheiser HD280Pro
[10] Datasheet of data acquisition card NI 6008USB

Transmission of the Acoustic Signal through the Middle Ear – An Experimental Study

M. Kwacz[1,*], M. Mrówka[2], and J. Wysocki[2]

[1] Warsaw University of Technology / Faculty of Mechatronics,
ul. sw. A. Boboli 8, Warsaw, 02-525, Poland
[2] Institute of Physiology and Pathology of Hearing MCSM,
ul. Mokra 17, Kajetany k. Nadarzyna, 05-830, Poland

Abstract. The transmission of the acoustic signal through the middle ear structures was determined on the basis broad-band measurements (400 Hz to 10 kHz) of the stapes, the round window and the Teflon piston stapes prosthesis displacements. The measurements were made in 4 fresh cadaver temporal bones with AC stimulation at 90 dB SPL applied to the external auditory canal. Two conditions in the same specimen were measured: the physiologial and the implanted state. A PSV 400 Scanning Laser Doppler Vibrometer system (Polytec GmbH, Waldbronn, Germany) was used to determine the effect of the Teflon piston stapes prosthesis implantation on transmission of the acoustic signal through the middle ear. It was shown that the cochlear stimulation after stapes piston prosthesis implantation, in comparison with a physiological specimen was reduced several times.

1 Introduction

The acoustic function of the middle ear (ME) is to match sound passing from the low impedance of air to the high impedance of cochlear fluid. The acoustic signal enters the ear is transmitted through the external ear canal (EC) and converted to vibrations by the tympanic membrane (TM). These vibrations are, via the middle ear ossicles, transmitted to the stapes footplate (SF) where they are converted to a sound pressure in the cochlear fluid. Since the perilymph, filling the inner ear, is incompressible, oscillations of the stapes cause the round window (RW) membrane interaction. In order to determine the degree of stimulation of the cochlea the vibration amplitude of the RW membrane can be used. This is especially crucial in a situation in which the real stimulation of the cochlea is difficult to predict, e.g. after implanting various types of prostheses which transmit vibration energy to the cochlea instead of the immobilized stapes footplate (stapedotomy or stapedectomy procedures).

* Corresponding author.

The aim of this study is to estimate the effect of stapedotomy on the related changes in the sound transmission from the TM, through the ME, and into the cochlea.

2 Methods

2.1 Temporal Bone Preparation

Four temporal bones were used for these experiments. The temporal bones were harvested from human cadavers selected at the Forensic Medicine Department, Warsaw Medical University, within 2 days of death, in accordance with the standard methodology authored by Schuknecht. After harvesting, the bones were kept in a saline solution at 5°C and frozen immediately upon extraction. The dissection was performed at the Head and Neck Clinical Anatomy Laboratory, International Center of Hearing and Speech, Kajetany, Poland. Literature sources [1-3] show, that the functioning of middle ear structures in fresh cadaver temporal bones is the same as their functioning in the physiological condition. Our procedure of preparing the physiological specimen required the following steps to be taken: (1) harvesting a segment of the temporal bone within 48 hours of death, (2) microscopic inspection of the specimen, (3) reaming until the tympanic membrane (the bony rim around the tympanic annulus was left intact), (4) conducting a wide posterior tympanotomy and making the round window and stapes visible, (5) gluing ER3-14A (Etymotic Reasearch, USA) foam eartips onto the remaining bony rim around the tympanic annulus, (6) placing the ER7-14C (Etymotic Research, USA) microphone tube approx. 2 mm away from the tympanic membrane and an ER3-04 (Etymotic Reasearch, USA) loudspeaker adapter approx. 4-5 mm away in the earmold, (7) periodically dipping the specimen in a saline solution to maintain the proper hydration of the structure.

2.2 Measurement System

Acoustic system: The acoustic signal introduced into the external auditory canal of the temporal bone specimen induces the tympanic membrane vibrations, and causes vibration of the ossicular chain. A sound is produced by a probe loudspeaker connected to an adapter. In our study an input sound signal generated by a computer system (VIBSOFT, Polytec PI) and amplified by an acoustic amplifier (Revox A78) was supplied to the earmold placed into the specimen. The procedure of acoustic stimulation involves: (1) calibrating the VIBSOFT system, the loudspeaker and probe microphone, (2) setting the sound level intensity of the stimulating signal (90dB SPL), (3) selecting and setting the frequency of the input signal (sounds with a center frequency of successive one-third octave bands), (4) transmitting sounds into the external auditory canal with the aid of a loudspeaker (ER-2, Etymotic Research) attached to an adapter (ER3-04, Etymotic Research), (5) controlling the sound intensity level of the stimulating sound with a probe microphone (ER-7C, Etymotic Research) placed in the microphone tube approx. 2 mm away from the TM.

Optical system: The optical measurement system was created from a commercially available SLDV PSV 400 scanning laser Doppler vibrometer produced by Polytec GmbH, Waldbronn, Germany. A detailed description of the measuring system as well as its parameters is supplied in [4]. Figure 1 shows the scheme of the experimental measurements in a physiological specimen and a partial view of the measurement system.

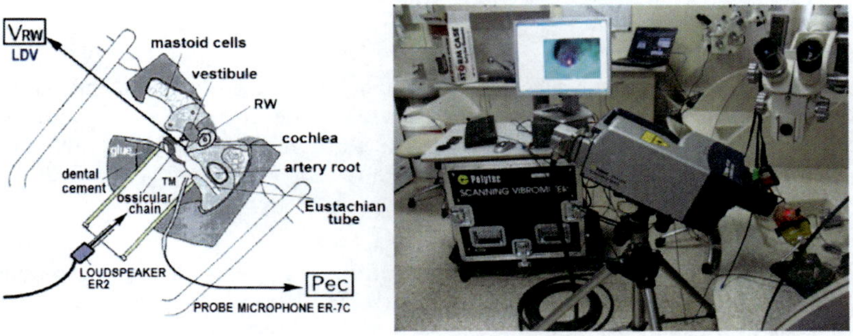

Fig. 1. Scheme of the experimental measurements (stage 1) and a partial view of the measurement system

2.3 Measurement Methodology

In order to prevent individual physiological differences among individual specimens from having impact on the results, the experiment was conducted in two stages: (1) stage I – measurements taken in a physiological specimen (the ossicular chain and tympanic membrane were left intact, the middle and inner ear structures were properly hydrated and the tympanic cavity properly ventilated), (2) stage II – vibration measurements in the same specimen after implanting a stapes prosthesis (retail 0.47mm diameter Teflon piston stapes prosthesis was used, fastened appropriately to the long crus of the incus by a platinum ribbon).

3 Results

The results of original empirical research into the parameters of the human ear RW membrane vibration in four physiologically fresh temporal bone specimens, conducted in accordance with the SLDV method were obtained for AC stimulation at a calibrated sound intensity level of 90 dB and with frequencies ranging from 400 Hz up to 10 kHz in the external auditory canal. The characteristics of the RW membrane vibrations were determined based on measurements of displacement in measurement grid nodes. The amplitude-frequency characteristics of specimen Preparat 1 before and after stapes prosthesis implantation for targets located on the surface of the RW membrane are shown on Figure 2.

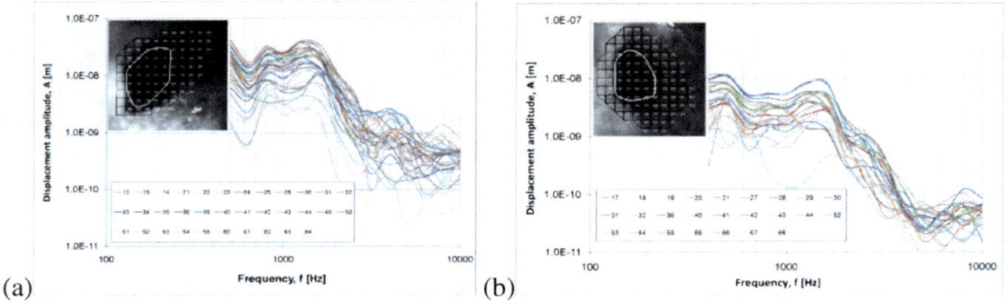

Fig. 2. (a) Magnitude of the displacement amplitude of 34 measurement points on the RW membrane – stage I (physiological), (c) Magnitude of the displacement amplitude of 25 measurement points on the RW membrane – stage II (implanted). The results are from specimen Preparat 1 stimulated with an AC when the sound pressure is 90 dB SPL in the external auditory canal.

The displacement amplitude for all measurement points in low-frequency range (0.5–2 kHz) is 10-15 times greater than in the high frequency range (2–10 kHz). Characteristic resonant frequencies of the middle ear are noticeable. The decrease in displacement amplitude of vibrations for frequencies above 2 kHz is related with different vibration phases in each measurement points on the RW membrane.

4 Discussion

The measurement results for the human ear RW membrane vibrations in four fresh cadaver temporal bone specimens showed that the maximum displacement amplitude for 1 kHz frequency in the central area of the RW membrane averaged 25 nm, whereas for 2 kHz, 4 kHz and 8 kHz frequencies it averaged 10 nm, 1.6 nm and 0.6 nm, respectively. The dispersion of the maximum vibration amplitude was related with different shapes and sizes of the RW membrane in each specimen, which is a characteristic trait of biological objects demonstrating individual variability. Based on detailed iso-amplitude chart analysis it was found, that the vibrations of measurement points spread across the entire surface of the RW membrane for all examined specimens in low-frequency ranges were single-phase vibrations. Above 1250 Hz – 2000 Hz frequencies, the phase of vibration for points placed in various parts of the RW membrane was different.

Analysis of the measurements results revealed, that the vibration amplitude of the RW membrane after stapes Teflon piston prosthesis implantation, in comparison with the vibration amplitude in a physiological specimen was reduced several times. Therefore one can assume, that as a result of conducting a standard implanting procedure, a significant change in biomechanical parameters of the middle ear conductive apparatus takes place, which causes a significant changes in the input impedance of the cochlea and a significant decrease in perilymph stimulation levels. The result of a decrease in stimulation of the perilymph in post-implantation condition, in comparison with its physiological condition, in case of otologic surgery conducted in-vivo, could be the incomplete closure of the air-bone gap resulting in hearing outcomes showing signs of conductive hearing loss.

5 Conclusion

Results of conducted empirical research, shows the character of vibrations of the RW membrane in fresh human temporal bone specimens in their physiological condition. The obtained characteristics form a basis for differentiating hearing results achieved after surgical ossicular chain reconstruction. Findings presented in this paper will be of practical use in the development of a new type of stapes prosthesis.

Acknowledgements

This work was supported by Polish Ministry of Science and Higher Education (Research Project Nr N N518 377637).

References

[1] Chien, W., Rosowski, R., Merchant, S.: Otol. Neurotol. 28, 782 (2007)
[2] Nakajima, H., et al.: The Laryngoscope 115, 147–154 (2005)
[3] Stenfelt, S., Hato, N., Goode, R.: J. Acousc. Soc. Am. 115, 797 (2004)
[4] Kwacz, M., Wysocki, J., Mrowka, M.: PAK 57, 25–29 (2011)

Simulation of Human Circulatory System with Coronary Circulation and Ventricular Assist Device

A. Siewnicka, B. Fajdek, and K. Janiszowski

Warsaw University of Technology Faculty of Mechatronics,
ul. A. Boboli 8, Warsaw, 02-525, Poland

Abstract. This paper presents results of modeling of human circulatory system together with the coronary circulation and the possibility of parallel heart assistance. The developed numerical model is a useful tool for studying behavior of individual components of the human circulatory system as well as the influence of the support application. It allows for simulation of the proper and the pathology circulation conditions, such as left or right heart failure. The addition of coronary circulation allows for coronary vessels occlusion simulation and enables verification of influence of left ventricle assistance on the coronary flow conditions. The whole model was implemented as a part of PExSim application. A short description of it is included and applied methods are presented.

1 Introduction

The development of technology enabled for frequent use of its achievements in medical applications. One of the biggest bioengineering projects in Poland is the Polish Artificial Heart program, whose aim is to develop of the construction and control algorithms for the heart assist devices. Applying the developed solutions requires accurate testing. For this purpose the modeling methods are widely used. In recent years some different models of the human circulatory system were developed, both numerical and physical ones, for example electrical or hydraulic. All of them are widely used to reproduce hemodynamic conditions of circulation system. Besides this, they can be applied for the testing of medical devices such as the blood pumps or assist devices. In our case, the main aim of development was to create a research platform for general purpose that could be easily adopted to solve different problems. For example it could be used for determination and testing of a Polish Ventricular Assist Device (POLVAD) control algorithms.

2 Model of the Circulatory System

The developed platform is a numerical mathematical model of the circulatory system composed of different individual components. It is based on the description proposed by Ferrari [1] and implemented as a part of the PExSim application [2]. This software consists of predefined function blocks that represent basic mathematical and logic relationships, dynamic and static elements or support an input and output operations. The possibility of easy extension with user-written objects makes it a flexible tool that can be used for emulation of complex dynamic system. For the simple and clear presentation of circulation system and to ensure the ability of easy parameters changes, each of the blocks is responsible for reproducing behavior of different part of human circulatory system. They are grouped in *Human Circulatory System* library, which consists such elements as: left and right ventricle (LH, RH), systemic arterial and venous circulation (SAC, SVC), pulmonary arterial and venous circulation systems (PAC, PVC) and lately developed, here described, model of the coronary circulation. Also the simple model of ventricular assist device (VAD), based on the ventricle description, was included. The mathematical equations in the circulation blocks were change [3] to allow the parallel ventricle support joining. The proper connection of these elements creates a complete model of circulatory system. A full description of structure and function of the blocks one can find in [3].

3 The Coronary Circulation Model

As the extension of functionality of the developed platform, the coronary circulation model was included. In fact, the dysfunction of it is a usual reason of some problems with circulatory system. The mathematical description is the combination of models proposed in [4] and [5]. The schematic representation of this, as an analogy to circuit diagrams, is shown in Fig. 1. The driving pressure for the coronary circulation is taken as a proportional to left ventricle pressure. According to this, the input values for the model are pressure values in aorta, left ventricle and right atrium (P_{il}, P_{lv}, P_{ra}). Output values are the following flows: the coronary arterial flow (Q_{art}), the flow supplying right atrium (Q_{ven}), the capillaries input and output flows (Q_{lcx1}, Q_{lcx2}, Q_{lad1}, Q_{lad2}) and extraordinary venous outflow (Q_{ex}).

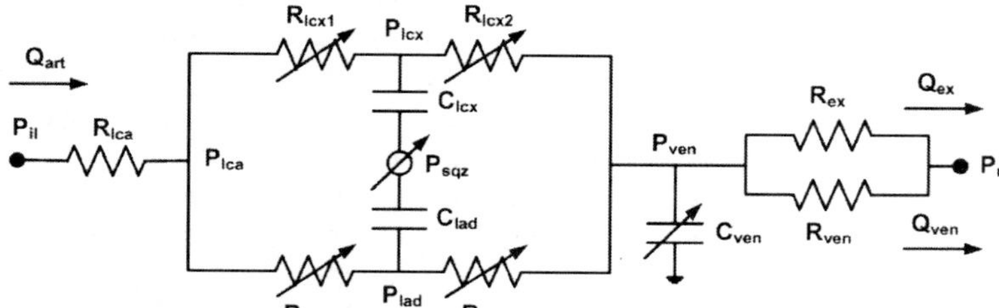

Fig. 1. The electric analogue of the coronary circulation system

The model was added to the PExSim application as a new Coronary Circulation (CC) function block. In order to join the coronary circulation function block to the main model of the circulation system, the modifications in equations for systemic arterial and venous circulation models had to be implemented. In the SAC element the input for coronary arterial flow (Q_{art}) was added and the description was modified. In the SVC element two inputs were added, for the coronary venous and extraordinary flows.

4 The POLVAD Modeling

The next step for creating a complete simulation platform was to create a model of the real assist device based on measurement results. For this purpose, an identification experiment of an extracorporeal ventricular assist device (POLVAD) was carried out at the Foundation of Cardiac Surgery Development in Zabrze (Poland). The Polish Cardiac Assist System POLCAST undergoing testing consists of a pneumatic driving unit POLPDU-402 and pneumatic drive pulse pump with disc valves and two chambers separated by a flexible membrane (Fig. 2).

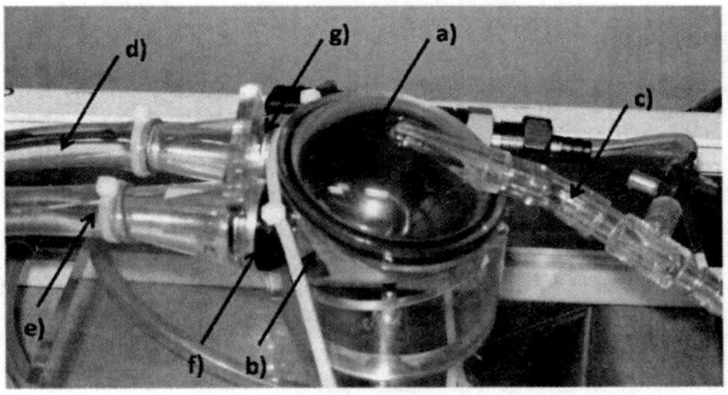

Fig. 2. The extracorporeal heart suport ventricle POLVAD: a) pneumatic ventricle, b) blood ventricle, c) pneumatic pipe, d) input cannula, e) output cannula, g), f) disc valves

The blood flow is a result of membrane movement, which is caused by air pressure changes in the pneumatic chamber. For the *in vivo* experiments, instead of blood, a fluid with similar properties was used. The assist device was connected to a hybrid circulatory system model developed at the Institute of Biocybernetics and Biomedical Engineering of the Polish Academy of Sciences. The following variables were measured: a supply pressure (Ppn) and pressures and flows in input and output cannula (Pin, Pout, Qin, Qout). As a result of the experiment, data series for different supply and load conditions were obtained. They were used as a basis for modeling the dynamics of the ventricular assist device. The modeling of the ventricular assist device is not a trivial task. The change in parameters strongly affects the nature of the process. Moreover, a model of mechanical activity of valves must be represented. The next problem is a strong dependence of the process conditions from appearing of full filling and emptying of the blood chamber. One from the approach to the description of the dynamics of the assist device was an attempt to determine the model using the fundamental principles of hydraulics and pneumatics. First, the pressure in the blood chamber was determined as a function of the control pressure. An algorithm was obtained to detect changes in the control pressure phases, which facilitated the modeling the opening and closing of the valves. The output flow was calculated on the basis of the pressure balance and changes in the valve resistance [6]. The parameters of the process were selected manually. Another method was the fuzzy modeling [7]. As parameterized variables were used two signals: output flow value and a set pressure different. For the output flow, the fuzzyfication was depended on the state of flow: positive flow (ejection), backward flow (closing of the valve) and the close valve phase (no flow). For the second variable the parameterization was divided in three areas: low, medium and high values of pressure. In both cases obtained fairly good results of modeling which are presented in [6, 7].

5 Summary

The paper presents new results of systematically developed numerical tool for modeling and simulation of human circulatory system with the extension of coronary circulation model and the possibility of the parallel assist device connection. The coronary model addition allowed simulation and verification of the influence of the left ventricle assistance on the coronary flow conditions. The whole package was implemented as a part of PExSim application. As a result a useful tool was developed, which gives the opportunity to study functions of individual components of the human circulatory system as well as the system as whole. It allows for simulation of the proper and the pathology circulation conditions, such as left or right heart failure. Implemented model of the ventricle assist device gives the opportunity for modeling the influence of the heart assistance on the hemodynamic conditions.

Acknowledgment

This work was partially supported by the National Centre for Research and Development (NCBiR) under Project "Development of metrology, information and telecommunications technology for the prosthetic heart" as part of the multiannual "Polish Artificial Heart" Program (task 2.1 - Developing a system of automatic control and supervision of work for extracorporeal cardiac prosthesis).

References

[1] Ferrari, G.: Study of Artero-ventricular Interaction as an Approach to the Analysis of Circulatory Physiopathology: Methods, Tools and Applications. Ph. D. dissertation, Consiglio Nazionale delle Ricerche, Rome, Italy

[2] Janiszowski, K., Wnuk, P.: A novel approach to the problem of the investigation of complex dynamic systems in an industrial environment. Maintenance Problems 4, 17–36 (2006)

[3] Fajdek, B., Golnik, A.: Modelling and simulation of human circulatory system. Methods and Models in Automation and Robotics (MMAR) 15, 399–404 (2010)

[4] Shreiner, W., Neumann, F., Mohl, W.: The Role Of Intramyocardial Pressure During Coronary Sinus Interventions: A Computer Model Study. IEEE Transactions on Biomedical Engineering 37, 956–967 (1990)

[5] Lim, K.M., Kim, I.S., Choi, S.W., Min, B.G., Won, Y.S., Kim, H.Y., Shim, E.B.: Computational analysis of the effect of the type of LVAD flow on coronary perfusion and ventricular afterload. The Journal of Physiological Sciences 59, 307–316 (2009)

[6] Siewnicka, A., Fajdek, B., Janiszowski, K.: Application of a PExSim for modeling a POLVAD artificial heart and the human circulatory system with left ventricle assistance. Polish Journal of Medical Physics and Engineering 16(2), 107–124 (2010)

[7] Janiszowski, K.: Fuzzy identification of dynamic systems used for modeling of hearth assisting device POLVAD., Pomiary Automatyka Robotyka, 90–95 (November 2010)

Novel FTIR Spectrometer for the Biological Agent Detection

G. Szymański[1,*], R. Jóźwicki[1], L. Wawrzyniuk[1], M. Rataj[2], and M. Józwik[1]

[1] Warsaw University of Technology, Institute of Micromechanics and Photonics, 8 Sw. A. Boboli St., Warsaw, 02-525, Poland

[2] Space Research Center of Polish Academy of Science, 18A Bartycka St. Warsaw, 00-716, Poland

Abstract. This paper presents the work carried to investigate the possible usage of Fourier Transform Infrared Spectroscopy (FTS) to remotely detect biological agents. We present a novel design of a Fourier transform spectrometer for biological agent detection. After presenting the opto-mechanical design of the device, a brief interpretation of selected measurements acquired by the device will be discussed. The spectrometer itself was tested on field tests in May 2011.

1 Introduction

Fourier Transform Infrared spectroscopy (FTS) is a known method for the pollutant detection of the atmosphere [1]. It concerns especially for both chemical [2-3] and biological [4-6] hazards. The recognition of chemical agents is relatively easy due to their linear spectra. On the contrary the spectra of biological agents are blurred, so till now their detection is uncertain. In this paper we present a new design and construction of FTIR spectrometer for making possible more accurate investigation of the biological agent detection. The project was realized within a framework of the project nr PBZ-MNiSW-DBO-03/I/2007 "Standoff detection and identification of biological agents with advanced opto-electronical methods" sponsored by Ministry of Science and Higher Education of Republic of Poland.

2 Requirements

Dynamic FTIR spectrometers with the purpose of standoff detection can be characterized by a number of parameters. The need for quick detection of possible life threatening agents dictates high speeds of the scan. Furthermore, some of the spectrometers [2] are used with the addition of scanning heads for passive scanning of large portions of space which also increases the requirement for high speeds. The typical resolution for such spectrometers is in the range of 1 to 8 cm^{-1}. This translates to required optical path difference (OPD) of order of several mm. To

* Corresponding author.

achieve high number of scans and to maintain desired resolution a special drive for moving mirror had to be implemented.

FTIR spectrometers for outside laboratory applications require detectors of high detectivity and usually use amplifiers to enhance the signal. The detectors should be cooled to reduce thermal noise.

The presented spectrometer is able to perform two scans per second with resolution of 1 cm^{-1} in two IR spectral windows: form 3 to 5 μm and from 8 to 12 μm. The speed of scans can be improved by reducing the maximum optical path difference in the spectrometer's interferometer. This results in lower resolution of instrument.

3 Description of Construction

The designed spectrometer is based on the Michelson interferometer. In order to simultaneously measure in both spectral ranges a novel construction has been proposed. The interferometer consists of two interferometers, which share a double-sided mirror mounted on the moving drive. The optical setup of the device is presented in Fig. 1. As for moving mirror drive, an ultrasonic piezo stage was proposed. The translation range of such stage is limited to 19 mm. The stage can achieve speeds of hundreds of millimeters per second. The speed, acceleration and range or translation are selected by user. Therefore, the desired speed of scans and resolution can be obtained.

Having measurement channels separated from each other, it was possible to use two different sets of beam-splitters and compensators. For both channels, this means for 3-5 and 8-12 μm calcium fluoride (CaF_2) and zinc selenide (ZnSe) were used, respectively. Due to the separation, both pairs of elements could be finely tuned for desired ranges. Also, detectors optimized to a given range could be used. Both detectors used in our construction are based on gallium arsenide (GaAs) and are cooled by multiple step Peltier coolers.

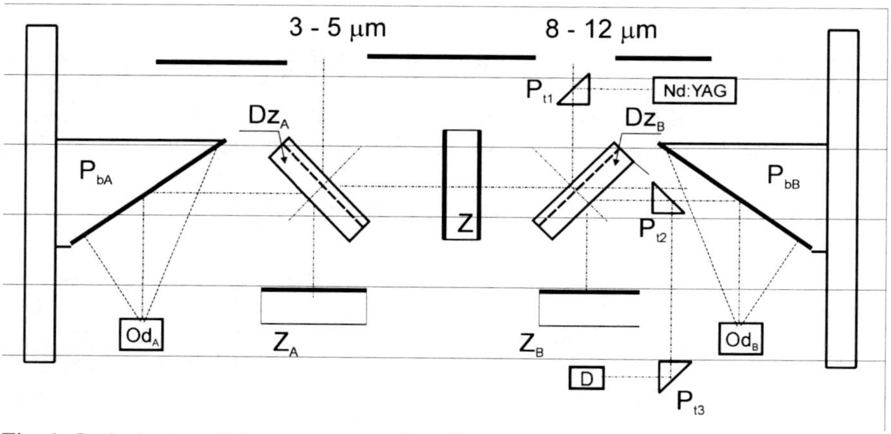

Fig. 1. Optical setup of the spectrometer. P_{bA}, P_{bB} – paraboloid mirrors, Dz_A, Dz_B – beamsplitters and compensators, Z_A, Z_B – flat mirrors, Z – double-sided mirror mounted on translation stage, Od_A, Od_B – detectors, Nd:Yag – Nd:Yag laser, D – detector for Nd:Yag laser, P_{t1}, P_{t2}, P_{t3} – prisms

In order to obtain spectrum without additional noise FTIR spectrometers require signal sampling in equal space intervals which is necessary for Fourier transform. Typically it is done by using additional coherent reference light source, i.e. a helium-neon laser. An additional interferometer is incorporated, be it additional mechanic setup or as a part of main interferometer. Signal sampling is generated on each positive or negative slope of an interferogram of the reference source. For our spectrometer two different options for obtaining equal space sampling were proposed. First is when Nd:Yag laser is incorporated in interferometer (as shown in Fig.1). Second option is to use internal optical encoder built into the ultrasonic piezo stage. The encoder produces pulses every 400 nm of translation, which corresponds to every 800 nm of OPD.

The entrance pupil of the device is equal to 23 mm, its field of view is equal to 1.2 deg x 1.2 deg. For field applications a dedicated mounting was designed to mount the spectrometer on a tripod. In order to aim on desired object the spectrometer a hunting scope was mounted onto the top of the device.

A complete setup with the Nd:Yag laser without the instrument enclosure is presented in Fig. 2.

Fig. 2. FTIR spectrometer with a scope for precise aiming without enclosure

Taking all in consideration, the described design possesses very high standards. The proposed construction is a subject of patent application in Patent Office of Republic of Poland [7].

4 Measurements

The demonstrator was field tested on Military University of Technology. A special measurement chamber was available in which biological agents were released. Our device was aligned about 50 meters away from the chamber. A series of different substances were then released, both in water aerosol and in powder state. For each agent series of measurements were obtained. For the data interpretation, the obtained interferograms were phase corrected and averaged. A sample result is presented in Fig. 3.

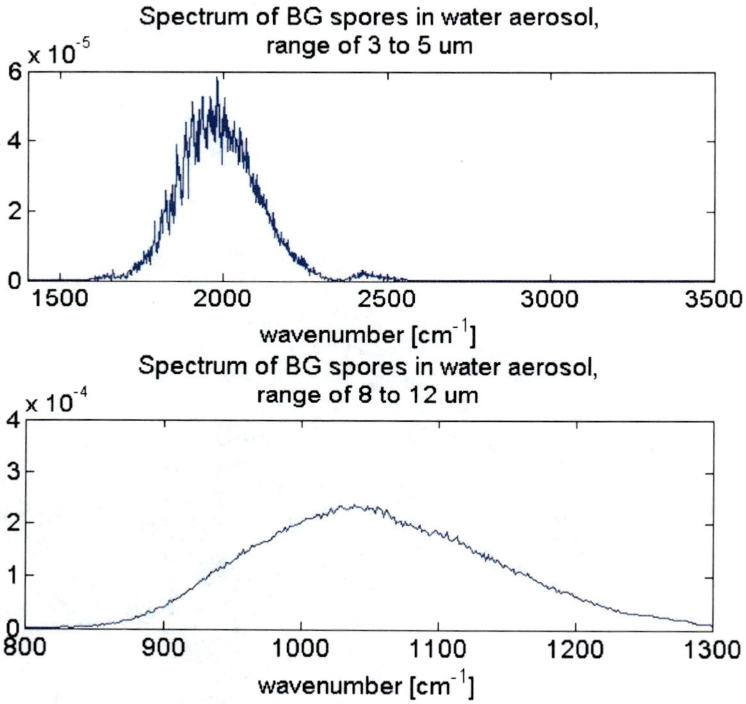

Fig. 3. Sample results obtained by FTIR spectrometer by 3 to 5 μm channel and 8 to 12 μm. The measured substance was BG spores in water aerosol

5 Conclusions

The presented FTIR spectrometer demonstrated the capability to measure spectra in desired regions. It is characterized by high specs, which are exceptional in comparison to similar devices. Its capability for biological detection and identification require further tests and data interpretation. The preliminary results show the presence of some distinctive features in acquired spectra that can be selected as indicators of presence or absence of biological agents. A base of different biological agent spectra must be composed as a reference material.

Acknowledgements

The authors would like to thank Space Research Center of Polish Academy of Science, Military University of Technology, Military Institute of Hygiene and Epidemiology and Military Institute of Chemistry and Radiometry for cooperation, assistance and for possibility to field test designed spectrometer.

References

[1] Harig, R., Matz, G.: Toxic Cloud Imaging by Infrared Spectrometry: A Scanning FTIR System for Identification and Visualization. Field Analytical Chemistry and Technology 5(1-2), 75–90 (2001)
[2] Harig, R., Matz, G., Rusch, P., Gerhard, H.-H., Gerhard, J.-H., Schlabs, V.: Infrared Remote Sensing of Hazardous Vapours: Surveillance of Public Areas during the FIFA Football World Cup 2006. In: Proc. of SPIE, vol. 6538 (2007)
[3] Farley, V., Chamberland, M., Lagueux, P., Vallières, A., Villemaire, A., Giroux, J.: Chemical agent detection and identification with a hyperspectral imaging infrared sensor. In: Proc. of SPIE, vol. 6661 (2007)
[4] Ben-David, A., Ren, H.: Detection, Identification, and Estimation of Biological Aerosols and Vapors with a Fourier-Transform Infrared Spectrometer. Appl. Opt. 42, 4887–4900 (2003)
[5] Thériault, J.-M., Puckrin, E., Jensen, J.O.: Passive Standoff Detection of Bacillus Subtilis Aerosol by Fourier-Transform Infrared Radiometry. Appl. Opt. 42, 6696–6703 (2003)
[6] Gittins, C.M., Piper, L.G., Rawlins, W.T., Marinelli, W.J., Jensen, J.O., Akinyemi, A.N.: Passive and active standoff infrared detection of bio-aerosols. Field Analytical Chemistry & Technology 3, 274–282 (1999)
[7] Rataj, M., Jóźwicki, R., Wawrzyniuk, L.: Dwupasmowy spektrometr fourierowski, patent application nr P-393118 (2.12.2010)

Selected Mechanical Properties of the Implant-Bone Joint

M. Zaczyk* and D.J. Choromańska*

Institute of Micromechanics and Photonics,
Warsaw University of Technology, ul. A. Boboli 8, Warsaw, 02-525, Poland

Abstract. The following experimental studies shows the results of mounting forces at the interface between implant and bone tissue as well as the quality of its mounting. On the basis of test results is possible to evaluate the proposed implant patient, depending on who is elected individually.

Using of artificial joints makes the patient after an operation of a endoprosthesis can recover the proper functioning of the motor system. During search for new methods of mounting an artificial joint within the bone tissue should be taken into account the characteristics of the unit. Experimental studies are required to show the quality of the results of mounting the implant mandrel within bone tissue. The results of research are related of the endoprosthesis and mandrels which are made of bioactive materials [6,7].

1 Introduction

A big issue of modern man is a sedentary lifestyle which causes high loads of motor system. During the year, ca.10% of bone mass is converted as a result of the restructuring process. Deteriorating state of the motor system is the cause of the increase of this organ injuries. Damages to the motor system are the cause for the search for new ways of treating organs and improve existing ones [2,3,5,8]. Research objective is to create a system to help determine the impact of implant mounted within the bone tissue on bone tissue. Currently it is not possible to clearly determine the problem.

2 Methodology of the Studies

In the experiments the impact of mounting the implant within the bone tissue was set way to define the relationship between an applied load and displacement of the mandrel within the bone tissue due to tissue density.

In experimental studies started checking the possibilities of measuring the parameters defining influence of the mounted implant mandrel on the bone tissue

* Corresponding author.

within the implant-bone joint. Studies have also shown the ability to monitor displacements of mandrel mounted in the bone tissue in relation to the acts in the direction of the load. During the measurements also observed displacement perpendicular to the direction in order to monitor and assess the deviation of the mandrel sinking into the bone tissue under the load forces. The experiment involved the technical weights, which have been subjected to loaded the sample. Load resulted from the forces generated by the weight of 80 kg. This type of load presents a picture of behavior of bone tissue which is loaded to the limit of elasticity without the effect of damaged tissue by the mandrel mounted within it.[7].

The results are recorded electronically used sensors which allow the registration of the results to be achieved in the time of the axes x, y, z. The results were recorded at two levels of the sample. A way of recording the displacements is shown in Fig.1.

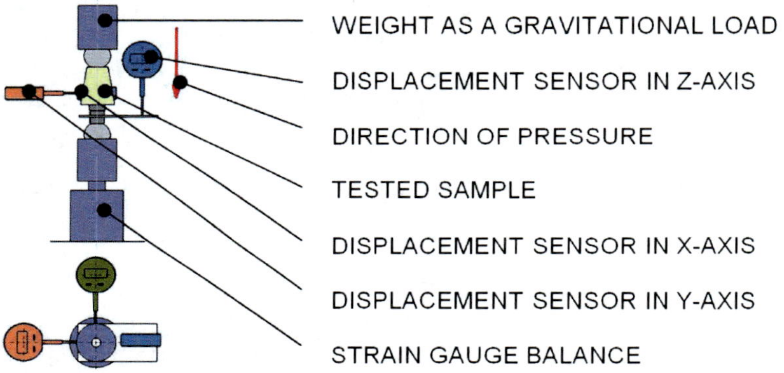

Fig. 1. Schematic of realization of the measurements of displacements along particular axes

Constructed test eliminated the undesirable situation of the component in case of force at a desired eccentric load. In the asymmetrical structure of the bones should be used with spherical spaces that automatically reject the forces that could cause a typical compression samples.

3 Object of the Studies

The studied object was mounting of a cylindrical frame which was made out of cobalt alloy (ISO 5832-12 standard), filled with a material resorbable in the osseous environment [1]. The material it was a synthetic osseous substitute by the name of Bio-Oss. Mounting of the cylindrical frame filled within a fragment and the closer shaft of the thigh bone was carried out with high accuracy and is suitable for all tested samples. While mounting it was used the same method as for the technical mounting during the implantation of the endoprosthesis during surgery.[4] (Fig.2.)

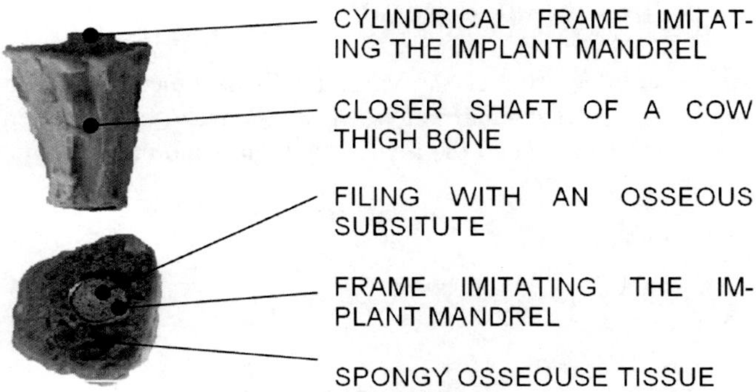

Fig. 2. Structure of the studied object presenting a technique of mounting a mandrel with a layer of resorbable material

4 Results of the Studies

The research shows possibility to create the characteristics of the displacements of mandrel mounted in relation to the bone, which was placed. During the experimental research on the test samples could be obtained characteristic for an implant-bone joint. Studies have shown that the reaction of bone tissue is a function of applied load. The table below (Tab.1) presents the average values of displacement of the mounted mandrel in relation to the bone which the mandrel was placed.

Table 1. Averaging values of the displacements for the third solution of the mounting, calculated on the basis of three tests

o	Y O H O J O	O	
		( )	(mm)
	Y O H H W J O		

5 Analysis of the Obtained Results

Analysis of the results determines the value of displacements in the direction of the applied load. (Fig. 3). The chart below shows the trend in movements that that can be seen between the bone and the mandrel which is mounted within it.

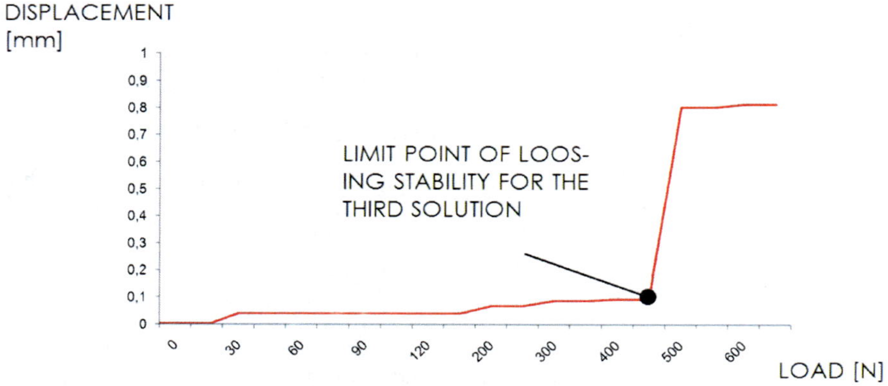

Fig. 3. Results of measurements of the displacements of the mounted mandrel with respect to the osseous tissue as dependent on the applied load

6 Discussion and Conclusions

During creating a mathematical model of the phenomenon one of the sources of information are needed, despite its imperfections, are biomechanical testing. Although experimental studies take a lot of time, enables the analysis of the results arising from clinical tests on the patient. Obtained results during the experimental studies of the implant-bone joint shows the pressure impacts in the area of the implant and bone tissue in which it is mounted.

References

[1] Marciniak, J.: Biomateriały w chirurgii kostnej, Wydawnictwo Politechniki Śląskiej, Gliwice (1992) (in Polish)

[2] Łaskawiec, J., Michalik, R.: Zagadnienia teoretyczne i aplikacyjne w implantach, Wydawnictwo Politechniki Śląskiej, Gliwice (2002) (in Polish)

[3] Morloock, M., Schneider, E., Bluhn, A., Vollmer, M., Bergmann, G., Muller, B., Honl, M.: Duration and frequency of every day activities in total hip patients. Journal of Biomechanics 34, 873–881 (2001)

[4] Pietruska, M.D.: A comparative study on the use of Bio-Oss and enamel matrix derivative (Emdogain®) in the treatment of periodontal bone defects. European Journal of Oral Science 109(3), 178–181 (2001)

[5] Marczynski, W.: Leczenie zaburzeń zrostu i ubytków tkanki kostnej" Wydawnictwo Bellona Warszawa (1995) (in Polish)
[6] Zaczyk, M., Jasińska-Choromańska, D., Kołodziej, D.: Selected mechanical Propierties of the Endoprosthesis-Bone Joint. Machine Dynamics Research 34(1), 118–127 (2010)
[7] Zaczyk, M., Jasińska-Choromańska, D., Miśkiewicz, A.: Stanowiska do badań jakości osadzenia trzpienia implantu w kości. In: Materiały konferencyjne Majówki Młodych Biomechaników", Ustroń (2011) (in Polish)
[8] Będziński, R.: Biomechanika Inżynierska, zagadnienia wybrane,Oficyna Wydawnicza Politechniki Wrocławskiej Wrocław (1997) (in Polish)

Rehabilitation Device Based on Unconventional Actuator

K. Židek*, O. Líška, and V. Maxim

Technical University of Kosice, Letná 9, Košice, 042 00, Slovakia

Abstract. In paper the rehabilitation device for upper arm based on artificial muscles is introduced. Presented automated rehabilitation device has three degrees of freedom: 2 DOF in arm and 1 DOF in elbow that provides almost all basic rehabilitation exercises. Artificial pneumatics muscles will be tested in connection with spring and antagonistic connection. This system provides lifting and falling of arm construction based on patient force. Possibility to generate help force during rehabilitation or opposite load is there. Artificial muscles are controlled through pneumatic valve terminal from micro computer based on MCU. Higher level control system provides artificial intelligence implementation based on neural network for prediction and change of load according sensor values history (incremental sensor, force sensor). For prototype testing there is described usability of industrial robot to test load and precision of trajectories during rehabilitation. This device will help to reduce therapeutics work with patient automate and improve rehabilitation process.

1 Introduction

Automated rehabilitation is nowadays in fast development in physical therapy. [3]. Automated rehabilitation is a special branch of rehabilitation medicine focused on devices that can be used by people to recover from physical trauma. The first results in this area are described for example in these articles [5], [6]. Within the area of rehabilitation automatized machines are more likely to be used. They replace manual procedures by autonomous excercises. There are three main areas of physical therapy: cardiopulmonary, neurological, and musculoskeletal. Though automated rehabilitation has applications in all three areas of physical therapy, most of the work and development is focused on musculoskeletal uses. Musculoskeletal therapy assists in strengthening and restoring functionality in the muscle groups and the skeleton, and in improving coordination. In the current paradigm of physical therapy, many therapists often work with one patient, especially at the early stages of therapy. Automatized rehabilitation allows rehabilitation to occur with only one therapist, or none with adequate results. Automated systems allow more consistent training program with automated tracking patient's progress and shifting the stress level accordingly, or making recommendations to the human therapist. In the future automated rehabilitation promises effective results. As the technology develops and prices decrease, rehabilitation systems will be available in everyday life.

---

* Corresponding author.

2 Construction

Construction of prototype device is mainly based on aluminum profiles and rotary joints. All actuators are based on pneumatics artificial muscles. Artificial muscles are suitable for these devices because of their flexibility especially in end positions. Presented automated rehabilitation device has three degrees of freedom: 2 DOF in arm and 1 DOF in elbow that provides almost all basic rehabilitation exercises as it was described by [1]. Artificial pneumatics muscles will be tested in connection with spring and antagonistic connection according design [4].

This system provides lifting and falling of arm construction. Possibility to generate help force during rehabilitation or opposite load is there. Artificial muscles are controlled through pneumatic valve terminal from micro computer based on MCU. Higher level control system provides artificial intelligence based on neural network for prediction and change of load according sensor values history. Schematic and design of rehabilitation device is displayed in Fig. 1.

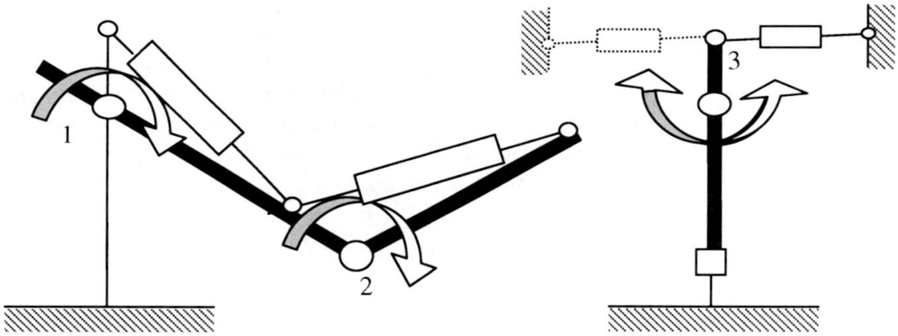

Fig. 1. Kinematics scheme of rehabilitation device

The mechanism is fixed to chair for rehabilitation in a comfortable sitting position. Rehabilitation system is designed for both arm (left, right), but not in same time. The patient must change chair for adequate arm.

3 Control System

The main control part is 8bit MCU (ATMEGA128L microcontroller) which control pneumatics artificial muscle and cooperate with sensors detailed showed in Fig. 2. The main output part for switching the electromagnetic valves is integrated transistor array, which is directly connected to the microcontroller output. Device is equipped by display and keypad for monitoring of rehabilitation process and practices selection. Microcontroller communicates with a PC by serial link (USART). There is possibility to connect device to mobile PC trough USART to USB transducer.

The basic control algorithm for the control of the rehabilitation process consists of three parts:

a) Regulatory part
b) Protective part
c) User part

The regulatory part of the algorithm ensures that the rehabilitation device copying required trajectories. An important feature for controlling the rehabilitation is pressure sensing of arm rehabilitation from limbs. Based on this property we can achieve a suitable speed of shoulder rehabilitation practices in the prescribed mode.

The protective part of the algorithm is designed to ensure safety of the patient during rehabilitation exercises, where for example: in case of detecting of acceleration level over the certain threshold device has to stop the movement of limb within a few milliseconds. The important elements for detection is included acceleration sensor, gyroscope and temperature sensor of human body (to monitor of the muscles during practice).

The user part of the algorithm ensures communication between the user and the microcontroller. By means of eight buttons it is possible to choose several types of rehabilitation practices with various parameters. All data during practice are displayed on the display unit. Since the display that is not able to display all values at once, so individual information rotated cyclically in a time loop [7].

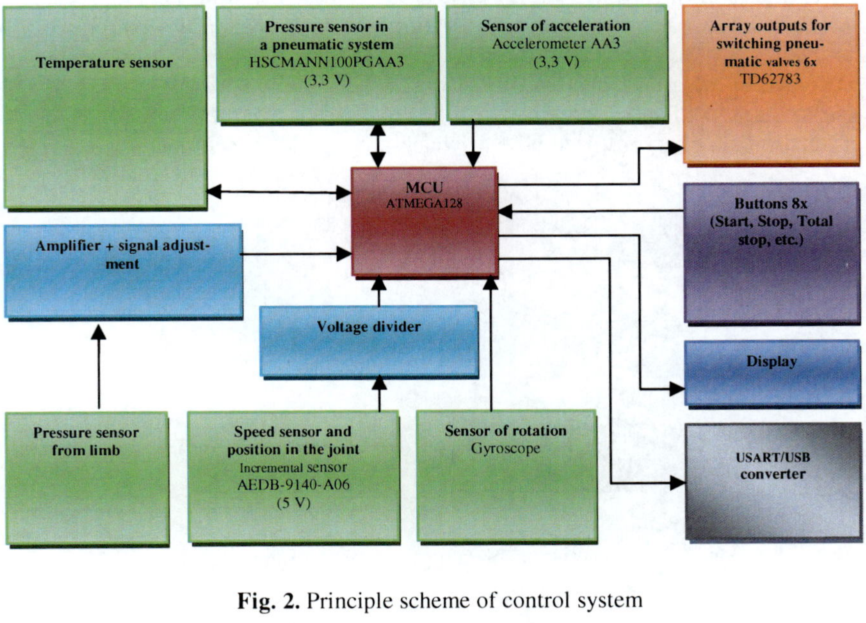

Fig. 2. Principle scheme of control system

Figure 3 is displayed design of PCB control board with user interface LCD and user buttons.

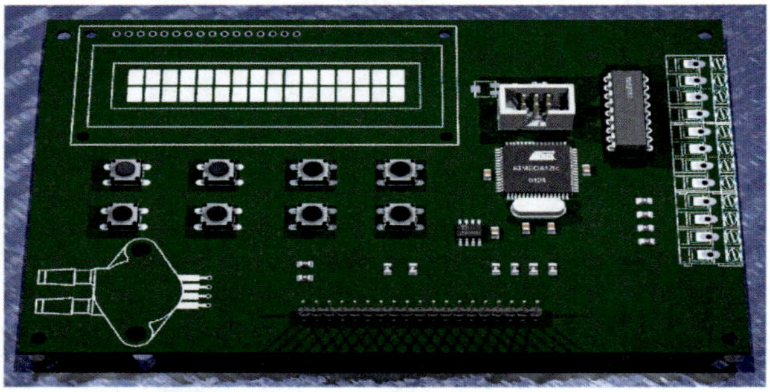

Fig. 3. Printed circuit board of control module

4 Neural Network Implementation

Utilization of artificial intelligence is widely applied in present. There are many experiments with various algorithms, methods and their combination e. g.: neural networks, theory of learning machines (machine learning), fuzzy logic, genetic algorithms, experts systems etc. As it was mentioned above the pneumatic artificial muscle is now unused mostly in reason of complicated control because of there is high non-linearity. Standard types of regulator fail what is main reason of using neural networks. Sequence of operation is visible in the Fig. 4 on the left. It describes operation of rehabilitation facility. In diagram is rehabilitation facility represented by operation system. In the Fig. 3, right there is displayed neural network with assigned specific values of four inputs to one output important for correct function NS. There is used NS with back propagation teaching.

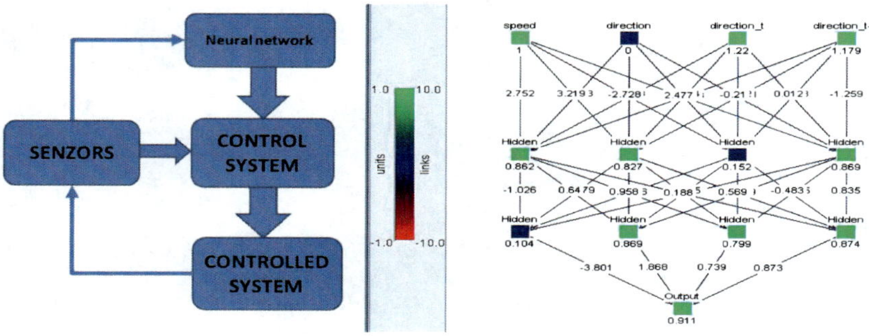

Fig. 4. Neural network and control scheme of implementation

5 Testing System

Testing platform is based on articulated robot with 5 DOF Mitsubishi RV-2AJ [2]. The robot is controlled from external C# application through serial port. Rehabilitation device is connected to end of robot efector through flexible coupling. We can reach any position in 3D robot workspace to define testing trajectory easy in drawing area. Testing device can help check safety of rehabilitation device before testing with life patient. Testing device is displayed in Fig. 5.

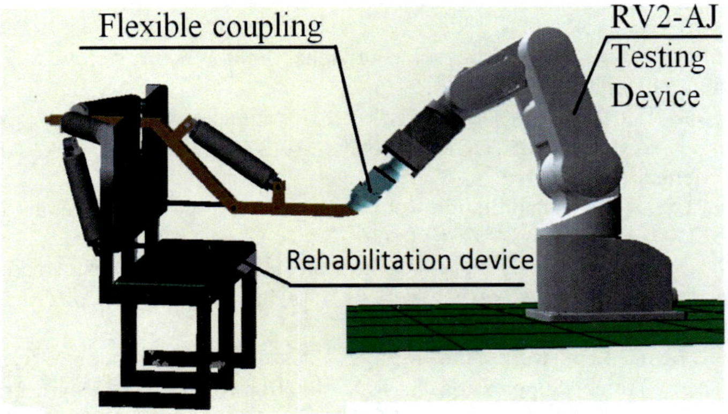

Fig. 5. Simulation of testing device and rehabilitation system

6 Conclusion

The project is using artificial muscle as joint actuator because of silent operation and flexibility during movement, start and end position. The developed automated rehabilitation device will save therapeutist capacity, provide improving in prediction of increasing and decreasing load during rehabilitation excersises according patient progress. System is monitored by many sensors during operation together with low level safety circuit. Prediction of load is based on integrated neural network algorithm. There is designed prototype testing system based on industrial robot. Next works after successful testing process will be development of mobile version without chair as outhouses (exoskeleton) for direct rehabilitation in patient household.

The research work is supported by the Project of the Structural Funds of the EU, Operational Program Research and Development, Measure 2.2 Transfer of knowledge and technology from research and development into practice: Title of the project: Research and development of the intelligent non-conventional actuators based on artificial muscles ITMS code: 26220220103.

We are support research activities in Slovakia / Project is cofounded from sources of ES.

References

[1] Cuccurullo Sara, J.: Physical medicine and rehabilitation board review, p. 938. Demos Medical Publishing (2010)

[2] Hopen, J.M., Hosovsky, A.: The servo robustification of the industrial robot. In: Annals of DAAAM for 2005, Opatija, October 19-22, pp. 161–162. DAAAM International, Vienna (2005)

[3] Kommu, I.S., et al.: Rehabilitation Robotics, p. 638. I-Tech Education and Publishing, Vienna (2007) ISBN 978-3-902613-01-1

[4] Piteľ, J., Balara, M., Boržíková, J.: Control of the Actuator with Pneumatic Artificial Muscles in Antagonistic Connection. VŠB – Technical University of Ostrava LIII(2), 101–106 (2007) ISSN 1210-0471

[5] Pons, J.L., Rocon, E., Ruiz, A.F., Moreno, J.C.: Upper-Limb Robotic Rehabilitation Exoskeleton: Tremor Suppression. In: Rehabilitation Robotics, Bioengineering Group, Instituto de Automática Industrial - CSIC, Spain, pp. 453–470 (2007) ISBN: 978-3-902613-04-2

[6] Sarakoglou, I., Kousidou, S., Nikolaos, G., Tsagarakis, C.D.G.: Italy Exoskeleton-Based Exercisers for the Disabilities of the Upper Arm and Hand in Rehabilitation Robotics, pp. 499–522. Genoa University of Salford1, UK (2007)

[7] Sun, J., Yu, Y., Ge, Y., Chen, F.: Research on the Multi-Sensors Perceptual System of a Wearable Power Assist Leg Based on CANBUS. In: Proceedings of the 2007 International Conference on Information Acquisition 2007 (2007)

Part VII

MEMS and Nanotechnology

Specific Measurements of Tilt with MEMS Accelerometers

S. Łuczak

Institute of Micromechanics and Photonics, Warsaw
University of Technology, ul. A. Boboli 8, Warsaw, 02-525, Poland

Abstract. The paper describes a specific kind of tilt measurements, where pitch of an axis is determined. The measurements are realized by means of MEMS accelerometers. Various mathematical formulas for computing the tilt are presented and results of their application, as well as the relevant requirements, are discussed. Experimental verification of the theoretical considerations is also presented, including values of an evaluated accuracy of the measurements performed by means of a tilt sensor built of two commercial MEMS accelerometers.

1 Introduction

One of numerous applications of accelerometers are tilt measurements [1]. If the employed accelerometers are of small dimensions, like their MEMS versions, they can be applied for sensing tilt of various devices having miniature size, like micro-robots [2].

Conventional component tilt angles are presented in Fig. 1 along with components of the gravitational acceleration. Let us assume that axes x, y, z are attached to an object, whose angular orientation must be determined, and axes x_0, y_0, z_0 are an immobile reference.

In typical measuring cases, both component tilt angles (pitch α and roll β) are determined (dual axis measurements), or only one angle (single axis measurements). However, there are cases when it is the most convenient to determine the arbitrarily oriented tilt angle φ (Fig. 1) instead. It should be noted that value of this angle does not change while the object rotates around z axis (neither value of the respective component acceleration).

Practical cases where such tilt measurements are interesting are for instance tilt of the rotation axis of a drill (e.g. in geodetic boreholes [3]). Then, the rotation angle of the drill (or of the drilling machine) around its symmetry axis is insignificant – only the tilt of the axis matters. If a standard two-axial tilt sensor were applied in the considered case (fixed e.g. to the chassis of the drilling machine), both its indications (corresponding to the pitch and the roll) would vary with rotation of the drilling machine around the symmetry axis of the drill, yet at a constant tilt of the axis.

This effect can be explained using an example of the leaning tower of Pisa. Even though its leaning angle is constant, an observer going up its stairs (and thus rotating around the symmetry axis of the tower at the same time) will be getting tilted to the right and to the left alternatively (as well as backward and forward – yet such tilting will be more difficult for him to observe in these directions because of balancing his body involuntarily).

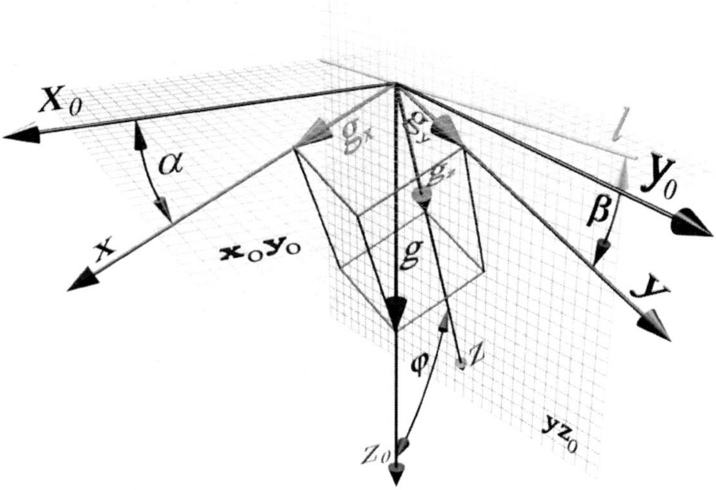

Fig. 1. Component tilt angles vs. the gravity acceleration

Quantities presented in Fig. 1 have the following meaning:

g_x, g_y, g_z – components of the gravitational acceleration indicated by accelerometers with sensitive axes arranged into the Cartesian x, y, z system,
g – gravitational acceleration,
φ – tilt angle,
α – pitch,
β – roll.

The tilt angle φ (corresponding to the leaning angle of the tower in our example) could be of course calculated on the basis of the pitch α and the roll β, however it would result in a decrease of the related accuracy, as well as in lengthening and complication of the data processing.

So, the most advantageous way of determining the tilt angle φ is to calculate its value on the basis of appropriate components of the gravitational acceleration.

2 Mathematical Relations

As resulted from Fig. 1, the specified tilt angles can be computed according to the following basic formulas [4]:

$$\alpha = \arcsin \frac{g_x}{g} \tag{1}$$

$$\beta = \arcsin \frac{g_y}{g} \tag{2}$$

$$\varphi_1 = \arccos \frac{g_z}{g} \tag{3}$$

The tilt angle φ has been subscripted as there are other ways of determining this angle:

$$g_{xy} = \sqrt{g_x^2 + g_y^2} \tag{4}$$

$$\varphi_2 = \arcsin \frac{g_{xy}}{g} \tag{5}$$

$$\varphi_3 = \arctan \frac{g_{xy}}{g_z} \tag{6}$$

The author proposed still other ways of determining tilt angles. The first consists in combining Eq. (3) with (5) as follows [4]:

$$\left\{ \begin{matrix} \varphi_1, \varphi_2 < 45^\circ \Rightarrow \varphi_4 = \varphi_2 \\ \varphi_1, \varphi_2 > 45^\circ \Rightarrow \varphi_4 = \varphi_1 \end{matrix} \right\} \tag{7}$$

The second idea is to combine Eq. (3) with (5) in a more sophisticated way – as a weighted average with variable weight coefficients [5], what ensures an even better effect (see Fig. 2) [6]:

$$\varphi_5 = \varphi_1 \sin^2 \varphi_4 + \varphi_2 \cos^2 \varphi_4 \tag{8}$$

Eq. (7) and (8) are based on a rational assumption that accuracies of the measured component accelerations are equal.

3 Sensitivity of the Measurements

The most important difference in using the presented mathematical relations is the sensitivity of determining the tilt angle. For the particular formulas, the respective sensitivity will be [4]:

$$k_1 = g \sin \varphi_1 \tag{9}$$

$$k_2 = g \cos \varphi_2 \tag{10}$$

$$k_3 = g = const \tag{11}$$

In the case of Eq. (7) and (8), the resultant sensitivities can be calculated respectively as:

$$\left\{ \begin{matrix} \varphi_1, \varphi_2 < 45° \Rightarrow k_4 = k_2 \\ \varphi_1, \varphi_2 > 45° \Rightarrow k_4 = k_1 \end{matrix} \right\} \tag{12}$$

$$k_5 \approx g = const \tag{13}$$

Courses of the sensitivity for each case are illustrated in Fig. 2. As can be clearly observed, the most advantageous course is taken by sensitivity k_3 and k_5, i.e. while using the arc tangent function – Eq. (5), or the weighted average – Eq. (8).

Course of sensitivity k_1 and k_2 justifies the idea of combining Eq. (3) with Eq. (5) – in this way the resultant sensitivity k_4 decreases only by ca. 30% (instead of decreasing down to 0 like k_1 and k_2).

So, generally, as far as the sensitivity of the considered tilt measurements is concerned (and the accuracy analogously), the most advantageous way of determining the tilt is to use the arc tangent formula (practical implementation of the weighted average is quite complicated [6]). The other formulas may be convenient in some particular cases.

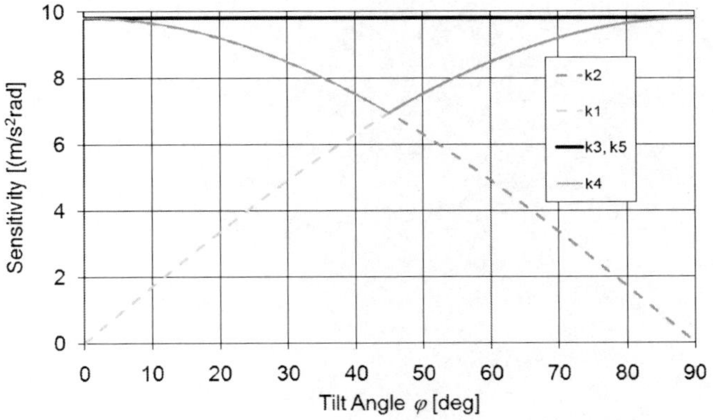

Fig. 2. Sensitivity of tilt measurements for various mathematical formulas

The employed functions expressed by Eq. (3), (5) and (6) have a limited domain: (0°, 180°) in the case of arc cosine, whereas (-90°, 90°) in the case of arc sine and arc tangent. So, in order to determine the tilt over the full angle, additional computation must be applied.

The simplest approach is to observe the sign of components g_x or g_y, and on this basis define the ultimate value of the tilt.

4 Experimental Verification

In order to experimentally verify the theoretical considerations presented above, appropriate tests were carried out. The test station used for this purpose has been described by the author in [7]. The tilt sensor was built of two dual-axis MEMS accelerometers ADXL 202E manufactured by Analog Devices Inc. [8].

The experimental verification consisted in applying a desired tilt of the tested sensor by means of the station, and then reading the analog output signals of the sensor. Comparison of the tilt θ applied by means of the test station and the tilt φ_i calculated on the basis of the sensor indications made it possible to determine values of measurement errors for particular mathematical formulas. The error e_i was defined as follows:

$$e_i = |\theta - \varphi_i| (i = 1, 2, 3, 4, 5) \tag{14}$$

Its courses for particular formulas are presented in Fig. 3.

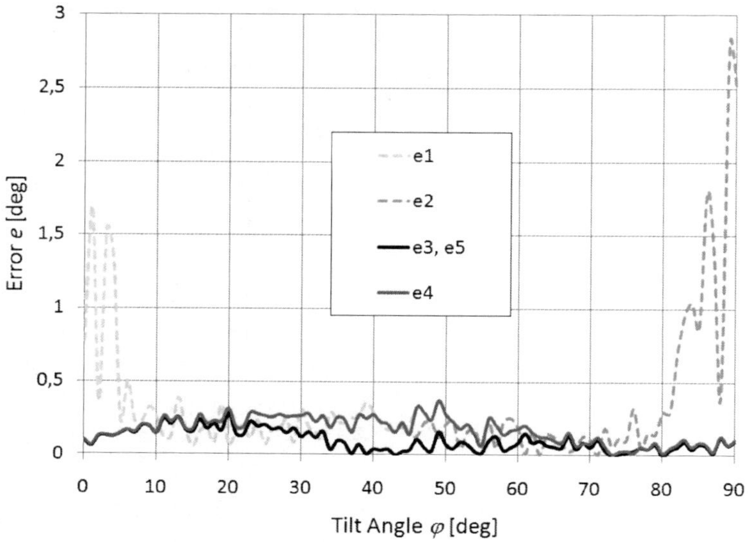

Fig. 3. Results of the experimental studies

As can be observed, the obtained results are consistent with the theoretical assumptions. Courses presented in Fig. 3 fully correspond to the variations of the respective sensitivities illustrated in Fig. 2.

Where the sensitivities k_1 and k_2 decrease, the respective errors e_1 and e_2 considerably increase. Courses of errors e_3 and e_5 are variable only due to the occurring random errors, as the amplitude of their variations is approximately constant over the whole angular range. While comparing errors e_3 and e_4 it can be stated that they are equal at the boundaries of the angular range, while slightly different in the middle, just as in the case of the respective sensitivities. Within the presented interval, value of a cumulated error for e_3/e_5 was of 9.3°, and 15.3° for e_4 (difference of ca. 86%).

5 Conclusions

The presented mathematical formulas for determining the considered tilt angle are analogous to respective equations for pitch and roll, presented in many various publications, e.g. [3,4,6,8-10]. It must be realized that application of particular formula results in different requirements pertaining to the employed accelerometer (or accelerometers).

First, there is involved a different number of sensitive axes of the accelerometer: 1 for Eq. (3), 2 for Eq. (5), 3 for Eq. (6)-(8).

Second, the measurement range of the accelerometer must be at least of ca. ±0,7 g while using Eq. (7), whereas it is to be slightly bigger, at least of ±1 g, while using the other formulas. This fact can be used when the tilt must be measured under vibrations, which increase the minimal value of the accelerometer measurement range.

Third, complication of the respective data processing is also different for various formulas. It is the most simple for Eq. (3), and for (5) - (8) gradually more and more difficult, what may be important while determining tilt in real-time.

In order to realize the discussed measurements by means of MEMS accelerometers with analog output signals, one can adapt for instance the algorithm proposed by the author in [10].

So, while choosing particular formula for determining a tilt angle of an axis, the following requirements must be considered: first, sensitivity of the measurements, then parameters of the applied accelerometer (number of sensitive axes, measuring range), and at last, the complication of the related computations.

References

[1] Wilson, J.: Sensor Technology Handbook. Newnes (2005)

[2] Fatikow, S., Rembold, U.: Microsystem Technology and Microrobotics. Springer, Heidelberg (1997)

[3] Qian, J., Fang, B., Yang, W., Luan, X., Nan, H.: Accurate Tilt Sensing with Linear Model. IEEE Sensors J. (to be printed)

[4] Łuczak, S., Oleksiuk, W., Bodnicki, M.: IEEE Sensors J. 6(6), 1669 (2006)

[5] Grepl, R.: Eng. Mech. 16(2), 141 (2009)

[6] Łuczak, S., Oleksiuk, W.: Eng. Mech. 14(3), 143 (2007)

[7] Łuczak, S.: Dual-Axis Test Station for MEMS Tilt Sensors. Metrology and Measurement Systems (to be printed)

[8] Low Cost ±2g Dual Axis Accelerometer with Duty Cycle Output, ADXL 202E, Analog Devices Inc., Norwood (2000)

[9] Tilt Sensing with Kionix MEMS Accelerometers, Kionix, Ithaca (2005)

[10] Łuczak, S.: Recent Advances in Mechatronics, p. 511. Springer, Heidelberg (2007)

Low Cost Inkjet Printing System for Organic Electronic Applications

K. Futera[1,2,*], M. Jakubowska[2], G. Kozioł[1], A. Araźna[1], and K. Janeczek[1]

[1] Tele and Radio Research Institute, ul. Ratuszowa 11, Warszawa, 03-450, Polska
[2] Warsaw University of Technology, Faculty of Mechatronics,
ul. Św. Andrzeja Boboli 8, 02-525 Warszawa, Polska

Abstract. This work shows in details the design and performance of precise ink jet printing system which has been constructed for organic electronic technology analysis. The printing system was designed for laboratory investigation of inks and substrates compatibility. Printing system has been tested by its precision and abilities by fabrication electronic elements. PEDOT: PSS, Sun Tronic U6415, Nano silver based inks were tested. Glass, alumina ceramic, PEN foil and paper were tested as substrates. Printer was design in order to solve disadvantages of commercial systems. It has improved software with user friendly graphic interface, improved accuracy and precision. New drop watch solution has been develop. In the investigation, compatibility of materials and inks was tested. Methods of making inks and substrates compatible, by setting the substrate temperature, ink jet printhead voltage and geometry, were studied. Printed lines after sintering process become conductive. Ohmic resistance of lines was measured and their quality was evaluated.

1 Introduction

Printed electronics belongs to one of the most important emerging technologies. It is also called flexible, organic or plastic electronics. It brings new approach of producing electronic components which are made nowadays mostly from silicon. This material allows to manufacture a highly complicated electronic circuits used in high-tech systems. However, a lot of applications can be found which requires cheap products used commonly. It does not pay off to produce them from silicon because of a high unit cost. Their price can be decreased through applying of printed electronics which is an alternative for traditional production process of electronic circuits. Printed electronics makes possible to produce electronic components on flexible substrates (foils, paper or fabrics). In the consequence, printed items can be bent repeatedly or even rolled. This causes no changes in their properties. It is not possible in the traditional products made from silicon which are rigid and their bending or rolling leads to irreparable damages. Moreover, used substrate materials are cheap. In the connection with it, an unit price of printed

[*] Corresponding author.

components can be decreased comparing to traditional one. Electronic components, like RFID (Radio Frequency Identification) tags, can be printed directly on a product like barcodes. This is also a big advantage of printed electronics. Another important feature of printed electronics is a possibility to reduce chemical waste left after etching process used commonly i.a. in a production of printed circuit boards. Printing is an additive process and a functional material is deposited directly to a substrate without applying chemical substances. It means that printed electronics is environmentally friendly production process of electronic components. Printed components can be manufactured using techniques used so far in the printing industry. Among them screen printing, ink jet, flexography and offset lithography play an important role. These techniques allow to produce components in a high throughput process like newspaper. Each of them characterizes with printing resolution, accuracy and requirements regarding to applied substrate and ink. For research purposes screen printing or ink jet are mostly used. These techniques requires a small amount of substrate and ink what is their big advantage. It causes that it is commonly used in prototyping. In screen printing a stencil is applied. It consists of a polyester or steel mesh tightened on an aluminum frame. An ink is squeezed on a substrate though this stencil using a squeegee drawing across the frame. After printing process a printed layer is dried by evaporation in an oven or by UV. Inks used in screen printing should have a viscosity in the range of 0.1 ÷ 10 Pa.s. Printed layers after drying are 20 nm to 100 μm thick. Comparing to screen printing ink jet is a relatively new techniques. Now it is developing rapidly, especially drop-on-demand (DOD) ink jet. In this process a single droplet is jetted through a nozzle when a pressure in a reservoir increases as a result of a vibration of piezo element. An ink should be around 10 mPa.s viscous and is drying also by evaporation or UV. Because of small DOD droplet's size (a few pico liters) an excellent resolution can be achieved, a few times higher than in screen printing. In the case of large area production flexography, offset lithography or gravure is used. These are roll-on-roll printing techniques. An ink is deposited on a substrate by a roller rolled on it. In the result a high throughput production process can be achieved. Printed electronics can be used in a variety of applications. The most important are: organic light emitting diodes (OLED), RFID tags, memories, sensors and photovoltaic cells. OLED may be used in the future to produce backlight in dash-boards or flexible low-cost screen which are durable to bending or rolling. Printed RFID tags will find an application in supply chain management, ticketing or brand protection. These tags can be integrated with various sensor allowing to identify object's exposure on temperature, humidity or UV. Comparing to currently applied indicators RFID-enabled sensors are read wirelessly using radio waves mostly in HF 13,56 MHz or UHF 860 ÷ 960 MHz. This is their significant advantage because there is no need to unpack a transported or stored product to check its state. These sensors will be used to create smart object fitted with an ability to react for changing environmental condition. Furthermore, printed RFID biosensors which monitor a chosen parameter of human organism (i.e. temperature) will find an application in hospitals. According to analytics prediction printed electronics will be developed rapidly in the next few years. It will play a crucial role in the future in a production of cheap, flexible consumer products. Each printing techniques brings variety of advantages, by using large scale flexography the price of final product can be lowered down more than two orders of magnitude according to single scale production. By combining layers printed with different inks, conductive, semi conductive and dielectric, complicated electronic structures can be manufactured. Organic electronic will never compete with silicon based electronic on field of efficiency or integration scale, this is not the way it's evolves into. Organic electronic elements can fill the market niche with flexible, low cost simple electronic devices. Main targets are "smart clothes" – sensors integrated clothes for firefighters, one time usage low cost biosensor – paper glucose sensor for diabetics and other basic but very useful devices. [1] Low cost RFID antennas can bring more wireless applications of sensors to new areas of applications. [2] Screen printing can be found very useful for material science, flexibility of process variables gives an opportunity to optimize paste properties easily. Variety of substrates material can be tested using screen printing, there are no limitations to flexibility or hardness as in flexography are. In ITME and ITR screen printed organic electronic investigation are being carried. Applying this method in ITR are being printed UHF RFID antennas on flexible substrates using organic and nanoparticle pastes[3]. Investigation of using CNT [4] based screen printable pastes as conductive material for flexible, transparent electrode applications have been also carried in ITME [5]. Third mentioned technique Inkjet is widely described in following article. Basic advantage of this technique is that additional tools are not needed for printing different patterns. It is very useful technique for rapid prototyping of electronic devices. [6] Inkjet printing is a simplest way to print the pattern in small series or even only one time. This method bases on jetting single, pico litter volume drop of an ink "on demand". Jetting single, round drop on demand is a critical variable of inkjet printing process.[7] Most common inkjet print heads are based on piezoelectric ceramic, electric pulse, sent from the driver, deform piezoelectric element which generates acoustic wave inside print head. Reaching nozzle edge wave generates drop.[8] There are print heads based on other that piezoelectric actuators, acoustic wave can be generated by small heater. Heater rapidly boils ink inside of the print head for a very short period of time, vapor bubble rapidly gain volume and rapidly disappears. This rapid volume change squeeze out drop through the nozzle.

2 The Design of Inkjet Printer

In this paper the design of piezoelectric print head based inkjet printer is described. It has been designed and built in Tele & Radio Research Institute (ITR). It is a small and low cost inkjet printer. The issue was to design flexible, low cost printing system to support material science investigation, technology optimization and new technology development. New printing system should have printing abilities no lower than commercial printers and low cost couldn't be an excuse. Commercial available printing systems has already been found inaccurate for some task realized in R&D units or in small laboratories. Those disadvantages, commonly described by the users were supposed to be removed from new designed printing system. Built printing system, is based on MicroDrop, piezoceramic PZT print head system [9]. This specific system gives the highest spectrum of compatible inks. It is possible to use inks with pH in the range 1-12, viscosity 1-12 cPs. Nozzle diameters in this printer are 50 μm and 100 μm, drop volume is in range from 35 pl to 85 pl. Printhead has a heater, build in, that allows to heat up ink inside of the nozzle up to 100°C. Printhead is controlled by attached driver unit. Generated electric pulse deforms piezoelectric actuator inside of the printhead. Operator controls the right pulse shape to get single "drop on demand" (DoD). Amplitude, length and delay of the pulse can be also set according to optimize drop shape. Amplitude can be set in range 0-250 V, length from 10 – 300 ms, and delay from 1000 to 9999 μs. Control unit generates strobe signal for the strobe light. This gives the ability to observe drops in still – "frozen" position using simple CCD camera. This system is also called "Drop Watcher". Delay of the strobe light signal can be adjust to freeze the right moment of the drop forming process. Commercial drop watch systems does not allow to freeze the picture Therefore, adjusting and optimization is less accurate then. Drop forming sequence taken by the strobe light CCD camera in different time delays is shown in Fig. 1.

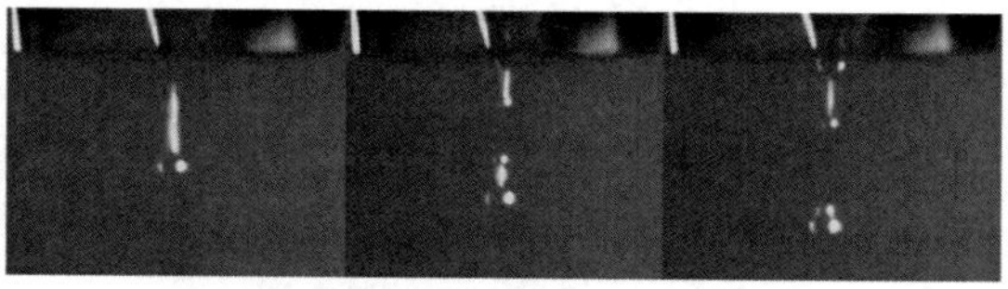

Fig. 1. Drop forming sequence of PEDOT:PSS taken by the CCD camera. Time offset was 20μs, 30μs, 40μs.

To perform printing patterns, the printhead must cooperate with highly precise positioning system. For new system designed in ITR ThorLabs LNR50S motorized translation stage[10] precise positioning system was used. Two linear positioning stages was combined into two axis positioning system. Scheme of printer is shown in fig 2.

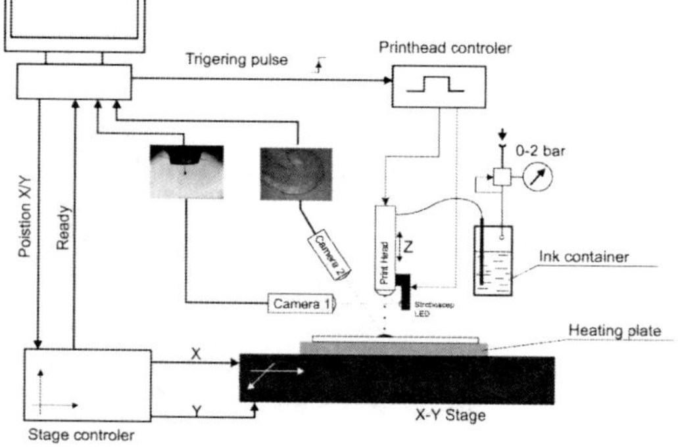

Fig. 2. Schematic of designed printing system.

Printing accuracy is lower than 5μm, due to advanced acceleration control software. Because of the dynamic of the printing process, speed and acceleration control had been improved by developing new software. Software uses acceleration control based on maximum speed calculation algorithm. To avoid "loosing steps" by the step engines, which is typically caused by the mass inert ion of the unit, algorithm calculates maximum possible speed of motion. This operation is being performed each time the move is to begin. Motion speed is optimized to maximum value, but the critical acceleration is never achieved. Printing system is equipped with two CCD TELI TOSHIBA cameras [11]. Camera 1 (see Figure 1.) is a part of Drop Watch System. Camera 2 controls the position of substrate stage by reference points. Reference points position control sets the stage position to pattern axis zero point, so that multilayer structures can be easily printed and putting out unfinished structure from the stage, for curing or drying, and putting it back did not cause position errors. Inkjet system has thermal conditioned vacuum stage, achievable temperature can be 10 °C lower and 80 °C higher than the surroundings. Thermal conditioning is solved by the Peltier cells[12] set and specially designed stage. Construction of the stage gives thermal stability below 0,1 °C even on paper or ceramic substrates. Thermal stability was measured using FLIR Infrared Camera. [13] Final build of new, low cost, inkjet printing system is shown in Fig. 3.

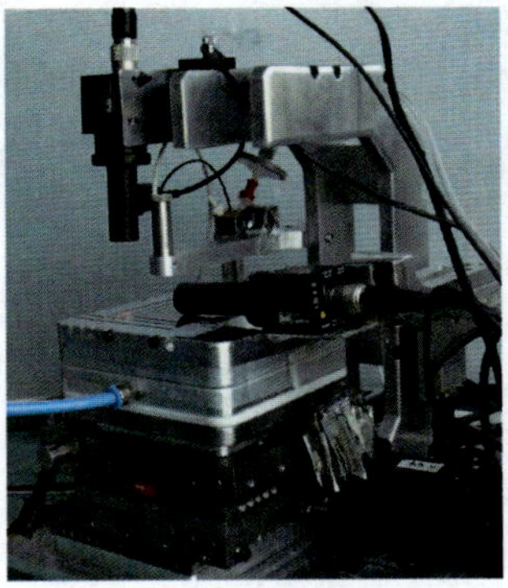

Fig. 3. Inkjet printing system build in Tele & Radio Research Institute.

However to perform technology investigation, graphic user interface software was developed. Software was designed to give maximum pattern design efficiency, so its graphic interface is similar to common basic CAD software. It is a simple, vector graphic editor with printer parameters control module. Software enables importing, and exporting pattern data to and from .CSV and .dxf files. That gives the user a choice, were the pattern will be created. Software calculate specific data such as stage position, stage speed and its trigger the drop on demand. This makes our printer a complete inkjet printing system designed for printing organic electronic. Our system provides performance of commercial printing systems with low cost and even higher flexibility. We have add new features and abilities to make inkjet printing more user friendly and give higher precision prints.

3 Printing Tests

Printer abilities has been tested by printing patterns on most use substrates and by using variety of inks. To test precision and accuracy specific pattern was designs. An array of single dots, dot lines, horizontal and vertical. For system tests nano silver based ink was used, this ink is a commercial ink provide by Amepox[14]. Dot spacing was 0,15 mm, dot diameter 100 μm. Printhead impulse was set to 110 V and 50 μs. Substrate temperature was 65 °C. As a substrate glass was used. 10 exact patterns has been printed. Specimen were measured by using optical microscopy (see fig. 4). Diameter, spacing and position of each dot was measured and analyzed. Dots diameter was 100 μm ± 4 μm, horizontal spacing 150 μm ± 2 μm, vertical spacing 150 μm ± 4 μm, position error 5 μm (as a distance from theoretic position to printed dot center), shape error 11 μm (as o standard deviations from ideal round shape).

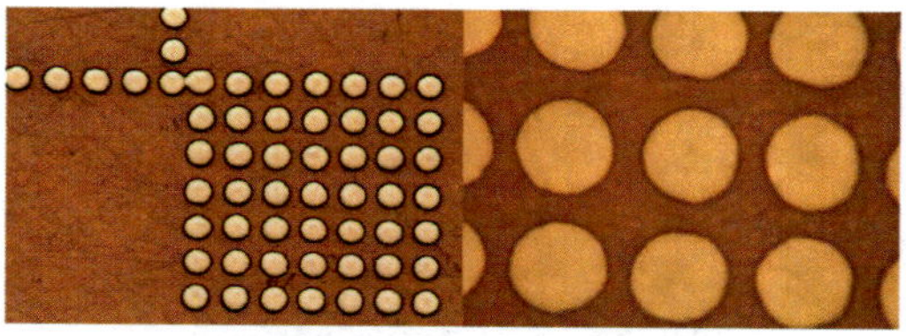

Fig. 4. Printed test pattern and shape precision.

Shape precision and jetting accuracy was also investigated. As a substrate for this test, glass was used. Single drop was jetted one more time on the printed pattern. As it can be seen, dots from the second layer are deposited precise on centers of first layers dots. This proves very high printing precision. Second layer dots exhibit smaller shape error, less than 3μm from model circular shape. However second layer dots have smaller diameter, the average diameter is 92 μm. This is caused by different surface tension on clean substrate surface and printed dots surface.

4 Line Printing Investigation

Line printing ability was investigated. Using built printing system conductive lines were printed on ceramic substrate. As ink, nano silver based ink was used. Line printing tests were divided into four categories by quantity of layers. Lines were printed on substrate heated to four temperatures, 40°C, 60°C, 80°C, 100°C. The issue was to find optimal parameters for printing conductive lines using nano silver based ink. Results of this test can be seen in tab. 1.

Table 1. Different substrate temperature and layers quantity printing results.

layers /temp C	1	2	3	4
40		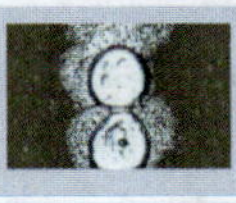	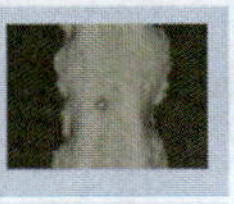	
60		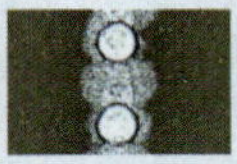		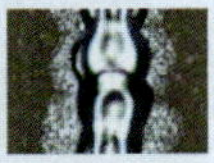
80			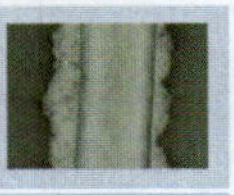	
100			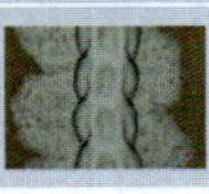	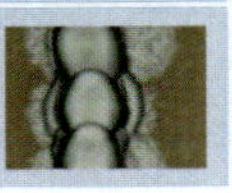

Blur edges of prints are caused by the structure of ceramic substrate. First layer of ink soaked in the porous substrate and deform the edge. Second layer of ink forms a line, solvents evaporates fast enough, to prevent dots from spilling on the first layer. Different line structure, width and morphology can be achieved by using different combination of parameters. Due to the test, best results can be obtained by printing on ceramic substrate heated to 80 °C with three layers. All printed patterns were sintered in 350 °C for one hour. In this process organic ink ingredients evaporates, and lines became conductive. After sintering resistance of lines was measured. The lowest resistance has lines printed on substrate heated to 60 °C with four layers. Resistance of this line is 0,063 Ω, lines with five layers and more, has higher resistance. It is caused by surface cracks, lines with four layers and more has long, wide surface cracks. Resistance value for line with five layers is 0,083 Ω, six 0,097 Ω . (see fig. 5).

Fig. 5. Surface cracks on line printed on substrate heated to 60 °C with five layers. Sintered in 350 °C for one hour.

Lines after sintering process can be seen in table 2. The "Coffee Ring" effect is strongly present. With line with three layers "Stack Coins" effect is observed. For four layer line none of these effects is present, but "Halo" effect shows up. Halo effect is caused by porous structure of surface of alumina ceramic substrate that was used in this experiment.

Table 2. Printed conductive lines, after sintering by firing in 350 °C for one hour.

Layers	2	3	4
			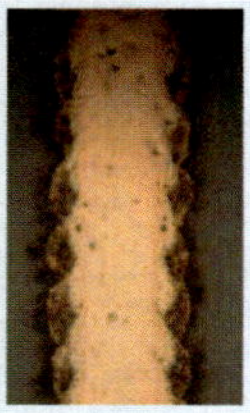
Resistance	1,29 Ω	0,85 Ω	0,06 Ω

5 Conclusions

Achieved results show, that precise inkjet printing system for organic electronic has been build. The advantage of this system is that it is low budget. Parameters of the new built system meet the commercial Inkjet printing systems, but it has been reached with ten times smaller budget. Developed new software makes pattern design easer for user, and graphic interface is more clear and friendly. Accuracy of the system was improved by using optimized maximum speed calculation algorithm. Printing precision of the printer is lower than 5μm and dot shape precision is not higher than 4μm. 100 μm ±5 μm lines can be printed using this system, this meets most requirements for manufacturing simple electronic devices. Moreover substrate stage design gives higher standard of temperature stability. It improves heat transfer and even distribution on substrate surface for low heat conductive materials like ceramic or paper. New drop watch solution adopted in this system uses CCD camera directed at printhead nozzle, makes drops observation during printing possible. This gives drops control feedback while printing process available. All these features sums up in high flexibility of printing system which opens new areas of possible usage of inkjet printers in electronic industry. Printed nano silver lines become conductive after sintering process. Sintering was executed by firing specimens in 350 °C for one hour. It is planned to investigate resistance sintering, as low temperature process, ready to print conductive lines on substrates fragile for high temperature firing. Printed conductive lines after firing have resistance 0,063 Ω. This values and precision makes this system ready for printing Radio Frequency Identification antennas, sensors, precise MEMS devices [17], conductive lines and pads or transistor electrodes. It is planned to manufacture organic electronic circuits using this system, OFET (Organic Field Effect Transistor) based logic structures and OLED / PLED screens. System has now entered the final test stage were its abilities will be investigated in small scale production line. It is ready to fabricate elements on semi commercial scale, and new materials for new applications are being developed.

References

[1] Sibinski, M., Jakubowska, M., Sloma, M.: Flexible Temperature Sensors on Fibers. Sensors 10(9), 7934–7946 (2010)

[2] White Paper, OE-A Roadmap for Organic and Printed Electronics. Organic Electronic Association (2009)

[3] Janeczek, K., Młożniak, A., Kozioł, G., Araźna, A., Jakubowska, M., Bajurko, P.: Screen printed UHF antennas on flexible substrates. In: Proceeding SPIE, vol. 7745, p. 77451B (2010)

[4] Słoma, M., Jakubowska, M., Młożniak, A.: Multiwalled carbon nanotubes deposition in thick film silver conductor. In: Proc. of SPIE, vol. 6937 (2007)

[5] Ptak, J., Cież, M., Zaraska, K., Słoma, M., Jakubowska, M.: A comparison of electrooptical characteristics of the electroluminescent lamps with transparent electrodes based on ITO and CNT. In: Large-area Organic and Printed Electronics Convention, June 23-25 (2009)

[6] Sitek, J., Futera, K., Janeczek, K., Bukat, K., Stęplewski, W., Koscielski, M., Jakubowska, M.: Investigation of inkjet technology for printed organic electronics. Elektronika LII Nr 3/2011, s. 112 (2011)

[7] Mohebi, M.M., Evans, J.R.G.: A Drop-on-Demand Ink-Jet Printer for Combina-torial Libraries and Functionally Graded Ceramics (2002)

[8] Wijshoff, H.: The dynamics of the piezo inkjet printhead operation. Physics Reports (2010)

[9] Producer datasheet, http://www.Microdrop.de

[10] Thorlabs GmbH, http://www.thorlabs.de – Motorized 2 LNR datasheet

[11] Toshiba-Teli Corporation, http://WWW.toshiba-teli.com prpduces datasheet.

[12] Choi, J., Yamaguchi, J., Morales, S., Horowitz, R., Zhao, Y., Majumdar, A.: Design and control of a thermal stabilizing system for a MEMS optome-chanical uncooled infrared imaging camera. Elsevier Science, Amsterdam (2003)

[13] FLIR SC645 data, http://www.flir.com/themrmography

[14] Nanosilver ink datasheet, http://www.amepox-mc.com

Determining the Influence of Load Near Machine's Component, Caused by Reshaped Material

M. Jonák

Brno University of Technology, Faculty of Mechanical Engineering,
Institute of Automotive Engineering, Technická 2896/2, Brno, 616 69, Czech Republic

Abstract. This article provides a short description of stress and deformation states in non-homogeneous material; it also deals with three coordinate systems that can be identified in a system with non-homogeneous material, the relation between these coordinate systems, transformation of the final acceleration vector and computing and transformation of normal and shear stresses vector between individual coordinate systems. Computing model, facilitating solution of the aforementioned problem in a 3-dimensional space, has been programmed in software Maple.

1 Introduction

Various kinds of non-homogeneous materials are utilized in many branches of mechanical and process industries. Design and construction of machines working with non-homogeneous materials requires the knowledge of its behaviour in various operating states (transport, stirring, pouring etc.). The better knowledge is available of the material's behaviour and its influence on machine components, the more accurate design or machine control will be possible.

With a closer look it is apparent that, regarding its behaviour, non-homogeneous material is a complex system and an accurate account of non-homogeneous material's behaviour is not available so far. This is the reason why the undermentioned theory is based on the assumption that actual non-homogeneous material can be substituted with a "suitable continuum" with defined properties; more precisely, "piecewise continuous continuum" with unstable properties, including size.

The calculation is based on physical-mathematical relations, which can be used – taking into account the aforementioned simplification – to describe the system with non-homogeneous material. Individual mathematical operations have been programmed in Maple, forming a compact computing model, which makes it possible to study the problem in a 3-dimensional space. For better illustration, the results are presented here especially in a graphical form.

2 Short Description of Tension

Tension in a 3-dimensional space can be described as a Cartesian tensor of second-order. If the tension exceeds a certain boundary value, its particles, possibly whole blocks, start moving towards each other or towards the machine. This is referred to as a limit tension state and the non-homogeneous material is reshaped.

The set of aforementioned values of limit normal stresses or shear stresses constitutes so-called limit line. In our case, the limit line is created from experimentally obtained values $[\sigma_i, \tau_i]$, which are connected with a cubic spline.

Elements of non-homogeneous material move with respect to each other or with respect to the machine following so-called shear curves. A shear curve can be defined as a spatial curve within the material, which satisfies the conditions of the limit tension state of particulate matter, i.e. $\sigma = \sigma_f \Lambda \tau = \tau_f$ [1]. In this case, considering further calculations, the shear curve is replaced with parametric cubic spline, which was created in Maple and its final form in the coordinate system x, y, z is presented in Figure 1.

3 Determining the Influence of Load Near Machine's Component Using the Maple Software

For an arbitrary point Z lying on a shear curve, a final acceleration vector a_v is defined in the dextrorotatory coordinate system x, y, z. In this case, this vector is coincide with gravity acceleration g and it causes the hydrostatic stress $h*\rho*g$ to emerge.

To find tension tensor it is necessary to create a tangent, a normal and a binormal in the point Z (see Figure 1); these define the axes of the second coordinate system t, n, b. Vector a_v is transformed into this system through directional cosines. It needs to be added that in all calculations, which use directional cosines, all vectors and directions have the form of unit vectors.

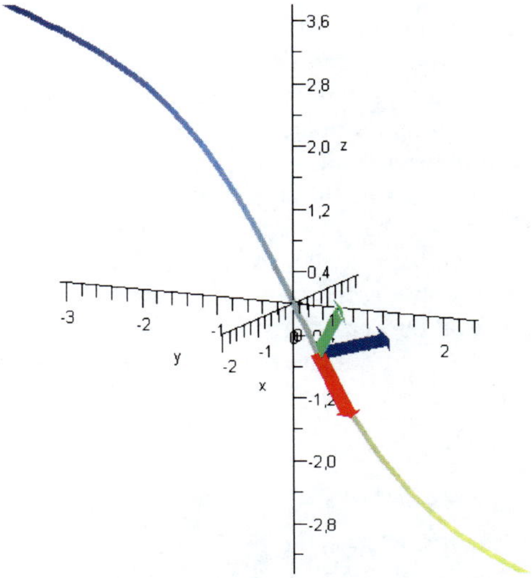

Fig. 1. Parametric cubic spline with the *t, n, b* coordinate system, where red, blue and green arrows is the tangent, normal and binormal, respectively.

If the limit line is available, it is possible to determine the directions and sizes of the main stresses σ_1, σ_2 and σ_3, which will be obtained from their functional dependence on the limit normal stress σ_f and the limit shear stress $\tau_f = f(\sigma_f)$.

Consequently, using directional cosines of the unit vector of final acceleration a_v which are known in all three coordinate systems, directional cosines of unit vectors σ_{01}, σ_{02} of the directions of main stresses σ_1, σ_2 are obtained. With respect to tangent and normal, axes of directions of main stresses σ_1 and σ_2 are rotated around the binormal on an angle as defined by (1) [2].

$$\pm\left(\frac{\pi}{4} - \frac{arctgf'(\sigma_f)}{2}\right) \tag{1}$$

Directional cosines of the unit vector σ_{03} of the direction of the main stress σ_3 coincide with the directional cosines of the unit vector of the binormal.

If point Z of the shear curve is very near the surface of the machine's component, determine in point Z directional cosines of the normal to the surface of machine's component n_{Z0} with respect to the coordinate system x, y, z [1]. Consequently, this normal may be transformed to the coordinate system defined by the directions of main stresses σ_1, σ_2 a σ_3 using its directional cosines belonging to the system x, y, z.

Mathematical equations for calculating of the size of normal stress acting through the point Z in the direction of the normal to the surface of the machine component n_{Z0}, the size of the shear stress τ_Z acting through the point Z perpendicularly to the normal to surface of the machine's component n_{Z0} and formulas give directional cosines of the shear stress τ_Z in the coordinate system x, y, z can be found in [1].

Now, we have determined the directions and sizes of normal and shear stresses in the point *Z*. Figure 2 compare the calculated values to the photograph taken during the experiment. Shear curve is in the form of plane curve and it lies in the plane parallel to the photograph's plane. For illustration purposes the coordinate system *x, y, z* has a general orientation.

The experiment was performed through moving the paddles in the particulate material which is affected by the gravity acceleration *g* and which bypasses the paddle following the shear curves, its direction coinciding with the edge of the paddle. The shear curve (cubic spline) and the point *Z*, the subject of the above-mentioned calculation, was chosen in a suitable fashion. The following material was used as the experimental particulate material: M-COLOR white 00052 PE/PP; spec. density 1,9418 g.cm-3; bulk density 1,12g.cm-3; POLYMER INSTITUTE Brno, CZ.

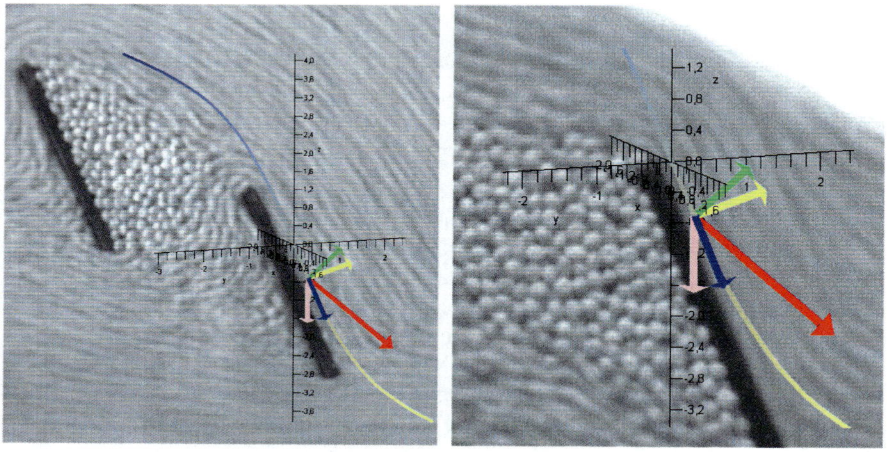

Fig. 2. Comparison of the calculated data with the experiment. Yellow and blue arrows are directions of the normal and shear stresses, respectively. Pink arrow represents of gravity, red arrow is the direction of the main stress σ_1 and green arrows are direction of the main stresses σ_2 and σ_3 (not seeing).

From the comparison presented in Figure 2 it is apparent that the directions of the normal and shear stresses vectors correspond to those obtained experimentally.

4 Conclusion

Using the Maple software, we have carried out the computation of the sizes and directions of the normal and shear stresses acting in a given point of the shear curve. With respect to the number of dimensions, the calculation was performed in a real, i.e. 3-dimensional space. This is one of the possible ways of description of the above-mentioned problematic, which is acceptable for low velocities, when the inertial forces can be ignored.

Acknowledgements

The published results have been achieved with the help of the project No. FSI-S-11-8 granted by specific university research of Brno University of Technology.

References

[1] Malášek, J.: Identifikace napjatosti, deformace a proudění v nehomogenních materiálech (2008) ISBN 978-80-214-3738-8
[2] Malášek, J.: Orientation of stress tensors in transformed matters. In: Engineering Mechanics 2006, pp. 224–225 (2006) ISBN 80-86246-27-2

MEMS Accelerometers Usability for Dangerous Tilt with Kalman Filter Implementation

K. Zidek*, M. Dovica, and O. Líška

Technical University of Kosice, Letná 9, Košice, 042 00, Slovakia

Abstract. This article contains description of mems accelerometers implementation to device which is able to measure danger tilt. We can find out actual tilt in two basic axis X and Y from -90° to +90°. Z Axis can only detect fall of device or in vehicle system very fast downhill grade during movement. For testing of solution we select small mobile robotic carriage. There is described hardware and software part of solution. Because data from sensor are in raw format from analog MEMS Accelerometer, we use free C# library with Kalman Filter implementation to remove signal error. We can acquire next information from sensor data for example movement's trajectory in X/Y axis (Cartesian system) and actual speed in all three axes. Fast alarm is provided by RGB diode or piezo buzzer. Integrated LCD display provides accurate information about actual device tilt.

1 MEMS Basic Information

Shortcut MEMS means micro electromechanical systems, marks mechanical and electromechanical construction of very small dimensions, and technologies used for their preparation too. MEMS technology is based on many tools and methods, which are used for creating very small structure with dimension of couple micrometers. An important part of technology was takeover from production of Integrated circuit (IC technology). Almost all of these devices are based on silicon substrate. MEMS structures are realized from thin layer. There are produced by photo lithographic methods. There are exists some other methods, that aren't derivate straight from technology of IC. There are three based step of operation in MEMS technology for layer applying to silicon material to substrate. Process of MEMS is usually structured sequence of this operation for creating real application. Real device then contains central unit for processing of data (microprocessor) and some other mechanical part which compose unit named micro sensor too. [4]

* Corresponding author.

2 MEMS Accelerometers

One of usual application for MEMS is sensor for measuring of acceleration. This MEMS sensor is usually named Accelerometer. They are divided to one-, two- or three Axes. Measuring of acceleration is possible to use in electronics and robotics for measuring: acceleration, deceleration, tilt, rotation, vibration, collision (crash) or gravitation. Accelerometers are used in many devices, special equipment and personal electrotechnics, for example:

- Robots and automated devices with balancing function (segway)
- Controls with tilt measuring
- Auto pilots of aero planes
- Car alarm systems
- Car crash detection (used in airbag system)
- Monitoring of people movements (virtual reality gloves)

Principle of MEMS microstructure sensor is displayed on figure 1., on the right side is displayed reference model.

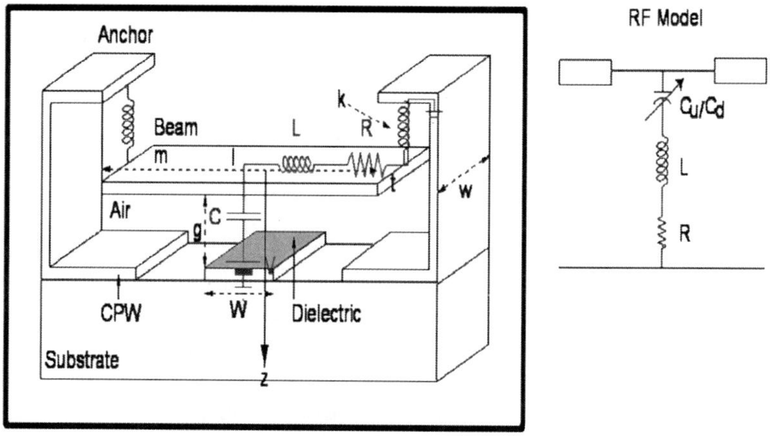

Fig. 1. MEMS sensor principle scheme and reference model

Older Accelerometers had big dimension and was very expensive. The construction was created from standard metal parts, springs and PCB. That was reason why wasn't used in electrotechnics nor robotics. This changed thanks to progress in MEMS technology. MEMS technologies do lowering the price, energetic consumption and dimension. Main usability is measuring of acceleration in three Axes: X - forward/backward, Y - left/right, Z - up/down. For mobile robotics we can use this sensor for measuring acceleration or deceleration by movement front and back, second Axis for change direction of movement right or left and third

Axis for fall detection of device. Second method of usability is measuring of device tilt based on simple mathematics. Picture 2 is shown MEMS Axis configuration and principle of tilt measuring with this sensor.

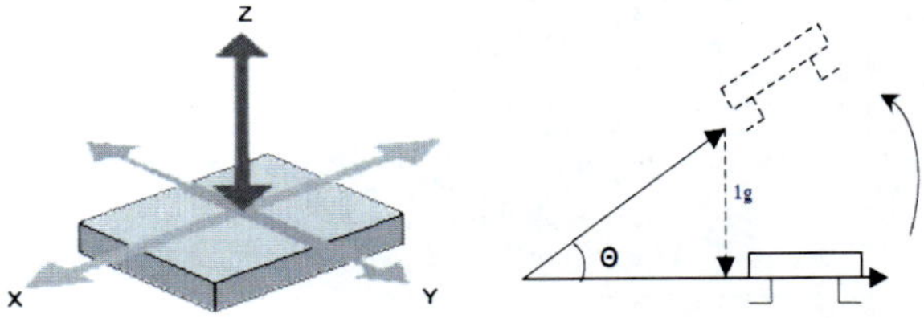

Fig. 2. MEMS sensor: Axis configuration, principle of tilt measuring

Output information from accelerometer is voltage which depends on movement or tilt of sensor in space. A static characteristic of sensor is not exactly linear. For common application we can this nonlinearity omit. The acceleration is usually in MEMS application measured in G unit. Expression "1 g = 9,80665 m/s2" means, that for every second, which passed the speed change will be 9,80665 meters per second. That is approximately speed 35.30394 km / h. The three Axis accelerometer can get null G on every Axis, if is in ballistic trajectory known as inertial or free fall. If we turn the accelerometer to 90 ° the output from one Axis will be exactly +1 g. In this situation, accelerometer measuring gravitation Force and can be in static position. Described characteristics for analogue MEMS sensor is displayed on figure 3. [1]

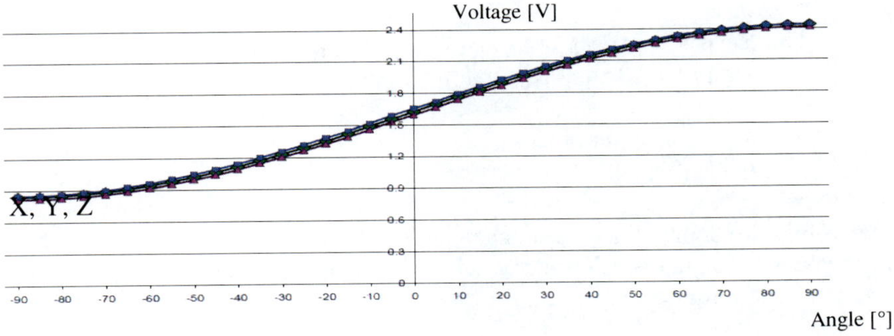

Fig. 3. Static characteristics of MEMS accelerometer sensor [2]

Example of block scheme sensor connection to user application is displayed on the figure 4. Additional LCD display, RGB led and piezo buzzer is connected straight to microcontroller enabling testing of application without Computer necessity.

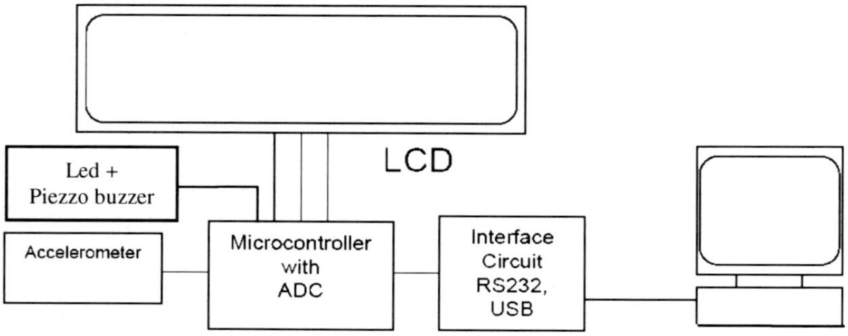

Fig. 4. Block scheme of application based on MEMS accelerometer sensor

Sensitivity of measured values depends on sensor G range. Most precise we acquire if sensor is set to ±1g. A disadvantage is that we cannot measure higher values of acceleration. Common sensors are produced to ±5g and it is possible to switch between ranges during application activity. Computations of tilt angles θ are realized thru basic mathematics and goniometric function. Vout is actual value of voltage, Voffset is voltage by 0g. Sensitivity of sensor is defined by technical documentation. In math is necessary find out positive or negative acceleration according to offset value. Datasheet math count according this:

$$V\_OUT = V\_OFF + \Delta V/\Delta g * 1g * \sin\theta \quad (1)$$

$$\theta = \arcsin(V\_OUT - V\_OFF)/(\Delta V/\Delta g) \quad (2)$$

where:

V_OUT = output of accelerometer (V) from ADC
VOFF = acceleration 0 g offset
$\Delta V/\Delta g$ = sensitivity
1 g = world gravitation
θ = tilt angle

Our values are counted according changed math, because we don't know max and min values for actual accelerometer. This maths gets extreme values during accelerometer operation.

$$incr = 180 * (H\_max - H\_min) \quad (3)$$

$$\theta = incr * (H\_nam - H\_min) \quad (4)$$

Where:

H_max, H_min – initial value of accelerometer extreme
H_nam – actual accelerometer value

3 Tested Hardware Platform

Introduced solution was tested on mobile computer with open source application in programming language C#. A prototype board contains Akcelerometer MMA7341L (analog) and accelerometer MMA7455 (digital) from Freescale. Currently there is active only analog Accelerometer. Microcontroller computes values of voltage for all Sensor Axis with help of three 9 bits ADC converters. Data are coded to frames (9 bytes as string $XXYYZZ1310). Every axis has value coded to two bytes (Low and High 8 bites).

First method of accelerometer communication is only for debug the application. Sensor is connected straight to PC. Data are sent thru serial line to serial port of PC.

For implementation to mobile robot is used USART interface without UART/RS232 Transducer and communicate straight with High Level control system based on AT91 control board with Linux Embedded OS. These serial data are transferred to TCP packet thru ser2net command line application. Data are sent next thru wife interface to C# application. Block diagram of testing debug solution and mobile control system implementation is displayed on figure 5.

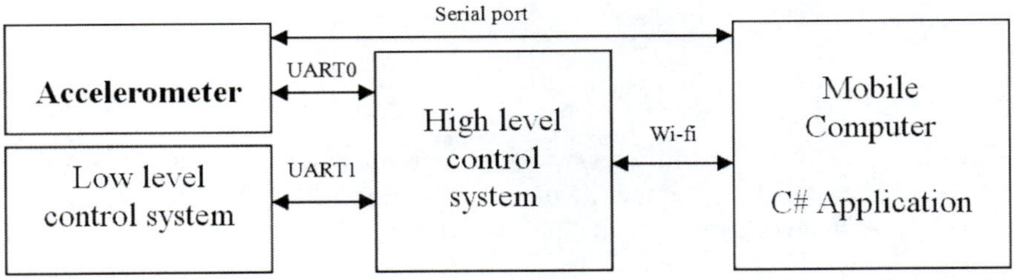

Fig. 5. Block diagram of connection sensor to testing mobile control system

Figure 6. is displaing PCB design on the left side, next is first prototype of sensor without RS232/USB transducer.

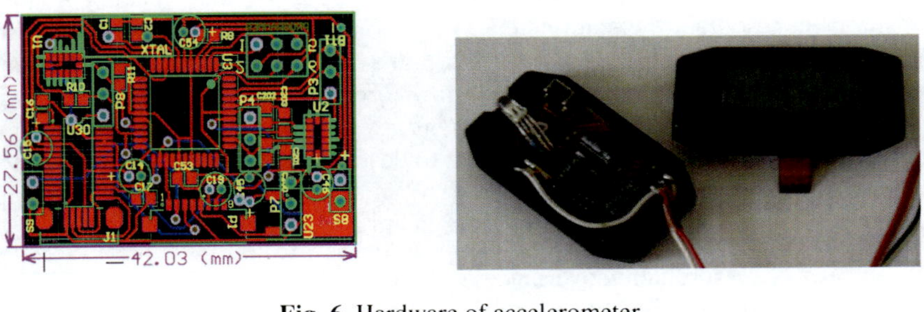

Fig. 6. Hardware of accelerometer

4 Software Platform Implementation

Software solution is based on open source C# application, which is currently implemented to mobile solution Graphical Interface of solution is displayed on figure 7. left is displayed 2D graphics, tilt in x-Axis. All values of real time tilt are displayed in graphical interface in text edit boxes.

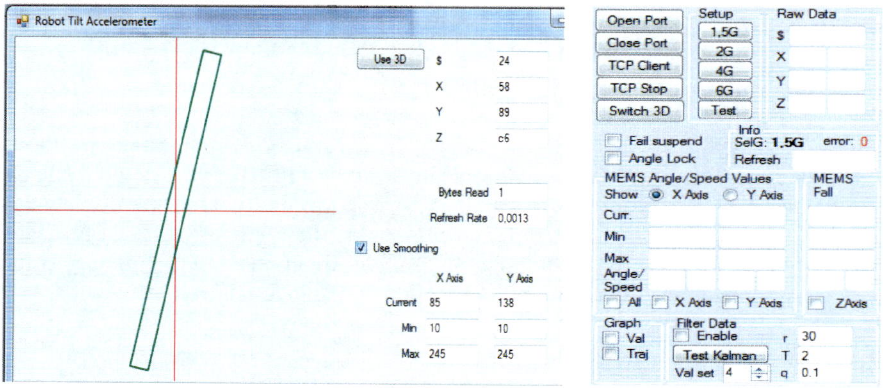

Fig. 7. C# application, left 2D tilt in X-Axis, configuration panel

Basic value of danger tilt is set to value bigger than 40°. This value starts critical routine and block movement of mobile device to actual direction. Danger tilt value can be changed thru graphical interface from 0-90°.

5 Kalman Filter Implementation to Smooth Raw Data

The Kalman filter is an efficient recursive filter that estimates the internal state of a linear dynamic system from a series of noisy measurements.

The Kalman filter is used in sensor fusion and data fusion. Typically real time systems produce multiple sequential measurements rather than making a single measurement to obtain the state of the system. These multiple measurements are then combined mathematically to generate the system's state at that time instant. Acquired data from MEMS sensor are in raw form with many disturbances, white noise etc. For testing solution we implement free C# Math.NET Neodym (Signal Processing)[8] with Kalman filter function to desktop application. Graphical interface provide settings of three basic values of Kalman filtering r, T, q which are necessary for customizing filter for real application.

r - Measurement covariance modifier
T- Time interval between measurements
q - Plant noise constant

Discrete Kalman Filter consist of two parts prediction and update.

Prediction:

x(k|k-1) = F(k-1) * x(k-1|k-1) (5)

P(k|k-1) = F(k-1)*P(k-1|k-1)*F(k-1) + G(k-1)*Q(k-1)*G'(k-1) (6)

Update:

S(k) = H(k)*P(k|k-1)*H'(k) + R(k) (7)

K(k) = P(k|k-1)*H'(k)*S^(-1)(k) (8)

P(k|k) = (I-K(k)*H(k))*P(k|k-1) (9)

x(k|k)= x(k|k-1)+K(k)*(z(k)-H(k)*x(k|k-1)) (10)

where:

S - Measurement covariance, K - Kalman Gain, P - Covariance update,
x - State update, F - State transition matrix, G - Noise coupling matrix,
Q - Plant noise covariance matrix, H - Measurement model,
R - Covariance of measurements, I – Matrix identify, z - Measurements of the system

Figure 8 displayed graph of actual values when MEMS sensor is stand statically on the ground (blue plotline). Black plotline shown filtered value cleared from errors and noise from ADC transduction. There is used for testing application only 1D Kalman filter for filtering only actual acceleration value. Next extension will be implementation of 2D or 3D filter for all three Axes.

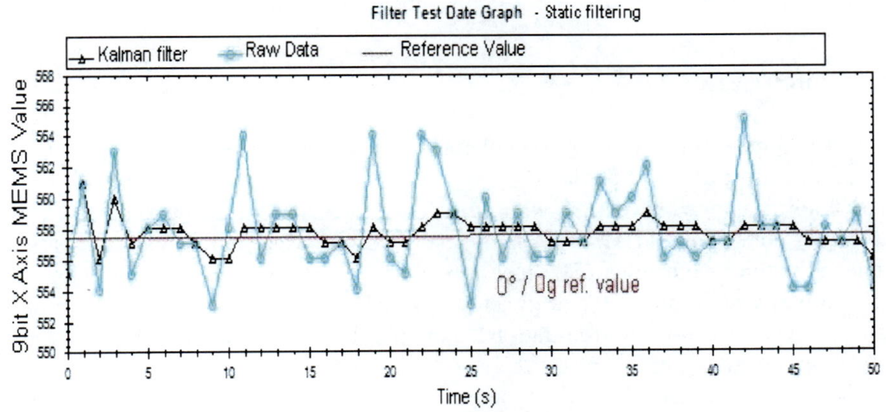

Fig. 8. Kalman filtering for raw Accelerometer data in static position

On the figure 9 is displayed data from accelerometer during tilt to 90° to one side, next to static position and then tilt to opposite side. Reference signal is red plotline. Black line is Kalman filtered value.

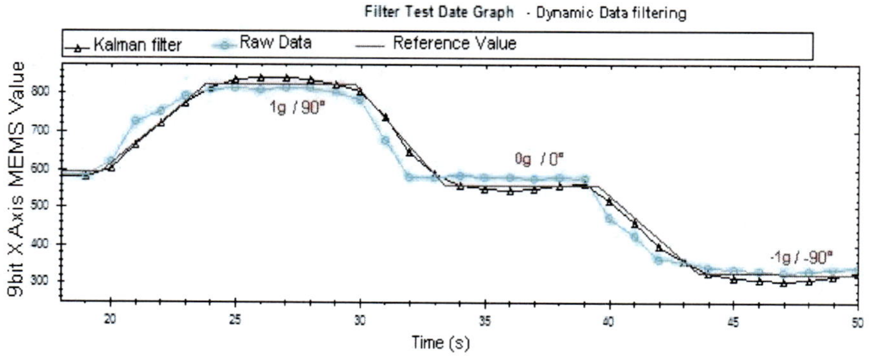

Fig. 9. Dynamic data from MEMS Accelerometer sensor with kalman filtering

We experimentally find out constants for kalman filter with compromise of minimal displace during dynamic and static operation.

$$r = 30.0, T = 2.0, q = 0.1$$

There is one problem in setup filter, when there is very fast acceleration and deceleration. This situation can occur when the real device fall or crash to the obstacle. We can avoid this situation by setting adequate value to danger alarm tilt and implementation of obstacles sensor detection (infra or ultrasonic) to mobile solution.

6 Conclusion

Introduced measuring solution is implemented to mobile device. Actual possibilities are measuring of tilt device -90° to 90°. You can select bound angle for start indication of danger device tilt with next visual or sound alarm. We can improve precision data from MEMS sensor by using 12 bit ADC but then is necessary change the microcontroller. Next idea can be change of Accelerometer with digital I2C output, which remove error generated by ADC conversion. We are computing next values from acquired data for example: trajectory, deceleration, average and actual speed.

Next works on this solution will be implementation of Kalman filter to program of MCU firmware and display actual angle value and alarm on LCD display. This remove testing mobile computer from actual solution and application will be small and compact device. This researched accelerometer device will be used in rehabilitation system as safety circuit to monitor extreme acceleration and deceleration for fast reaction of device stop. Another example of usability is in the figure 10.

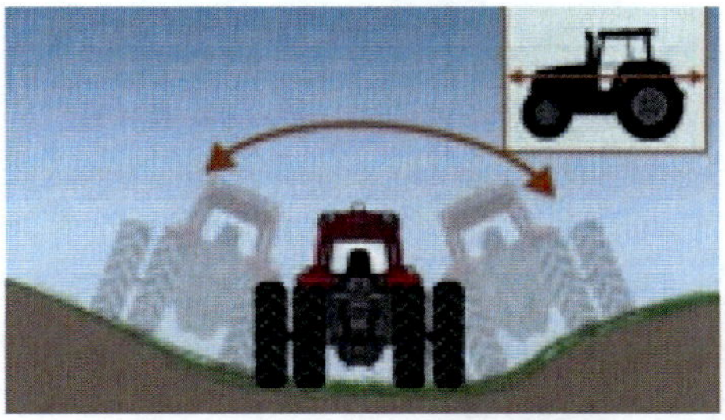

Fig. 10. Example of usability in agricultural equipment

The research work is supported by the Project of the Structural Funds of the EU, Operational Program Research and Development, Measure 2.2 Transfer of knowledge and technology from research and development into practice: Title of the project: Research and development of the intelligent non-conventional actuators based on artificial muscles ITMS code: 26220220103

Európska únia

We are support research activities in Slovakia / Project is cofounded from sources of ES.

References

[1] Tuck, K.: Tilt Sensing Using Linear Accelerometers, Application Note, AN3461, Rev 2, (June 2007), http://www.freescale.com/files/sensors/doc/app_note/AN3461.pdf

[2] Clifford, M., Gomez, L.: Measuring Tilt with Low-g Accelerometers, Application Note, AN3107, Rev 0 (May 2005), http://www.freescale.com/files/sensors/doc/app_note/AN3107.pdf

[3] What is MEMS Technology?, https://www.memsnet.org/mems/what-is.html (Online, cit. 8.2.2010)

[4] Johnson, C.D.: Accelerometer Principles. Process Control Instrumentation Technology, April 14 (2009), 0-13-441305-9

[5] Zidek, K.: Open robotics control system, Technical University Kosice, http://www.orcs.sebsoft.com

[6] Saloky, T., Piteľ, J., Vojtko, I.: Control systems design with reliability defined in advance. In: Proceedings of the 1st IFAC Workshop on New Trends, Design of Control Systems, Smolenice, Slovakia, September 7-10, pp. 404–407 (1994)

[7] Zidek, K.: MEMS Accelerometer SVN, Google code (2010), http://code.google.com/p/orcs/source/browse/#svn/MEMS_Accelerometer

[8] Rüegg, C.: Math.NET Neodym 2008 February Release, v2008.2.2.364 (2008), http://www.mathdotnet.com/downloads/Neodym-2008-2-2-364.ashx?From=NeodymCurrentRelease

Modelled and Experimental Analysis of Electrode Wear in Micro Electro Discharge Machining with Carbon Fibres

Anna Trych*

Division for Precision and Electronic Products Technology,
Faculty of Mechatronics, Warsaw University of Technology,
A. Boboli 8, 02-525 Warszawa, Poland

Abstract. This paper presents an analysis of electrode wear in a micro electro discharge machining with carbon fibres as tool electrodes. Experimental data are discussed and compared with theoretical calculations. Both approaches are focusing on a frequency of discharges impact on the process. The thermal model is presented because melting and evaporation are main phenomena occurring in EDM process. Basing on certain assumptions and simplifications the approximated depth of single discharge crater can be estimated. The experimental data indicating a strong dependence between electrode wear parameter and energy. It also presents a tendency of this parameter in comparison with a frequency of discharges.

1 Introduction

Current trends and applications require a constant investigation and research on development of micro- and nano-manufacturing technologies. Micro EDM, which is one of key technologies for precision devices, has a good established position as one of such. However, as authors in [1] indicate, it is becoming to lose its importance in a research environment, even though an industry points it as one of current and future field of strong interest. As a crucial issue from an industrial viewpoint it should be developed. Other point, which is to determine, is the direction where the development of this technology and research should be led. In this paper, as in previous ones [2, 3], the emphasis is on novel electrodes. The thorough examination of possible usage of micro EDM process with carbon fibre electrodes underlies this research. The paper is focused on a wear of electrodes. As one of important parameters during manufacturing processes, analysis of tool wear is highly desired from industrial viewpoint. It determines the material policy, production planning and also a measure of technology efficiency.

* Corresponding author.

2 Carbon Fibres as Tool Electrodes

Carbon fibres are well known material used in many branches of engineering. Their properties, however, can also be utilized in micro electro discharge machining. One of the most important issues when searching for a material to apply in this manufacturing technique is conductivity. A material for electrodes should be conductive. Carbon fibres can differ in properties because of producers, manufacturing technology and material used as precursor to obtain them. As taking into account micro machining the next problem to solve is a dimension of electrodes. Fig. 1. represents comparison of single fibre with human hair. Such fibres can be as small as 5, 10 μm in diameter. This is very interesting from an industrial viewpoint, because obtaining such dimensions with other materials would greatly increase the cost of electrodes and also of the end product. Furthermore, the cost per single filament is very low and fibres are widely available on the market. Comparing it to operations that must be performed to achieve similar electrodes in such quantities, carbon fibres seem to be reasonable solution.

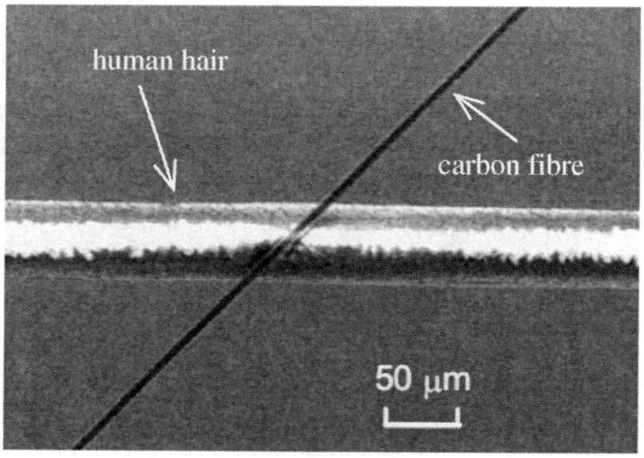

Fig. 1. Comparison of human hair with single carbon fibre

As EDM is mostly consider as thermal process good thermal conductivity is also desired. It is at acceptable level of 15-20[W/(K·m)]. These key properties mentioned before are well fulfilled by carbon fibres and so they can by recognised as a potential novel material for electrodes. Further analysis of carbon fibres can be consulted in [2].

3 Model of Single Discharge Thermal Penetration Depth

A model of thermal heat penetration into carbon fibre electrodes is proposed. Basing on parameters used during experiments, it is possible to calculate theoretical depth on which a single discharge can reach through material penetrating it thermally.

Considering a carbon fibre as a semi-infinite body with a step increase in surface temperature representing a single discharge, the temperature distribution can be described by equation:

$$\frac{\partial T}{\partial t} = a\frac{\partial^2 T}{\partial g^2} \tag{1}$$

were t - time, T - temperature, g – depth, a - temperature diffusivity (material constant). Fig. 2. presents considered situation.

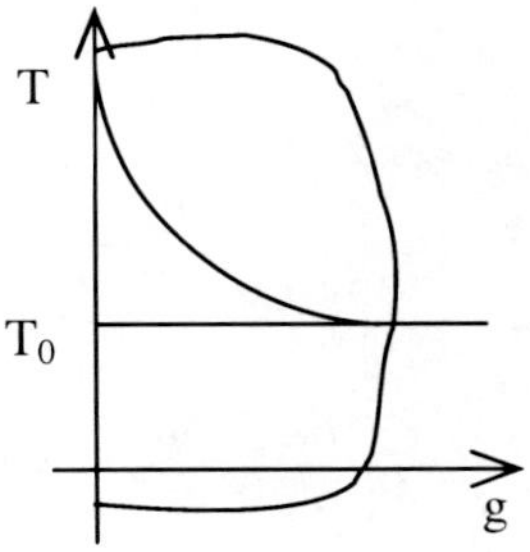

Fig. 2. Carbon fibre as a semi-infinite body heated with a step increase in surface temperature (adapted from [4])

Assuming boundary conditions as follows:

$$T(0,t) = T_s \tag{2}$$

for the depth 0 the Ts - surface temperature is delivered,

$$T(\infty,t) = T_0 \tag{3}$$

for infinite depth the material obtains T_0 - ambient temperature.

Defining initial condition as:

$$T(g,0) = T_0 \tag{4}$$

the initial temperature was also T_0,

the following equation is obtained:

$$\frac{T(x,t) - T_S}{T_0 - T_s} = erf\left(\frac{g}{2\sqrt{at}}\right) \tag{5}$$

were g is penetration depth, a, T_0 and Ts are as mentioned before. Detailed derivation of equation (5) can be found e.g. in handbook [4]. In considered case T_0 is assumed to be an ambient temperature 20°C and Ts is the temperature delivered during the discharge assumed as 5700°C.

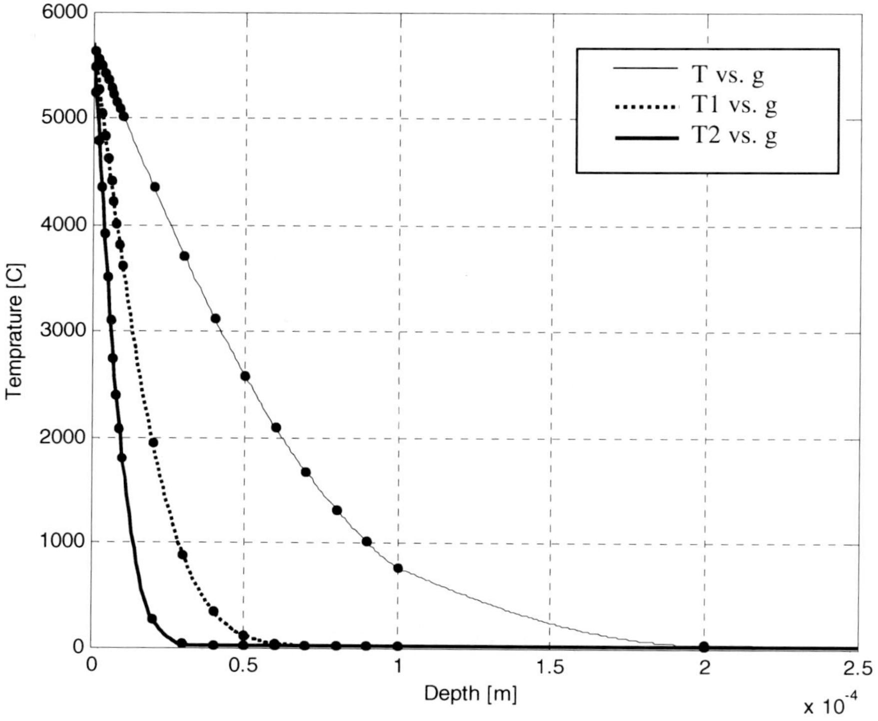

Fig. 3. Temperature at different depths

According to literature, temperature occurring in EDM process can vary between 4000K to 8000K depending on machining conditions [5, 6].

Solving equation (5) for t equal 0.7, 3 and 30μs, the curves in Fig. 3. are obtained. The time taken into consideration for this solution corresponds to frequency from performed experiments 1490, 328 and 33 kHz. In an RC- generator, which was used in experiments, the frequency depends on resistance R and capacitance C values. The time needed for charging the capacitor and than discharging it, so that an actual discharge can occur between the workpiece and the tool electrode, can be considered as a circuit constant RC (it has a value of time). Thus, it is assumed that for higher frequency which is inversely proportional to RC, less time is needed for single discharge. However, shorter time is compensated by more discharges.

The solutions of equation (5) are presented in Fig. 3. T, T1 and T2 correspond with 30, 3 and 0.7 μs respectively. The longer time of discharge gives the longer time of high temperature to penetrate carbon fibre. Thus, the temperature drop is slower for the longer time – Fig. 3.

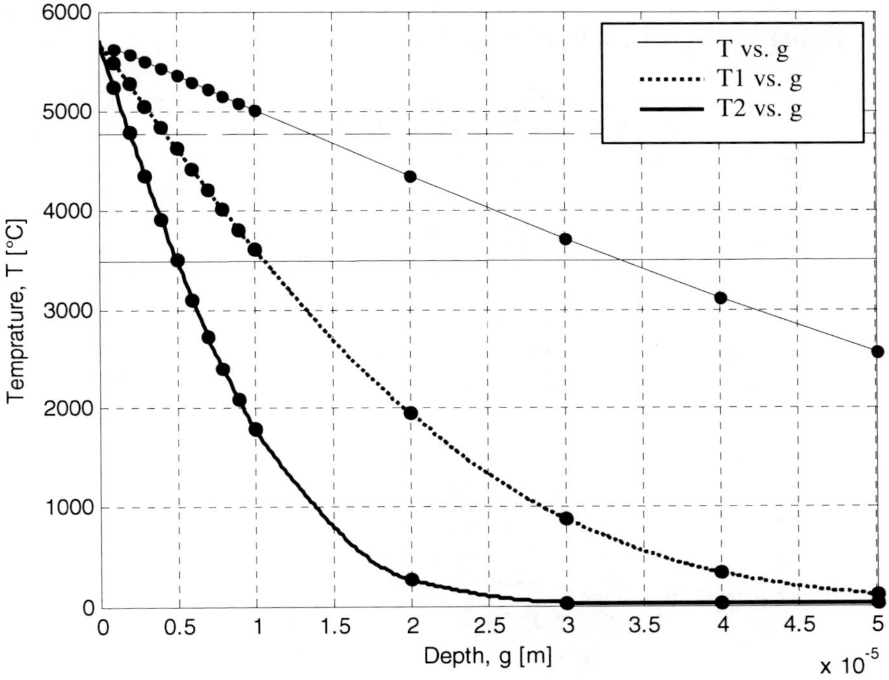

Fig. 4. Temperature at different depths (magnification of selected area)

Furthermore, the temperature at any depth can be shown, indicating e.g. where melting point was exceeded. Fig. 4. represents the magnification of the selected area form Fig. 3. Horizontal continuous and dashed lines represent melting temperature and boiling temperature for carbon respectively. It can be assumed that above boiling point the material will be evaporated. For temperatures above melting point analogically, it can be viewed as a liquid phase. However, determination how much material will be evacuated in this form is difficult to estimate. According to [7] only 50% of melted material is removed during the discharge. Since, this value is given for the workpiece material it can only estimated that for the electrode would be similar situation of the partial material removal. EDM process is directional which means that one electrode is being machined more than the other. In most cases the workpiece is removed more than the tool electrode. However, for carbon fibre electrodes, it was noted [2] that in certain conditions the tool electrode is being worn rather than the material machined. Thus, estimation about the volume of material removed in the single discharge from presented model can only be coarse, indicating only an approximate depth of penetration. For example from Fig. 4. the depth of craters modelled in such a way would be about 5, 11, 34 μm for curves T, T1 and T2 respectively with the assumption of the entire material removal.

4 Experimental Procedures and Tests

EDM with carbon fibres as tool electrodes, as it was mentioned before, can take high values of the electrode wear rate (EWR). Thus, experiments to determine this parameter were performed. The series of EWR parameters are shown in Fig. 5.

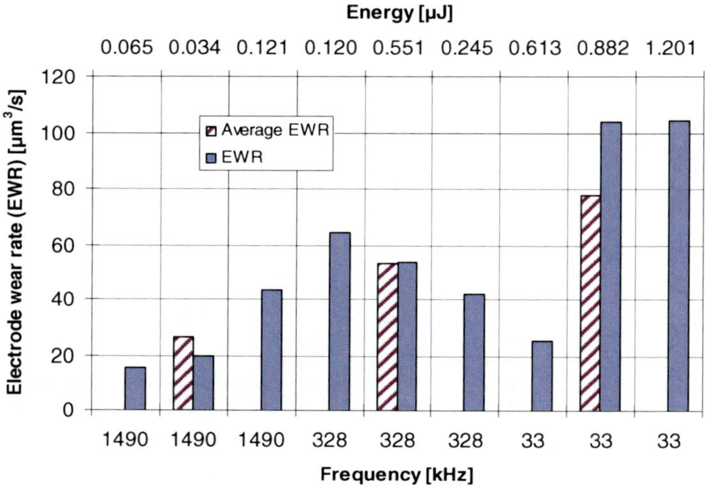

Fig. 5. Electrode wear rate (EWR) vs. frequency and energy

Each test was performed using RC generator with the constant R equals 22kΩ and different values of voltage and capacitance. The dependence of EWR on the frequency is not clear due to the fact that each trial was performed with a different value of energy. Thus, for each of 3 tests with the same parameter of R and C an average was also shown. This presents the tendency that for lower frequencies the higher value of EWR is obtained. According to the previous theoretical model for these frequencies the time of a single discharge was considered to be longer, so the depth of crater would be deeper and as a resulting parameter the volume would be bigger.

However, more trials should be performed because the calculated energy for each of them can sometimes differ by one order of magnitude and can strongly affect the influence of frequency. The good example of such effect is a slight difference of EWR values between first and seventh bar (trial) whereas the energy is 65 and 613 nJ respectively – Fig.5.

It is also difficult to consider values of EWR as accurate because the time measured during experiments was a summary time. Short circuits and other breaks that could occur during trials were neglected. It was impossible to estimate the time consumed by such pauses and furthermore, they were appearing more frequently in certain conditions than others. Thus, some tests might indicate more affected data.

5 Conclusions

Carbon fibre as the novel material for micro EDM has been considered. The model analysis of electrode wear caused by a single discharge was proposed. It provides the possibility of measuring of heat penetration into the material. The determined value of depth in melting temperature can serve as a tool for setting the radius of discharge crater. When considering it in such a way, it is possible to calculate the EWR parameter of a single discharge. Assuming an appropriate partial removal of material from a crater it can indicate, in certain amount of time, the actual EWR value for a model experiment. Such approach can serve as an estimation tool of usage of carbon fibres in planning of research experiments. From industrial viewpoint such a tool can provide the more reasonable material policy for a manufacturer. The experimental trials showed certain tendencies when considering material removal rate. However, testing procedures should be adapted to eliminate one parameter influence on the other to highlight the actual relations between discussed factors.

References

[1] Dimov, S.S., Matthews, C.W., Glanfield, A., Dorrington, P.: A roadmapping study in Multi-Material Micro Manufacture. 4M Network of Excellence in Multi-Material Micro Manufacture (2006)

[2] Kudła, L., Trych, A.: Characterisation and Experimental Testing of the Properties of Carbon Fibres in respect of their Application as Tool Electrodes for Micro EDM. In: Proceedings of the 16th International Symposium on Electromachining, Shanghai, Chiny, April 19-23 (2010)

[3] Kudła, L., Trych, A.: Testing of carbon fibres as tool electrodes in micro electrical discharge machining. Journal of Automation, Mobile Robotics & Inteligent Systems 3(4), 153–156 (2009)

[4] Taler, J., Duda, P.: Solving Direct and Inverse Heat Conduction Problems. Springer, Heidelberg (2006)

[5] Natsu, W., et al.: Temperature Distribution Measurement in EDM Arc Plasma Using Spectroscope. JSME Int. Journal. Ser. C. Mech. Systems, Mach. Elem. Manuf., 384–390 (2004)

[6] Ruszaj, A.: Non-conventional manufacturing methods of machine components and tools, Krakow (1999) (in Polish)

[7] Nowicki, B., Dmowska, A., Podolak-Lejtas, A.: A new method of investigating crater and flash made by individual discharge using scanning profilometers. Wear 270, 121–126 (2011)

Part VIII

Photonics

Improved Data Processing for an Embedded Stereo Vision System of an Inspection Robot

Przemysław Łabęcki

Institute of Control and Information Engineering, Poznan University of Technology, ul. Piotrowo 3a, Poznań, 60-965, Poland

Abstract. The task of mapping the surrounding terrain is crucial for inspection robots, if they have to operate autonomously. A STereo-On-a-Chip (STOC) camera from Videre Design is a good sensor for the task if low payload and computational capabilities are a factor. This paper presents a method of building a grid map of the terrain using the STOC camera. Different geometric configurations of the sensing system are analyzed and an optimal one is chosen. A post-processing algorithm for the noisy range data acquired from the camera is proposed and compared to the filtering implemented in the camera hardware. Experimental results are presented.

1 Introduction

Mobile inspection robots need a model of the surrounding terrain to operate autonomously. Such a model is necessary for motion planning and, in case of walking robots used for inspection or search tasks, also for foothold selection [1]. Because walking robots impose tight constraints as to mass and size of hardware components, passive stereo vision systems seem to be a good choice for such robots, due to low power consumption and small size. Typically, calculating a dense depth map using a stereo system has high computational requirements, however advanced intelligent sensors, like the Videre Design STOC cameras, can provide such measurements in real time, thanks to data processing embedded in hardware. This paper focuses on the usage of a STOC camera for rough terrain mapping, considering various limitations of such system.

There has been some work related to the Videre Design STOC cameras and the use of stereovision in terrain mapping. In [2], a STOC camera was considered as a sensor for mobile robots in search and rescue tasks. However, the authors did not take into account the false measurements produced by the camera. Rusu et al. [3] used stereovision for terrain mapping with a walking robot, but none of the data processing was done onboard the robot. Moreover the compliant legs of the RHex robot allowed to limit the usage of the terrain map only to path planning. In [4], the LittleDog legged robot used stereo vision to build a terrain map for navigation and foothold selection, but also all data processing was performed off-board.

2 Distance Measurements Using the Videre Design STOC Camera

2.1 The Built-In Stereo Processing Algorithm

The STOC cameras implement the Small Vision System (SVS) algorithm [5]. As any stereovision algorithm, it needs to find stereo correspondency in order to calculate depth. The stereo matching is achieved by means of area correlation. To do this, both left and right images are first transformed to a form where epipolar lines correspond to image rows. Then the correlation matching is performed row-by-row. This process produces a disparity image, which holds information about how position of corresponding pixels differs from left to right image. This image can be easily transformed to a depth image. All depth measurements provided by the STOC system are in the reference frame of the left camera.

The stereo processing often produces incorrect matches. Such errors occur as a result of insufficient image texture (featureless areas) and around depth boundaries. False measurements are also produced when objects are located outside the horopter (i.e. too close to the camera) and on reflective surfaces. These artifacts are discussed in detail in section 3.1. The SVS system incorporates three kinds of filters to reduce incorrect matches. One of them is the uniqueness filter, which removes pixels if the correlation value is too similar to other matches. The confidence filter removes matches appearing on weak textures. False measurements caused by image noise are removed by the speckle filter, which imposes a region size constraint. However, these filters are generally too restrictive, as they also tend to remove useful information. Moreover, they can't handle errors produced by out-of-horopter objects or reflective surfaces. Fig. 1 shows the disparity image of a rough terrain mockup filtered with two different threshold sets. Brighter colors in disparity images indicate points closer to the camera, while black color means no measurement at the given pixel.

Generally, the speckle filter is always set to default values, as it doesn't remove useful information. The other two filters, when working at default settings, remove much of useful information, while still retaining some false measurements, mostly on the edges of objects.

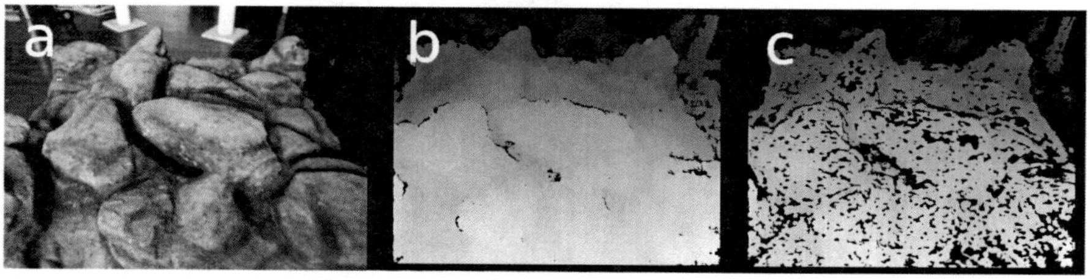

Fig. 1. The application of filters built-in to the SVS system. (a) Image from the left camera, (b) disparity image with low threshold values, (c) disparity image with default threshold values

2.2 Measurement Uncertainty

The uncertainty of the measurements generally rises with the distance from the camera. The range resolution is expressed by the following equation:

$$\Delta r = \frac{r^2}{b \cdot f} \Delta d\,, \qquad (1)$$

where Δr is the resolution at the given distance r, b is the stereo baseline, f is the focal length and Δd is the minimal disparity that the system can distinguish (which is by default 1/16 pixel).

The limited depth resolution is not the only source of measurement uncertainty. Also calibration errors (of both stereo and intrinsic parameters) can decrease measurement accuracy. It has been observed, that the quality of the measurements decreases near the edges of the disparity image.

2.3 Configuration of the Camera for Terrain Model Acquisition

Some factors have to be considered when choosing the geometric configuration of the sensing system. The location of the camera above ground and the tilt of the camera can affect such parameters as the field of view (and maximal range), the number of object appearing closer than the minimal range and the scale of occlusions. Fig. 2 illustrates measurements of a terrain mockup performed in three different configurations and visualized as an elevation map, where height is coded by color. It is apparent, that increasing the camera tilt reduces occlusions, as does placing the camera higher above ground. However, the distance of the camera from the ground is limited by the decreasing resolution of the measurements and the physical dimensions of the whole system. The camera tilt also cannot be too high, as it limits the range of the system, reducing its capability to acquire maps useful for path planning.

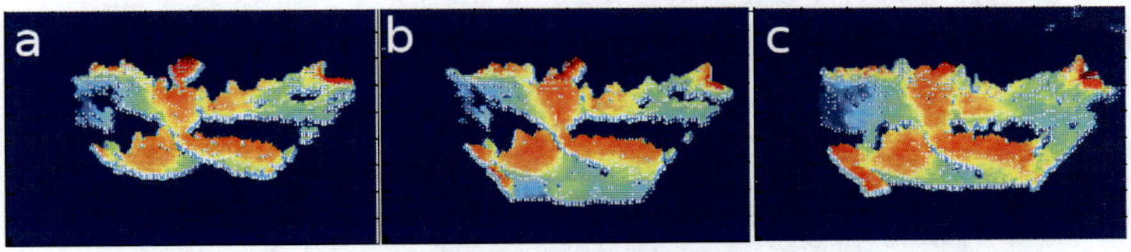

Fig. 2. Measurements performed in three different configurations of the system. (a) Tilt angle α=15°, elevation of the camera h=60cm, (b) tilt angle α=35°, elevation of the camera h=60cm, (c) tilt angle α=35°, elevation of the camera h=85cm.

3 Rough Terrain Mapping

3.1 Measurement Post-processing

Because the filters built into the camera hardware don't remove all of the artifacts, there is a need for measurement post-processing. The artifacts that are most resilient to the built-in filters are those on the borders of objects with disparity values much different from the background. These erroneous measurements manifest as halos stretched from the foreground to the background object (fig. 3d). The second group of erroneous measurements is formed by out-of-horopter artifacts, and artifacts appearing on reflective surfaces and textureless areas. Such measurements often show as groups of points near the camera (fig. 3c).

Both groups of measurements can be removed by analyzing the depth image. The depth image can be acquired by projecting the point cloud provided by the system onto an image or directly from the disparity image, using the following equation:

$$r = \frac{bf}{d}, \quad (2)$$

where r is the depth, b is the stereo baseline, f is the focal length and d is the disparity at the given pixel. At pixels where the disparity is invalid (no matching was determined or the built-in filters removed the measurement), the depth is set to 0.

At every nonzero pixel of the depth image, two coefficients are calculated: the point density and the depth span of the neighboring pixels. Pixels of low density are likely to be a part of the halos while pixels with large depth span in the neighborhood are likely to be either a part of the halos or a part of an artifact from the second group.

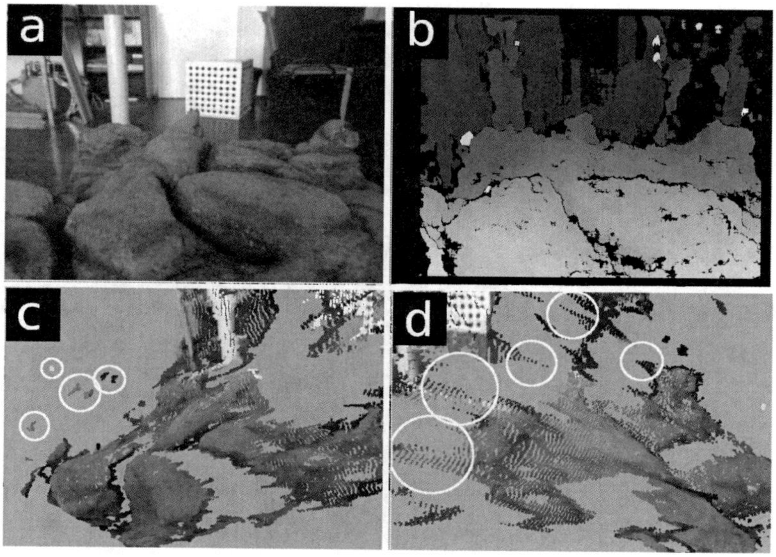

Fig. 3. Common measurement errors. (a) Grayscale image, (b) disparity image, (c) artifacts appearing out of horopter or on reflective surfaces (marked with circles), (d) erroneous measurements appearing on the edges of objects (marked with circles)

Both coefficients are calculated in a window based at the given pixel. Because the size of the window is large (25x25), only every other pixel in the widow was used.

The pixel density is calculated as follows. At the point in 3D space corresponding to the given pixel, a cube is defined. The length of the cube sides is determined by the size of the window: in perspective projection, at a given distance from the camera, the width of the image window corresponds to a specific length in 3D space. The number of pixels in the window that are contained within the cube is used to calculate the density. This number is each time scaled to the total number of pixels in the window. This way, a normalized density in range of 0 to 1 is calculated for each detected pixel of the depth image.

The calculation of the depth span needs some more post-processing. All zero-pixels in the depth image must be first filled-in to provide neighborhood for the pixels surrounded by areas of invalid disparity (which is often the case with the out-of-horopter or insufficient texture artifacts). To do this, the zero-pixels are recursively replaced by the average of the surrounding nonzero pixels. This filling-in is independently done twice, starting from the left and from the right side of the image. Then, at a given pixel, the greater of both depth values is chosen.

Once the zero-values are filled-in, the depth span is simply calculated as the depth difference between the closest and the farthest pixel in the window.

Normally, at the edges of the objects, the depth span would be larger and the density smaller than in the center of the objects. This could lead to removal of actual edges of the objects. To avoid this, the objects are detected in the grayscale image, and the window is each time limited to fit within an object. The detection of objects is performed by means of segmentation of the grayscale image. In this work, watershed segmentation was used. Generally, undersegmentation is preferred over oversegmentation. Additional segments could limit the window to a small area, reducing the depth span and increasing density at the given pixel, and subsequently prevent the algorithm from removing an artifact.

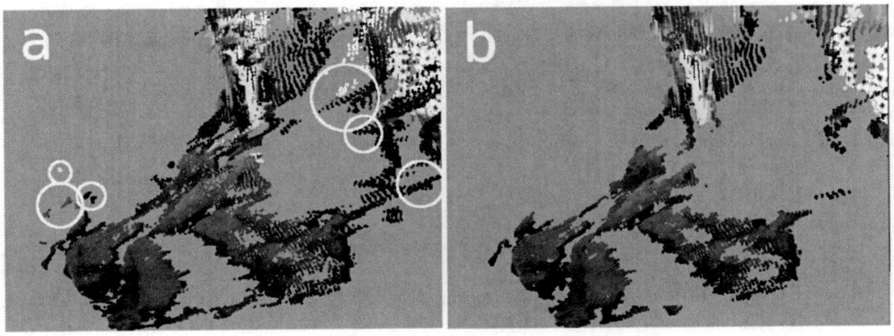

Fig. 4. Application of the data post-processing. (a) Cloud of points with erroneous measurements of both types (marked in circles), (b) cloud of points with the artifacts removed.

The values of the calculated coefficients are used to discard artifacts. Pixels with the following coefficient values are treated as faulty measurements:

- density lower than 0.3,
- depth span greater than 0.8m,
- density lower than 0.7 and depth span greater than 0.5m.

Figure 4 illustrates the application of the range data post-processing. Both types of artifacts are removed, with little loss of correct measurements. The post-processing was implemented in *OpenCV* to maintain the properties of a real-time system.

3.2 Map Building Algorithm

For the purpose of mobile robot navigation and foothold selection, a 2.5D grid representation of the environment was used. Such representation is easy to update and can be directly used in foothold selection and navigation [1]. The environment model consists of an elevation map, where each cell holds the height of the object at that cell, and a certainty map of equal size, where each cell represents the measurement certainty in the corresponding cell of the elevation map [6].

Each time new measurements are available, the measured points should be merged with the existing map. Every map cell that contains new points, can be updated with a new elevation value, if only the measurements are of better quality. To calculate the new elevation value, the space corresponding to the given cell is divided into vertically arranged bins of equal size. The bin that contains the most points from the new measurement is selected to calculate the new elevation and certainty values. The certainty value is calculated as the average certainty of all points from the selected bin, while the elevation is calculated as a weighted average of the vertical coordinates of the points accumulated in this bin, with certainty values of the points used as weights.

The certainty value of a single point is based on the depth resolution of the measurement, given by equation 1. Moreover, as the quality of the measurements degrades near the edges of the disparity image, an additional weighting factor is introduced:

$$w = 1 - \frac{\sqrt{(u - C_x)^2 + (v - C_y)^2}}{I_w / 2}, \tag{3}$$

where u, v are the horizontal and vertical image coordinates of the given point, C_x and C_y determine the position of the principal point of the camera and I_w is the width of the image. It is easy to notice, that the fraction part of the weighting factor is in fact a normalized distance from the image center. The weighting factor equals 1 in the center of the image and is zero near the left and right edges of the image. Finally, the certainty of the measured point is calculated as:

$$c = \frac{1}{\Delta r} w . \tag{4}$$

The new values based on the points from the selected bin are used to update the map at the given cell, if the new certainty value is greater than the old one.

One last step in the map updating algorithm is the evaluation of the visibility constraint. Sometimes new measurements contain gaps, in which no new elevation values are available. Old but erroneous measurements can be retained in these gaps. Such measurements can be removed by checking the visibility constraint. Each cell in a gap is checked whether it occludes newly updated cells. If this is the case, a conflict arises, as the newly updated cells wouldn't have been observed if the old measurements in the gaps were correct. Therefore, if the certainty of the cells in the gap is lesser than of the occluded cells, the measurements in the gap are invalidated.

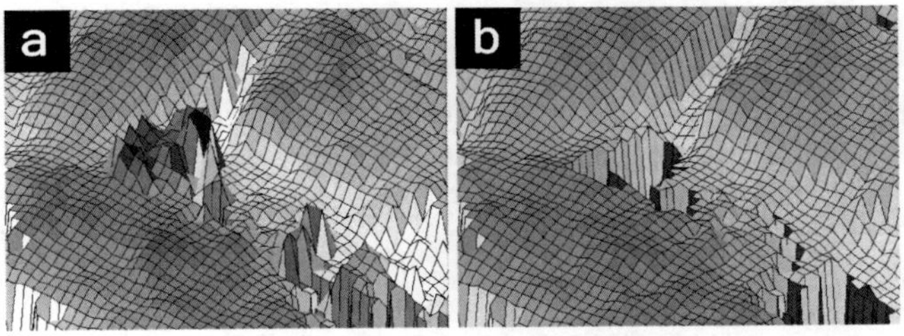

Fig. 5. Applying the visibility constraint. (a) Fragment of a map built without checking for visibility, (b) fragment of a map with the constraint applied.

4 Experimental Results

Tests of the mapping system were performed using a terrain mockup of size 150x150 cm. The camera was 80cm above ground level and the tilt angle was 35°. The camera was moved manually in equal intervals towards the mockup, and in each position, measurements with both low filter values (minimal filtering) and default filter values were made. The map visible in fig. 6b was created from data acquired with minimal filtering, and the map visible in fig. 6c was created with the additional data post-processing.

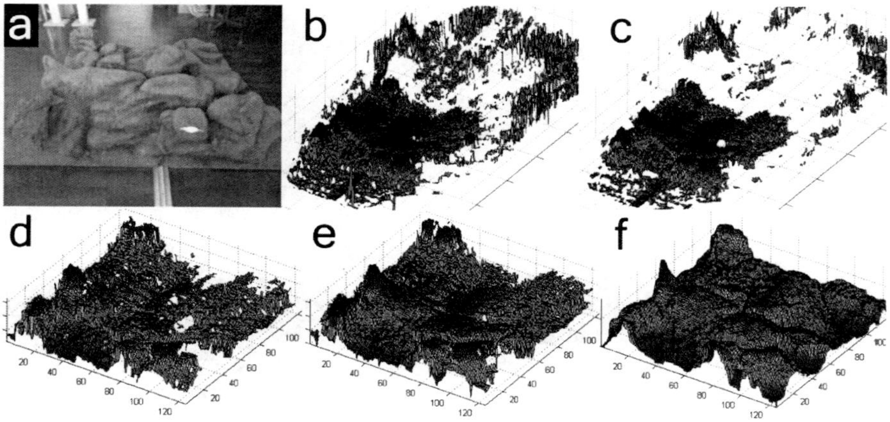

Fig. 6. Terrain maps. (a) Picture of the mockup, (b) created map – low filter values, (c) created map – post-processing applied, (d) fragment of the map containing the mockup – default filter values, (e) fragment of the map containing the mockup – low filter values and data post-processing, (f) ground truth map of the mockup

To compare the presented post-processing approach with the default filters in a quantitative way, a version of the Performance Index (*PI*), introduced in [6] was used. Because the original index does not account for missing measurements, a modification was needed:

$$PI = \frac{\sum_{i=1}^{n}\sum_{j=1}^{m} o_f(i,j)\cdot\left(h_f(i,j)-h_t(i,j)\right)^2}{\sum_{i=1}^{n}\sum_{j=1}^{m} o_m(i,j)\cdot\left(h_m(i,j)-h_t(i,j)\right)^2}\cdot\frac{N_{om}}{N_{of}}. \tag{5}$$

In the above equation, i and j are the coordinates of the point in the grid map. The above formula compares the elevation values $h_f(i,j)$ of the map built from corrected range data: filtered with default filter values or post-processed using the proposed approach, with the elevation values $h_m(i,j)$ obtained with minimal filtering, and the ground truth map elevations $h_t(i,j)$. N_{om} and N_{of} are the numbers of the observed cells in the maps build from data filtered with default filter values/post-processed and with minimal filtering, and $o_f(i,j)$ and $o_m(i,j)$ can be either zero or one and determine if the given cell in the respective map has been observed. The ground truth (fig. 6f) needed to calculate the *PI* value was acquired by scanning the terrain mockup with a Hokuyo URG-04LX 2D scanning laser rangefinder attached to a precisely moved arm of an industrial robot.

The *PI* value tells how much the investigated approach reduces mapping errors (defined as the sum of squared elevation differences in all observed cells, divided by the number of these cells) with regard to a map build only with minimal filtering. Lower *PI* values mean better reduction of the mapping error, while a *PI* value equal 1 means no error reduction at all.

The *PI* value for the map built with default filter values was equal 0.991, which means that most of the removed data was correct. This can be seen in fig. 6d: there are holes even in the observed areas. The proposed post-processing algorithm scored a *PI* value of 0.917, while removing most of the artifacts in greater distance, that are visible in fig. 6b.

5 Conclusions

This work presents a new method for terrain map building with a hardware-embedded stereo camera. A data post-processing algorithm was proposed, which removes more artifacts than the built-in filters, while retaining more of useful information. The drawback of the proposed post-processing algorithm is the added computational cost. So far the mapping algorithm was tested in a controlled environment using a terrain mockup. Outdoor tests with the Messor walking robot are in preparation.

References

[1] Belter, D., Skrzypczyński, P.: Rough Terrain Mapping and Classification for Foothold Selection in a Walking Robot. In: IEEE Works. on Security, Search and Rescue Robots, Bremen, CD-ROM (2010)

[2] Poppinga, J., Birk, A., Pathak, K.: A characterization of 3D sensors for response robots. In: Baltes, J., Lagoudakis, M.G., Naruse, T., Ghidary, S.S. (eds.) RoboCup 2009. LNCS, vol. 5949, pp. 264–275. Springer, Heidelberg (2010)

[3] Rusu, R.B., Sundaresan, A., Morisset, B., Hauser, K., Agrawal, M., Latombe, J.-C., Beetz, M.: Leaving Flatland: Efficient Real-Time 3D Perception and Motion Planning. Journal of Field Robotics 26(10), 841–862 (2009)

[4] Kolter, J.Z., Kim, Y., Ng, A.Y.: Stereo Vision and Terrain Modeling for Quadruped Robots. In: IEEE International Conference on Robotics and Automation, pp. 1557–1564 (2009)

[5] Konolige, K.: Small vision systems: hardware and implementation. In: Eighth International Symposium on Robotics Research, pp. 111–116 (1997)

[6] Ye, C., Borenstein, J.: A Novel Filter for Terrain Mapping with Laser Rangefinders. IEEE Transactions on Robotics and Automation, 913–921 (2003)

Multi-channel Laser Interferometer for Parallel Characterization of MEMS Microstructures

K. Liżewski[1], A. Styk[1], M. Józwik[1], M. Kujawińska[1],
R. Paris[2], and S. Beer[3]

[1] Institute of Micromechanics and Photonics,
Warsaw University of Technology, 8 Sw. A. Boboli St.,
Pl-02-525 Warsaw, Poland
[2] Institut für Mikroelektronik- und Mechatronik-Systeme Gemeinnützige GmbH,
Ehrenbergstr. 27, D - 98693 Ilmenau, Germany
[3] CSEM, Technoparkstrasse 1, 8005 Zürich, Switzerland

Abstract. In the paper the new approach towards microsystems characterization at wafer level, developed under EU project "SMART InspEction system for High Speed and multifunctional testing of MEMS and MOEMS" (SMARTIEHS) is presented. The goal of the project is to provide fast, cost effective and flexible optical inspection system. Two different interferometer approaches are pursued in the project's development process: a refractive optics based Mirau type low coherence interferometer and a diffractive optics based Twyman-Green laser interferometer. In the paper the design and the tests of the laser interferometer are presented. The multifunctional approach of the interferometer measurement concept allows inspecting MEMS shape and deformation as well as characterizing spatial distribution of vibration amplitude at a resonance frequency within one instrument.

1 Introduction

Development of micro technology and increasing number of systems utilizing M(O)EMS devices in many field imposes strict conditions on manufacturers [1]. In general during different steps of production process contact and non-contact measurement methods applied to investigate substrates, coatings or single M(O)EMS devices are used [2]. Such approach makes production expensive as well as time consuming. To increase the efficiency of measurement the parallel measurement scheme has been proposed. The main aim of this scheme is to measure several units at once. The realization of the proposed scheme came with SMARTIEHS system (created under EU project SMARTIEHS "SMART InspEction system for High Speed and multifunctional testing of MEMS and MOEMS") [3]. The goal of the project is to provide fast, cost effective and flexible optical inspection system utilizing two types of interferometer arrays and hence allowing inspection of several objects at the time.

This paper focuses on the laser interferometer array designed and manufactured for SMARTIEHS system as one of two interferometer types. In the system laser interferometer is mainly responsible for objects' dynamic properties characterization (resonant frequency and mode shape evaluation) but it also allows for static properties measurement.

2 Inspection System Mechanical Design

The design of the SAMRTIEHS instrument platform is motivated by industrial requirements on one hand and by a modular design and flexibility on the other hand. The mechanical design of the platform is determined by the three following substructures:

- A commercial prober system is used as system platform. The prober system performs a chuck movement in x, y, z and rotation around z as well as a scope movement in x, y and z. These features are used to realize a semiautomatic test system.
- The optical system consists of interferometer wafers, light sources with the corresponding lenses, beam splitters and a camera unit. The mechanical mounting of these components has to realize the required optical path lengths including various adjustment possibilities to handle tolerances.
- A high precision z-drive is needed for focusing the measurement interferometer towards the MEMS devices. Furthermore the z-drive has to realize a highly uniform movement and stable positioning of the optical system in order to perform measurement tasks.

The mechanical unit was manufactured and is assembled at SUSS probe station (Suss prober PA 200 [4]). The main aim of the unit is to assure mechanical stability and protection for optical measurement head. The optical head consists of the two interferometer types. For each interferometer special modules holding an array of interferometers and light sources as well as beamsplitters were designed. CAD model of system is presented on Figure 1.

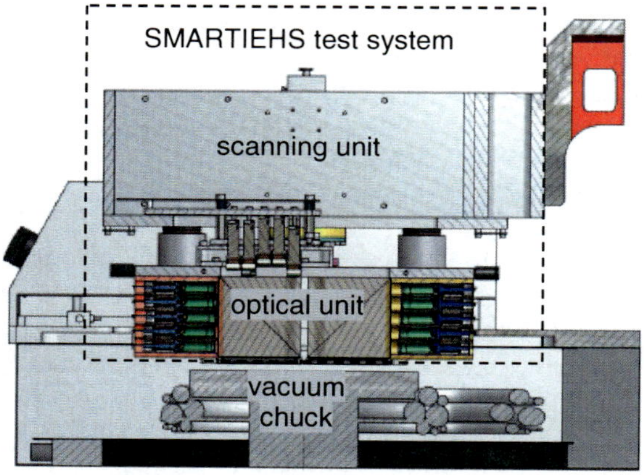

Fig. 1. Cross section of SMARTIEHS system attached to the prober

In order to perform measurements each interferometer type requires scanning or step-like changing of the optical path difference. For this purpose the high precision drive consisting of three voice coil drives is realized (see Figure 2). The selection of the voice coils is motivated by the large required travel range of 1 mm with a linearity of better than 1%.

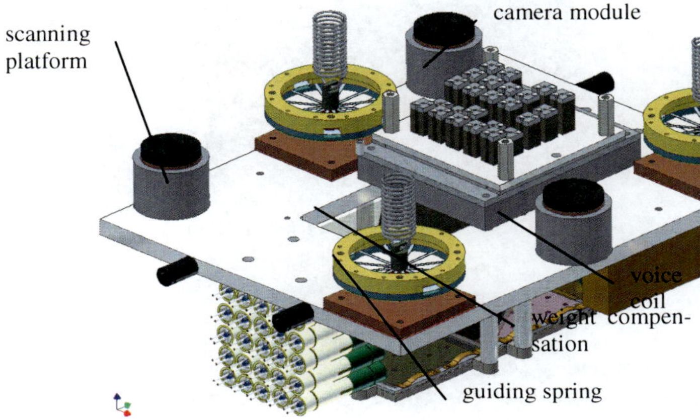

Fig. 2. Concept of the z-scan platform

The position signals in z-axis (scanning axis) are measured by three commercial interferometers by *SIOS* and their measuring heads are fixed at the carrier frame. Based on the *SIOS* interferometer signal resolution of 3 nm and after an interpolation the controller deliver the required positioning accuracy of 10 nm.

Besides the z-scan the platform enables pitch and roll motions (rx- and ry- direction) for the parallel alignment of the interferometer array to the MEMS-wafer. To realize a straight-lined and uniform z-motion the platform is weight compensated by pull-springs and guided with star-shaped leaf springs which provide a ratio of horizontal to vertical stiffness of over 10000. The nonlinearity of the leaf springs stiffness is compensated by feed forward control. Figure 3 presents the photography of the assembled SMARTIEHS system mounted on probe station.

Fig. 3. The picture of SMARTIEHS system

3 Laser Interferometer Unit

Compact designed of optical wafer and large scale of integration allows extending single channel to multi-channel concept. Figure 4 shows the scheme of optical wafer with array of laser interferometers. A pitch between individual channels is adapted to distance between measurement structures on MEMS wafer under test. In SMARTIEHS project pitch is defined as 13.46 mm and final optical wafer consists an array of 5x5 independent interferometer channels.

The presented interferometer configuration is an adaptation of Twyman- Green interferometer. This type of interferometer is often used to non- contact measurements of reflecting MOEMS devices featuring low gradient of shape and continuity surface. In the project classical optical elements were replaced by special designed and optimized diffractive optical elements etched on glass substrates. The picture of the manufactured interferometer array is presented on Figure 5.

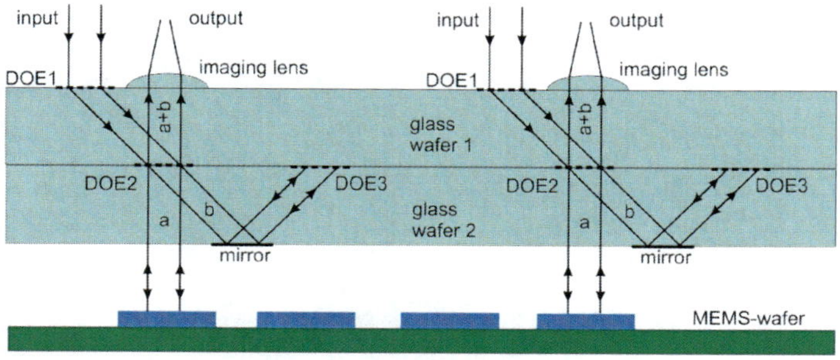

Fig. 4. Scheme of multi-channel interferometer

Fig. 5. Manufactured array of laser interferometers

The beam radiated by the laser diode placed in cartridges (light module in the system, see Figure 1.) is collimated by lens (f = 4,51 mm) in order to obtain plane wavefront. Illumination is then directed down to optical wafer by the beam splitter. Optical wafer is placed under cube beam splitter in distance 2 mm. Optical axis is deflected 90° and the beam illuminates perpendicularly first diffractive structure DOE 1. Diffractive structures (DOE) used in the system are gratings with precisely defined parameters and different profiles etched on two optical glass substrates. The substrates are bonded together by 35μm thick elastomer layer deposited on the backside of the upper grating substrate. Diffraction structures geometrical parameters are shown on Figure 6.

DOE 1 grating diffracts the incident collimated beam in order to illuminate the beam splitting diffraction structure – DOE 2. A key task is to optimize power efficiency of first order grating up to 70 % of entire incident beam. Second optical wafer consists of two DOEs etched on top surface and one mirror coated on bottom side.

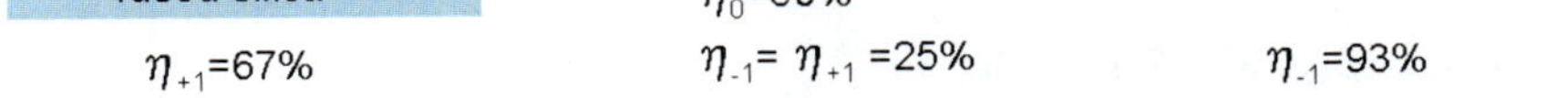

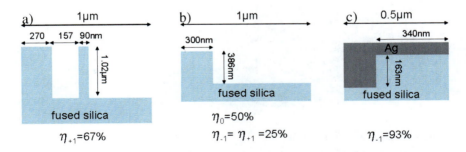

Fig. 6. Designs of the diffraction grating: a) DOE 1, b) DOE 2, and c) DOE 3

DOE 2 works as a beam splitter in the interferometer unit and divides beam into reference and object beam and recombine them after reflections. Transmitted light of DOE 2 creates reference beam that after reflections on mirror and DOE 3 goes back to DOE 2. DOE 3 is a reference element in this configuration. Diffracted light of DOE 2 is directed towards object surface and is called object beam. Both beams are joined by DOE 2 and give interference pattern. Diffractive structures were manufactured by electron lithography and reactive ion etching process. A period of first two structures was designed to be 1 μm while period of the third grating structure is 500 nm. After reflections from reference and object surface both beams are imaged on camera along the same optical axis by microlens placed on a top of first glass wafer. The diameter of lens is 3.2 mm while radius of curvature is 9.55 mm. Information about the investigated parameters of the object are obtained by analysis of acquired interferograms.

4 Signal Processing

In the project the specially designed Smart-Pixel Camera, consisting of the array of 5x5 imagers, is used. Every imager characterizes with the 140x140 smart-pixels that allows for the direct demodulation of the time dependent interferometric signal at the detection level with demodulation frequencies (ω_D) up to 100 kHz [5]. The working principle lies in direct demodulation implemented within every pixel creating two main signals: I (in-phase) and Q (quadrature-phase), see Figure 7. The interferometer employing the camera requires linear scanning (linear change of optical path difference) combined with synchronous camera demodulation frequency. The scan length in one demodulation (lock-in) period is λ/2. The optical signal collected and converted into the electrical signal is integrated and sampled in every pixel. Each pixel has background suppression circuit to avoid saturation and small input signal contrast.

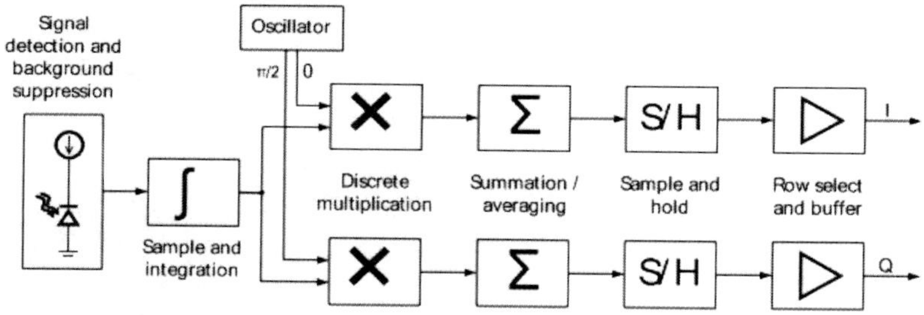

Fig. 7. Smart-pixel camera working principle

The signal samples are further multiplied by a signal of the same frequency and averaged on two paths, phase shifted by π/2 with respect to each other. A sample and hold stage in both circuit branches allows simultaneous demodulation and read-out, which means that while the stored values are read out of the imager pixel field, the input signal is demodulated and generate next values. These two branches create the *I* and *Q* channels of the imager.

This unique camera is especially useful for the vibration measurement technique that was developed within described project [6]. Since the project assumes that measured vibration amplitude of the objects will not exceed 80 nm, and the measurements have to be full-field, standard measurement methods for resonant vibration characterization, such as time averaging interferometry can not be used. The proposed method id based on classical two beam interferometry, with additional sinusoidal modulation of the light source. Due to the light modulation and additional small changes of the intensity caused by the measured vibration, the time dependent signal in each pixel of the camera is obtained. This signal can be further processed by the camera in the demodulation mode since it oscillates with the frequency specially matched to the camera demodulation frequency. The frequency match is realized when the light modulation frequency (ω_L) and the object excitation frequency (ω_V) differs of the value equal to the demodulation frequency i.e. $\omega_L = \omega_V - \omega_D$. The interferometric signal is then described by:

$$I(t) = I_0 \begin{bmatrix} 1 \\ + \gamma \cos(\varphi_0 + \delta) \\ + \frac{2\gamma\chi\pi A_V}{\lambda} \sin(\varphi_0 + \delta)\cos(\omega_D t + \varphi_V - \varphi_L) \end{bmatrix}, \tag{1}$$

where I_0 is the average of the interferogram, γ is the contrast of the interferogram, χ is the laser source modulation depth, φ_0 is the phase of the interferogram (depends on the shape of the object), φ_L and φ_V are the initial phases of light source and excitation control signals respectively and δ is an additional phase shift that may be introduced intentionally.

The first two terms of the above equation are constant. The third term, however, oscillates directly with camera lock-in frequency. The observed signal is a classical two beam interferogram with the vibration amplitude encoded in its contrast/modulation distribution. Two Smart Pixel camera output signals (I and Q, respectively) are:

$$I = \left(\frac{2\pi\gamma\chi A_V}{\lambda}\right) I_0 \sin(\varphi_0 + \delta)\cos(\varphi_V - \varphi_L) \tag{2}$$

$$Q = \left(\frac{2\pi\gamma\chi A_V}{\lambda}\right) I_0 \sin(\varphi_0 + \delta)\sin(\varphi_V - \varphi_L) \tag{3}$$

From the above signals one may calculate expressions containing information on amplitude vibration:

$$MD = \sqrt{(I)^2 + (Q)^2} = \left| \left(\frac{2\pi\gamma\chi A_V}{\lambda}\right) I_0 \sin(\varphi_0 + \delta) \right|. \tag{4}$$

In order to find the vibration characteristic of measured objects, first resonant frequencies need to be found and then distribution of the amplitude (vibration mode shape) at resonant frequency needs to be measured.

5 Experimental Results

In the SMARTIEHS system laser interferometer array is used to investigate the static and dynamic properties of measured MEMS devices. The static properties such as topography and deformations are measured using conventional Temporal Phase shifting technique for fringe pattern processing. For static measurements the IR sensor of the lateral size of 600 μm was used as an object. Figure 8 presents the evaluated wrapped phase (mod 2π) proportional to the topography of the investigated device in 10 channels measured parallel (at single measurement cycle).

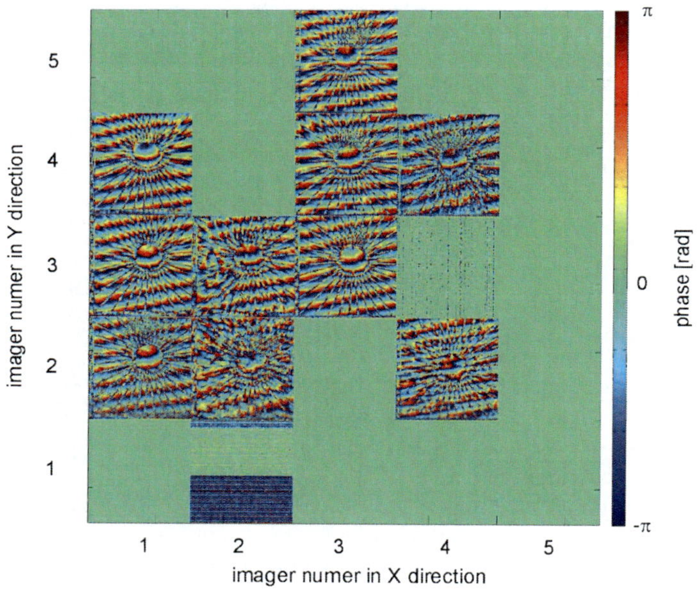

Fig. 8. Evaluated mod 2π phase distribution measured parallel at 10 different channels

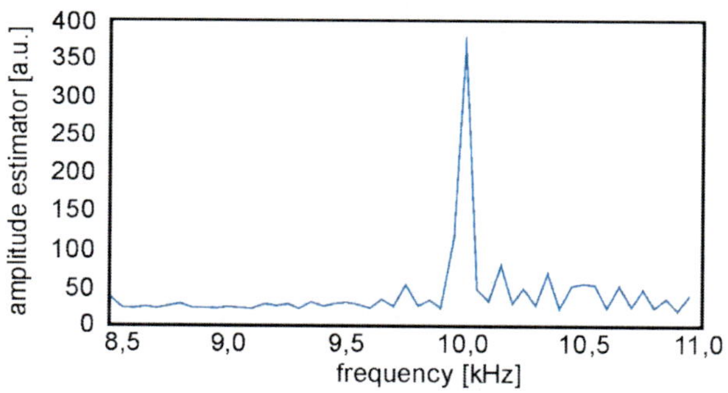

Fig. 9. Frequency spectrum for vibrating object measured in (1,3) channel

The dynamic properties of the objects were measured with proposed quasi heterodyned method. As an object the special calibrated piezoelectric transducer with the silicon mirror glued at the top was used. The object was excited by controlled voltage signal with changeable frequency and amplitude. That approach allows for selective measurements process. An exact vibration frequency could be set and investigated. The measurements were performed for selected frequencies from the 10 – 70 kHz range. The smart-pixel camera demodulation frequency was set to 1 kHz. Figure 9 shows the results of resonant frequency search within the range of 8,5 – 11 kHz in (1,3) channel. It is clearly visible that the measured object vibrated at 10 kHz – the exact frequency of applied excitation signal.

6 Conclusions

The system concept introduced in this paper is a novel solution in industry measurements. It is based on a parallel inspection approach. An array of micro optical sensors integrated on a probing wafer is adapted to the M(O)EMS wafer under test and is thus able to inspect more than 100 structures within one measurement cycle on an 8 inch wafer. The developed inspection system can be improved with regards to the spatial resolution, the numbers of channels, the range of measured parameters and the measurement speed. The main limitation for a smaller footprint of the channels is the size of the smart pixel imager. The footprint of the micro optical interferometers is only a few mm^2. The next generation smart pixel cameras will be smaller and have increased spatial resolution and measurement speed.

Acknowledgements

SMARTIEHS is a collaborative project funded under the Grant Agreement 223935 to the 7th Framework Program Objective 2007-3.6.

This work was co-supported by statutory founds of IMiF PW.

The authors want to thank all involved people for their valuable contribution in SMARHIES project realization.

References

[1] NEXUS MST/MEMS Market Analysis III 2005-2009 report (2009)
[2] Osten, W. (ed.): Optical inspection of Microsystems. Taylor and Francis, New York (2006)
[3] SMARTIEHS - SMART InspEction system for High Speed and multifunctional testing of MEMS and MOEMS, `http://www.ict-smartiehs.eu`
[4] Cascade Microtech, Inc., `http://www.cmicro.com/`
[5] Beer, S.: Real-time photon-noise limited optical coherence tomography based on pixel-level analog signal processing, PhD dissertation, University of Neuchâtel, Neuchâtel (2006)
[6] Styk, A., Lambelet, P., Røyset, A., et al.: SPIE, vol. 7387, p. 73870M (2010)

2D and 3D Digital Image Correlation Method and Its Application for Building Structural Elements' Investigation

D. Szczepanek, M. Malesa, and M. Kujawińska

Institute of Micromechanics and Photonics, Warsaw University of Technology, Sw. A. Boboli St., 02-525 Warsaw, Poland

Abstract. This paper aims to showcase some applications of Digital Image Correlation obtained using IMiF's own DIC software, and to present it as a general-purpose displacement measurement method, which can in many cases be a relatively cheap and simple, yet accurate replacement for other, more demanding methods.

1 Introduction

In the domain of experimental mechanics and vision-based optical metrology, Digital Image Correlation is a well-established, tried and tested measurement method for determining displacements (full-field or pointwise) in a wide array of test objects. Since its inception in the early 1980s, the basic algorithm went through a series of significant modifications which substantially widened its area of application and improved its robustness, but it wasn't until mass-market, affordable computer hardware became powerful enough to handle the computational demands of calculating correlation over ever-growing, high-res images that it actually gained enough momentum to slowly start pushing against other measurements methods, which were previously considered weapons-of-choice for particular applications. The experimental setup for a DIC measurement is extremely simple, requiring hardly any equipment save for image acquisition hardware, and possibly a computer system for synchronization and data storage. This means that the entire complexity of the method is contained in the software, and that software is constantly evolving to enable new functionality and overcome remaining limitations thanks to on-going research (see for example [1]) and peer information exchange through scientific publications which are eagerly followed by those using DIC in their work. IMiF at the Warsaw University of Technology is no exception, and this paper presents some results of DIC application, obtained using our homegrown correlation software, that the authors felt were quite interesting and worth sharing with the research community to illustrate the wide area of DIC's applicability.

2 2D DIC

The two-dimensional variant of DIC is the basic implementation that also lies at the core of 3D measurements. The basic idea is deceptively simple. Regular white-light images of a test object undergoing some sort of deformation or displacement are recorded in a sequence. The rest of the measurement is performed by the software, which divides the region of interest in a reference image into discrete blocks or subsets. Those are subsequently matched against similar subsets in an another (object) image from the sequence, using a correlation criterion. The difference in position between a reference subset and its matched object equivalent defines the displacement vector at the subset's origin. Repeating this process over the area of interest yields a displacement map, that is sampled as defined by the subset size and the separation of subsequent subsets selected for analysis (step size). Of course the denser the displacement map, the higher the computational cost.

While this seems quite easy in theory, there are a few difficulties to overcome. If the deformation is not purely translational, the subsets change shape between frames in the image sequence and their detectability is severely impaired, proportionally to the amount of shape deformation. Extra steps need to be taken to account for these deformations (usually approximated by an affine transform) to find subsets in images with complex displacements. Also, it's generally desirable to extract the information with sub-pixel accuracy, which in combination with the aforementioned shape deformation complexity, introduces the need to compute many variants of subset interpolations, which grow in number when the requested area of interest is more densely sampled. The algorithm's implementation must be „smart" enough to know where and how to look to avoid unnecessary computations without sacrificing accuracy. Changing lighting conditions between exposures also provide a challenge which is generally targeted using a specific form of the correlation criterion (ZNSSD, ZNCC) which was proven to be insensitive to lighting variations.

The most distinctive characteristic of DIC is the object surface preparation (or sometimes lack thereof). Theoretically, an optimal pattern of an object's surface for DIC analysis is a random speckle pattern – that way each subset is sure to be unique and can be unambiguously identified. However, in practice it is sometimes difficult or altogether impractical to apply a random patter (through spray-painting or sticking a template onto the object) and an object's natural surface pattern (texture, rust, etc.) can be used successfully, in combination with appropriate values of specific parameters of the algorithm (e.g. the subset size).

Using appropriate imaging optics, the images for correlation analysis can be obtained at virtually any scale – microscopic DIC is also a possibility [2]. An extra calibration step is required to convert the calculated displacements from pixel to real-world (e.g. mm) values. It is usually accomplished using a detail with a known dimension in the field of view, but geometrical camera calibration (using a calibration target, for example with a chessboard or dot pattern) may be also used to minimize the influence of image distortions caused by optical imperfections, account for the object's angular tilt (non-perpendicularity) in relation to the imager, or pixel-milimeter conversion if the precise distance between the imager and the object is known.

3 3D DIC

While two-dimensional DIC is generally capable of determining displacements in the image plane only, the 3D variant enables it to find all three (spatial) components of the displacement vector by adding an extra camera to the measurement setup. The details remain essentially the same, with the one exception that this time geometrical camera calibration is a sine qua non requirement. It establishes the relation between the two cameras and allows to backproject detected positions in the imager's pixel coordinates into real-world 3D point coordinates. The stereo-correspondence problem is also solved using correlation (optionally taking the perspective transform between the cameras into account). Thus, a shape map is obtained for one (reference) frame, which can be compared to another (object) in order to determine three-dimensional displacement vectors of the object's area of interest. 3D DIC enables the measurement of non-planar test objects and out-of-plane motion. While inevitably more computationally demanding than 2D DIC, it yields much more information about the displacement characteristics of a yet wider class of test objects and enjoys a great range of applications in many diverse fields of research. A flowchart of 3D DIC has been presented in Figure 1.

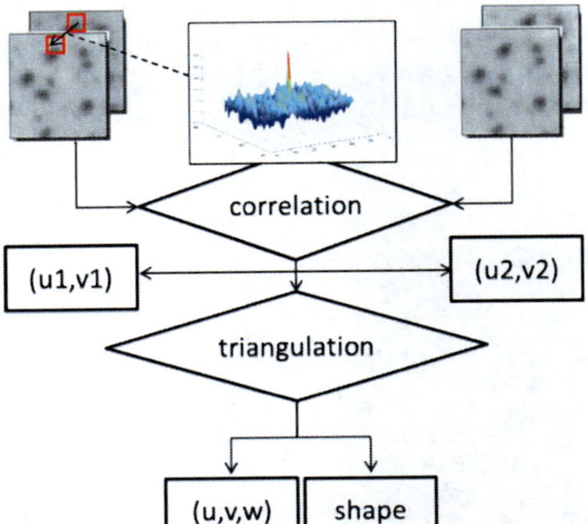

Fig. 1. 3D DIC flowchart

All of the example applications presented below were carried out in co-operation with the Building Research Institute.

4 Concrete Lintels Shearing

The aim of the measurements was to investigate the overall resistance of concrete lintels in regard to shearing. The lintels were 4 meters long and were subjected to load in a range of 0÷60kN. Displacements were expected to be predominantly in-plane and therefore a 2D DIC system was used. For measurements, a 2Mpx AVT Stingray camera was used. The data acquisition frequency was 10 frames per second. In Figure 2, ε_{yy} strain maps calculated for 40kN, 50kN and 60kN load values have been presented.

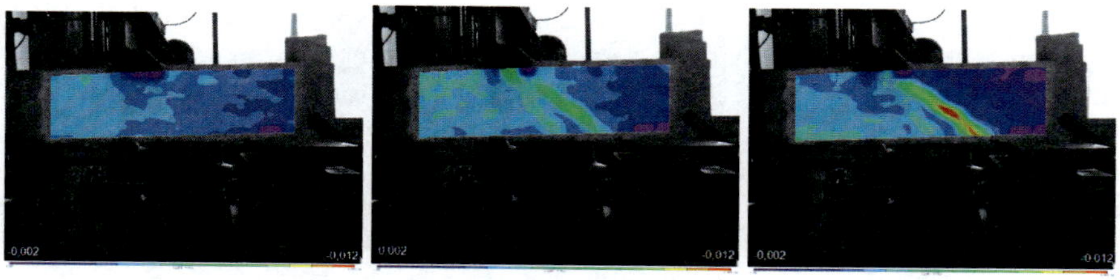

Fig. 2. ε_{yy} strain maps, calculated for 40kN, 50kN and 60kN

A substantial gradient of U and V displacements, measured for loads higher than 40kN, caused high ε_{yy} strains concentration in the middle of the FOV(Fig. 2) and subsequently lead to a crack propagation. With the support of DIC it was possible to predict the crack localization before it actually occurred. Presented in Figure 3, is the crack propagation graph.

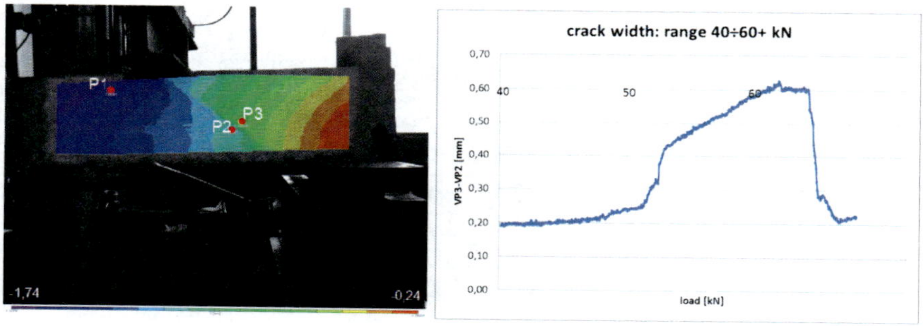

Fig. 3. Crack propagation graph

The width of the crack was calculated as a distance between two points localized respectively to the right (P3) and to the left (P2) of the crack.

The DIC technique is a powerful tool for custom-requirement, very specific and detailed analysis of a variety of effects. This example shows the capacity of DIC to predict and possibly to prevent a failure process in various materials and constructional elements.

5 Brick Walls Compression

The measured objects were 1,2m high brick walls subjected to compression in a range of 0÷2400kN. The aim of the tests was to compare different kind of welds in aspect of their resistance to compression and the homogeneity of transferred loads. The 2D DIC setup with a 18Mpx camera installed on a tripod and a measured object with sprayed-on speckle pattern are presented in Figure 4.

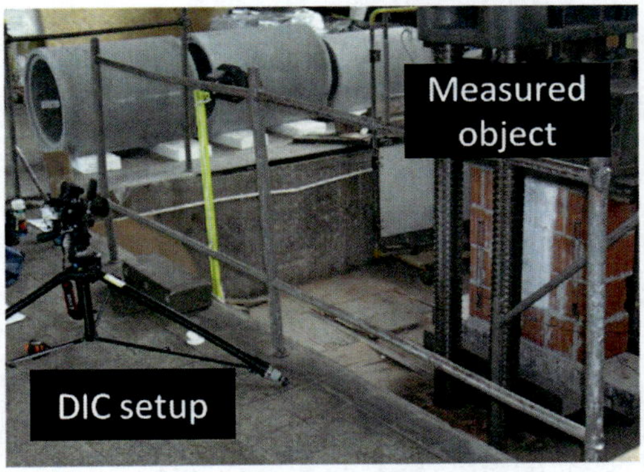

Fig. 4. Measurement setup

To validate the DIC measurements, two strain gauges were installed on the object. Although they were the principal method for determining the critical load, they could not be used for the evaluation of load transfer homogeneity. DIC, which is a full-field technique was perfect for both tasks. Presented in Figure 5 is an example V displacement map.

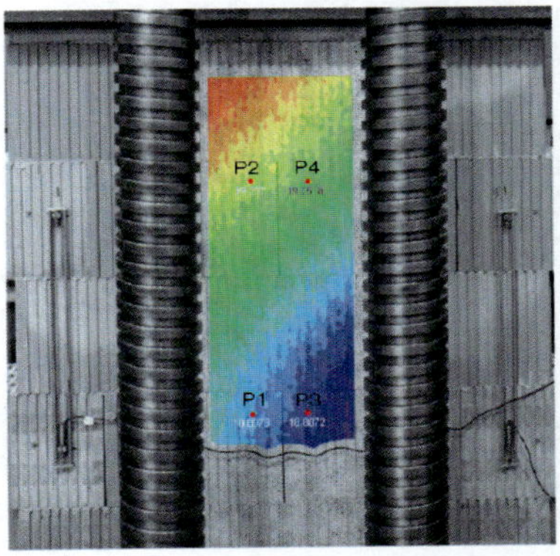

Fig. 5. Example V displacement map captured in an early stage of the test

The four points overlaid on the displacement map (points P1, P2, P3 and P4 in Fig. 5) were used to calculate Ɛyy strains, which were subsequently compared with the strain gauges' results. The corresponding points (P1, P2 and P3, P4) were set apart by 0.5m, which corresponds to the strain gauges' base length.

An example strain map has been presented in Figure 6.

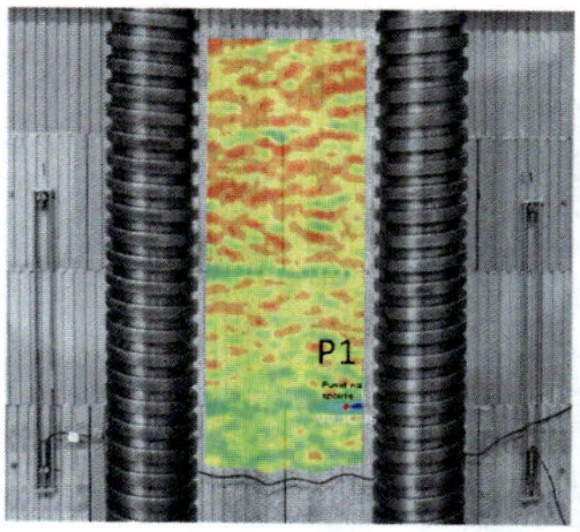

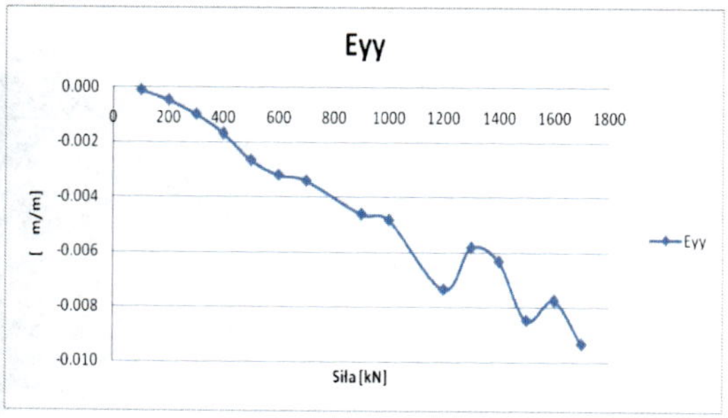

Fig. 6. Example strain map for 1200kN load and strains against load

In Figure 6 higher strains' concentration in welds can be clearly observed. In the bottom right of the field of view an inhomogeneity has been detected. A plot of Ɛyy strain calculated in point P1 as function of load has has been also presented in Figure 6.

The brick walls' tests were conducted in accordance with ISO standards. The utilization of the DIC method extends the range of drawn conclusions. The capacity of homogenous load transfer by a different kinds of welds could be easily evaluated.

6 Window Pressure Resistance

The third example comes from pressure tests of windows. The aim of this standard test is to evaluate a window's resistance to high pressure, for example as induced by wind. A standard approach would be to measure the out-of-plane displacement in only one, central point of the object with an inductive displacement sensor. The application of 3D DIC increases the amount of information, which can be obtained from the test.

The measured object was subjected to pressures in range of ±2000Pa. The 3D DIC system consisted of two AVT Stingray cameras equipped with 28mm lenses. Presented in Figure 7 is the shape of the measured window subjected to maximum pressure.

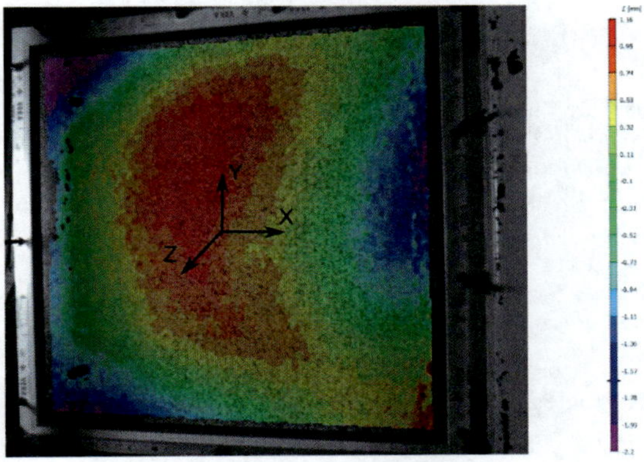

Fig. 7. Shape of the measured window

The result of the measurements were U, V and W displacement maps. Although the W displacement map is a dominant factor for overall window deformation, U and V displacement maps brings information about possible asymmetry of the object. Significantly larger U and V displacements close to the frame are noticeable.

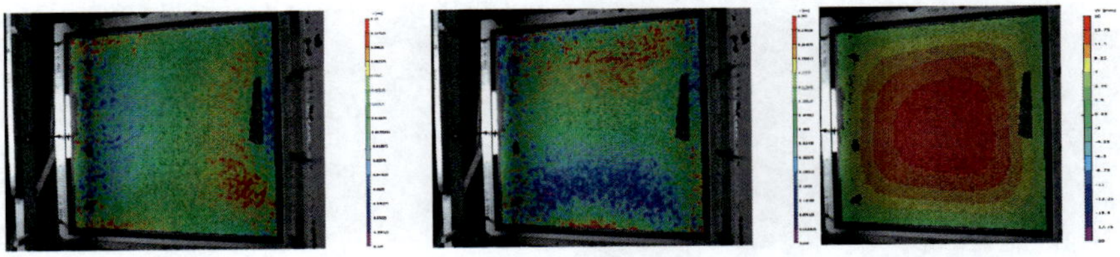

Fig. 8. U, V and W displacements maps

It is also worth noticing, that using 3D DIC, the point of the actual maximum displacement can be indicated in course of the analysis. Predefining the measurement point in standard, inductive sensor measurement techniques can lead to inaccurate or distorted results.

7 Conclusions

The DIC method has been successfully used for standard measurements and evaluation of building elements. In addition to very accurate and comparable standard results, it proved to be a useful tool for the comprehensive analysis of side-effects, which can occur during tests. Additional data and information obtained from the analysis can be useful for designers of new building materials and elements.

Acknowledgements

The financial support from the project "Health Monitoring and Lifetime Assessment of Structures" – MONIT – POIG.0101.02-00-013/08-00 from the EU Structural Funds in Poland is gratefully acknowledged.

The authors wish to thank Mr. Krzysztof Malowany from the Institute of Micromechanics and Photonics for valuable input during the measurements.

References

[1] Pan, B.: Recent Progress in Digital Image Correlation. Experimental Mechanics (2010) (published online)

[2] Srinivasan, V., Radhakrishnan, S., Zhang, X., Subbarayan, G., Baughn, T., Nguyen, L.: High resolution characterization of materials used in packages through digital image correlation. In: Proceedings of IPACK 2005, San Francisco, CA, USA (2005)

[3] Shmidt, H., Sablik, J.: Phys. Rev. Lett. 81, 1268 (1996)

Low Coherence Interferometric System for MEMS/MOEMS Testing

A. Pakula* and L. Salbut

Institute of Micromechanics and Photonics,
Warsaw University of Technology,
8 Sw. A. Boboli Str. Warsaw, 02-525, Poland

Abstract. Low Coherence Interferometry (LCI) known also as White Light Interferometry (WLI) is a useful tool for vast range of specimen characterization especially for MEMS/MOEMS testing [1,2]. In the paper specially design laboratory interferometric setup, together with examples of various objects characterized by it, is presented.

1 Introduction

The main power of low coherence interferometry is its vast use for mainly shape and surface characteristic examination. The commercially available setups are not always able to measure required especially when the object is complicated or has to be specially mounted, instead of being just lied on the table. To overcome those problems in the Institute of Micromechanics and Photonics, Warsaw University of Technology, the laboratory setup of an interferometer was build. Its construction enables introduction of necessary changes in order to obtain the best results or adjusting to uniqueness of measured object.

2 Principle of Measurement Method

The phenomenon of partially time coherent light interference is used as a base for LCI. For Twyman – Green interferometer, as shown in figure 1, the intensity distribution, in detector plane, for light beam of coherence length of lc, is given by equation 1:

$$I = 2I_0\left\{1+\exp\left[-\left(\frac{2OPD}{lc}\right)^2\right]\cos\left[\frac{2\pi OPD}{\lambda_0}\right]\right\} \tag{1}$$

where OPD – optical patch difference.

* Corresponding author.

The zero optical path difference is indicated by the fringe of the highest intensity (constructive interference) or the lowest intensity (destructive interference) – the zero order fringe. The intensity of other fringes drop rapidly. Those points (areas) which are optically conjugated with reference surface are detected by the detection of zero order fringe. Measured object is scanned layer by layer by changing the position of a reference surface and detecting new position of zero order fringe.

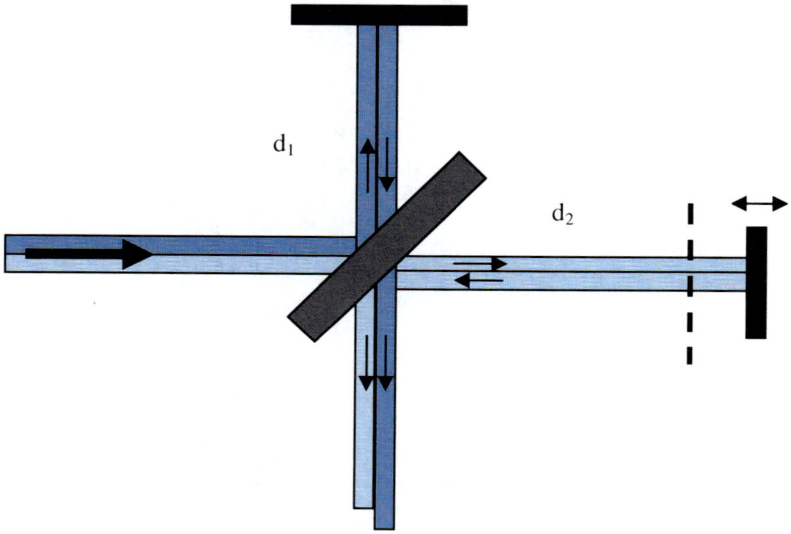

Fig. 1. Scheme of Twyman – Green interferometer

The scanning range in LCI is limited only by the length of scanning motion. This is not questionable advantage while comparing with similar methods like spectral optical coherence tomography.

3 LCI Laboratory Setup

As a base for the interefrometric setup the Twyman – Green configuration was used. Separation of reference arm of the interferometer from the object one enables easy reference surface change without introducing much change in the setup. Moreover, the possibility of changing the scanning move, from moving the reference surface to moving an object, depending on the height of the inspected specimen allows maintaining the object in being in focus. It is especially important with relatively high objects, but also requires scanning device to have higher load capacity.

Setup's structure is divided into modules: illuminating module, interferometric module, imagining module. All options of each module are described below. The scheme of one of possible combinations of those modules is shown in figure 2, while the setups picture is showed in figure 3.

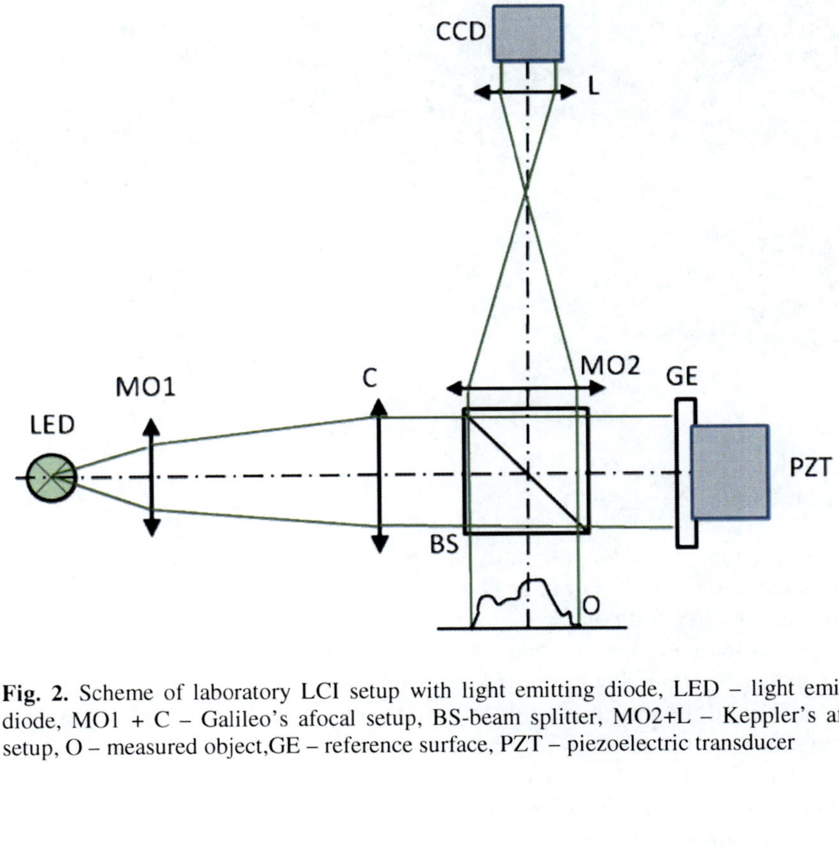

Fig. 2. Scheme of laboratory LCI setup with light emitting diode, LED – light emitting diode, MO1 + C – Galileo's afocal setup, BS-beam splitter, MO2+L – Keppler's afocal setup, O – measured object,GE – reference surface, PZT – piezoelectric transducer

Fig. 3. Laboratory LCI setup with organic light emitting diode

3.1 Illumination Module

Laboratory setup grants the possibility to alter the illumination module in order to adjust to measured object. For objects with high reflectivity and those with diffuse surfaces different conditions are required. Additionally, used numerical algorithm [3] require different scanning steps for different illumination and vertical resolution is limited by spectral characteristic of a light source [4].

Depending on the measurement conditions possible light sources, used in illuminating module, are as follow: single light emitting diode (LED), superluminescent pigtailed LED, white LED with three separated sources, organic light emitting diode (OLED). The last one is currently being tested.

For each light source different optics is needed in order to obtain the plane wave front which enters the interferometric module. In the table 1 the light sources parameters and types of accompanying optical setup is shown.

Table. 1. Optional light sources with a type of an accompanying optical setup

Light source	Main wavelength [nm]	FWHM [nm]	Optical setup type
LED (green)	523	30	Galileo's afocal setup
White LED	630	15.1	Galileo's afocal setup
	515	33.6	
	476	20.3	
Pigtailed LED (green)	525	30	Collimator
Pigtailed LED (white)	4000K	250	Collimator
OLED	610	30	Collimator, Keppler's afocal setup

Choice of light source and its accompanying optical setup is made in order to obtain the best illuminating conditions for different measurement objects classes. For measurements of objects with small height discontinuities shorter PZT transducer is used and SEST (squared-envelope function estimation by sampling theory) algorithm is applied [3] – which requires Gaussian envelope spectrum light source. Scanning step is adjusted specially to the light source – in this case the green diodes are used. For SEST algorithm the steps are relatively high, in a range of tens of micrometers. As a result of scanning the maximum of the coherence envelope function is calculated and found.

White light LEDs are used for obtaining of higher contrast of the fringes [5]. Spectral characteristic of such diodes is more complex and may be represented as a sum of two or more Gaussian sources. Because of it the usage of SEST algorithm is impossible. With those light sources a maximum intensity detection algorithm is applied together with dense sampling – tens to hundreds of nanometers.

For diffusing or objects with low reflectability superluminescent pigtailed diodes are required, because of the large amount of intensity losses in the setup.

3.2 Interferometric Module

Twyman – Green interferometric module consists of cubic beam splitter, designed to work with light from visible spectrum, reference surface and measured object. Scanning motion is introduced by one of transducers: PZT transducer with working range of 60 μm or step motor with working range up to 25 mm. PZT transducer is used for inspection of objects with low step heights and may be used for moving reference surface or light objects, while step motor is suitable for examination of larger objects which may weight up to 9 kg.

Choice for reference surface depends on measured objects. For specimen with metallic, polished surface as a reference flat silver or aluminum covered mirror is used, for glass objects, those with low reflectance or diffusing flat glass is used. Changing the reference surface allows to match the beam intensities in both interferometer's arms.

In one measurement it is possible to cover only the area of 19 mm x 14.25 mm because of the size of used beam splitter. Usually lower fields of view are used in order to obtain higher lateral resolution with used detector.

3.3 Imagining Module

Imaging module depends highly on the measured object size and needed lateral resolution. Optical setup has to enlarge or shrink the measured area to the size of a CCD matrix (Point Grey GRAS-50S5M, μEye Imaging Developement System UI-2230-M). In order to obtain desired magnification and field of view a single lens or, more often, afocal Keppler's setup is used. For building afocal setups achromatic doublets are used.

3.4 Technical Parameters

Described interferometric system is characterized by following parameters:

- Field of view: from 1.02 mm x 0.77 mm up to 19 mm x 14.25 mm;
- Lateral resolution: up to 1 μm/pixel;
- Measurement resolution: up to 0.1 μm;
- Scanning range: up to 60 μm, or up to 25 mm;
- Measurement time: depending on the scanning range and numerical algorithm, from 10 s to quasi – real time.

4 Exemplary Results of Measured Objects

Described optical setup was used for examination various objects from category of MEMS/MOEMS. In the figure 4 only some of them together with exemplary results are shown. The collection was made just to show the flexibility of laboratory LCI setup.

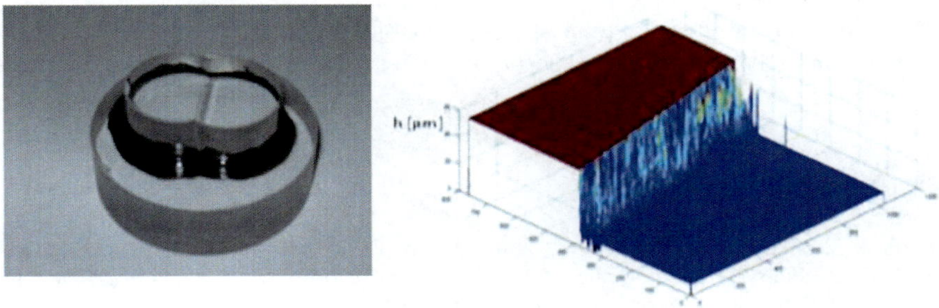

a) Glass step height, h=47.6μm

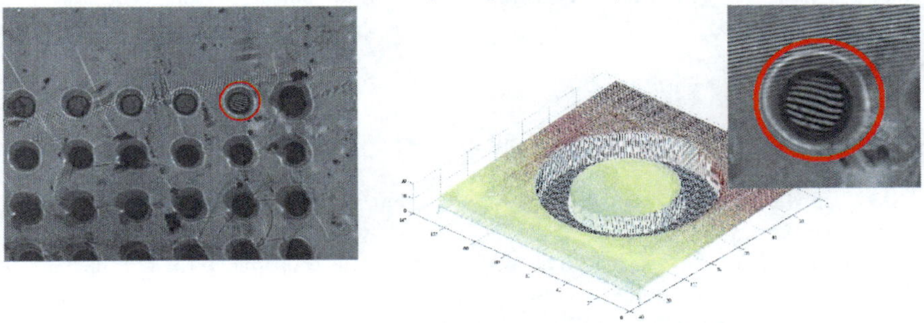

b) Fiber protrusion in a fiber connector [6]

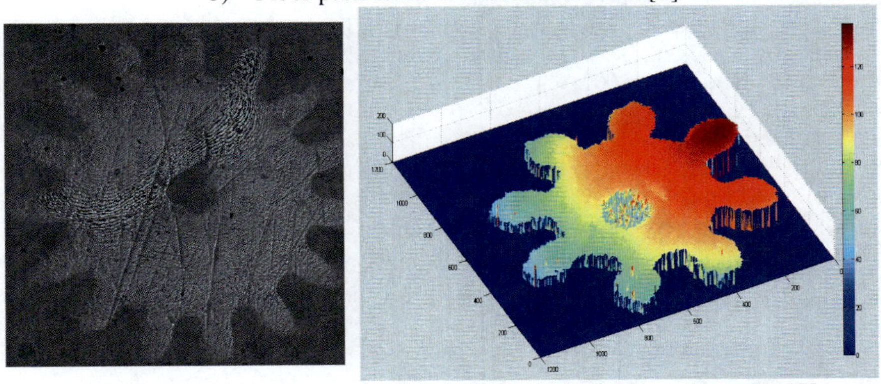

c) Characteristic parameters of small cogged wheels and moulds [7]

Fig. 4. Exemplary objects measured with laboratory LCI setup

Exemplary objects measured by laboratory LCI setup, shown in figure 4, are: a) step height made of glass, used as the standard for this setup – it posses a certificate given by Central Office of Measures of Poland, and is used as an reference object while comparing results obtained on described setup and on commercial setup; b) fiber facet protruding in a new type of a fiber connector – measurement done while inserting the fiber – also shows the quality of measured fiber's facet [6]; c) small cogged wheel and its mould – measured parameters: feet diameter and heads diameter, shrinkage due to fabrication process [7].

Designed interferometric setup is very flexible and the variety of measured objects is limited only by the opportunity to receive new, interesting object to examine.

Acknowledgements

Authors would like to thank Dr. Jurgen van Erps from Vrije Universiteit Brussel and M.Sc. Eng. Andrzej Skalski from Institute of Metrology and Biomedical Engineering Warsaw University of Technology for granting their specimens for examination.

The work was supported by the Ministry of Science and Higher Education within the grant no NN202 354136 and partially within statutory work realized in IMiF WUT.

References

[1] Born, M., Wolf, E.: Principles of optics, 7th edn. Cambridge University Press, Cambridge (1999)

[2] Wayant, J.C., Creath, K.: Advances in interferometric optical profiling. Int. J. Mach. Tools Manufact. 32, 5–10 (1992)

[3] Hirabayashi, A., Ogawa, H., Kitagawa, K.: Fast surface profiler by white-light interferometry by use of a new algorithm based on sampling theory. Appl. Opt. 41(23), 4876–4883 (2002)

[4] Goodman, J.W.: Statistical optics. John Wiley & Sons, Chichester (1985)

[5] Pakuła, A., Salbut, L.: Fringe contrast improving in low coherence interferometry by white light emitting diodes spectrum shaping. In: Fringe 2009, pp. s. 358–s. 363. Springer, Heidelberg (2009)

[6] Van Erps, J., Pakula, A., Tomczewski, S., Salbut, L., Vervaeke, M., Thienpont, H.: In Situ Interferometric Monitoring of Fiber Insertion in Fiber Connector Components. IEEE Photonics Technology Letters 22(1), 60–62 (2010)

[7] Pakula, A., Tomczewski, S., Skalski, A., Biało, D., Salbut, L.: Measurements of characteristic parameters of extremely small cogged wheels with low module by means of low-coherence interferometry. In: Photonics Europe, April 12-14. Proc. of SPIE, vol. 7718-0M (2010)

Mobile Optical Coherence Tomography (OCT) System with Dynamic Focusing and Smart-Pixel Camera

S. Tomczewski* and L. Salbut

Institute of Micromechanics and Photonics, Warsaw University of Technology, 8 Sw. A. Boboli St., Pl-02-525 Warsaw, Poland

Abstract. The new mobile system based on optical coherence tomography (OCT) for measurement of surface layers structure is presented. Due to application of special type detection matrix (called smart-pixel camera) it can be used for very fast measurements also in outdoor conditions. Additionally, the dynamic focusing mechanism causes that the surface of zero optical path difference is always sustained within depth of focus of imaging system. The concept and design of the mobile system is described and exemplary results of its application are presented.

1 Introduction

Optical coherence tomography called OCT is a technique that allows three dimensional structural imaging of investigated sample with the micrometer-scale resolution. Developed in early nineties [1] OCT was originally intended for measurements of a biological tissues in 800nm wavelength region. Later the technical materials were also introduced as a samples in OCT measurements. For example till now we have seen tomography images of the composite materials [2], polymer and polyester foil used in food industry, laminate floor panels [3], GFRP skin of helicopter rotor blade [4], subsurface defects in ceramic materials [5].

OCT systems are based on measurements of light backscattered from investigated sample. This measurement is done by means of low coherence interferometry. The first incarnation of OCT systems is called time domain optical coherence tomography (TD-OCT). Those systems are usually based on fiber optic Michelson's interferometer. The low coherent light emitted from the source propagates through optical fiber and is divided into reference and sample arm at the fiber coupler. Light in sample arm illuminates investigated specimen and is then backscattered to optical fiber. In the reference arm light is reflected from mirror and goes back to the optical fiber. The mirror in the reference arm is moved and the changes of light intensity are registered at photo detector.

The registered intensity changes are called A-scan and they contain information about reflectivity profile of measured sample at the point of measurement. Line of

* Corresponding author.

combined A-scans is called B-scan, and then by combining a number of B-scans whole volumetric image is acquired.

Few years later the different method for OCT measurements was developed. This method is called spectral domain optical coherence tomography and it's main advantage over TD-OCT is that the mechanical scanning is not needed for acquiring an A-scan data. There are two types of spectral OCT, first called FD-OCT and uses a spectrometer at the output of Michelson interferometer for acquiring A-scan data. Second type of spectral domain OCT is called SS-OCT (swept source OCT) and in its basic form it uses swept laser source and the point single detector. Detector registers intensity changes that result from wavelength sweeping process. In both this cases the Fourier transform must be performed for calculation of A-scan data. As reference scanning is not required spectral domain methods present huge advantages in sensitivity and signal to noise ratio over TD-OCT.

There is also Full Field and Wide Field optical coherence tomography. Those methods use CCD or CMOS cameras for acquiring en-face interferometric images instead of obtaining group of A-scan data. By combining those en-face images volumetric data can be built. With those methods scanning in reference arm is required as in TD-OCT, but there is no need for transversal scanning.

The imaging depth of OCT depends on spectral characteristics of light source and is limited by investigated object characteristics (absorption and scattering). The axial resolution of OCT system is determined by spectral characteristics of light source. If the source has Gaussian spectral characteristics then the axial resolution can be calculated as [6] (1):

$$\Delta l = \frac{2ln2}{\pi}\frac{\lambda_0^2}{\Delta\lambda} \tag{1}$$

Where λ_0 represents the central wavelength of the considered source and $\Delta\lambda$ is the full-width at half-maximum (FWHM).

The transversal resolution of an OCT systems is analogue to this in conventional microscopy and it's inversely proportional to imaging system's depth of focus. To overcome limited depth of focus in measurements of larger structures we present modified wide-field OCT system based on Twyman-Green's interferometer with scanning module in measurement arm.

2 Mobile OCT System

The idea of this system was first proposed at Speckle 2010 conference [7]. Since then the whole system was developed and built.

The designed setup is based on Twyman-Green's interferometer geometry and it's schematic is presented at Fig. 1. The light emitted from the source propagates through optical fiber, is collimated by collimator (C) and divided at the beam splitter into sample arm and reference arms. The light in the reference arm is then directed by the mirror (M1) and travels to the reference glass surface (RM) which position is fixed. After reflecting from the reference surface the light propagates back to the beam splitter. In the sample arm light from beam splitter travels to the scanning module (SM) which consists of two mirrors (M2 and M3) installed at the linear stage. After leaving the scanning module light is directed by the mirror (M4) and illuminates measured sample (O). After reflecting from the investigated sample the light goes back to the beam splitter over the same path. The light reflected from the reference surface and light reflected from the sample are then combined at beam splitter. Then light travels over the imaging lens system (L1+L2) and its intensity is registered by the smart-pixel camera which performs in-pixel signal demodulation [8]. The change in optical path difference is induced by movement of mirror system (M2 and M3) installed on the linear stage. The stage is equipped with encoder and is set to move with constant velocity. Velocity must be held constant for smart-pixel camera to perform acceptable signal demodulation.

a)

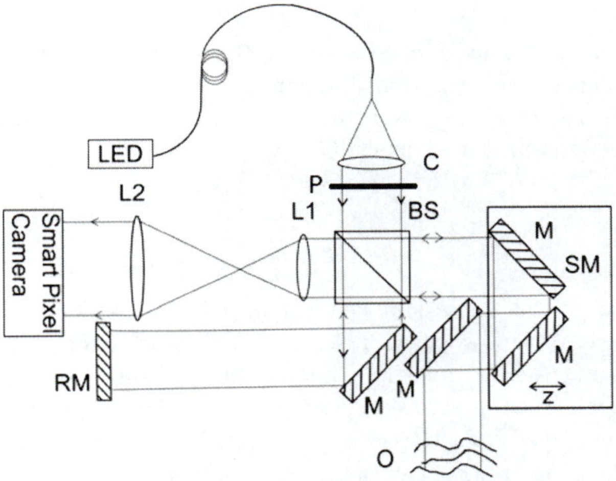

b)

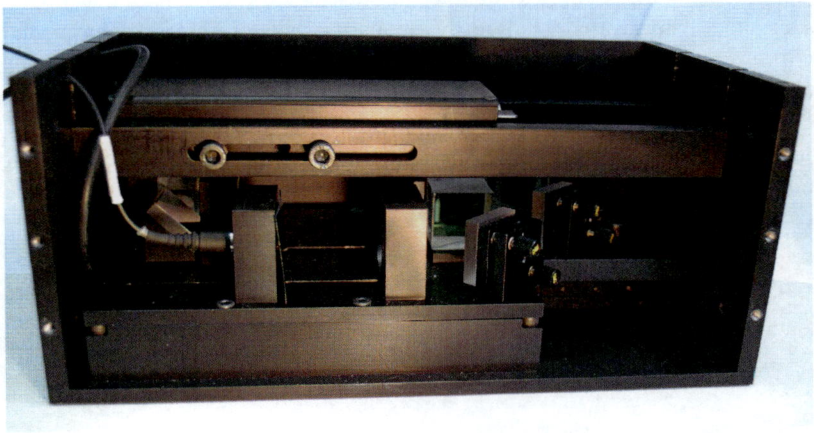

Fig. 1. The a) scheme; b) photo of mobile tomography system: LED – light emitting diode, C-collimator, BS-beam splitter, L1+L2 – imaging lens system, O – measured object, RM–reference surface, M-mirrors, SM – scanning module, P –polarizer

The use of the scanning module allows movement of imaging plane in the examined object during scanning process so the plane of zero optical path difference will always be in an imaging system's depth of focus. There won't be decrease in transversal resolution resulting from theoretically limited depth of focus of imaging system, even if the object's position is fixed with respect to the interferometric system. This advantage is crucial in measurements of the objects that can't be put on scanning stage because of being part of bigger structures.

Insertion of polarizer (see Fig. 1a) with axis perpendicular to the incidence plane highly increased contrast of interference fringes. The increase of fringes contrast would also occur if polarizer is inserted with axis parallel to incidence plane.

The size of measurement area is about 10mm. The scanning range is about 13mm. The dimensions of designed system are 265x150x130 mm. The Heliotis heliCam C3 smart-pixel camera, with resolution of 292x280 pixels is used. The camera is spectrally limited to about 900nm.

3 Results

For now some tests of our system were conducted in visible light, we present two of them. The movement velocity of linear stage with scanning module was set to 1.7mm/s in both cases. To confirm possibility of using this system outside laboratories, all of the measurements were conducted on ordinary computer desk without vibration isolation. The LED (λ_0=625nm; FWHM=20nm) was used as a light source. Axial resolution was calculated as 8,6μm in air for this light source.

First of them was a detection of flat glass surface. The aim of this test was verification of system's ability to conduct proper signal demodulation with selected linear stage. The results acquired by smart-pixel camera are presented at Fig. 2. The presence of the glass surface is clearly visible at presented pictures (Fig. 2a). The signal was also appropriately demodulated by smart-pixel camera.

Second of presented examples is detection of a glue layer on polymer tape. Horizontal sections obtained from measurements are presented at Fig. 3. Reflections from both layers are also clearly visible. The distance between adjacent layers (1:3 and 4:6 at Fig. 3) was calculated as 2,5μm.

a)

b)

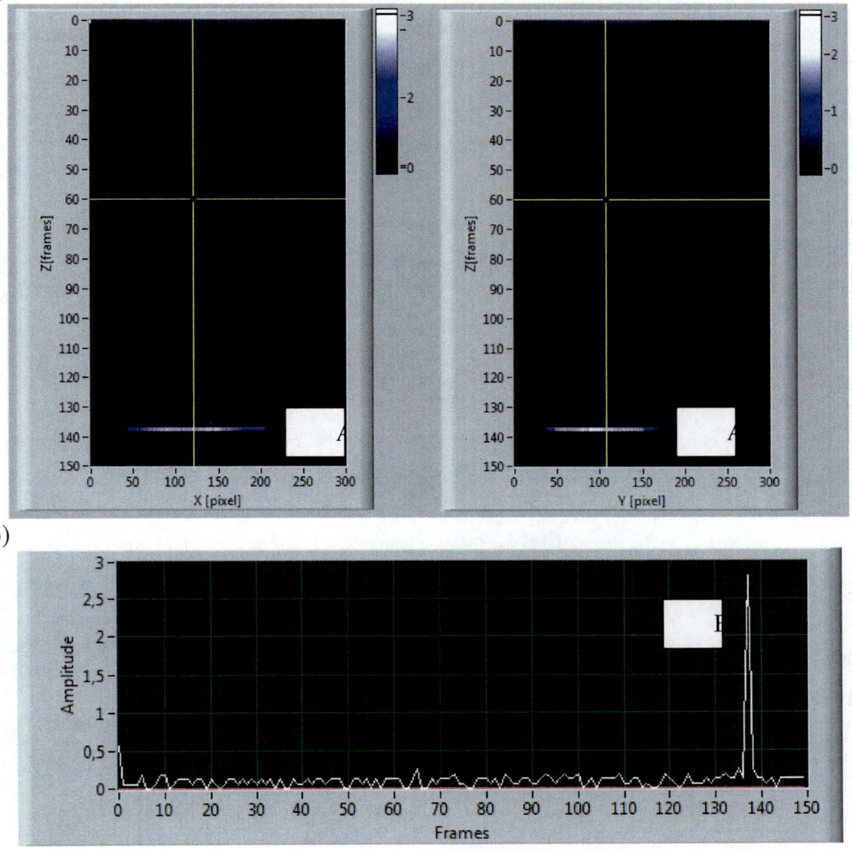

Fig. 2. Results of reflective surface detection a) vertical sections (reflective surface – A); b) exemplary demodulated signal with peak from reflective surface (B)

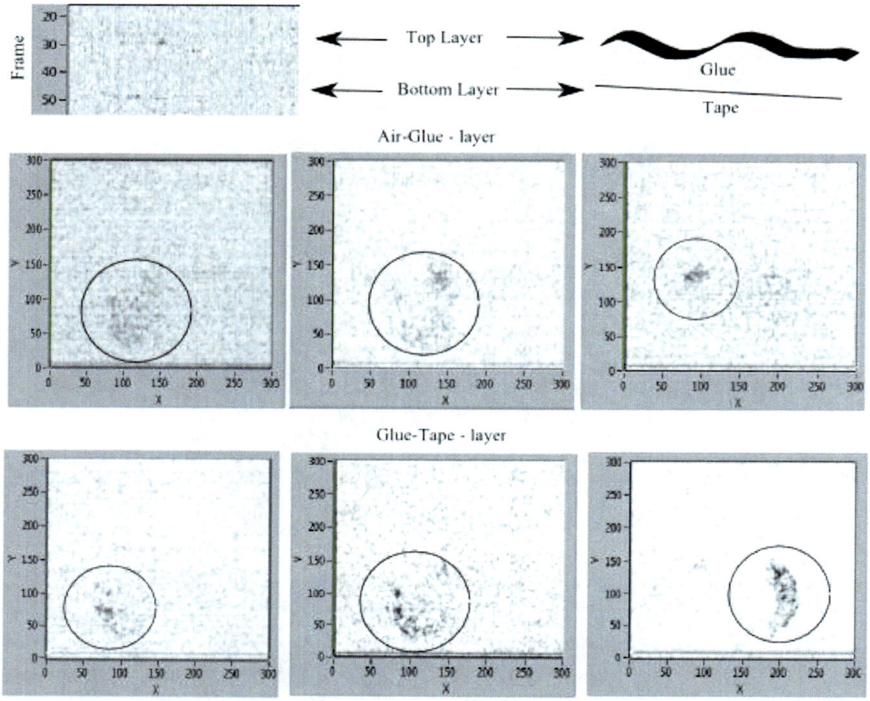

Fig. 3. Detection of glue layer on polymer tape

4 Conclusion

The new mobile system based on optical coherence tomography (OCT) was presented. Due to application of smart-pixel camera and specially designed dynamic focusing mechanism the system can be used for testing of surface layers structure and nondestructive defects detection in laboratory, workshop and outdoor conditions. In close future we are going to conduct tests and measurements in near infrared with a different light source more suitable for this kind of measurements (λ_0=850nm), this will highly increase system's applications range by decrease of light scattering and decrease of light absorption. There will be possibility to conduct measurements in wider range of materials, especially non transparent in visible light.

Acknowledgments

Authors acknowledge the financial support through MONIT project no. WND-POIG.01.01.02-00-013/08.

References

[1] Huang, D., Swanson, E.A., Lin, C.P., Schuman, J.S., Stinson, W.G., Chang, W., Hee, M.R., Flotte, T., Gregory, K., Puliafito, C.A., Fujimoto, J.G.: Science 254, 1178–1181 (1991)

[2] Dunkersa, J., Parnasa, R., Zimbaa, C., Petersona, R., Flynna, K., Fujimoto, J., Bouma, B.: Composites Part A: Applied Science and Manufacturing 31(12), 1345–1353 (2000)

[3] Wiesauer, K., Pircher, M., Gotzinger, E., Hitzenberger, C.K., Oster, R., Stifter, D.: Composites Science and Technology 67(15-16), 3051–3058 (2007)

[4] Wiesauer, K., Pircher, M., Götzinger, E., Bauer, S., Engelke, R., Ahrens, G., Grützner, G., Hitzenberger, C.K., Stifter, D.: Optics Express 13(3), 1015–1024 (2005)

[5] Bashkansky, M., Lewis, D., Pujari, V., Reintjes, J., Yu, H.Y.: NDT & E International 34, 547–555 (2001)

[6] Brezinski, M.E., Fujimoto, J.G.: IEEE Journal of Selected Topics in Quantum Electronics 5(4), 1185–1192 (1999)

[7] Salbut, L., Pakuła, A., Tomczewski, S., Styk, A.: Proc. SPIE 7387, p. 738714 (2010)

[8] Beer, S., Zeller, P., Blanc, N., Lustenberger, F., Seitz, P.: Smart Pixels for Real-time Optical Coherence Tomography. Swiss Center for Electronics and Microtechnology (CSEM SA), Badenerstrasse 569, CH-8048 Zurich, Switzerland

Author Index